CLIMATE AND CIRCULATION OF THE TROPICS

ATMOSPHERIC SCIENCES LIBRARY

STEFAN HASTENRATH

University of Wisconsin, Madison

Climate and circulation of the tropics

D. Reidel Publishing Company

A MEMBER OF THE KLUWER ACADEMIC PUBLISHERS GROUP

Dordrecht / Boston / Lancaster / Tokyo

Library of Congress Cataloging-in-Publication Data

Hastenrath, S.
    Climate and circulation of the tropics.

    (Atmospheric sciences library)
    Bibliography: p.
    Includes indexes.
    1. Tropics — Climate. 2. Atmospheric circulation —
Tropics. I. Title. II. Series.
QC993.5.H37   1985   551.6913   85—19651

ISBN-13: 978-94-010-8878-7   e-ISBN-13: 978-94-009-5388-8
DOI: 10.1007/978-94-009-5388-8

Published by D. Reidel Publishing Company,
P.O. Box 17, 3300 AA Dordrecht, Holland.

Sold and distributed in the U.S.A. and Canada
by Kluwer Academic Publishers,
190 Old Derby Street, Hingham, MA 02043, U.S.A.

In all other countries, sold and distributed
by Kluwer Academic Publishers Group,
P.O. Box 322, 3300 AH Dordrecht, Holland.

to the memory of Hellmut Berg

PREFACE

Tropical atmosphere and ocean are receiving increased attention in relation to the functioning of the global climate system, the remarkable climatic variability in low latitudes, and the associated manifold environmental and societal consequences. Beyond the traditional emphasis of meteorology on weather analysis and forecasting, there is a growing interest in the climate and large-scale circulation of the tropics. This book may serve as a text for graduate and upper-division undergraduate students in meteorology, and is also intended as a reference work for practicing meteorologists, and researchers in the atmospheric, oceanic, and other environmental sciences.

I began writing this book in 1979, but the roots reach further back. Early experiences in North Africa fuelled my curiosity about the low latitudes. In 1960 I seized the opportunity to work in the National Meteorological Service of El Salvador in Central America. My interest in the tropics continued after joining the University of Wisconsin in 1963. Field research brought me to the equatorial Pacific, and many times to the tropical Americas and Africa. This involved visits and correspondence with many weather services. My acquaintance with Australasia and South Asia is limited to short study visits, but includes continuous contacts with colleagues at key research institutions in India, namely the India Meteorological Department, the Indian Institute of Tropical Meteorology, and Andhra University. A guest semester at the University of the Witwatersrand in 1971 and related travels provided a perspective on the problems of Southern Africa. My work with the World Meteorological Organization included in 1973–74 an affiliation with the University of Nairobi, Kenya. I have since maintained contact with both the University of Nairobi and the Kenya Meteorological Department. Concerning South America, I followed invitations to the Universidad de los Andes in Mérida, Venezuela, in 1978 and 1981, to the División de Glaciología y Seguridad de Lagunas of Electroperu in the Cordillera Blanca of Peru in 1978 and 1982, and to the Instituto Nacional de Pesquisas Espaciais in São José dos Campos, S.P., Brazil, variously since 1979. I acknowledge the stimulation received at numerous conferences and workshops in the United States and overseas. Surely I also learned through osmosis from my academic colleagues and students at the University of Wisconsin. The U.S. National Science Foundation supported our work on tropical problems over the past seventeen years. All these experiences and contacts shaped my perception of the tropics.

For permission to reproduce illustrations, I thank R. T. Barber, Robert W. Burpee, Karl W. Butzer, Toby N. Carlson, Pao-Shin Chu, Lawrence Coy, Jean Dettwiller, Clive Dorman, John Findlater, John Flenley, Neil L. Frank, T. Fujita, William M. Gray, John Griffiths, John Horel, R. A. Houze, John Imbrie, A. P. Kershaw, T. N. Krishnamurti, Ernest C. Kung, Wilhelm Lauer, Marcel Leroux, Edward Lorenz, Syukuro Manabe, Andrew McIntyre, Robert L. Molinari, Walter Munk, Gerhard Neumann, Reginald Newell, Chester W. Newton, Neville Nicholls, Albert Pallmann, George Philander, Stephen Pond, Warren Prell, Colin S. Ramage, Eugene M. Rasmusson, Richard Reed, Herbert Riehl, H. U. Roll, James C. Sadler, Jagadish Shukla, Joanne S. Simpson, Henry Stommel, Alayne F. Street-Perrott, I. Subbaramayya, John C. Swallow, Paul Tchernia, Kevin

Trenberth, Thomas Van der Hammen, E. M. Van Zinderen Bakker, George Veronis, John M. Wallace, Ming-Chin Wu, Klaus Wyrtki; Academic Press, Akademie der Wissenschaften der DDR, Akademie der Wissenschaften und der Literatur Mainz, American Association for the Advancement of Science, American Geophysical Union, American Meteorological Society, American Scientist, ASECNA, Balkema Publishers, Birkhäuser Verlag, Blackwell Scientific Publications, British Meteorological Office, Edward Arnold Publishers, Elsevier Publishing Company, Enslow Publishers, Geological Society of America, Goddard Space Flight Center, Her Brittanic Majesty's Stationery Office, India Meteorological Department, John Wiley Publishers, Koninklijk Nederlands Meteorologisch Instituut, La Recherche, Macmillan Journals Limited, Marine Technology Society, McGraw-Hill Publishing Company, Meteorological Society of Japan, MIT Press, New Zealand Meteorological Service, Pergamon Press, Prentice-Hall, Quaternary Research Center of University of Washington, Inc., Rockefeller Institute Press, Royal Meteorological Society, Springer Verlag Wien, Tellus, University of California Press, University Press of Hawaii, World Meteorological Organization.

As with my earlier books, Eva Singer typed the numerous generations of manuscript drafts and updates. Without her patience and her superb mastery of the word processor, this book would have remained a pile of illegible hand-scribbled sheets. Doug Stenz assisted me with the graphics and Klaus Wolter with the indexes and the bibliography of Chapter 12.

Although, in writing this book, I consulted thousands of publications in seven languages from all around the tropics and extending from the last century to the middle of 1985, I indubitably omitted important references by subjective selection or mere oversight, and I apologize.

I was fortunate to have various colleagues read and comment on all or specific chapters of a draft manuscript. I feel grateful to Colin Ramage, University of Hawaii; Peter Lamb, Illinois State Water Survey; Jay McCreary, Nova University; Hermann Flohn, University of Bonn; Dick Grove, University of Cambridge; John Imbrie, Brown University; Vernon Kousky, Climate Analysis Center of NOAA; Abraham Oort, Geophysical Fluid Dynamics Laboratory of NOAA; Dick Reed, University of Washington. In the publication of this book, I appreciate the effective cooperation with David Larner of D. Reidel Publishing Company. Finally, I am indebted to my old teacher Hellmut Berg, who inspired us, who 25 years ago encouraged me to take a job in the tropics, and to whose memory this book is dedicated.

Madison, July 1985   STEFAN HASTENRATH

TABLE OF CONTENTS

LIST OF FIGURES

LIST OF TABLES

CHAPTER 1

INTRODUCTION

The functioning of the global climate system can be understood only upon proper appreciation of processes in the tropics. In view of the large time scales involved in climate processes, investigation of the World ocean is indispensable. It is therefore not surprising that international and national efforts at global climate dynamics have recognized studies of climate mechanisms in low latitudes and of the role of the hydrosphere as tasks of high priority (National Academy of Sciences, 1969, pp. 63–81; 1975, pp. 62–83; World Meteorological Organization – ISCU, 1975, pp. 26–39; 1983, vol. 1., pp. 1–20; World Meteorological Organization, 1980, pp. 12–15, 38–44; Federal Coordinating Council, 1977, pp. 11, 19–20, 23, 35, 41, 43; National Oceanic and Atmospheric Administration, 1977, pp. 11–13, 17–18, 20–29, 41–45; 1980, pp. 36–68; Department of Commerce, 1980, pp. 8, 13, 24).

The latitude domain of the tropics is here taken in a generous sense. Delimitation by the tropics of cancer and capricorn has astronomical, but little meteorological meaning. Regarding the atmospheric circulation, the subtropical high pressure cells and associated anticyclonic axes, while shifting seasonally, offer a natural demarcation. The tropical atmosphere is characterized by high temperature and abundant moisture content. However, bodies of air with tropical properties and origin may penetrate deep into middle latitudes during the respective summers. Considering the net radiation at the top of the planet Earth, the low latitudes are a region of heat gain for the atmosphere-hydrosphere-lithosphere system, but the lines of zero net radiation also undergo considerable latitude displacements in the course of the year. A further elementary climatic distinction between the lower and higher latitudes relates to the relative roles of radiational forcings: the diurnal cycle is more important than the annual march in the lower as opposed to the higher latitudes, although transition is gradual. Furthermore, the latitude variation of the Coriolis parameter entails a gradual changeover in the dynamics of atmospheric and oceanic flow. In accordance with these considerations and for the present purposes, 30°N and 30°S are here suggested as approximate boundaries of the tropics. These parallels would divide the globe into a tropical half and the extratropical caps. However, no rigid limits are implied. In fact, the connection between the tropics and the higher latitudes is of particular interest.

Research on the tropics in the past decade has developed an increasing appreciation of the large time and space scales, and has turned its focus on the dynamics of climate, as compared to the emphasis on weather events and smaller scale processes which prevailed throughout the earlier part of this century. Associated with the evolving interest in the functioning of the climate system are an enhanced awareness of the role of the oceans, greater attention to past climatic regimes, and a growing concern for the economic and societal implications of climate. Accordingly, the objectives of the present book differ from the various earlier treatises on the meteorology of the low latitudes (e.g. Ramage, 1971; Carlson and Lee, 1978; Riehl, 1979; Krishnamurti, 1979) not only in the inclusion of more recent research results, but also in its thrust directed to the dynamics of climate. Work over a century provided an essential basis for the important advances in low latitude meteorology during the past 10–20 years. The recent evolution in research on the climate and circulation of the tropics encompasses a wide range of contributions from diverse fields and calls for a broad scope. Our own work at the University of Wisconsin is included in this book, to the extent that it forms part of this evolution.

It seems appropriate to begin this book with consideration of the diurnal forcings and local circulations which are so strongly developed in the low latitudes (Chapter 2). The daily march of insolation and atmospheric tides are particularly pronounced in the tropics, and as a result day-periodic processes and circulations are much more vigorous than in the higher latitudes. In many tropical regions, the annual average climatic conditions can be understood only from the interaction between diurnal factors, local circulations, and the large-scale flow. Accordingly, day-periodic processes are considered as relevant to various subjects discussed later on in the book, where emphasis is on the large space and time scales.

The planetary scale circulation is reviewed (Chapters 3–5) as background reference for subsequent more special topics with regional focus. Concerning the atmosphere (Chapter 3), the reviews by Lorenz (1967) and Palmén and Newton (1969), and the analyses of Oort and Rasmusson (1971) and Newell *et al.* (1972, 1974) remain important references, and are complemented by the updated data evaluation of Oort (1983). The planetary scale atmospheric circulation of the tropics is dominated by an easterly lower-tropospheric wind regime and the two thermally direct Hadley cells. The latter perform most of the poleward transport of westerly absolute angular momentum in low latitudes, while eddy mechanisms become most effective beyond the fringe of the tropics. The kinetic energy budget of the tropics during the respective winter seasons is characterized by a large production associated with the Hadley circulation, small dissipation rates, and an export to the extratropical caps, while during summer generation and export rates are much smaller. In mechanical terms, the tropical atmosphere thus plays a vital role in the maintenance of the global circulation.

While the planetary scale atmospheric circulation appears now reasonably well explored, the tropical ocean circulation (Chapter 4) is the subject of much current research and controversy. The surface circulation of the low-latitude Pacific and Atlantic is characterized by quasi-permanent anticyclonic gyres centered in the subtropics and predominantly zonal flow regimes in the equatorial region, whereas ocean currents in the Indian Ocean are subject to the reversals between the Northern winter and summer monsoon wind regimes. Annual and interannual variations in the strength of the surface winds over the equatorial oceans are in particular reflected in changes of the ocean circulation, the fast travelling equatorially trapped waves being of special interest in the origin of certain climate anomalies.

In the study of the heat and water budget, treatment of the coupled atmosphere-ocean system (Chapter 5) is imperative. Fundamental to the heat budget of the planet Earth is the net radiation at the top of the atmosphere, which can now in principle be measured from satellites. The annual mean heat budget of the hydrosphere is evaluated through calculation of net heat gain through the surface on the basis of long-term ship observations. Estimates of meridional heat transport in the ocean are obtained from the study of the surface heat budget and direct evaluation of hydrographic sections. The atmospheric heat and moisture budget is evaluated from aerological and surface information. On this basis, the relative role of atmosphere and hydrosphere in the total poleward transport of heat can be appreciated. The ocean accounts for about half of the total poleward transport around 30°N and 20°S. Error tolerances are considerable for the determination of atmospheric and hydrospheric heat transport divergences, as well as for estimates of net radiation at the top of the atmosphere.

Significant progress has recently been made in the detailed quantitative analysis of quasi-permanent large-scale circulation systems (Chapter 6). Emanating from the subtropical high pressure belts of either hemisphere, the lower-tropospheric trade winds meet within a band of highest surface temperature and low pressure trough near the Equator. Embedded in the low pressure trough is a belt of maximum convergence-cloudiness-rainfall. Seasonal variations of heat low processes are a major factor for the annual latitude migration of the equatorial trough. The monsoon area of the World is delineated in terms of the complete annual reversal of wind regimes, encompassing the Indian Ocean sector and much of tropical Africa. The most spectacular development is reached by the Northern summer Southwest monsoon in the Indian Ocean sector, which is related to the establishment of a heat-low induced monsoon trough over South Asia. On the Indian Subcontinent and adjacent regions, the bulk of the annual precipitation falls during the Southwest monsoon. Various jet stream systems are developed at different times of year, namely the upper-tropospheric Subtropical Westerly Jet, the likewise upper-tropospheric Tropical Easterly Jet, the West African Mid-tropospheric Jet, and the East African Low Level Jet. Zonal circulations along the Equator are caused by East–West differences in surface and tropospheric temperatures, most prominent being the 'Walker Circulation' over the Pacific.

Weather systems and phenomena on the synoptic time scale have received up-to-date treatment in Ramage (1971), Carlson and Lee (1978), Riehl (1979), Krishnamurti (1979), and Anthes (1982). Although none of these texts is comprehensive, they collectively provide a useful account of the subject matter. Chapter 7 concentrates on the climatology of weather systems, in particular their structure, large-scale environmental setting, regional distribution, and seasonal characteristics. In comparison with the higher latitudes, there is a vast diversity of weather systems in the tropics. Specifically discussed are Hurricanes, Waves in the Easterlies, Squall Lines, Dust Storms of the Sudan, Monsoon Depressions, Subtropical Cyclones, Temporales of Pacific Central America, and wintertime cold surges. Various of these weather systems, through the associated rainfall, play an essential role in the regional climate. Their variability from year to year forms part of the circulation and climate anomalies in the tropics.

Climatic disasters in various tropical regions during the recent past and their severe socio-economic consequences have awakened public awareness to the important interannual variability of the atmosphere-ocean system (Chapter 8). Marked interannual variations are an intrinsic part of tropical climate. Various regional climate anomalies are related to the surface pressure seesaw of the Southern Oscillation, on a time scale of 2–10 years, and with dipoles over the Eastern South Pacific and the greater Indonesian – Australasian region, but spanning the global tropics. While these near-global pressure variations were known since the 19th century, their causal relation to the El Niño events on the West coast of South America has been recognized only since the 1970's. Variations of the upper-air circulation are an essential part of the El Niño Southern Oscillation (ENSO) phenomenon. Regional climate anomalies discussed specifically include the vagaries of the Indian monsoon, the droughts of Northeast Brazil, rainfall variations in the Central American – Caribbean region, droughts in Subsaharan Africa, rainfall and sea surface temperature anomalies at the Angola coast, and precipitation departures in the Zaïre (Congo) basin. Interannual variability appears concentrated in various preferred time scales.

Because of the considerable human impact of interannual variability in key tropical regions, there has long been a keen interest in the possibility of predicting climatic hazards of great magnitude well in advance (Chapter 9). Thus attempts at climate prediction have a long tradition in some densely populated agricultural lands of the tropics, in particular India and Indonesia. Other problems considered include El Niño, Northeast Brazil droughts, rainfall in Subsaharan Africa and Southern Africa, North Atlantic Hurricanes, and Hong Kong summer rainfall. Work during the 1980's offers the prospect for climate prediction based on a combination of extensive diagnostic general circulation studies with statistical methods; verification on an independent data set being essential. In fact, various studies indicate that empirically based climate prediction is possible for *certain* tropical climate problems.

Natural climatic disasters such as discussed in Chapters 8 and 9 may have severe economic and societal implications, but at the same time human activities may seriously affect the climatic environment. These two kinds of impact are the subject of Chapter 10. Human interference with the natural setting includes the deliberate deforestation throughout extensive regions of the tropics, as well as certain land use patterns that alter surface albedo and dust input into the atmosphere. Fish life concentrated in a few coastal regions of the tropical oceans is vulnerable to both the natural interannual variability and excessive fishing. The human impact of natural climatic diasters is discussed for India, Northeast Brazil, and Subsaharan Africa, in particular. Climate prediction can only be useful where the economic, social, and political systems are flexible enough to respond effectively to predictions of such natural diasters.

Tropical glaciers are of interest in the study of climate (Chapter 11) on both the paleo and modern time scales. Glacier and climate variations are documented, in varying detail, from 500 000 years ago to the present. A general ice recession which began in the 19th century in all tropical high mountain regions is continuing. Concerning the recent glaciation, two methods are being developed to infer climate variations from glacier observations, namely the numerical modelling of the complete causality chain from climatic forcing to terminus response, and climatic ice core studies.

The final Chapter 12 of this book combines vegetation, lake, glacier evidence, and deep-sea core analyses, to produce a preliminary synopsis of climate and circulation changes in the tropics during the past 30 000 years. Large regions specifically considered include Pacific and Australasia, the Indian Ocean and surrounding continents, Africa and the adjacent tropical Atlantic, and the tropical Americas. Numerical simulations of the atmospheric circulation during certain episodes of the geological past complement the evaluation of field evidence.

Many of the issues addressed in the present attempt at a synthesis are central to ongoing research. Highlighting the gaps in our knowledge and an appraisal of recent research results on tropical circulation and climate are the objectives of this book.

References

Anthes, R. A., 1982: *Tropical cyclones, their evolution, structure, and effects*. Meteorological Monographs, vol. 19, no. 41, Amer. Meteor. Soc., Boston, 208 pp.

Carlson, T. N., Lee, J. D., 1978: *Tropical meteorology*, Independent Study by Correspondence, Pennsylvania State University, University Park, Pa., 390 pp.

Department of Commerce, 1980: 'National climate program', annual report 1979. Washington, D.C., 49 pp.

Federal Coordinating Council for Science, Engineering and Technology, 1977: 'A United States climate program'. ICAS 20b-FY 1977, Rockville, Maryland, 82 pp.

Krishnamurti, T. N., 1979: *Tropical meteorology*. Compendium of Meteorology, vol. 2, part 4 (ed. A. Wiin-Nielsen), WMO No. 364, World Meteorological Organization, Geneva, 428 pp.

Lorenz, E. N., 1967: *The nature and theory of the general circulation of the atmosphere*. World Meteorological Organization, WMO No. 218, TP. 115, Geneva, 161 pp.

National Academy of Sciences, 1969: *An oceanic quest: The International Decade of Ocean Exploration*. Washington, D.C., 115 pp.

National Academy of Sciences, 1975: *Understanding climatic change: a program for action*. Washington, D.C., 239 pp.

National Oceanic and Atmospheric Administration, 1977: 'NOAA Climate Program'. Rockville, Maryland, 94 pp.

National Oceanic and Atmospheric Administration, 1980: 'National climate program, five-year plan'. Washington, D.C., 102 pp.

Newell, R. E., Kidson, J. W., Vincent, D. G., Boer, G. J., 1972, 1974: *The general circulation of the tropical atmosphere*. MIT Press, Cambridge, Mass, 258 and 371 pp.

Oort, A. H., 1983: 'Global atmospheric circulation statistics, 1958–1973'. NOAA Professional Paper 14, U.S. Government Printing Office, Washington, D.C., 180 pp.

Oort, A. H., Rasmusson, E. M., 1971: 'Atmospheric circulation statistics'. NOAA Professional Paper 5, Rockville, Maryland, 323 pp.

Palmén, E., Newton, C. W., 1969: *Atmospheric circulation systems*. Academic Press, New York and London, 603 pp.

Ramage, C. S., 1971: *Monsoon meteorology*. Academic Press, New York and London, 296 pp.

Riehl, H., 1979: *Climate and weather in the tropics*. Academic Press, London, New York, San Francisco, 611 pp.

World Meteorological Organization, 1980: 'Outline plan and basis for the World Climate Programme, 1980–83'. Geneva, WMO – No. 540, 64 pp.

World Meteorological Organization – ICSU, 1975: *The physical basis of climate and climate modelling*. GARP Publication Series, No. 16, 265 pp.

World Meteorological Organization – ICSU, 1983: *Large-scale oceanographic experiments in the WCRP*. WCRP Publication series, No. 1, 2 vols., 121 and 544 pp.

CHAPTER 2

DIURNAL FORCINGS AND LOCAL CIRCULATIONS

Day-periodic processes and circulations on the local to meso-scale dominate the regional climate and weather in the tropics, in contrast to the higher latitudes where the diurnal is much smaller than the annual cycle. Forcings on the time scale of a day or less result from horizontal contrasts in the surface heat budget and thermal and gravitational atmospheric tides; both heat budget and tidal forcings being particularly pronounced in the low latitudes. Differential heat budget forcings lead to vigorous topographic and meso-scale circulations, which in turn dominate the spatial pattern and diurnal cycle of cloudiness and rainfall. Day-periodic processes must accordingly be recognized as important factors for the regional climates, the latent heating distribution, and consequently the large-scale circulation. The role of diurnal and local forcings for the long-term and large-scale functioning of the tropical atmosphere does not seem to have received the attention it deserves. The present chapter is intended as a preliminary review of such diurnal and local processes, and as reference for subjects discussed in later chapters of this book.

2.1. Insolation and Heat Budget Forcing

Radiation geometry (Sellers, 1965, pp. 11–39; Budyko, 1974, pp. 1–11) accounts for the following characteristics of the tropics as compared to the higher latitudes: the annual total of solar radiation is larger, while the annual range is less; accordingly, the diurnal variation greatly exceeds the annual range. An elementary consequence of these features of solar radiation is illustrated in the thermoisopleth diagram, Fig. 2.1:1. In the higher latitudes, temperature varies little throughout the day, but greatly in the course of the year; thus isopleths in Fig. 2.1:1, part a (station Berlin, Germany), run broadly parallel to the 'hour' axis and perpendicular to the 'month' axis. By contrast, in the tropics the annual range of daily mean temperature is typically only 1–2°C, with the diurnal range being about an order of magnitude larger; accordingly, the orientation of isopleths in Fig. 2.1:1, part b (station Quito, Ecuador) is approximately orthogonal to the higher latitudes. Fig. 2.1:1 illustrates the role of insolation geometry in the predominantly diurnal surface temperature regime of the tropics. Great diurnal range and reduced seasonality are the hallmarks of the temperature regime throughout the tropics, even in high mountain locations where the annual mean temperature may approach that of higher latitudes. The station Quito shown in Fig. 2.1:1, part b has such a highland location. However, tropical lowland stations (see Lauer, 1975) not displayed here show a similar isopleth pattern, albeit with larger absolute values of temperature. The pronounced diurnal march of insolation together with the large amounts of solar radiation are readily visualized as controlling factors for the development in the course of the day of spatial contrasts in surface heat budget and temperature, given spatial differences in surface properties (heat capacity, albedo), and topography. The spatial temperature contrasts in turn largely determine the intensity and dimensions of local circulation systems on the diurnal time scale.

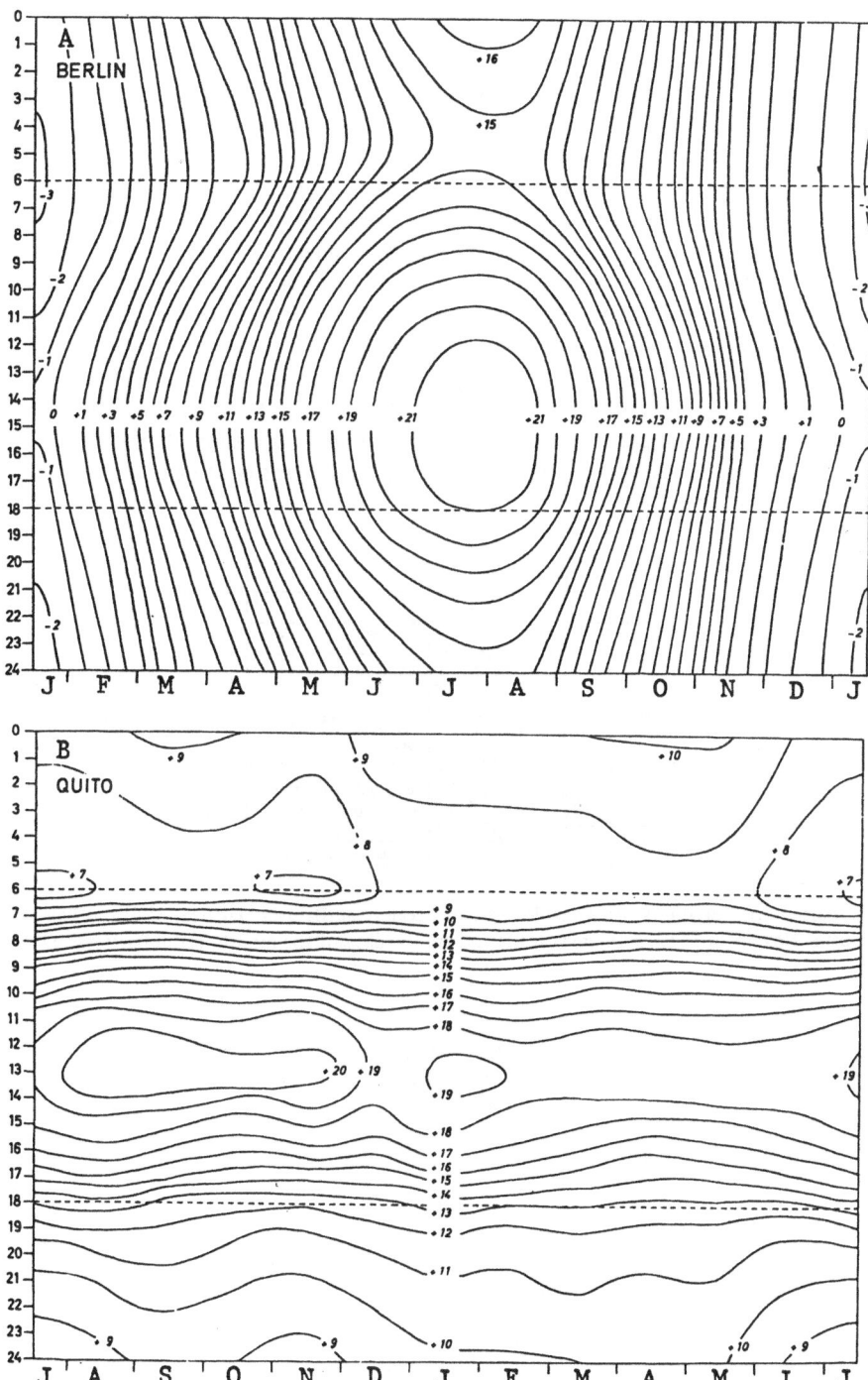

Fig. 2.1:1. Annual and diurnal temperature regimes. Horizontal scale represents months and vertical scale hours in local time. Thermoisopleths are in °C. (a) top, mid-latitudes (station Berlin, Germany, 52°31′N, 13°22′E, 35 m). (b) bottom, tropics (station Quito, Ecuador, 0°14′S, 78°32′W, 2850 m). From Lauer (1975).

2.2. Atmospheric Tides

Atmospheric tides have in the higher latitudes been isolated only through careful statistical analysis of large data volumes (Bartels, 1928, 1932; Chapman and Lindzen, 1970, pp. 24–105), whereas in the tropics the periodic pressure variations are conspicuously apparent on the barograph trace of any single day. This diversity between the higher and lower latitudes is illustrated in the classical graph of Bartels (1928) reproduced as Fig. 2.2:1. For a review of the various tides and their thermal and gravitational nature refer to Chapman and Lindzen (1970), as well as to Defant and Defant (1958, pp. 492–518), and Mitra (1952, pp. 31–37). Of interest here are the solar diurnal (24-hourly) tide S_1, the solar semi-diurnal (12-hourly) tide S_2, and the solar ter-diurnal (8-hourly) tide S_3. Their global patterns have been studied by Bartels (1932) and are depicted in Fig. 2.2.:2.

The solar diurnal tide S_1 is subject to large regional diversity, but its spatial pattern shows above all the influence of the continents (Chapman and Lindzen, 1970, pp. 38–40). At the period of 12 solar hours, two components are distinguished which both show definite global patterns. A standing wave (Fig. 2.2:2, part b) has nodal lines at 35 degrees North and South, and at 11.5 h universal time a maximum in the high latitudes and a minimum in the tropics, amplitudes being

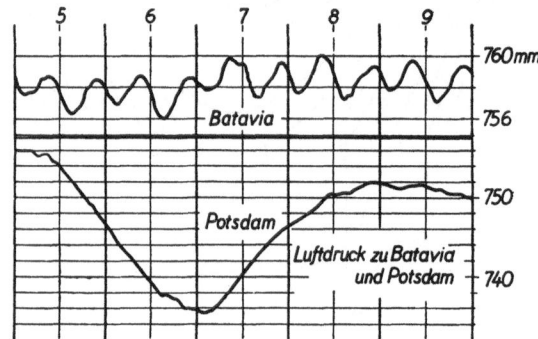

Fig. 2.2:1. Barometric variations (on twofold different scales) at Batavia (Jakarta, 6°S) and Potsdam (52°N) during November 1919. Numbers on abscissa are the days of the month. From Bartels (1928).

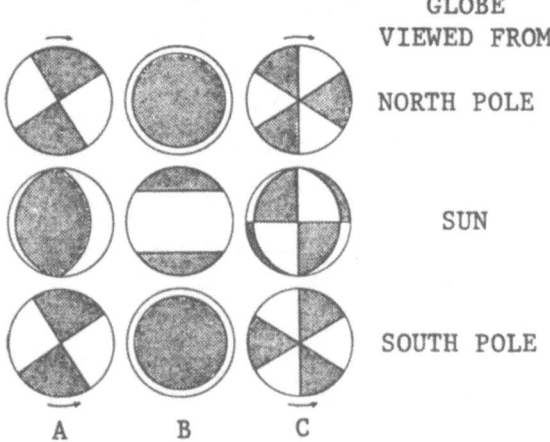

Fig. 2.2:2. Global patterns of atmospheric tides. (a) solar semi-diurnal progressive wave; (b) solar semi-diurnal standing wave; (c) solar ter-diurnal progressive wave. Shading denotes positive, and blank area negative pressure departures. From Bartels (1932).

about 0.1 mb at the poles and 0.05 mb at the Equator. Thus this wave interferes with other tides in the higher latitudes but is negligible in the tropics. Most important for the low latitudes is the 12-hourly progressive pressure wave (Fig. 2.2:2, part a). This has a largest amplitude of over 1 mb at the Equator with a decrease poleward into either hemisphere, and travels from East to West in such a way that at any location maxima are experienced around 10 and 22 h and minima around 4 and 16 h local time. The solar 8-hourly wave (Fig. 2.2:2, part c) is antisymmetric about the Equator, possesses a maximum amplitude of about 0.2 mb at latitudes 30 degrees North and South, also progresses from East to West, and has one of the maxima around 06 h local time in the respective winter hemisphere. The spatial distribution of the solar 8-hourly pressure wave is rather regular, and the phase opposition between hemispheres and the reversal between the extreme seasons are thought to be related to a similar behavior of the solar 8-hourly harmonic of temperature (Chapman and Lindzen, 1970, p. 43). Of the S_1, S_2, and S_3 tides, it is the 12-hourly travelling wave that dominates the daily march of pressure in the tropics, although the superposition of the 24- and 8-hourly waves result in a generally larger morning than evening maximum, as well as some variation in the course of the year.

The various pressure waves require for mass continuity periodic wind variations. The wind departure patterns postulated from theory for the 12-hourly and the 8-hourly travelling waves are depicted in Fig. 2.2:3. Because of its larger magnitude the 12-hourly wave is of primary interest here. Fig. 2.2:4 schematically summarizes the relation of wind and pressure variations for the 12-hourly wave. In either hemisphere, departure easterlies are largest at the time of the pressure maximum, while largest westerly departures coincide with the pressure minima. In either hemisphere, largest equatorward wind departures follow the pressure minima and precede the pressure maxima, the reverse being the case for the poleward wind departure. At the Equator, the meridional wind departures vanish. This wind departure pattern corresponds to maximum convergence at the Equator after the pressure minima and before the pressure maxima, and similarly to largest divergence after the pressure maxima and before the pressure minima.

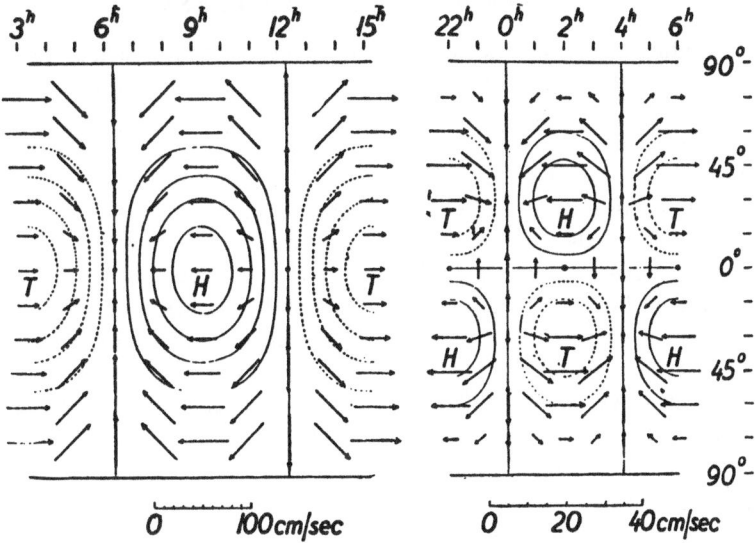

Fig. 2.2:3. Map of wind departures associated with tidal pressure variations. (a) left, solar semi-diurnal progressive pressure wave, isobar spacing 0.2 mm Hg. (b) right, solar ter-diurnal anti-symmetric progressive pressure wave in Northern winter, isobar spacing 0.05 mm Hg. Source: Defant and Defant (1958), and Bartels (1932).

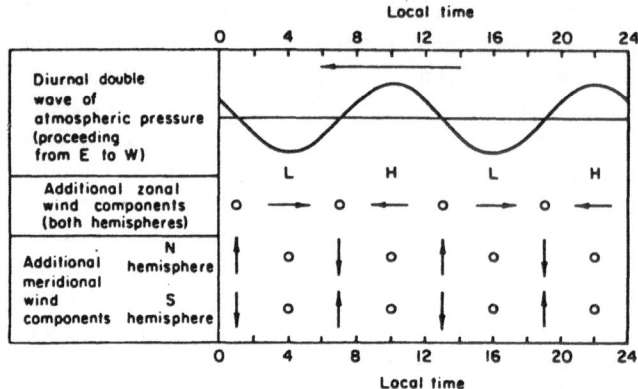

Fig. 2.2:4. Diurnal double wave of atmospheric pressure and corresponding wind components according to theory. From Roll (1965), and Kuhlbrodt and Reger (1938).

While the 12-hourly pressure variation is readily apparent on any single day, the corresponding wind variations seem greatly obliterated by noise. The more regular appearance of the pressure wave may result from the integration of wind variations and associated divergence/convergence over a deep atmospheric layer. However, wind variations associated with the 12-hourly pressure wave and consistent with theory have been demonstrated by appropriate statistical analysis not only for the surface but also for the free atmosphere (review in Hastenrath, 1972).

The variations of surface wind speed and divergence/convergence associated with the solar semi-diurnal tide are of interest in relation to the daily march of sea-air exchange, cloudiness, and rainfall. These implications are discussed in Section 2.4.

2.3. Circulations on the Local and Meso-Scale

Day-periodic circulations in mountainous terrain and across coastlines, such as they are also found in the higher latitudes, are as a rule more vigorously developed in the tropics. Various prominent examples are considered in this section.

It seems appropriate to begin with Leopold's (1949) classical study on the interaction of sea breeze and trade winds in the Hawaiian Islands. Fig. 2.3:1 illustrates various flow configurations. In the case of small islands (10—80 km wide) of modest relief (highest elevations around 1000 m), such as Lanai and Molokai, the Northeast trades blow over the island and during daytime meet a sea breeze on the leeward side, where convergence and cloudiness develop. At night the trades would oppose the land breeze on the windward side and combine with it on the leeward side. Maui is the example of a larger island with a 3000 m high volcano. The trades flow around this mountain obstacle and largest rainfall is found on the windward side. Daytime convection on the leeward side develops cloud streets at the edges between the sea breeze and trade wind systems. On large mountainous islands (Mauna Kea) topographic obstacles cannot be entirely circumvented. There is cloudiness and rain on the windward side. Strong trade winds blow through mountain passes. At night the land breeze on the windward side flows counter to the trades resulting in convergence and convection. During daytime, a convergence between sea breeze and trade winds is characteristic for the leeward side (Kona type). The convergence and convection resulting from the interaction of land breeze, sea breeze, and trade winds are then similar to the small islands (Lanai type) but more vigorous.

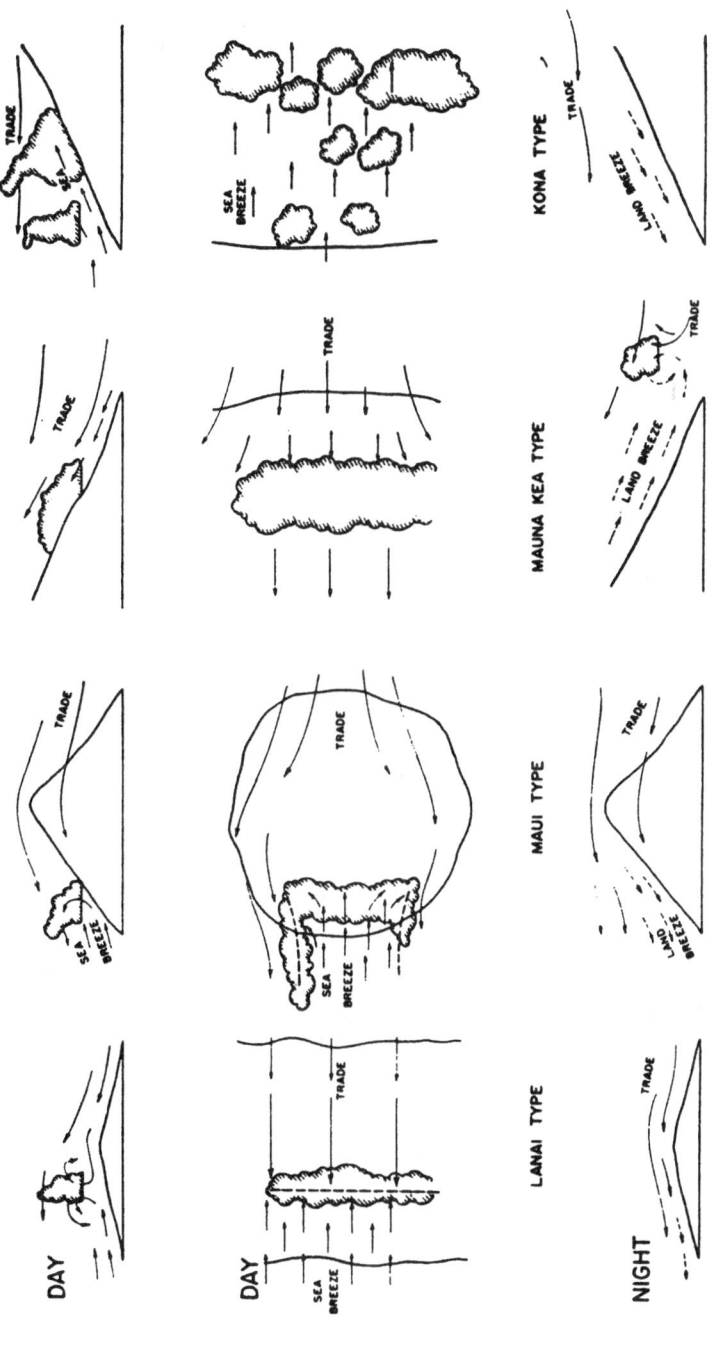

Fig. 2.3:1. Schematic representation of four types of interaction of sea and land breezes and trade wind over the Hawaiian Islands. Daytime conditions (top and middle) shown in vertical and horizontal sections, nocturnal conditions (bottom) in vertical section only. From Leopold (1949) (*Journal of Meteorology*, American Meteorological Society).

Since Leopold's (1949) classical paper, the local wind systems of the Hawaiian Islands have been studied more extensively (Ramage, 1978). These investigations have documented the role of vertical momentum exchange in the diurnal cycle of wind speed and have shown that high volcanoes penetrating the trade inversion (Section 6.5) force lower-tropospheric flow around them.

Such effects as described above for Hawaii are not limited to islands. At the East coast of Brazil, rainfall is concentrated in the late night to early morning hours, in remarkable contrast to locations farther inland. Kousky (1980) studied this phenomenon on the basis of satellite imagery and surface observations, and describes the development of a convergence between a land breeze and the Southeast trades during the late night to early morning. Leroux (1983, vol. 1, pp. 82–85) presents a similar example for the Ivory Coast.

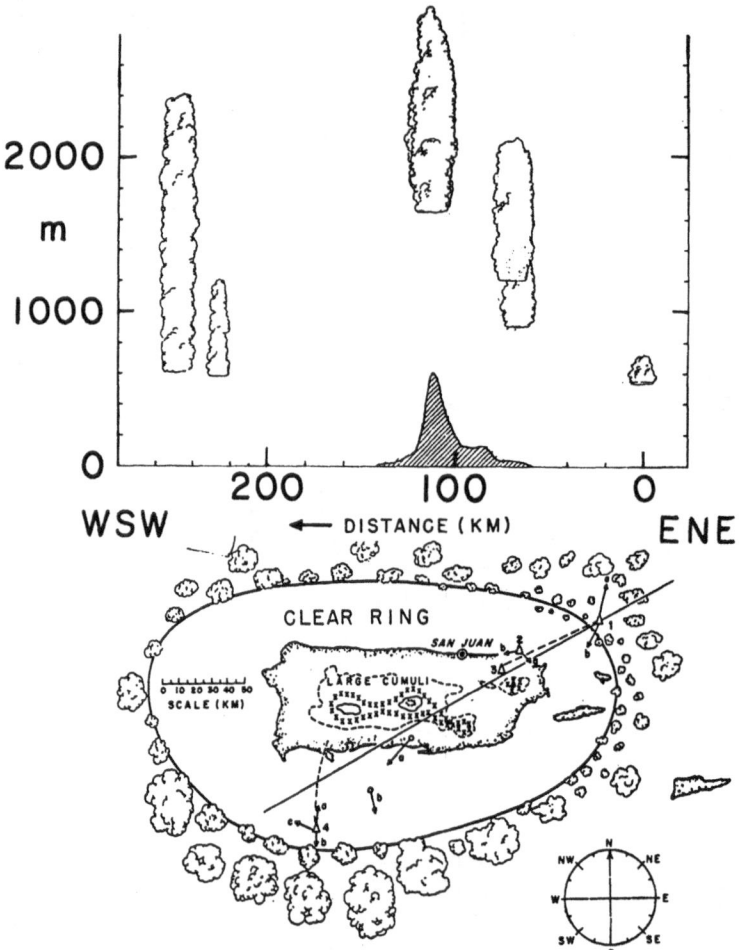

Fig. 2.3:2. Patterns of daytime cloudiness over and around the island of Puerto Rico. Plan view sketch at bottom shows development of a clear ring around the island; thin straight solid line indicates orientation of vertical transect at top of graph. This shows convection over mountainous interior of island and cloud-free areas. Source: Malkus (1955).

Fig. 2.3:2 illustrates the development of the daytime cloudiness pattern around the large island of Puerto Rico. In the course of the day, strong convection develops over the mountains of the interior, which can be considered as an elevated heat source, and this entails a lateral inflow of air. A compensating subsidence seems to take place around the island, manifesting itself in a clear ring, and trade cumuli confined to larger distances from the island. The clear ring increases in width until the middle of the afternoon, but with the decay of convection over the central mountains the ring contracts again towards the evening. Related phenomena can be observed in the interior of the continents. Near the Indian Ocean coast of Tanzania in East Africa, Kilimanjaro looms from extensive plains near sea level abruptly to almost 6000 m. Convection and cloudiness over the mountain during daytime is also associated with the tendency for the development of a wide cloud-scarce ring round it, presumably again a reflection of compensating subsidence.

Similar compensatory vertical motion patterns are commonly found in many tropical mountain regions, though with geometrically less regular configuration. An example is the Mount Kenya area of East Africa as illustrated in Fig. 2.3:3. At night, surface winds are directed down the slopes and valleys and the peak regions clear with preference. At the same time airstreams form a confluence over the basins, where cloudiness develops. In the morning the topographically induced circulations reverse. Surface winds are directed up the slopes and valleys, the mountain clouds up from about noon onward, while the basins clear. In the late afternoon, the downslope winds take over again.

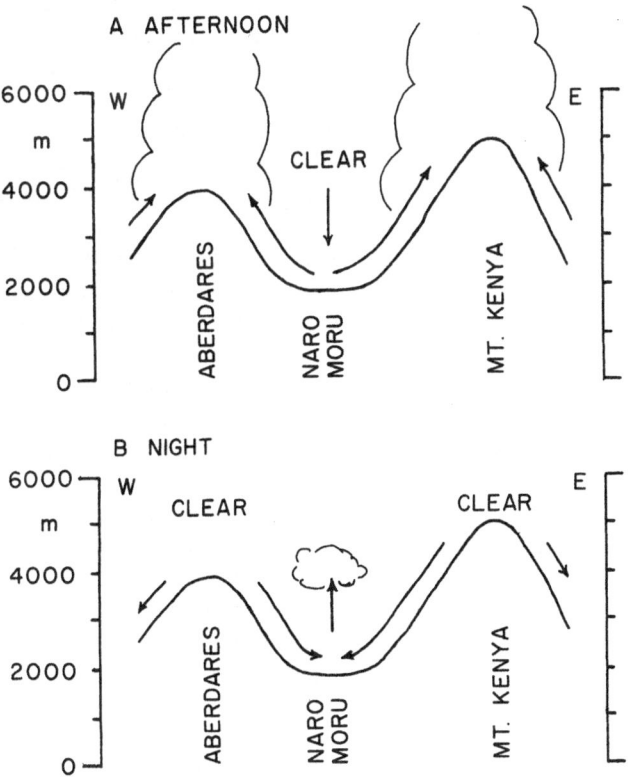

Fig. 2.3:3. Diurnal circulation and cloudiness patterns in the Mount Kenya area, as synthesized from numerous visual observations. (a) afternoon, (b) night. Vertical exaggeration about 1:25.

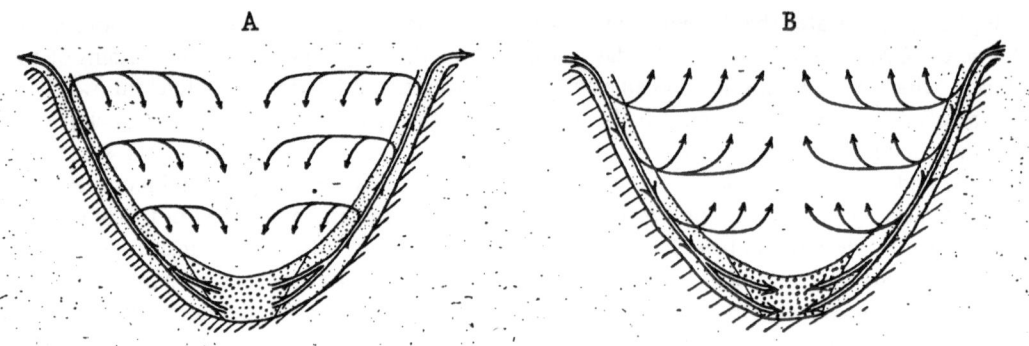

Fig. 2.3:4. Cross-circulations associated with local wind systems in broad valleys. (a) left, up-valley directed wind during daytime, featuring upslope motion on the valley sides and subsidence over the longitudinal valley axis. (b) right, down-valley directed wind at night, with cross-circulation opposite to daytime conditions. From Wagner (1932).

Clearing may set in an hour or two before nightfall – a circumstance of some interest to mountain climbers.

Mountain and valley winds tend to be particularly regular and vigorous in the high mountain regions of low latitudes. In addition to the flow along the longitudinal valley axis, cross circulations may arise. Thus Troll (1952) and Schweinfurth (1956) hypothesized for the daytime valley winds upslope motion at the valley walls and subsidence over the longitudinal axis of the valley, as illustrated in Fig. 2.3:4. The vegetation pattern in certain broad valleys of the Southern Himalayas and Eastern Andes suggests particularly arid conditions at the valley bottom, passing to a more humid small-scale environment at the valley walls. In view of the general thermal stability at night, the daytime circulation is considered to be of greater consequence on the small-scale climate and vegetation pattern.

Land- and sea-breeze circulations in the higher latitudes have a typical horizontal extent of 10–20 km, major controlling factors presumably being the Coriolis effect and the magnitude of horizontal contrasts in surface heating. In the tropics, however, diurnal wind reversals extend some 100 km inland. In many instances, the afternoon airstream may extend continuously from the ocean shore over a coastal mountain range to basins in the interior, as is the case at the Pacific coast of El Salvador or the North coast of Venezuela. The diurnal wind systems of such large proportions may not be sea breezes in the classical sense, but result from a combination of various local circulations.

That day-periodic mass exchanges between the atmospheric columns over land and sea and the associated variations in the wind field may have truly continental and oceanic proportions is indicated by the work of Wallace and Hartranft (1969), who demonstrated diurnal variations in the upper-air flow patterns organized in large gyres. This notion of the very large spatial scale of diurnal wind variations is further supported by Minnis and Harrison's (1984a, b, c) evaluation of satellite measurements for the month of November 1978, which suggests a large-scale diurnally modulated circulation between the Amazon basin and the adjacent oceans.

Thermally induced circulations between the air columns over the large tropical inland lakes and the surrounding land areas are particularly noteworthy for their intensity. Flohn and Fraedrich (1966), and Fraedrich (1968, 1971) studied the day- and nighttime circulations over Lake Victoria in East Africa. During the day, air flows out from over the lake towards the surrounding land areas and subsidence and clear skies prevail over the lake. At night, airflow from over the land converges

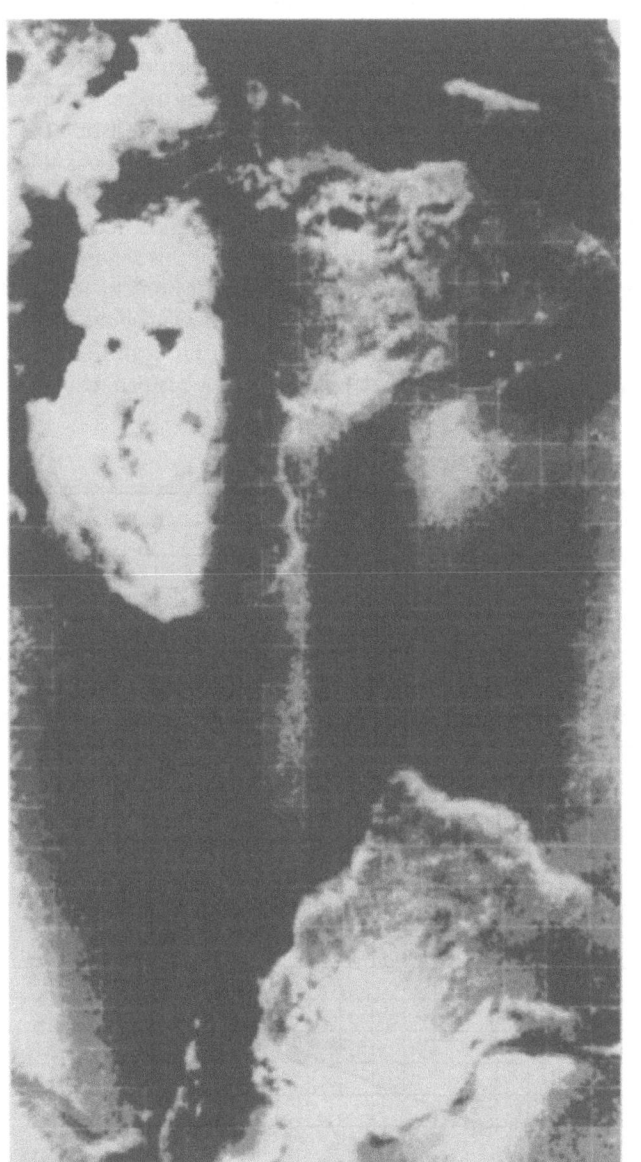

Fig. 2.3:5. Satellite-derived daytime cloudiness over Africa and South America, 1400 local time, 0–2 oktas, annual. Note that large inland lakes stand out as dark, such as Lake Victoria, East African Rift lakes, Lake Chad in Africa, and Lake Titicaca in South America. From U.S. Department of Commerce and U.S. Air Force (1971).

over the lake, resulting in abundant cloudiness and rainfall. Similar diurnal circulation systems are found at Lake Titicaca on the South American Altiplano (Kessler and Monheim, 1968) and probably also at Lake Chad and the Great Rift lakes of Eastern Africa. The chart of satellite-derived daytime cloudiness (Fig. 2.3:5) clearly demonstrates the scarcity of clouds over all of these large tropical inland lakes. The daily marches of cloudiness and rainfall associated with these local to meso-scale circulation systems are further considered in Section 2.4.

Less uniquely related to local wind systems is the vertical distribution of rainfall on tropical mountains. While in the higher latitudes precipitation generally increases upward to the peak regions of the mountains, a different altitudinal variation is characteristic of the tropics. In his classical treatise of Java, Junghuhn (1854, vol. 1, pp. 145–154, 166–174, 369–404, 471–488, 555–570) already noted that rainfall tends to increase from the lowland plains to some altitude zone on the mountain slopes, from which further upward precipitation decreases so that the higher regions typically remain dry. In more than a century since then, this general pattern of altitudinal zonation has been repeatedly confirmed for many mountain regions in low latitudes, a typical elevation of the belt of maximum precipitation being about 1000 m above the surrounding plains (review in Hastenrath, 1967). By way of example, Fig. 2.3:6 illustrates the vertical distribution of rainfall at the Pacific slope of the Guatemalan highlands. A major factor in the origin of the

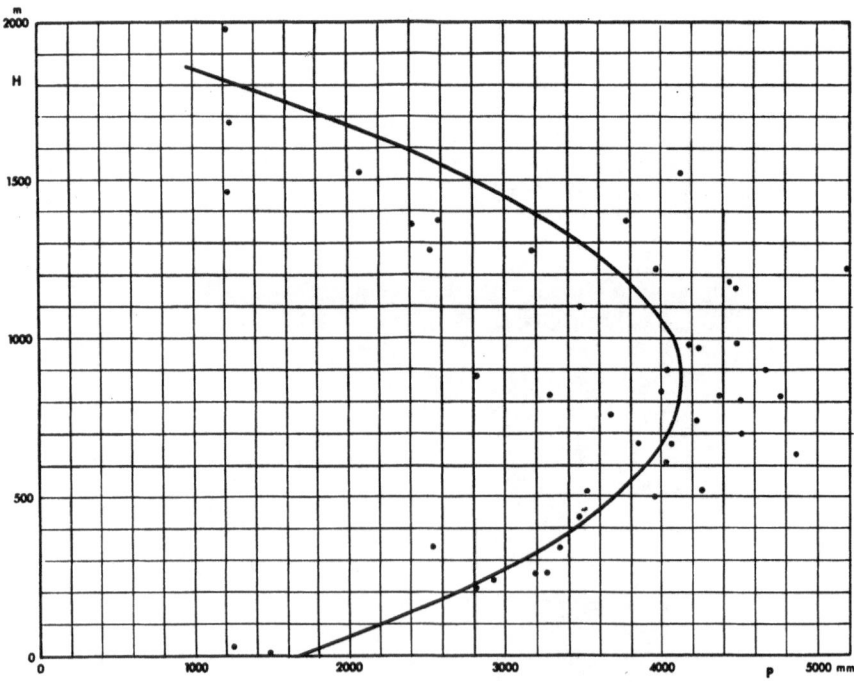

Fig. 2.3:6. Vertical distribution of annual rainfall at the Pacific coast of Guatemala, Central America. Dots denote station values and solid line is least-square fit parabola. From Hastenrath (1967).

altitudinal belt of maximum precipitation is the predominantly convective nature of tropical rainfall. The lowlands benefit less from orographic uplift and are affected by evaporation of rain below the cloud base. Proceeding beyond the belt of maximum precipitation, the moisture content of the atmospheric column at large and the precipitable water in the convective cloud in particular diminish upward, thus adversely affecting the precipitation potential of the higher mountain regions. By comparison, stratiform cloudiness enveloping the mountains as a whole, is associated with a spatially more even precipitation pattern, a situation prevalent in the higher latitudes. However, even in the European Alps, precipitation tends to decrease from an intermediate altitude band towards the peak regions during summertime in weather situations with predominantly convective cloudiness (Hastenrath, 1967). Conversely in the tropics the altitudinal belt of maximum rainfall tends to vanish during weather situations with predominantly stratiform cloudiness and persistent non-convective precipitation (Hastenrath, 1968). Weischet (1965) points out that the typically rather low elevation of the belt of maximum rainfall has unfortunate consequences for the generation of hydroelectric power in the tropics.

That mountainous terrain may entail large differences in precipitation and hence sharp climatic gradients over small distances is widely recognized. More surprising are the marked contrasts in rainfall and vegetation encountered at certain *flat* tropical coasts. This problem was first treated theoretically by Bryson and Kuhn (1961). A somewhat different approach is taken here. The vorticity equation can in an approximate form be written

$$\frac{d(\zeta + f)}{dt} + (\zeta + f)\nabla \cdot \mathbf{V} = \left(\frac{\partial F_y}{\partial x} - \frac{\partial F_x}{\partial y} \right), \tag{2.3:1}$$

where ζ is relative vorticity, f Coriolis parameter, $\nabla \cdot \mathbf{V}$ divergence of horizontal wind, and F_x and F_y the zonal and meridional components of the frictional acceleration. While the solenoidal and twisting terms are commonly neglected, the frictional term (right-hand side of Eq. 2.3:1) is here retained because application is specifically intended for the surface layer. Moreover, disregard the first left hand term. The simplified vorticity equation consisting of the two remaining terms is strictly valid for barotropic conditions without vertical shear and with absolute vorticity invariant following the motion. More generally, we can consider the effect of the right-hand term on the second left-hand term, noting that f largely determines the sign of absolute vorticity ζ_a and that ζ_a is essentially positive in the Northern and negative in the Southern hemisphere. Some examples are as follows.

Consider easterly flow along a zonally oriented coast in the Northern hemisphere. The West to East directed component of the frictional acceleration F_x decreases northward, F_y is zero, and ζ_a positive, so that sign convention calls for divergence, and because of mass continuity for subsidence. This is the situation along the dry North coast of Venezuela during much of the year. The cold surface waters along the North coast of South America (Hastenrath and Lamb, 1977) are not to be regarded as primary factor of the aridity, but are themselves the result of mechanical forcing by the lower atmosphere. Further details on the dynamic climatology of the Caribbean coast of South America are found in Lahey (1958). As another example consider northerly flow along a meridionally oriented coast in the Northern hemisphere. Sign convention requires convergence, and by implication ascending motion. This is the case at the Nicaraguan East coast in winter, when abundant rain falls while dry season conditions prevail in much of Central America and the Caribbean (Hastenrath, 1967). Bryson and Kuhn (1961) offer various other examples from both hemispheres. Note from Eq. (2.3:1) that the stress-differential induced divergence effect is scaled according to ζ_a, so that effects would be largest in the low latitudes.

2.4. Diurnal Marches of Sea-Air Exchange, Cloudiness, and Precipitation

The thermal forcings and resultant diurnal circulations on the local to meso-scale, as described in Sections 2.1 and 2.3 provide a powerful control for the diurnal marches of vertical motion, cloudiness, and precipitation. Of further interest are the oscillations in wind speed and vergence associated with atmospheric tides (Section 2.2), as well as variations in sensible and latent heat flux at the surface that may affect stability and fuel convection. For details of cloud physics and dynamics related to tropical convection refer to Krishnamurti (1979, pp. 89–142).

Fig. 2.4:1 summarizes the diurnal cycles of pertinent atmospheric processes over the open ocean. Roll (1965, pp. 238, 368) has evaluated the observations of the Meteor Expedition (Kuhlbrodt and Reger, 1938) and other data. For sea surface temperature T_o, the diurnal range is less than 0.3°C, with a minimum at 4 and a maximum at 15 h local time. The measured diurnal range of air temperature T_a is slightly larger, with a similar timing of the extrema. The resultant sea minus air temperature difference $(T_o - T_a)$ has a range of a few tenths of degree centigrade (Fig. 2.4:1, part a). $(T_o - T_a)$ is largest positive (or smallest negative) in the late night to early morning hours, and largest negative (or smallest positive) around local noon; details depending on the sign of the daily average difference $(T_o - T_a)$ in a given region. The plot of $(T_o - T_a)$ in Fig. 2.4:1, part a, is thus consistent with various observational studies (Roll, 1965, pp. 238, 368). It should be noted, however, that ship heating may bias measurements of T_a and thus make estimates of $(T_o - T_a)$ spurious. This would further affect inferences based on this temperature difference. Thus, $(T_o - T_a)$ is related to the stability of the surface layer. Moreover, in terms of the bulk-aerodynamic equations (Roll, 1965, p. 252), the difference $(T_o - T_a)$ and the wind speed are directly proportional to the sensible heat flux across the sea-air interface. In an easterly wind regime, as in the trade wind regions, the solar semi-diurnal tidal wind variations tend to combine with the basic flow so as to produce wind speed maxima around 10 and 22 and minima around 4 and 16 h local time (Figs. 2.2:4 and 2.4:1, part d). However, a considerable diversity of diurnal wind speed variations can be found over various regions of the tropical oceans. The variations of wind speed and of the interface temperature difference $(T_o - T_a)$ would produce variations of sensible heat flux Q_s. Such resultant variations are illustrated in Fig. 2.4:1, part b, but this curve may be compromised by the aforementioned shortcomings in estimating $(T_o - T_a)$. Concerning the latent heat flux Q_e and evaporation, Hoeber (1969) finds in the equatorial Atlantic a semi-diurnal variation due to the semi-diurnal variation of wind speed. In addition to the cycle of stability in the surface layer, the small semi-diurnal tidal variation of convergence (Figs. 2.2:4 and 2.4:1, part e) has been considered as a factor in producing the diurnal marches of cloudiness and rainfall over sea (Malkus, 1964; Brier and Simpson, 1969).

A different approach to the problem of diurnal march of the oceanic heat budget was taken by Delinore (1972), but limited to a 12-day period during the 1969 Barbados Oceanographic and Meteorological Experiment (BOMEX). Combining subsurface temperature soundings for the top 50 m of the ocean with surface radiation measurements, he calculated the latent and sensible heat fluxes as residual. From this method he obtained largest evaporation rates around local sunset, and smallest values around midnight and during midmorning, but notes a disagreement with results of the bulk-aerodynamic approach.

For the open ocean, various observational studies (Kuhlbrodt and Reger, 1938; reviews of later work in Malkus, 1964, and Hastenrath, 1976) have examined the diurnal march of rainfall. There is general agreement regarding the main maximum in the late night to early morning hours. A minimum in the middle of the day and a secondary maximum in the evening (Fig. 2.4:1, part c)

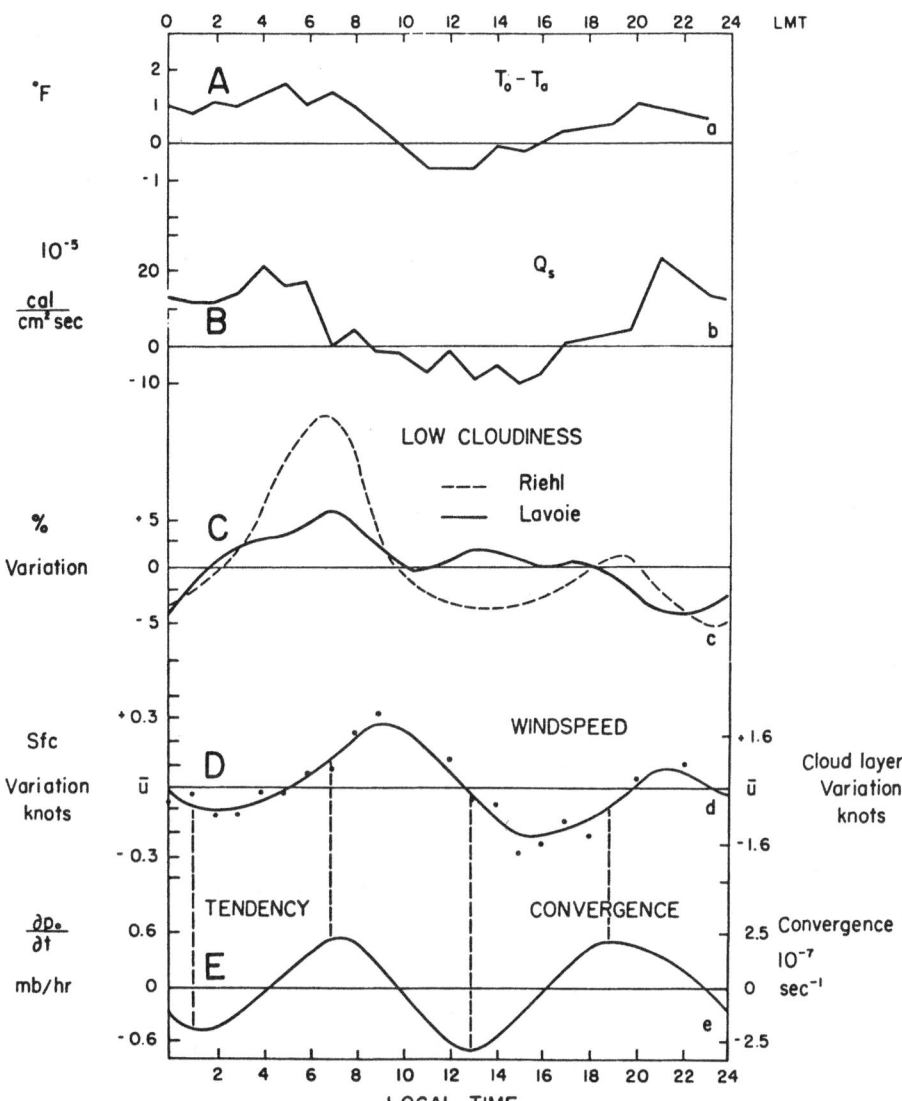

Fig. 2.4:1. Diurnal cycles as a function of local time. From top to bottom: (a) temperature difference sea (T_O) minus air (T_a) in °F; (b) sensible heat transfer from sea to air, in 10^{-5} cal cm^{-2} s^{-1}; (c) low cloudiness, variation about the mean in percent; (d) wind speed variation in knots; (e) pressure tendency in mb hr^{-1} and convergence in 10^{-7} s^{-1}. From Malkus (1964).

appear in some but not all data sets and are therefore observationally less well established. Concerning the prominent night maximum of precipitation over the mid-latitude oceans, Kraus (1963) emphasizes the effects of absorbed solar radiation which through evaporation would reduce liquid water content and cloud thickness. In addition Malkus (1964) calls attention to the enhanced sea minus air temperature difference ($T_O - T_a$) and sensible heat transfer across the sea surface Q_s during the late night to early morning hours and the semi-diurnal tidal vergence variations (Fig. 2.4:1, part e) as possible factors in the diurnal cycle of cloudiness over sea, although the wind vergence departures associated with the solar semi-diurnal tide are recognized to be small. Gray and Jacobson (1977) and McBride and Gray (1980) also note the prevalence of a morning maximum of rainfall in various regions of the tropical oceans, and hypothesize that the diurnal cycle in deep convection with a morning maximum is associated with organized weather disturbances. In particular, they suggest that the upper layered clouds of organized weather systems reduce nighttime infrared cooling in the layers underneath, as compared to cloud-free areas; by contrast during the day solar energy acts to increase the temperature of the clear regions throughout the troposphere, while in the disturbances temperature increase is primarily confined to the upper cloud decks. These radiative-convective heating contrasts between disturbances and their surroundings are thought to affect pressure gradients and hence cause large diurnal mass convergence differences. In contrast to Gray and Jacobson's (1977) results, Augustine (1984) finds for August 1979 over part of the tropical Pacific inferred rainfall maxima near dawn and in midafternoon. Beyond the above considerations of potential contributing factors, the causes of the diurnal marches of cloudiness and rainfall over the open ocean are not conclusively established.

Over the tropical continents, the diurnal marches of cloudiness and precipitation are dominated by the circulations on the local to meso-scale discussed in Section 2.3. In analogy to the oceans an afternoon convective maximum of cloudiness and rainfall appears plausible, and is in fact found over extensive tropical land areas, such as the Western Ghats of India (Iyer and Dass, 1946), Northern Ghana (Gbeckor-Kove and Dankwa, 1966), and the interior of Southern Africa (review in Hastenrath, 1970). However, there is a puzzling diversity of diurnal regimes, presumably as a result of diurnal vertical motion patterns related to compensating mass exchanges (Section 2.3). Only some examples but no exhaustive review are presented in the following.

An afternoon maximum of cloudiness and rainfall is particularly well developed in the higher portions of tropical mountains, as explained in Section 2.3. This leads to an azimuth asymmetry of the surface heat budget and consequently also of other environmental conditions. Refer to Fig. 2.4:2. In the predominantly cloud-free morning, the eastward facing slopes receive abundant insolation. By contrast, the westward facing slopes depend on the direct solar radiation in the afternoon, when the azimuth of the sun is appropriate but the direct rays are blocked by the then common cloud cover (Section 2.3). Among the more striking consequences of these effects of the local circulations is the characteristic zonal asymmetry in the extent of glaciers (Section 11.1). Thus on the Kibo cone of Kilimanjaro in East Africa (Hastenrath, 1984, pp. 63–92), the eastern crater rim is now free of ice, while glaciers extend to much lower elevations on the western slopes. Fossil moraine morphology indicates that these West-East contrasts in ice distribution, that are caused by local circulations, already existed in Pleistocene times (Hastenrath, 1984, pp. 50–53). In the vicinity of the Equator, these zonal asymmetries can be more pronounced than the more straightforward North-South contrasts. Similar azimuthal asymmetries in the ice distribution and other environmental features can be recognized on Mount Kenya, the Mexican volcanoes, and many other mountains of the tropics (Hastenrath, 1981, pp. 52–53; 1984, pp. 293–296). In regions with near perennial cloud cover and weak local circulations, these zonal asymmetries in ice distribution fade out.

The interaction of land breeze and large-scale flow producing a coastal convergence in the latter part of the night has been discussed in Section 2.3, with reference to the studies of Leopold (1949) for the Hawaiian Islands, and of Kousky (1980) for the East coast of Brazil. The largest cloudiness and rainfall in the latter part of the night at the Indian Ocean coast of Southern Africa appear to be related to a similar mechanism (Monteiro Correia, 1970; Preston-Whyte, 1970).

As discussed in Section 2.3, lower-tropospheric convergence during the night is characteristic of the large inland lakes of the tropics. This is reflected in the nighttime maxima of cloudiness and

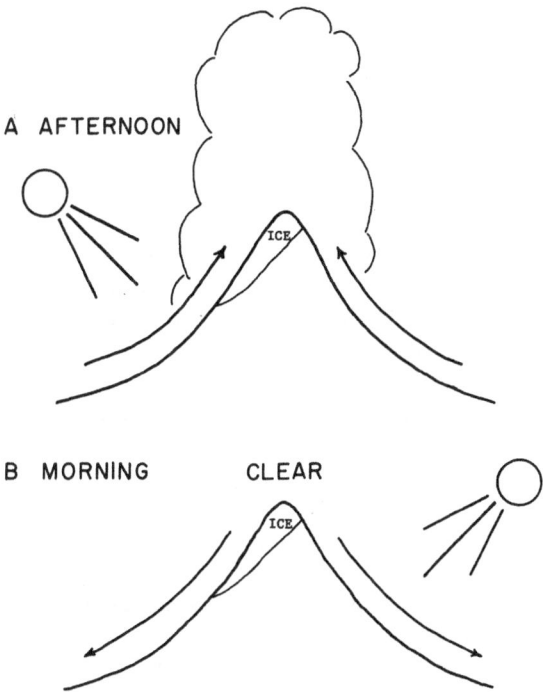

Fig. 2.4:2. Scheme of diurnal circulations, cloudiness, and zonal asymmetries on tropical mountains, as synthesized from numerous visual observations in the high mountains of Eastern Africa and South America. (a) afternoon, (b) morning.

rainfall over Lake Victoria (Flohn and Fraedrich, 1966; Fraedrich, 1968, 1971), the lakes of the Great Rift of Eastern Africa, Lake Chad, as well as over Lake Titicaca on the South American Altiplano (Kessler and Monheim, 1968). Lower-tropospheric divergence, subsidence, clearing of skies, and absence of rain is typical of the daytime conditions, as is also strikingly apparent on satellite imagery (Fig. 2.3:5). Passing from the nighttime cloudiness-rainfall maximum over the lakes towards the daytime peak characteristic of extensive land areas, a vast variety of transitional regimes are found (Thompson, 1957, 1968; Hastenrath, 1970).

Departures from the conventional picture of afternoon convective rainfall can encompass extensive land areas without pronounced topographic relief. Delorme (1963) documents for Subsaharan West Africa a prevalence of late night to early morning rainfall. Pedgley (1969) finds for the Sudan an afternoon to evening rainfall maximum only at places distant from the Ethiopian highlands, while elsewhere precipitation is more evenly distributed throughout the day, and an early morning maximum occurs widely. An interaction on the scale of hundreds of km and on a daily basis is suggested between the lowlands of the Sudan in the North and the Ethiopian highlands to the South.

For the land areas surrounding the Red Sea, Flohn (1965) recognizes a veritable mosaic of diurnal rainfall regimes, all related to circulation systems on the local to meso-scale. Watts (1955) and Ramage (1964) call attention to the great diversity of wind systems and diurnal rainfall regimes in Malaya. Similarly complex patterns, while as a rule not analyzed, are found in regions of complicated coastal configuration and mountain topography in other parts of the tropics.

2.5. Synthesis

In the tropics, day-periodic processes and local to meso-scale circulations are much more vigorous than in the higher latitudes. As a result of radiation geometry, the amplitude of the diurnal cycle of insolation and temperature exceeds the annual cycle by about an order of magnitude. Of the tidal forcings the solar semi-diurnal tide is most important. Associated with the prominent 12-hourly pressure oscillation are periodic variations of wind speed and vergence, that affect the diurnal marches of sea-air exchange and cloudiness.

Over the open ocean, the cloudiness-rainfall maximum occurs commonly in the late night to early morning hours. Over the tropical land areas, land- and sea-breeze and mountain circulations dominate the diurnal march of cloudiness, rainfall, and weather. Anabatic flows lead to an afternoon cloudiness-precipitation maximum on the mountains, with compensating subsidence and clearing over the basins. Rainfall in the basins occurs with some preference at night. At the coasts, a land breeze may interact with the large-scale flow to produce a convergence and rainfall maximum in the latter part of the night. Large inland lakes experience lower-tropospheric convergence, cloud cover and rainfall during the night, and subsidence and clear skies during the day. Diurnal mass exchanges between highlands and adjacent plains on the scale of hundreds of km may be responsible for the suppression of daytime convection and the origin of a nighttime cloudiness-rainfall maximum over the lowlands.

The marked diurnal and local controls in the tropics, as sketched in this chapter, must be considered as important factors for the regional climates and the large-scale circulation. Thus, in many tropical regions, the annual or seasonal rainfall distribution can be appreciated only from the interaction between diurnal factors, local circulations, and the large-scale flow; the resultant spatial pattern of latent heating is in turn essential in the driving of the circulation on a planetary scale. The realism of numerical simulation experiments may well depend on the proper formulation of the diurnal cycle. The long-term and large-scale implications of diurnal and local forcings in the tropics have received remarkably little attention, and the pertinent literature references are accordingly limited. However, day-periodic processes are expected to be relevant to various subjects discussed later on in this book, in particular to the quasi-permanent large-scale circulation (Chapter 6), weather systems (Chapter 7), interannual climate variability (Chapter 8), as well as to inadvertent or intentional climate modification (Chapter 10).

References

Augustine, J. A., 1984: 'The diurnal variation of large-scale inferred rainfall over the tropical Pacific Ocean during August 1979'. *Mon. Wea. Rev.*, 112, 1745–1751.

Bartels, J., 1928: 'Gezeitenschwingungen der Atmosphäre'. *Handbuch der Experimentalphysik*, 25 (Geophysik), 163–210.

Bartels, J., 1932: 'Tides in the atmosphere'. *Scientific Monthly*, 35, 110–130.

Brier, G. W., Simpson, J., 1969: 'Tropical cloudiness and rainfall related to pressure and tidal variations'. *Quart. J. Roy. Meteor. Soc.*, 95, 120–147.

Bryson, R. A., Kuhn, P. M., 1961: 'Stress-differential induced divergence with application to littoral precipitation'. *Erdkunde*, 15, 187–294.

Budyko, M. I., 1974: *Climate and life*. Academic Press, New York, San Francisco, London, 508 pp.

Chapman, S., Lindzen, R. S., 1970: *Atmospheric tides, thermal and gravitational*. Gordon and Breach, New York, 200 pp.

Defant, A., Defant, F., 1958: *Physikalische Dynamik der Atmosphäre*. Akademische Verlagsgesellschaft, Frankfurt, 527 pp.

Delinore, V. E., 1972: 'Diurnal variation of temperature and energy budget for the oceanic mixed layer during BOMEX'. *J. Phys. Oceanogr.*, 2, 239–247.

Delorme, G. A., 1963: 'Repartition et durée des precipitations en Afrique Occidentale'. Monographies de la Météorologie Nationale, No. 28, Paris, 27 pp.

Flohn, H., 1965: 'Klimaprobleme am Roten Meer'. Erdkunde, 19, 179–191.

Flohn, H., Fraedrich, K., 1966: 'Tagesperiodische Zirkulation und Niederschlagsverteilung am Viktoria-See'. Meteorol. Rundschau, 19, 157–165.

Fraedrich, K., 1968: 'Das Land und Seewindsystem des Viktoria-Sees nach aerologischen Daten'. Archiv Meteor. Geophys. Bioklim., Ser. A., 17, 186–206.

Fraedrich, K., 1971: 'Modell einer lokalen atmosphärischen Zirkulation mit Anwendung auf den Viktoria-See'. Beiträge zur Physik der Atmosphäre, 44, 95–114.

Gbeckor-Kove, N. A., Dankwa, J. B., 1966: 'Diurnal variation of rainfall in Ghana'. Ghana Meteorological Services Department, Departmental Note No. 15, Accra 1966, 17 pp.

Gray, W. M., Jacobson, R. W., 1977: 'Diurnal variation of deep cumulus convection'. Mon. Wea. Rev., 105, 1171–1188.

Hastenrath, S., 1967: 'Rainfall distribution and regime in Central America'. Archiv Meteor. Geophys. Bioklim., Ser. B, 15, 201–241.

Hastenrath, S., 1968: 'Zur Vertikalverteilung des Niederschlags in den Tropen'. Meteorol. Rundschau, 21, 113–116.

Hastenrath, S., 1970: 'Diurnal variation of rainfall over Southern Africa'. Notos, 19, 85–94.

Hastenrath, S., 1972: 'Daily wind, pressure, and temperature variation up to 30 km over the tropical Western Pacific'. Quart. J. Roy. Meteor. Soc., 98, 48–59.

Hastenrath, S., 1976: 'Daily variation of the atmospheric mass and moisture budget over the tropical Western Pacific'. J. Meteor. Soc. Japan, 54, 226–232.

Hastenrath, S., 1981: The glaciation of the Ecuadorian Andes. Balkema, Rotterdam, 159 pp.

Hastenrath, S., 1984: The glaciers of equatorial East Africa. Reidel, Dordrecht, Boston, Lancaster, 353 pp.

Hastenrath, S., Lamb, P. J., 1977: Climatic atlas of the tropical Atlantic and Eastern Pacific Oceans. University of Wisconsin Press, 112 pp.

Hoeber, H., 1969: 'Wind-, Temperatur- und Feuchteprofile in der wassernahen Luftschicht über dem äquatorialen Atlantik'. "Meteor" Forschungsergebnisse, Series B, No. 3, p. 1–26, Borntraeger, Berlin, Stuttgart.

Iyer, V. D., Dass, L., 1946: 'Diurnal variation of rainfall at Mahalabaleshwar'. India Meteor. Dept. Sci. Notes, 9, No. 105, 33–36.

Junghuhn, F., 1854: Java, zijne gedaante, zijn plantentooi, en inwendige bouw. C. W. Mieling, 's Gravenhage, second edition, 4 vols.: 671, 538, 914, 498 pp.

Kessler, A., Monheim, F., 1968: 'Der Wasserhaushalt des Titicacasees nach neueren Messergebnissen'. Erdkunde, 22, 275–281.

Kousky, V. E., 1980: 'Diurnal rainfall variation in Northeast Brazil'. Mon. Wea. Rev., 108, 488–498.

Kraus, E. B., 1963: 'The diurnal precipitation change over sea'. J. Atmos. Sci., 20, 551–556.

Krishnamurti, T. N., 1979: Tropical meteorology. Compendium of meteorology, vol. 2, part 4, editor A. Wiin-Nielsen, WMO No. 364, World Meteorological Organization, Geneva, 428 pp.

Kuhlbrodt, E., Reger, J., 1938: 'Die meteorologischen Beobachtungen; Methoden, Beobachtungsmaterial und Ergebnisse'. Wissenschaftliche Ergebnisse der Deutschen Atlantischen Expedition "Meteor", 1925–27, vol. 14, Parts A and B, Walter de Gruyter and Co., Berlin-Leipzig, 392 and 212 pp.

Lahey, J. F., 1958: 'On the origin of the dry climate in Northern South America and the Southern Caribbean'. University of Wisconsin, Department of Meteorology, Scientific Report No. 10, 290 pp.

Lauer, W., 1975: 'Vom Wesen der Tropen'. Akademie der Wissenschaften und der Literatur Mainz, Abhandlungen der Mathematisch-Naturwissenschaftlichen Klasse, No. 3, 52 pp. plus 20 plates.

Leopold, L. B., 1949: 'The interaction of trade wind and sea breeze, Hawaii'. J. Meteor., 6, 312–320.

Leroux, M., 1983: Le climat de l'Afrique tropicale. Champion, Paris, 2 vols., 633 pp. and 24 pp. plus 250 maps.

Malkus, J. S., 1955: 'The effect of a large island upon the trade-wind airstream'. Quart. J. Roy. Meteor. Soc., 81, 538–550.

Malkus, J. S., 1964: 'Tropical convection: progress and outlook'. pp. 247–277 in Proceedings of the WMO-IUGG Symposium on Tropical Meteorology, 5–13 Nov. 1963, Rotorua, New Zealand, Wellington, New Zealand, 737 pp.

McBride, J. L., Gray, W. M., 1980: 'Mass divergence in tropical weather systems, paper 1: diurnal variation'. Quart. J. Roy. Meteor. Soc., 106, 501–516.

Minnis, P., Harrison, E. F., 1984a: 'Diurnal variability of regional cloud and clear-sky radiative parameters derived from GOES data. Part 1: Analysis method'. J. Climate Appl. Meteor., 23, 993–1011.

Minnis, P., Harrison, E. F., 1984b: 'Diurnal variability of regional cloud and clear-sky radiative parameters derived from GOES data. Part 2: November 1978 cloud distributions'. *J. Climate Appl. Meteor.*, **23**, 1012–1031.

Minnis, P., Harrison, E. F., 1984c: 'Diurnal variability of regional cloud and clear-sky radiative parameters derived from GOES data. Part 3: November 1978 radiative parameters'. *J. Climate Appl. Meteor.*, **23**, 1032–1051.

Mitra, S. K., 1952: *The upper atmosphere*. The Asiatic Society, Calcutta, 713 pp.

Monteiro Correia, M., 1970: 'Variação diurna da precipitação na Beira'. Serviço Meteorológico de Moçambique, Lourenço Marques, SMM 67, MEM 62, 10 pp.

Pedgley, D. E., 1969: 'Diurnal variation of the incidence of monsoon rainfall over the Sudan, Parts I and II'. *Meteor. Mag.*, **98**, 97–107, and 129–134.

Preston-Whyte, R. A., 1970: 'Land breezes and rainfall on the Natal coast'. *South African Geogr. J.*, **52**, 38–43.

Ramage, C. S., 1964: 'Diurnal variation of summer rainfall over Malaya'. *J. Trop. Geogr.*, **19**, 62–68.

Ramage, C. S., 1978: 'Effect of the Hawaiian Islands on the trade winds'. pp. 62–67, in: Preprint volume of conference on climate and energy: climatological aspects and industrial operations, May 1978, Asheville, N.C., Amer. Meteor. Soc., Boston, Mass.

Roll, H. A., 1965: *Physics of the marine atmosphere*. Academic Press, New York and London, 426 pp.

Schweinfurth, U., 1956: 'Über klimatische Trockentäler im Himalaya. Erdkunde', **10**, 297–302.

Sellers, W. D., 1965: *Physical climatology*. University of Chicago Press, 272 pp.

Thompson, B. W., 1957: 'The diurnal variation of precipitation in British East Africa'. East African Meteorological Department, Technical Memorandum No. 8, Nairobi, 70 pp.

Thompson, B. W., 1968: 'Tables showing the diurnal variation of precipitation in East Africa and Seychelles'. East African Meteorological Department, Technical Memorandum No. 10, Nairobi, 49 pp.

Troll, C., 1952: 'Die Lokalwinde der Tropengebirge und ihr Einfluss auf Niederschlag und Vegetation'. *Bonner Geogr. Abhandl.*, **9**, 124–182.

U.S. Department of Commerce and U.S. Air Force, 1971: *Global atlas of relative cloud cover 1967–70*. Washington, D.C., 237 pp.

Wagner, A., 1932: 'Neue Theorie des Berg- und Talwindes'. *Meteorol. Zeitschrift*, **49**, 329–341.

Wallace, J. M., Hartranft, F. R., 1969: 'Diurnal wind variations surface to 30 km'. *Mon. Wea. Rev.*, **97**, 446–459.

Watts, I. E. M., 1955: 'Rainfall of Singapore Island'. *J. Trop. Geogr.*, **7**, 71 pp.

Weischet, W., 1965: 'Der tropisch-konvektive und der aussertropisch-advektive Typ der vertikalen Niederschlagsverteilung'. *Erdkunde*, **19**, 6–14.

CHAPTER 3

PLANETARY SCALE ATMOSPHERIC CIRCULATION

Atmospheric circulation characteristics on a planetary scale can largely, but not completely be described by the distribution of properties in meridional-vertical cross-sections. Thus the distribution of temperature and humidity is discussed in Section 3.1. The temperature and humidity distribution is, through the thickness pattern and the slope of constant pressure surfaces, related to the zonal wind regime, as reviewed in Section 3.2. The mean meridional circulation is likewise considered in Section 3.3. By contrast, the maintenance of the circulation in terms of angular momentum and kinetic energy, as discussed in Sections 3.4 and 3.5, involves eddy mechanisms and cannot be accounted for by mean property distributions in meridional-vertical planes. Since kinetic energy amounts only to a small fraction of the total energy content, it is here considered separate from the budgets of heat and moisture addressed in Chapter 5.

3.1. Mean Meridional Distribution of Temperature and Humidity

Latitude-mean temperature is shown in Figs. 3.1:1a and b by meridional-vertical cross-sections for the extreme seasons. Throughout the year tropospheric temperature is horizontally rather uniform within the tropics, with poleward temperature decrease concentrated in the midlatitudes. The inverse temperature gradients are characteristic of the stratosphere. A temperature minimum reflects the tropical tropopause near 100 mb.

During the Northern summer (Fig. 3.1:1a), the zone of highest lower-tropospheric temperature is found well to the North of the Equator and meridional temperature gradients in the Northern hemisphere are comparatively slack. During Southern summer (Fig. 3.1:1b), the maximum of lower-tropospheric temperature is located near the Equator, the meridional gradient in the Northern hemisphere midlatitudes is especially steep, while the Southern hemisphere extratropical cap shows meridional contrasts only moderately weaker than in winter.

Specific humidity is depicted in Figs. 3.1:2a and b by meridional-vertical cross-sections analogous to Figs. 3.1:1a and b. Annual mean conditions are also described in Peixoto and Oort (1984). The pattern of specific humidity appears largely determined by the temperature distribution. Thus the largest values are found in the lower troposphere of the equatorial zone, with a drastic decrease upward and towards the higher latitudes of either hemisphere. Furthermore, the lower-tropospheric maximum of specific humidity is found well in the Northern hemisphere during Northern summer (Fig. 3.1:2a), while seasonal differences are less pronounced in the extratropical cap of the Southern hemisphere. The pattern of relative humidity, not depicted here, is characterized by mid-tropospheric minima in the subtropics (Lorenz, 1967, p. 43). These may reflect the prevailing subsidence (ref. Section 3.3).

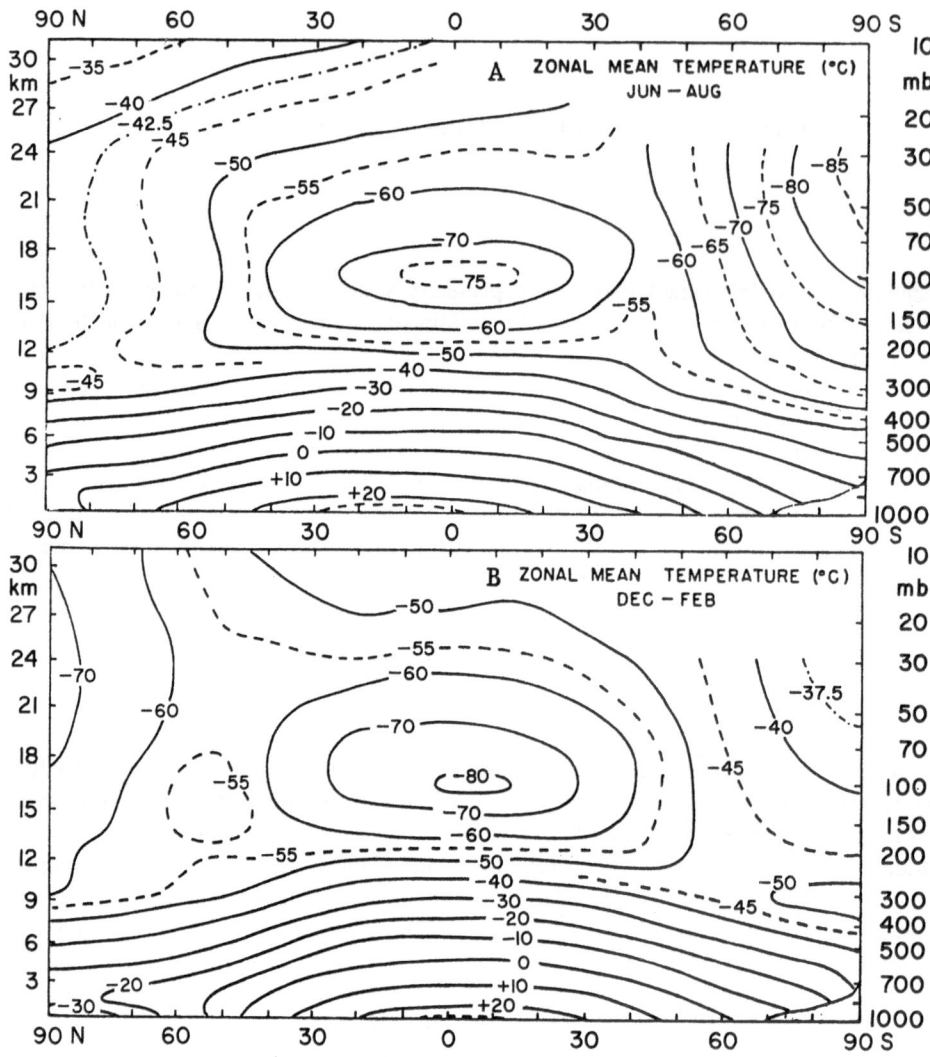

Fig. 3.1:1. Meridional-vertical cross-sections of temperature for (a) June–July–August, (b) December–January–February. Source: Newell *et al.*, 1972, 1974, vol. 1. (*General circulation of the tropical atmosphere*, MIT Press, copyright 1972 by the Massachusetts Institute of Technology.)

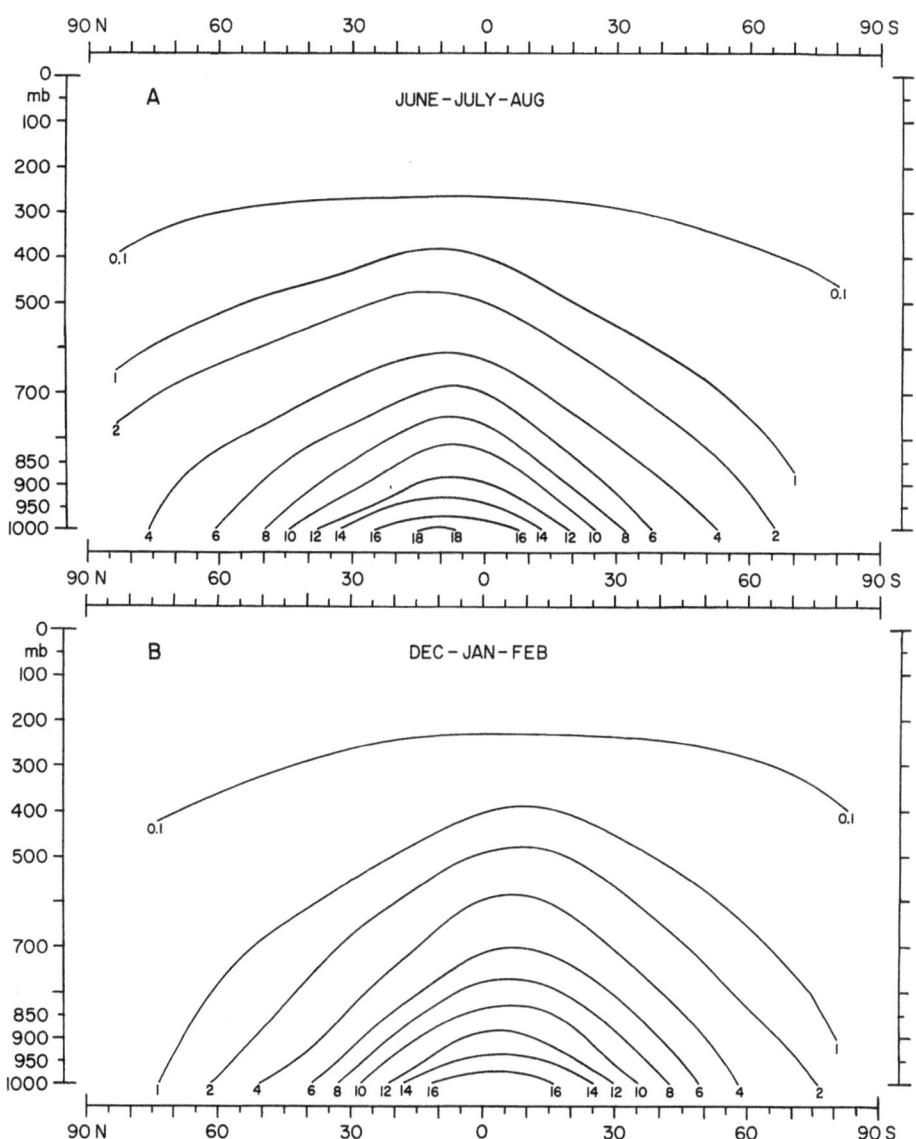

Fig. 3.1:2. Meridional-vertical cross-sections of specific humidity for (a) June–July–August, (b) December–January–February. Source: Oort, 1983.

3.2. Zonal Wind Regime

The temperature distribution as reviewed in Section 3.1 provides — by virtue of the thickness pattern and the thermal wind relationship — a major control for the zonal wind regime. This is illustrated by meridional-vertical cross-sections for the extreme seasons in Figs. 3.2:1a and b. Annual mean conditions are also described in Peixoto and Oort (1984).

The meridional profile of surface pressure characterized by ridges in the subtropics is commensurate with lower-tropospheric easterlies in the tropics and westerlies in midlatitudes. Consistent with the meridional temperature gradient and its hydrostatic implications, the boundary between temperate-latitude westerlies and the tropical easterlies slopes upward towards the lower latitudes. Strongest westerlies occupy broadly the zones of most pronounced meridional temperature contrasts.

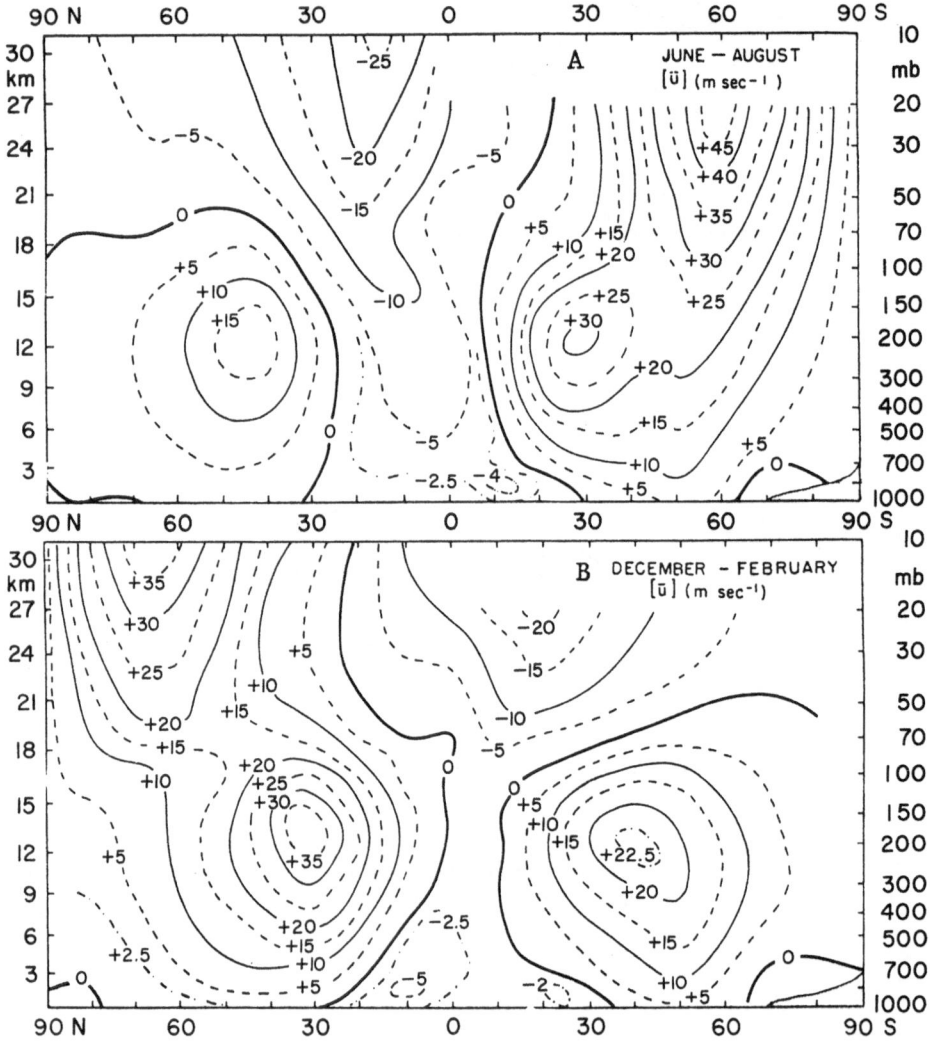

Fig. 3.2:1. Meridional-vertical cross-sections of zonal wind speed for (a) June–July–August, (b) December–January–February. From Newell *et al.*, 1972, 1974, vol. 1. (*General circulation of the tropical atmosphere*, MIT Press, copyright 1972 by the Massachusetts Institute of Technology.)

Seasonal differences in the zonal wind regime are related to changes in the thickness pattern. During the respective winter seasons, the temperate-latitude westerlies are strongest and in the upper troposphere they extend deep into the tropics. By contrast, during summer the anticyclonic axes are steep and in the tropics easterlies extend from the surface into the upper troposphere. It should be noted that zonal averaging (Fig. 3.2:1) obliterates longitudinal contrasts, such as the occurrence of monsoonal westerlies over the Northern Indian Ocean in Northern summer, and of near-equatorial westerlies over the eastern Pacific and Atlantic, as compared to the prevalence of easterlies over the central and western parts of these oceans (Section 6.1).

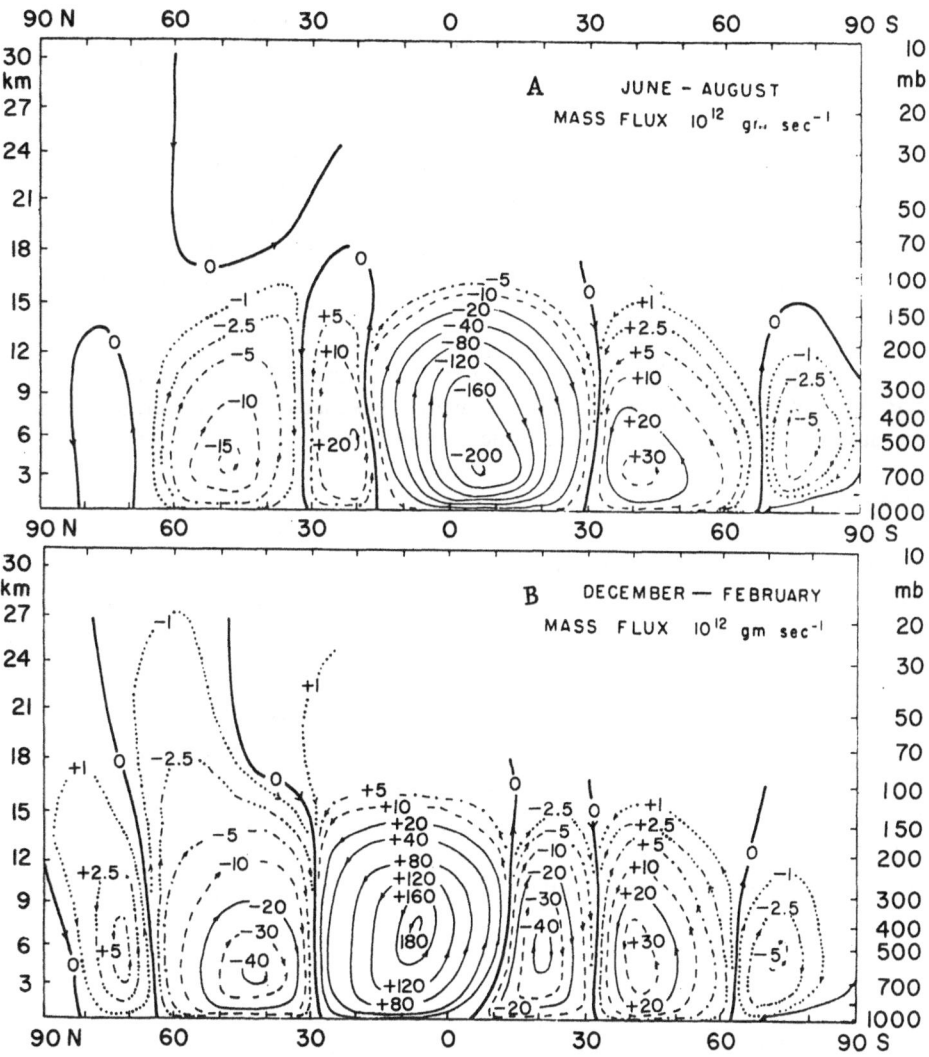

Fig. 3.3:1. Meridional-vertical cross-sections of mean meridional circulation for (a) June–July–August, (b) December–January–February. From Newell *et al.*, 1972, 1974, vol. 1. (*General circulation of the tropical atmosphere*, MIT Press, copyright 1972 by the Massachusetts Institute of Technology.)

3.3. Mean Meridional Circulation

The latitude-mean meridional wind is ageostrophic in contrast to the predominantly geostrophic mean zonal circulation. Furthermore, the meridional is much smaller than the zonal wind component, and estimates of latitude-mean values are fraught with considerable relative error. Unlike the zonal component, however, a useful constraint is available in that for reasons of continuity the total vertically integrated mass flux across a latitude circle must vanish in the long-term annual mean. This postulate appears also plausible for the extreme seasons, although to a lesser extent for the transition seasons. Observed winds are to be adjusted accordingly. From meridional-vertical transects of the meridional wind analogous to those of the zonal component shown in Figs. 3.2:1a and b, it is then possible to construct cross-sections of the circulation in the mean meridional-vertical plane, such as presented in Figs. 3.3:1a and b for the extreme seasons.

Figs. 3.3:1a and b illustrate the major mean meridional circulation cells during the extreme seasons. Annual mean conditions are also described in Peixoto and Oort (1984). In the tropics, two cells are apparent, with ascending motion in the warmer equatorial region and subsidence at their poleward extremities. These are the thermally direct Hadley cells. The middle latitudes of either hemisphere are occupied by cells with subsidence on their equatorward and ascending motion on their poleward sides. These are the thermally indirect Ferrel cells.

During the Northern summer (Fig. 3.3:1a), the Northern hemisphere Hadley cell is comparatively weak, is confined to a narrow latitude band, and possesses a total mass circulation of only about 25×10^9 kg s^{-1}. At the same time, the Southern hemisphere Hadley cell is most vigorously developed: its mass circulation amounts to about 200×10^9 kg s^{-1} and it extends well into the Northern hemisphere.

During Northern winter (Fig. 3.3:1b), the Northern hemisphere Hadley cell reaches an intensity about an order of magnitude larger than in summer, it is situated further southward, and in fact extends into the Southern hemisphere. Concurrently, the Southern hemisphere Hadley cell is weak.

The extension of the Southern hemisphere Hadley cell across the Equator during Northern summer is especially conspicuous, thus reflecting in particular, but not exclusively, the contributions of the monsoons of South Asia and Africa (Sections 6.8.3 and 6.8.2). The tropical Hadley cells play a paramount role in the maintenance of the global circulation, as will be discussed in the subsequent Sections 3.4 and 3.5.

3.4. Maintenance of the Global Circulation: Angular Momentum

The total angular momentum of the earth—atmosphere system must remain invariant with time. Since the rotation rate of the earth is nearly constant, the atmosphere must also conserve its angular momentum. The distribution of sources and sinks entails an exchange of momentum in the atmosphere between the lower and higher latitudes.

The absolute angular momentum per unit mass

$$m = (\Omega a^2 \cos^2 \phi + a \cos \phi u), \qquad (3.4:1)$$

where Ω is the angular velocity of the earth's rotation, ϕ latitude, a the earth radius, and u the eastward wind component. The terms on the right-hand side are known as the 'Ω momentum' and the 'u momentum', respectively.

According to Newton's second law, the absolute angular momentum per unit mass can be changed by the torques due to the zonal pressure gradient force $-(1/\rho)\,(\partial p/\partial x)$ and the frictional force, here approximated by $+(1/\rho)\,(\partial \tau/\partial z)$, where τ is the stress. This principle can be expressed as (Holton, 1972, p. 225)

$$\frac{\mathrm{d}m}{\mathrm{d}t} = -a\cos\phi\,\frac{1}{\rho}\,\frac{\partial p}{\partial x} + a\cos\phi\,\frac{1}{\rho}\,\frac{\partial \tau}{\partial z}\,. \tag{3.4:2}$$

Elementary transformation using the continuity equation and integration in height and zonally around a latitude circle (Holton, 1972, p. 225) yields the following expression for the rate of change of absolute angular momentum for a zonal ring of air of unit meridional width at latitude ϕ:

$$\frac{\partial}{\partial t}\int_0^\infty \int_0^{2\pi} \rho m a \cos\phi\,\mathrm{d}\lambda\,\mathrm{d}z$$

$$= -a^2\cos^2\phi\int_0^{2\pi} \tau_0\,\mathrm{d}\lambda - \int_0^\infty \int_0^{2\pi} a^2\cos^2\phi\,\frac{\partial p}{\partial x}\,\mathrm{d}\lambda\,\mathrm{d}z -$$

$$- \int_0^\infty \int_0^{2\pi} \frac{\partial}{\partial y}\,(\rho m v)\,a\cos\phi\,\mathrm{d}\lambda\,\mathrm{d}z, \tag{3.4:3}$$

where τ_0 is the stress at the surface and v the meridional wind component. For the long-term mean conditions considered here, the left-hand term vanishes, so that the three terms on the right must balance. The first term represents the exchange of angular momentum between earth and atmosphere due to the surface frictional torque. The second term is the pressure torque. This would be zero for a zonal ring of atmosphere on a flat earth. However, this term provides a net contribution if there are pressure differences between the western and eastern sides of mountains. The third term is the convergence of absolute angular momentum associated with the meridional transport. These contributions are discussed in the following.

The earliest evaluations from observational data (Priestley, 1951) were undertaken with reference to the first right-hand term of Eq. (3.4:3). Surface easterlies occupy the tropical half of the Earth, while westerlies prevail in the extratropical caps (Figs. 3.2:1a, b). By means of the surface friction, the earth exerts an eastward torque upon the atmosphere above in the tropics, which is tantamount to a flux of westerly angular momentum from earth to atmosphere. Similarly, the opposite frictional torque exerted by the surface westerlies in the extratropical caps corresponds to a flux of angular momentum from atmosphere to earth. Since the total angular momentum of the entire atmosphere must be conserved in the long-term average, the upward flux in the tropical half of the earth must equal the downward flux in the extratropical caps. Accordingly, angular momentum must be transferred poleward within the atmosphere, with a maximum flux at the latitudes separating the domains of surface easterlies and westerlies.

The vertical exchange of angular momentum through friction in the tropical half of the earth and the extratropical caps, respectively, can in principle be evaluated from surface wind information. With latitude ϕ, earth radius a, and mean eastward surface stress at a given latitude $\bar{\tau}_x$, the torque per unit surface exerted with reference to the earth's rotational axis

$$t = a\cos\phi\,\bar{\tau}_x, \tag{3.4:4}$$

with the dimension of length (say of a leverage arm) times force per surface. Area integration yields the total torque contributed by a latitude band

$$T = 2\pi a^3 \int \bar{\tau}_x \cos^2 \phi \, d\phi, \tag{3.4:5}$$

with the dimension of length times force, or angular momentum per time. In fact, extending the area integration over the latitude domains of the surface easterlies and westerlies, respectively, yields the upward and downward transfers of angular momentum between earth and atmosphere.

Following these considerations, Priestley (1951) evaluated the mean zonal surface stress over the World oceans from ship observations. Difficulties of this approach are related to the choice of representative drag coefficients describing the surface roughness. Results are reproduced in Fig. 3.4:1.

The mountain torque has been evaluated from pressure charts and profiles of the principal mountain ranges (Lorenz, 1967, pp. 51–53). This task entails quantitative estimates of pressure differences between the eastern and western side of the mountains. Results are also included in Fig. 3.4:1. In the Northern hemisphere temperate zone, the mountain and the frictional torques have the same sign, but the former is smaller.

From the evaluation of frictional and pressure torques such as illustrated in Fig. 3.4:1, part a, it is possible to calculate the poleward transport that must be performed within the atmosphere. Results are depicted by the broken line in Fig. 3.4:1, part b.

Independently, the poleward transport of absolute angular momentum in the atmosphere can be evaluated directly from aerological data. Of particular interest are the mechanisms by which these transports are performed within the atmosphere. Refer to Eq. (3.4:1), consider meridional transports, and denote latitude means by square brackets, and departures from such means by a prime and a subscript x. Then

$$[mv] = a \cos \phi \, (\Omega a \cos \phi \, [v] + [u] \, [v] + [u_x' v_x']). \tag{3.4:6}$$

A time average shall be denoted by an overbar. The time-averaged transport of absolute angular momentum across a wall extending around an entire latitude circle and over the entire depth of the atmosphere

$$M = 2\pi a^2 \cos^2 \phi \int_0^{p_{\text{sfc}}} (\Omega a \cos \phi \, [\bar{v}] + [\bar{u}] \, [\bar{v}] + [\overline{u_x' v_x'}]) \frac{dp}{g}. \tag{3.4:7}$$

With integration extending over the entire depth of the atmosphere, the contribution of the first term in the parenthesis vanishes for all practical purposes. $[\bar{u}] \, [\bar{v}]$ is called 'circulation flux' or 'drift' and is related to the mean meridional motion $[\bar{v}]$, and $[\overline{u_x' v_x'}]$ is an 'eddy flux'.

Results of the direct evaluation of angular momentum transport from aerological data are plotted in Fig. 3.4:1, part b. Transports required by the observed surface torques are also entered for comparison. The curves agree in the major features in that they illustrate the supply of westerly angular momentum from the tropical atmosphere to higher latitudes. The circulation flux plays a role in the realm of the tropical Hadley cells, while eddies account for most of the transport further poleward. The asymmetric upper-air waves associated with travelling midlatitude disturbances are known to be the main agents in the poleward eddy flux.

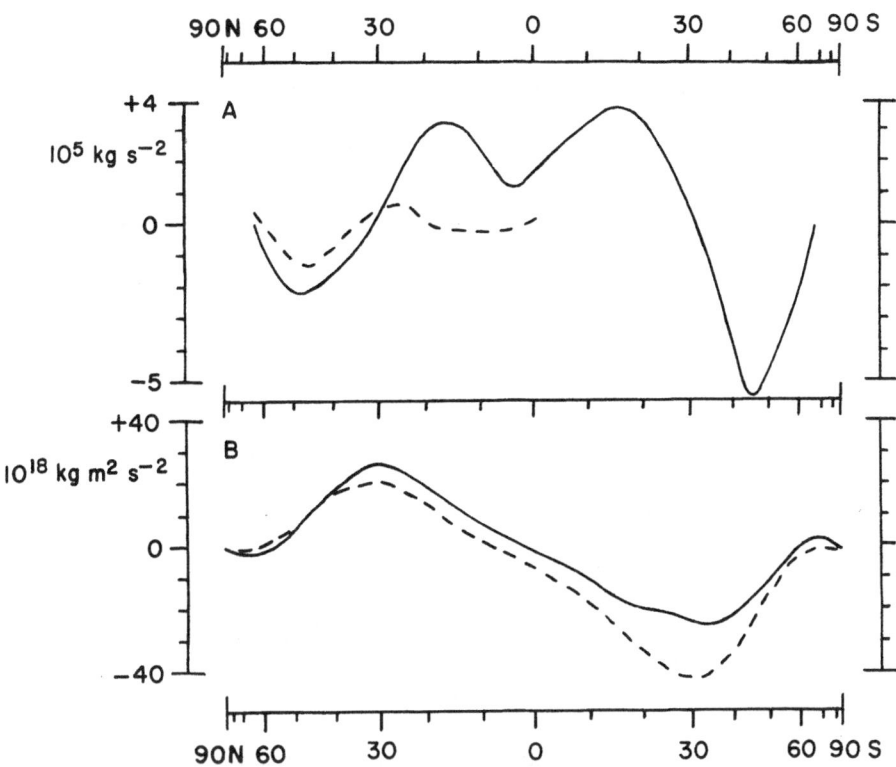

Fig. 3.4:1. Torque and absolute angular momentum. (a) average eastward torque per unit area exerted upon the atmosphere by surface friction, solid, and by mountains, broken line, in units of 10^5 kg s^{-2}. (b) northward transport of westerly absolute angular momentum in the atmosphere: observed transport, solid; transport required by observed surface torques, broken line; in 10^{26} kg m^2 s^{-2}. Sources: Lorenz, 1967, Holton, 1972.

3.5. Maintenance of the Global Circulation: Kinetic Energy

The present section focuses on the role of the tropics in the kinetic energy budget of the global atmosphere. This issue was first addressed by Palmén (1959). The budget equation of horizontal kinetic energy for a volume of the atmosphere extending from the surface to the top of the atmosphere, or to a level with vanishing vertical motion, can be written (Palmén, 1959)

$$\frac{\partial}{\partial t} \int_\nu \rho K \, d\nu = - \int_\nu \nabla \cdot (\rho \mathbf{V} K) \, d\nu - \int_\nu \mathbf{V} \cdot \nabla p \, d\nu + \int_\nu \rho \mathbf{V} \cdot \mathbf{F} \, d\nu. \qquad (3.5:1)$$

$$\qquad\quad S \qquad\qquad\qquad A \qquad\qquad\quad G \qquad\qquad\quad D$$

Here ρ is the density of the air, $K = V^2/2$ the horizontal kinetic energy per unit mass, $\mathbf{V}$ the horizontal wind vector, $-\nabla p$ the horizontal pressure gradient, $\mathbf{F}$, the horizontal frictional force per unit mass and ν volume.

We shall here consider the domain 0–30°N and the Northern hemisphere extratropical cap in particular, in accordance with the available data. Discussion shall be limited to the extreme seasons, so that the storage of kinetic energy S essentially disappears and a balance must result from the three right-hand terms. A represents the lateral import, G the generation, and D the frictional dissipation of kinetic energy inside the volume.

Term A contains the product VK. In the following, the implications of space and time averaging shall be considered for the example of the meridional with component v. A bar signifies a time mean and a bracket a space mean, primes denote deviations from the means and the subscripts x and t specify the deviation from a space and time mean, respectively. The space-time mean of the flux of kinetic energy at a given level can then be expanded in the 'mixed space-time domain' (Oort, 1964) and expressed as

$$[\overline{vK}] = [\overline{v}] \; [\overline{K}] + [\overline{v'_x K'_x}] + [\overline{v'_t K'_t}] \tag{3.5:2}$$

The first right-hand term is a flux associated with a non-zero mean rate of flow ($\overline{v}$) and is called advective flux in Priestley's (1949) terminology. This term can be computed from time and space averaged values of v and K. The second right-hand term represents the flux associated with standing eddies, and results from the correlation in space between the time means of v and K; in other words it is due to the semi-permanent features of the atmosphere. The standing eddy flux can be computed from time mean values of v and K at fixed locations. The third right-hand term has to be determined from instantaneous and simultaneous values of v and K. This so-called transient eddy flux represents the effect of travelling disturbances. An expansion analogous to Eq. (3.5:2) applies to the product $\mathbf{V} \cdot \nabla p$ in term G. Evaluation of observations with a breakdown as in Eq. (3.5:2) serves to identify the contribution of various mechanisms.

Table 3.5:1 summarizes the evaluation of the kinetic energy budget from time-averaged aerological data, while Table 3.5:2 represents computations of meridional transports based on daily soundings. These transports are performed largely by transient eddies. Frictional dissipation values given in Table 3.5:1 refer to the surface layer only. Dissipation in the free atmosphere is difficult to assess, but is estimated to be somewhat smaller than the surface dissipation (Palmén, 1959).

TABLE 3.5:1

Kinetic energy budget of the Northern hemisphere troposphere, 1000–100 mb in 10^{13} W, during winter (December–February) and summer (July–August).

(G) generation by mean meridional circulation and standing eddies;
(A) divergence of kinetic energy transport by mean meridional circulation and standing eddies;
(D) dissipation of kinetic energy in the surface layer (sources: Palmén, 1959; Kung, 1963; Hastenrath, 1969).

		0–30°N	30–60°N
	winter		
(G)		+ 26	− 4
(A)		+ 4	− 4
(D)		6	27
	summer		
(G)		+ 9	+ 2
(A)		+ 1	− 0
(D)		3	8

TABLE 3.5:2

Poleward transport of kinetic energy in 10^{13} W, according to various sources.

Kao (1954): 1010−100 mb, January 1949;
Pisharoty (1955): surface−150 mb, January−February 1949;
Mintz (1955): 1910−200 mb, winter and summer 1949.

winter	30°N (maximum)	60°N
Kao	24	4
Pisharoty	20	–
Mintz	17	7
summer	40°N (maximum)	60°N
Mintz	2	2

During Northern winter the generation of kinetic energy by the mean meridional circulation and standing eddies in the domains 0−30°N is from Table 3.5:1 found to exceed the local dissipation by an amount comparable to the poleward transport across 30°N given in Table 3.5:2. Transports across the Equator may be small by comparison. Concerning the extratropical cap poleward of 30°N, frictional dissipation can from Table 3.5:1 be estimated to be of the order of 60×10^{13} W (Palmén, 1959). Accordingly, during winter the tropics may supply about one third of the kinetic energy requirements in the extratropical cap.

During the Northern summer the kinetic energy generation by the tropical Hadley cell approximately balances the local dissipation (Table 3.5:1) leaving at best a small amount for export. This is consistent with the very small poleward transport shown in Table 3.5:2. However, the frictional dissipation in the extratropical cap is small during summer and can approximately be met by the local generation of kinetic energy.

This very approximate estimation of budget terms indicates the important role of the tropical mean meridional circulation for the maintenance of circulation in the extratropical cap during winter.

Recent evaluations of the atmospheric energy cycle such as summarized in Holton (1972, pp. 213−223) consider in particular the conversions between zonal and eddy available potential energy and zonal and eddy kinetic energy. Because of the scarcity of aerological observations in the tropics and the Southern hemisphere, computer simulations (Otto-Bliesner, 1980) of the energy cycle have considerable merit. Numerical experiments with this frame of reference seem to offer much promise for elucidating the role of the tropics in meeting the kinetic energy demands of higher latitudes.

3.6. Synthesis

The zonally averaged planetary scale circulation of the tropics consists of an easterly wind regime bounded by the anticyclonic axes of the subtropical highs, and the two thermally direct Hadley cells.

Concerning the maintenance of the global circulation in mechanical terms, the tropical atmosphere represents a source region of westerly absolute angular momentum. This is given back to the solid earth in the higher latitudes. The required poleward transports in low latitudes are performed primarily by the Hadley circulations, while eddy mechanisms assume the dominant role beyond the fringe of the tropics.

The kinetic energy budget during the respective winter seasons is characterized by a large production associated with the mean meridional circulation and comparatively small dissipation rates. Thus a substantial surplus of kinetic energy is available for export to the extratropics. This is accomplished by the Hadley cell within the tropics, and transient and standing eddies further poleward. During the respective summer seasons, kinetic energy generation by the then much weaker Hadley circulation is small, and is approximately offset by the dissipation within the tropics. However, in the summertime extratropics kinetic energy generation is comparatively small and is approximately met by local generation.

In mechanical terms, the tropical atmosphere thus is instrumental in the maintenance of the global circulation. Further important constraints of the general circulation relate to the thermodynamics and the budgets of heat and moisture. In contrast to the topics of angular momentum and kinetic energy, however, these can only be adequately treated for the atmosphere-hydrosphere system as a whole. Accordingly, following the synopsis of ocean circulation in Chapter 4, the heat and water budgets of atmosphere and hydrosphere are discussed in Chapter 5.

References

Hastenrath, S., 1969: 'On the role of meridional circulations in the kinetic energy budget of the Northern hemisphere'. *Archiv Meteor. Geophys. Bioklim., Ser. A*, 18, 1–16.

Holton, J. R., 1972: *An introduction to dynamic meteorology.* Academic Press, New York and London, 319 pp.

Kao, S. K., 1954: 'The meridional transport of kinetic energy in the atmosphere'. *J. Meteor.*, 11, 351–361.

Kung, E. C., 1963: 'Climatology of aerodynamic roughness parameter and energy dissipation in the planetary boundary layer of the Northern hemisphere'. Dept. Meteorology, University of Wisconsin, Annual report 1963, Contract DA–36–039–AMC–00878, Task 1–A–6–11001–B–021–08.

Lorenz, E. N., 1967: 'The nature and theory of the general circulation of the atmosphere'. WMO No. 218, TP. 115, Geneva, 161 pp.

Mintz, Y., 1955: 'The total energy budget of the atmosphere'. Dept. Meteorology, UCLA, Final Report, General Circulation Project, Paper No. 13, Contract AF 19(122)–48.

Newell, R. E., Kidson, J. W., Vincent, D. G., Boer, G. J., 1972, 1975: *The general circulation of the tropical atmosphere*, vols. 1 and 2. MIT Press, Cambridge, Mass., 258 and 321 pp.

Oort, A. H., 1964: 'On estimates of the atmospheric energy cycle'. *Mon. Wea. Rev.*, 11, 483–493.

Oort, A. H., 1983: 'Global atmospheric circulation statistics, 1958–1973'. NOAA Professional Paper 14, U.S. Government Printing Office, Washington, D.C., 180 pp.

Otto-Bliesner, B. L., 1980: 'The dynamics of seasonal change of the long waves as deduced from a lower-order general circulation model'. Ph.D. Dissertation, Department of Meteorology, University of Wisconsin, Madison, 223 pp.

Palmén, E., 1959: 'On the maintenance of kinetic energy in the atmosphere'. pp. 212–224 in B. Bolin, ed: *The atmosphere and the sea in motion*, Rossby memorial volume, Rockefeller Institute Press, New York, 509 pp.

Peixoto, J. P., Oort, A. H., 1984: 'Physics of climate'. *Rev. Modern Phys.*, 56, 365–429.

Pisharoty, P. R., 1955: 'The kinetic energy of the atmosphere'. Dept. Meteorology, UCLA, Scientific Report No. 6, Contract AF 19(122)–48.

Priestley, C. H. B., 1949: 'Heat transport and zonal stress between latitudes'. *Quart. J. Roy. Meteor. Soc.*, 75, 28–40.

Priestley, C. H. B., 1951: 'A survey of the stress between the ocean and the atmosphere'. *Australian J. Sci. Res.*, A4, 315–318.

OCEAN CIRCULATION

The tropical oceans are an essential component of the global climate system. Air-sea coupling plays an important role in the global heat and moisture budget (Chapter 5), in the quasi-permanent circulation systems of the lower atmosphere (Chapter 6), in the mechanisms of interannual variability and regional climate anomalies (Chapters 8–9), and in the characteristics of past climates (Chapters 11–12). As a background for these topics, the basics of ocean circulation are reviewed here, largely following the presentations in standard texts. However the dynamics of the equatorial ocean in particular are still at the focus of current research and new perspectives are imminent.

The surface circulation of the World ocean is closely related to the wind systems of the lower atmosphere. As discussed by Defant (1961, vol. 1, pp. 544–555), under stationary conditions all forces such as wind stress, pressure gradient and Coriolis accelerations, are in equilibrium and the mass distribution is adapted accordingly; there being usually a mutual adjustment between the internal field of force and the existing current. If non-linear terms are also included, the presence of eddies precludes strict stationarity. Even for the stationary conditions considered by Defant (1961, vol. 1, pp. 544–553) it is not possible to distinguish between cause and effect. A change in the field of force entails a subsequent change in the current, and conversely a change in the current is associated with a rearrangement of the field of force. These circumstances pertain to the way in which wind influences density currents. It is with this perspective that Fig. 4:1 should be viewed, which reproduces Munk's (1950) scheme of circulation in a rectangular ocean resulting from a simplified mean surface wind distribution.

The circulation in Fig. 4:1 is divided into gyres equivalent in scale to climatic belts. A gyre is composed of zonal currents to the North and South, an intense current narrowly concentrated along the western boundary, and a compensating very broad drift in the eastern portion. The outstanding features in low latitudes are the subtropical gyres and the predominantly zonally arranged currents in the equatorial belt. These current systems are discussed in Sections 4.2 and 4.3. The surface circulation of the tropical Indian Ocean, characterized by strong seasonal reversals and related to the monsoon wind systems is considered in Section 4.3. The review of deep circulation in Section 4.5 and the summary in Section 4.6 complete the synopsis of low-latitude ocean circulation.

4.1. Wind Stress and Motion Field in the Upper Ocean

The relation of wind stress to oceanic mass transport and vertical motion is discussed extensively in Stommel (1976, pp. 154–156), McLellan (1965, pp. 82–87), Pond and Pickard (1983, pp. 118–144), and Gill (1982, pp. 319–322, 482–491). The basic idea is that the wind stress directly forces a thin surface layer (the Ekman layer), which in turn drives currents in the deep ocean. The following is a brief summary of the theory.

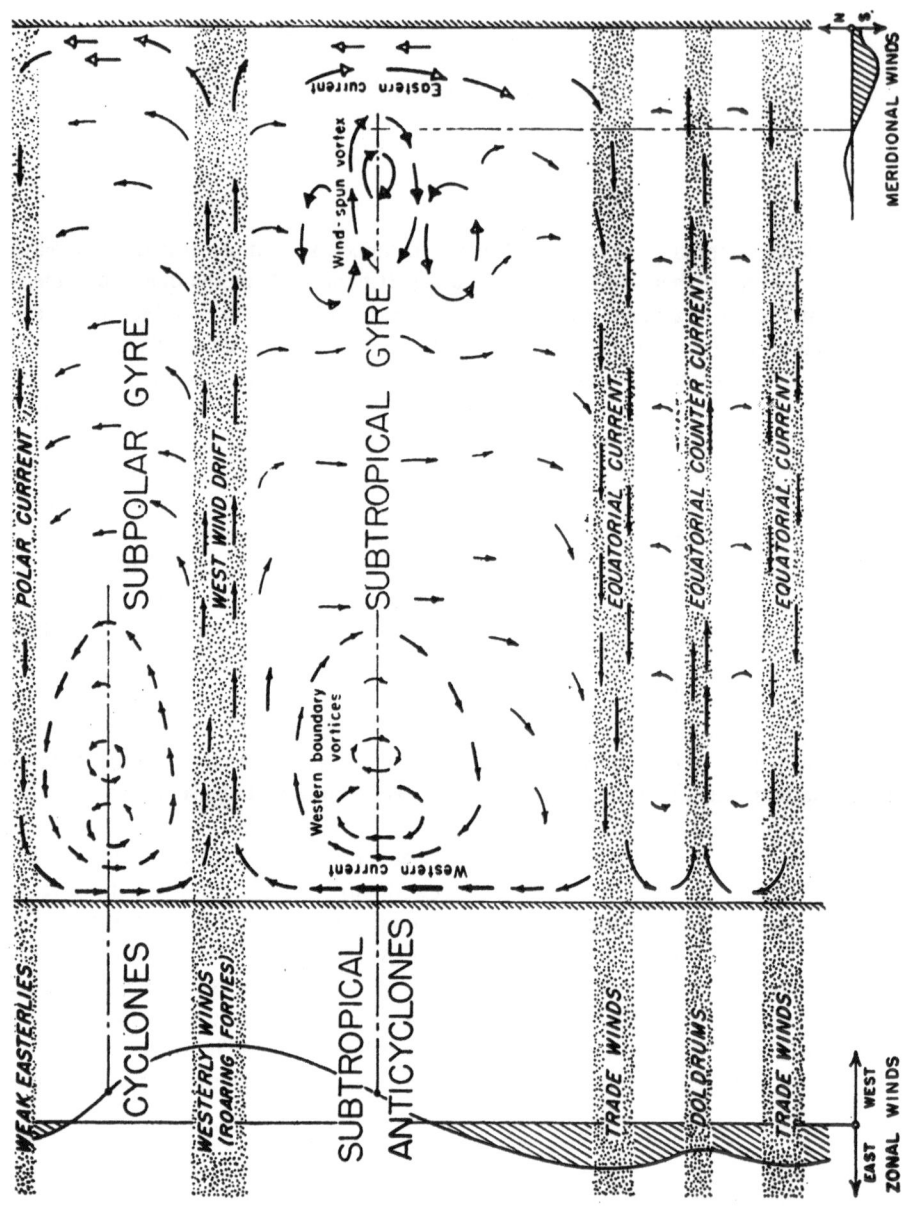

Fig. 4:1. Schematic presentation of circulation in a rectangular ocean resulting from zonal winds (filled arrowheads), meridional winds (open arrowheads), or both (half-filled arrowheads). The width of the arrows is an indication of the strength of the currents. The nomenclature applies to either hemisphere, but in the Southern hemisphere the subpolar gyre is replaced largely by the Antarctic Circumpolar Current (West wind drift) flowing around the World. From Munk, 1950 (Journal of Meteorology, American Meteorological Society).

The equations of motion assuming negligible accelerations and friction from horizontal gradients are (Pond and Pickard (1983, p. 119))

$$-f\rho v = -\frac{\partial p}{\partial x} + \frac{\partial \tau_x}{\partial z},$$

$$+f\rho u = -\frac{\partial p}{\partial y} + \frac{\partial \tau_y}{\partial z}.$$

(4.1:1)

Conservation of mass, assuming that water is incompressible, implies

$$\frac{\partial u}{\partial x} + \frac{\partial v}{\partial y} + \frac{\partial w}{\partial z} = 0.$$

(4.1:2)

The x, y, and z coordinates are counted positive to the East, North and upwards, respectively. The velocity components u and v, density ρ, pressure p, and vertical shearing stress components τ_x and τ_y vary with x, y, z, while the Coriolis parameter $f = 2\Omega \sin \phi$ depends only on y.

Following Gill (1982, pp. 320–321), the velocities on the left-hand sides of Eqs. (4.1:1) can be separated into a part driven by pressure gradients and a part driven by the stress. The components driven by pressure gradients are

$$v_p = +\frac{1}{\rho f}\frac{\partial p}{\partial x},$$

$$u_p = -\frac{1}{\rho f}\frac{\partial p}{\partial y}.$$

(4.1:3)

These currents extend deep into the water column and are geostrophic. They are directed at right angles to the right (left) of the pressure gradient in the Northern (Southern) hemisphere. The part driven by the stress, the Ekman velocity, is given by

$$v_E = -\frac{1}{\rho f}\frac{\partial \tau_x}{\partial z},$$

$$u_E = +\frac{1}{\rho f}\frac{\partial \tau_y}{\partial z}.$$

(4.1:4)

This part of the motion field is confined to a shallow friction layer. It should be noted that these simple equations (4.1:3) and (4.1:4) break down near the Equator, where f approaches zero.

Vertical integration of Eqs. (4.1:4) yields the equations of Ekman mass transport

$$M_{yE} = -\frac{\tau_{xo}}{f},$$

$$M_{xE} = +\frac{\tau_{yo}}{f},$$

(4.1:5)

where the subscript o refers to the surface, and so τ_{xo} and τ_{yo} are the components of wind stress at the ocean surface. The corresponding components of the geostrophic mass transport are M_{yp} and M_{xp}.

As the wind stress varies spatially, so does the Ekman transport. The associated horizontal divergence and convergence produces compensatory vertical motion in the boundary layer, or 'Ekman pumping'. For $w = 0$ at the surface, application of Eq. (4.1:2) to the boundary layer gives

$$\rho w_E = \frac{\partial M_{xE}}{\partial x} + \frac{\partial M_{yE}}{\partial y}. \tag{4.1:6}$$

Combining Eqs. (4.1:6) and (4.1:5) yields the vertical mass flux across the bottom of the oceanic boundary layer

$$\rho w_E = \frac{\partial}{\partial x}\left(\frac{\tau_{yo}}{f}\right) - \frac{\partial}{\partial y}\left(\frac{\tau_{xo}}{f}\right). \tag{4.1:7}$$

Gill (1982, p. 327) calls this quantity 'Ekman pumping velocity' and points out that it has the same sign in the atmosphere as in the ocean, or in other terms (Gill, 1982; p. 483) Ekman convergence in the atmospheric boundary layer is associated with Ekman divergence in the ocean.

Cross-differentiation of the vertically integrated form of Eq. (4.1:1) yields the curl equation of meridional mass transport (Stommel, 1976, p. 155; Pond and Pickard, 1983, p. 120)

$$M_y = M_{yE} + M_{yp} = \frac{\operatorname{curl}\tau_o}{\beta}, \tag{4.1:8}$$

where $\beta = \partial f/\partial y$, and

$$\operatorname{curl}\tau_o = \frac{\partial\tau_{yo}}{\partial x} - \frac{\partial\tau_{xo}}{\partial y}. \tag{4.1:9}$$

The wind stress also leads to mass adjustments and the creation of pressure gradients, associated with which there are in turn pressure-gradient driven velocities v_p and u_p (Eq. 4.1:3) and mass transports M_{yp} and M_{xp}. Substituting from Eq. (4.1:8) into Eq. (4.1:7) yields

$$\rho w_E = \frac{\beta(M_y - M_{yE})}{f} = \frac{\beta M_{yp}}{f}. \tag{4.1:10}$$

This shows that the Ekman pumping forces a deep geostrophic meridional current. Upward (downward) directed Ekman pumping velocity entails poleward (equatorward) directed geostropic mass transport, which as a consequence of the poleward convergence of the meridians is convergent (divergent). Divergence vanishes for the sum of the vertically integrated Ekman plus geostrophic transports.

4.2. The Subtropical Gyres

Most extensively studied are the strong subtropical gyres of the North Atlantic and North Pacific basins. Regarding the Southern hemisphere, counterparts are well developed in the South Atlantic and South Pacific Oceans, while the subtropical gyre of the South Indian Ocean is less distinct and subject to substantial seasonal changes. Stommel's (1976) treatise on the Gulf Stream not only deals with the surface circulation of the greater North Atlantic, but is also pertinent to the dynamics of the subtropical gyres in general. Accordingly, Stommel's (1976) monograph serves as a major basis for the present section. More recent work on the North Atlantic subtropical gyre is discussed in Armi and Stommel (1983).

The map scheme Fig. 4.2:1, part a, illustrates the anticyclonic wind pattern over the North Atlantic, which is essentially symmetric in the zonal direction. The pattern is repeated in the map Fig. 4.2:1, part b. Intuitively, this anticyclonic atmospheric circulation and the associated wind stress pattern appear conducive to an anticyclonic gyre in the surface circulation of the ocean. However, instead of a zonally symmetric gyre as in the atmosphere such as sketched in Fig. 4.2:1, part a, the observed oceanic pattern is markedly asymmetric, featuring an intense poleward flow at the western boundary, compensated by a weak southward drift spread over the eastern part of the gyre (Fig. 4.2:1, part b). The westward intensification and asymmetry are dominant characteristics of the subtropical gyres, and their causes are discussed in the following.

The vorticity equation for the oceanic flow can in approximate form be written

$$\frac{d\zeta}{dt} = -\frac{df}{dt} + \left[\frac{\partial F_{yi}}{\partial x} - \frac{\partial F_{xi}}{\partial y} \right] + \left[\frac{\partial F_{yo}}{\partial x} - \frac{\partial F_{xo}}{\partial y} \right] \qquad (4.2:1)$$

$$R \;\; = - \; P \;\; + \qquad\qquad I \qquad\qquad + \qquad\qquad W$$

Here ζ is relative vorticity and f Coriolis parameter ($2\Omega \sin \phi$). F signifies the frictional force per unit mass exerted on the water body, with the subscripts x and y denoting the eastward and northward components, respectively; the subscript o refers to the ocean surface, and the subscript i indicates effects in the interior and at the bottom and sides of the water body. Thus the term R is the individual time rate of change of relative vorticity following the motion, P is the individual time rate of change of planetary vorticity (Coriolis parameter) following the motion, I the rate of change of vorticity due to friction in the interior and at the bottom and sides of the water body, and W the rate of change of vorticity due to wind stress at the ocean surface. Divergence, solenoidal, and twisting terms, which can be important in atmospheric flow, are not considered in the above formulation. All terms in Eq. (4.2:1) would have units of s^{-2}.

The conditions described in Fig. 4.2:1 represent steady state, so that the total vorticity over the whole oceanic gyre must be constant; as a particle travels through a complete circuit around the gyre, it must arrive back at its starting point with no net change in vorticity. With reference to Eq. (4.2:1), Table 4.2:1 compares the vorticity tendencies in northward and southward flows for symmetric and asymmetric gyres. The symmetric and anticyclonic wind gyre (Fig. 4.2:1) makes for uniform negative vorticity tendencies (W) on the western and eastern sides of the ocean for both the symmetric and the asymmetric case of ocean circulation. The other terms of Eq. (4.2:1) differ between the cases of symmetric and asymmetric ocean gyres. The vorticity tendency associated with the frictional effects in the interior and at the bottom and sides of the water body (I) is positive in both cases, as these frictional effects would act opposite to the anticyclonic sense of oceanic flow. In the symmetric ocean case, these tendencies (I) would be of the same magnitude at the western and eastern side of the gyre, while for the asymmetric ocean circulation frictional

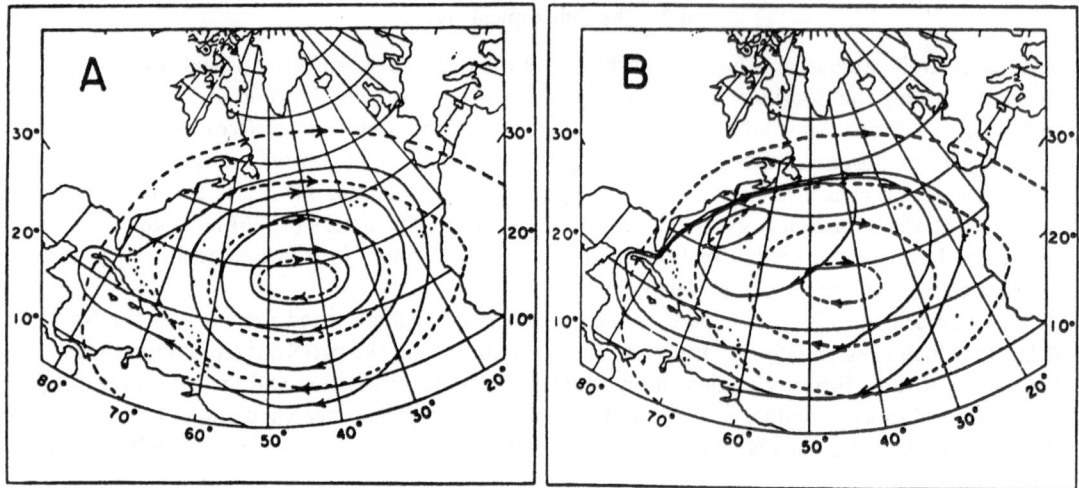

Fig. 4.2:1. Winds and surface circulation in the North Atlantic. (a) schematic wind system (broken) and currents (solid) which would occur were there no asymmetry in the circulation; (b) schematic relation of wind (broken) and current (solid) with asymmetry. From Stommel (1976).

TABLE 4.2:1

Vorticity tendencies related to terms in Eq. (4.2:1) for (i) symmetric, and (ii) asymmetric ocean circulation, with arbitrary units. Source: Stommel (1976).

	(i) symmetric		(ii) asymmetric	
	N flowing, W side of ocean	S flowing, E side of ocean	N flowing W edge	S flowing remainder of ocean
W	−1.0	−1.0	− 1.0	−1.0
I	+ 0.1	+ 0.1	+ 10.0	+ 0.1
−P	−1.0	+ 1.0	− 9.0	+ 0.9
total	−1.9	+ 0.1	0.0	0.0

effects and the positive vorticity tendency (I) would be greatly enhanced in the fast northward flow at the western edge of the oceanic gyre. Finally, the planetary vorticity tendency ($-P$) is negative in the northward flow on the western side and positive in the southward stream on the eastern side of the oceanic gyre, for both the symmetric and asymmetric cases. In the symmetric case these tendencies of opposing sign are of the same magnitude, while in the asymmetric case the negative tendency in the fast northward flow on the western side is much larger than the positive tendency in the southward stream on the eastern side of the gyre. Table 4.2:1 shows that under the circumstances steady state would not be obtained for the symmetric case, while an asymmetric ocean circulation would allow for a vorticity balance in the northward and southward flowing branches and thereby also in the gyre as a whole. Calculations by Stommel (1976, pp. 87–93) demonstrate that an asymmetric ocean circulation is obtained with the Coriolis parameter varying in latitude, but not for an ocean rotating uniformly with latitude or non-rotating.

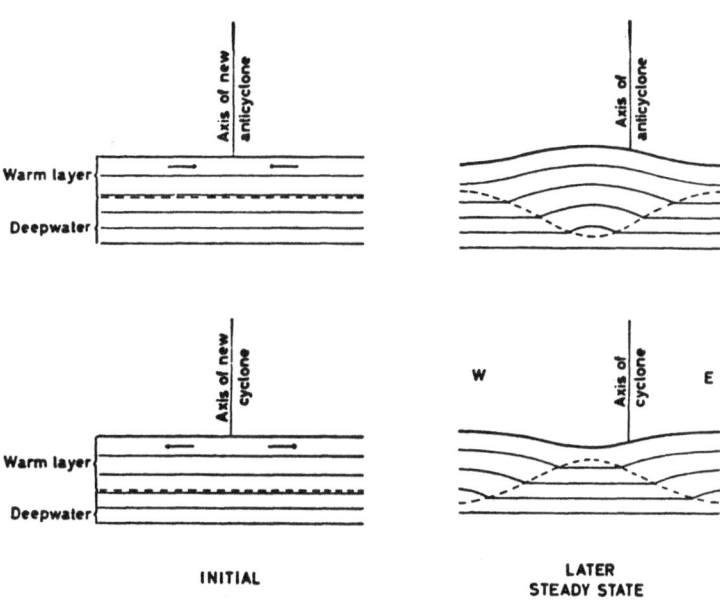

Fig. 4.2:2. Zonal vertical profiles showing schematically the creation and maintenance of maximum thickness of the warm oceanic surface layer under the influence of anticyclonic wind stress, and minimum thickness of same layer under cyclonic wind stress. Full lines: sea surface and interior isobaric surfaces. Dashed lines: density discontinuity surface. From Bjerknes (1959) (in *The atmosphere and the sea in motion*, Bolin, ed., copyright 1959, The Rockefeller Institute Press, p. 70).

The effect of anticyclonic and cyclonic surface wind patterns on the thermal structure of the upper ocean is schematically shown in Fig. 4.2:2. Refer to Eq. (4.1:5). Under the influence of an anticyclonic wind stress pattern, the Ekman mass transport is directed towards the central portion of the atmospheric anticyclone. This, along with the fact that wind tends to be weaker in the central portion than in the periphery of the anticyclone, leads to Ekman convergence and a downward directed Ekman pumping velocity (Eqs. (4.1:6) and (4.1:7)) under the atmospheric anticyclone. The Ekman mass transport affects the warm surface layer in particular, so that warm waters are piled up under the atmospheric anticyclone, resulting in a deep thermocline and an upward bulge of the free ocean surface. Fig. 4.2:2 shows a maximum thickness of the warm oceanic surface layer somewhat to the West of the atmospheric anticyclonic axis. The upper part of Fig. 4.2:2 exemplifies conditions for subtropical ocean gyres. The response of the upper ocean to an atmospheric cyclone should be understood analogously with reference to Eqs. (4.1:5), (4.1:6), and (4.1:7). Fig. 4.2:2 shows a shallow warm oceanic surface layer below a cyclonic wind pattern, with a minimum thickness again somewhat to the West of the atmospheric cyclone.

Fig. 4.2:2 and Eqs. (4.1:5) to (4.1:7) are also interesting in relation to the origin of currents in the upper ocean. The anticyclonic wind stress pattern leads to an upward bulging of constant pressure topographies and of the free ocean surface under the atmospheric anticyclone. A pressure pattern is thus created which allows for currents driven by pressure gradients, in accordance with Eqs. (4.1:3). It is then interesting to note that the large anticyclonic ocean gyres are directly related to a maximum of constant pressure topography in the central portion of the ocean, but only indirectly to the anticyclonic wind pattern.

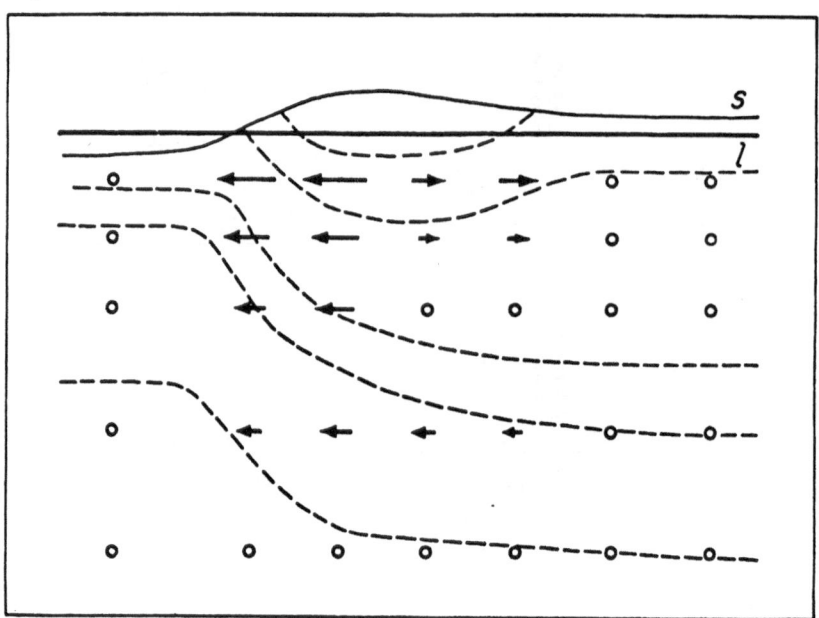

Fig. 4.2:3. Schematic cross section of the Gulf Stream. The arrows indicate direction of horizontal pressure gradients, and the little circles vanishing horizontal pressure gradient. The current is flowing perpendicular to the plane of the page. The broken lines are contours of equal water density. The line l is a level surface, and the line s the actual sea surface. From Stommel (1976).

The relation of pressure, density, and flow fields in the subtropical gyre is further schematically illustrated in Fig. 4.2:3 for the Gulf Stream. The current would be directed approximately perpendicularly to the right of the pressure gradient, except that near the surface the wind stress contributes substantially to the balance of forces, thus allowing for a flow component across the surface contours, i.e. there is Ekman drift. At some sufficiently deep level, pressure surfaces are nearly horizontal, friction is negligible and accordingly flow vanishes. This entails denser fluid to the left of the core of the surface flow where the sea surface is lower, and less dense fluid towards the right, or the center of the gyre, where the free sea surface is highest. The diagram Fig. 4.2:3 extends to a depth of about 1000 m. The free ocean surface, as well as density gradients are much steeper on the western than on the eastern side of the gyre, consistent with the fast western boundary currents.

The water transports by western boundary currents have been estimated from observation and theoretical computations. Fig. 4.2:4 shows the results of Munk's (1950) calculations for the North Pacific Ocean based on the mean zonal wind stress shown at the left of the graph. For the Kuroshio and the Gulf Stream systems values of about 4 and 7×10^7 m^3 s^{-1} are obtained from wind stress data and direct current measurements, respectively. By contrast, the transport of the Brazil Current is less than 1×10^7 m^3 s^{-1}.

A major control on the surface climate exerted by the subtropical gyres consists of the equatorward advection of cold waters in the eastern part of the oceans (Canary, Benguela, California, and Humboldt currents) and the intense poleward directed warm currents concentrated at the western boundaries (Gulf Stream, Kuroshio, Brazil currents), that decisively mitigate the thermal regime of midlatitude coasts.

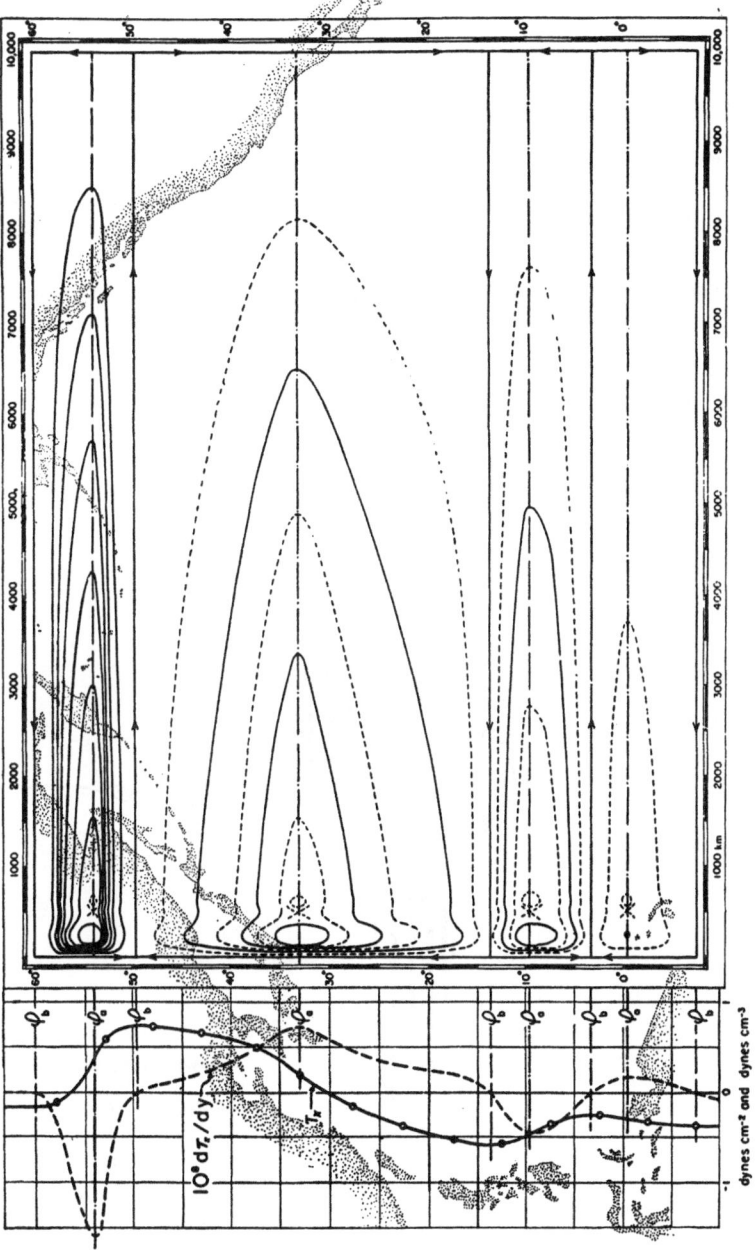

Fig. 4.2:4. The mean annual zonal wind stress $\tau_X(y)$ over the Pacific and its curl $\partial\tau_X/\partial y$ are plotted on the left. Shown on the right are mass transport stream-lines, transport between adjacent solid lines being 10^7 m^3 s^{-1}. From Munk (1950) (*Journal of Meteorology*, American Meteorological Society).

4.3. Equatorial Current Systems

The currents discussed in this section occupy a large portion of the low-latitude oceans. In contrast to the large horizontal gyres that dominate the ocean circulation from the outer tropics to mid-latitudes (Section 4.2), the equatorial zone is characterized by largely zonally oriented current systems.

4.3.1. OVERVIEW OF SURFACE CIRCULATION

Throughout the year, three major surface currents are recognized in the Atlantic and Pacific Oceans (Figs. 4:1, 4.2:4, and 4.3.1:1): the westward flowing North Equatorial Current (NEC) poleward of about 10°N, the likewise westward directed South Equatorial Current (SEC) extending from the Southern hemisphere to around 5°N, and sandwiched between the former two streams the eastward flowing Equatorial Countercurrent (ECC). Expanding on the schemes shown in Figs. 4.1 and 4.2:4, Leetmaa *et al.* (1981) quote evidence for a South Equatorial Countercurrent

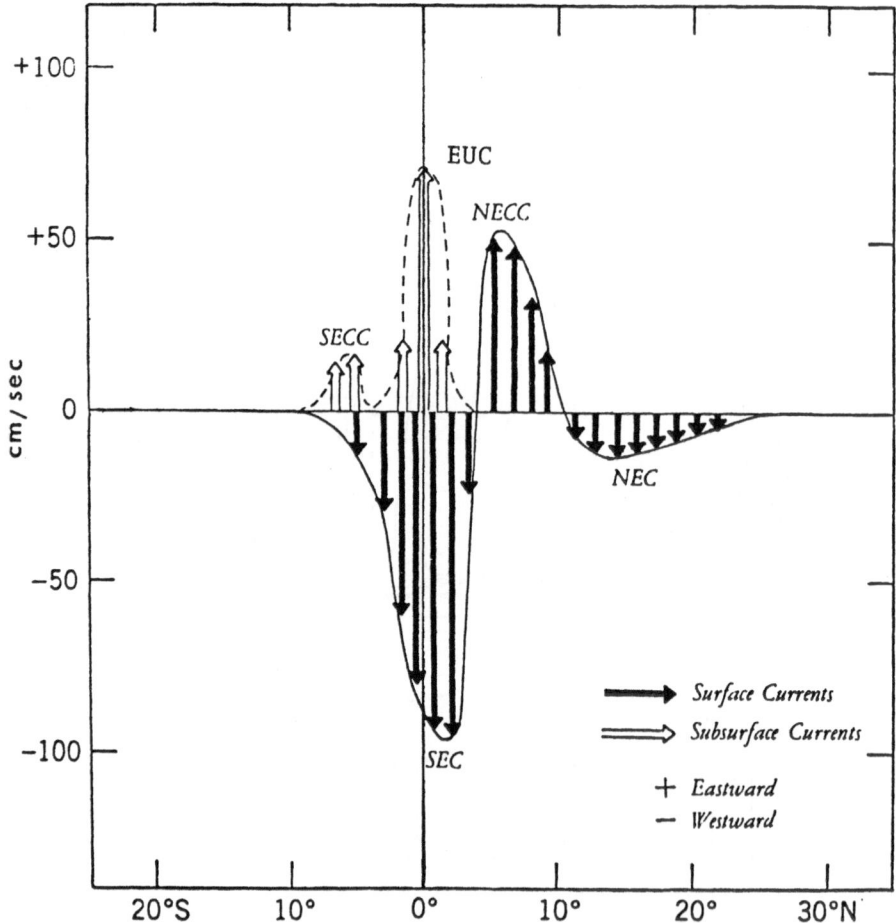

Fig. 4.3.1:1. Sketch of equatorial Atlantic surface currents along 28°W during GATE: South Equatorial Counter-current SECC, South Equatorial Current SEC, North Equatorial Countercurrent NECC, North Equatorial Current NEC, Equatorial Undercurrent EUC. Source: Philander (1977).

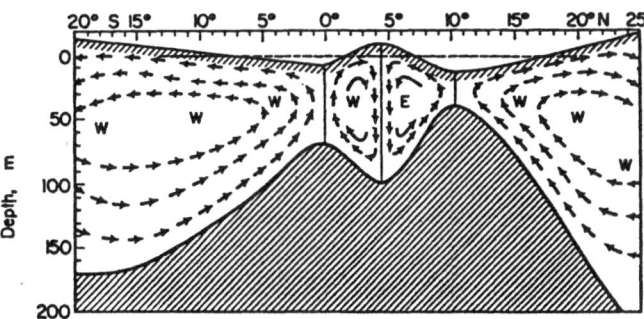

Fig. 4.3.1:2. Schematic representation of the zonal and meridional velocity components of the tropospheric circulation in the Atlantic Ocean (the topography of the thermocline is exaggerated in the vertical scale by about 1:1 million; that of the physical sea surface even more); W current towards West, E current towards East. From Defant (1961), vol. 1 (reprinted with permission from "A. Defant: *Physical Oceanography*", copyright 1961, Pergamon Press).

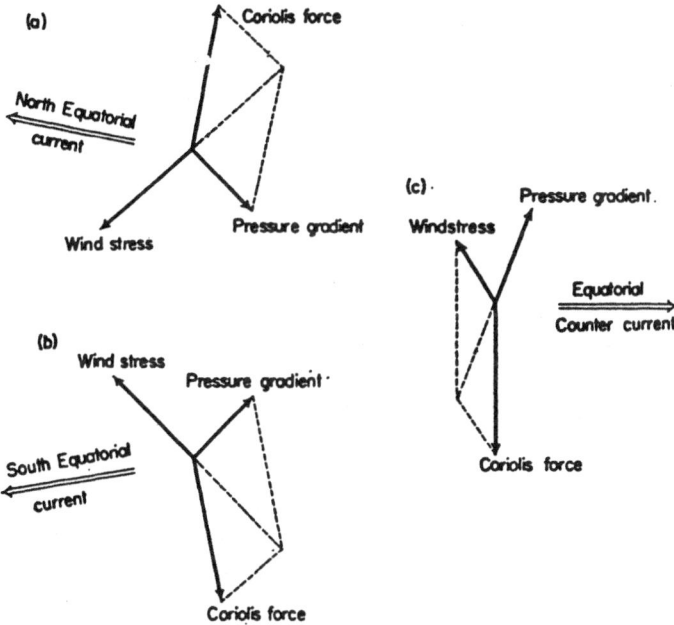

Fig. 4.3.1:3. Diagrams of forces: (a) for the North Equatorial Current; (b) for Southern hemispheric domain of the South Equatorial Current; (c) for the Equatorial Countercurrent. From Defant (1961), vol. 1 (reprinted with permission from "A. Defant: *Physical Oceanography*", copyright 1961, Pergamon Press).

in both oceans at about 5–10°S, albeit less well developed than its Northern hemisphere counterparts. In the Indian Ocean, both the wind systems (Section 6.8) and surface currents reverse seasonally, as will be discussed in Section 4.4. The westward flowing North and South Equatorial Currents seem most immediately plausible in relation to the Northeast and Southeast trades of the two hemispheres. Conditions appear less straightforward for the Equatorial Countercurrent which is directed opposite to the surface winds.

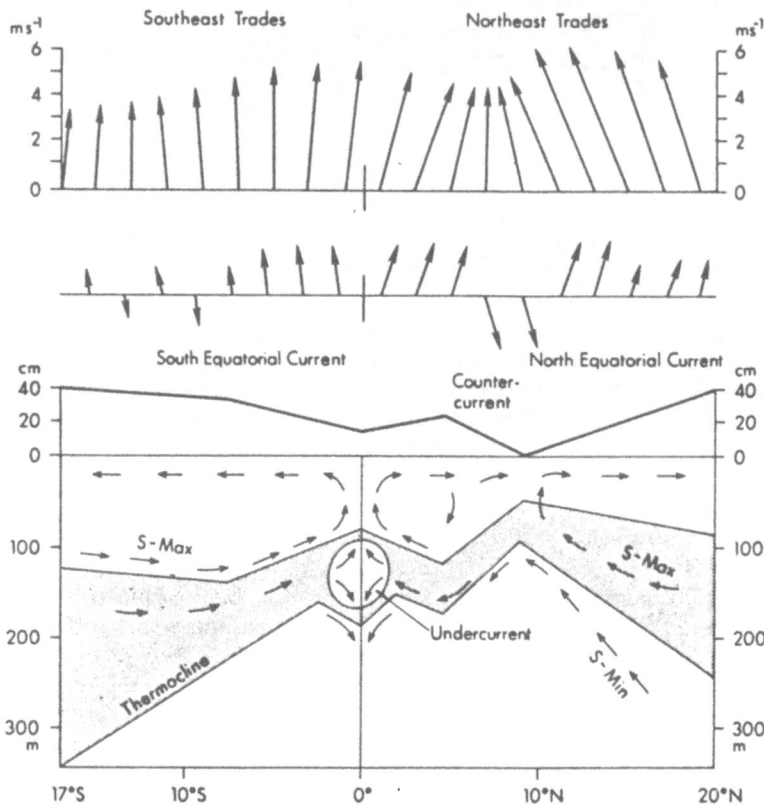

Fig. 4.3.1:4. Oceanographic conditions in the central equatorial Pacific between 150 and 160°W, showing winds, surface currents, dynamic height, thermal structure and meridional circulation. From Wyrtki and Kilonsky (1984) (*Journal of Physical Oceanography*, American Meteorological Society).

The triple structure of equatorial ocean surface circulation is presented in Fig. 4.3.1:1 for the Atlantic. The relation between zonal currents and the topography of the free ocean surface is illustrated in Fig. 4.3.1:2, depicting conditions in the low-latitude Atlantic, but pertaining also to the Pacific. Calendar monthly maps of the surface dynamic topography of the tropical Atlantic have recently been produced by Merle and Arnault (1985). Complementing Fig. 4.3.1:2, the scheme Fig. 4.3.1:3 illustrates the arrangement of forces in the equatorial surface current systems to be discussed in Section 4.3.2. Fig. 4.3.1:4 partly duplicates the information content of Fig. 4.3.1:2 except for a sketch of surface winds and ocean currents and a subsurface feature to be discussed further below. Fig. 4.3.1:2 shows minima of sea level near the Equator and 10°N, and a maximum around 5°N. The corresponding slopes of the free sea surface are commensurate with westward flow in the North Equatorial Current, to the North of 10°N. Likewise, the rise of the free sea surface from the Equator into the Southern hemisphere and from the Equator to 5°N is consistent with the westward flow in both the Southern and Northern hemispheric portions of the South Equatorial Current. Finally, the drop of the free sea surface from 5 to 10°N corresponds to the eastward flowing Equatorial Countercurrent.

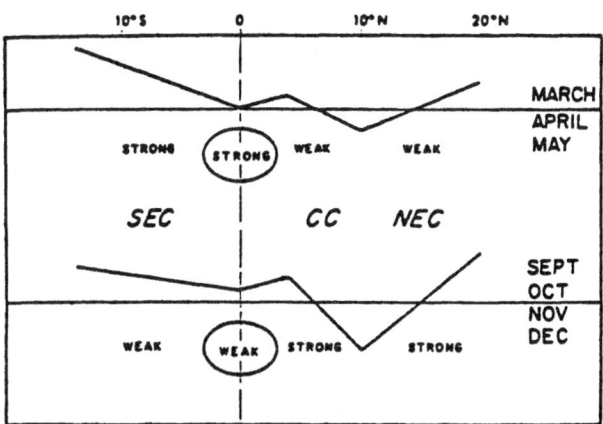

Fig. 4.3.1:5. Schematic diagram of the meridional profile of sea level and the strength of the equatorial currents in the Pacific in Spring and Fall. Closed line below the surface at the Equator denotes the Equatorial Undercurrent. From Wyrtki (1974a) (*Journal of Physical Oceanography*, American Meterological Society).

The surface wind regime undergoes a marked annual cycle. In response to this there are variations of sea surface topography and of currents in the course of the year. These are schematically shown in Fig. 4.3.1:5. This indicates for the Pacific smallest meridional gradients of surface topography in the realm of the North Equatorial Current and Equatorial Countercurrent during March–May, and steepest slopes in September–December. Similarly, Katz (1981) finds for the Atlantic that the mean differences in height between the countercurrent trough, the equatorial ridge, and the South equatorial ridge during February–April are only about half as large as during July–September. Seasonal variations of the Atlantic North Equatorial Countercurrent have been studied by Garzoli and Katz (1983) and Richardson and McKee (1984).

4.3.2. BALANCE OF FORCES

Fig. 4.3.1:3 depicting the balance of forces in the equatorial surface current systems should be considered with reference to Fig. 4.3.1:2. The Equatorial Countercurrent is drawn flowing due East, while the North Equatorial Current and the Southern hemispheric domain of the South Equatorial Current are shown with some poleward components. The pressure gradient vectors in Fig. 4.3.1:3 conform to the meridional slopes of the sea surface sketched in the North-South cross-section, Fig. 4.3.1:2, but are all shown with some eastward component, indicating an overall rise of the free sea surface from the eastern towards the western side of the low-latitude Pacific. This reflects the piling up of waters resulting from the westward wind stress over the vast expanses of the tropical Pacific. The wind stress vectors indicate the Northeast trades over the North Equatorial Current (Fig. 4.3.1:3, part a), the prevalence of the Southeast trades over the Southern hemisphere portion of the South Equatorial Current (Fig. 4.3.1:3, part b), and the cross-equatorial airflow from the Southern hemisphere extending to the realm of the Equatorial Countercurrent (Fig. 4.3.1:3, part c). The Coriolis acceleration appears at right angles to the direction of the surface currents, being directed towards the right in the two Northern hemispheric regions (Fig. 4.3.1:3, parts a and c) and towards the left in the Southern hemisphere (part b of Fig. 4.3.1:3). Fig. 4.3.1:4 also illustrates the pattern of surface winds and ocean currents.

In the realm of North Equatorial Current (Fig. 4.3.1:3, part a), the resultant of Coriolis and pressure gradient accelerations counteract the wind stress effect of the Northeast trades.

The South Equatorial Current straddles the Equator (Fig. 4.3.1:2). In the Southern hemisphere portion of the South Equatorial Current (Fig. 4.3.1:3, part b), the wind stress of the Southeast trades is similarly counterbalanced by the resultant of Coriolis and pressure gradient accelerations. The arrangement of forces is substantially different in the Northern hemispheric portion of the South Equatorial Current (not depicted in Fig. 4.3.1:3). The wind stress is similar in direction but weaker than for the Southern hemispheric portion of the South Equatorial Current (Fig. 4.3.1:3, part b). The Coriolis acceleration can be visualized as being rather smaller and opposite in direction. The pressure gradient is also reversed in direction, as is illustrated in Fig. 4.3.1:2. This constellation of forces would again allow for a balance.

In the Equatorial Countercurrent (Fig. 4.3.1:3, part c), the wind stress associated with the cross-equatorial airstream from the Southern hemisphere is similar in direction but weaker than for the two more southerly domains just considered. This is counteracted by the resultant of Coriolis and pressure gradient accelerations, either of which is oriented broadly opposite to the conditions described for the North Equatorial Current (Fig. 4.3.1:3, part a).

The arrangement of forces illustrated in Fig. 4.3.1:3 is thus directly related to the direction of the large currents and the slope of the free ocean surface depicted in the meridional cross-section Fig. 4.3.1:2. It should be realized, however, that Figs. 4.3.1:2 and 4.3.1:3 provide merely a description of equatorial surface current systems, but no explanations of their origin. This issue is addressed in Sections 4.3.3 and 4.3.4.

4.3.3. THE NORTH EQUATORIAL COUNTERCURRENT

Sverdrup (1947) pioneered an understanding of the dynamics of the North Equatorial Counter-current with a linear steady-state baroclinic theory relating the zonal current systems of the low-latitude Pacific to the field of surface wind stress. The essence of this approach is summarized in the following (Pond and Pickard, 1983, pp. 121–124).

Note from mass continuity that

$$\frac{\partial M_x}{\partial x} + \frac{\partial M_y}{\partial y} = 0. \tag{4.3.3:1}$$

Neglect τ_{yo} and $\partial \tau_{yo}/\partial x$ for the predominantly zonal regimes of the trade wind zones. Note that $y = R \, d\phi$, $\beta = df/dy = d(2\Omega \sin \phi)/R \, d\phi = 2\Omega \cos \phi/R$, where R = earth radius, and $f = 2\Omega \sin \phi$ Coriolis parameter. Then differentiating Eq. (4.1:7) with respect to y gives

$$\frac{\partial M_x}{\partial x} = -\frac{\partial M_y}{\partial y} = \frac{1}{2\Omega \cos \phi} \left(R \frac{\partial^2 \bar{\tau}_{xo}}{\partial y^2} + \frac{\partial \bar{\tau}_{xo}}{\partial y} \tan \phi \right). \tag{4.3.3:2}$$

Integration from the coast, where $x = 0$, and $M_x = 0$, yields

$$M_x = \frac{x}{2\Omega \cos \phi} \left(\frac{\partial \bar{\tau}_{xo}}{\partial y} \tan \phi + \frac{\partial^2 \bar{\tau}_{xo}}{\partial y^2} R \right) \tag{4.3.3:3}$$

Here x (negative) is the distance from a West coast westward to a specific point P in the ocean. The bars over the stress terms indicate averaging over the distance x, and the value of M_x is for the point P.

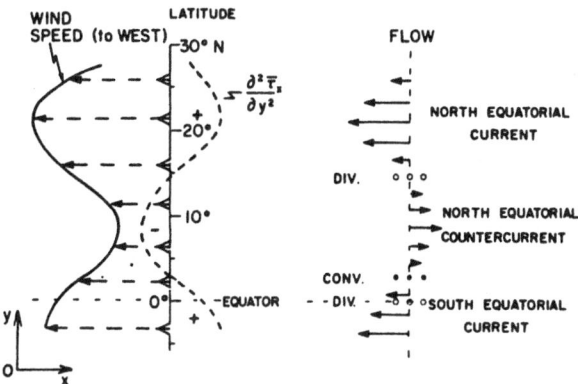

Fig. 4.3.3:1. Smoothed representation of the zonal components of wind speed and stress terms, and related currents at low latitudes in the eastern Pacific Ocean. DIV indicates divergence and CONV convergence. From Pond and Pickard (1983) (reprinted with permission from "S. Pond and G. L. Pickard: *Introductory dynamical oceanography*", copyright 1983, Pergamon Press).

Pond and Pickard (1983, pp. 122–123) point out that in the trade wind and equatorial zones the second term in the bracket in Eq. (4.3.3:3) is dominant. Fig. 4.3.3:1 for the eastern Pacific illustrates schematically meridional profiles of three quantities, namely the zonal wind component, the curvature of the meridional profile of zonal surface stress $\partial^2 \tau_{xo}/\partial y^2$ (second term in the bracket in Eq. (4.3.3:3)), and the zonal ocean surface currents. The following features stand out. North of about 15°N and South of about 2°N $\partial^2 \tau_{xo}/\partial y^2$ is positive, so that with a negative value of x Eq. (4.3.3:3) yields a negative value for M_x, that is westward directed flow, corresponding to the North Equatorial Current and the South Equatorial Current. By contrast, between about 2 and 15°N $\partial^2 \tau_{xo}/\partial y^2$ is negative, so that with a negative value of x the zonal mass transport M_x becomes positive or is directed eastward; this would correspond to the North Equatorial Countercurrent. Thus Fig. 4.3.3:1 shows qualitatively how Sverdrup's (1947) solution explains the existence of the westward flowing North Equatorial and South Equatorial Currents sandwiching the eastward directed North Equatorial Countercurrent.

Sverdrup's (1947) analysis was expanded by Reid (1948). Stommel (1948) and Munk (1950) included friction and added the treatment of western boundary currents. Munk's (1950) results are illustrated in Figs. 4:1 and 4.2:4. Essential to these studies is the meridional profile of zonal wind stress which, while being westward throughout, features a minimum to the North of the Equator (see left-hand side of Figs. 4:1, 4.2:4, and 4.3.3:1).

Pond and Pickard (1983, pp. 126–127) call attention to the limitations of Sverdrup's (1947) results. The theory should be limited to the eastern part of the ocean, because x in Eq. (4.3.3:3) lets M_x increase westward faster than observed. As a reason for this discrepancy Pond and Pickard (1983, pp. 126–127) consider lateral friction between the currents, which would increase with the intensity of the flows. Inasmuch as the equations are integrated from a deep level to the surface, the solution gives the information about the vertical structure of the motion field or subsurface currents such as discussed in Sections 4.3.5 and 4.3.7.

Leetmaa *et al.* (1981) expressed reservations about the apparent success of Sverdrup's (1947) steady-state theory. Mindful of the marked annual and interannual variability of the wind field, they repeated Sverdrup's (1947) calculations using more recent observations including oceanographic sections during 1967–68. They found some similarity to Sverdrup's results but also seasonal variations in intensity of the three currents. With reference to a level of no motion of only 500–1000 m and considering the time variability of wind stress, it remains open whether barotropic and baroclinic adjustments can take place rapidly enough to warrant application of steady-state theory. Leetmaa *et al.* (1981) conclude that it cannot be said whether the Sverdrup relationship provides an accurate description of the currents in the eastern tropical Pacific, and that the test of Sverdrup's theory hinges on the assumed level of no motion.

4.3.4. WIND STRESS, VERTICAL MOTION, THERMOCLINE AND SURFACE TOPOGRAPHY

Associated with the triple structure of current systems discussed in Sections 4.3.1 to 4.3.3, the vertical cross-sections Figs. 4.3.1:2 and 4.3.1:4 show characteristic meridional profiles of free ocean surface, thermocline depth, and vertical motion. The extrema of ocean surface topography are qualitatively consistent with the direction of three three surface currents (ref. Section 4.3.2), and largest upwelling tends to coincide with minimum surface topography and mixed-layer depth, and downwelling with high sea surface and deep mixed layer. It is interesting to consider the meridional profiles of free ocean surface, mixed-layer depth, and vertical motion in the upper ocean, in relation to the surface wind stress pattern.

Gill (1982, pp. 482–497) discusses the North Equatorial Countercurrent in terms of the vorticity budget of atmosphere and ocean. As a consequence of the easterly surface winds, the oceanic Ekman transport is according to Eq. (4.1:5) directed away from the Equator in most of the tropical zone, producing divergence and upwelling at the Equator. However, as is seen from Eqs. (4.1:5) and (4.1:7), because of the poleward increase of the Coriolis parameter, along with the decrease of zonal wind stress, the Ekman transport decreases and the resulting convergence produces downwelling. This combination is found on the equatorward side of the wind minimum, which broadly corresponds to the confluence zone between Northeast trades and cross-equatorial flow from the Southern hemisphere. Proceeding from this latitude poleward towards the core of the trade winds, both zonal wind stress and Coriolis parameter increase, which according to Eq. (4.1:5) would have opposing effects on the latitude variation of the meridional Ekman transport. At any rate, the poleward increasing zonal wind stress would be conducive to divergence, as indicated in Fig. 4.3.3:1, as well as to upwelling, low ocean surface, and shallow thermocline, as shown in Fig. 4.3.1:4. The meridional profiles of ocean surface topography in turn appear relevant to the zonal surface current systems.

The above discussion is pertinent to the predominantly zonal wind regime over the more central portions of the Pacific and Atlantic Oceans. However, somewhat different wind stress mechanisms may be operative in the realm of the clockwise turning cross-equatorial boundary layer flow over the Eastern Equatorial Pacific and Atlantic (Section 6.7.3). Thus, according to Eq. (4.1:5) the Ekman transport in the Southern hemisphere would be directed away from the Equator, involving upwelling to the South of it. Regarding the Northern hemisphere, the recurvature of the surface winds from southeasterly to southwesterly near 5°N (Section 6.7.3) would be conducive to a confluence of Ekman transport around 5°N and hence downwelling. Somewhat similar arguments have been previously proposed by Cromwell (1953).

4.3.5. THE EQUATORIAL UNDERCURRENT

The major features of surface circulation in the low-latitude Pacific and Atlantic Oceans as described above were well explored by the middle of the 20th century. As a surprise came the discovery, in the early 1950's, of an intense eastward flowing current at a depth of about 100 m along the Pacific Equator (Cromwell *et al.*, 1954), by direct current observations. The Equatorial Undercurrent in the Pacific was named Cromwell Current, after one of the discoverers killed in an airplane crash *en route* to another oceanographic expedition. Subsequently Neumann (1960) pointed out that an Equatorial Undercurrent was also inherent in earlier dynamic height determinations in the Equatorial Atlantic. Moreover, Taft and Knauss (1967) quote observational evidence of an Atlantic Equatorial Undercurrent dating back to 1886. The Equatorial Undercurrent has since been confirmed as a quasi-permanent feature of equatorial Atlantic circulation (Metcalf *et al.*, 1962; Katz *et al.*, 1980). The Atlantic Equatorial Undercurrent is also shown schematically in Fig. 4.3.1:1. In the Indian Ocean an Equatorial Undercurrent is a seasonal feature (Swallow, 1964; Bruce, 1973; Luyten and Swallow, 1976; Leetmaa and Stommel, 1980; McPhaden, 1982a, b), as are other current systems in that monsoon ocean.

A series of early studies on the Cromwell Current are published together in an issue of Deep-Sea Research (Wooster, 1960). Further useful summaries are found in Knauss (1963), Moore and Philander (1977), Wyrtki (1979a), Lukas and Firing (1984), and McCreary (1985). The Cromwell Current is symmetrical about the Equator, some 300 km wide and 200 m thick, as defined by the 0.25 m s^{-1} contour, with a core speed of more than 1 m s^{-1} and a transport of about 40×10^6 m^3 s^{-1}. By comparison, the transport of the Atlantic Equatorial Undercurrent is only about 10×10^6 m^3 s^{-1} (Katz *et al.*, 1980). The core of the Cromwell Current rises from more than 100 m depth in the West to less than 50 m depth in the East, as does the thermocline with which it is associated. Lukas (1986) discusses the termination of the Equatorial Undercurrent in the Eastern Pacific. The subsurface features pertinent to the Undercurrent are illustrated in Fig. 4.3.1:4, but not Fig. 4.3.1:2 which predates the discovery of the Undercurrent.

Essential for the origin of the Undercurrent (Fofonoff and Montgomery, 1955; Veronis, 1960; Wyrtki, 1975b) are easterly winds along the Equator (Figs. 4.3.5:1 and 4.3.5:2) and a bounded wind stress (McCreary, 1985). In a non-rotating bounded basin, these winds will pile up water on the western side, and thus give rise to an eastward pressure gradient, which in turn drives the subsurface water back toward the East (Fig. 4.3.5:1). In regions away from the Equator, the Coriolis acceleration gains importance in the flow dynamics, limiting the westward pile-up of waters and ultimately setting latitudinal bounds to the undercurrent phenomenon. To the extent that the horizontal pressure field at some appropriately deep level remains unaffected by the westward rise of the free ocean surface, the higher sea surface to the West must be compensated by a thicker mixed layer with less dense waters. Accordingly, the rise of the sea level from the eastern to the western extremity of the equatorial Pacific of the order of 40 cm is accompanied by a drop of the thermocline of about 150 m (Fig. 4.3.5:2); values for the Atlantic and Indian Oceans being somewhat smaller.

Fofonoff and Montgomery (1955) and Cane (1980) called attention to a further factor considered conducive to the maintenance of the Undercurrent. Easterly wind produces poleward Ekman flow and divergence in the surface layer at the Equator. This is compensated by meridional convergence at depth. Vorticity conservation in the equatorward directed flow is further conducive to a subsurface eastward flowing current at the Equator, but provides by itself no adequate explanation for the origin of the undercurrent.

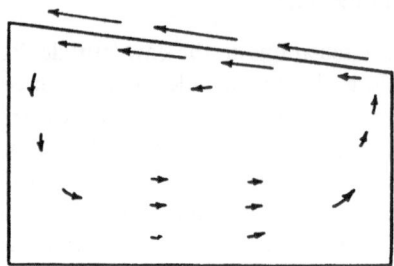

Fig. 4.3.5:1. Sketch of the overturning cellular motion in a confined non-rotating basin when the surface is subjected to a tangential shear stress acting to the left. The pile-up of water towards the left forces a slow return circulation in the deeper parts of the basin. From Veronis (1960) (reprinted with permission from "*Deep-Sea Research*, vol. 6; G. Veronis: *An approximate theoretical analysis of the equatorial undercurrent*", copyright 1960, Pergamon Press).

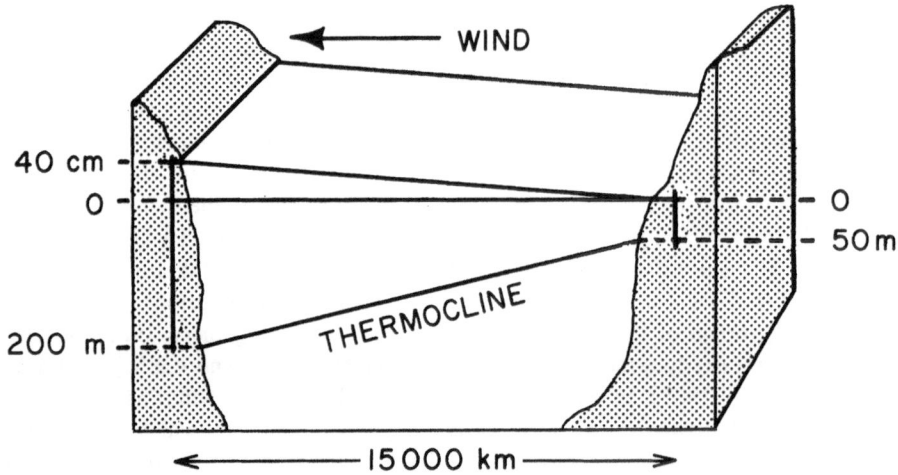

Fig. 4.3.5:2. Schematic diagram of wind, sea surface and thermocline slopes along the Pacific Equator. Source: Wyrtki (1979b).

Regarding adjustments of the mass field, the Equatorial Undercurrent is considered to be in geostrophic balance. Thus the temperature above the core increases with latitude, the pattern being inverse below the core. Such a thickness pattern corresponds to a speed maximum at the thermocline, with both thermocline and current core shoaling eastward (Figs. 4.3.1:4, 4.3.1:5, 4.3.5:2).

Enhancement/reduction of the easterlies along the Equator (Fig. 4.3.5:2) entails increase/ decrease of the slopes of both sea surface and thermocline, and strengthening/weakening of the Undercurrent. Such variations occur as part of the annual cycle (Fig. 4.3.1:5). A complete disappearance of the Pacific Equatorial Undercurrent has been reported during the 1982–83 El Niño (Firing *et al.*, 1983).

4.3.6. EQUATORIAL WAVES AND REMOTE FORCING

Wind stress forcing and oceanic response are recognized as instrumental in the interannual variability of the coupled atmosphere-hydrosphere system. These processes have first been studied by Wyrtki (1973a, 1974a, b, 1975a, b, 1978, 1979a, b), Hurlburt *et al.* (1976), and McCreary (1976). Certain propagating wave phenomena (Kelvin and Rossby waves) of climatic consequence are confined to the vicinity of the Equator, the so-called 'equatorial wave guide' (Gill, 1982, pp. 440–444; Cane and Sarachik, 1976, 1977, 1979; Schopf and Harrison, 1983; Knox and Anderson, 1985). Typical wave speeds for the eastward travelling Kelvin waves are of the order of $1-3$ m s^{-1} (Moore and Philander, 1977; Cane and Moore, 1981), corresponding to time spans of about 2–3 and 1–2 months to cross the Pacific and Atlantic basins, respectively. Thus, these waves are of interest because they can transmit information rapidly along the Equator from one boundary of the ocean basin to the other. As a result, remote forcing is important. Various recent observational studies (Eriksen *et al.*, 1983; Voorhis *et al.*, 1984; Hayes and Halpern, 1984) are concerned with wind-stress induced Kelvin waves in the equatorial Pacific Ocean. The Pacific El Niño phenomenon (Section 8.3), an example of remote forcing, is in fact still at the focus of ongoing research in tropical oceanography. It is already apparent, however, that the effect of surface winds on the current systems of the equatorial oceans is essential in mechanisms of interannual climate variability.

4.3.7. RECENT DISCOVERIES OF SUBSURFACE CURRENTS

Recent cross sections across the current systems of the equatorial Pacific have been published by Eldin (1983), Wyrtki and Kilonsky (1984), and Leetmaa and Molinari (1984). Fig. 4.3.7:1 illustrates the complicated current structure in the equatorial central Pacific. Wyrtki and Kilonsky (1984) evaluated the speed and transport of the various currents. The westward flowing North Equatorial Current transports 23 and the eastward directed North Equatorial Countercurrent 20×10^6 m^3 s^{-1}. The westward flowing South Equatorial Current consists of three branches totalling a transport of 55×10^6 m^3 s^{-1}. The transport by the Equatorial Undercurrent is given as 32×10^6 m^3 s^{-1}. The South Equatorial Countercurrent is weak and transports only about $1-5 \times 10^6$ m^3 s^{-1}. Two eastward flowing subsurface countercurrents were identified by Tsuchiya (1975) in the eastern Pacific, located symmetrically to the Equator on both sides of the equatorial thermostad. These are also apparent in Fig. 4.3.7:1. The Northern Subsurface Countercurrent carries 9×10^6 m^3 s^{-1} and the Southern Subsurface Countercurrent, which consists of two branches, 4×10^6 m^3 s^{-1}. The Equatorial intermediate current varies considerably in depth, and its transport is estimated at about $12-19 \times 10^6$ m^3 s^{-1}. Subsurface Equatorial Countercurrents have also been reported for the Atlantic (Cochrane *et al.*, 1979). McPhaden (1984) studied the dynamics of this phenomenon.

The discoveries of a westward flowing Intermediate Equatorial Undercurrent situated below the Equatorial Undercurrent in the Pacific (Moore and Philander, 1977), of the thermostad (layer of uniform temperature) and associated subsurface countercurrents, and of multiple jet structures at the Equator (Leetmaa *et al.*, 1981) still await comprehensive theoretical explanation. Periodic reviews, workshop proceedings, expedition monographs, newsletters (National Science Foundation, National Oceanic and Atmospheric Administration, 1977; O'Brien, 1979; Cane and Sarachik, 1983; Anonymous, 1978; McCreary *et al.*, 1981; Düing, 1970; Siedler and Woods, 1980; Halpern, 1980–85; Weisberg, 1984; Knox and Anderson, 1985; McCreary, 1985), atlases (Fuglister, 1960; Deutsches Hydrographisches Institut, 1960; U.S. Navy Hydrographic Office, 1966; Robinson, 1976; Robinson *et al.*, 1979; Hastenrath and Lamb, 1977, 1978, 1979; Merle, 1978), and journal publications attest to the increased attention to problems of the equatorial oceans.

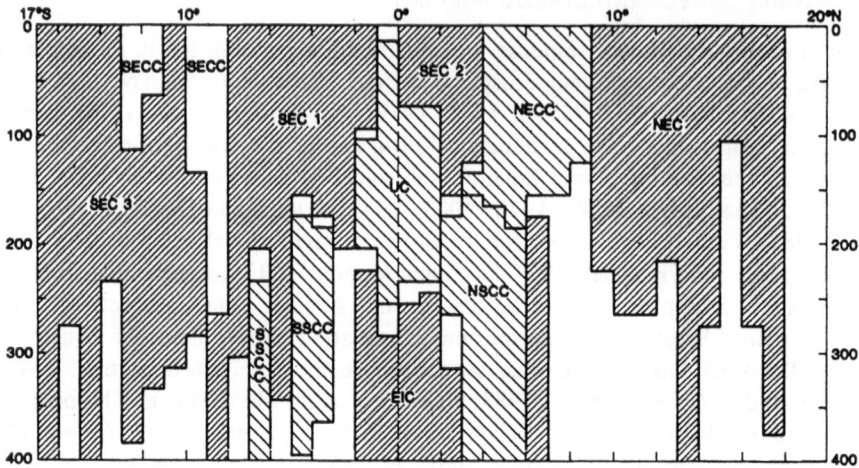

Fig. 4.3.7:1. Areas occupied by the main zonal currents between Hawaii and Tahiti in the upper 400 m. Dark shading is westward flow, light shading eastward flow, and blank areas have zonal speeds less than 2 cm s^{-1}. Letter codes for the various currents are as follows: North Equatorial Current NEC; North Equatorial Countercurrent NECC; South Equatorial Current, three branches, SEC 1, 2, 3; South Equatorial Countercurrent SECC; Equatorial Undercurrent UC; Northern Subsurface Countercurrent NSCC; Southern Subsurface Countercurrent, SSCC; Equatorial Intermediate Current EIC; From Wyrtki and Kilonsky (1984) (*Journal of Physical Oceanography*, American Meteorological Society).

4.4. The Monsoon Ocean

Nowhere else on the globe is the annual reversal of the surface wind regime as spectacular as over the Indian Ocean sector (Section 6.8). The change from the Northern winter Northeast monsoon to the summer Southwest monsoon winds (Figs. 4.4:1 to 4.4:3) is associated with a similarly radical alternation of the ocean surface circulation over the North and equatorial Indian Ocean, while in the Southern tropical Indian Ocean the pattern of surface currents varies only moderately in the course of the year (Figs. 4.4.4a and b). An atlas of surface currents by ten day intervals has just been published by Cutler and Swallow (1984).

In the Southern tropical Indian Ocean an anticyclonic gyre prevails, comparable to the Northern and Southern hemispheric portions of the Pacific and Atlantic (Section 4.2). It features equatorward flow on the eastern side along the coast of Australia, and the poleward directed Agulhas Current in the Mozambique Channel between Madagascar and the African mainland. The westward directed flow on the equatorward side of the gyre is a counterpart to the South Equatorial Current in the Pacific and Atlantic Oceans. The relation of these surface currents to the wind pattern is in broad terms as discussed in Sections 4.2 and 4.3 for the other two tropical oceans.

It is in the Equatorial and North Indian Ocean where surface circulation regimes differ substantially from the Pacific and Atlantic. An arrangement of westward flowing South and North Equatorial Currents enclosing an eastward directed Equatorial Countercurrent is found during the Northern winter (Fig. 4.4:4, part a). However, in contrast to the Pacific and Atlantic (Section 4.3), the Equatorial Countercurrent is situated in the Southern hemisphere; the boundary against the North Equatorial Current being at about 0–3°S, and the limit to the South Equatorial Current at

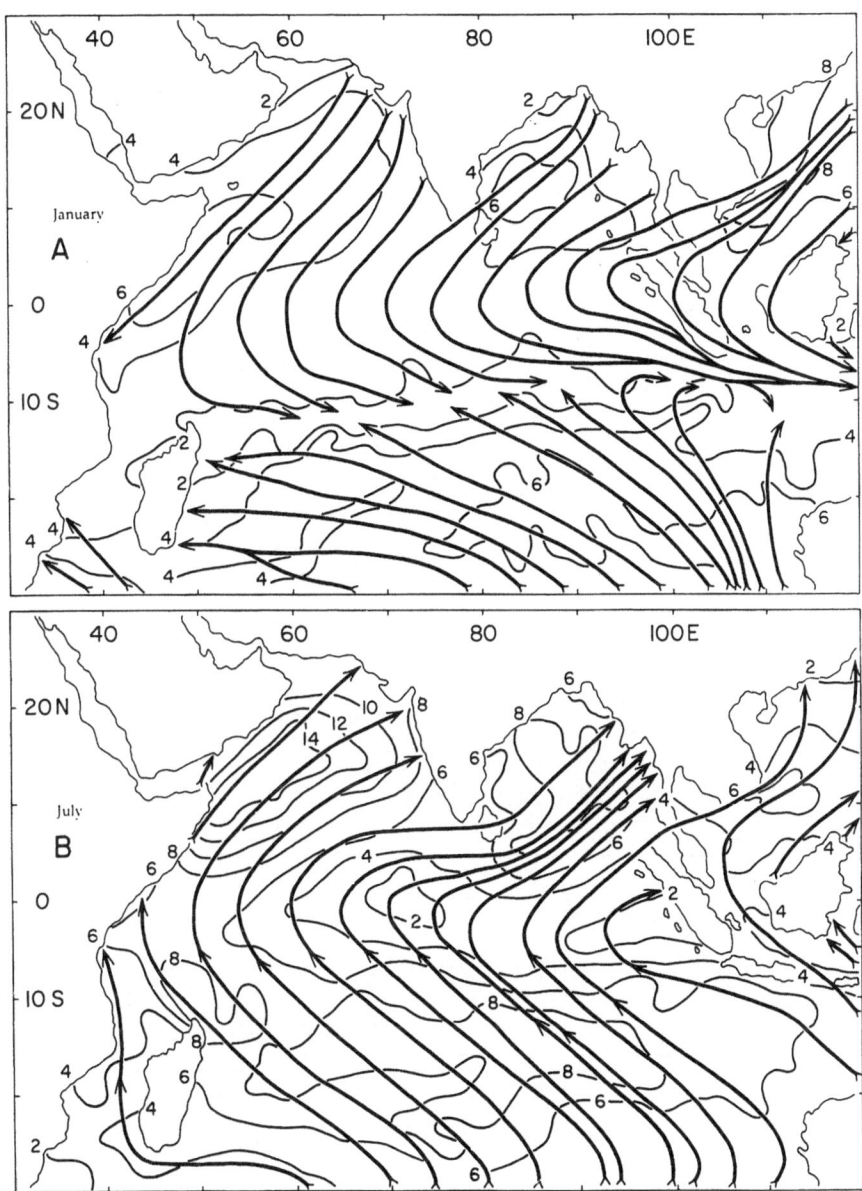

Fig. 4.4:1. Surface resultant wind field over the tropical Indian Ocean in (a) January and (b) July; streamlines and isotachs in m s^{-1}; Source: Hastenrath and Lamb (1980).

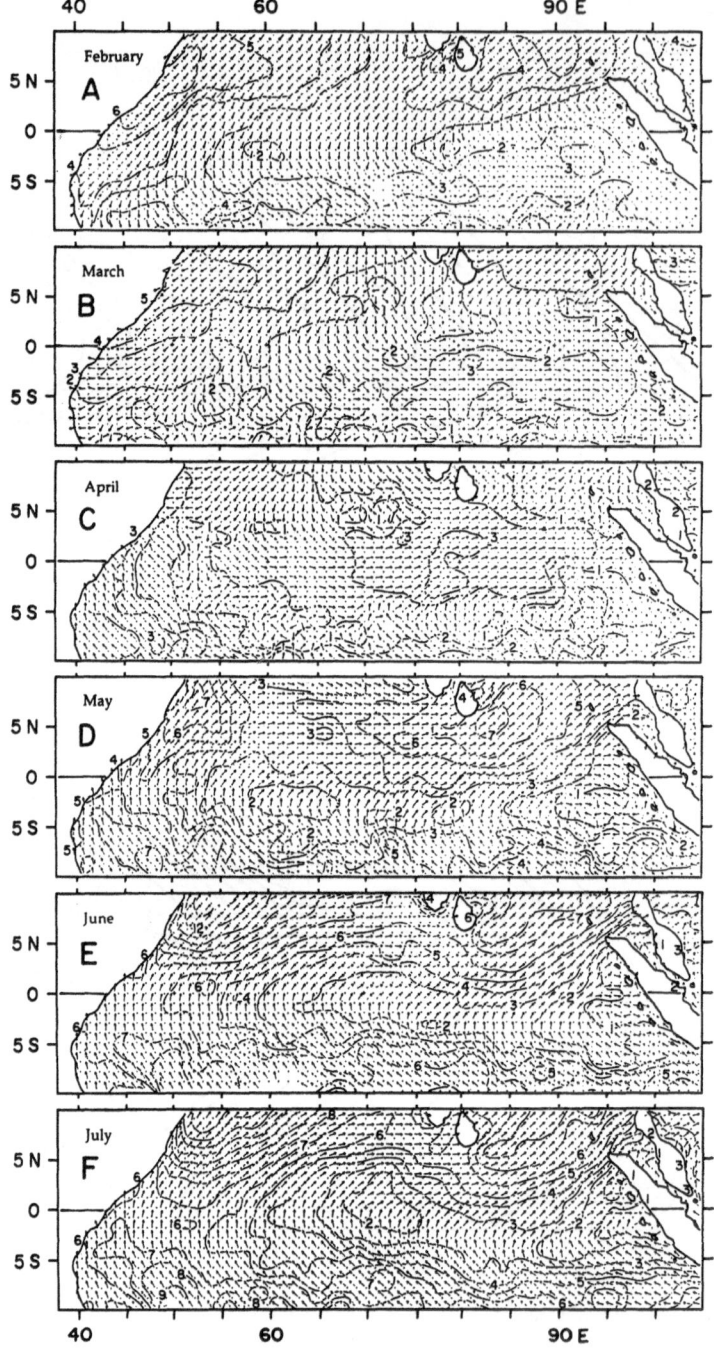

Fig. 4.4:2. Surface resultant wind field over the equatorial Indian Ocean during February (a) through July (b). Isotachs are in m s^{-1}. Source: Hastenrath and Lamb (1979).

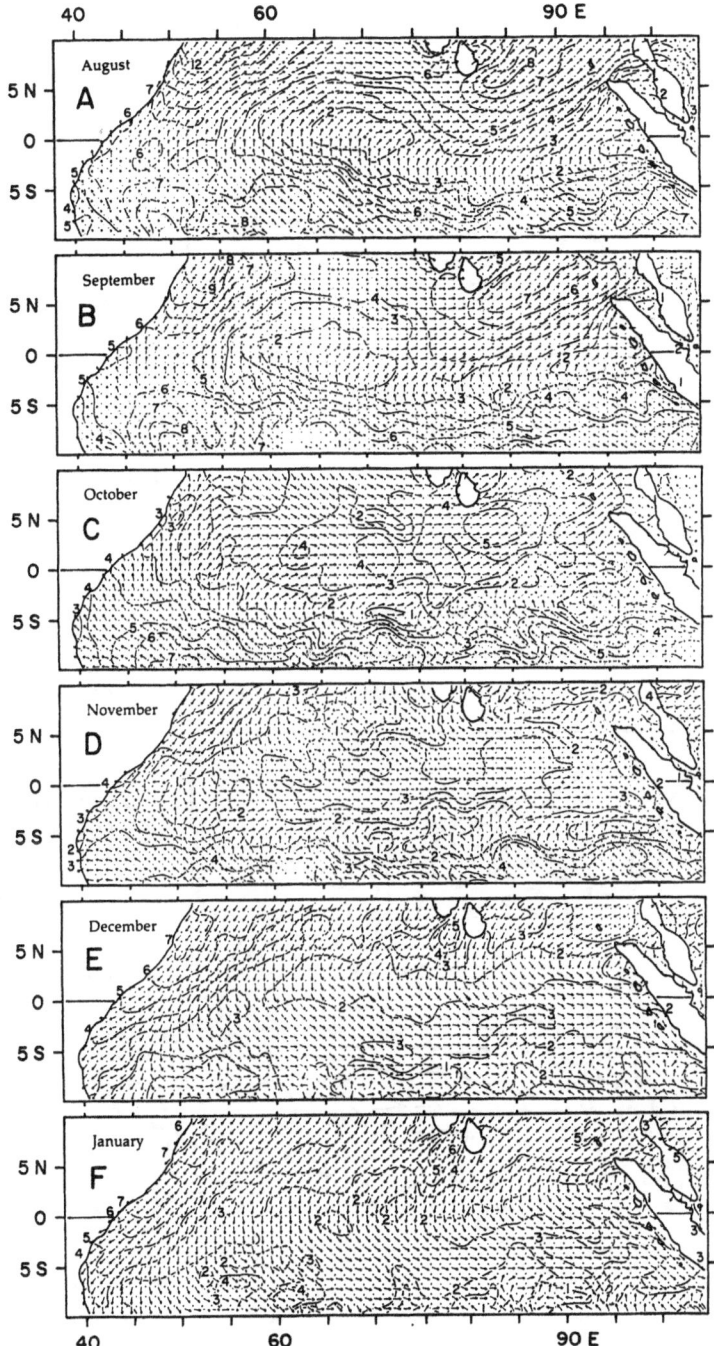

Fig. 4.4:3. Surface resultant wind field over the equatorial Indian Ocean during August (a) through January (b). Isotachs are in m s^{-1}; Source: Hastenrath and Lamb (1979).

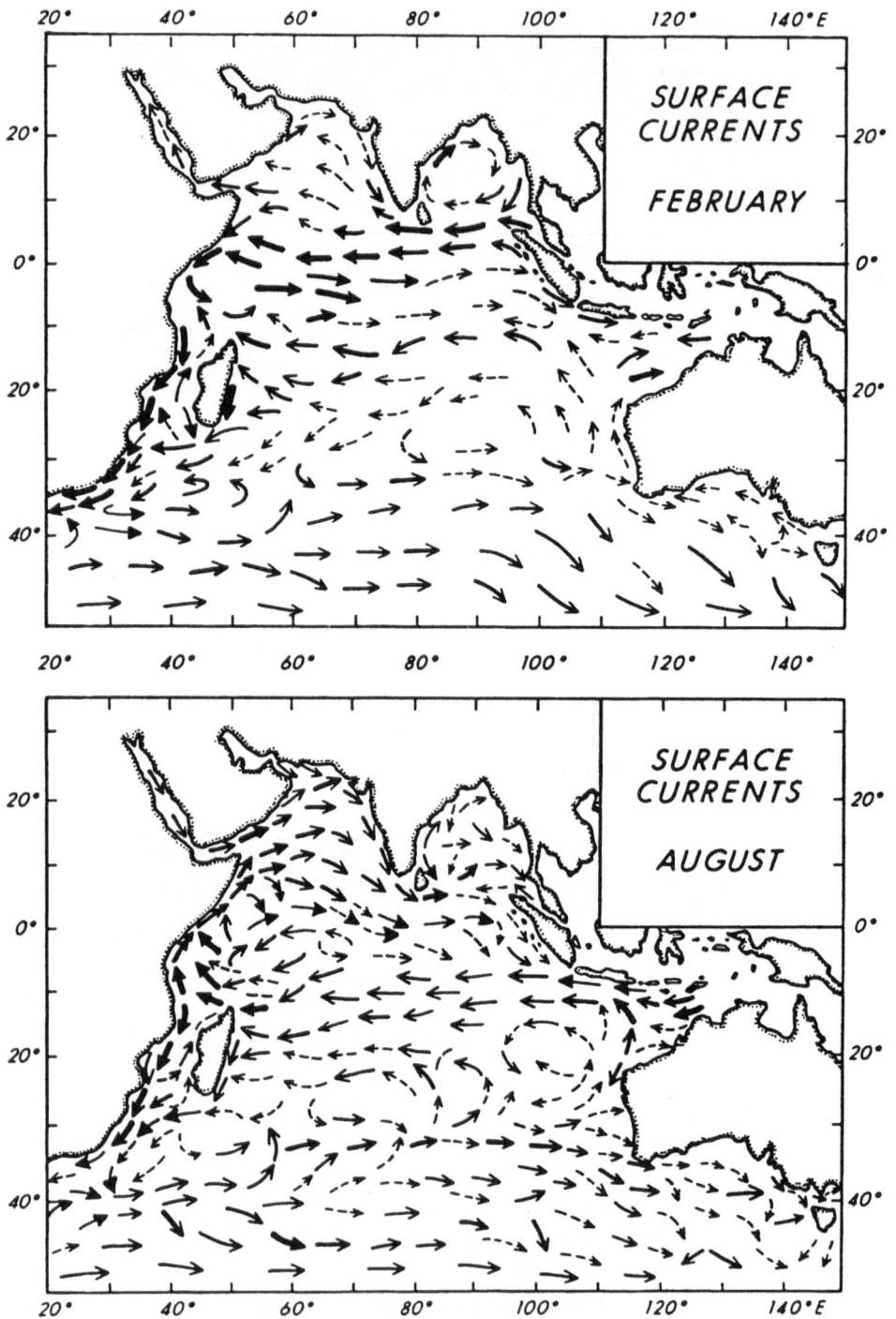

Fig. 4.4:4. Surface currents in the Indian Ocean. (a) Northern hemisphere winter; (b) Northern hemisphere summer. From Düing (1970) (reprinted by permission of University of Hawaii Press, from *The monsoon regime of the currents in the Indian Ocean*, by Walter Düing, copyright 1970 by East–West Center Press).

about 8°S. This appears broadly consistent with the surface wind pattern during Northern winter, when the Northeast monsoon from Southern Asia crosses the Equator extending into a Southern hemispheric trough (Sections 6.7.3, 6.8.3). That is – unlike the Pacific and Atlantic – the belt of weakest winds enclosed between the cores of the Northern and Southern hemispheric trades is located to the South of the Equator. Also, the near-equatorial westerlies of the Southern Indian Ocean are not duplicated in the other two oceans.

The ocean surface circulation changes (Fig. 4.4:4b) with the onset of the Southwest monsoon winds. The Southern hemispheric trades cross the Equator eventually penetrating into Southern Asia, so that the trough of weak winds between the Southern and Northern hemispheric trades is eliminated (Sections 6.7.3, and 6.8.3). Consistent with this is the disappearance of the Equatorial Countercurrent in the South Equatorial Indian Ocean. The South Equatorial Current at the coast of Africa still feeds into the southward flowing Agulhas Current, which in this season becomes weaker. More importantly, however, the South Equatorial Current now develops a continuation into an intense boundary current along the coast of Eastern Africa, the Somali Current. The evolution of this current regime in the Western Indian Ocean has been extensively studied during recent expeditions (Düing and Schott, 1978; Düing et al., 1980: Schott and Quadfasel, 1980, 1982; Swallow, 1981; Leetmaa et al., 1980, 1982; Quadfasel and Schott, 1982). The northward mass transport by the Somali current is estimated at about 10×10^6 m^3 s^{-1} (Düing and Leetmaa, 1980). The development of this intense poleward flowing boundary current in the Western Indian Ocean is associated with marked changes in the sea surface temperature pattern (Hastenrath and Lamb, 1979), and in the spatial distribution of mixed layer depth (Wyrtki, 1971).

The evolution of surface flow conditions during 1979 in particular has been studied by Düing et al. (1980). The northeastward flow of the Somali Current progressed from the Equator in April to 4°N in August. The separation of the current from the coast at its poleward boundary shifted northward in distinct steps. Between 6 and 10°N, that is poleward of the separation latitude, a clockwise gyre with northeastward flow along the coast developed during June. Fig. 4.4:5 illustrates the surface flow conditions during June–July 1979. In a numerical model experiment Luther and O'Brien (1985) succeeded in reproducing many of the observed features of ocean circulation in the Western Arabian Sea, such as the two-gyre pattern, and the timing and displacement of circulation features.

Lighthill (1969) and Wunsch (1977) were the first to model the wind-driven circulation of the Indian Ocean basin. More recent modelling studies of the Indian Ocean include the works of Gent et al. (1983) and Luther et al. (1984). Luyten and Roemmich (1982) find that moored records of zonal velocity in the western equatorial Indian Ocean are dominated by motion of semi-annual period, with an eastward propagating equatorial Kelvin wave and a westward long equatorial Rossby wave. McPhaden (1982a) analyzed a 29 months record at the Island of Gan in the central equatorial Indian Ocean and concludes from the amplitude and phase of semi-annual variations below the mixed layer that energy propagates downward in the form of equatorial Kelvin and non-dispersive Rossby waves. McPhaden (1982a) also calls attention to a wind-forced response at periods between 30 and 60 days. McPhaden (1982b) further suggests that in the equatorial Indian Ocean westerlies suppress upwelling by depressing the thermocline and driving eastward currents with relatively little vertical shear.

In the months of transition between the two monsoon wind systems, the ocean circulation near the Equator becomes complicated and variable. The strong westerly winds along the Equator in the transition between the two monsoons (Hastenrath and Lamb, 1979, charts 16–18, 23–25; and present Fig. 4.4:2, part c, and Fig. 4.4:3, part c) drive an eastward equatorial surface jet, with waters accumulating to the East and then returning through the North and South Equatorial

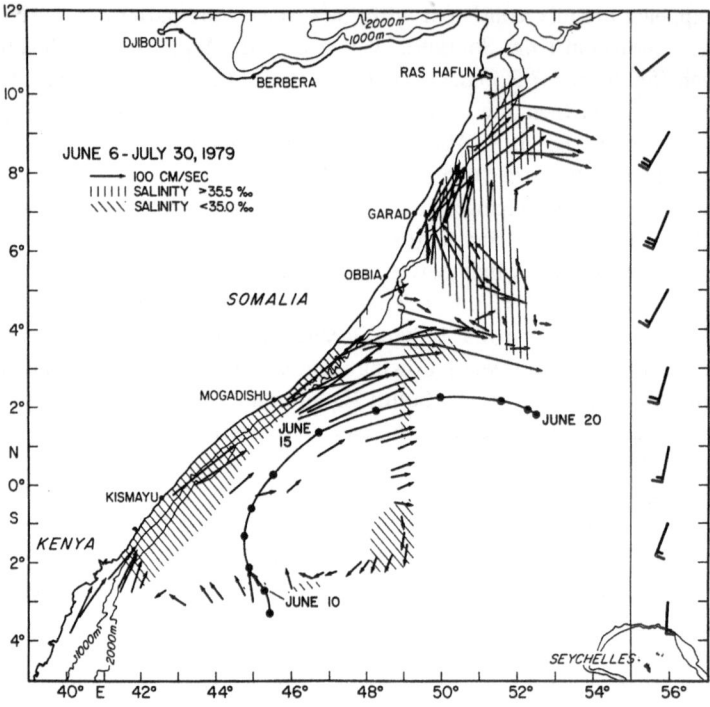

Fig. 4.4:5. Distribution of surface currents in the Western Arabian Sea during June—July 1979. Current arrows are centered on the observation point. Wind data averaged over two degree latitude bands are plotted to the right of the map, with head of wind barbs on the center latitude of each band; one half-feather on a barb representing 5 knots. Satellite-tracked buoy trajectories are shown by the lines connecting closed circles and representing days. From Düing et al., 1980 (Science, 209, 588—590; copyright by American Association for the Advancement of Science).

TABLE 4.4:1

Observations of Equatorial Undercurrent in the Indian Ocean

Source	Date	Longitude E	Core speed cm s^{-1}	Remarks
Taft and Knauss (1967)	March 63	61°	30—80	centered on Equator
	Apr 63	92°	″	″
	March—Apr 63	53°—92°	″	″
Swallow (1964)	18—23 March 64	58°	120	stronger in S hemisphere
	26 Apr—1 May 64	67°30′	120	″
Bruce (1973)	28 Aug—1 Sept 64	55°	50	axis to S of Equator
Leetmaa and Stommel (1980)	Feb—June 75	55°30′	80	in May—June EUC
	Feb—June 76	55°30′	80	moved southward
McPhaden (1982a) (record Jan 73—May 75)	Feb—March 73 Feb—March 75 not in 74	73°10′	50	

Currents (Wyrtki, 1973b; O'Brien and Hurlburt, 1974; Knox, 1976; Eriksen, 1979; Luyten *et al.*, 1980; Leetmaa and Stommel, 1980). The development of the jet is associated with a rise of thermocline to the West and a depression to the East (Wyrtki, 1973b; 1971, charts 330 and 332; Robinson *et al.*, 1979, charts 70 and 140). The westerly wind implies equatorward Ekman flow in both hemispheres, and vorticity conservation implies enhanced eastward flow at the Equator. The convergent Ekman drift results in downwelling at the Equator, thus favoring the downward transport of westerly momentum, which may be a cause for the deep vertical extent of the current (Cane, 1980). The Equatorial Westerly Jet has a speed of about 1 m s^{-1}, performs an eastward transport of about $20 \times 10^6 \text{ m}^3 \text{ s}^{-1}$, and while short-lived is considered instrumental in changing the mass structure in the ocean (Wyrtki, 1973b; Eriksen, 1979).

Corrollaries to the Pacific and Atlantic Equatorial Undercurrent have been observed in the Indian Ocean in the wake of the Northern winter monsoon, in particular around March–April (Swallow, 1964; Taft and Knauss, 1967; Luyten and Swallow, 1976; Leetmaa and Stommel, 1980; McPhaden, 1982a) and around the end of the summer monsoon in late August to early September (Bruce, 1973), with a core around 75 m depth, a speed of the order of 1 m s^{-1}, and an eastward mass transport of $5–25 \times 10^6 \text{ m}^{-3} \text{ s}^{-1}$. The published observations are summarized in Table 4.4:1. It is noted that in contrast to the Pacific and Atlantic, the Equatorial Undercurrent in the Indian Ocean has the tendency to be asymmetric, being stronger to the South than to the North of the Equator.

The factors considered instrumental for the origin of the Pacific and Atlantic Equatorial Undercurrent are reviewed in Section 4.3, indispensable being easterlies near the Equator, where the Coriolis effect vanishes. The map sequences Fig. 4.4:2, parts a to f, and Fig. 4.4:3, parts a to f, are to be examined with this perspective.

Regarding the transition from the Northern winter to the summer monsoon, Table 4.4:1 shows the months of February to April to be of particular interest. The Undercurrent observations during this season are all from longitudes eastward of 55°E. The maps for February to April, Fig. 4.4:2, parts a to c, reveal substantial easterly wind components only to the West of about 55°E where they are, however, also present at the height of the Northern winter monsoon. East of 55°E, winds in March are from the North, changing to westerly from April onward. The thermocline depth at the Equator diminishes from the African coast eastward during March–April (Wyrtki, 1971, chart 329; Robinson *et al.*, 1979, charts 42 and 56), again similar to conditions at the height of the winter monsoon (Wyrtki, 1971, chart 328; Robinson *et al.*, 1979, charts 14 and 28).

Table 4.4:1 contains one published report of an Equatorial Undercurrent at the end of the Northern summer monsoon season, being from 55°E. The map sequence Fig. 4.4:3, parts a to f, gives little indication for easterly wind components along the Equator at this time of year, over the Western Indian Ocean. Both at the height and at the end of the Northern summer monsoon, the thermocline depth at the Equator decreases from the African coast eastward (Wyrtki, 1971, chart 321; Robinson *et al.*, 1979, charts 112, 126).

The sequence of wind maps, Figs. 4.4:2 and 4.4:3 shows that during the limited seasons of monsoon transitions, when an Equatorial Undercurrent exists in the Indian Ocean, conditions for its formation are less straightforward than in the Pacific and Atlantic. Easterlies which would make for a westward pile-up of waters and deepening of the mixed layer are then not prominent. However, the observed zonal slopes of the thermocline along the Equator do appear conducive to Equatorial Undercurrent development. It therefore seems that understanding the causes for the annual cycle of the zonal thermocline profile along the Equator would shed light on the origin of the short-lived Equatorial Undercurrent in the Indian Ocean.

The causes for a hemispheric asymmetry of the Equatorial Undercurrent have been considered by Charney (1960), Charney and Spiegel (1971), Cane (1980), and Gill (1982, p. 483). A northward wind forces the surface layers downwind, thus lowering the thermocline and raising the surface to the North. The resulting pressure gradient amounts to a reduction of the current to the North and a speed increase to the South of the Equator. This would explain the asymmetry at the end of the Northern summer monsoon (Table 4.4:1, Fig. 4.4:3, parts a and b), while conditions may be more complicated at the end of the winter monsoon (Table 4.4:1, Fig. 4.4:2, parts a to c).

An important accomplishment of the International Indian Ocean Expedition in the 1960's was the comprehensive mapping of the subsurface distribution of pertinent variables (Wyrtki, 1971). The oceanographic programs of the 1970's, in particular INDEX (1975–76) and MONEX (1978–79), have produced a wealth of further information. This work is expected to result in an updated documentation and understanding of Indian Ocean circulation (Knox and Anderson, 1985) and provides the basis for the planning of future research (Anonymous, 1984).

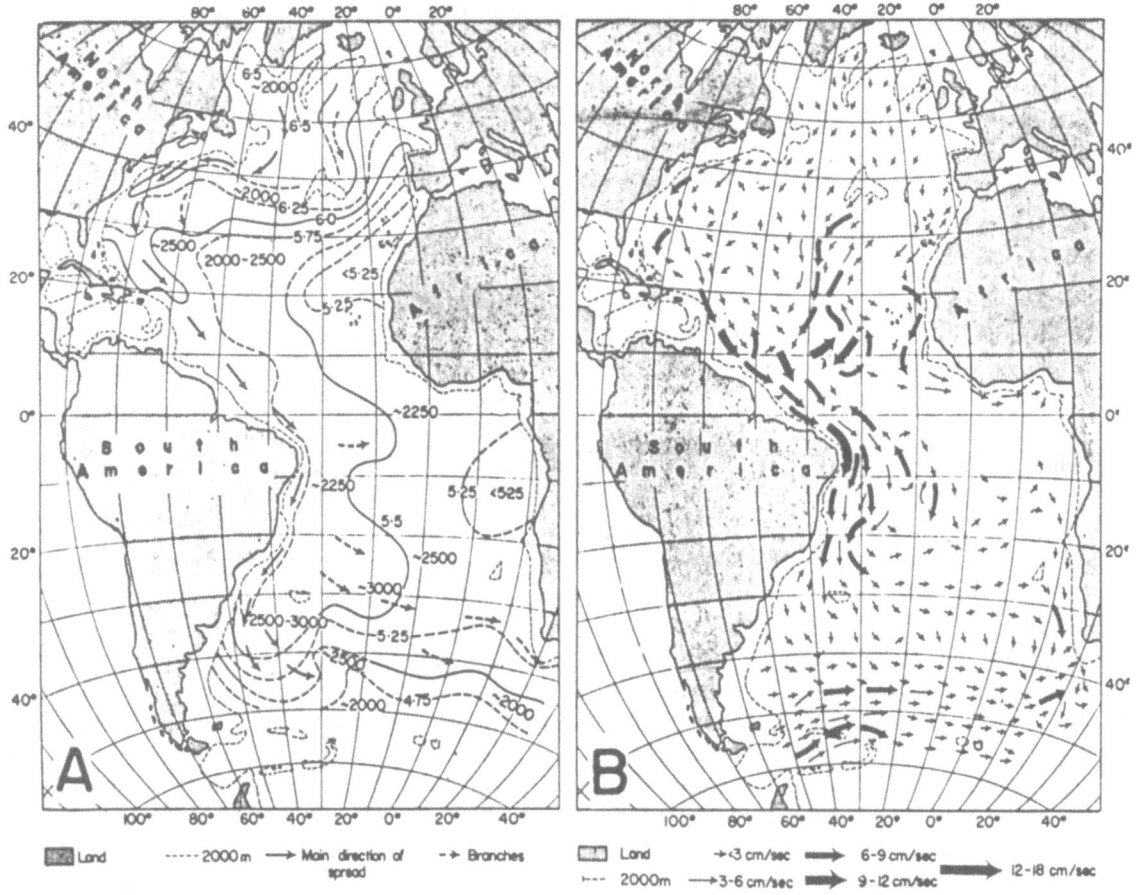

Fig. 4.5:1. Deep circulation of the Atlantic. (a) spreading of the core masses of the mean North Atlantic Deep Water; O_2 content in the layer of the intermediate oxygen maximum in a depth of about 2000–3000 m; numbers larger than 100 indicate depth of this maximum; (b) current field at depth of 2000 m, computed from absolute topography of 2000 decibar surface. From Defant (1961), vol. 1 (reprinted with permission from "A. Defant: *Physical Oceanography*", copyright 1961, Pergamon Press).

4.5. Deep Circulation

Unlike the surface flow pattern, the deep circulation of the World ocean is not readily amenable to direct current measurements. Insight has been gained from dynamic topographies, the distribution of water mass properties, and theoretical considerations.

The high latitudes, in particular the northern North Atlantic and the Weddell Sea sector of the Antarctic, are thought to be the major source regions of deep water. Through interaction with the atmosphere cold and dense water masses are formed, which sink on their travel equatorward. Streams of water masses of different origin are located at intermediate layers.

The picture of the deep circulation is somewhat less sketchy for the Atlantic than for the other ocean basins. Fig. 4.5:1 depicts the deep circulation of the Atlantic deduced from both water mass properties and dynamic topographies. Both approaches indicate a concentration of southward flow along the coast of the Americas from the North Atlantic across the Equator into the South Atlantic. Fig. 4.5:2 similarly shows a southward flow in the bottom layer concentrated in the westernmost part of the Atlantic, and thus directed broadly opposite to the flow in the surface layer. The maps Figs. 4.5:1 and 4.5:2 are complemented by the scheme of circulation in a meridional-vertical plane in the Atlantic, Fig. 4.5:3. Concerning the World ocean, only Stommel's (1958) global view can be offered in Fig. 4.5:4, the most prominent features being the North Atlantic and Weddell Sea source regions and deep western boundary currents in all three oceans. Although little explored, the deep ocean circulation must be regarded as instrumental in the functioning of the global climate system on the long time scales. Important insight can be expected from recent efforts in numerical modelling of the global ocean circulation.

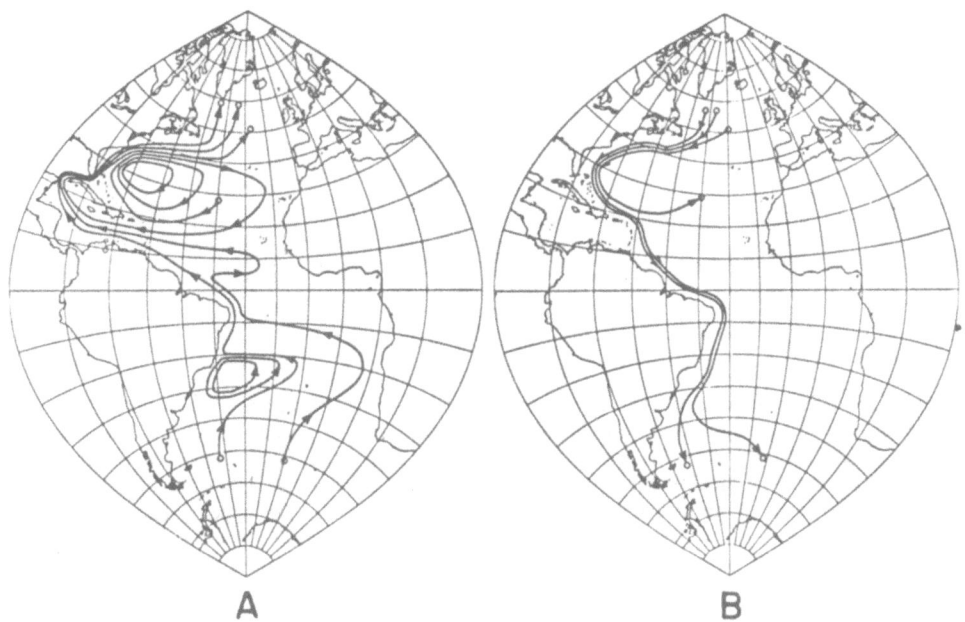

A   B

Fig. 4.5:2. Atlantic Ocean circulation; (a) surface layer, and (b) bottom layer. From Stommel (1957, 1976).

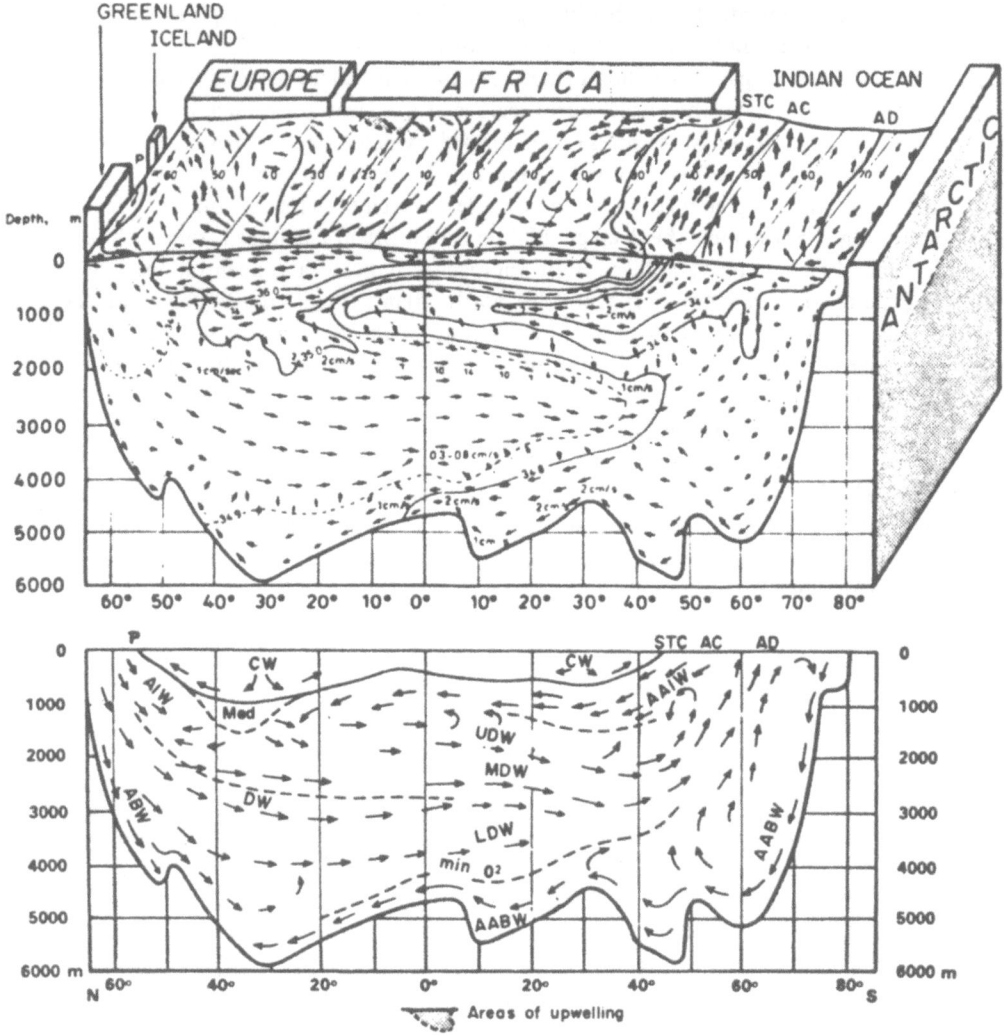

Fig. 4.5:3. Circulation in a meridional-vertical plane in the Atlantic. Notations are as follows: ABW Arctic bottom water, AABW Antarctic bottom water, AAIW Antarctic intermediate water, AIW Arctic intermediate water, CW central water, DW Atlantic deep water, UDW upper deep water, MDW mean deep water, LDW lower deep water, Med water of Mediterranean origin, AC Antarctic Convergence, AD Antarctic Divergence, STC Subtropical Convergence, P Arctic Polar Front. From Tchernia (1980). (The upper portion of the figure is from Wüst, 1949, 1950; the lower portion is from: Gerhard Neumann, Willard J. Pierson, *Principles of physical oceanography*, 1966, p. 466, reprinted by permission of Prentice-Hall Inc., Englewood Cliffs, N.J.)

Fig. 4.5:4. Deep circulation of the World ocean. Dots indicate the North Atlantic and Weddell Sea source regions. From Stommel (1958) (reprinted with permission from *"Deep-Sea Research*, vol. 5; H. Stommel: *The abyssal circulation"*, copyright 1958, Pergamon Press).

4.6. Synthesis

The surface circulation of the low-latitude oceans is characterized by anticyclonic gyres centered in the subtropics and predominantly zonal flow regimes in the equatorial region. The subtropical gyres are related to the surface wind pattern, but feature a concentration of poleward flow in western boundary currents due to the Coriolis parameter varying with latitude.

In the equatorial zone of the Pacific and Atlantic, a westward flowing South Equatorial Current occupies a domain extending from the Southern hemisphere to about 5°N, the westward directed North Equatorial Current has a southern limit of about 10°N, and enclosed between these two westward directed streams the Equatorial Countercurrent flows eastward, that is opposite to the surface winds. These three currents all owe their existence to the meridional variation of the zonal wind stress. An eastward flowing Equatorial Countercurrent at a depth of about 100 m centered along the Equator results from the westward pile-up of waters due to the surface wind stress. Annual and interannual variations in the strength of the easterlies along the Equator are reflected in changes of equatorial ocean circulation, of particular interest being the fast travelling equatorially trapped waves.

The surface circulation of the Indian Ocean is subject to the reversals between the Northern winter and summer monsoon wind regimes. During Northern winter, the westward flowing North and South Equatorial Currents enclose an eastward directed Equatorial Countercurrent situated to the South of the Equator, consistent with a belt of weakest winds. During the Northern summer Southwest monsoon, winds sweep from the Southern hemisphere across the Equator into Southern

Asia. As a consequence, the Equatorial Countercurrent as well as the North Equatorial Current are not apparent during that season. The South Equatorial Current persists and feeds into an intense northward directed boundary current hugging the coast of Eastern Africa. During the transition seasons, the circulation of the equatorial Indian Ocean becomes complicated and variable.

Concerning the deep circulation, the Northern North Atlantic and the Weddell Sea sector of the Antarctic are considered as the major source regions of deep water. In the Atlantic, a southward flow in the bottom layer is concentrated along the coast of the Americas, that is broadly opposite to the surface flow pattern. The picture of deep circulation for the other ocean basins is more sketchy. However, circulations in a vertical plane merit particular attention for the meridional heat and freshwater transports in the World ocean, a subject to be discussed in Chapter 5.

References

Anonymous, 1978: *Review papers of equatorial oceanography*. FINE Workshop Proceedings, June–Aug. 1977, SIO, La Jolla, Calif. 477 pp.

Anonymous, 1984: *Proceedings of a workshop on the oceanography of the Indian Ocean, Nov. 1984*, University of Miami, Miami, Fla., 135 pp.

Armi, L., Stommel, H., 1983: 'Four views of a portion of the North Atlantic subtropical gyre'. *J. Phys. Oceanogr.*, 13, 828–857.

Bjerknes, J., 1959: 'The recent warming of the North Atlantic'. pp. 65–73 in B. Bolin, ed.: *The atmosphere and the sea in motion*, Rossby Memorial Volume, Rockefeller Institute Press, New York, 509 pp.

Bruce, J. G., 1973: 'The equatorial undercurrent in the Western Indian Ocean during the Southwest monsoon'. *J. Geophys. Res.*, 32, 419–423.

Cane, M. A., 1980: 'On the dynamics of equatorial currents, with application to the Indian Ocean'. *Deep Sea Res.*, 27A, 525–544.

Cane, M. A., Moore, D. W., 1981: 'A note on low-frequency equatorial basin modes'. *J. Phys. Oceanogr.*, 11, 1578–1584.

Cane, M. A., Sarachik, E. S., 1976: 'Forced baroclinic ocean motions, I: The linear equatorial unbounded case'. *J. Mar. Res.*, 34, 629–665.

Cane, M. A., Sarachik, E. S., 1977: 'Forced baroclinic ocean motions, II: The linear equatorial bounded case'. *J. Mar. Res.*, 35, 395–432.

Cane, M. A., Sarachik, E. S., 1979: 'Forced baroclinic ocean motions, III: The linear equatorial basin case'. *J. Mar. Res.*, 37, 253–299.

Cane, M. A., Sarachik, E. S., 1983: 'Equatorial oceanography'. *Rev. Geophys. and Space Phys.*, 21, 1137–1148.

Charney, J. G., 1960: 'Non-linear theory of a wind-driven homogeneous layer near the Equator'. *Deep Sea Res.*, 6, 303–310.

Charney, J. G., Spiegel, S. L., 1971: 'The structure of wind-driven currents in homogneous oceans'. *J. Phys. Oceanogr.*, 1, 149–160.

Cochrane, J. D., Kelly, F. J., Olling, C. R., 1979: 'Subthermocline countercurrents in the western equatorial Atlantic Ocean'. *J. Phys. Oceanogr.*, 9, 724–738.

Cromwell, T., 1953: 'Circulation in a meridional plane in the central equatorial Pacific'. *J. Mar. Res.*, 12, 196–213.

Cromwell, T., Montgomery, R. B., Stroup, E. D., 1954: 'Equatorial undercurrent in Pacific Ocean revealed by new method'. *Science*, 119, 648–649.

Cutler, A. N., Swallow, J. C., 1984: 'Surface currents of the Indian Ocean (to 25°S, 100°E): compiled from historical data archived by the Meteorological Office, Bracknell, U.K.' Institute of Oceanographic Sciences, Wormley, U.K., Report No. 187, 42 pp.

Defant, A., 1961: *Physical Oceanography*. 2 vols. Pergamon Press, Oxford, New York, 727 and 598 pp.

Deutsches Hydrographisches Institut, 1960: 'Monatskarten für den Indischen Ozean'. Publ. No. 2422, third edition, Hamburg, 50 pp.

Düing, W., 1970: 'The monsoon regime of the currents in the Indian Ocean'. International Indian Ocean Expedition, Oceanographic Monograph No. 1, University Press of Hawaii, Honolulu, 68 pp.

Düing, W., Leetmaa, A., 1980: 'Arabian Sea cooling: a preliminary heat budget'. *J. Phys. Oceanogr.*, 10, 307–312.

Düing, W., Molinari, R. L., Swallow, J. C., 1980: 'Somali current: evolution of surface flow'. *Science,* 209, 588–589.

Düing, W., Schott, F., 1978: 'Measurements in the source region of the Somali Current during the monsoon reversal'. *J. Phys. Oceanogr.,* 8, 278–289.

Eldin, G., 1983: 'Eastward flows of the South equatorial central Pacific'. *J. Phys. Oceanogr.,* 13, 1461–1467.

Eriksen, C. C., 1979: 'An equatorial transect of the Indian Ocean'. *J. Mar. Res.,* 37, 215–232.

Eriksen, C. C., Blumenthal, M. B., Hayes, S. P., Ripa, P., 1983: 'Wind-generated equatorial Kelvin waves observed across the Pacific Ocean'. *J. Phys. Oceanogr.,* 13, 1622–1640.

Firing, E., Lukas, R., Sadler, J., Wyrtki, K., 1983: 'Equatorial Undercurrent disappears during 1982–1983 El Niño'. *Science,* 222, 1121–1123.

Fofonoff, N. P., Montgomery, R. B., 1955: 'The Equatorial Undercurrent in the light of the vorticity equation'. *Tellus,* 7, 518–521.

Fuglister, F. C., 1960: *Atlantic Ocean atlas of temperature and salinity profiles and data from the International Geophysical Year of 1957–58.* Woods Hole Oceanographic Institution, Atlas series, vol. 1, Woods Hole, Mass., 209 pp.

Garzoli, S. L., Katz, E. J., 1983: 'The forced annual reversal of the Atlantic North Equatorial Countercurrent'. *J. Phys. Oceanogr.,* 13, 2082–2090.

Gent, P. R., O'Neill, K., Cane, M. A., 1983: 'A model of the semiannual oscillation in the equatorial Indian Ocean'. *J. Phys. Oceanogr.,* 13, 2148–2160.

Gill, A., 1982: *Atmosphere-ocean dynamics.* Academic Press, New York, London, Paris, San Diego, San Francisco, Saõ Paulo, Sidney, Tokyo, Toronto, 662 pp.

Halpern, D., ed., 1980–85: 'Tropical Ocean-Atmosphere Newsletter'. NOAA Pacific Marine Environmental Laboratory, Seattle, nos. 1–35.

Hastenrath, S., Lamb, P. J., 1977: *Climatic atlas of the tropical Atlantic and Eastern Pacific Oceans.* University of Wisconsin Press, 112 pp.

Hastenrath, S., Lamb, P. J., 1978: *Heat budget atlas of the tropical Atlantic and Eastern Pacific Oceans.* University of Wisconsin Press, 103 pp.

Hastenrath, S., Lamb, P. J., 1979: *Climatic atlas of the Indian Ocean.* Part 1. Surface climate and atmospheric circulation. Part 2. The oceanic heat budget. University of Wisconsin Press, 116 and 110 pp.

Hastenrath, S., Lamb, P. J., 1980: 'On the heat budget of hydrosphere and atmosphere in the Indian Ocean'. *J. Phys. Oceanogr.,* 10, 694–708.

Hayes, S. P., Halpern, D., 1984: 'Correlation of current and sea level in the eastern equatorial Pacific'. *J. Phys. Oceanogr.,* 14, 811–824.

Hurlburt, H. E., Kindle, J., O'Brien, J. J., 1976: 'A Numerical simulation of the onset of El Niño'. *J. Phys. Oceanogr.,* 6, 621–631.

Katz, E. J., 1981: 'Dynamic topography of the sea surface in the equatorial Atlantic'. *J. Mar. Res.,* 39, 53–63.

Katz, E. J., Bruce, J. G., Petrie, B. D., 1980: 'Salt and mass flux in the Atlantic Equatorial Undercurrent'. *Deep Sea Res., Part A,* vol. 26, Suppl. 1, 137–160.

Knauss, J. A., 1963: 'Equatorial current systems'. pp. 235–252, in M. N. Hill, ed.: *The Sea,* vol. 2, Interscience Publishers, New York, London, 554 pp.

Knox, R. A., 1976: 'On a long series of measurement of Indian Ocean equatorial currents near Addu Atoll'. *Deep Sea Res.,* 23, 211–221.

Knox, R., Anderson, D. L. T., 1985: 'Recent advances in the study of the low-latitude ocean circulation'. *Progr. Oceanogr.,* 14, 259–317.

Leetmaa, A., McCreary, J. P., Moore, D. W., 1981: 'Equatorial currents; observations and theory'. pp. 184–196 in Warren, B. A., Wunsch, C., eds.: *Evolution of physical oceanography,* MIT Press, 623 pp.

Leetmaa, A., Molinari, R. L., 1984: 'Two cross-equatorial sections at 110°W'. *J. Phys. Oceanogr.,* 14, 255–263.

Leetmaa, A., Quadfasel, D. R., Wilson, D., 1982: 'Development of the flow field during the onset of the Somali Current, 1979'. *J. Phys. Oceanogr.,* 12, 1325–1342.

Leetmaa, A., Rossby, H. T., Saunders, P. M., Wilson, P., 1980: 'Subsurface circulation in the Somall Current'. *Science,* 209, 590–592.

Leetmaa, A., Stommel, H., 1980: 'Equatorial current observations in the Western Indian Ocean in 1975 and 1976'. *J. Phys. Oceanogr.,* 10, 258–269.

Lighthill, M. J., 1969: 'Dynamic response of the Indian Ocean to onset of the Southwest monsoon'. *Phil. Trans. Roy. Soc. London,* A265, 45–92.

Lukas, R., 1986: 'The termination of the Equatorial Undercurrent in the eastern Pacific'. *Progr. Oceanogr.,* submitted.

Lukas, R., Firing, E., 1984: 'The geostrophic balance of the Pacific equatorial undercurrent'. *Deep-Sea Res.,* 31, 61–66.

Luther, M. E., O'Brien, J. J., 1985: 'A model of the seasonal circulation in the Arabian Sea forced by observed winds'. *Progr. Oceanogr.*, 14, 353–385.

Luther, M. E., O'Brien, J. J., Meng, A. H., 1984: 'Morphology of the Somali Current system during the Southwest monsoon'. Proceedings of the JSC/CCCO 16th International Liège Colloquium 'Coupled atmosphere-ocean models'.

Luyten, J. R., Fieux, M., Gonella, J., 1980: 'Equatorial currents in the western Indian Ocean'. *Science*, 209, 600–602.

Luyten, J. R., Roemmich, D. H., 1982: 'Equatorial currents of semi-annual period in the Indian Ocean'. *J. Phys. Oceanogr.*, 12, 406–413.

Luyten, J. R., Swallow, J. C., 1976: 'Equatorial undercurrents'. *Deep Sea Res.*, 23, 999–1001.

McCreary, J. P., 1976: 'Eastern tropical ocean response to changing wind systems: with application to El Niño'. *J. Phys. Oceanogr.*, 6, 632–645.

McCreary, J. P., 1985: 'Modeling equatorial ocean circulation'. *Ann. Rev. Fluid Mech.*, 17, 359–409.

McCreary, J. P., Moore, D. W., Witte, J. M., eds., 1981: *Recent progress in equatorial oceanography*. Report of the final meeting of SCOR Working Group 47, Venice, Italy, April 1981. Nova University, N.Y.I.T. Press, 466 pp.

McLellan, H. J., 1965: *Elements of physical oceanography*. Pergamon Press, Oxford, London, Edinburgh, New York, Paris, Frankfurt, 180 pp.

McPhaden, M. J., 1982a: 'Variability in the central equatorial Indian Ocean. Part 1: Ocean dynamics'. *J. Mar. Res.*, 40, 157–176.

McPhaden, M. J., 1982b: 'Variability in the central equatorial Indian Ocean. Part 2: Oceanic heat and turbulent energy balances'. *J. Mar. Res.*, 40, 403–419.

McPhaden, M. J., 1984: 'On the dynamics of Equatorial Subsurface Countercurrents'. *J. Phys. Oceanogr.*, 14, 1216–1225.

Merle, J., 1978: 'Atlas hydrologique saisonnier de l'Océan Atlantique Intertropical'. Trav. Doc. ORSTOM, No. 82, 184 pp., 153 cartes.

Merle, J., Arnault, S., 1985: 'Seasonal variability of the surface dynamic topography in the tropical Atlantic Ocean'. *J. Mar. Res.*, 43, 267–288.

Metcalf, W. G., Voorhis, A. D., Stalcup, N. C., 1962: 'The Atlantic equatorial undercurrent'. *J. Geophys. Res.*, 67, 2499–2508.

Moore, D. W., Philander, S. G. H., 1977: 'Modelling of the tropical ocean circulation'. pp. 319–361, in Goldberg, E. D., McCave, I. N., O'Brien, J. J., Steele, J. H., eds.: *The Sea*, vol. 6, Inter-Science Publishers, New York, London, Sidney, Toronto, 1048 pp.

Munk, W. H., 1950: 'On the wind-driven ocean circulation'. *J. Meteor.*, 7, 79–93.

National Science Foundation, National Oceanic and Atmospheric Administration, 1977: *Report of the U.S. GATE Central Program Workshop, July–Aug 1977*, NCAR, Boulder, Colo., 723 pp.

Neumann, G., 1960: 'Evidence for an equatorial undercurrent in the Atlantic Ocean'. *Deep Sea Res.*, 6, 328–334.

Neumann, G., Pierson, W. J., 1966: *Principles of physical oceanography*. Prentice-Hall, Inc., Englewood Cliffs, N.J., 545 pp.

O'Brien, J. J., 1979: 'Equatorial oceanography'. *Rev. Geophys. Space Phys.*, 17, 1569–1575.

O'Brien, J. J., Hurlburt, H. E., 1974: 'Equatorial jet in the Indian Ocean: theory'. *Science*, 184, 1075–1077.

Philander, G., 1977: 'The large-scale oceanic circulation and its variability as observed during GATE'. pp. 79–89, in NSF–NOAA: *Report of the U.S. GATE Central Program Workshop, July–Aug 1977*, NCAR, Boulder, Colo., 723 pp.

Pond, S., Pickard, G. L., 1983: *Introductory dynamic oceanography*. Second edition, Pergamon Press, Oxford, New York, Toronto, Sidney, Paris, Frankfurt, 329 pp.

Quadfasel, D. R., Schott, F., 1982: 'Water-mass distributions at intermediate layers off the Somali coast during the onset of the Southwest monsoon'. *J. Phys. Oceanogr.*, 12, 1358–1372.

Reid, R. O., 1948: 'The equatorial currents of the Eastern Pacific as maintained by the stress of the wind'. *J. Mar. Res.*, 7, 74–99.

Richardson, P.L., McKee, T. K., 1984: 'Average seasonal variation of the Atlantic North Equatorial Counter-current from ship drift data'. *J. Phys. Oceanogr.*, 14, 1226–1238.

Robinson, M. K., 1976: *Atlas of North Pacific Ocean monthly mean temperatures and mean salinities of the surface layer*. NAVOCEANO, Washington, D.C.

Robinson, M. K., Bauer, R. A., Schroeder, E. H., 1979: *Atlas of North Atlantic – Indian Ocean monthly mean temperatures and mean salinities of the surface layer*. Naval Oceanographic Office, NOO RP-18, NSTL Station, Bay St. Louis, Mississippi, 234 pp.

Schopf, P. S., Harrison, D. E., 1983: 'On equatorial waves and El Niño. I: influence of initial states on wave-induced currents and warming'. *J. Phys. Oceanogr.*, 13, 936–948.

Schott, F., Quadfasel, D. R., 1980: 'Development of the subsurface currents of the Northern Somali current gyre from March to July 1979'. *Science*, 209, 593–595.

Schott, F., Quadfasel, D. R., 1982: 'Variability of the Somali Current system during the onset of the Southwest monsoon, 1979'. *J. Phys. Oceanogr.*, 12, 1343–1357.

Siedler, G., Woods, J. D., 1980: 'Oceanography and surface layer meteorology in the B/C – scale; GATE vol. 1'. Suppl. 1. to *Deep-Sea Research Part A*, vol. 26, Pergamon Press, 294 pp.

Stommel, H. M., 1948: 'The westward intensification of wind-driven ocean currents'. *Trans. Amer. Geophys. Union*, 29, 202–206.

Stommel, H., 1957: 'A survey of ocean current theory'. *Deep Sea Res.*, 4, 149–184.

Stommel, H., 1958: 'The abyssal circulation'. *Deep Sea Res.*, 5, 80–82.

Stommel, H. M., 1976: *The Gulf Stream, a physical and dynamical description*. Second edition, University of California Press, Berkeley, 248 pp.

Sverdrup, H. U., 1947: 'Wind-driven currents in a baroclinic ocean; with application to the equatorial currents of the Eastern Pacific'. *Proc. Nat. Acad. Sci.*, 33, 318–326.

Swallow, J. C., 1964: 'Equatorial undercurrent in the Western Indian Ocean'. *Nature*, 204, 436–437.

Swallow, J. C., 1981: 'Observations of the Somali current and its relationship to the monsoon winds'. pp. 444–452 in: Lighthill, J., Pearce, R. P., eds.: *Monsoon dynamics*, Cambridge University Press, 735 pp.

Taft, B. A., Knauss, J. A., 1967: 'The equatorial undercurrent of the Indian Ocean as observed by the Lusiad Expedition'. *Bull. Scripps Institution of Oceanography*, 9, 163 pp.

Tchernia, P., 1980: *Descriptive regional oceanography*. Pergamon Press, Oxford, New York, Toronto, Sidney, Paris, Frankfurt, 253 pp.

Tsuchiya, M., 1975: 'Subsurface countercurrents in the eastern equatorial Pacific'. *J. Mar. Res.*, 33 (Suppl.), 145–175.

U.S. Navy Hydrographic Office, 1966: 'Atlas of surface currents, Indian Ocean'. U.S. Navy Hydrogr. Off. Publ. 566, Washington, D.C., 12 pp.

Veronis, G., 1960: 'An approximate theoretical analysis of the equatorial undercurrent'. *Deep Sea Res.*, 6, 318–327.

Voorhis, A., Luyten, J. R., Needell, G., Thompson, J., 1984: 'Wind-forced variability of upper ocean dynamics in the central equatorial Pacific during PEQUOD'. *J. Phys. Oceanogr.*, 14, 615–622.

Weisberg, R. H., ed., 1984: 'SEQUAL/FOCAL: first year results on the circulation in the equatorial Atlantic'. *Geophys. Res. Lett.*, 11, no. 8, 713–804.

Wooster, W. S., ed., 1960: 'The Cromwell Current (a series of articles)'. *Deep Sea Res.*, 6, 263–334.

Wunsch, C., 1977: 'Response of an equatorial ocean to a periodic monsoon'. *J. Phys. Oceanogr.*, 7, 497–511.

Wüst, G., 1949: 'Die Kreisläufe der atlantischen Wassermassen, ein neuer Versuch räumlicher Darstellung'. *Forschungen und Fortschritte*, 25, 285–289.

Wüst, G., 1950: 'Blockdiagramme der Atlantischen Zirkulation auf Grund der "Meteor"-Ergebnisse'. *Kieler Meeresforschungen*, 7, no. 1, pp. 24–34.

Wyrtki, K., 1971: *Oceanographic atlas of the International Indian Ocean Expedition*. National Science Foundation, Washington, D.C., 531 pp.

Wyrtki, K., 1973a: 'Teleconnections in the Equatorial Pacific Ocean'. *Science*, 180, 66–68.

Wyrtki, K., 1973b: 'An equatorial jet in the Indian Ocean'. *Science*, 181, 262–264.

Wyrtki, K., 1974a: 'Sea level and the seasonal fluctuations of the equatorial currents in the Western Pacific Ocean'. *J. Phys. Oceanogr.*, 4, 91–103.

Wyrtki, K., 1974b: Equatorial currents in the Pacific 1950 to 1970 and their relations to the trade winds. *J. Phys. Oceanogr.*, 4, 372–380.

Wyrtki, K., 1975a: 'Fluctuations of the dynamic topography in the Pacific Ocean'. *J. Phys. Oceanogr.*, 5, 450–459.

Wyrtki, K., 1975b: 'El Niño – The dynamic response of the Equatorial Pacific Ocean to atmospheric forcing'. *J. Phys. Oceanogr.*, 5, 572–584.

Wyrtki, K., 1978: 'Monitoring the strength of equatorial currents from XBT sections and sea level'. *J. Geophys. Res.*, 83, 1935–1940.

Wyrtki, K., 1979a: 'Comments on the variability of the tropical ocean'. *Dyn. Atmos. Oceans*, 3, 209–212.

Wyrtki, K., 1979b: 'El Niño'. *La Recherche*, 10, 1212–1220.

Wyrtki, K., Kilonsky, B., 1984: 'Mean water and current structure during the Hawaii-to-Tahiti Shuttle Experiment'. *J. Phys. Oceanogr.*, 14, 242–254.

HEAT AND WATER BUDGETS

Net radiation at the upper boundary of the planet Earth amounts to a heat gain in the tropics and a loss in higher latitudes. Redistribution of energy over the continents takes place exclusively in the atmosphere, whereas in the realm of the vast tropical oceans both atmosphere and hydrosphere cooperate in the heat export to other parts of the globe. The partitioning of atmospheric versus oceanic heat transport is of fundamental interest in global energetics. Sea-air interaction processes play an important role in this context. Because of the latent heats involved in phase changes, the continuity of water is directly related to that of energy.

The review of theory in Section 5.1 is to serve as reference for the subsequent treatment of the various heat and water budget components of the atmosphere-ocean-land system. The discussion of the hydrospheric and atmospheric domains in Sections 5.2 and 5.3 is based on satellite measurements of net radiation at the top of the atmosphere and computations of the oceanic heat budget from long-term ship observations (Hastenrath, 1977a, b, 1980, 1982, 1984a, b). This permits inference of the relative roles of atmosphere and ocean in the poleward heat transport (Section 5.6). In addition to the inferred atmospheric contribution, direct evaluations of heat and water budget processes in the atmosphere are presented in Sections 5.4 and 5.5. A synthesis is attempted in Section 5.7.

5.1. Basic Theory

A heat budget scheme for the atmosphere-ocean-land system is illustrated in Fig. 5.1:1. The heat budget equation for the system can be written

$$\text{SWLW}\uparrow\downarrow_{\text{top}} = Q_{va} + Q_{ta} + Q_{vo} + Q_{to}. \tag{5.1:1}$$

The left-hand term signifies the net radiation at the top of the atmosphere; in the right-hand terms, the subscripts v and t denote divergence of heat transport and storage, respectively, and the subscripts a and o refer to atmosphere and ocean-land.

For the long-term mean annual conditions considered here, the second and fourth right-hand terms of Eq. (5.1:1) vanish. Moreover, for land areas the third right-hand term does not arise, and hence $\text{SWLW}\uparrow\downarrow_{\text{top,land}} = Q_{va,\text{land}}$. On the other hand, for ocean areas, $\text{SWLW}\uparrow\downarrow_{\text{top,ocean}} = Q_{va,\text{ocean}} + Q_{vo}$. For the heat budget of ocean-land

$$\text{SWLW}\uparrow\downarrow_{\text{sfc}} = Q_e + Q_s + Q_{vo} + Q_{va}. \tag{5.1:2}$$

The left-hand term being the net radiation at the ocean and/or land surface. The third right-hand term is zero for land surfaces as in Eq. (5.1:1). Likewise, the terms Q_{to} and Q_{ta} in Eq. (5.1:2) vanish as in Eq. (5.1:1) for long-term mean annual conditions.

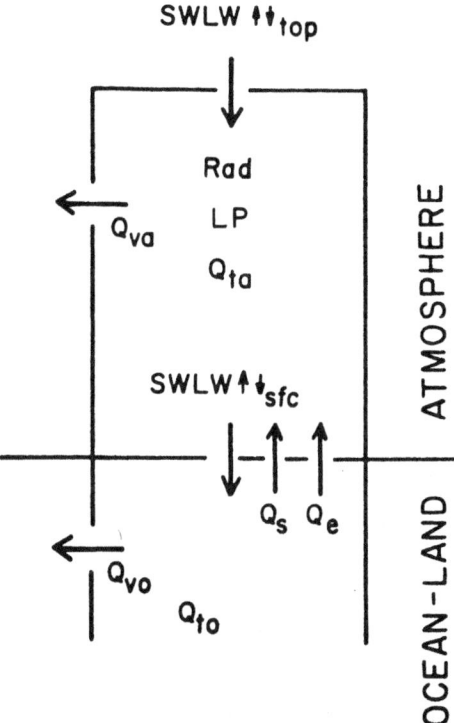

Fig. 5.1:1. Heat budget scheme for atmosphere-ocean-land system. SWLW↑↓top and SWLW↑↓sfc denote net radiation at the top of the atmosphere and at the ocean and land surface, respectively; Rad = SWLW↑↓top − SWLW↑↓sfc; Q_v divergence of horizontal heat transport, Q_t storage, subscripts a and o referring to atmosphere and ocean-land; Q_s, Q_e, sensible and latent heat flux at the surface.

The heat budget equation for the atmospheric column reads

$$Q_{va} + Q_{ta} = [\text{SWLW}\uparrow\downarrow_{\text{top}} - \text{SWLW}\uparrow\downarrow_{\text{sfc}}] + Q_e + Q_s, \tag{5.1:3}$$

where Rad = $[\text{SWLW}\uparrow\downarrow_{\text{top}} - \text{SWLW}\uparrow\downarrow_{\text{sfc}}]$ is the net radiative cooling of the atmospheric column, and Q_e and Q_s are the latent and sensible heat transfer at the ocean and/or land surface, respectively.

In the direct evaluation of the atmospheric heat export separate budget equations are used for geopotential energy plus sensible heat $H = (gz + c_p T)$ and for latent heat Lq, respectively (Riehl and Malkus, 1958; Hastenrath, 1966; Hastenrath and Lamb, 1980); the contribution of kinetic energy $V^2/2$ is small by comparison.

The budget equation for potential energy and sensible heat reads

$$\text{div } H + H_t = Q_s + LP + \text{Rad}. \tag{5.1:4}$$

The left-hand terms signify, respectively, the divergence of the total vertically integrated transport and the storage of $(gz + c_p T)$ in the atmospheric column. The first and third terms on the right-hand side have been explained before, and the second right-hand term denotes the latent heat release through precipitation. The second left-hand term vanishes for long-term mean conditions.

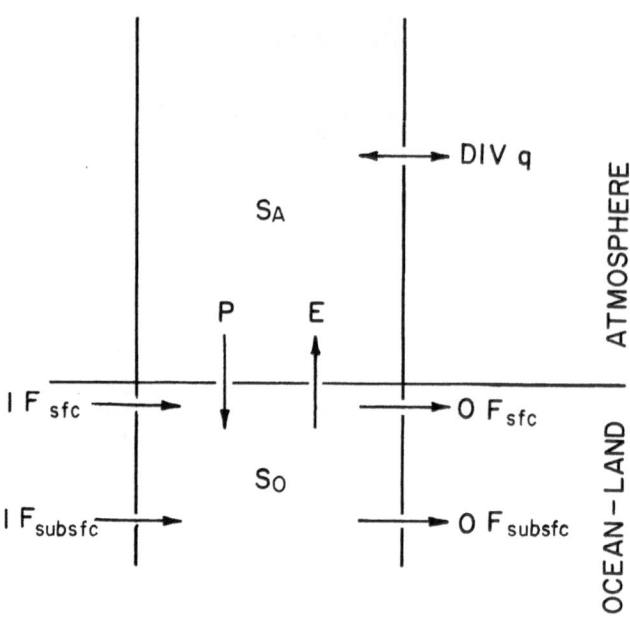

Fig. 5.1:2. Water budget scheme for the atmosphere-lithosphere-hydrosphere system. P denotes precipitation, E evaporation from land or water surfaces, div q divergence of the total vertically integrated water vapor flux in the atmosphere, q_t rate of change of precipitable water, and S_O storage in the lithosphere-hydrosphere. The terms IF_{sfc}, OF_{sfc}, IF_{subsfc}, OF_{subsfc} signify, respectively, surface inflow and outflow, and subsurface inflow and outflow.

In a form analogous to Eq. (5.1:4), the budget equation of latent heat can be written

$$\text{div } Lq + Lq_t = Q_e - LP. \tag{5.1:5}$$

In similarity to Eq. (5.1:4), the terms on the left-hand side represent the divergence of the total vertically integrated transport and storage of Lq in the atmospheric column, the latter being small in the long-term mean.

Note that

$$Q_{va} = \text{div } H + \text{div } Lq \tag{5.1:6}$$

and

$$Q_t = H_t + Lq_t. \tag{5.1:7}$$

Adding Eqs. (5.1:4) and (5.1:5) and noting Eqs. (5.1:6) and (5.1:7) yields the total heat budget equation of the atmospheric column, Eq. (5.1:3). Furthermore, adding the heat budget equations for the atmosphere Eq. (5.1:3) to that for ocean-land Eq. (5.1:2), yields the budget equation for the atmosphere-hydrosphere-lithosphere system as a whole, Eq. (5.1:1).

Concerning the water budget of the atmosphere-ocean-land system, refer to Fig. 5.1:2. The water budget equation for the combined system can be written

$$\text{div } q + q_t = IF_{sfc} - OF_{sfc} + IF_{subsfc} - OF_{subsfc} - S_O \tag{5.1:8}$$

The terms on the left refer to processes in the atmosphere, and the terms on the right to the lithosphere-hydrosphere. The terms on the left denote, respectively, the divergence of the total vertically integrated transport of water, primarily in vapor form, and the rate of change of precipitable water in the atmospheric column. The top of the water vapor atmosphere is not well defined, but the 500 or 300 mb levels are good approximations. The right-hand terms are the surface inflow and outflow, the subsurface inflow and outflow (seepage), and the soil moisture and ground water storage, respectively.

For the atmospheric column

$$\text{div } q + q_t = E - P \tag{5.1:9}$$

where $E = Q_e/L$ is evaporation and P precipitation. Multiplication of Eq. (5.1:9) by the latent heat of evaporation L yields the latent heat budget equation (5.1:5).

For the lithosphere-hydrosphere

$$P - E = IF_{\text{sfc}} + OF_{\text{sfc}} - IF_{\text{subsfc}} + OF_{\text{subsfc}} + S_o \tag{5.1:10}$$

Adding Eqs. (5.1:9) and (5.1:10) yields the water budget equation for the atmosphere-lithosphere-hydrosphere system as a whole.

5.2. Net Radiation at the Top of the Atmosphere

The annual mean spatial patterns of net radiation at the top of the atmosphere, and of heat export within ocean and atmosphere are mapped in Fig. 5.2:1. The major cloud bands and the sea surface temperature distribution during the extreme seasons are sketched in Fig. 5.2:2 for reference. Mean meridional profiles of net radiation at the top of the atmosphere are also included in Figs. 5.2:3 and 5.2:4 and Table 5.2:1. Table 5.2:2 lists the area of the three oceans and four continents by latitude zones.

Fig. 5.2:1a. constructed from data by Vonder Haar and Ellis (1974), shows a heat gain for the entire map area except for the deserts of Northern Africa, largest amounts in the equatorial belt, and a decrease towards the subtropics. Spatial details, such as the maxima in the Northern equatorial Pacific and Atlantic, the Northwest to Southeast orientation of isopleths over South America, and the smaller values to the North as compared to the South of the Equator in Africa, are broadly consistent with the cloudiness pattern (Atkinson and Sadler, 1970). These features are reflected in hemispheric asymmetries over the individual oceans and continents in Table 5.2:1, as well as in the graphs of meridional profiles (Figs. 5.2:3 and 5.2:4) as derived from the same data source.

In contrast to the estimates of Q_{vo} and Q_{va}, an integral constraint is available for $SWLW\uparrow\downarrow_{\text{top}}$, in that an approximate balance can be postulated for the multi-annual mean for the globe as a whole. The eight published and partly overlapping data sets listed in Table 5.2:3 differ in such features as spectral bands used, Equator crossing times, solar constant, and other assumptions in processing (Stephens et al., 1981). It is not possible here to resolve this diversity and to single out the *real* from the *apparent* interannual variability. The annual and global average of planetary net radiation resulting for the various data sets is listed in Table 5.2:3, and ranges from about +10 to beyond −10 W m⁻². The *apparent* imbalances differ considerably even between sets with predominantly overlapping time periods, thus pointing to deficiencies in sampling, sensing, and processing. An attempt at separating signal and noise has been undertaken by Short and Cahalan

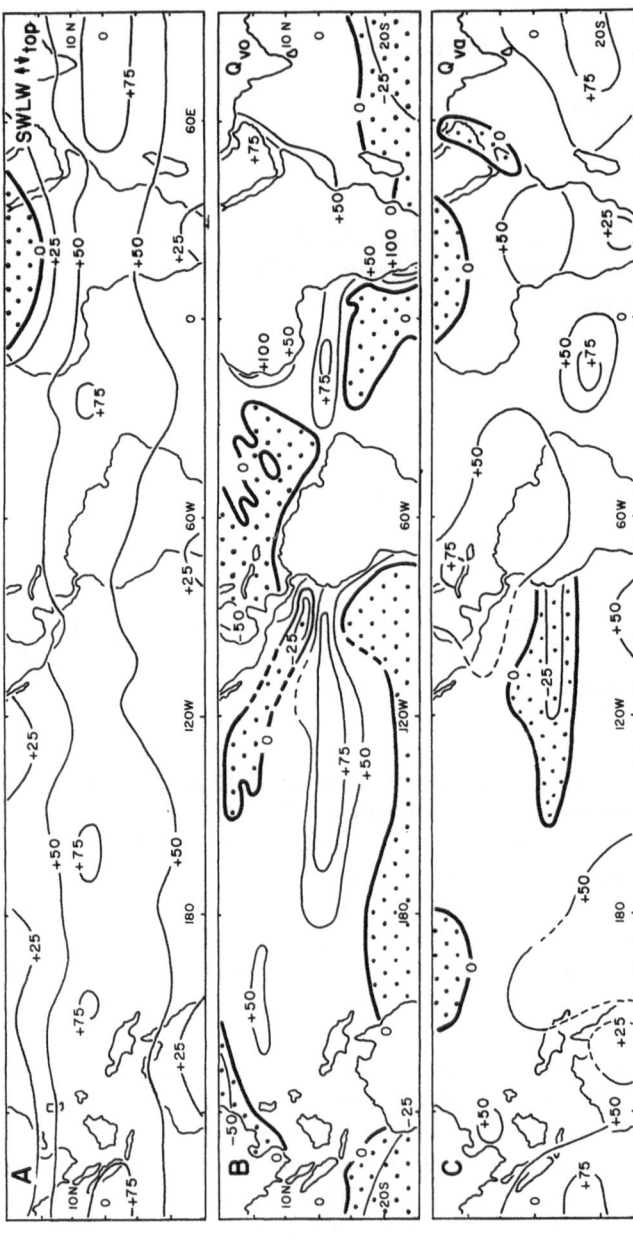

Fig. 5.2:1. Annual mean maps of heat budget components. (a) Net radiation at the top of the atmosphere SWLW↑↓ (source: Vonder Haar and Ellis, 1974). (b) Heat export (positive) within the ocean Q_{vo}, (sources: Hastenrath and Lamb, 1980, for the Indian Ocean; Hastenrath and Lamb, 1978b, for the Atlantic; and Wyrtki, 1966, Hastenrath and Lamb, 1978b, and Budyko, 1963, for the Pacific). (c) Heat export (positive) within the atmosphere, Q_{va}. (sources: as for a and b). Spatial resolution corresponds to five degree squares for map (b), and to ten degrees for map (b), and (c). In maps (b) and (c) isopleths of 25 and 75 are drawn at extrema only, and interpolated isopleths are denoted by broken lines. Negative areas are shown by dot raster. Units in W m⁻². From Hastenrath (1980).

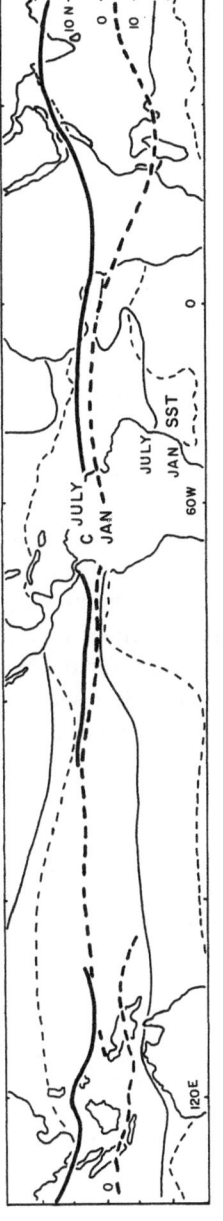

Fig. 5.2:2. Major cloud bands heavy, and 25 °C isotherm of sea surface temperature thin lines, during January (broken) and July (solid). Cloud information is taken from Atkinson and Sadler (1970), and sea surface temperature data from Hastenrath and Lamb (1979) for the Indian Ocean, from Hastenrath and Lamb (1977) for the Atlantic, and from U.S. Navy Hydrographic Office (1944) and Hastenrath and Lamb (1977) for the Pacific Ocean. From Hastenrath (1980).

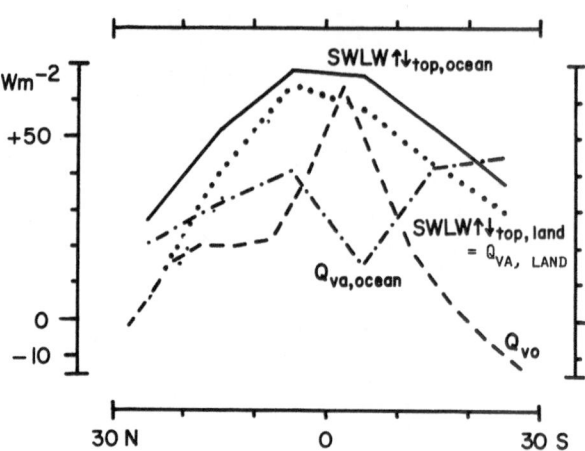

Fig. 5.2:3. Annual mean meridional profiles for the oceanic and continental portions of the global tropics. Net radiation at the top of the atmosphere over the ocean SWLW↑↓top,ocean is denoted by solid, and over land, SWLW↑↓top,land (= Q_{va}, land), by dotted line. Heat export within the ocean, Q_{vo}, is entered as broken, and residual heat export within the atmosphere, Q_{va}, ocean) a dash-dotted line; in W m^{-2} (source: Fig. 5.2:1). From Hastenrath (1980).

(1983), but limited to the outgoing longwave radiation at the top of the atmosphere and three sample areas of the North Pacific.

An adjustment uniform with latitude was applied to the eight data sets so as to achieve a zero global annual mean, and then the respective meridional heat transports were calculated. As maxima were found at 30–40°N and °S, only the values for these latitudes and for the Equator are listed in Table 5.2:4. Also included are the results of London's (1957) study, which predates the satellite era and contains a zero annual average for the Northern hemisphere. The diagrams Figs. 5.2:3 and 5.2:4 show values only of Vonder Haar and Ellis' (1974) data set. Although the various data sets agree on the most fundamental features, namely small cross-equatorial fluxes and largest poleward transports around 30–40°N and 30–40°S, Table 5.2:4 reveals conspicuous differences in the magnitude of the required meridional heat transport. For the latitude of largest transport in the Southern hemisphere, 30–40°S, transport values from the various data sets have a range of about 100 × 10^{13} W, or some 20% of the average of all data sets. By contrast, for the Northern hemisphere, the range is about 200 × 10^{13} W, or about 40% of the mean value. It is remarkable that the results of London's (1957) study of a quarter century ago and well before the dawn of the satellite era compare comfortably with the diversity of satellite results, although they are somewhat smaller.

Various calculations have also been presented by Hastenrath (1980) and Carissimo et al. (1985) with adjustments to satellite radiation data differing between the tropics and the extratropical caps. These lead to different values for the maximum required poleward heat transport in the combined atmosphere-ocean system.

The satellite measurements summarized in Tables 5.2:3 and 5.2:4 yield a large apparent interannual variability of heat storage and transport in the combined hydrosphere-atmosphere system. Order-of-magnitude comparisons (Hastenrath, 1984a) serve to gauge real interannual variability against deficiencies in sampling, sensing, and processing. Storage in the atmosphere can be disregarded because of its small heat capacity. Concerning the ocean, the currently available observations of surface and subsurface temperatures are inadequate to estimate the year-to-year variations of heat content in the top layer of the global ocean. Barring a much more extensive data base for the global hydrosphere, the possibility cannot be discarded that the interannual variability of

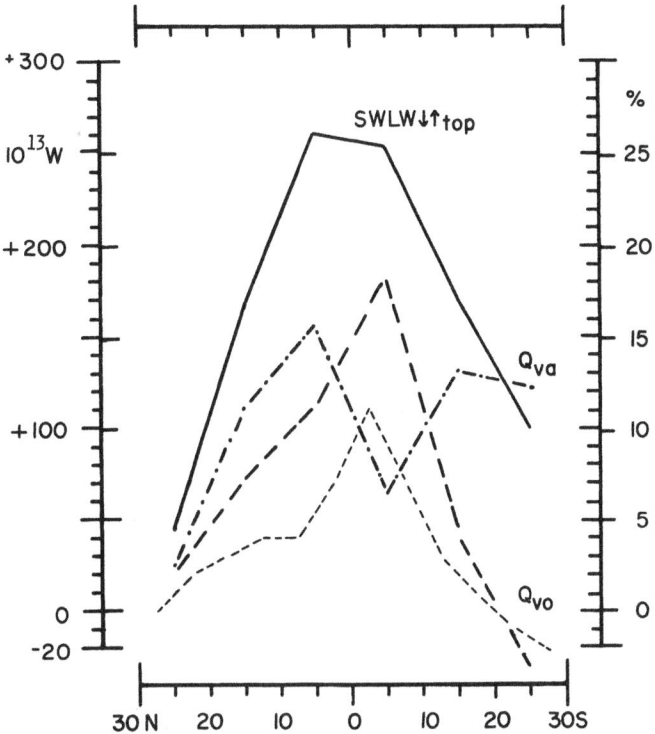

Fig. 5.2:4. Annual mean meridional profiles for the global tropics, heat budget terms for land and ocean combined. Net radiation at the top of the atmosphere, SWLW↑↓$_{top}$, solid; heat export within the atmosphere, Q_{va}, dash-dotted; heat export within the ocean, Q_{vo}, broken; with thin line referring to five and thick line to ten degree bands. Scales are in 10^{13} W and in percent of SWLW↑↓$_{top}$ for $30°$ N $- 30°$S $= 1008 \times 10^{13}$ W $= 100\%$. Vonder Haar and Ellis' (1974) values of SWLW↑↓$_{top}$ were adjusted uniformly with latitude so as to achieve a zero global balance. (source: Tables 5.2:1 and 5.2:2 and Figs. 5.2:1 and 5.3:1). From Hastenrath (1980).

global ocean heat content may reflect a substantial portion of the *apparent* interannual variability of global-mean net radiation at the top of the atmosphere.

Turning to the year-to-year variations in meridional heat transport, Oort (1977, pp. 53–60) has tabulated the interannual standard deviation of meridional heat transport in the atmosphere, separately for the contributions by the mean meridional circulation, standing and transient eddies, and the three energy forms latent heat, sensible heat, and geopotential energy. No values are available for the resultant of these transport mechanisms and energy forms, and it is realized that there tends to be a compensation between the various mechanisms. If all contributions to interannual variability were in the same sense, Oort's (1977, pp. 53–60) tabulations would yield interannual standard deviations of meridional heat transport in the atmosphere of about 68 and 40×10^{13} W at latitudes 30 and $45°$N, respectively. However, these numbers represent extreme upper bounds, and compensation effects may yield considerably smaller values for the resultant of transport mechanisms and energy forms. Then the overall interannual standard deviation of atmospheric meridional heat transports may be much smaller than the *apparent* interannual standard deviation of meridional heat transport in the combined hydrosphere-atmosphere system as resulting from Table 5.2:4. Thus the ocean *may* account for a substantial interannual variability in poleward heat transport.

TABLE 5.2:1

Annual mean net radiation at the top of the atmosphere, $SWLW\uparrow\downarrow_{top}$, in W m^{-2}. From Hastenrath (1980).

	Pacific	Atlantic	Indian	Americas	Africa	S. Asia	Australia
30–20°N	+24	+35	–	+35	– 7	+13	–
20–10	+51	+54	+56	+52	+32	+49	–
10– 0	+65	+72	+78	+59	+65	+70	–
0–10	+61	+66	+78	+54	+63	+63	–
10–20	+48	+52	+61	+42	+45	–	+47
20–30°S	+37	+36	+34	+35	+21	–	+21

TABLE 5.2:2

Total area of latitude zones, in 10^9 m^2, and percentage contribution of various oceans and continents (in part from List, 1968, Table 164). From Hastenrath (1980).

	Total area 10^9 m^2	Pacific	Atlantic	Indian	Americas	Africa	S. Asia	Australia
30–25°N	19 699	35	23	1	5	13	23	0
25–20	20 499	37	22	6	3	15	17	0
20–15	21 145	39	20	13	3	15	10	0
15–10	21 663	42	18	17	2	16	5	0
10– 5	21 960	44	13	18	6	17	2	0
5– 0	22 124	45	17	17	8	10	3	0
0– 5	22 124	42	15	19	11	8	5	0
5–10	21 960	42	13	22	12	8	3	0
10–15	21 633	44	14	22	11	8	0	1
15–20	21 145	41	14	21	9	9	0	6
20–25	20 499	39	16	21	7	7	0	10
25–30°S	19 699	38	18	23	6	4	0	11

These considerations suggest that part of the *apparent* interannual variability in heat input to the hydrosphere-atmosphere system as a whole (net allwave radiation at the top of the atmosphere, $SWLW\uparrow\downarrow_{top}$) *may* be accounted for by year-to-year changes of temperature in the top layer of the World ocean. Tables 5.2:3 and 5.2:4 underline the need for a quantitative intercomparison of the published estimates of net radiation at the top of the atmosphere, in terms of sensing, sampling, and processing, with the aim of identifying the *real* portion of the *apparent* interannual variability – a task well beyond the scope of the present review. The interannual variability of oceanic heat transport *may* be substantial. Inasmuch as the net allwave radiation at the top of the atmosphere $SWLW\uparrow\downarrow_{top}$ is a small difference between the downward and upward directed shortwave and the upward directed longwave radiation, efforts at substantially improving over a 6–10 W m^{-2} error in $SWLW\uparrow\downarrow_{top}$ may involve a formidable task in the determination of component fluxes. The pertinent ongoing endeavors are in part described in Gruber *et al.* (1983), Kandel (1983), and Office of Space Science and Applications, NASA (1983). This brief review of evidence cautions against optimistic estimates of uncertainties in the presently available satellite-derived net radiation at the top of the atmosphere.

TABLE 5.2:3

Annual global average of net radiation at the top of the atmosphere for eight partly overlapping data sets, in W m^{-2}. From Hastenrath (1984a).

	Source	Satellite	Period	Net radiation [W m^{-2}]
I.	Gruber (1978)	NOAA operational polar orbiting, two-channel SR	June 74–May 75	−10
II.	Vonder Haar and Ellis (1974)	TIROS 4, TIROS 7, NIMBUS 2, NIMBUS 3, ESSA 3, Experimental,	51 months during	+ 8
III.	Ellis and Vonder Haar (1976)	NIMBUS 2, NIMBUS 3, ESSA 3, Experimental, NOAA 1, ITOS 1	29 months during July 64–May 71	− 0.01
IV.	Jacobowitz et al. (1979)	NIMBUS 6 ERB	July 75–June 76	+ 6
V.	Stephens et al. (1981)	ESSA 3, Experimental, ESSA 7, NIMBUS 6	48 months during July 64–June 77	+ 9
VI.	Winston et al. (1979)	NOAA operational polar orbiting two-channel SR	June 74–May 75	−20
VII.	Winston et al. (1979)	NOAA operational polar orbiting two-channel SR	June 75–May 76	−14
VIII.	Winston et al. (1979	NOAA operational polar orbiting two-channel SR	June 76–May 77	− 9

TABLE 5.2:4

Required meridional heat transport in the hydrosphere-atmosphere system, computed for the various data sets after adjustment for zero annual global average of net radiation at the top of the atmosphere, as shown in Table 5.2:2, in 10^{13} W, positive northward. From Hastenrath (1984a).

		40°N	30°N	0	30°S	40°S
I.	Gruber (1978)	+518	+510	0	−560	−585
II.	Vonder Haar and Ellis (1974)	+547	+562	+ 61	−495	−531
III.	Ellis and Vonder Haar (1976)	+477	+477	−40	−537	−537
IV.	Jacobowitz et al. (1979)	+656	+673	+65	−560	−572
V.	Stephens et al. (1981)	+548	+558	+18	−534	−545
VI.	Winston et al. (1979)	+564	+547	−26	−620	−633
VII.	Winston et al. (1979),	+571	+563	+15	−572	−586
VIII.	Winston et al. (1979),	+569	+548	0	−580	−604
average of I through VIII		+553	+555	+11	−459	−585
IX.	London (1957)	+459	+423	0	−	−

5.3. Oceanic Heat Budget

The map of mean annual heat gain by the oceanic water body, Fig. 5.2:1b., shows heat export for much of the low latitude oceans, but especially so for the cold water regions immediately to the South of the Equator in the Atlantic and the Eastern and Central Pacific, as well as off the West coasts of Southern and Northern Africa and South America (Figs. 5.2:2 and 6.1:1). Import is indicated for much of the Southern subtropical oceans, the warm current regions of the Western North Atlantic, and the westernmost North Pacific. Import or relatively small export is also characteristic of the regions of zonally oriented cloud bands (Fig. 5.2:2) immediately to the North of the Atlantic and Eastern Pacific Equator.

Complementing the assessment of annual mean conditions, Lamb (1981) and Lamb and Bunker (1982) studied the annual march of the heat budget of the North and tropical Atlantic using subsurface observations. Such analyses for other regions of the tropical oceans are in order for the understanding of seasonal heat budget and transports in the tropical hydrosphere.

Mean meridional profiles for the three oceans are depicted in Fig. 5.3:1. Both the Atlantic and Pacific show the major maximum of Q_{vo} in the latitude band immediately to the South of the Equator. Export is indicated for most latitude bands. The Indian Ocean shows a contrastingly different pattern in that export, i.e. positive Q_{vo}, is largest in the Northern Indian Ocean, decreases southward, and changes to import in the outer tropics of the Southern hemisphere. The most remarkable feature in the latitudinal distribution of Q_{vo} for all oceans combined (Figs. 5.2:3 and 5.2:4) is the hemispheric asymmetry with a prominent maximum to the South of the Equator. Thus the meridional pattern for the tropical hydrosphere as a whole is dominated by the Pacific and Atlantic Oceans.

A major cause for the differing meridional heat budget patterns in the various oceans can be appreciated from the profiles of sea surface temperature plotted in Fig. 5.3:1. In the Atlantic and Pacific, the zone immediately to the South is much colder than the band to the North of the Equator (Figs. 5.2:2 and 5.3:1). This is a feature of the Eastern Atlantic and Pacific in particular (Hastenrath, 1977b, 1978; Hastenrath and Lamb, 1977, 1979; Wyrtki, 1966; U.S. Navy Hydrographic Office, 1944). These are also regions of rather small sensible and latent heat transfer (Hastenrath, 1978; Hastenrath and Lamb, 1978b, 1979; Wyrtki, 1965, 1966). In the Indian Ocean, the sensible and latent heat flux increases from the Northern towards the Southern hemisphere, without a minimum near the Equator (Hastenrath and Lamb, 1979, 1980). A near-equatorial minimum of sea surface temperature is likewise lacking. This examination shows that the prominent maximum of Q_{vo} immediately to the South of the Equator, which characterizes the tropical hydrosphere as a whole (Figs. 5.2:3 to 5.2:4), results from the contribution of the reduced heat transfer in the cold water tongues of the Eastern Atlantic and Pacific Oceans in particular. The cold water surface is conducive to small latent and sensible heat expenditures and hence a large oceanic energy gain. However, the latter is not used locally to raise the sea surface temperature, as a result of the hydrospheric vertical motion pattern, which in turn is related to the dynamics of the coupled ocean-atmosphere system (Hastenrath, 1978; Hastenrath and Lamb, 1978a).

The error in Q_{vo} in the sense of a 'total uncertainty' has been estimated to be of the order of 20–30 W m^{-2} (Hastenrath, 1977a). A sizeable portion of this figure accounts for a possible systematic error in Q_{va}, as it would result from systematic uncertainties primarily in the net longwave radiation. Truly random errors would be of little consequence by comparison.

The meridional transport of heat within the ocean can be obtained by assumption of an appropriate boundary condition and integration of Q_{vo} (Fig. 5.3:1) over latitude bands. Such curves have been constructed by Bryan (1962) and Emig (1967) for individual oceans on the basis of Budyko's (1963) atlas.

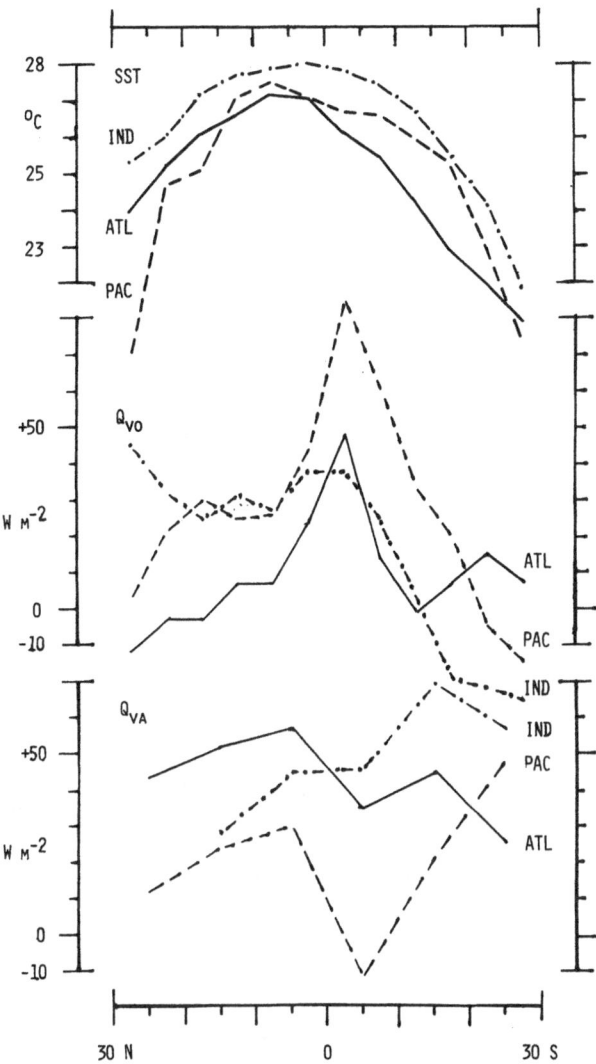

Fig. 5.3:1. Annual mean meridional profiles for sea surface temperature, and of divergence of heat transport within ocean Q_{vo}, and atmosphere Q_{va}. Atlantic solid, Pacific broken, and Indian Ocean dash-dotted lines; in degree C and W m^{-2} '(source: Figs. 5.2:1a, b for Q_{vo} and Q_{va}; Hastenrath and Lamb, 1977, 1979, U.S. Navy Hydrographic Office, 1944, for T). From Hastenrath (1980).

Vonder Haar and Oort (1973) and Oort and Vonder Haar (1976) inferred the oceanic heat transport in the Northern hemisphere from satellite-derived net radiation at the top of the atmosphere, and calculations of the atmospheric heat transport. The same approach was applied by Trenberth (1979) to the Southern hemisphere and by Carissimo *et al.* (1985) to the globe as a whole. In these studies oceanic heat transport is computed as the residual between two quantities which themselves are obtained as differences between large terms, so that sampling errors are sizeable.

Meridional heat transports were evaluated from hydrographic sections (Wüst and Defant, 1936; Fuglister, 1960; Niiler and Richardson, 1973) by Bryden and Hall (1980), Wunsch (1980), and

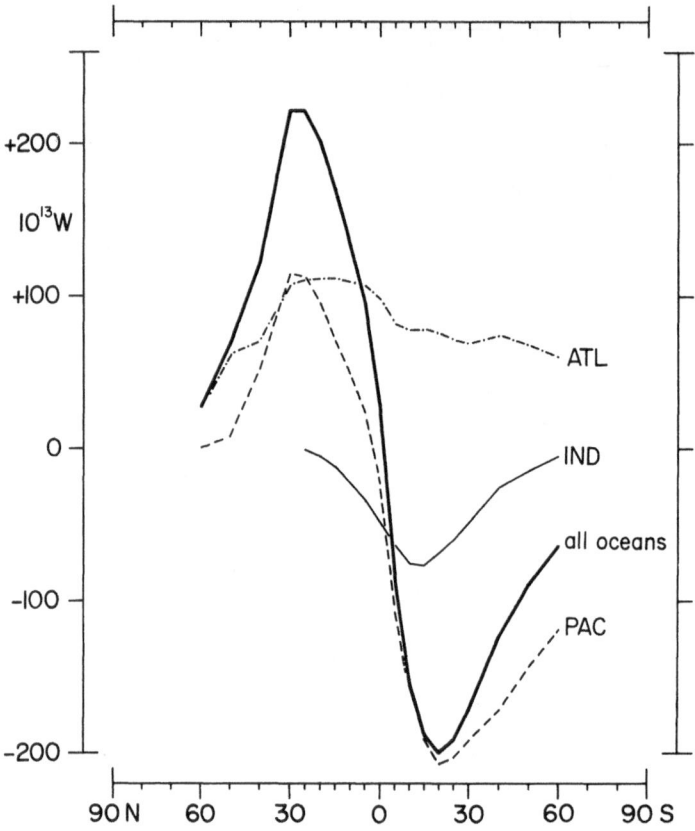

Fig. 5.3:2. Annual mean meridional heat transport within the oceans. Pacific broken, Atlantic dash-dotted, Indian Ocean thin solid, and all oceans combined heavy solid line. Northward transport positive, units in 10^{13} W (source: Fig. 5.3:1 and Table 5.1:1 for 30°N – 30°S; Budyko, 1963, for 30–66°N and 30–60°S; Aagaard and Greisman, 1975, for 66–90°N; and Lamb, 1981, and Lamb and Bunker, 1982, for 30–66°N in the Atlantic). From Hastenrath (1982).

Roemmich (1980) for the North Atlantic, and by Bennett (1978) and Wunsch *et al.* (1983) for the Southern oceans. Meridional heat transports within the individual oceans were calculated (Hastenrath, 1980, 1982) from the values of Q_{vo} for latitude bands between 30N – 30°S, Fig. 5.3:1. For purposes of a northern boundary condition, Aagaard and Greisman's (1975) estimates for the Arctic basin were used. Furthermore, Budyko's (1963) atlas was available for the zones 66–30°N and 30–60°S. For the North Atlantic, the studies of Bunker (1976), Bunker and Worthington (1976), Lamb (1981), and Lamb and Bunker (1982) are furthermore pertinent. Integration proceeded from North to South. In the high latitudes of the Southern hemisphere, 70°W, 20°E, and 120°E were used as divisions between the Pacific, Atlantic, and Indian Oceans. Emig (1967) chose similar boundaries. Results are presented in profile form in Fig. 5.3:2, and further illustrated in the map scheme, Fig. 5.3:3.

The characteristics of meridional heat transport differ remarkably for the three oceans. In the Pacific sector, heat exchange across the Bering Straits is small (Aagaard and Greisman, 1975), as is the poleward transport around 60°N. The largest values are indicated for 30°N, and transport changes to southward somewhat North of the Equator. Thus the large heat gain in the band 0–5°S (Figs. 5.3:1 and 5.3:3) is carried southward in its entirety. The largest southward transport is

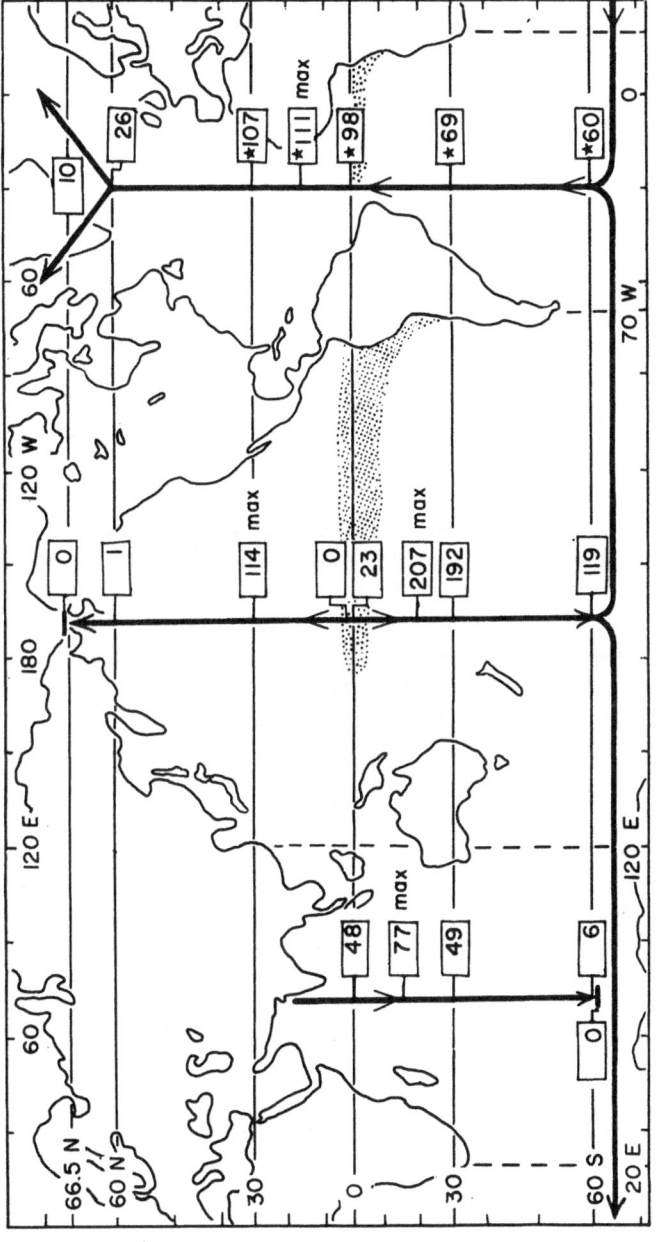

Fig. 5.3:3. Map scheme of annual mean meridional heat transport within the oceans (from Fig. 5.3:2). Heavy cross-bar denotes latitude of zero, and 'max' that of maximum meridional transport, with numbers indicating amounts in units of 10^{13} W. Stippling marks areas with $Q_{vo} > +50$ W m^{-2} in Fig. 5.2:1b, and broken lines show the meridians used as boundaries between oceans in the high Southern latitudes. From Hastenrath (1982).

CHAPTER 5

TABLE 5.3:1

Estimates of northward heat transport in the Atlantic Ocean, in 10^{13} W. Derived (a) from surface heat budget, (b) from hydrographic sections. From Hastenrath (1982).

		60°N	36	30	24/25	0	8	15/16	21	24	28	30	32	60°S
Hastenrath (1980)	a	26		155	157	144						115		106
Lamb (1981)	a	28		107	113	102								
Bryden and Hall (1980)	a				111									
	b				110									
Roemmich (1980)	b		80		120									
Wunsch (1980)	b		75		120									
Bennett (1978)	b									65			68	
Fu (1981, Meteor)	b						42	86	54		83			
Fu (1981, IGY)	b						18	73		54			77	
Hastenrath (1982)	a	26		107	110	98						69		60

found near 20°S. Various sources of heat budget estimates for the tropical Pacific are compared in Hastenrath (1984b) and Talley (1984). By virtue of its large area, the Pacific dominates the pattern for all oceans combined.

In the Atlantic, the oceanic heat transport is directed northward, from 60°S all the way to the Arctic. The large oceanic heat surplus in the band 0–5°S (Figs. 5.3:1 and 5.3:3) is reflected in a marked increase of northward heat transport. Values remain high throughout the Northern hemisphere tropics, reach a maximum at 15°N, and drop off sharply poleward of 30°N. The prominent transport characteristics in the Atlantic, as compared to the Pacific, are related to the open communication with the Arctic basin and the concomitant large heat deficit in high Northern latitudes, and to the large oceanic heat gain in the Southern hemisphere tropics associated with the comparatively low sea surface temperature (Figs. 5.3:1 and 5.2:2).

Estimates of meridional heat transport derived from hydrographic sections in the Atlantic are of interest in relation to Figs. 5.3:2 and 5.3:3. Results are compared in Table 5.3:1.

Bryden and Hall (1980) evaluated the oceanic heat transport across 25°N both from Bunker's (1976) assessment of the surface heat budget and directly from hydrographic sections (Fuglister, 1960; Niiler and Richardson, 1973). As shown in Table 5.3:1, they obtained almost identical numbers from the two independent methods. Wunsch (1980) calls this agreement fortuitous, and gives 24°N as latitude of the section. Roemmich (1980) determined the oceanic heat transport at 24 and 36°N from hydrographic sections, as did Wunsch (1980) for these and other latitudes. They obtained similar values, which are also included in Table 5.3:1.

For the realm 30–25°N a remarkable agreement is emerging from Table 5.3:1 between the estimates based on surface heat budget and hydrographic sections. It is furthermore noted that Behringer and Stommel's (1981) independent assessment of oceanic heat gain in the tropical North Atlantic based on subsurface data is consistent with evaluations of the surface heat budget (Hastenrath and Lamb, 1978b). Concerning the South Atlantic, the value of Hastenrath (1982) for 30°S agrees closely with Bennett's (1978) results for 24 and 32°S based on the evaluation of hydrographic sections (Fuglister, 1980; Wüst and Defant, 1936).

Fu (1981) applied the inverse method used by Wunsch (1980) and Roemmich (1980) to the observations of the Meteor and IGY Expeditions in the South Atlantic (Fuglister, 1960; Wüst and Defant, 1936). He evaluated the meridional heat transports across various latitudes, with reference to the 2000 and 4000 db surfaces. The reference level makes a difference for the IGY

TABLE 5.3:2
Estimates of southward heat transport across 60°S in the
World ocean as a whole, in 10^{13} W. From Hastenrath (1982).

Hastenrath (1980)	13
Zillman (1972)	10–20
Trenberth (1979)	100
Gordon (1981)	54
Hastenrath (1982)	65

observations, but matters little for the Meteor data set. The arithmetic mean of his results for the two reference levels is also listed in Table 5.3:1. Fu's numbers for 32 and 28°S agree reasonably well with Hastenrath's (1982) and Bennett's (1978) results. Fu's relative minimum at 24 and 21°S appears startling, and his values for 8°S are much smaller than expected from the surface heat budget. Fu (1981) considers his errors as largest for this section.

As a result of the continent configuration, meridional heat transport in the Indian Ocean (Hastenrath and Lamb, 1980) is directed southward, with a maximum around 10–15°S and a decrease to near nil at 60°S. Note that consideration of heat exchange to the North of Australia may alter the partitioning of meridional transports between the Pacific and Indian Oceans. Talley (1984) considers the error tolerances in estimating this component. These are in part related to uncertainties in the volume transport westward through the Indonesian archipelago, which is estimated at 2, 5, 10, and 14 Sv (1 Sv = 10^6 m³ s⁻¹), respectively, by Wyrtki (1961, p. 136), Fine (1985), Godfrey and Golding (1981), and Piola and Gordon (1984). At any rate, heat deficits in the higher Southern latitudes of the Indian and Atlantic Ocean sectors must be compensated by contributions from the Pacific domain. Furthermore, heat input from the Pacific helps balance the heat budget of the Atlantic sector at large.

For all oceans combined, maximum northward and southward heat transport are found at 30°N and 20°S, respectively, and a net northward transport is indicated at the Equator. A closure of Fig. 5.3:3 is desirable in terms of a comparison with the heat budget of the Antarctic Ocean poleward of 60°S. Various estimates of southward transport across 60°S are compared in Table 5.3:2. The value of Fig. 5.3:3. (Hastenrath, 1982) is in good agreement with Gordon's (1981) recent estimate derived from the heat budget of the South Polar Ocean. Zilman's (1972, Fig. 1.10(e), p. 114) heat budget maps for the Southern Ocean have a less satisfactory data base although they broadly agree with the earlier estimate of Hastenrath (1980). Trenberth's (1979) value far exceeds all other estimates; the method being fraught with the combined uncertainties in planetary net radiation SWLW↑↓$_{top}$ and the atmospheric heat export Q_{va}.

Referring to the early analysis of a hydrographic section by Bryan (1962), Wunsch (1980) suggests the possibility of an equatorward heat transport at 32°N in the Pacific Ocean. For the World ocean as a whole, this would entail a considerably larger southward heat transport across 60°S than obtained above. However, in the light of the large oceanic surface heat losses of the northern North Pacific, a southward heat transport across 32°N does not appear plausible. Also, the scale and configuration of the Pacific may be unfavorable for the evaluation of meridional transports from hydrographic sections.

Errors in transport estimates derived from the surface heat budget are discussed in Hastenrath (1980) and Lamb (1981). Concerning errors in the evaluation from hydrographic sections, Bryden and Hall (1980) propose a tolerance of ±30 × 10^{13} W for their meridional heat transport value at 25°N of 110 × 10^{13} W. However, the error margin does evidently not account for variability on the seasonal, interannual, and longer time scales. The reservation regarding the temporal sampling applies to the hydrographic section approach in general (Bennett, 1978; Wunsch, 1980; Roemmich, 1980; Fu, 1981; Hall and Bryden, 1982; Wunsch et al., 1983; Bryan, 1983).

Meridional heat transports were here mostly arrived at indirectly. The mechanisms by which the oceans perform these transports are of considerable interest, in that the surface circulation does not seem sufficient to explain the required transports. In fact, meridional overturnings appear to be instrumental in the merdional heat transports. Empirical results are being substantiated by recent computer simulations (Meehl et al., 1982; Miller et al., 1983; Pares-Sierra et al., 1985). In view of the paucity of direct observations numerical models seem to deserve particular attention.

For the World Ocean as a whole, Emig (1967) found similar latitudes of the extrema as in Fig. 5.3:2, but smaller and larger poleward transports in the Northern and Southern hemispheres, respectively. Accordingly, her tabulations show a southward rather than a northward directed net transport across the Equator. The atlas available to Emig (1967) for the tropics is inferior to the present data sources. Vonder Haar and Oort (1973), Oort and Vonder Haar (1976), Trenberth (1979), and Carissimo et al. (1985), used a novel approach in that they calculated Q_{vo} as a residual in Eq. (5.1:1). The values of meridional heat transport in the World ocean from the three former studies are entered in Fig. 5.6:1 for comparison. Moreover, Carissimo et al. (1985) show that their results differ little from Oort and Vonder Haar (1976) and Trenberth (1979). In the light of the considerable uncertainties in the satellite-derived net radiation at the top of the atmosphere borne out by Tables 5.2:3, and 5.2:4, as well as the assessment of the atmospheric heat transports, the agreement between the independent methods is remarkable.

The effect of errors in $SW\uparrow\downarrow_{top}$ and in Q_{vo}, as estimated further above, on the calculation of heat transports is considered in the following. A systematic error of 10 W m^{-2} in Q_{vo} would lead to an error of the order of 100×10^{13} W in the oceanic heat transport around 30°N. Similar considerations apply to $SWLW\uparrow\downarrow_{top}$. It is suggested that errors in the oceanic transport calculated by Vonder Haar and Oort (1973), Oort and Vonder Haar (1976), and Carissimo et al. (1985), as a residual from satellite-derived net radiation at the top of the atmosphere and computations of the atmospheric heat budget are at least of comparable magnitude. Carissimo et al. (1985) propose uncertainties of 100 and 50×10^{13} W for the satellite-derived required transport in the combined atmosphere ocean system and in the atmosphere in midlatitudes, respectively, and an uncertainty of 150×10^{13} W for the oceanic transport calculated in their method as residual. Their residual method does not seem to be less deficient than the calculations based on the surface heat budget. Within the plausibly broad error bands, the rather larger oceanic transport estimates by Vonder Haar and Oort (1973), Oort and Vonder Haar (1976), Trenberth (1979), and Carissimo et al. (1985), must be considered as compatible with the present ones, Figs. 5.3:2, 5.3:3, 5.6:1, and Table 5.3:2.

Throughout this section, the oceanic heat budget was considered for annual mean conditions only. Inasmuch as heat storage Q_{to} in the average over many years is small compared to other terms in Eqs. (5.1:1) and (5.1:2), the treatment is then greatly simplified. Little is presently known about the annual cycle of heat content in the upper hydrosphere. Assessment of seasonal storage and depletion of heat in the tropical oceans would allow the estimate, not only of annual, but also of seasonal heat transports from evaluation of the surface heat budget, Eqs. (5.1:1) and (5.1:2). The works of Lamb (1981) and Lamb and Bunker (1982) for the Atlantic and the recent global analysis by Levitus (1984) are important steps in this direction. Among his most important results is a maximum in the annual range of oceanic heat content around 10°N and 75–100 m depth in the Atlantic and Pacific, which appears related to large-scale redistributions of heat and displacements of the thermocline. In fact, mechanical forcing and ocean dynamics, rather than merely surface net radiation, must be regarded of foremost importance for the heat budget of the low-latitude oceans.

Taking advantage of the extensive data bank compiled by Levitus (1982, 1984), Carissimo *et al.* (1985) extended Oort and Vonder Haar's (1976) earlier Northern hemispheric analysis to a study of seasonal heat transports in the global ocean. They note a strong annual cycle in poleward oceanic heat transport, with a maximum of about 600×10^{13} W at 25–30 degrees latitude in the winter hemisphere, and weak transports during summer, except in the inner tropics, 10°S – 10°N, where a strong transport is directed towards the winter pole. The bulk of the transport takes place above the 225 m level.

5.4. Atmospheric Heat Budget

The appraisal of the atmospheric heat budget is here centered on the quantitative evaluation of Eq. (5.1:4) for annual mean conditions (Tables 5.4:1, 5.4:2, 5.4:3). Since the net allwave radiation at the surface SWLW↑↓$_{sfc}$ as a rule exceeds that at the top of the atmosphere SWLW↑↓$_{top}$, the net radiative flux divergence R_a = SWLW↑↓$_{top}$ − SWLW↑↓$_{sfc}$ amounts to a cooling of the atmospheric column at all latitudes. This loss is approximately, but not exactly balanced by the heating due to precipitation LP. The latter term will further be of interest in the discussion of the water budget in Section 5.6. The sensible heat flux across the surface Q_S also results in a heat gain

TABLE 5.4:1

Terms of the atmospheric heat budget, Eq. (5.1:4), by latitude bands in W m^{-2}. Net radiation at the top of the atmosphere SWLW↑↓$_{top}$, and at the Earth surface SWLW↑↓$_{sfc}$, net radiative flux divergence in the atmospheric column R_a = SWLW↑↓$_{top}$ − SWLW↑↓$_{sfc}$, precipitational heating LP, sensible heat flux at the surface Q_S and residually determined lateral export of sensible heat plus potential energy div H. Sources: (a) Hastenrath (1982); (b) Budyko (1974, p. 219), Sellers (1965, p. 5, 103).

	SWLW↑↓$_{top}$ a	SWLW↑↓$_{sfc}$ b	R_a	LP b	Q_S b	div H
90–80°N	−110	− 12	− 98	10	−13	−101
80–70	−106	+ 1	−107	14	− 1	− 94
70–60	− 81	+ 26	−107	37	+13	− 57
60–50	− 45	+ 39	− 84	56	+18	− 10
50–40	− 29	+ 63	− 92	62	+22	− 8
40–30	− 1	+ 97	− 98	60	+31	− 7
30–20	+ 18	+126	−108	55	+31	− 22
20–10	+ 44	+138	− 94	92	+21	+ 19
10– 0	+ 62	+137	− 75	151	+14	+ 90
0–10°S	+ 59	+137	− 78	118	+13	+ 53
10–20	+ 45	+136	− 91	97	+14	+ 20
20–30	+ 26	+123	− 97	67	+21	− 9
30–40	+ 5	+105	−100	72	+14	− 14
40–50	− 26	+ 73	− 99	80	+13	− 6
50–60	− 57	+ 37	− 94	76	+14	− 61
60–70	− 91	− 17	−108	33	−14	− 61
70–80	− 97	− 3	−100	7	− 5	− 98
80–90°S	−105	− 14	−119	2	−14	−131

TABLE 5.4:2

Observed annual mean northward transports of sensible heat (c_pT), geopotential energy (gz), latent heat (Lq), and water vapor (q) in the layer 1012.5–75 mb, by transient eddies (TE), standing eddies (SE), and the mean meridional circulation (MMC). Units for c_pT, gz, and Lq are 10^{13} W, and for q 10^7 kg s^{-1}. Source: Oort (1983), and Oort and Rasmusson (1971).

	c_pT			gz			Lq			q		
	TE	SE	MMC	TE	SE	MMC	TE	SE	MMC	TE	SE	MMC
90°N	−	−	7	−	−	0	0	0	0	0		0
80	+ 25	+ 2	− 7	−	−	+ 21	+ 3	+ 0	0	+ 1	+ 0	0
70	+ 82	+14	− 14	−4	−0	+ 41	+ 17	+ 1	− 13	+ 7	+ 0	− 5
60	+191	+64	0	−4	+4	− 20	+ 46	+ 6	+ 20	+18	+ 2	+ 8
50	+ 294	+65	+ 77	−7	+2	−155	+ 83	+ 4	+ 26	+33	+ 2	+10
40	+338	+18	+ 92	−3	+0	−184	+117	+ 5	+ 61	+47	+ 2	+24
30	+ 201	+31	+104	−7	−0	−139	+113	+27	+ 35	+45	+11	+14
20	+ 57	+41	−226	−7	+0	+377	+ 71	+37	−113	+28	+15	−45
10°N	+ 28	+ 4	−513	−0	+0	+789	+ 46	+ 1	−237	+19	+ 0	−95
0	0	− 8	+ 80	+0	+0	−160	0	− 6	+ 80	0	− 2	+32
10°S	− 43	−12	+355	+4	+0	−553	− 46	−11	+198	−19	− 4	+79
20	− 98	−23	+188			−339	− 78	−18	+113	−31	− 7	+45
30	−233	−21	+ 35			− 69	−109	− 9	+ 34	−43	− 3	+14
40	−350	− 3	−123			+215	−113	+ 1	− 61	−45	+ 0	−24
50	−305	− 3	−103			+181	− 85	+ 1	− 26	−34	+ 0	−10
60	−127	−10	− 40			+ 80	− 43	− 0	− 20	−17	− 0	− 8
70	− 11	−11	+ 14			− 14	− 12	− 1	0	− 5	− 0	0
80	− 7	− 6	+ 28			− 42	− 2	− 1	0	− 1	− 0	0
90°S	0	0	0			0	0	0	0	0	0	0

by the atmosphere, except for very high latitudes. However, Q_s is much smaller than either R_a or LP. The sum of the three terms, R_a, LP, and Q_s, represents the amount of geopotential energy plus sensible heat that must be exported or imported by processes within the atmosphere.

Net radiation at the top of the atmosphere SWLW↑↓$_{top}$ decreases from the equatorial belt towards the higher latitudes of either hemisphere (Table 5.4:1), as was shown in Section 5.2. The net radiation at the Earth surface SWLW↑↓$_{sfc}$ exhibits a similar pattern. Accordingly, the latitudinal contrasts in R_a are comparatively small. The meridional profile of precipitation is characterized by a decrease from the equatorial zone towards the poles, with secondary minima in the subtropics. The sensible heat flux Q_s amounts to a heat gain to the atmosphere in the tropics, but a loss in higher latitudes. From the latter three terms, the required export of sensible heat plus geopotential energy $(c_pT + gz)$ in the atmosphere, div H, is obtained as a residual. Integration with respect to latitude yields the required poleward transport of $(c_pT + gz)$ in the atmosphere, as listed in Table 5.4:3.

TABLE 5.4:3

Annual mean northward transport of sensible heat plus geopotential energy ($c_pT +$ gz) within the atmosphere, in 10^{13} W. (a) required transport in the entire atmospheric column according to Table 5.4:1, and (b) observed transport in the layer 1012.5 to 75 mb as listed in Table 5.4:2.

	Required	Observed
90°N	0	0
80	+ 38	+ 41
70	+ 146	+ 119
60	+ 253	+ 235
50	+ 278	+ 276
40	+ 302	+ 261
30	+ 326	+ 190
20	+ 413	+ 242
10°N	+ 331	+ 308
0	− 67	+ 88
10°S	−302	−249
20	−389	−272
30	−354	−288
40	−304	−261
50	−286	−230
60	−277	− 97
70	−163	− 22
80	− 50	− 27
90°S	0	0

In addition to the determination as residual (Table 5.4:1), direct assessment of energy transports from aerological data is available in the publications of Oort (1983), Oort and Rasmusson (1971), and Peixoto and Oort (1984). Table 5.4:2 summarizes their results for long-term annual mean conditions, with a breakdown by contributions of transient and standing eddies and the mean meridional circulation, as defined in Section 3.5. The northward transport of gz by the mean meridional circulation is pronounced in the realm of the wintertime Hadley cell, but the annual mean pattern of the total transport of ($c_pT + gz$) shows another maximum in the zone of the travelling disturbances in midlatitudes where the transient eddy transport of c_pT is pronounced.

The observed transport of sensible heat plus geopotential energy ($c_pT + gz$) by all three mechanisms combined (Table 5.4:2) is listed in Table 5.4:3, for comparison with the residual estimates from Table 5.4:1. The observed transports agree broadly with the required values for the Northern hemisphere midlatitudes where transports are largest and the radiosonde network is most plentiful. It is recalled that the direct evaluation pertains to the layer 1012.5 to 75 mb only. Tables 5.4:1 and 5.4:2 are further of interest in relation to Sections 5.5, 5.6, and 5.7.

5.5. Water Budget

The water budget of the atmosphere-lithosphere-hydrosphere system is here discussed with reference to the scheme Fig. 5.1:4 and Eqs. 5.1:8 to 5.1:10. As interest is focused on annual mean conditions, storage terms are not considered. The budgets of water and latent heat shall first be presented in terms of zonally averaged conditions, in a way analogous to the treatment of sensible heat and geopotential energy in Section 5.4. This shall be complemented by an analysis of the regional water budget of the tropical oceans and continents.

Latitude mean values of the three most important terms of the atmospheric water budget Eq. (5.1:9) are listed in Table 5.5:1. The precipitation P and evaporation E are compiled from surface information, and the required atmospheric water vapor flux divergence div q is calculated as a residual. Precipitation broadly decreases from the lower to the higher latitudes, but with secondary minima in the subtropics of either hemisphere. Rainfall is largest in the equatorial belt, with a marked maximum to the North of the Equator. Evaporation also decreases from the lower towards the higher latitudes, but a weak minimum is found in the equatorial belt to the North of the Equator. The latitude mean pattern of the atmospheric water vapor flux divergence div q calculated as difference between the two aforementioned terms is characterized by vapor export from the outer tropics and subtropics and import into the equatorial belt as well as into the higher latitudes. The required meridional transport of water vapor in the atmosphere calculated from the third column in Table 5.5:1 is listed in Table 5.5:3.

TABLE 5.5:1

Terms of the annual mean atmospheric water budget, Eq. (5.1:9), by latitude bands, in 10^{-7} kg m^{-2} s^{-1}. Precipitation P, evaporation E, residually determined lateral export of water vapor div q = $E-P$. Sources: Budyko (1974, p. 219), Sellers (1965, p. 5, 103).

	P	E	div q
90–80°N	40	16	− 24
80–70	64	47	− 17
70–60	148	105	− 43
60–50	224	147	− 77
50–40	248	199	− 49
40–30	240	309	+ 69
30–20	368	424	+ 56
20–10	368	424	+ 56
10– 0	604	377	−227
0–10°S	472	398	− 74
10–20	388	472	+ 84
20–30	268	435	+167
30–40	288	388	+100
40–50	320	278	− 42
50–60	304	163	−141
60–70	132	52	− 80
70–80	28	16	− 12
80–90°S	8	0	− 8

TABLE 5.5:2

Terms of the annual mean atmospheric latent heat budget, Eq. (5.1:5), by latitude
bands, in W m^{-2}. Precipitational heating LP, latent heat flux across surface LE,
residually determined lateral export of latent heat div $Lq - LE - LP$. Sources as for
Table 5.5:1.

	LP	LE	div Lq
90–80°N	10	4	– 6
80–70	14	12	– 2
70–60	37	26	–11
60–50	56	37	–19
50–40	62	50	–12
40–30	60	77	+ 17
30–20	55	96	+ 41
20–10	92	106	+ 14
10– 0	151	94	–57
0–10°S	118	100	–18
10–20	97	118	+ 21
20–30	67	109	+ 42
30–40	72	97	+ 25
40–50	80	69	–11
50–60	76	41	–35
60–70	33	13	–20
70–80	7	4	– 3
80–90°S	2	0	– 2

The required water vapor transport is in Table 5.5:3 compared with the observed values as
obtained from the aerological method and detailed in Table 5.4:2. Table 5.5:3 shows agreement
in the major features, namely transport from the Northern hemisphere subtropics towards the
Equator, as well as poleward transport in midlatitudes. Table 5.4:2 indicates that the mean
meridional circulation, the Hadley cell, primarily effects the equatorward transport in the tropics,
whereas transient eddies account for most of the poleward transport in midlatitudes. The standing
eddy contributions are largest in the latitudes of the subtropical highs and subpolar lows.

The same considerations as presented above for the water budget also pertain to the latent
heat budget (Eq. 5.1:5) summarized in Tables 5.5:1, 5.5:3, and 5.4:2. The meridional profiles of
atmospheric water vapor transport shown in Tables 5.4:2 and 5.5:1 to 5.5:3 are broadly consistent
with the classical studies of Starr and White (1955), Starr et al. (1958), Starr and Peixoto (1958,
1960, 1964). Rosen et al. (1979) document the interannual variability of the atmospheric water
vapor transport.

The divergence of the atmospheric water vapor transport, or the difference of evaporation
minus precipitation, has a direct bearing on the salinity conditions of the upper ocean, in that
rainfall dilutes and evaporation enhances the salt concentration. Wüst (1954) first called attention
to the close causal relationship illustrated in Fig. 5.5:1. The meridional profiles of P, E, and $E-P$
in Fig. 5.5:1 are broadly consistent with Table 5.6:1. $E-P$ is negative in a zone immediately to

TABLE 5.5:3

Annual mean northward transport of water vapor (in 10^7 kg s^{-1}) and of latent heat (in 10^{13} W) within the atmosphere. (a) required transport in the entire atmospheric column according to Tables 5.5:1 and 5.5:2, and (b) observed transport in the layer 1012.5 to 75 mb as listed in Table 5.4:2.

	Water vapor		Latent heat	
	Required	Observed	Required	Observed
90°N	0	0	0	0
80	+ 1	+ 1	+ 3	+ 3
70	+ 4	+ 3	+ 5	+ 5
60	+ 12	+ 28	+ 27	+ 72
50	+ 33	+ 47	+ 79	+ 113
40	+ 49	+ 76	+ 118	+ 183
30	+ 25	+ 70	+ 58	+ 175
20	− 41	− 2	− 105	− 5
10°N	− 64	− 75	− 163	− 190
0	+ 37	+ 30	+ 90	+ 74
10°S	+ 71	+ 56	+ 170	+ 141
20	+ 36	+ 7	+ 82	+ 17
30	− 30	− 32	− 75	− 84
40	− 65	− 68	− 175	− 173
50	− 53	− 43	− 138	− 110
60	− 16	− 25	− 46	− 63
70	− 1	− 6	− 6	− 13
80	− 0	− 2	− 1	− 3
90°S	0	0	0	0

the North of the Equator and reaches largest positive values in the subtropics of either hemisphere. Surface salinity S shows a very similar latitudinal distribution, namely smallest values immediately to the North of the Equator and largest amounts in the subtropics. Fig. 5.5:1 highlights the important linkage between the atmospheric water vapor transport and the salt budget of the World ocean.

Discussion so far was confined to the atmospheric branch of the global water and latent heat budget, Eqs. (5.1:9) and (5.1:5). The lithosphere-hydrosphere portion of the water budget, Eq. (5.1:10), is examined in the following for large continental and oceanic domains and with reference to Fig. 5.5:2.

The Atlantic basin as a whole is characterized by evaporation substantially exceeding precipitation. This deficit is met by lateral inflow from adjacent ocean sectors and river discharge from the continents. The Arctic basin receives the waters from the large rivers of Siberia and Northern North America. Moreover, precipitation exceeds evaporation over the Arctic Ocean. Accordingly, a large surplus is available for net southward transport into the Atlantic. Nearly all of Europe and much of North America drains into the North Atlantic. The major portion of the water surplus of tropical Africa flows to the Atlantic, mainly through the Nile and Zaïre rivers. The contributions of the Senegal and Niger are substantially smaller. Discharge to the Indian Ocean is small by

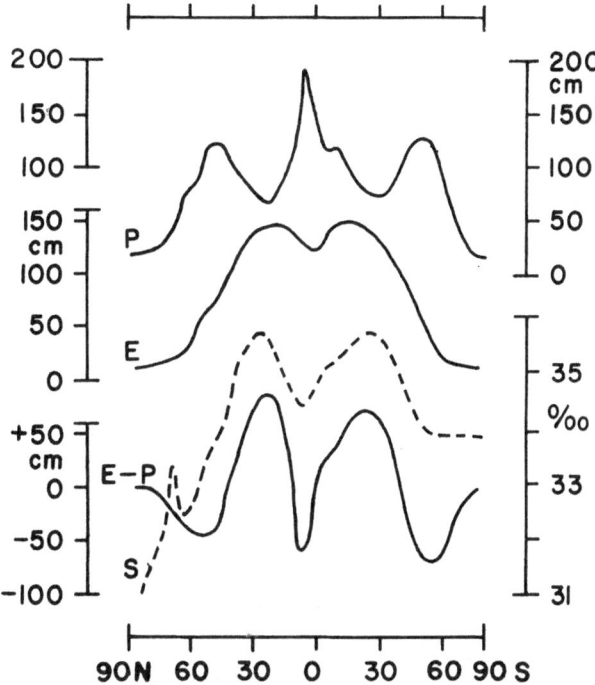

Fig. 5.5:1. Latitude distribution of precipitation P, evaporation E, the difference $E-P$, and surface salinity S in the World Ocean. From Wüst (1954).

comparison (Zambezi). Tropical South America likewise drains primarily towards the Atlantic. The largest discharge stems from the Amazon, with further substantial contributions by the Magdalena and Orinoco rivers that flow into the Caribbean Sea. In addition to all these contributions from the Arctic Ocean and the discharge from the continents, the Atlantic requires a further water import from the Southern Ocean to meet its water balance.

The Indian Ocean as a whole also evaporates more water than it receives through precipitation – commensurate with its role as major water vapor source for the rainfall over South Asia during the Northern summer Southwest monsoon. The Indian Ocean receives freshwater input primarily from the rivers of monsoon Asia (Indus, Ganges, Brahmaputra), with smaller contributions from Australasia and Eastern Africa. In addition to these contributions by continental runoff, a further water input from the Southern Ocean is required to make up for the deficit.

An excess of evaporation over precipitation is also indicated for the Pacific Ocean. This deficit is only in part compensated by river discharge from the surrounding continents, so that a substantial import of water is required from the Southern Ocean.

Turning to the closure of the budget sketched in Fig. 5.5:2, precipitation far exceeds evaporation in all three sectors of the Southern Ocean. This excess together with the ice discharge across the coast of Antarctica then compensates for the remaining deficits in the Atlantic, Indian, and Pacific Oceans. Although the source used (UNESCO, 1978) is the most comprehensive presently available, Fig. 5.5:2 can provide only a preliminary orientation.

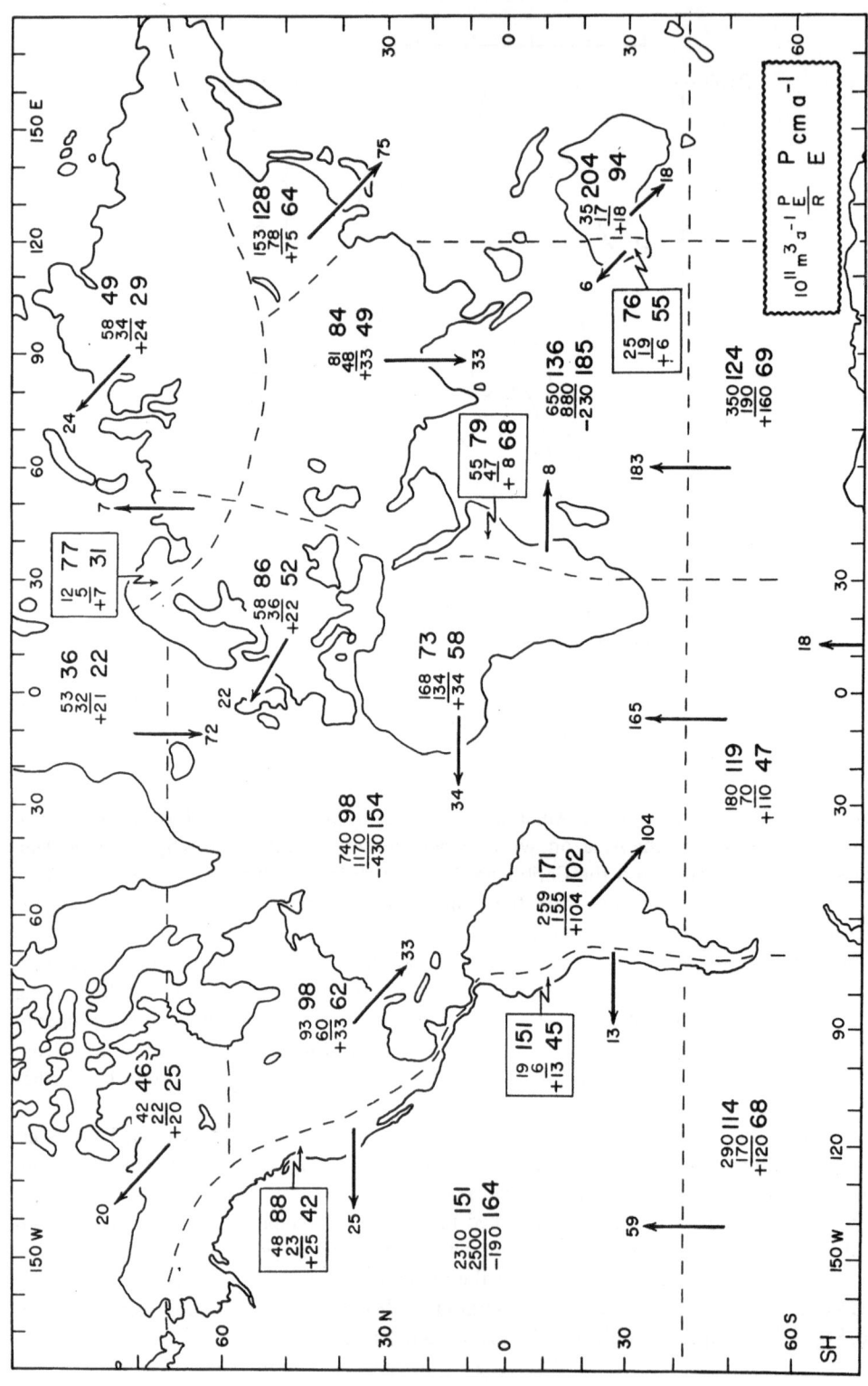

Fig. 5.5:2. Schematic map of the water budget of continents and oceans. Broken lines delineate continental drainage basins and sectors of the ocean. The large numbers in boxes denote precipitation, P (top) and evaporation E (bottom) in cm a^{-1}. The small numbers signify, from top to bottom, P, E, and $P-E$, in 10^{11} m^3 a^{-1}. Arrows indicate continental discharge and net water exchange between segments of the ocean, with the same units. Source: UNESCO (1978).

5.6. Relative Roles of Oceanic Versus Atmospheric Heat Transports

The assessment of net allwave radiation at the top of the atmosphere $SWLW\uparrow\downarrow_{top}$ in Section 5.2 and of the net oceanic heat gain Q_v in Section 5.3 allows inference of the heat export within the atmosphere Q_{va}, and thus an appraisal of the relative roles of hydrosphere and atmosphere in the poleward heat transport.

The map of heat export within the atmosphere Q_{va}, Fig. 5.2:1c, is the difference between the patterns of net radiation at the top of the atmosphere $SWLW\uparrow\downarrow_{top}$ and oceanic heat export Q_{vo} depicted in Figs. 5.2:1a and 5.2:1b, respectively. Fig. 5.2:1c shows atmospheric heat export for most of the low latitudes, but especially the Southern Indian Ocean, the South Atlantic, the western North Atlantic, and equatorial Africa. Heat import is indicated for the deserts of Northern Africa, the cold water regions along the coasts of Eastern Africa and Arabia, and particularly a band immediately to the South of the Equator in the Eastern and Central Pacific. Similarly, the atmospheric heat export is comparatively small to the South of the Atlantic Equator. The latter features are a direct result of the maxima of oceanic heat export Q_{vo} to the South of the Pacific and Atlantic Equator and off the Somali-Arabian coasts, and the rather uniform horizontal pattern of net radiation at the top of the atmosphere $SWLW\uparrow\downarrow_{top}$.

Mean meridional profiles of Q_{va} are shown in Fig. 5.3:1 for the three oceans and in Table 5.2:1 also for the four continents. The profiles for the three individual oceans illustrate the existence of a minimum of Q_{va} to the South of the Equator in the Atlantic and Pacific, and its absence in the Indian Ocean. This is a direct consequence of the meridional profiles of Q_{vo} in the three oceans; the cold water conditions during the Northern summer monsoon contributing towards the large Q_{vo} in the Northern Indian Ocean. For all oceans combined (Figs. 5.2:3 and 5.2:4), the prominent maximum of oceanic heat export Q_{vo} immediately to the South of the Equator along with the more nearly hemispherically symmetric pattern of net radiation at the top of the atmosphere $SWLW\uparrow\downarrow_{top}$ result in a conspicuous minimum of atmospheric heat export Q_{va} in a latitude band immediately to the South of the Equator. This feature is strongly dictated by the patterns of the Pacific and Atlantic Oceans.

It is recalled that net atmospheric heat export Q_{va} typically results from import of latent heat Lq in the lower troposphere, and export of geopotential energy and sensible heat $(gz + c_pT)$ in the upper troposphere, and vice versa (Riehl and Malkus, 1958; Hastenrath, 1966; Hastenrath and Lamb, 1980); this being related to the characteristic vertical distribution of the various atmospheric energy forms and the intensity and direction of the vertical circulation. The transports of geopotential energy and sensible heat $(gz + c_pT)$ and of latent heat Lq were explicitly considered in Sections 5.4 and 5.5.

From the information contained in Tables 5.2:1 and 5.2:2 and Figs. 5.2:1 and 5.3:1 the meridional profiles of $SW\uparrow\downarrow_{top}$, Q_{va}, and Q_{vo} in Fig. 5.2:4 were constructed. In addition to absolute energy quantities by latitude belts, the terms are also expressed as percentages of the total annual radiative heat gain at the top of the atmosphere in the tropics, $30°N - 30°S$. Consistent with Fig. 5.2:2, the profiles in Fig. 5.2:4 show approximate hemispheric symmetry for $SWLW\uparrow\downarrow_{top}$, with the maximum to the North of the Equator. The maximum of oceanic heat export Q_{vo} immediately to the South of the Equator is concomitant with a conspicuous minimum of atmospheric heat export Q_{va}. The largest amounts of atmospheric heat export Q_{va} are found in a band immediately to the North of the Equator. However, despite the comparatively large share of land in the total surface of the band $0-10°N$, the entire atmospheric heat export from this belt is little larger than the oceanic heat export Q_{vo} from the band $0-10°S$. In fact, the oceanic heat export effected by the band $0-5°S$ is more substantial by comparison.

Classical concepts of the global heat budget have emphasized the role of upper-tropospheric export from the realm of the quasi-planetary near-equatorial band of maximum convergence ('ITCZ'). A 'hot-tower' mechanism of undiluted cumulonimbus cores was invoked to carry heat from the surface layer through the mid-tropospheric energy minimum into the upper troposphere, from where it would be exported laterally to higher latitudes (Riehl and Malkus, 1958). The analysis summarized in Fig. 5.2:4 invites a reappraisal of the relative roles of the tropical atmosphere and hydrosphere in the heat budget of the globe.

The planetary cloud-band ('ITCZ') is throughout the year confined to the latitude band 0–10°N, except for the longitudes 50–100°E of the Indian Ocean sector, where the maximum cloud zone is displaced farther North during Northern summer (Fig. 5.2:2). In the following discussion, the magnitude of heat budget terms is compared to the net allwave radiation at the top of the atmosphere SWLW↑↓$_{top}$ in the zone 30°N to 30°S as reference (1008×10^{13} W = 100 units, caption to Fig. 5.2:4). The annual atmospheric heat export Q_{va} from the latitude belt 0–10°N amounts to only 13 units (Fig. 5.2:4). If the band 10–20°N is substituted for the longitudes 50–100°E, a substantially smaller number is obtained (Fig. 5.2:1c, Table 5.2:2). By comparison, the oceanic export Q_{vo} from the zone 0–10°S is 18 units, and 11 of these units stem from the belt 0–5°S alone. In fact in this strip immediately to the South of the Equator some 90% of the net radiation at the top of the atmosphere is disposed of by export within the oceanic water body Q_{vo}. Viewed in perspective, heat export from the realm of the 'ITCZ' by atmospheric motions, Q_{va}, plays a modest *relative* role, while oceanic export Q_{vo} from the cold water zones immediately to the South of the Equator in the Eastern Atlantic and Eastern and Central Pacific is an at least comparably important factor in the heat export to the extratropics. In absolute terms, however, the atmospheric heat export Q_{va} from the band 0–10°N as given in Fig. 5.2:4 is about twice as large as the classical estimates (Riehl and Malkus, 1958) that minimize the role of the oceans but use the much smaller values of SWLW↑↓$_{top}$ calculated by London (1957) as input. Thus Fig. 5.2:4 shows for the band 0–10°N values of SWLW↑↓$_{top}$ = 26 and Q_{va} = 15, as compared to London's (1957) annual SWLW↑↓$_{top}$ = 8 units, all in 10^{14} W.

Tables 5.2:1 to 5.2:4 and Figs. 5.2:1 and 5.6:1 illustrate the relative importance of the ocean in disposing of the heat surplus from the tropics. The radiative heat gain at the top of the atmosphere over land is rather smaller than over sea, and is in its totality disposed of within the atmospheric column. For the tropical oceans, the ratio of hydrospheric to total heat export follows a pattern similar to Q_{vo}, because of the comparatively modest spatial variation of net radiation at the top of the atmosphere.

The partitioning of total heat export into the atmospheric and hydrospheric contributions requires satellite-derived estimates of planetary net radiation, that are consistent with a balance for the globe as a whole. In Fig. 5.2:4 imbalances are adjusted uniformly with latitude. On this premise, of the total annual heat gain in the belt 30°N – 30°S, only 20 units are exported by the atmosphere over land (Tables 5.2:1, 5.2:2, Fig. 5.2:4). For the tropical oceans at large, exports by atmosphere and water body amount to 41 and 39 units.

The poleward heat transport for all oceans combined, Fig. 5.3:2, is in Fig. 5.6:1 compared with the total transport in the ocean-atmosphere system. For the latter parameter, the average of data sets I–IV identified in Table 5.2:3 is used. This is within drafting accuracy almost identical to a curve constructed from the data set V. The difference between the curve of total required transport in the ocean-atmosphere system and the plot of oceanic heat transport represents an estimate of the heat transport within the atmosphere. The curve of atmospheric heat transport differs from that in Fig. 11 of Hastenrath (1980), because of altered satellite input and oceanic transport figures. In Fig. 5.6:1 the hydrosphere accounts for 40 and 46% of the total poleward

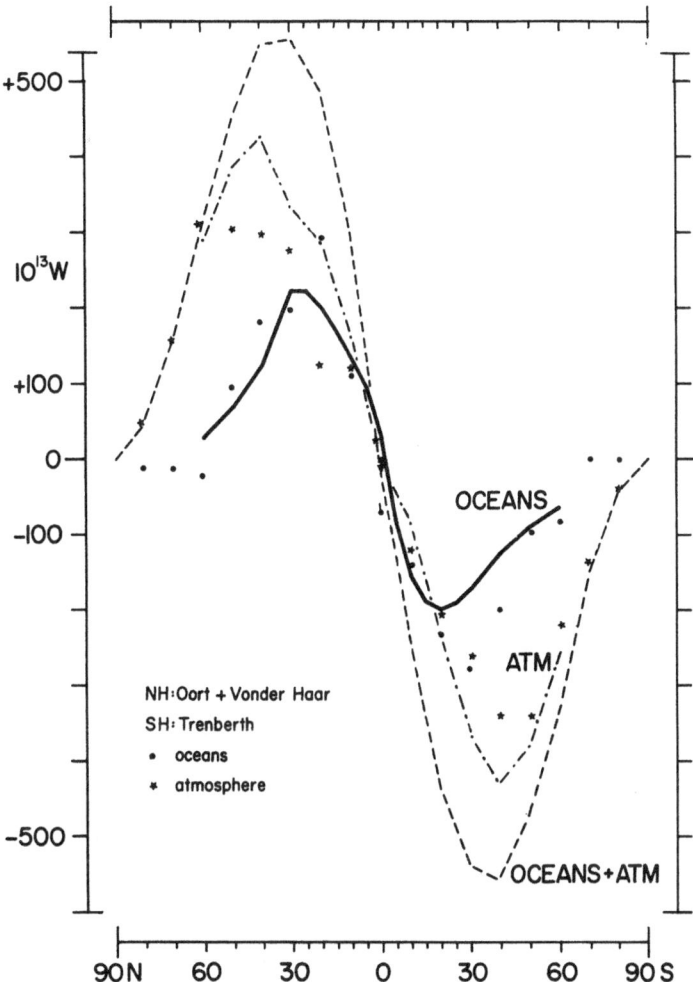

Fig. 5.6:1. Annual mean meridional transport within the atmosphere-ocean system broken (source: Gruber, 1978), within the oceans solid (from Fig. 5.3:2), and within the atmosphere (residual) dash-dotted line. Estimates by Oort and Vonder Haar (1976) for the Northern and by Trenberth (1979) for the Southern hemisphere are entered as dots and stars for ocean and atmosphere, respectively. Northward transport positive, units in 10^{13} W. From Hastenrath (1982).

transport at 30°N and 20°S, respectively. The corresponding shares in Hastenrath (1980, Fig. 11) are 53 and 35%.

Estimates of Oort and Vonder Haar (1976) and Trenberth (1979) are also entered in Fig. 5.6:1 for comparison. Vonder Haar and Oort (1973) concluded that the hydrosphere performed 74 and 47% of the total poleward heat transport at latitudes 20 and 30°N, respectively. The corresponding numbers in Oort and Vonder Haar's (1976) study are 73 and 47%. Trenberth's (1979) estimate of maximum oceanic transport at 30°S amounts to 52% of the total required poleward transport. This partitioning hinges critically on representative values of the net radiation at the top of the atmosphere.

Meridional heat transport within the atmosphere is largest in midlatitudes where it indeed effects the bulk of the total transport for the ocean-atmosphere system. As shown in Table 5.4:2,

TABLE 5.6:1

Annual mean total northward heat transport ($c_p T + gz + Lq$) within the atmosphere, in 10^{13} W. O+R 71: Oort and Rasmusson (1971, p. 127–135, layer 1012.5 to 75 mb); O+V 76: Oort and Vonder Haar (1976, layer 1012.5 to 75 mb); T 79: Trenberth (1979); O 83: Oort (1983, 1012.5 to 75 mb); Fig. 5.6:1: of this chapter; H 82: Hastenrath (1982); computed from net radiation at the top of the atmosphere SWLW↑↓$_{top}$ in Fig. 5.6:1, and values of surface net radiation SWLW↑↓$_{sfc}$, and sensible Q_s and latent heat flux Q_e at the surface published by Budyko (1974, p. 219) and Sellers (1965, p. 5, 103).

	O+V 76	T 79	O+R 71	O 83	Fig. 5.6:1	H 82
90°N	0			0		0
80	+ 50			+ 44		+ 41
70	+160		+136	+124		+151
60	+310		+277	+307	+292	+280
50	+310		+310	+389	+390	+357
40	+300		+344	+444	+428	+420
30	+280		+282	+365	+334	+384
20	+110		+137	+237	+283	+308
10°N	+100		+105	+118	+162	+168
0	+ 20	− 17	+ 89	− 14	− 3	+ 23
10°S		−123	−190	−108	− 6	−132
20		−209		−255	−234	−307
30		−258		−372	−367	−439
40		−335		−434	−433	−479
50		−344		−340	−384	−424
60		−201		−160	−262	−323
70		−145	− 35	− 35		−169
80		− 48		− 30		− 51
90°S	0			0		0

a substantial portion of the poleward heat transport in the atmosphere of the Northern hemisphere extra-tropics takes place in the form of latent heat (see also Oort and Rasmusson, 1971; Rosen et al., 1979; Oort, 1983).

The annual mean meridional heat transport in the atmosphere obtained as a residual between the two other curves in Fig. 5.6:1, is in Table 5.6:1 compared with various other sources. The columns O + V 76, O + R 71, T 79, and O 83 are based on aerological observations. The column H 82 was obtained from evaluation of the heat budget equation (5.1:3) of an atmospheric column, using the net radiation at the top of the atmosphere SWLW↑↓$_{top}$ from Fig. 5.6:1, and values of surface net radiation SWLW↑↓$_{sfc}$, sensible Q_s, and latent heat flux Q_e at the surface published by Budyko (1974, p. 219) and Sellers (1965, p. 5, 103).

Table 5.6:1 shows that the present results, Fig. 5.6:1, agree broadly with the calculations of Oort and Rasmusson (1971, pp. 127–135) for the Northern hemisphere midlatitudes, where transports are largest and the radiosonde network is most plentiful. In addition, the atmospheric transport curve of Fig. 5.6:1 is in good agreement with the column H 82 for most latitudes of both hemispheres. Given the values of the three right-hand terms of Eq. (5.1:3) from Budyko (1974) and Sellers (1965), the required atmospheric meridional heat transport would depend only on the latitudinal pattern of SWLW↑↓$_{top}$. Thus the indirect assessment of atmospheric meridional heat transport presented in Fig. 5.6:1 is compatible with the results of two other independent approaches.

Errors in atmospheric transport related to interannual variability (Oort, 1977) have been estimated by Vonder Haar and Oort (1973) and Oort and Vonder Haar (1976). These figures are small, in fact the corresponding probable error in the annual atmospheric heat export Q_{va} from the zone 0–30°N would amount to only about 2 W m^{-2}. It is also appreciated that appropriate adjustments have ensured internal consistency in the MIT General Circulation Project data. However, the lack of actual upper-air observations in large areas of the tropics (Oort, 1978) cannot be made up by analysis techniques. Among the most critical data-sparse areas are the vast expanses of the Pacific, the realm of jet-like wind concentrations over the Western Indian Ocean, and others. Furthermore, an evaluation of the annual mean meridional energy transports tabulated by Oort and Rasmusson (1971) according to energy components, layers, and latitude belts, illustrates that the net export results are usually a small difference between large terms of opposite sign. It is here proposed that the error in the net atmospheric energy export Q_{va} from the belt 0–30°N, as related to the effect of small differences, may well be of the order of 5 W m^{-2}. More importantly, internal data consistency does not preclude sizeable systematic errors in Q_{va}, the magnitude of which is not known.

In order to appreciate error tolerances in heat transport estimates, consider zero heat transport across the Equator and a systematic error in, say, the satellite-derived net radiation at the top of the atmosphere of about 10 W m^{-2}. This would correspond to an error in the heat transport across 30°N of the order of 100×10^{13} W. Similar considerations apply for the atmospheric and hydrospheric domains of the system separately.

In conclusion, integration of oceanic heat export indicates poleward heat transport from the tropics into either hemisphere in the Pacific; a northward directed transport from high Southern latitudes all the way to the Arctic in the Atlantic; and southward transport in the Indian Ocean. Oceanic heat surplus in the Pacific compensates deficits in the higher Southern latitudes of the Atlantic and Indian Ocean sectors, as well as in the Atlantic at large. For all oceans combined, heat transport is largest around 30°N and 20°S, where it accounts for 40 and 46% of the total meridional transport in the ocean-atmosphere system. For the globe as a whole, the sign of the net cross-equatorial oceanic heat transport is uncertain. Atmospheric transport is largest, and effects the bulk of the total transport in mid-latitudes.

The relative partitioning proposed here is contingent upon the representativeness of satellite-measured net radiation at the top of the atmosphere and the computation of the oceanic heat budget. Appreciably different estimates of SWLW↑↓$_{\text{top}}$, Q_{vo}, and Q_{va}, must be considered as compatible within the broad error limits indicated for all three terms.

5.7. Synthesis

The most elementary requirement in general circulation theory is the continuity of mass of the atmosphere (Section 3.3). As this is satisfied, the maintenance of the atmospheric circulation in terms of angular momentum and kinetic energy (Sections 3.4 and 3.5) can be considered, essentially in isolation from the ocean. Turning to the continuity of heat and water, however, treatment of the coupled atmosphere-ocean system (Chapters 4 and 5) is imperative. In this context, the role of kinetic energy in the overall energy budget of the atmosphere can be disregarded, because it is of much smaller magnitude than sensible heat, geopotential energy, and latent heat.

Fundamental to the heat budget of the planet Earth is the net radiation at the top of the atmosphere. This can now in principle be measured from satellites. An integral constraint is available in that this quantity should be approximately zero in the global multi-annual mean. Measured meridional profiles of planetary net radiation must be adjusted accordingly. Even so the results for various years and from different satellites differ considerably. The crossover from positive to negative net radiation values is found somewhat poleward of 30 degrees latitude. Uncertainties are related to a combination of real interannual variability and deficiencies in sampling, sensing, and processing. The uncertainties concerning a representative mean meridional profile of net radiation at the top of the atmosphere hamper the partitioning of total poleward heat transports into the contributions by the atmosphere and hydrosphere.

The annual mean heat budget of the hydrosphere is evaluated through calculation of latent and sensible heat fluxes and net radiation at the surface on the basis of long-term ship observations. The net oceanic heat gain is then obtained as a residual. Integration with latitude yields the annual mean meridional heat transports in the Atlantic, Indian, and Pacific Oceans. For the Atlantic, these estimates are complemented by the direct evaluation of hydrographic sections.

The heat budget of the Atlantic is characterized by the open communication with the Arctic basin and the concomitant large heat deficit in high Northern latitudes, and the large oceanic heat gain in the Southern hemisphere tropics associated with the comparatively low sea surface temperature, especially in a band immediately to the South of the Equator. Therefore, a heat import from other parts of the World ocean is required, and the oceanic heat transport in the Atlantic is directed northward from 60°S all the way to the Arctic. As a result of the continent configuration, meridional heat transport in the Indian Ocean is directed southward, with a maximum around 10–15°S and a decrease to near zero at 60 S. In the Pacific oceanic heat transports are directed from the equatorial belt poleward into either hemisphere. Contributions from the Pacific domain make up for the heat deficits in the higher Southern latitudes of the Indian and Atlantic Ocean sectors, and the Atlantic and Arctic Oceans at large. For all oceans combined, maximum poleward heat transports are found around 30°N and 20°S.

The continuity of sensible heat plus geopotential energy in the atmosphere is largely controlled by the meridional distribution of precipitational heating, net radiative cooling of the atmospheric column, and sensible heat flux from the surface. The required transports calculated from these terms are compared with those obtained directly from aerological information. Close agreement is found in the Northern hemisphere midlatitudes, where transports are largest and the radiosonde network is most plentiful. The northward transport of geopotential energy by the mean meridional circulation is pronounced in the realm of the wintertime Hadley cell, but the annual mean pattern of the total transport of sensible heat plus geopotential energy is dominated by the transient eddy transport of sensible heat, which has a maximum in the zone of the travelling disturbances in midlatitudes.

The transport of water vapor and latent heat in the atmosphere is directly related to the meridional pattern of precipitation and evaporation from the surface. Both are large in low latitudes, but rainfall has a maximum in a zone immediately to the North of the Equator and minima in the subtropics, while evaporation shows a weak minimum in the zone of maximum rainfall to the North of the Equator. Required meridional transports of water vapor according to the latitude distribution of precipitation and evaporation are again compared with direct evaluation from aerological information. Agreement is best for the Northern hemisphere midlatitudes, where transports are largest and the radiosonde network is most nearly adequate. The mean meridional circulation effects the equatorward transport of water vapor and latent heat in the tropics, while transient eddies account for most of the poleward transports in midlatitudes.

The lithosphere-hydrosphere portion of the water budget can be appraised for large continental and ocean domains. Over all three tropical oceans, evaporation exceeds precipitation, the deficit being made up by river discharge from the surrounding continents and net import from the Southern Ocean. The excess of evaporation over precipitation is particularly large for the Atlantic, which receives water import from the Arctic basin, and the freshwater inflow from all of Europe, much of North America, and most of the African and South American continents, in addition to the import from the Southern Ocean.

The total meridional heat transport in the atmosphere, including sensible heat, geopotential energy, and latent heat, can be estimated from the surface heat budget and measurements of net radiation at the top of the atmosphere, and independently from direct evaluation of aerological information. Results agree closely for the Northern hemisphere midlatitudes where transports are largest and the radiosonde network is most satisfactory. Similar values are obtained from the inference of atmospheric poleward heat transport as the difference between (a) the total required transport in the combined atmosphere-ocean system according to the net radiation at the top of the atmosphere and (b) the assessment of the hydrospheric transport.

On this basis, the relative roles of atmosphere and hydrosphere in the total poleward transport of heat can be appreciated. The ocean accounts for about half of the total poleward transports around 30°N and 20°S. Atmospheric transports are largest in midlatitudes. In general, the heat transport estimates from various independent methods agree remarkably closely, although error tolerances are large for all of these approaches.

References

Aagaard, K., Greisman, P., 1975: 'Towards new mass and heat budgets for the Arctic Basin'. *J. Geophys. Res.*, 80, 3821–3827.

Atkinson, G. D., Sadler, J. C., 1970: 'Mean-cloudiness and gradient-level-wind charts over the tropics'. U.S. Air Force, AWS Technical Report No. 215.

Behringer, D. W., Stommel, H., 1981: 'Annual heat gain of the tropical Atlantic from subsurface ocean data'. *J. Phys. Oceanogr.*, 11, 1393–1398.

Bennett, A. F., 1978: 'Poleward heat fluxes in Southern hemisphere oceans'. *J. Phys. Oceanogr.*, 8, 785–798.

Bryan, K., 1962: Measurements of meridional heat transport by ocean currents. *J. Geophys. Res.*, 67, 3403–3413.

Bryan, K., 1983: Poleward heat transport by the ocean. *Rev. Geophys. Space Phys.*, 21, 1131–1137.

Bryden, H. L., Hall, M. M., 1980: 'Heat transport by currents across 25°N latitude in the Atlantic Ocean'. *Science*, 207, 884–886.

Budyko, M. I., 1963: *Atlas of the heat balance of the Earth*. Kartfabrika Gosgeoltehizdata, Leningrad, 75 pp. (in Russian).

Budyko, M. I., 1974: *Climate and life*. International Geophysics Series, vol. 18. Academic Press, New York, San Francisco, London, 508 pp.

Bunker, A. F., 1976: 'Computations of surface energy flux and annual air-sea interaction cycles of the North Atlantic Ocean'. *Mon. Wea. Rev.*, 104, 1122–1140.

Bunker, A. F., Worthington, L. V., 1976: 'Energy exchange charts of the North Atlantic Ocean'. *Bull. Amer. Meteor. Soc.*, 57, 670–678.

Carissimo, B. C., Oort, A. H., Vonder Haar, T. H., 1985: 'On estimating the meridional energy transports in the atmosphere and ocean'. *J. Phys. Oceanogr.*, 15, 82–91.

Ellis, J. S., Vonder Haar, T. H., 1976: 'Zonal average earth radiation budget measurements from satellites for climate studies'. *Atmos. Sci. Paper*, No. 240, Colorado State University, Fort Collins, Colo., 58 pp.

Ellis, J. S., Vonder Haar, T. H., Levitus, S., Oort, A. H., 1978: 'The annual variation of the global heat balance of the Earth'. *J. Geophys. Res.*, 83, 1958–1962.

Emig, M., 1967: 'Heat transport by ocean currents'. *J. Geophys. Res.*, 72, 2519–2529.

Fine, R. A., 1985: 'Direct evidence using tritium data for through flow from the Pacific into the Indian Ocean'. *Nature*, 315, 478–480.

Fu, L.-L., 1981: 'The general circulation and meridional heat transport of the subtropical South Atlantic determined by inverse methods'. *J. Phys. Oceanogr.*, 11, 1171–1193.

Fuglister, F. C., 1960: *Atlantic Ocean atlas of temperature and salinity profiles and data from the International Geophysical Year of 1957–58*. Woods Hole Oceanographic Institution, Atlas series, vol. 1, Woods Hole, Mass., 209 pp.

Godfrey, J. S., Golding, T. J., 1981: 'The Sverdrup relation in the Indian Ocean, and the effect of Pacific-Indian Ocean throughflow on Indian Ocean circulation and on the East Australian Current'. *J. Phys. Oceanogr.*, 11, 771–779.

Gordon, A. L., 1981: 'Seasonality of Southern ocean sea ice'. *J. Geophys. Res.*, 86, 4193–4197.

Gruber, A., 1978: 'Determination of the earth-atmosphere radiation budget from NOAA satellite data'. NOAA Technical Report NESS 76, Washington, D.C., 28 pp.

Gruber, A., Ruff, A., Earnest, C. L., 1983: 'Determination of the planetary radiation budget from TIROS-N satellites'. NOAA Technical Report NESDIS 3, 12 pp.

Hall, M. M., Bryden, H. L., 1982: 'Direct estimates and mechanisms of ocean heat transport'. *Deep-Sea Res.*, 29, 339–360.

Hastenrath, S., 1966: 'On general circulation and energy budget in the area of the Central American Seas'. *J. Atmos. Sci.*, 23, 694–711.

Hastenrath, S., 1977a: 'Relative role of atmosphere and ocean in the global heat budget: tropical Atlantic and Eastern Pacific'. *Quart. J. Roy. Meteor. Soc.*, 103, 519–526.

Hastenrath, S., 1977b: 'Atmospheric and oceanic heat export in the tropical Pacific'. *J. Meteor. Soc. Japan*, 55, 494–497.

Hastenrath, S., 1978: 'Hemispheric asymmetry of the oceanic heat budget in the Equatorial Atlantic and Eastern Pacific'. *Tellus*, 29, 523–529.

Hastenrath, S., 1980: 'Heat budget of tropical ocean and atmosphere'. *J. Phys. Oceanogr.*, 10, 159–170.

Hastenrath, S., 1982: 'On meridional heat transports in the World ocean'. *J. Phys. Oceanogr.*, 12, 922–927.

Hastenrath, S., 1984a: 'On the interannual variability of poleward transport and storage of heat in the ocean-atmosphere system'. *Archiv. Meteor. Geophys. Bioklim.*, Ser. A, 33, 1–10.

Hastenrath, S., 1984b: 'On meridional transports of heat and freshwater in the Pacific Ocean'. *Archiv Meteor. Geophys. Bioklim.*, Ser A, 33, 91–99.

Hastenrath, S., Lamb, P. J., 1977: *Climatic atlas of the tropical Atlantic and Eastern Pacific Oceans*. University of Wisconsin Press, 112 pp.

Hastenrath, S., Lamb, P. J., 1978a: 'On the dynamics and climatology of surface flow over the equatorial oceans'. *Tellus*, 30, 436–448.

Hastenrath, S., Lamb, P. J., 1978b: *Heat budget atlas of the tropical Atlantic and Eastern Pacific Oceans*. University of Wisconsin Press, 103 pp.

Hastenrath, S., Lamb, P. J., 1979: *Climatic atlas of the Indian Ocean. Part 1. Surface circulation and climate. Part 2. The oceanic heat budget*. University of Wisconsin Press, 116 and 110 pp.

Hastenrath, S., Lamb, P. J., 1980: 'On the heat budget of hydrosphere and atmosphere in the Indian Ocean'. *J. Phys. Oceanogr.*, 10, 694–708.

Jacobowitz, H., Smith, W. L., Howell, H. B., Nagle, F. W., Hickey, J. R., 1979: 'The first eighteen months of planetary radiation budget measurements from the NIMBUS-6 ERB experiment'. *J. Atmos. Sci.*, 36, 501–507.

Kandel, R., 1983: 'Satellite observation of the Earth radiation budget'. *Beitr. Phys. Atmosph.*, 56, 322–340.

Lamb, P. J., 1981: 'Estimate of annual variation of Atlantic Ocean heat transport'. *Nature* 290, 766–768.

Lamb, P. J., Bunker, A. F., 1982: 'On the annual march of the heat budget of the North and tropical Atlantic Oceans'. *J. Phys. Oceanogr.*, 12, 1388–1410.

Levitus, S., 1982: *Climatological atlas of the World Ocean*. NOAA Professional Paper No. 13, U.S. Government Printing Office, Washington, D.C., 173 pp.

Levitus, S., 1984: 'Annual cycle of temperature and heat storage in the world ocean'. *J. Phys. Oceanogr.*, 14, 727–746.

List, R. J., 1968: *Smithsonian meteorological tables*. Smithsonian Institution Press, Washington, 6th ed., 527 pp.

London, J., 1957: 'A study of the atmospheric heat balance'. Final Report, Contract No. AF 19(122)–165, Res. Div., College of Engineering, New York University, 99 pp.

Meehl, G. A., Washington, W. M., Semtner, A. J. Jr., 1982: 'Experiments with a global ocean model driven by observed atmospheric forcing'. *J. Phys. Oceanogr.*, 12, 301–312.

Miller, J. R., Russell, G. L., Tsang, L.-C., 1983: 'Annual oceanic heat transports computed from an atmospheric model'. *Dyn. Atmos. Oceans*, 7, 95–109.

Niiler, P. P., Richardson, W. S., 1973: 'Seasonal variability of the Florida Current'. *J. Mar. Res.*, 31, 144–167.

Office of Space Science and Applications, NASA, 1983: 'Understanding climate: a strategy for solar and Earth radiation research (1984–1994)'. NASA, Washington, D.C., 72 pp.

Oort, A. H., 1977: 'The interannual variability of atmospheric circulation statistics'. NOAA Professional Paper No. 8, 76 pp., Rockville, Md.

Oort, A. H., 1978: 'Adequacy of the rawinsonde network for global circulation studies tested through numerical model input'. *Mon. Wea. Rev.*, 106, 174–195.

Oort, A. H., 1983: 'Global atmospheric circulation statistics, 1958–1973'. NOAA Professional Paper 14, U.S. Government Printing Office, Washington, D. C., 180 pp.

Oort, A. H., Rasmusson, E. M., 1971: 'Atmospheric circulation statistics'. NOAA Professional Paper No. 5, 323 pp., Rockville, Md.

Oort, A. H., Vonder Haar, T. H., 1976: 'On the observed annual cycle in the ocean-atmosphere heat balance over the Northern hemisphere'. *J. Phys. Oceanogr.*, 6, 781–799.

Pares-Sierra, A. F., Inoue, M., O'Brien, J. J., 1985: 'Estimates of oceanic horizontal heat transport in the tropical Pacific'. *J. Geophys. Res.*, 90, 3293–3303.

Peixoto, J. P., Oort, A. H., 1984: 'Physics of climate'. *Reviews Modern Phys.*, 56, 365–429.

Piola, A. R., Gordon, A. L., 1984: 'Pacific and Indian Ocean upper-layer salinity budget'. *J. Phys. Oceanogr.*, 14, 747–753.

Riehl, H., Malkus, J. S., 1958: 'On the heat balance of the equatorial trough zone'. *Geophysica*, 6, 503–538.

Roemmich, D., 1980: 'Estimation of meridional heat flux in the North Atlantic Ocean by inverse methods'. *J. Phys. Oceanogr.*, 10, 1972–1983.

Rosen, R. D., Salstein, D. A., Peixoto, J. P., 1979: 'Variability in the annual fields of large-scale atmospheric water vapor transport'. *Mon. Wea. Rev.*, 107, 26–37.

Sellers, W. D., 1965: *Physical climatology*. University of Chicago Press, 272 pp.

Short, D. A., Cahalan, R. F., 1983: 'Interannual variability and climatic noise in satellite observed outgoing longwave radiation'. *Mon. Wea. Rev.*, 111, 572–577.

Starr, V. P., Peixoto, J. P., 1958: 'On the global balance of water vapor and the hydrology of deserts'. *Tellus*, 10, 189–194.

Starr, V. P., Peixoto, J. P., 1960: 'On the zonal flux of water vapor in the Northern hemisphere'. *Pure Appl. Geophys.*, 47, 199–203.

Starr, V. P., Peixoto, J. P., 1964: 'The hemispheric eddy flux of water vapor and its implications for the mechanics of the general circulation'. *Archiv Meteor. Geophys. Bioklim.*, Ser. A, 14, 111–130.

Starr, V. P., Peixoto, J. P., Livadas, G. C., 1958: 'On the meridional flux of water vapor in the Northern hemisphere'. *Pure Appl. Geophys.*, 39, 174–185.

Starr, V. P., White, R. M., 1955: 'Direct measurement of the hemispheric poleward flux of water vapor'. *J. Mar. Res.*, 14, 217–225.

Stephens, G. L., Campbell, G. G., Vonder Haar, T. H., 1981: 'Earth radiation budgets'. *J. Geophys. Res.*, 86, 9739–9760.

Talley, L. D., 1984: 'Meridional heat transport in the Pacific Ocean'. *J. Phys. Oceanogr.*, 14, 231–241.

Trenberth, K. E., 1979: 'Mean annual poleward energy transport by the oceans in the Southern hemisphere'. *Dyn. Atmos. Oceans*, 4, 57–64.

UNESCO, 1978: *World water balance and water resources of the earth*. Paris-Leningrad, 663 pp.

U.S. Navy Hydrographic Office, 1944: 'World atlas of sea surface temperatures'. H.O. Publ. No. 225, 2nd ed., U.S. Government Printing Office.

Vonder Haar, T. H., Ellis, J. S., 1974: 'Atlas of radiation budget measurements from satellites'. Colorado State University, Department of Atmospheric Science, Colo. Atmospheric Science Paper No. 231, Front Collins, Colo., 180 pp.

Vonder Haar, T. H., Oort, A. H., 1973: 'New estimate of annual poleward energy transport by Northern hemisphere oceans'. *J. Phys. Oceanogr.*, 2, 169–172.

Winston, J. S., Gruber, A., Gray, T. I., Varnadore, M. S., Earnest, C. L., Manello, L. P., 1979: *Earth-atmosphere radiation budget analyses derived from NOAA satellite data, June 1974 – February 1978*, 2 vols., NOAA, Washington, D.C.

Wunsch, C., 1980: 'Meridional heat flux of the North Atlantic Ocean'. *Proc. Natl. Acad. Sci.*, 77, no. 9, 5043–4047.

Wunsch, C., Hu, D., Grant, B., 1983: 'Mass, heat, salt, and nutrient fluxes in the South Pacific Ocean'. *J. Phys. Oceanogr.*, 13, 725–753.

Wüst, G., 1954: 'Gesetzmässige Wechselbeziehungen zwischen Ozean und Atmosphäre in der zonalen Verteilung von Oberflächensalzgehalt, Verdunstung und Niederschlag'. *Archiv. Meteor. Geophys. Bioklim., Ser. A*, 7, 305–328.

Wüst, G., Defant, A., 1936: Atlas der Schichtung und Zirkulation des Atlantischen Ozeans. *Wissenschaftliche Ergebnisse der Deutschen Atlantischen Expedition Meteor, 1925–27*, vol. 6, atlas.

Wyrtki, K., 1961: 'Physical oceanography of the Southeast Asian waters', NAGA Report 2, Scientific Results of Marine Investigations of the South China Sea and the Gulf of Tailand 1959–61. Scripps Institution of Oceanography, 194 pp.

Wyrtki, K., 1965: 'The average annual heat balance of the North Pacific Ocean and its relation to ocean circulation'. *J. Geophys. Res.*, 70, 4547–4559.

Wyrtki, K., 1966: 'Seasonal variation of heat exchange and surface temperature in the North Pacific Ocean'. Hawaii Institute of Geophysics Report HIG–66–3, University of Hawaii, Honolulu, 8 pp. plus numerous figures and charts.

Zillman, J. W., 1972: *A study of some aspects of the radiation and heat budgets of the Southern hemisphere oceans*. Bureau of Meteorology, Meterological Study No. 26, Melbourne, 562 pp.

REGIONAL CIRCULATION SYSTEMS

The planetary scale atmospheric circulation of the tropics is dominated by the mean meridional Hadley cells of the two hemispheres (Section 3.3). The descending branches of these cells and the associated anticyclonic axes in the lower troposphere provide a natural delineation of the poleward extremities of the tropical atmosphere. Accordingly, the discussion of regional circulation systems in this chapter will in Sections 6.3 to 6.6 proceed from the realm of the subtropical anticyclones of both hemispheres, over the trade wind regimes, to the equatorial trough zone and associated confluence and convergence belts. The focus will be on the atmosphere, although it will on occasion become necessary to refer to the hydrospheric portion of the coupled system.

6.1. Overview of the Global Tropics

Patterns of circulation and climate in the global tropics are reviewed here as an orientation for the detailed discussion in later sections. The review is based on maps of surface temperature (Fig. 6.1:1, parts a and b), sea level pressure (Fig. 6.1:2, parts a and b), upper- and lower-tropospheric flow patterns (Fig. 6.1:3, parts a and b, and Fig. 6.1:4, parts a and b), and cloudiness (Fig. 6.1:5, parts a and b) during the extreme seasons, and of annual rainfall (Fig. 6.1:6). The maps reproduced here serve as basic background reference. The pertinent fields are documented in detail in various recent atlases (Atkinson and Sadler, 1970; Sadler, 1975b; Wyrtki, 1971; Ramage and Raman, 1972; Ramage et al., 1972; Robinson, 1976; Robinson et al., 1979; Hastenrath and Lamb, 1977a, 1978b, 1979; Merle, 1978; Godbole and Shukla, 1981). Likewise still useful is a series of upper-air atlases for the Southern hemisphere (Taljaard et al., 1969; Van Loon et al., 1971; Jenne et al., 1971; Crutcher et al., 1971). For state of the art analyses of upper- and lower-tropospheric flow patterns refer to Sadler (1975b) and Atkinson and Sadler (1970) in particular. The map reproductions Figs. 6.1:3 and 6.1:4 are to serve only as preliminary reference, within the scale limitations of the present book pages.

During Northern winter, the band of highest surface temperature (Fig. 6.1:1, part a) is located relatively far equatorward, as are various circulation features in the Northern hemisphere, such as the subtropical high pressure belt and anticyclonic axis and the equatorial trough of low pressure and associated confluence between the trade wind systems of the two hemispheres (Fig. 6.1:2, part a, and Fig. 6.1:3, part b). Concurrently, the subtropical high pressure belt and anticyclonic axis of the Southern hemisphere is located relatively far poleward. The subtropical anticyclonic axes of the two hemispheres and the aforementioned near-equatorial confluence zone stand out as regions of weak winds, while high speeds are found in the core of the trades of the two hemispheres (Fig. 6.1:3, part b).

The upper-tropospheric circulation over the Northern hemisphere during Northern winter (Fig. 6.1:3, part a) is dominated by the Subtropical Westerly Jet (Section 6.2), with westerlies extending deep into the tropics, while westerlies in the Southern hemisphere are less strongly

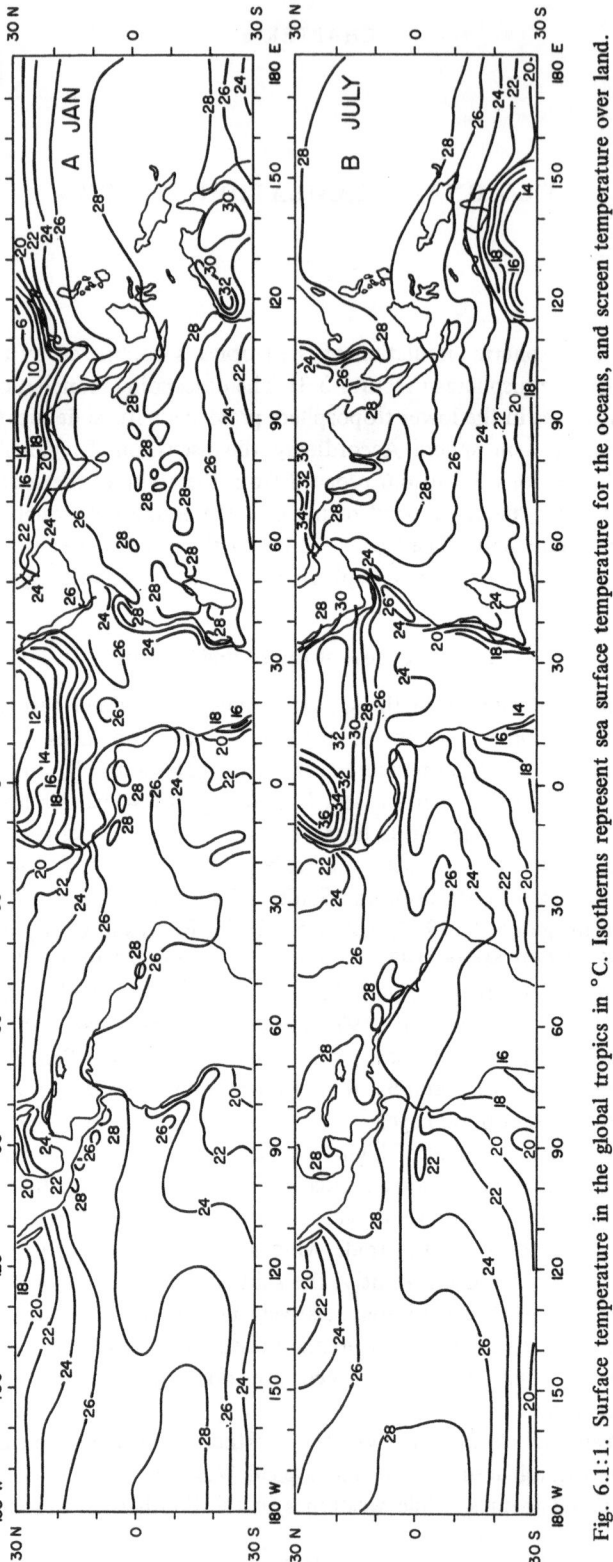

Fig. 6.1:1. Surface temperature in the global tropics in °C. Isotherms represent sea surface temperature for the oceans, and screen temperature over land. (a) January, (b) July. Sources: Hastenrath and Lamb (1977a, 1979), Wyrtki (1966), Weare et al. (1980), Ramage (1971), p. 12, 14, Scherhag (1969), p. 146, 152.

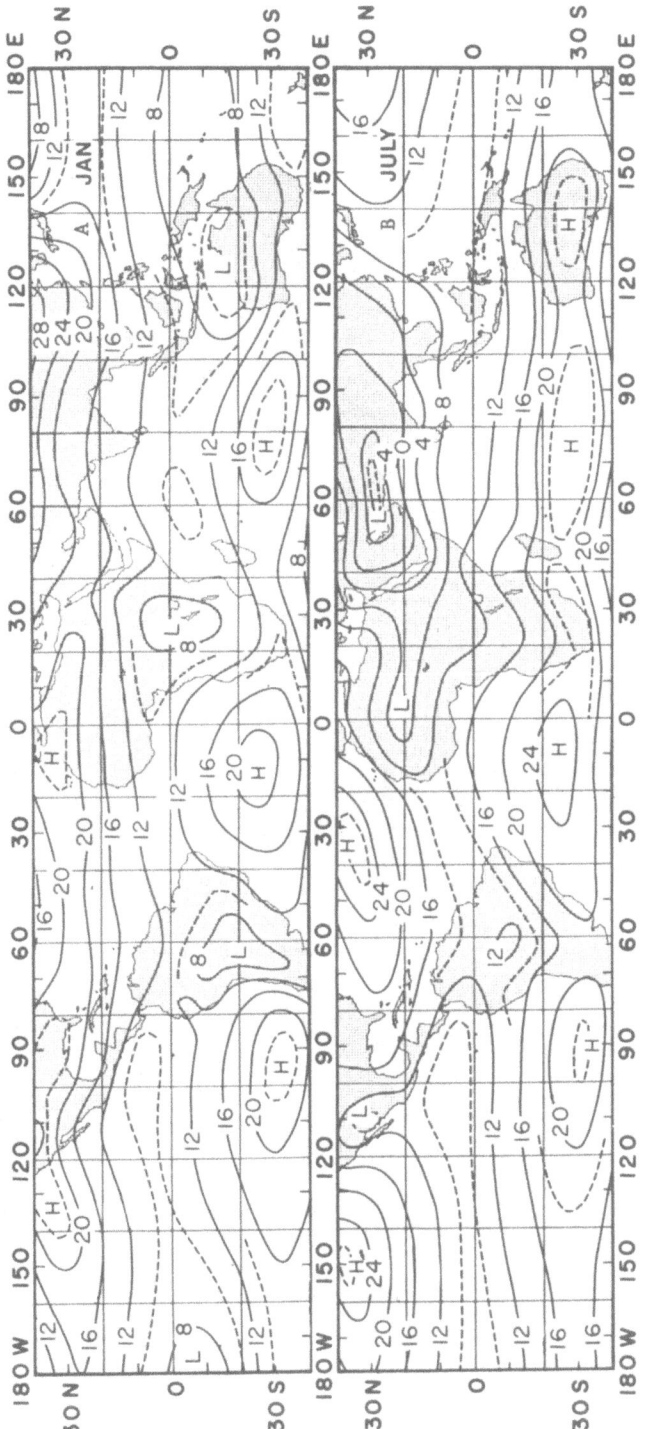

Fig. 6.1:2. Sea level pressure over the global tropics in 1000 + mb. (a) January, (b) July. Source: Godbole and Shukla (1981).

Fig. 6.1:3. Circulation over the global tropics in January. Resultant wind streamlines and isotachs in knots. (a) 200 mb, and (b) gradient wind level. Sources: Atkinson and Sadler (1970), Atkinson (1971), Sadler (1975b).

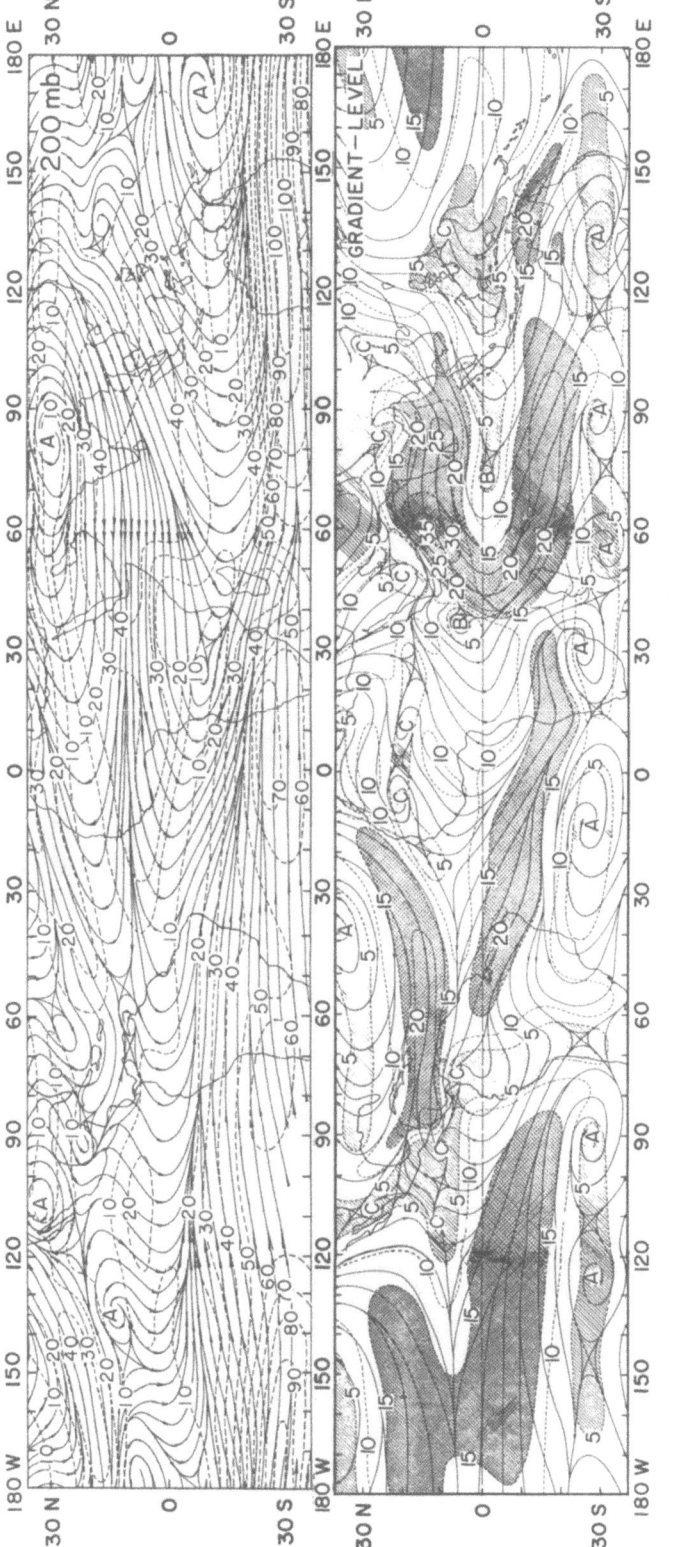

Fig. 6.1:4. Circulation over the global tropics in July. Resultant wind streamlines and isotachs in knots. (a) 200 mb, and (b) gradient wind level. Sources: Atkinson and Sadler (1970), Atkinson (1971), Sadler (1975b).

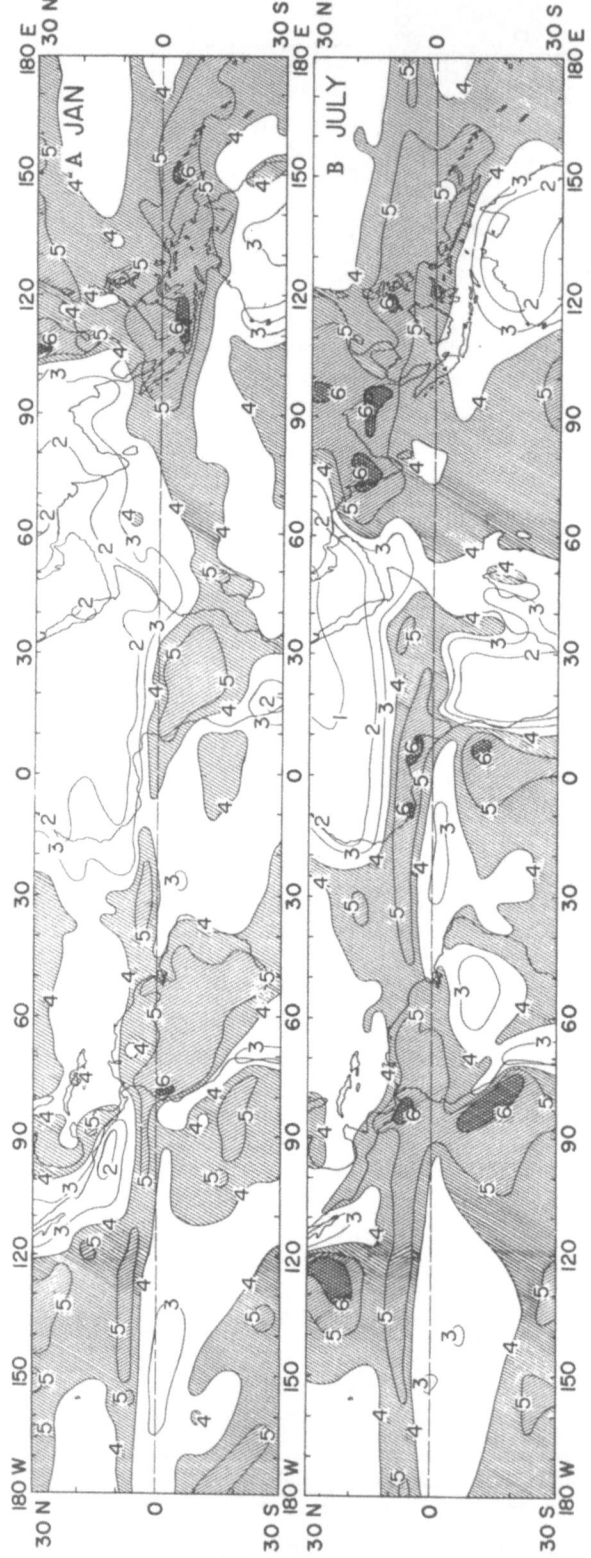

Fig. 6.1:5. Cloudiness over the global tropics, in oktas. (a) January, and (b) July. Source: Atkinson and Sadler (1970), and Atkinson (1971).

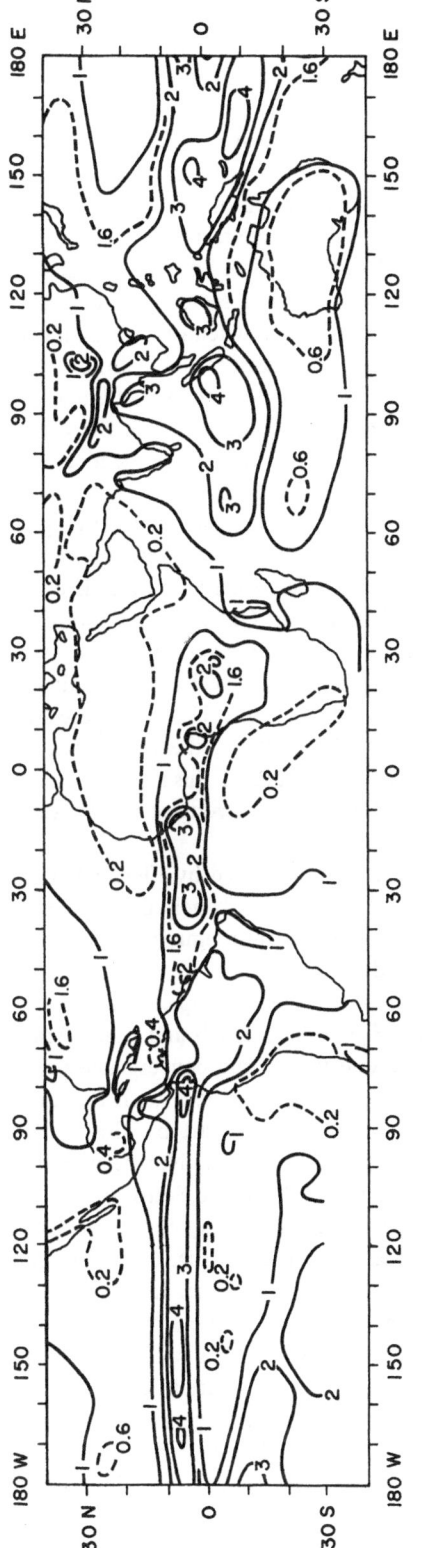

Fig. 6.1·6. Annual rainfall totals in the global tropics, in m. Sources: Solid lines over the oceans from Dorman and Bourke (1979, 1981) and Dorman (1982); broken lines over the continents from Baumgartner and Reichel (1975).

developed. Upper-tropospheric easterlies are found in the latitude band 0–10°S extending from the Central Pacific across the Indian Ocean into the African sector. 200 mb flow patterns during Northern winter are further detailed in Boyle and Chang (1984).

Circulation and climate patterns during Northern summer are characterized by a far poleward location of the band of highest surface temperature (Fig. 6.1:1, part b) and of quasi-permanent circulation features over the Northern hemisphere, such as the subtropical high pressure belt and anticyclonic axis, and the equatorial trough of low pressure and associated confluence zone between the Northeast trades and the cross-equatorial flow from the Southern hemisphere (Fig. 6.1:2, part b, and Fig. 6.1:4, part b). Airstreams of Southern hemispheric origin penetrate particularly far northward in the South Asia – Africa sector. At the same time, the subtropical high pressure belt and anticyclonic axis of the Southern hemisphere are located farther equatorward than during Northern winter. Again, winds are weakest in the realm of the anticyclonic axes of the two hemispheres and the confluence belt between Northeast trades and cross-equatorial flow from the Southern hemisphere, while the highest speeds are found in the core of the trades and the cross-equatorial monsoon flow in the Indian Ocean sector, embedded in which there is a jet stream system of interhemispheric proportions (Section 6.2).

The upper-tropospheric circulation during Northern summer (Fig. 6.1:4, part a) shows substantial differences from the winter chart. Over the Northern hemisphere tropics, the Subtropical Westerly Jet has disappeared at all longitudes. Anticyclonic cells dominate in the subtropics over North America and the Asia – Africa sector. On the equatorward side of the anticyclonic axis, the Tropical Easterly Jet (Section 6.2) extends from Southeast Asia across the Indian Ocean to Africa. In the longitudes between the anticyclonic cells over North America and the Asia – Africa sector 'tilted troughs' (oriented from Southwest to Northeast) are apparent at mid-ocean over the Pacific and Atlantic. In the Southern hemisphere, the anticyclonic axis is found closer to the Equator than in Northern winter, and poleward of this a strong Subtropical Westerly Jet (Section 6.2) extends around the globe; westerlies being strongest in the Australia – New Zealand sector (Ramage, 1971, p. 230).

The planetary-scale circulation features depicted by Figs. 6.1:3 and 6.1:4 are associated with distributions of vergence and vertical motion that largely determine the patterns of cloudiness and rainfall and thus the characteristics of the surface climate. An orientation for the global tropics is provided by the maps Figs. 6.1:5 and 6.1:6.

Cloudiness (Fig. 6.1:5) is smallest in the subtropical regions characterized by anticyclonic outflow in the lower troposphere (Figs. 6.1:3, part b and Fig. 6.1:4, part b), and largest in a belt to the North of the Equator where confluence and convergence prevail in the lower troposphere (Fig. 6.1:3, part b, and Fig. 6.1:4, part b). Consistent with the annual shift of circulation belts illustrated by Fig. 6.1:3, part b, and Fig. 6.1:4, part b), the near-equatorial belt of maximum cloudiness assumes a more northerly position during Northern summer. Particularly noteworthy is a band of minimum cloudiness immediately to the South of the Equator over the Atlantic and Eastern Pacific. In the Atlantic this feature is especially conspicuous during Northern summer.

The major features of the annual rainfall map, Fig. 6.1:6, are consistent with the cloudiness patterns during the extreme seasons (Fig. 6.1:5, parts a and b). Most prominent are the small precipitation amounts in the realm of the subtropical anticyclones, and the belt of maximum rainfall to the North of the Equator. This contrasts with small precipitation totals immediately to the South of the Equator especially over the Atlantic and Eastern Pacific, which contribute strongly to a hemispheric asymmetry for the average around the globe. It is over the tropical Eastern Atlantic and Eastern Pacific to the South of the Equator that rainfall and cloudiness patterns (Figs. 6.1:6, 6.1:5, parts a and b) most strikingly disagree: the extensive cloud cover in these regions consists largely of low cloud decks (Hastenrath and Lamb, 1977a, charts 70–97), which produce only small precipitation amounts.

6.2. Jet Streams

In the review of flow patterns over the global tropics (Section 6.1, Figs. 6.1:3 and 6.1:4) reference was made to jet-like concentrations of wind. The prominent jet stream systems of low latitudes will here be considered in context.

6.2.1. BASIC DYNAMICS

Jet streams — through the associated vergence and vertical motion patterns — exert a control on the surface weather and climate. The dynamics of jet streams are comprehensively treated in Reiter (1963), and only some basic features are considered here. An easterly jet is schematically shown in Fig. 6.2.1:1, which serves as reference for the following discussion of vergence pattern and (ageostropic) cross circulation in the entrance (upstream) and exit (downstream) portion of the jet.

The distribution of divergence and convergence at jet stream level can with reference to Fig. 6.2.1:1 be appreciated from the vorticity equation. This can in approximate form be written

$$\frac{d}{dt}(\zeta + f) = -(\zeta + f)\,\text{DIV}, \qquad (6.2.1:1)$$

where f is the Coriolis parameter, ζ relative vorticity, and DIV divergence. Without restriction of generality consider Northern hemispheric conditions. For the straight easterly flow shown in Fig. 6.2.1:1, $\zeta = -\partial u/\partial y$. From the isotach pattern in Fig. 6.2.1:1 it is seen that ζ is positive to the left of the jet axis, being largest at the longitude of the jet maximum. To the right of the jet axis ζ is negative, again with largest negative value at the longitude of the jet maximum. Accordingly,

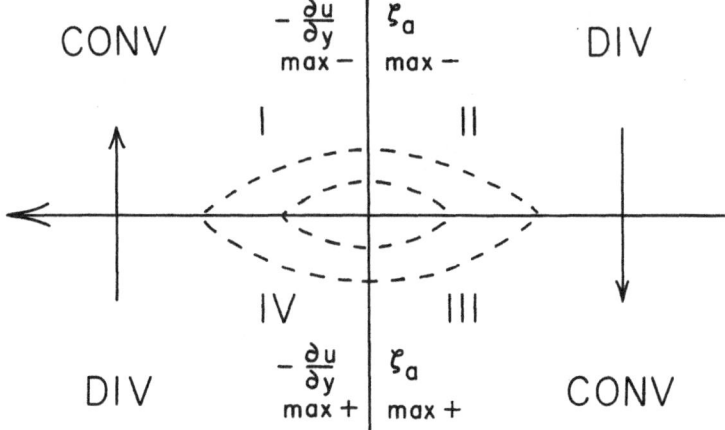

Fig. 6.2.1:1. Scheme of easterly jet stream. Arrow indicates easterly flow direction and jet axis, and broken lines are isotachs, with largest values at the origin of the rectangular coordinate systems. I, II, III, and IV denote, respectively, the left rear, right rear, left front, and right front quadrants.

the absolute vorticity $\zeta_a = (\zeta + f)$ is largest positive to the left of the jet maximum, and smallest positive, or even negative to its right. As the left-hand term of Eq. (6.2.1:1) denotes the rate of change of absolute vorticity following the motion, it is seen with reference to Fig. 6.2.1:1 that the sign convention for the right-hand side is satisfied with convergence in the left rear (III) and divergence in the left front (IV) quadrant. To the right of the jet axis ζ would have a sign opposite to f, making ζ_a less positive, so that divergence would be expected in the right rear (II), and convergence in the right front (I) quadrant.

Independent from these applications of the vorticity equation (6.2.1:1), consider the geostrophic departure (Haltiner and Martin, 1957, p. 191)

$$(\mathbf{V} - \mathbf{V}_g) = \frac{1}{f} \mathbf{k} \times \left(\frac{d\mathbf{V}}{dt} \right)_H \tag{6.2.1:2}$$

that is the vector difference from the geostrophic wind $\mathbf{V}_g$. In the entrance region of the jet (Fig. 6.2.1:1), wind increases in the direction of the flow, so that the geostrophic departure vector would be directed from the right rear (II) to the left rear (III) quadrant. Conversely, deceleration takes place in the exit region, corresponding to a geostrophic departure vector from the left front (IV) to the right front (I) quadrant. These cross-circulations in the entrance and exit regions are qualitatively consistent with the distribution of divergence and convergence in the four quadrants of Fig. 6.2.1:1 as discussed above.

6.2.2. SUBTROPICAL WESTERLY JET

In the overview over the global tropics (Section 6.1, Figs. 6.1:3 and 6.1:4) reference was made to the Subtropical Westerly Jet as dominating the wintertime upper-tropospheric circulation of the Northern hemisphere lower latitudes. More specifically, Figs. 6.2.2:1 and 6.2.2:2 illustrate the circumglobal extent of the jet at latitudes around 20–35°N. Fig. 6.2.2:1 shows speed maxima over the Americas, the Middle East, and East Asia, at which longitudes the jet axis has a far poleward position and where it has only a relatively small latitude separation from the Polar Front Jet (Fig. 6.2.2:2). It must be noted that Fig. 6.2.2:1 depicts the speed field relative to the jet core and not with respect to fixed geographic coordinates. For a presentation in the latter coordinate frame refer to Sadler (1975b, pp. 12–13; and Fig. 6.1:3, part a). As is apparent from Figs. 6.2.2:1 and 6.2.2:2, the wintertime Subtropical Westerly Jet is located in the latitude domain of the large Central Asian mountain massifs. Fig. 6.2.2:3 illustrates for the 500 mb region the preference of the jet axis to be situated either to the North or South of the mountains.

In the planetary perspective (Sections 3.2 and 3.3), the Subtropical Westerly Jet is located above the descending portion of the Hadley cell, and the jet is best developed in the season of the most intense mean meridional circulation. According to Palmén and Newton (1969, pp. 16–20), the Subtropical Westerly Jet owes its existence primarily to the convergence of the poleward transport of westerly absolute angular momentum in the upper poleward directed branch of the Hadley cell. In addition, the confluence between the upper-tropospheric branches of the Ferrel and Hadley cells (Palmén and Newton, 1969, pp. 112–113) leads to a concentration of property contrasts in the upper-tropospheric 'Subtropical Front' (Fig. 6.2.2:4), which is further conducive to the maintenance of the jet. However, as it overlies prevailing subsidence in the lower troposphere, the 'Subtropical Front' has no weather characteristics corresponding to the Polar Front (Palmén and Newton, 1969, p. 114). The Subtropical Westerly Jet may profoundly affect large-scale convection in the lower-troposphere (Palmén and Newton, 1969, p. 463).

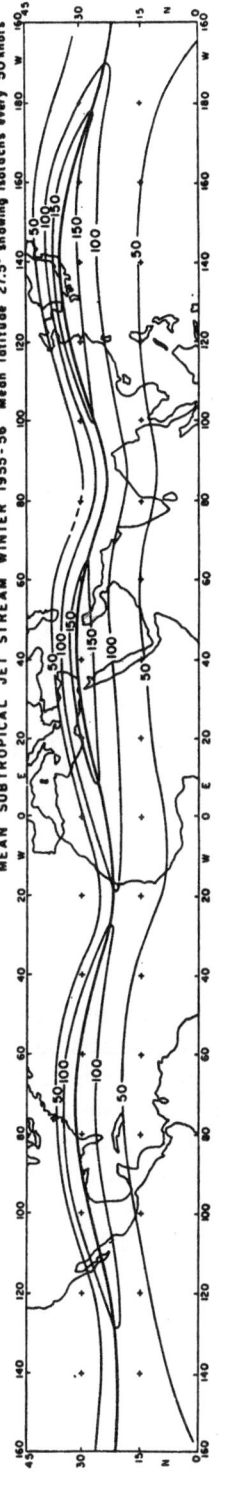

Fig. 6.2.2:1. Mean Subtropical Westerly Jet Stream for winter 1955—56. Isotachs in 50 knot interval at 200 mb surface. The mean latitude of the jet axis is 27.5°N. From Krishnamurti (1961) (*Journal of Meteorology*, American Meteorological Society).

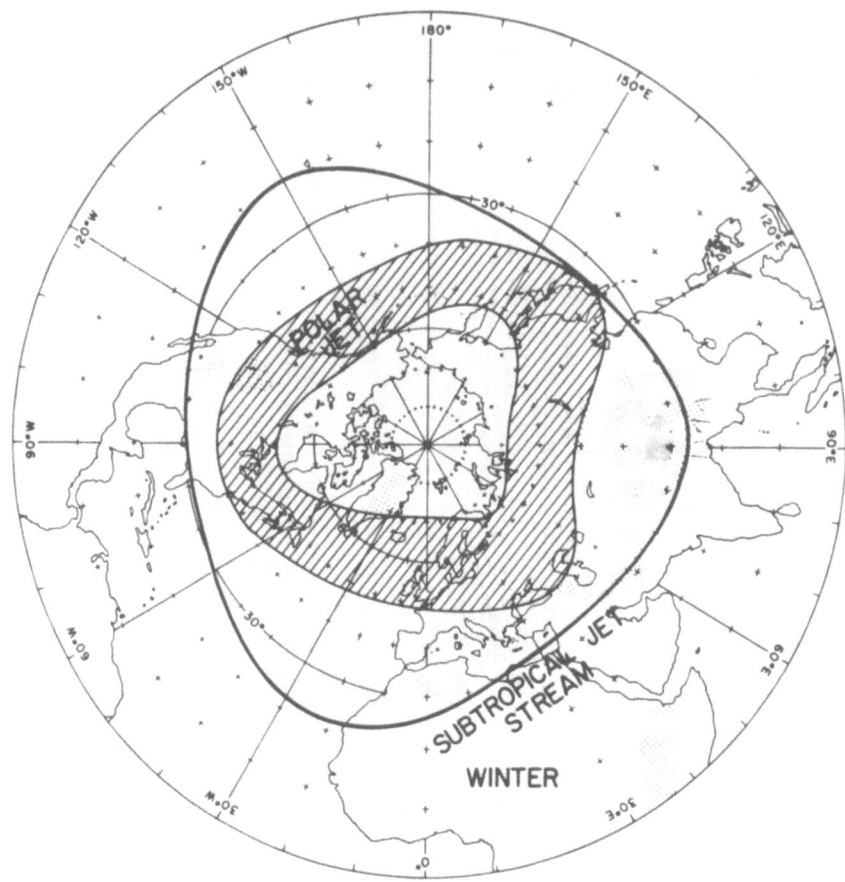

Fig. 6.2.2:2. Mean axis of Subtropical Westerly Jet Stream during winter, and area (shaded) of principal activity
of polar front jet stream. From Riehl (1962), Palmén and Newton (1969).

Although Palmén and Newton (1969, pp. 16–20, 112–114, 463) offer general suggestions,
Krishnamurti (1979, pp. 323–327) states that the precise mechanisms for the formation of the
wintertime Subtropical Westerly Jet are not well understood. He lists fifteen different factors that
may play a role in the origin of the jet. Large-scale meridional contrasts in surface heat budget,
temperature, and pressure, and the Central Asian mountain topography are considered particularly
important. As seen in Fig. 6.2.2:1, the jet speed is largest in the ridges and weaker in the troughs.
This pattern is explained from ageostrophic effects (Krishnamurti, 1979, p. 325): the southwesterly
flow upstream from the ridges has an ageostrophic component towards lower pressure, while the
northwesterly wind upstream from the troughs possesses an ageostrophic component towards
higher pressure; the concomitant accelerations and decelerations resulting in the speed extrema.
Krishnamurti (1979, pp. 325–326) also notes that the three ridges in the jet pattern are located
broadly at the longitudes of three equatorial regions with major rainfall activity during Northern
winter, namely South America, Central Africa, and Indonesia. Conversely the three jet troughs
broadly coincide with three mid-oceanic troughs in the Southern hemisphere during this season.

As is apparent from Figs. 6.1:3 and 6.1:4, the Subtropical Westerly Jet of the Southern hemisphere is best developed during the respective winter season, with strongest westerlies in the

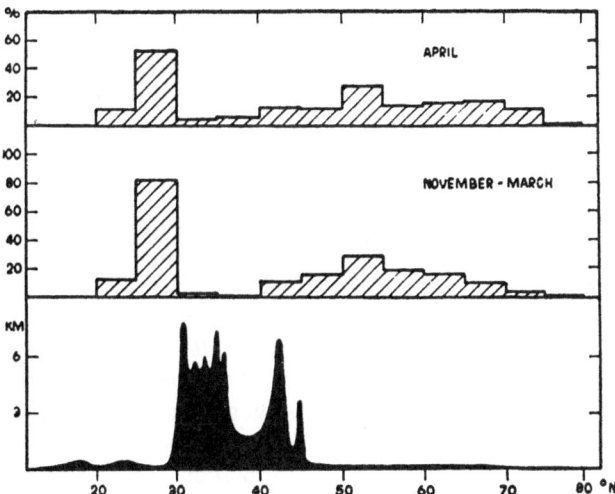

Fig. 6.2.2:3. Percentage frequency of jet stream axis in 5 degree latitude belts at 500 mb level in the cool season 1949–50, and the profile of topography, along longitude 80°E. From Ramage (1952) (*Journal of Meteorology*, American Meteorological Society).

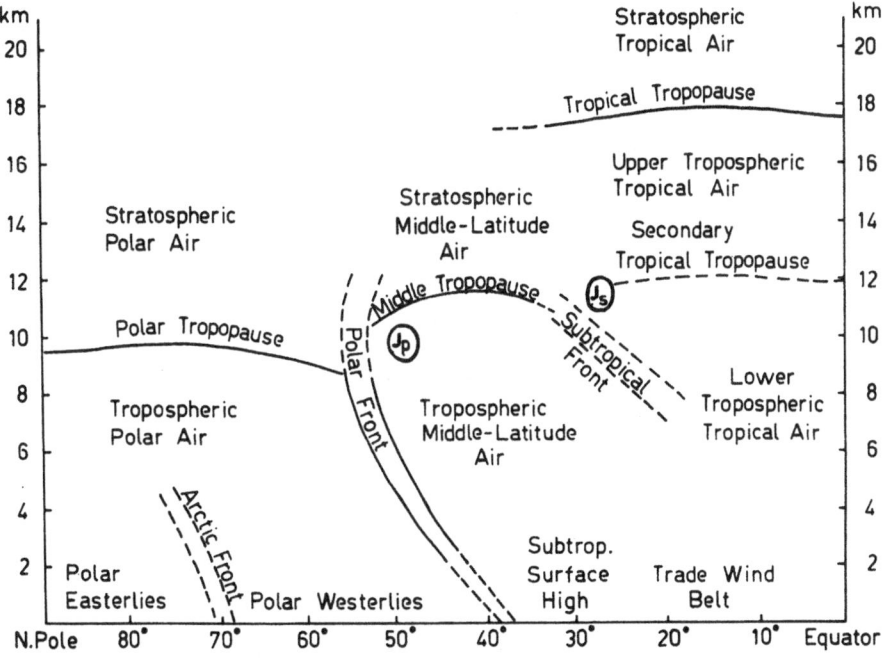

Fig. 6.2.2:4. The principal air masses, tropopauses and fronts, and jet streams in relation to the features of the low-level wind systems. From Palmén and Newton (1969).

Australia – New Zealand sector (Ramage, 1971, p. 230). However, the annual variation in jet development is smaller than for the Northern hemisphere, consistent with the lesser intensity variation of the Southern hemisphere Hadley cell (Palmén and Newton, 1969, p. 216).

6.2.3. TROPICAL EASTERLY JET

A prominent feature of the upper-tropospheric circulation apparent on the July 200 mb chart (Section 6.1, Fig. 6.1:4, part a) is a band of strong easterlies extending from Southeast Asia across the Indian Ocean and Africa to the Atlantic: the Tropical Easterly Jet. This jet is illustrated in more detail in Figs. 6.2.3:1 to 6.2.3:4.

The Tropical Easterly Jet is limited to the height of the summer circulation, lasting from late June into early September. From the development to the decay, the jet axis remains close to about 14°N, core speeds being about 40 m s^{-1}. The jet axis lies at 200–100 mb, sloping upward towards the North (Flohn, 1964; Osman and Hastenrath, 1969; Mishra and Tandon, 1983). The development of the Tropical Easterly Jet is related to the thermal wind pattern during Northern summer. In particular, the Tropical Easterly Jet owes its origin to the circumstance that the maximum heating and tropospheric thickness pattern in summer is located in the subtropics rather than near the Equator. This results from the radiation geometry and is aided by the presence of land surfaces in the subtropics of Eurasia and Africa, and especially the extensive elevated heating surfaces of the Central Asian mountain massifs. The evolution of the Tropical Easterly Jet parallels the build-up and decay of the upper-troposphere ridge in the subtropics.

Fig. 6.2.3:1 is a case study analysis of the Tropical Easterly Jet. For a climatic mean analysis of the upper-tropospheric flow field refer to Sadler (1975b, pp. 24–25) in particular. Fig. 6.2.3:1 shows over Southern Asia a jet axis with core speeds exceeding 40 m s^{-1} located to the South of

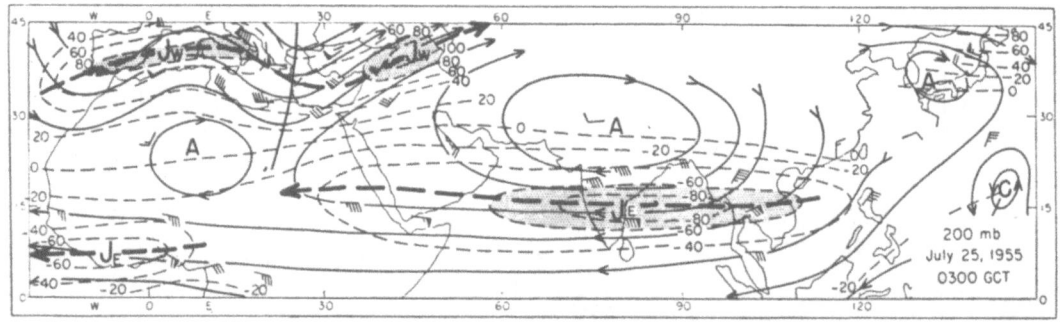

Fig. 6.2.3:1. Streamlines (solid lines) and isotachs (dashed, in knots) at the 200 mb level, 25 July 1955, 0300 GCT. JW = westerly jet maximum, JE = easterly jet maximum, A = anticyclones, C = cyclones. Heavy dashed lines with arrows indicate positions of jet axes. From Koteswaram (1958).

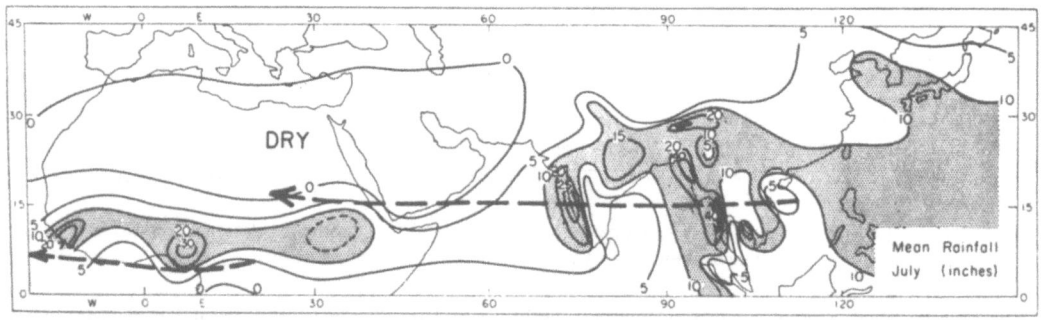

Fig. 6.2.3:2. Mean July precipitation (in inches) and position of Tropical Easterly Jet during August 1955. From Koteswaram (1958).

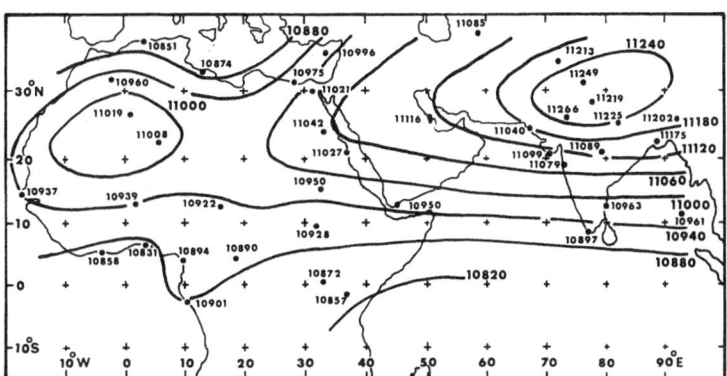

Fig. 6.2.3:3. July 200 − 850 mb layer thickness in gpm. Source: Osman and Hastenrath (1969).

an anticyclonic cell over the Tibetan Plateau. A separate jet entity is apparent over West Africa, associated with a separate anticyclonic cell over the Sahara. Related to this jet system is the rainfall distribution shown in Fig. 6.2.3:2, which is indicative of the vertical motion pattern in the lower troposphere. In the entrance region of the jet over Asia, abundant rainfall is found to the North of the jet axis, amounts being distinctly smaller to the South. For the exit region over West Africa, the reverse pattern is indicated.

The pioneering case study (Figs. 6.2.3:1 and 6.2.3:2) of Koteswaram (1958) is complemented by the climatic mean analyses of Flohn (1964) and the theoretical treatment of Mishra and Tandon (1983). Consistent with Fig. 6.2.3:1, the pattern of the climatic mean July 200−850 mb layer thickness (Fig. 6.2.3:3) shows largest values in the subtropics rather than the equatorial zone, and separate maxima over the Tibetan Plateau and the Sahara. The scheme Fig. 6.2.3:4 depicts the cross-circulation in the entrance and exit regions of the jet. Flohn (1964) points out that the cross-circulation in the entrance region is kinetic-energy producing, the summertime latent and sensible heating over the Tibetan Plateau providing a major source of energy. By contrast, the cross-circulation in the exit region re-converts kinetic into potential energy. Fig. 6.2.3:4 is consistent with the discussion in Section 6.2.1 (Eqs. (6.2.1:1) and (6.2.1:2); Fig. 6.2.1:1), and the vertical motion pattern to be inferred from the rainfall distribution, Fig. 6.2.3:2. Thus, in the entrance region of the jet, ascending motion would be conducive to rainfall over Southern Asia, in contrast to the Equatorial Indian Ocean. In the exit region, the Southern portion of West Africa with its abundant summer rainfall would be in the ascending branch of the cross-circulation, contrasting with the subsidence over the dry desert fringe to the North. The vertical motion field resulting from the cross-circulations would thus correspond to a checkerboard pattern in the rainfall distribution.

In the search for possible corollaries to the intense Tropical Easterly Jet phenomenon in the Asia − Africa sector, Flohn (1964) gives particular attention to the checkerboard arrangement in the rainfall distribution. Likewise, larger summertime heating and tropospheric thickness in the subtropics as compared to the equatorial zone is recognized as an essential characteristic of Tropical Easterly Jet corollary. Of interest in this context are the region from the Southern United States to Northern South America, as well as the three continents of the Southern hemisphere, where the fields in the checkerboard pattern would be inverse to the Northern hemisphere. It is noted that the powerful meridionally oriented mountain barrier may complicate matters for South

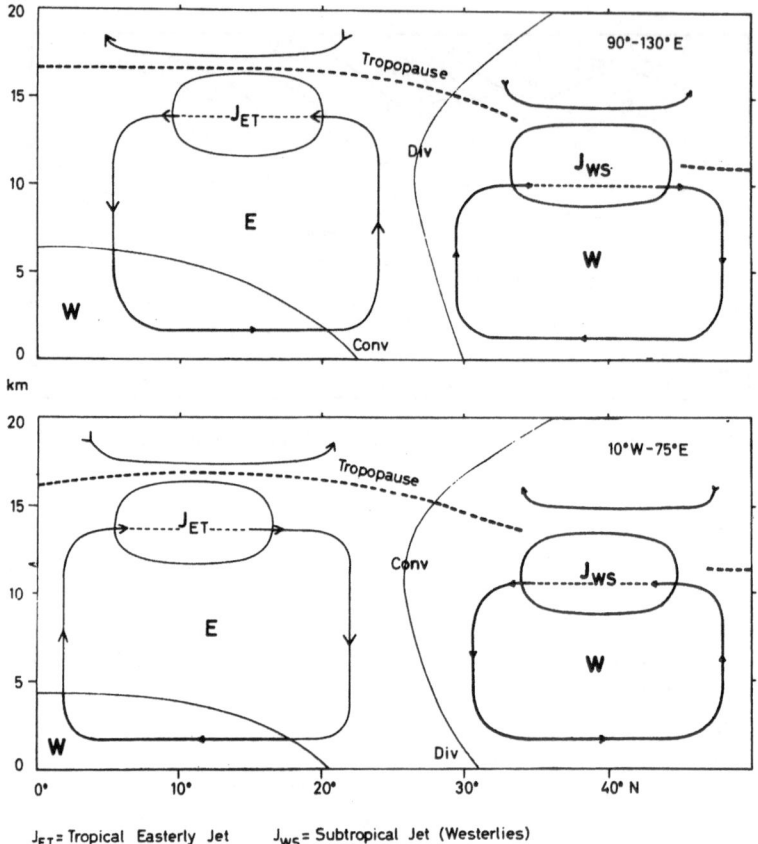

Fig. 6.2.3:4. Idealized cross-circulation at the entrance and exit regions of the Tropical Easterly Jet. From Flohn (1964).

America. At any rate, the various regions considered are at best weak counterparts to the intense Tropical Easterly Jet of the Asia — Africa sector, where the continent and mountain distribution is uniquely favorable for the poleward increase of heating and tropospheric thickness.

6.2.4. WEST AFRICAN MID-TROPOSPHERIC JET

Separate from the upper-tropospheric Tropical Easterly Jet discussed in Section 6.2.3, an easterly wind maximum is found in the middle troposphere over West Africa throughout the year, but the current becomes organized into a well-defined jet of more than 10 m s^{-1} from April to November, when it also attains its most poleward location (Burpee, 1972). The meridional and zonal cross-sections, Figs. 6.2.4:1 and 6.2.4:2, depict the jet at the height of the Northern summer. The jet has its core around 600 mb, is located at about 15°N, and extends from the Red Sea all the way to the Atlantic Ocean. The jet is related to the thermal wind pattern. The increase of easterlies from the surface upward reflects the juxtaposition of warmer air over the Sahara desert to the North and cooler maritime air to the South. Burpee (1972) suggests that the decrease of the easterly wind component from the jet level at 600 mb further upward is due to a reversal of the meridional temperature gradient (warmer air equatorward and cooler air poleward of the region), although this cannot be substantiated from observation.

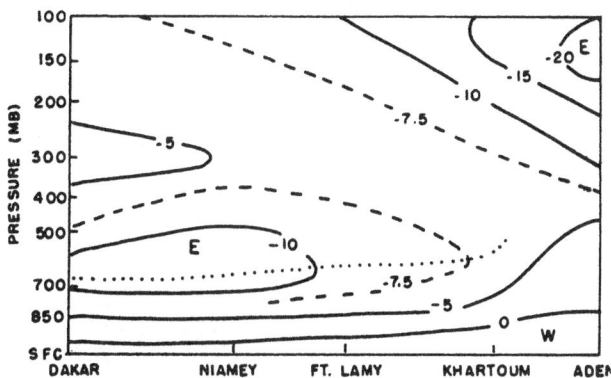

Fig. 6.2.4:1. August mean meridional-vertical cross-section of zonal wind component (m s^{-1}) along 13°N. From Burpee (1972) (*Journal of Atmospheric Sciences*, American Meteorological Society).

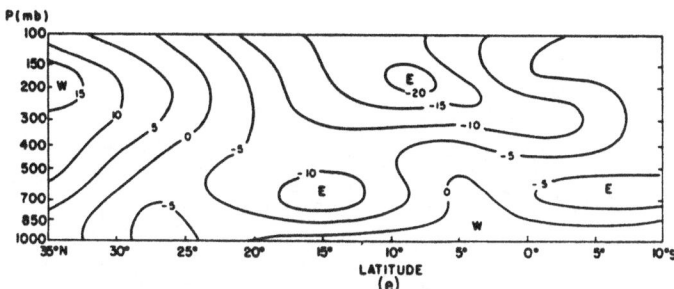

Fig. 6.2.4:2. August mean zonal-vertical cross-section of zonal wind component (m s^{-1}) along 5°E. From Burpee (1972) (*Journal of Atmospheric Sciences*, American Meteorological Society).

Krishnamurti (1979, pp. 180–185) suggests the existence of a corollary to the West African Mid-tropospheric Jet over Northwestern Australia in Southern summer. Lower-tropospheric thermal gradients on the equatorward side of the Australian desert are then similar to the Sahara in Northern summer. Meridional temperature contrasts are concentrated at 15°S, and the jet is near 600 mb, with a maximum speed around 15 m s^{-1}. Krishnamurti (1979, p. 180) considers the cyclonic wind shear on the equatorward side of the jet as instrumental in the lower-tropospheric wave activity.

6.2.5. EAST AFRICAN LOW LEVEL JET

As recently as the mid 1960's, John Findlater discovered the East African Low Level Jet, in the course of meteorological support for low-flying aircraft missions in the coastal region that were prompted by border incidents following the independence of Kenya. Subsequently, Findlater (1969a, b, 1972, 1977a, b) evaluated existing pilot balloon and aircraft observations and demonstrated that this jet system is a major lower-tropospheric circulation feature over the Western Indian Ocean during the Northern summer Southwest monsoon. This jet system then received much attention during the generously funded field programs of MONEX and in various numerical modelling studies (Krishnamurti et al., 1976, 1983; Hart, 1977; Hart et al., 1978; Bannon, 1979a, b; Krishnamurti and Wong, 1979). Synonyms found in the literature include 'Somali Jet' and 'Low Level Jet over the Western Indian Ocean.'

The large horizontal extent of the jet system over the Western Indian Ocean is shown in Fig. 6.2.5:1, its vertical structure in Fig. 6.2.5:2, and the month-to-month position of its axis in Fig. 6.2.5:3. The jet is centered around 1.5 km height, has core speeds of 12–15 m s^{-1} and strong horizontal and vertical shears, is a feature of the Northern summer, and must be regarded as an integral part of the Southwest monsoon current. The jet (Fig. 6.2.5:1) shares major characteristics with the surface circulation (Sections 4.4 and 6.8.3, Figs. 4.4:1 to 4.4:3), such as speed maxima

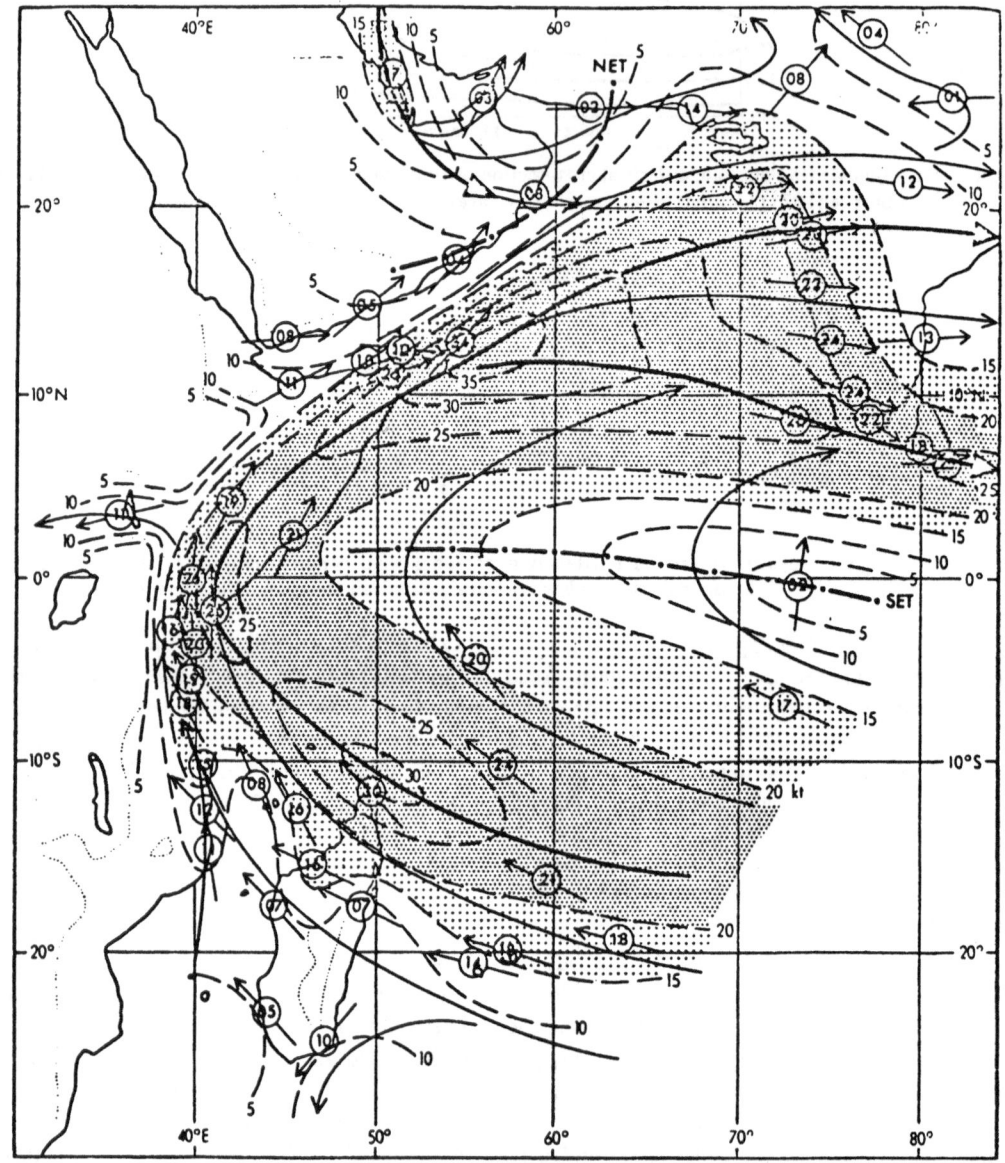

Fig. 6.2.5:1. Mean monthly airflow at 1 km in July. Major streamline axis of maximum flow is shown by horizontal line and arrows, isotachs at 5 knot intervals by thin solid and axis of minimum winds by heavy dash-dotted lines. From Findlater (1977b).

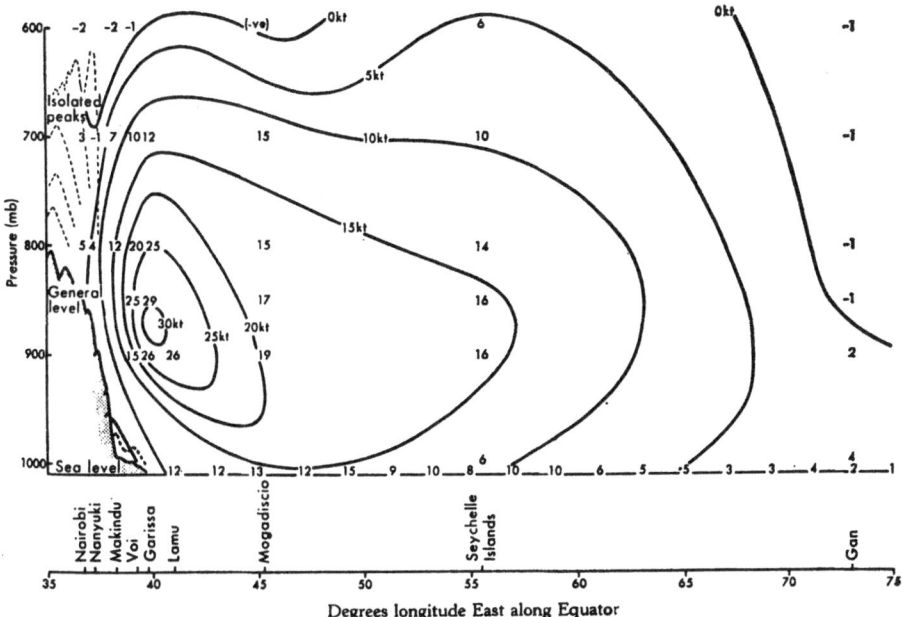

Fig. 6.2.5:2. Mean meridional airflow at the Equator in July. Isotachs are at intervals of 2.5 m s^{-1}, positive from South. From Findlater (1969b).

around 10°S off the Northern tip of Madagascar and over the Western Arabian Sea, as well as an apparent branching of the flow over the Eastern Arabian Sea. Moreover, in its seasonal march the jet evolves and decays in unison with the large-scale features of the surface circulation.

Saha (1974) points out an ancillary factor in the course of the jet development: wind stress causes oceanic upwelling at the East African coast, which in turn influences the atmospheric thickness pattern in the lower atmosphere and thus the jet. Moreover, Kinuthia and Asnani (1982) recognize an orographically channelled branch of the jet penetrating into the interior of the continent.

The aerological structure of the East African Low Level jet is illustrated in Fig. 6.2.5:4, the main features being as follows: (i) the air is coolest and moistest in the lower layers to the East and warmest and driest in the higher layers and to the West; (ii) the level of maximum wind slopes upward to the East, while the zone of maximum horizontal gradient of potential temperature slopes upward to the West, the intersection between these two axes being to the West of the jet core; (iii) the jet core is predominantly cloudy, with an axis of maximum cloudiness over land and to the West; (iv) over land and to the West of the jet axis a stable layer extends from about the jet level upward.

The aforementioned similarities in the wind pattern at jet level and surface are presumably related to common mechanisms. The dynamics of surface flow will be considered in Section 6.7.3. However, Fig. 6.2.5:1 shows a peculiarity not apparent in the surface wind pattern, namely a speed maximum at 1 km over the Equator at the coast of Kenya. This feature appears in Findlater (1971, 1977b), from where Fig. 6.2.5:1 is taken, but not in Findlater (1974, 1977a) for the same level, nor in Findlater (1977b) for the 600–2400 m layer.

126   **CHAPTER 6**

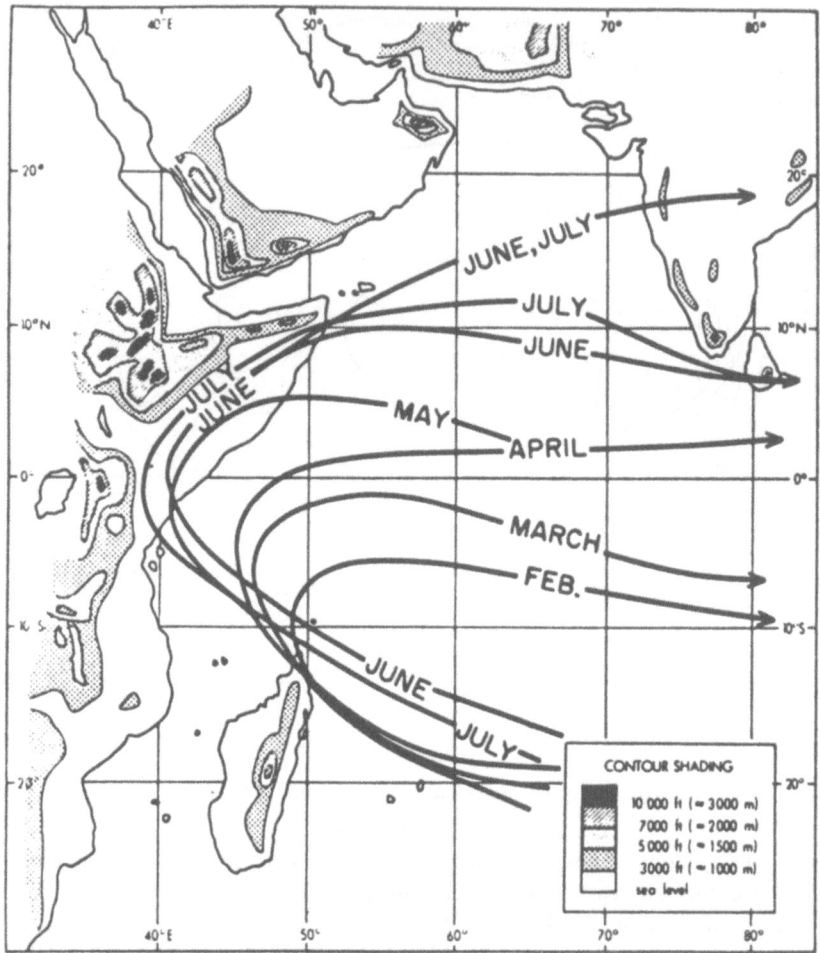

Fig. 6.2.5:3. Month-by-month progression of the axis of the East African Low-Level Jet. Source: Findlater (1971), National Research Council (1977) (reproduced with permission of Her Britannic Majesty's Stationary Office).

Various numerical studies are aimed at the dynamics of the jet (Washington and Daggupaty, 1975; Anderson, 1976; Krishnamurti *et al.*, 1976; Hart, 1977; Bannon, 1979a, b). Anderson (1976), Hart (1977), and Bannon (1979a, b) in particular call attention to analogies with western boundary currents in the ocean (Section 4.2). It is interesting to note that these simulations are evidently not geared to the same perceived jet wind pattern characteristics. By way of observational background, Washington and Daggupaty (1975) thus refer to a 1 km map of Findlater, but their bibliographic reference seems inconsistent. Their results differ substantially from the flow map ascribed to Findlater, not only in the isotachs but also in the streamline pattern. Anderson (1976) draws on Findlater's (1974) chart for the 1 km level, which does not contain a speed maximum (along the jet axis) over the Equator at the Kenya coast. Krishnamurti *et al.* (1976) use as reference the 1 km map in Findlater (1971) which originally contains the equatorial speed maximum at the Kenya coast, but in their reproduction they enter the label 'speed minimum' for about 1.5°N. Bannon (1979a) draws in his arguments on Findlater's (1977b) 1 km map with its

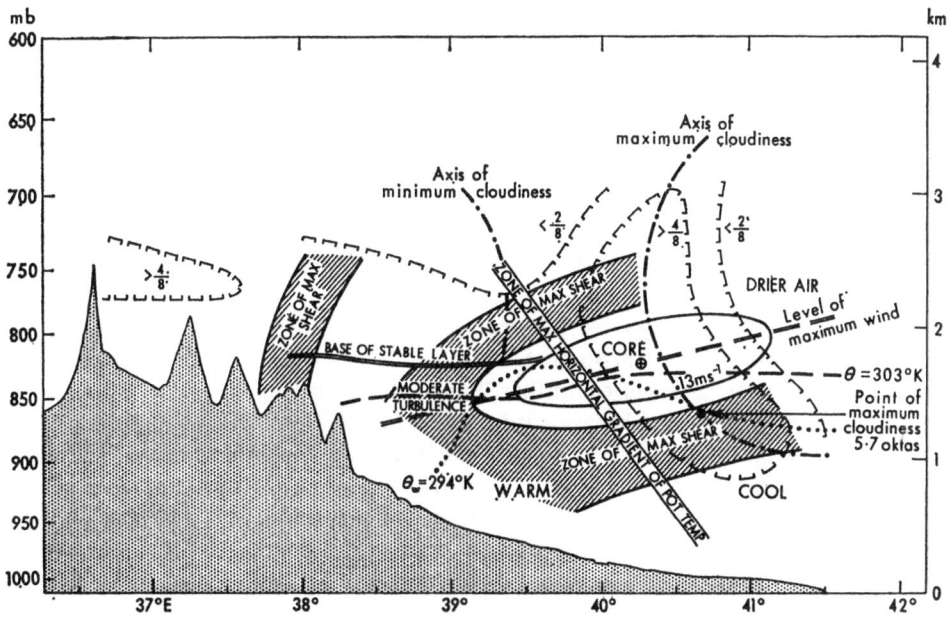

Fig. 6.2.5:4. Observed characteristics of the core of the monsoon current, based upon data from four flights. From Findlater (1972).

speed maximum over the Equator at the Kenya coast. In Hart (1977) no reference is found to the perceived presence or absence of an equatorial speed maximum (along the jet axis) at the Kenya coast. With such diversity of perceived atmospheric reality, the merits of the various numerical simulation studies are not readily compared.

Expanding on Krishnamurti *et al.* (1976), Krishnamurti and Wong (1979) and Krishnamurti *et al.* (1983) studied the vertical structure of the East African Low Level Jet through numerical simulation. Their results suggest that an advective boundary layer is dominant across the Equator; the jet core is located toward its poleward edge; further northward a transition takes place to a quasi-Ekman boundary layer.

The jet undergoes a marked diurnal variation (Findlater, 1977b; Hart *et al.*, 1978; Bannon, 1979b), with highest speeds in the early morning and a slowdown to the early afternoon. This diurnal march is presumably related to the increased convection during daytime and the concomitantly enhanced vertical exchange of momentum. Anderson (1984) applied an advective mixed-layer model to the diurnal cycle of the jet.

Findlater (1969b) estimates the cross-equatorial mass transport of the East African Low Level Jet in July at 7×10^{10} kg s^{-1}, which would amount to about half of the total global lower-tropospheric mass flow across the Equator. The jet likewise plays an effective role in the cross-equatorial water vapor transport (see also Rao *et al.*, 1978, 1981; Van de Boogaard and Rao, 1984). In the light of such magnitude assessments, it is not surprising that a positive linkage may exist between the intensity of the East African Low Level Jet and the monsoon rainfall over India, as suggested by Findlater's (1977a) preliminary studies. The role of the East African Low Level Jet within the greater context of the Southwest monsoon circulation will be considered further in Section 6.8.

6.3. Subtropical Highs

Within the subtropical high pressure belts of the two hemispheres, distinct cells are developed over the oceans, which are manifested as highs in the surface pressure field and reflected as anticyclones in the lower-tropospheric flow pattern (Fig. 6.1:3, part b, and Fig. 6.1:4, part b): the North Atlantic and North Pacific highs, and in the Southern hemisphere the South Atlantic, South Pacific, and South Indian Ocean highs. In the course of the year, these undergo variations of intensity and position. Being at the root of the trades (Section 6.4), the subtropical highs represent centers of action in the tropical circulation.

As shown by Fig. 6.3:1 and Table 6.3:1, the subtropical highs are closest to the Equator during the respective winter, and are displaced poleward in summer. In addition to the latitude displacements, the North Pacific, North and South Atlantic, and Indian Ocean highs are all located further East in Northern winter, and shifted further West in Northern summer (Fig. 6.3:1 and Table 6.3:1), while a more complicated behavior is indicated for the South Pacific high. In contrast to the latitudinal shift of the highs, the longitude displacements sympathetic between hemispheres are not well understood. Table 6.3:1 also shows that the meridional pressure gradient on the equatorward side of the subtropical highs tends to be steepest during the respective winter. This corresponds to stronger easterlies at lower latitudes, with a corresponding change in the wind stress pattern as compared to summer. Moreover, the vertical motion field related to the subtropical highs and its annual changes have a bearing on the surface weather and climate.

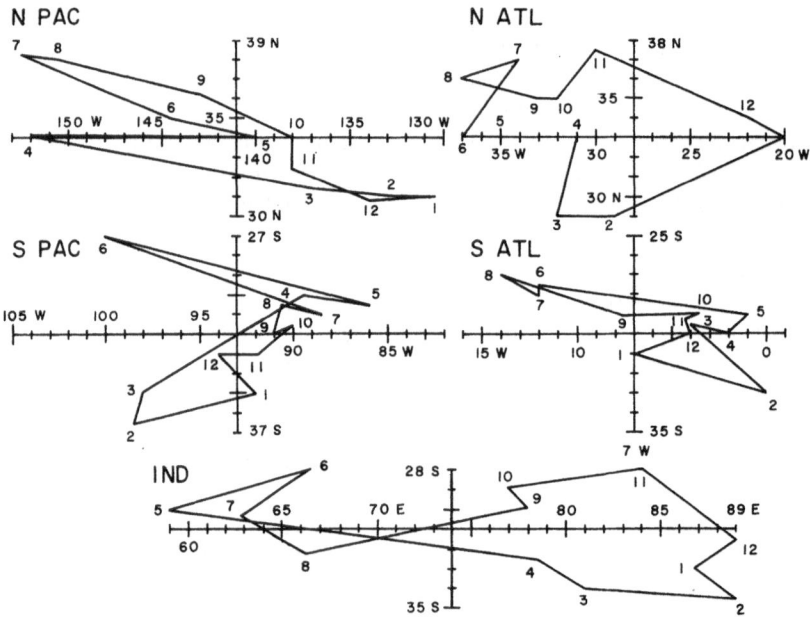

Fig. 6.3:1. Annual variation in the latitude and longitude position of the Subtropical high centers. The numbers 1 through 12 denote the calendar months. Source: U.S. Navy (1974, 1976, 1977, 1978, 1979).

TABLE 6.3:1

Annual cycle characteristics of the subtropical highs. $\Delta p/\Delta y$ (in 10^{-7} mb m^{-1}) is the meridional pressure gradient measured from the center of the high to the axis of the equatorial low pressure trough. Sources: U.S. Navy, 1974, 1976, 1977, 1978, 1979.

	N Atlantic	N Pacific	S Atlantic	S Pacific	S Indian Ocean
			northernmost position		
latitude	37.0°N	38.3°N	27.0°S	27.0°S	28.0°S
month	July (Nov)	July	Aug	June	June (Nov)
$\Delta p/\Delta y$	33	35	26	23	38
			southernmost position		
latitude	29.0°N	31.0°N	33.0°S	36.5°S	4.5°S
month	Feb–March	Dec–Jan–Feb	Feb	Feb	Feb
$\Delta p/\Delta y$	30	36	27	29	40
			westernmost position		
longitude	37°W	152.5°W	14.0°W	100.0°W	59.0°E
month	June, Aug	July	Aug	June	May
			easternmost position		
longitude	20°W	130.5°W	0.0°	86.0°W	89.0°E
month	Jan	Jan	Feb	May	Feb, Dec

The broad description of seasonal characteristics offered by Fig. 6.3:1 and Table 6.3:1 is for the North Atlantic and North Pacific highs complemented by Bryson and Lahey's (1958) analysis of singularities in the annual cycle. Their temporal resolution by pentades provides a more detailed description of the march of the seasons. The following singularities are of particular interest for the tropics. Bryson and Lahey (1958) date the breakdown of the Northern hemispheric winter circulation as around the end of March. A very abrupt northward shift and westward extension of the highs takes place at the end of June, the so-called 'mid-summer high jump'. The major southward retreat occurs in September and October. Broadly consistent with Bryson and Lahey's (1958) findings, Fig. 6.3:1 shows for the North Atlantic and North Pacific highs an abrupt shift from June to July, and more gradual changes during spring and fall.

These singularities in the annual cycle behavior of the Northern hemispheric subtropical highs correspond to prominent seasonal events in the surface climate, that will be considered in Sections 6.4, 6.7, and 6.8 in particular. Thus, in the Caribbean and Central America (Hastenrath, 1967a), the rainy season begins very gradually around the second half of April. The months of May and June bring copious precipitation, but the weather character tends to change abruptly around the end of June, with July and August being typically drier and less cloudy. The end of the rainy season around the end of October is a distinct event, much in contrast to the rather gradual onset. Likewise, over Subsaharan Africa (Sections 6.7.2:3 and 6.8:2), the northward progression of quasi-permanent circulation systems from winter to summer is more gradual than the southward movement. Furthermore, the Indian summer monsoon (Section 6.8), is known to have a gradual onset but a very distinct retreat (Ramage, 1971, p. v., 150–154; Reiter, 1963, pp. 398–404). In this perspective, the annual cycle behavior of the Northern hemispheric subtropical highs should be viewed not as an ultimate cause for seasonality in the tropics, but as part of season to season changes in the hemispheric circulation.

6.4. Trades

In the planetary perspective, the trades constitute the lower-tropospheric equatorward flowing branch of the tropical Hadley circulation (Section 3.3). The trades emanate from the eastern flanks of the subtropical highs (Section 6.3), flow westward and equatorward, and meet in confluence zones embedded within the equatorial trough of low pressure (Section 6.7). In their upstream portion, on the eastern equatorward side of the subtropical anticyclones, the trades are subsiding and divergent. Proceeding downstream, towards the Equator and the western side of the subtropical anticyclones, divergence gradually gives way to convergence and ascending motion. Surface climate and weather characteristics change accordingly along the trade wind trajectory.

The trades of the two hemispheres occupy nearly half of the surface of the globe and are among the steadiest large-scale wind systems on Earth. It is then not surprising that they attracted the imagination of the early general circulation theorists (Halley, 1686; Hadley, 1735). Two centuries elapsed between these early ideas and the beginning of modern trade wind research. Teisserenc de Bort and Rotch (1905, 1909) and Hergesell (1905a, b; 1906a, b; 1907) probed the Northeast trades of the Atlantic off the coasts of Northwest Africa around the turn of the 19th to the 20th centuries. Their expeditions included upper-air soundings from aboard ship. Sverdrup's (1917) doctoral dissertation, in part drawing on the aforementioned field observations, is a milestone in trade wind research. It presumably provided a stimulus for the Atlantic Meteor Expedition 1925–27, one of the best designed and most successful meteorological-oceanographic expeditions of the 20th century. Sverdrup's (1917) pioneering study, various publications stemming from the evaluation of the Meteor Expedition (von Ficker, 1936a, b; Kuhlbrodt, 1928; Laskowski, 1937; Schnapauff, 1937; Jaw, 1937), as well as later research (Kraus, 1959; Augstein, 1972; and others), emphasize the role of the trades as accumulators of latent and sensible heat picked up from the ocean surface and as exporters of heat to the equatorial trough zone. This energetic role of the trades is also of primary concern in Riehl and collaborators' (Riehl *et al.*, 1951; Riehl and Malkus, 1957) studies of the Northeast Pacific trades. Their work is complemented by the investigations of Neiburger (1960) and Neiburger *et al.* (1961). Important recent observations on the North Pacific trades includes systematic aircraft traverses during November 1977 to January 1978 (Ramage *et al.*, 1981).

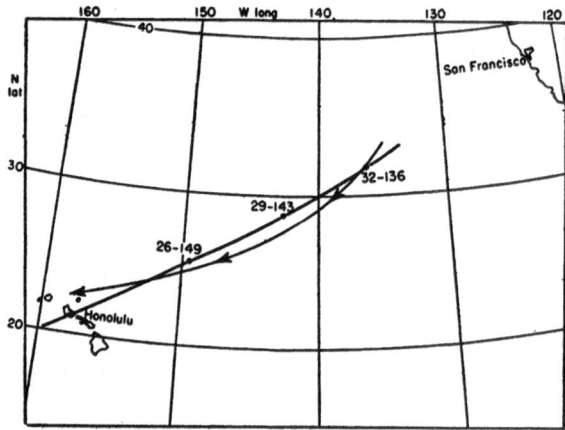

Fig. 6.4:1. Location of weather ships July–October 1945, and mean surface-air trajectory; the average clockwise turning of wind direction with height being only 6 degrees up to 3 km. From Riehl *et al.* (1951).

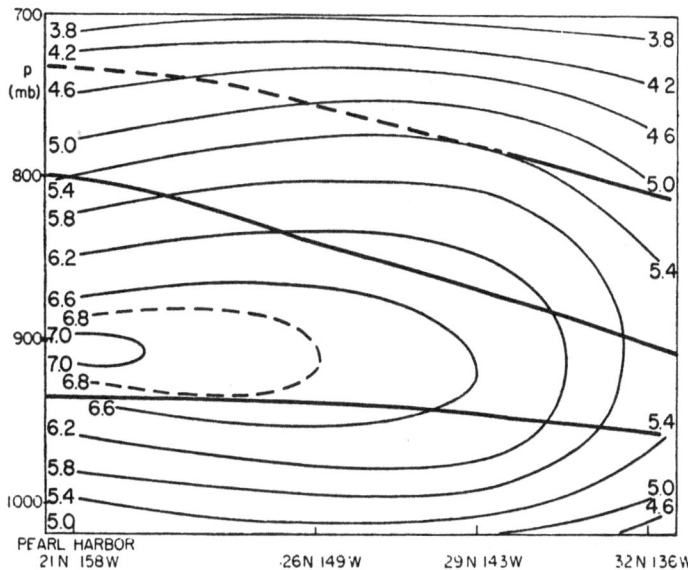

Fig. 6.4:2. Vertical cross-section of wind speed in m s⁻¹. Heavy lines mark the boundary of cloud layer and base and top of the trade inversion. From Riehl *et al.* (1951).

Figs. 6.4:1 through 6.4:10 show the major results of Riehl *et al.*'s (1951) classical study of the Northeast Pacific trades. The major observational basis for this investigation consists of time-averaged radio soundings taken from aboard stationary weather ships deployed along a great circle route from the California coast to Hawaii during July–October 1945 (Fig. 6.4:1). Analyses are presented in vertical cross-sections from surface to 700 mb and from the California coast to the Hawaiian Islands. Delineated in these cross-sections are, proceeding from the surface upward, a subcloud layer, a cloud layer, and above that a persistent stable layer, the trade inversion; the latter is to be discussed more extensively in Section 6.5.

The cross-section Fig. 6.4:2 shows largest wind speed in the lower layers but well above the surface, a decrease towards the upper portion of the diagram, and an increase downstream, from California towards Hawaii.

Fig. 6.4:3 depicts the directional steadiness of wind (ratio of vector mean wind speed to scalar mean). Steadiness is high in the lower layers and decreases rapidly upward across the trade inversion. Within the lower layers, steadiness increases downstream reaching values in excess of 95% in the region of the Hawaiian Islands.

The vertical profile of wind speed Fig. 6.4:4 is derived from the wind cross-section Fig. 6.4:2, and shows a maximum at somewhat below the 900 mb level or 1000 m height, and a marked decrease upward. The vertical profile of divergence derived from the wind cross-section shows largest divergence somewhat below 900 mb, a decrease to small divergence values near the surface, as well as a decrease to a region of non-divergence at the top of the cross-section near 700 mb. Accordingly, the profile of vertical motion is characterized by subsidence throughout, with largest values at the top of the section and a decrease to zero at the surface.

The seemingly small absolute amount of vertical motion corresponds to sizeable vertical displacements of slabs of air in the course of the long trade wind trajectory, as is illustrated in Fig. 6.4:5. The trade inversion rises downstream, while individual columns descend, shrink vertically,

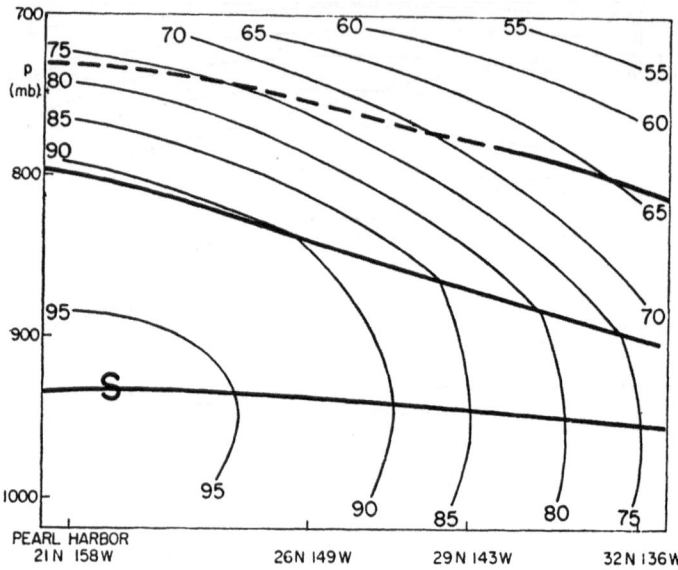

Fig. 6.4:3. Vertical cross-section of directional steadiness of wind, in percent. Heavy lines mark the boundary of cloud layer and base and top of trade inversion. From Riehl *et al.* (1951).

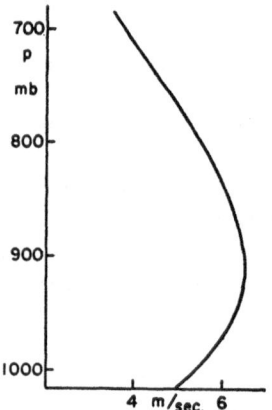

Fig. 6.4:4. Vertical profile of wind speed in m s^{-1}. From Riehl *et al.* (1951).

and spread horizontally. Trajectories cross inversion top and base. Accordingly, three masses of air are distinguished in the cross-section Fig. 6.4:6: air that has been below the inversion throughout the journey represented in the cross-section; air that has been incorporated into the cloud layer; and air that has been incorporated into the inversion layer.

While Figs. 6.4:2 to 6.4:6 describe the air circulation in a vertical section along the trade wind trajectory, Figs. 6.4:7 to 6.4:10 present the distribution of thermodynamic properties.

The temperature cross-section Fig. 6.4:7 shows isotherms running broadly parallel to the inversion. The inversion layer appears merely as a deep isothermal stratum rather than a region

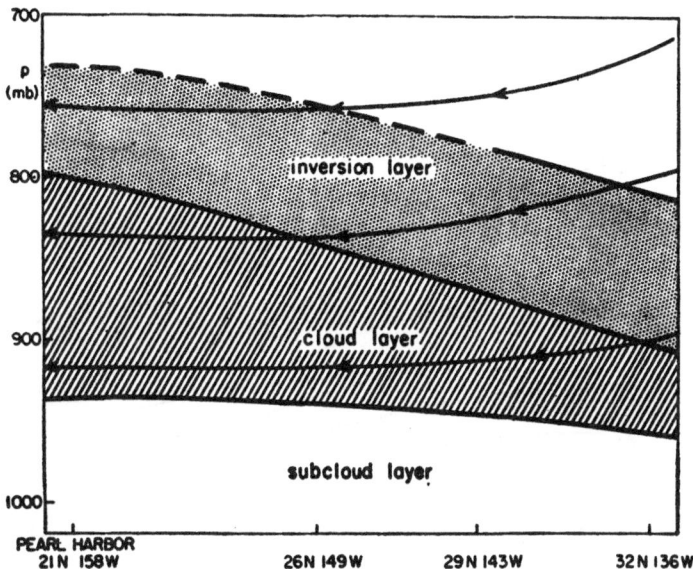

Fig. 6.4:5. Sample air trajectories in relation to atmospheric layers. From Riehl *et al.* (1951).

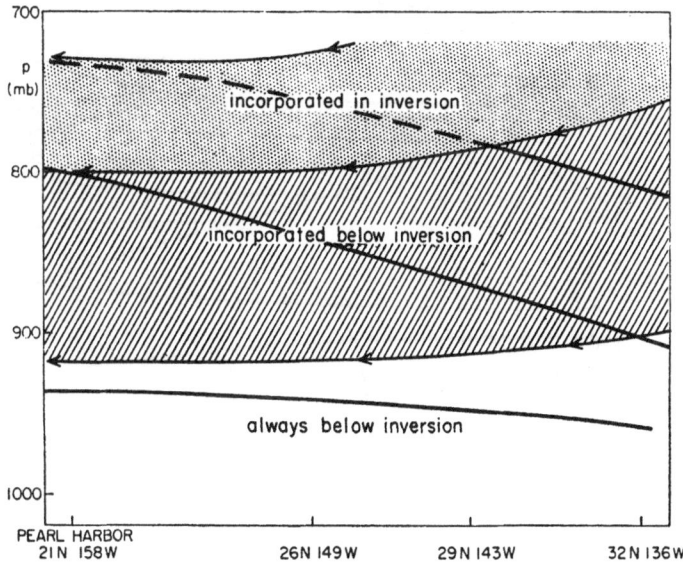

Fig. 6.4:6. Flow of mass through the trade inversion. From Riehl *et al.* (1951).

of upward temperature increase. This is a consequence of the analysis based on time-averaged measurements. Neiburger (1960) and Neiburger *et al.* (1961) analyzed individual soundings and obtained distinct inversion characteristics. Drawing on Neiburger *et al.*'s (1961) study, Riehl (1979, pp. 228–233) presented a modified version of the original (Riehl *et al.*, 1951) analysis. Further noteworthy in Fig. 6.4:7 is the downstream increase of temperature in the lower layers.

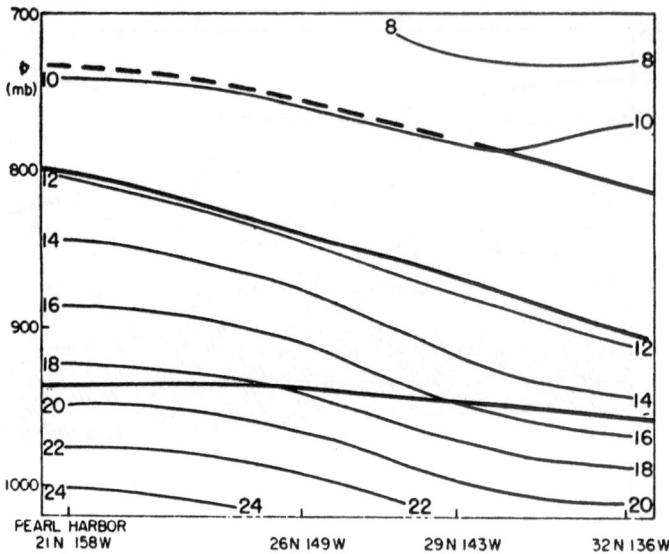

Fig. 6.4:7. Vertical cross-section of temperature in °C. From Riehl *et al.* (1951).

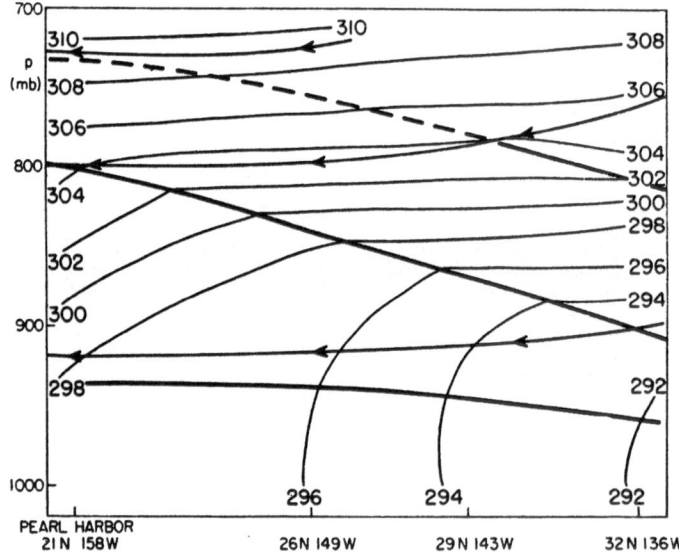

Fig. 6.4:8. Vertical cross-section of potential temperature in °K. From Riehl *et al.* (1951).

Fig. 6.4:7 is complemented by the cross-section of potential temperature, Fig. 6.4:8, which shows values increasing upward most strongly in the inversion layer, and in the surface layer an increase of potential temperature along the trajectory.

The humidity conditions are illustrated in the cross-sections Figs. 6.4:9 and 6.4:10. Specific humidity (Fig. 6.4:9) decreases upward, particularly abruptly across the inversion layer, while near the surface values increase along the trade wind trajectory. Relative humidity (Fig. 6.4:10) is largest in the layer below the inversion, across which it decreases abruptly upward.

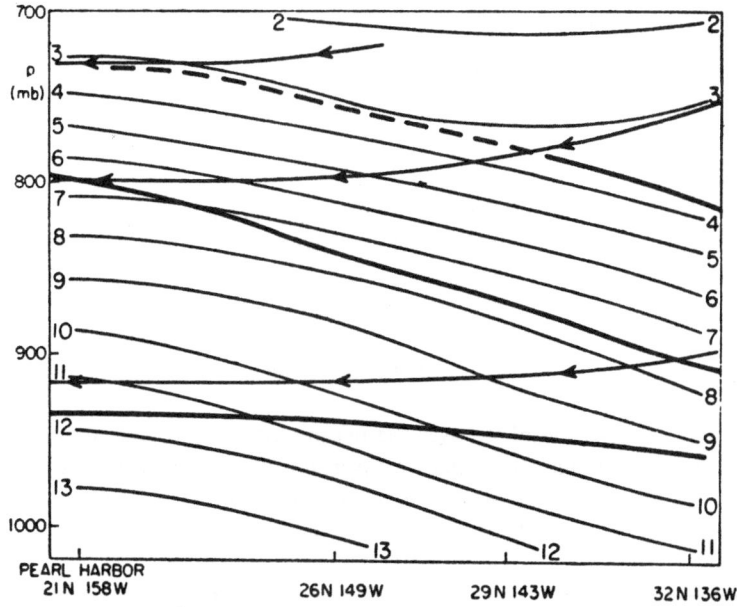

Fig. 6.4:9. Vertical cross-section of specific humidity in g kg⁻¹. From Riehl *et al.* (1951).

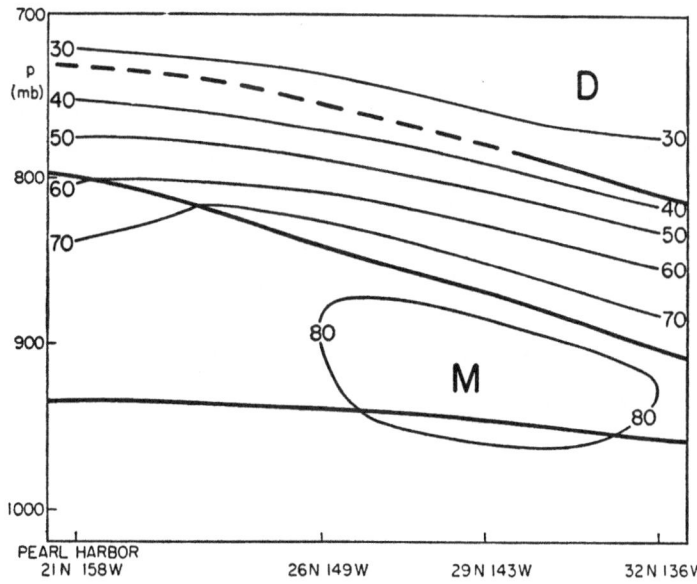

Fig. 6.4:10. Vertical cross-section of relative humidity in percent. From Riehl *et al.* (1951).

Figs. 6.4:8 and 6.4:9 are interesting concerning the way in which the trades pick up heat from the tropical oceans along their trajectory. From the downstream increase of properties it is seen that the trades act as accumulators of latent heat and to a much lesser extent of sensible heat. Riehl and Malkus (1957) point out that although the sensible heat pickup is of subordinate importance for the heat budget, it serves an important function in the maintenance of the trades; the heating along the trajectory would hydrostatically maintain a downstream directed pressure gradient in the surface layer and thus provide a self-driving mechanism for the trade wind circulation.

Thompson and Neiburger's (1953), Neiburger's (1960), and Neiburger *et al.*'s (1961) work of the lower atmosphere over the Northeast Pacific complement the study of Riehl *et al.* (1951) particularly in the following respects: it presents maps of spatial patterns and is based on the evaluation of individual soundings. The investigation is primarily aimed at the structure and maintenance of the trade inversion and is therefore more extensively discussed in the following Section 6.5.

6.5. Trade Inversion

The structure of the lower troposphere over the oceanic trade wind regions is characterized by a persistent, spatially continuous and extensive inversion. The trade inversion was mentioned in Section 6.4. Pattern characteristics of the trade inversion, processes operative in the origin and maintenance, and its role in the surface climate are discussed in the present section.

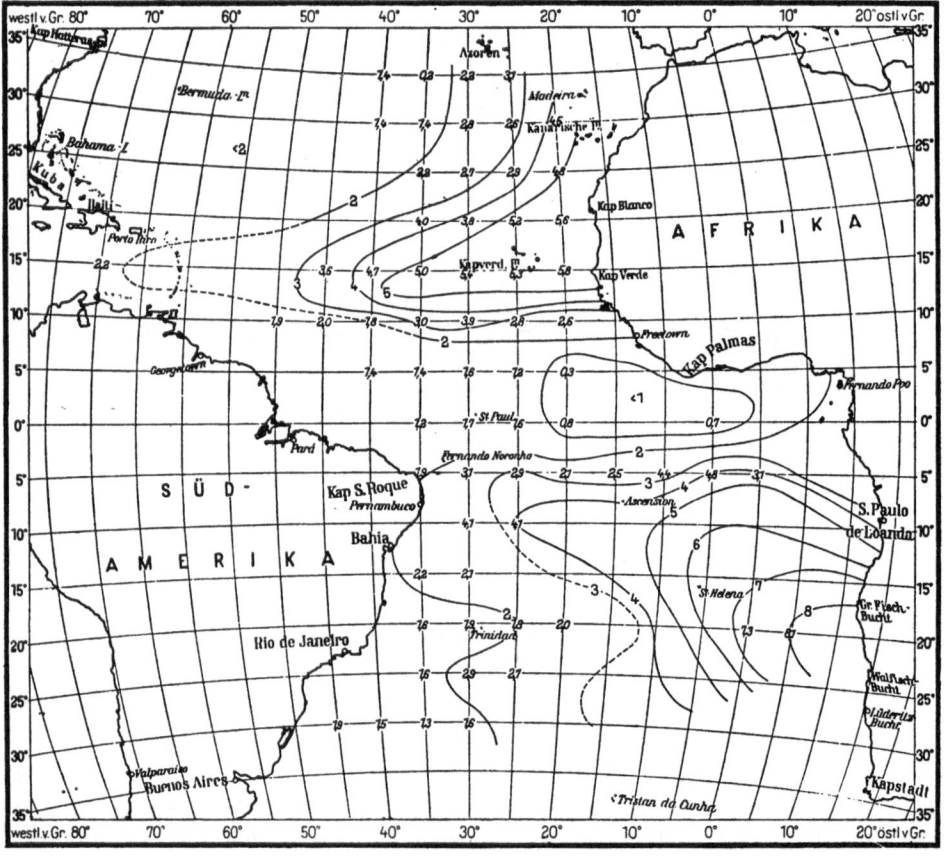

Fig. 6.5.1:1. Height of the base of the trade inversion over the Atlantic, in m. From von Ficker (1936a).

6.5.1. SPATIAL PATTERNS

The work of the Atlantic Meteor Expedition 1925–27 has remained the most comprehensive observational investigation of the trade inversion to date (von Ficker, 1936a, b; also Kuhlbrodt, 1928; Laskowski, 1937; Schnapauff, 1937; Jaw, 1937). The main results of von Ficker's (1936a) study are presented in a series of maps, Figs. 6.5.1:1 to 6.5.1:5.

Fig. 6.5.1:1 shows the spatial pattern of the inversion base. The base is lowest on the eastern equatorward side of the subtropical highs, values of less than 500 m being found off the coasts of Northwest and Southwest Africa. From these regions the inversion base rises both equatorward and towards the western side of the Atlantic Ocean; the rise being most pronounced over the first few 100 km off the coast, and more gradual thereafter. The inversion base is found at more than 1500 m over the western part of the Atlantic and at more than 2000 m in the equatorial zone.

The spatial pattern of inversion intensity mapped in Fig. 6.5.1:2 exhibits much similarity with the distribution of inversion base. Thus the largest temperature jump across the inversion is found again on the eastern equatorward side of the subtropical highs, the temperature difference off the coast of Northwest Africa being larger than 5°C, and off the coast of Southwest Africa even in excess of 8°C. The temperature differential decreases to less than 1°C in the equatorial region and to less than 2°C over the western part of the Atlantic.

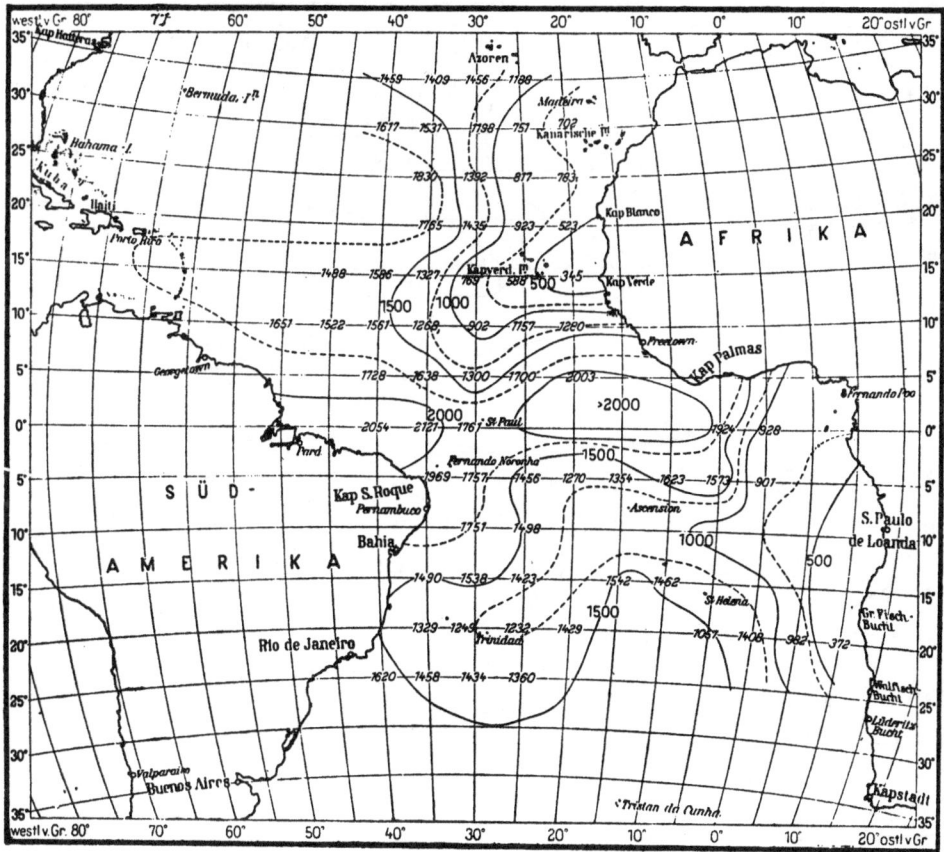

Fig. 6.5.1:2. Temperature increase from bottom to top of inversion over the Atlantic, in °C. From von Ficker (1936a).

The relative humidity decrease from bottom to top of the inversion (Fig. 6.5.1:3) is a further measure of inversion intensity. The large-scale pattern characteristics are again similar to those of temperature jump (Fig. 6.5.1:2) and inversion base (Fig. 6.5.1:1). Thus the largest values are found off the coasts of Northwest and Southwest Africa, where they exceed 60 and 70%, respectively.

Figs. 6.5.1:4 and 6.5.1:5 offer further details on the thermal structure of the lower atmosphere. The maximum inversion temperature (Fig. 6.5.1:4) again is largest off Northwest and Southwest Africa, where it amounts to more than 20°C, and decreases to less than 15°C in the equatorial zone and the western part of the Atlantic. The difference between surface air temperature and maximum inversion temperature (Fig. 6.5.1:5) is slightly negative off the coasts of Northwest and Southwest Africa, while positive values in excess of 10°C are found over the Western Atlantic and the equatorial region, where in fact the inversion is colder than the sea surface. Figs. 6.5.1:4 and 6.5.1:5 thus complement Figs. 6.5.1:1 to 6.5.1:3 in describing the spatial pattern of inversion characteristics.

Figs. 6.5.1:1 to 6.5.1:5 collectively show that the trade inversion tends to be lowest and best developed on the eastern equatorward side of the subtropical highs, and that it rises and weakens broadly following the trade wind airstream equatorward and towards the western part of the

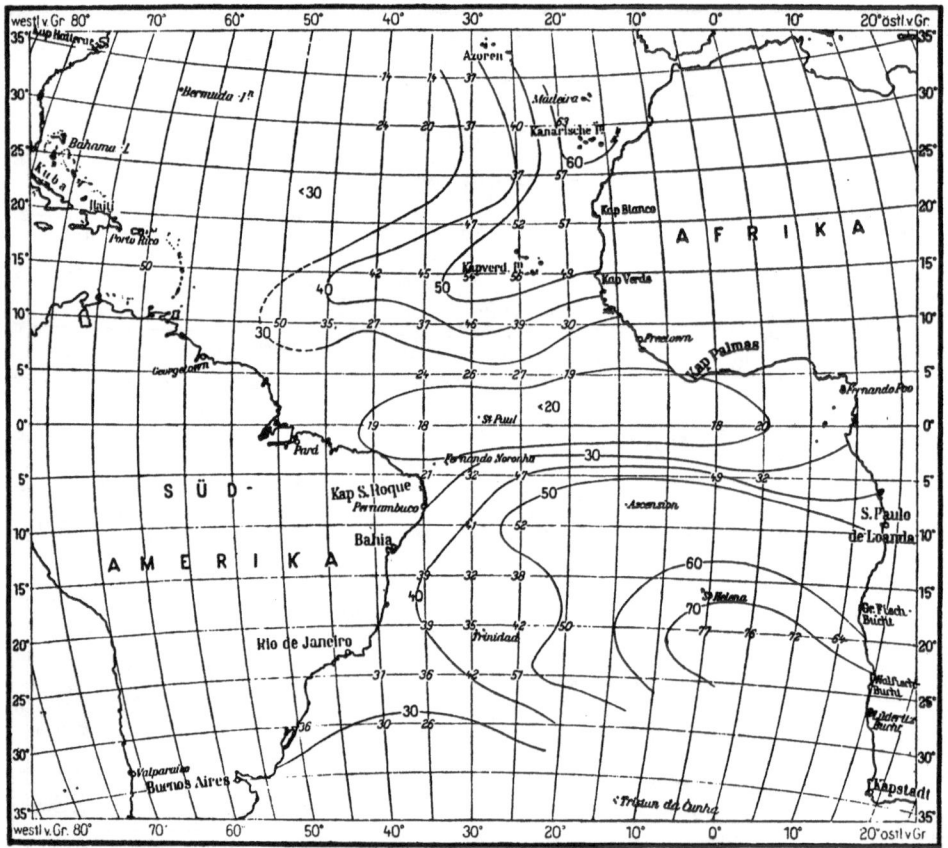

Fig. 6.5.1:3. Decrease of relative humidity from bottom to top of inversion over the Atlantic, in percent. From von Ficker (1936a).

ocean. The charts Figs. 6.5.1:1 to 6.5.1:5 are based on sampling limited in time and space during cruises spread over some years, and provide only a general orientation on inversion pattern characteristics. Annual cycle characteristics are not considered. Thus Riehl (1979, pp. 202–225) calls attention to the considerable day to day variations of inversion characteristics noted in recent field experiments. However, the Meteor investigations still represent the most comprehensive large-scale analysis of trade inversion characteristics over the Atlantic.

As can be appreciated from Figs. 6.5.1:1 to 6.5.1:5 the trade inversion tends to be rather less well developed over the Caribbean region, lying as it does on the western side of the North Atlantic subtropical high. The trade inversion over the Caribbean (Gutnick, 1958; Hastenrath, 1966) has a base at 2000 m or more, a temperature jump of the order of $1-2°C$, and is subject to pronounced variations in the course of the year. Fig. 6.5.1:6 illustrates the percentage frequency of lower-tropospheric inversions at stations along a meridional profile along 80°W extending from Southern Florida to the Panama Canal Zone. For the year as a whole, inversion frequency decreases from the latitude of the axis of the North Atlantic high towards the equatorial trough zone. Overall, inversions are most frequent towards the end of the winter half year, around March, and rarest during the summer half-year, which is the rainy season in much of the area.

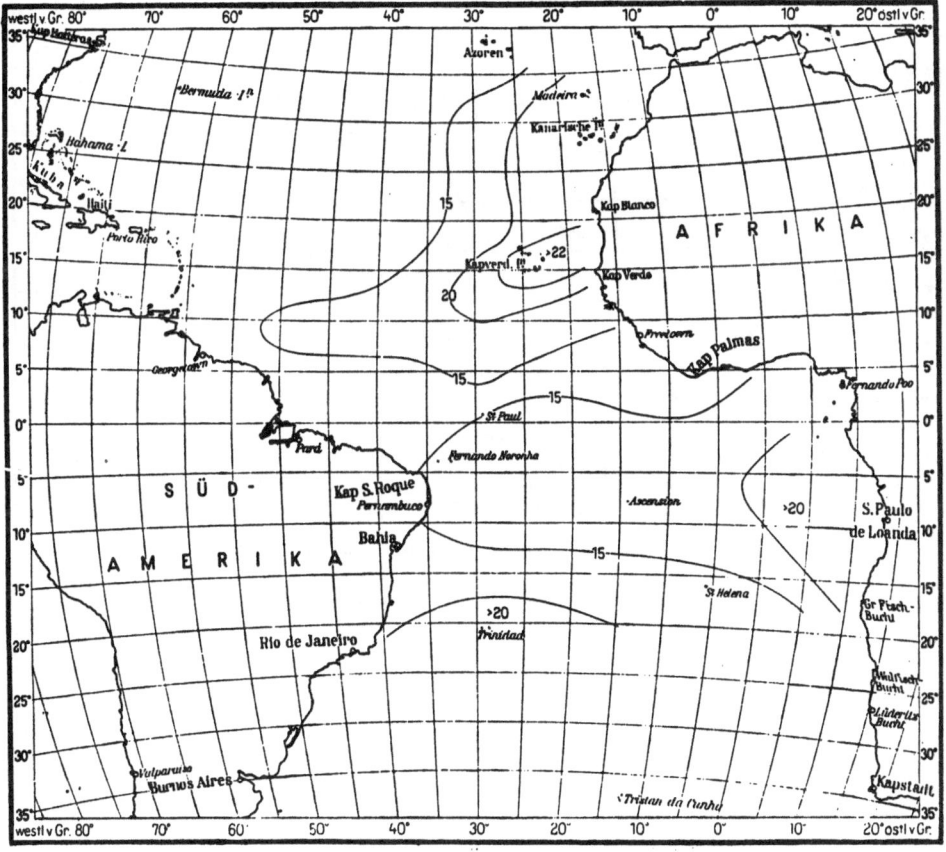

Fig. 6.5.1:4. Maximum inversion temperature over the Atlantic in °C. From von Ficker (1936a).

The inversion characteristics just described for the Atlantic are paralleled by investigations in the Northeast Pacific. A very preliminary picture of the inversion is provided by Riehl *et al.*'s (1951) paper, but a comprehensive investigation is due to Neiburger (1960) and Neiburger *et al.* (1961). Figs. 6.5.1:7 to 6.5.1:14 illustrate the major results. The surface circulation over the Northeast Pacific in July is described by Figs. 6.5.1:7 to 6.5.1:9. From the center of the Northeast Pacific high (Fig. 6.5.1:7), anticyclonic outflow (Fig. 6.5.1:8) takes place, in such a way that the airstreams emanating from the eastern sector of the high appear as the roots of the trade winds. It is on the eastern and southeastern sides of the high that strongest lower-tropospheric divergence and subsidence are found (Fig. 6.5.1:9). These general circulation analyses provide the essential background for the discussion of the spatial patterns of inversion characteristics.

The inversion base (Fig. 6.5.1:10) rises from about 400 m at the California coast to about 2000 m over the Hawaiian Islands. The temperature jump across the inversion (Fig. 6.5.1:11) amounts to more than 10°C at the California coast, from where it decreases broadly following the trade wind trajectory to about 2°C in the region of the Hawaiian Islands. The relative humidity jump across the inversion (Fig. 6.5.1:12) is likewise largest at the California coast, where it exceeds 60%, and decreases to about 40% over Hawaii.

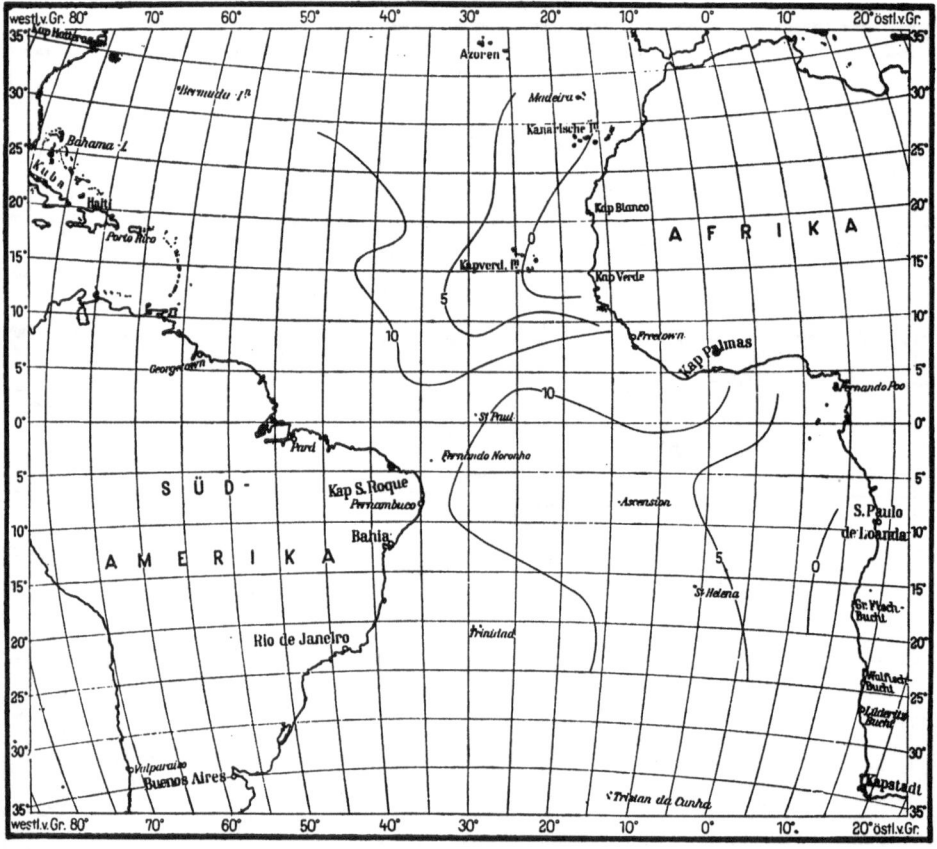

Fig. 6.5.1:5. Difference of surface air temperature minus maximum inversion temperature over the Atlantic in °C. From von Ficker (1936a).

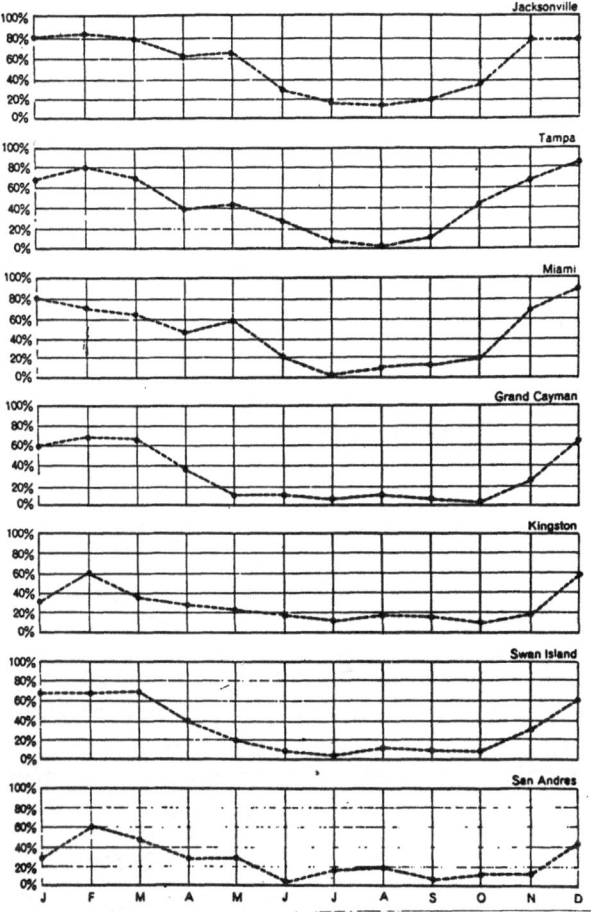

Fig. 6.5.1:6. Frequency of lower-tropospheric inversions at stations along a meridional profile at 80°W extending from Southern Florida to Panama, in percent of radio soundings during 1980. From Hastenrath (1966).

The maps of the percentage frequency of inversions, Fig. 6.5.1:13, shows that an inversion is found at all times in the region off the California coast, where the inversion is best developed. Proceeding towards Hawaii the inversion frequency gradually decreases. Streamlines (Fig. 6.5.1:8) cross the lines of equal inversion height (Fig. 6.5.1:10) at approximately right angle, but it is noteworthy that they intersect the isopleths of inversion frequency (Fig. 6.5.1:13) only at small angles. The structure and spatial variations of the inversion is illustrated in the vertical cross-section from the California coast to the Hawaiian Islands, Fig. 6.5.1:14. The cross-section shows the pronounced rise of the inversion over the first few 100 km off the California coast, and the more gradual change further towards the open ocean. The thickness of the inversion is about 400 m, with little systematic spatial variation. It is surmised that diurnal and day-to-day variations in inversion thickness may be rather larger than the average spatial differences apparent in Fig. 6.5.1:14. Neiburger's (1960) and Neiburger et al.'s (1961) investigations as summarized in Figs. 6.5.1:7 to 6.5.1:14 document for the Northeast Pacific inversion conditions that are broadly consistent with the results of the Meteor Expedition for the Atlantic. Ramage et al. (1981) report from their aircraft traverses in the Central Pacific during November 1977 to January 1978 little variation of the inversion height between the Hawaiian Islands and the equatorial region.

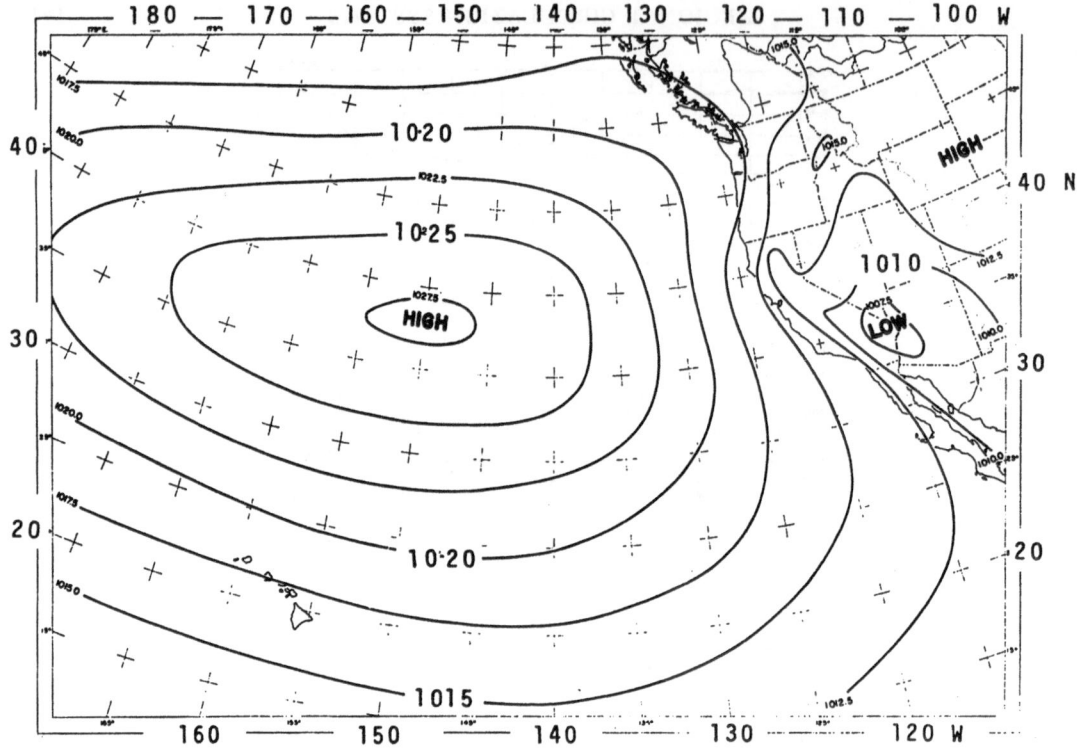

Fig. 6.5.1:7. Normal sea-level pressure over the Northeast Pacific in July, in mb. From Neiburger *et al.* (1961).

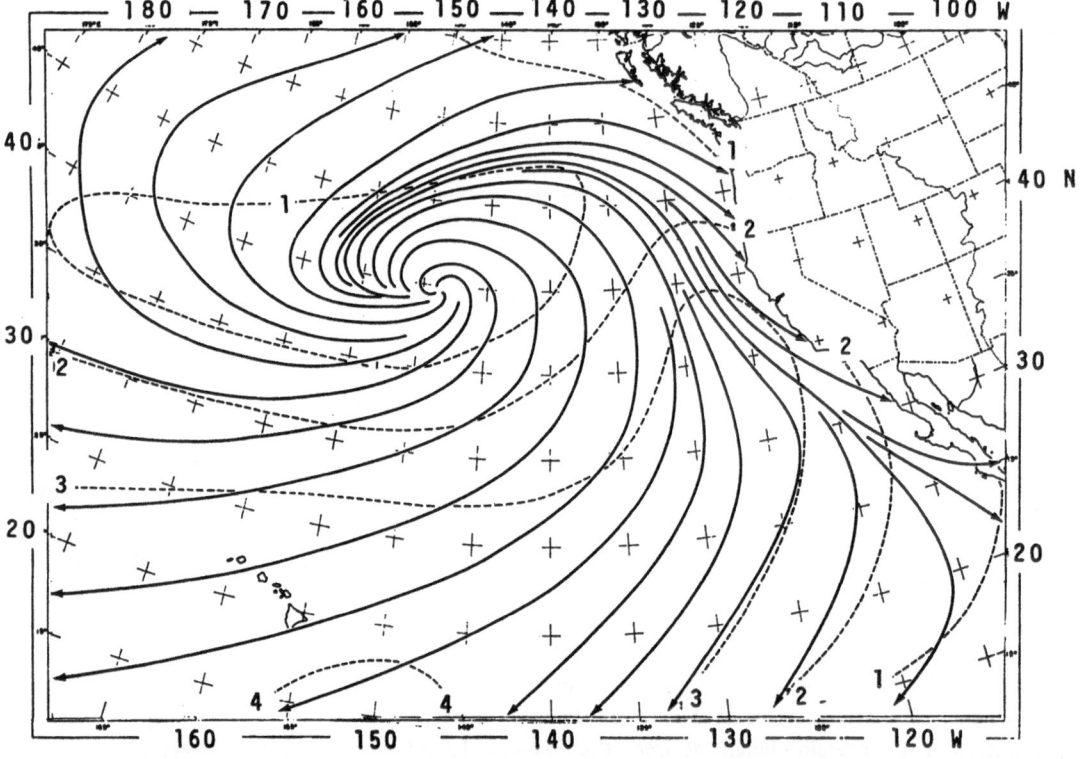

Fig. 6.5.1:8. Normal resultant wind streamlines and isotachs (m s⁻¹) over the Northeast Pacific in July. From Neiburger *et al.* (1961).

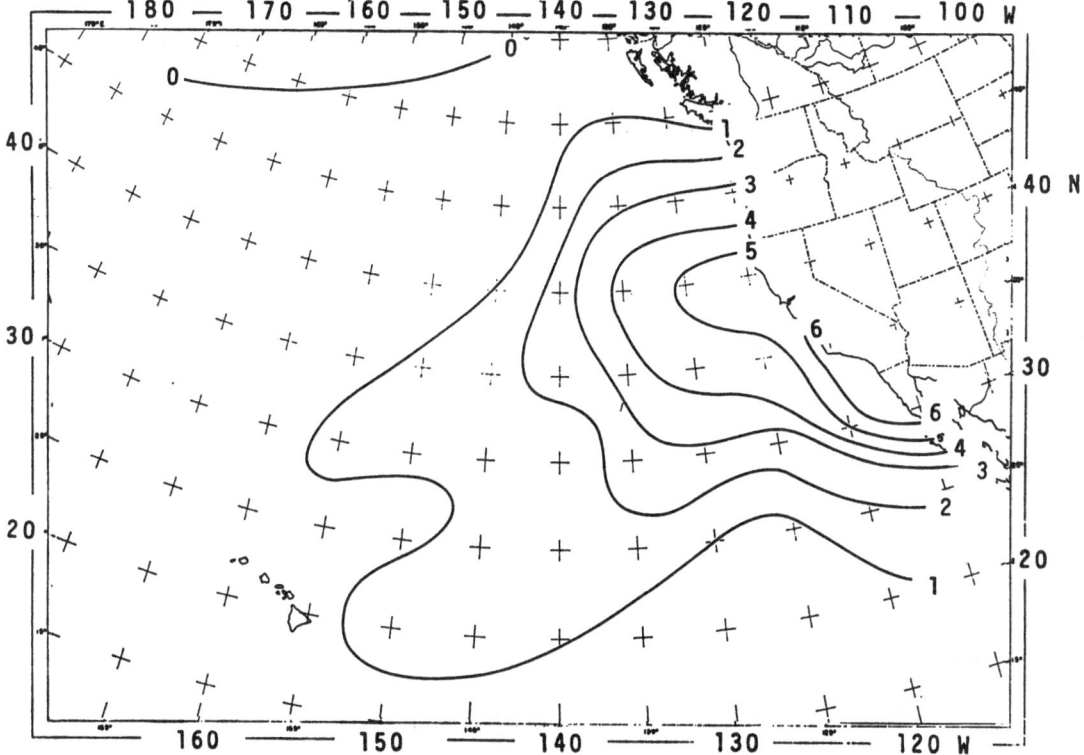

Fig. 6.5.1:9. Divergence of surface resultant wind over the Northeast Pacific in July, in 10^{-6} s^{-1}. From Neiburger *et al.* (1961).

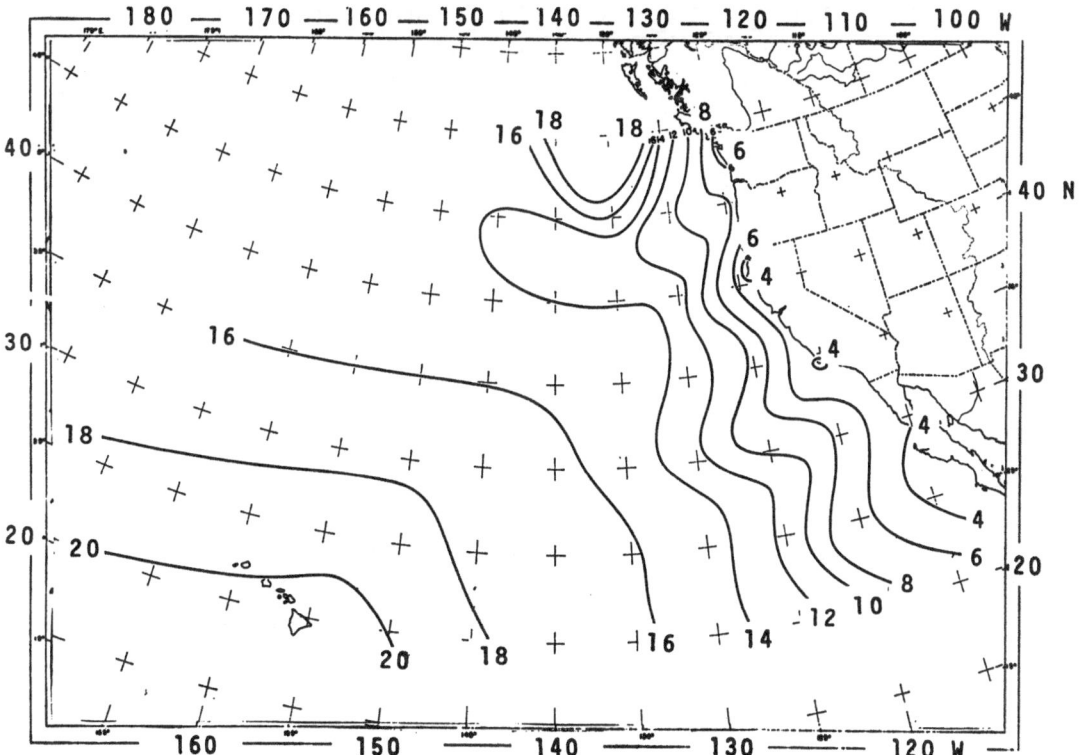

Fig. 6.5.1:10. Height of the base of the trade inversion over the Northeast Pacific, in 10^2 m. From Neiburger *et al.* (1961).

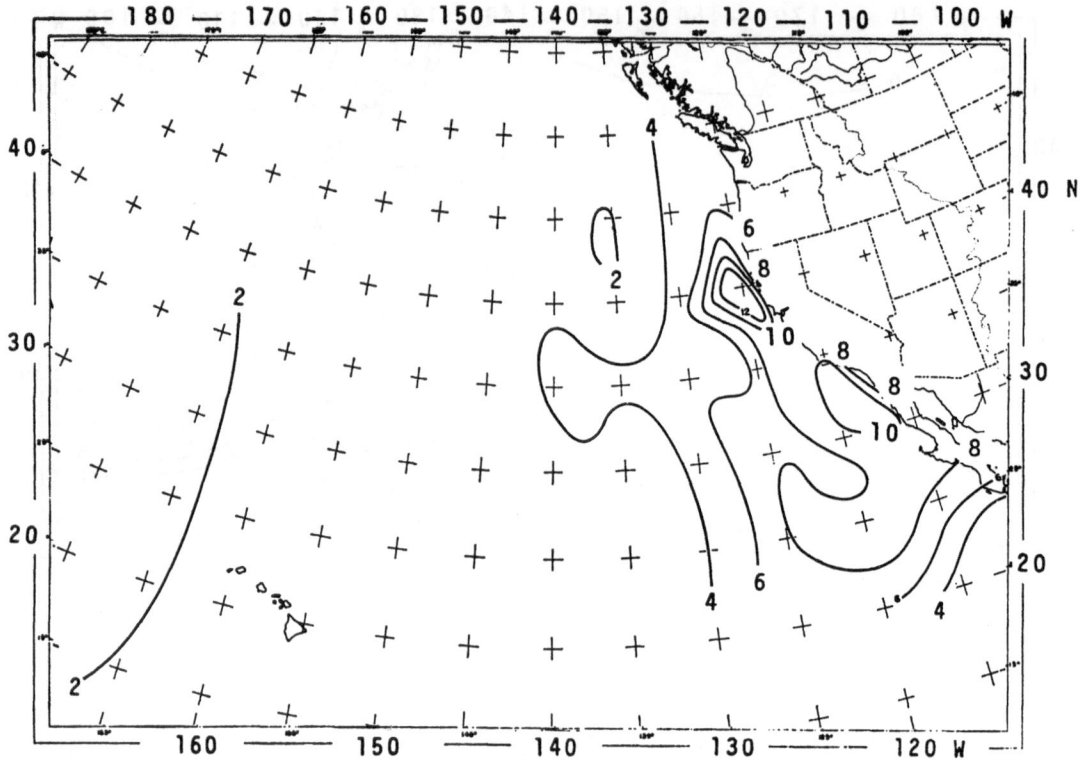

Fig. 6.5.1:11. Temperature increase from bottom to top of inversion over the Northeast Pacific, in °C. From Neiburger *et al.* (1961).

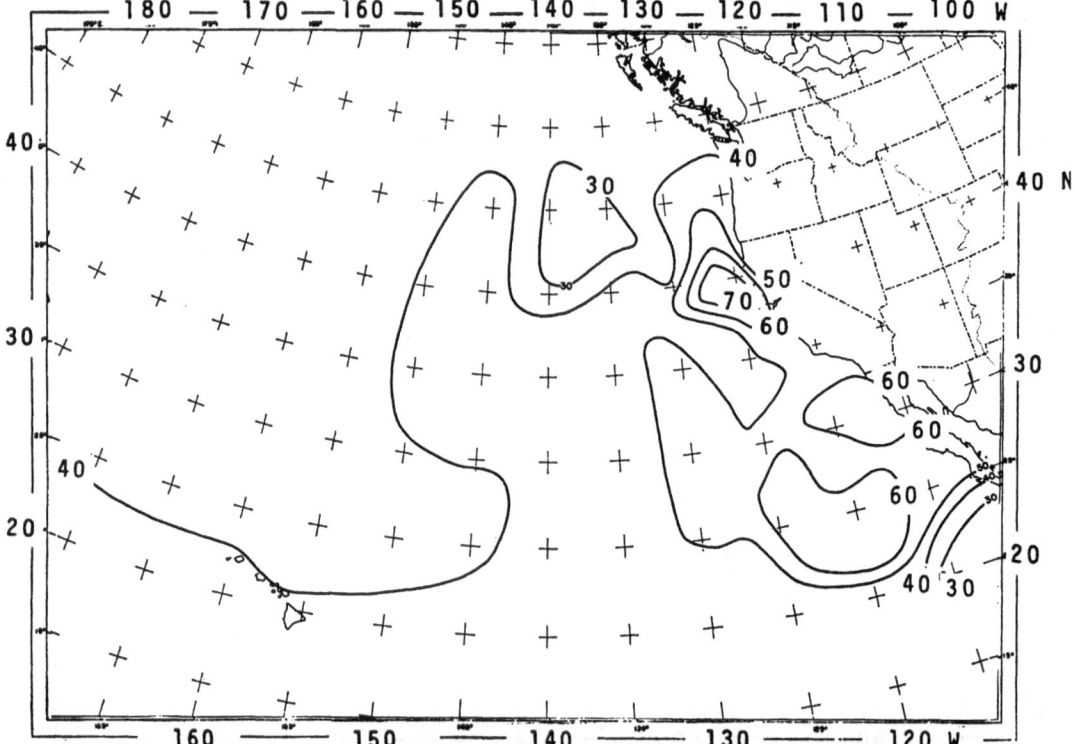

Fig. 6.5.1:12. Decrease of relative humidity from bottom to top of inversion over the Northeast Pacific, in percent. From Neiburger *et al.* (1961).

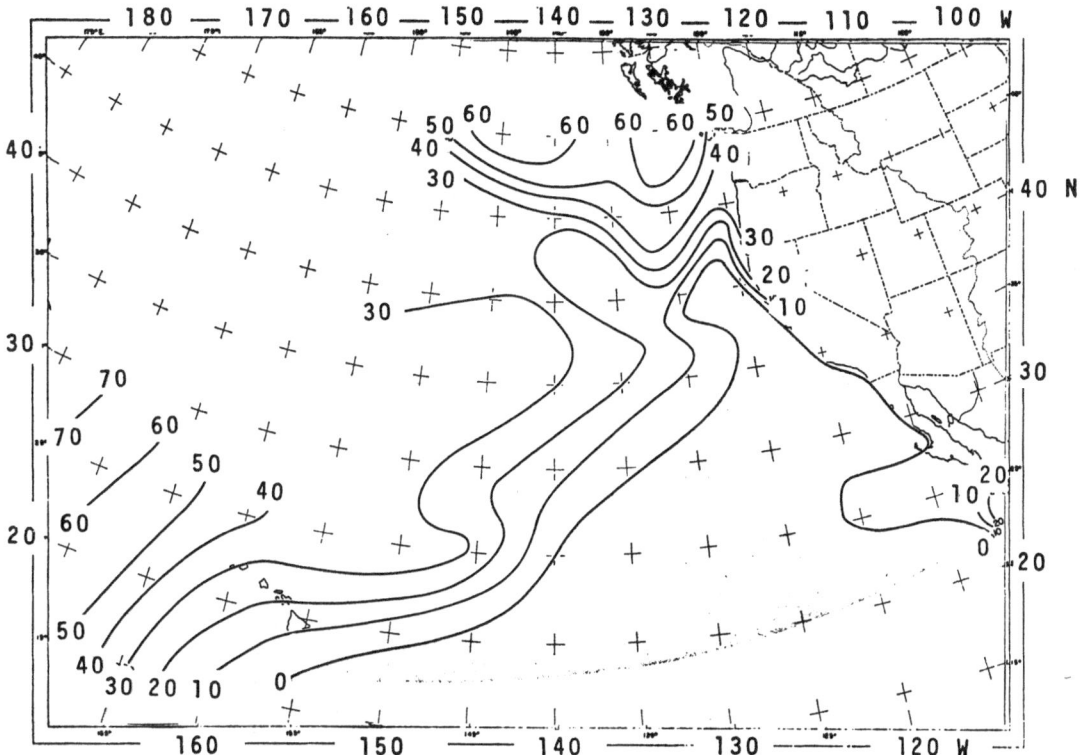

Fig. 6.5.1:13. Percentage frequency of no inversion over the Northeast Pacific. From Neiburger *et al.* (1961).

For the Eastern South Pacific, a useful sampling of inversion characteristics has been presented by Haraguchi (1968). In similarity to the Atlantic and North Pacific, the inversion is lowest and strongest on the eastern equatorward side of the Eastern South Pacific high. Proceeding from there equatorward and westward towards the open ocean, the inversion both rises and weakens; typical values of inversion base being about 500 m at the South American West coast and more than 2000 m over Tahiti.

Turning to the Indian Ocean, Taljaard (1953, 1955) and Émon (1948) studied the inversion conditions around Southern Africa. Concerning the Atlantic coast, Taljaard (1953) describes an inversion base around 500 m at Alexander Bay and a rise both northward and southward to values of about 1000 m at the mouth of the Zaïre (Congo) river (Kinshasa) and Cape Town, respectively. On the Indian Ocean side of Southern Africa, Émon (1948) finds for Southern winter a rise from about 1000 m at Durban and Maputo (Lourenço Marques) to more than 2500 m over Malagasy. During Southern summer the inversion is higher and less well developed; inversions being clearly defined at 1500 m over Durban, but less conspicuous over Malagasy. Taljaard (1955) considers the term 'trade-wind inversion' unfortunate in that inversions or stable layers extend continuously to either side of the subtropical high pressure belt.

With this more general perspective it seems appropriate to mention here also the discussions of lower-tropospheric inversion conditions over the Arabian Sea during Northern summer by Ramage (1966), Desai (1967), and Flohn *et al.* (1968). The controversy raised in these papers concerning the various processes conducive to the inversion phenomenon is in part related to the problem of origin and maintenance of the trade inversion.

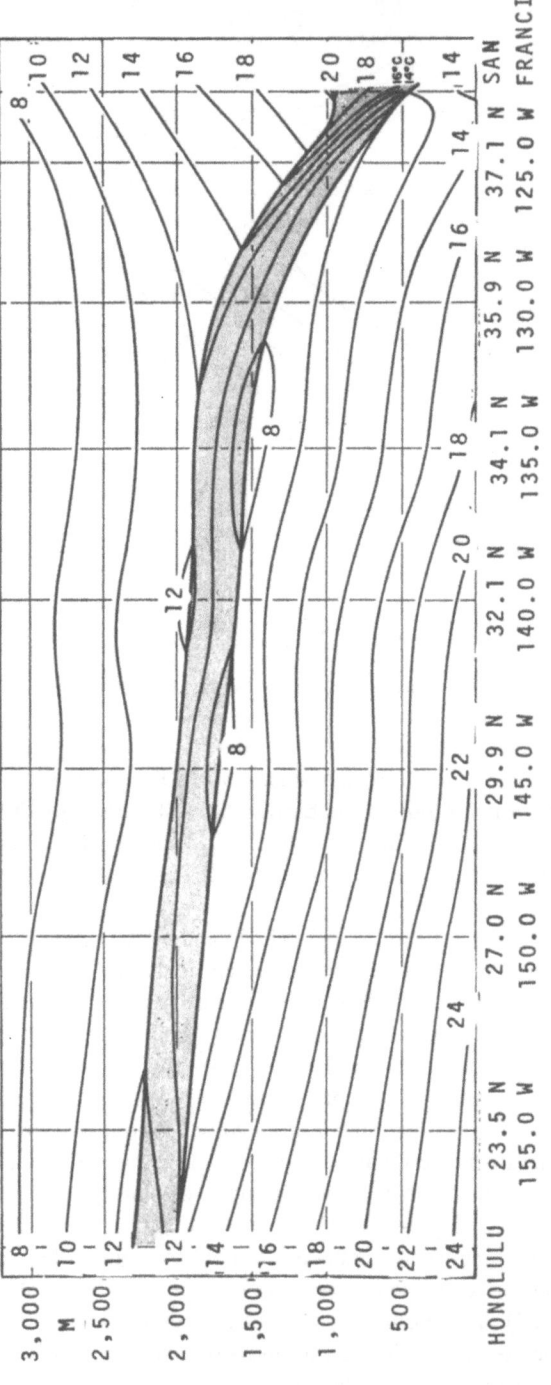

Fig. 6.5.1:14. Vertical cross-section of temperature (in °C) from the California coast to the Hawaiian Islands, showing downstream variation of trade inversion. From Neiburger *et al.* (1961).

6.5.2. ORIGIN AND MAINTENANCE

Mechanisms operative in the origin and maintenance of the trade inversion are to be considered in conjunction with the observed spatial patterns and annual cycle variations discussed in Section 6.5.1. In all generality, the following types of inversion formation are recognized: (i) turbulence; (ii) radiation/heat budget; (iii) advection; (iv) subsidence. While any of these processes may play some role for limited areas and periods, our primary concern here is to single out the most dominant processes.

Concerning mechanism (i) it is recalled that the regions of most persistent, strongest, and lowest inversion on the eastern equatorward side of the subtropical highs are among the areas of weakest wind and thus turbulence. Regarding mechanism (ii) it is interesting that the aforementioned regions of best inversion development are all characterized by cold equatorward flowing ocean currents: the Canary, Benguela, California, and Humboldt Currents (Section 4.2). While cold surface waters may hamper convection and thus be a factor in the especially strong inversion development in these particular regions, it is recognized that the trade inversion is not a surface inversion. Accordingly, mechanism (ii) cannot be the dominant cause of the trade inversion even in these regions. Mechanism (iii) was long considered a factor in the origin of the trade inversion, despite Piazzi-Smyth's (1858) early observations from his 1856 expedition to the Canary Islands, which demonstrated that the inversion is embedded in the trade wind circulation rather than forming its upper boundary. However, advection may contribute to the origin of the inversion at the coast of Northwest Africa, where hot dry air from the Sahara overruns a wedge of cooler and moister maritime air (Carlson and Prospero, 1972; Prospero and Carlson, 1972). Again, advection by itself does not explain the large-scale inversion pattern.

Mechanism (iv) is of particular interest. As to the spatial pattern, lower-layer divergence and subsidence is large at the subtropical anticyclonic axes and in the upstream portion of the trades on the eastern equatorward side of the subtropical highs – a distribution broadly coincident with the patterns of intensity, height, and persistence of the inversion as reviewed in Section 6.5.1. Regarding the annual march of inversion frequency and intensity, there is a striking parallelism with the latitude displacements of the subtropical high axis: as described in Section 6.5.1, the inversion over the Caribbean is best developed towards the end of the Northern winter, while along the coast of Southwest Africa inversions in Southern winter appear enhanced on the equatorward side of the South Atlantic high axis but reduced at the more poleward locations of Southern Africa.

Subsidence must indeed be considered as a major control in the maintenance of the trade inversion. It should be noted, however, that the inversion stays at some elevation above the surface, the subsidence effect presumably being counteracted by convection from below. The theoretical studies of Rouse and Dodu (1955) and Ball (1960) are relevant in this context. As columns of air travel downstream, convective clouds develop that are inhibited in their vertical development by the trade inversion. Some trade cumuli may penetrate the inversion, but quickly dissipate in the dry air above. The continued effect of these cloud developments is a gradual erosion, weakening and lifting of the inversion downstream. Concurrently, the air column on its travel downstream gets into regions with lesser divergence and subsidence. In conjunction these factors would be conducive to the weakening and rising of the inversion downstream. In addition to the interplay of these two major factors cold ocean currents and atmospheric advection may provide reinforcing factors in limited regions.

6.5.3. CLIMATIC IMPLICATIONS

Within the context of global energetics, and more specifically the heat transport in the equatorward directed lower-tropospheric branch of the Hadley cell (Section 3.3), the trade inversion serves an important function. Sensible heat and especially moisture picked up from the ocean by the surface air stream remains largely trapped below the inversion. As considered above, trade cumuli are impeded from carrying moisture to higher layers, and the stable stratification hampers convection and thus rainfall. As a consequence, evaporation tends to exceed precipitation in the trade wind region, and moisture is accumulated within the trade wind air stream and carried into the equatorial trough zone, where condensation is concentrated in the ascending branch of the Hadley cell. The release of latent heat in the equatorial trough zone, and the upward heat transport by high-reaching cumulonimbus towers (Riehl and Malkus, 1958), are considered instrumental in the functioning of the Hadley cell. Moreover, a well-developed mean meridional circulation is associated with strong subsidence in its poleward portion, which in turn largely controls the intensity of the trade inversion in the lower troposphere. Conversely, a weak Hadley cell is concomitant with less subsidence, a weaker trade inversion, less accumulation of latent and sensible heat in the surface layer of the trades, and hence less concentration of latent heat release in the equatorial trough zone. Kraus (1959) in particular has called attention to the role of feedback processes in the tropical ocean and trade wind system. Through the mechanisms sketched above, the trade inversion appears to play a role in the global heat budget and climate.

The trade inversion also has important implications for the regional climate. These are largely related to atmospheric stability hampering convection and vertical exchange. The roots of the trade inversion are at the fringes of the midlatitudes. Thus the lower-tropospheric inversion over Southern California finds its continuation in the trade winds of the tropical Northeast Pacific. The effect of aerosol production from the large urban concentrations of Southern California (Los Angeles) is compounded by the atmospheric stability. An analogous situation is encountered in the Santiago region of temperate latitude Chile where air pollution is accentuated by the persistent inversions, that apparently extend continuously out to the tropical Southeast Pacific. The pollution problems of the Peruvian capital city of Lima appear likewise due to a combination of anthropogenic factors and inversion effects.

The trade inversion enhances the humidity in the surface layer and extended stratiform cloudiness, and hampers rainfall. These characteristics are accentuated at certain low-latitude coastal regions, with marked environmental consequences. Most prominent among these regions is the West coast of tropical South America to the South of the Equator, to which Fig. 6.5.3:1 refers. At the latitude of Southern Peru, the terrain rises from the Pacific Ocean to a plateau around 1500 m and then to the Andes and the Altiplano at some 4000 m. In July 1964 thermometer and altimeter readings were taken on an automobile travel from the coast to the Altiplano. The values are plotted in Fig. 6.5.3:1. The profile is consistent with the broadly synchronous ascents at the coastal radiosonde stations Lima to the North and Antofagasta to the South. A marked inversion extends from about 800 to 1300 m, with a temperature differential of about 10°C. In the coastal region below the inversion atmospheric humidity is high, but there is little rainfall. This is reflected in the barren and desertic environment. The sparse vegetation may in part utilize water available from dew formation. This source may improve upward to below the inversion, where the extensive stratiform cloud deck over the coastal ocean touches the slope of the continent, and where fog is a persistent feature. Proceeding further upward to above the inversion, the atmospheric and terrestrial environment changes abruptly. The fog gives way to clear sky and bright sunshine, temperature increases, and vegetation is completely absent in the dune fields of the Pampa de la Joya (Lettau and Lettau, 1969, 1978; Hastenrath, 1967b; Stearns, 1969).

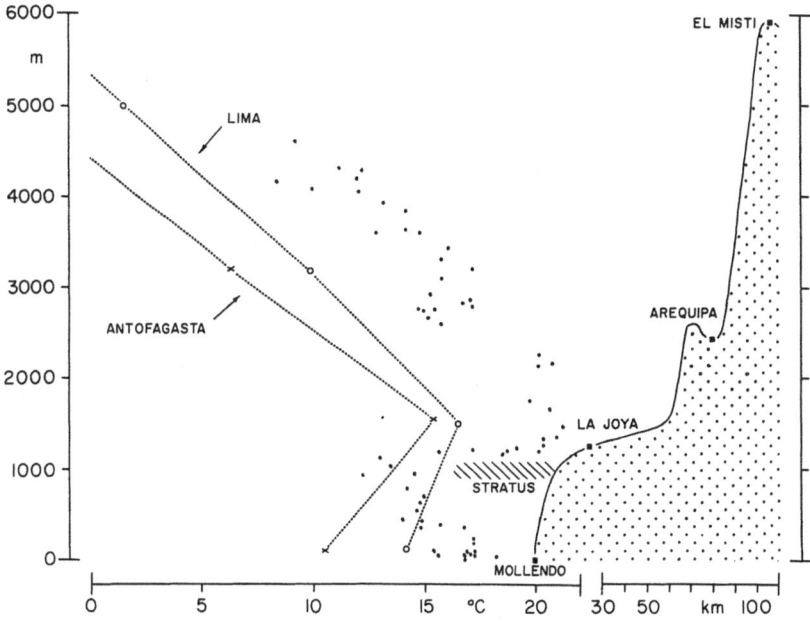

Fig. 6.5.3:1. Vertical temperature profiles at the West coast of South America. Dots denote measurements made by S. H. on a motor car travel from Mollendo to Puno on 16–17 July 1964, while crosses and open circles depict the 19 GMT, 16 July 64, radiosonde ascents at Antofagasta and Lima, respectively. Terrain topography from the Pacific Ocean to the Andes is shown to the right with a 40-fold vertical exaggeration.

Somewhat similar conditions are encountered at the coast of Southwest Africa, where the inversion touches the continental escarpment, separating the cool and overcast, albeit rainless environment of the coastal desert from the warm, dry, clear-sky conditions of the arid interior. Likewise, the inversion has profound environmental consequences at the flat coasts of Northwest Africa. The lower atmosphere is cool and humid and extensive stratiform cloudiness is common, but there is little rainfall. Rauh (1952, pp. 20–22, 36–47) has described the way in which the vegetation adapts to this peculiar atmospheric environment. The succulent thorny Euphorbia minimizes evaporative losses by its morphological design which is similar to the Cactus of the New World; dew formation being an essential factor. The Argania trees with their large hard leaves have a way of channeling water from dew formation down to the trunk and root region. Rainfall as such is of subordinate importance in the maintenance of these life forms.

Finally, the trade inversion merits attention for its potentialities as an elevated duct for transoceanic radio transmission. Katzin et al. (1960) discuss the engineering implications of lower-tropospheric inversions over the Arabian Sea and the South Atlantic, and recent studies in Chile (P. Aceituno, personal communication, 1983) likewise consider the radio communication potential of the trade inversion over the Eastern South Pacific.

6.6. Mid-Tropospheric Inversions

Less studied than the trade inversion are stable layers in the mid-troposphere at the lower latitudes. Taljaard (1953) calls attention to the tendency for inversions at 3—4 km over Southern Africa, especially during Southern winter, and relates them to large-scale subsidence processes. Over Northern hemispheric Africa mid-tropospheric inversions near 4 km are likewise common (Carlson and Prospero, 1972; Prospero and Carlson, 1972). These seem to have implications for the dust transport from the Sahara to the adjacent Atlantic (Morales, 1979), in that dust particles are impeded from being carried to higher layers and are thus trapped below the mid-tropospheric inversion. In fact, Prospero and Carlson (1972) point out that the dust of Saharan origin is concentrated in the layer comprised between the trade inversion below and the mid-tropospheric inversion above. Recent studies have been concerned with the size distribution (d'Almeida and Schütz, 1983), transport and removal processes (Lee, 1983), and the radiative effects (Guedalia et al., 1984) of Saharan dust. However, little is found in the current literature on the annual variation, spatial pattern, and origin of the mid-tropospheric inversion over Africa.

6.7. Equatorial Trough Zone

Enclosed between the subtropical high pressure belts of the two hemispheres, a zone of low pressure extends continuously around the globe near the Equator (Fig. 6.1:2). Within this trough of low pressure, the trade wind airstreams from either hemisphere meet, and ascending motion and copious precipitation prevail. In the planetary perspective, the equatorial trough zone constitutes the equatorward, ascending portions of the Hadley mean meridional circulation cells of both hemispheres (Section 3.3). The energetics are characterized by an import of water vapor concentrated in the trade wind layer, and an export of geopotential energy and sensible heat in the upper troposphere, resulting in a net atmospheric heat export from the equatorial trough zone to the higher latitudes (Sections 5.4 to 5.7). Regarding the regional climate, abundant weather and rainfall are characteristic. The regional circulation and climate features are the main objectives in the remainder of this section.

6.7.1. THE LARGE-SCALE SETTING

The circumglobal maps Figs. 6.1:1 to 6.1:6 offer a coarse picture of equatorial trough zone circulation. Very broadly, the zone of low surface pressure coincides with a band of highest surface temperature. Both are closest to the Equator in Northern winter, but shift well into the Northern hemisphere during Northern summer, attaining a particularly far poleward location in the sector of the monsoon regions from Western Africa to Southern Asia. The coincidence in annual march and spatial patterns of the surface pressure and temperature fields suggests that the equatorial low pressure trough is primarily thermally induced. The 'heat low' nature of the equatorial trough is furthermore indicated by the thickness pattern, in that the trough tends to weaken from the surface towards the higher layers, in part giving way to contour maxima. This tendency is, for example, implied by the flow pattern in the meridional-vertical cross-section Fig. 3.2:1 and by the comparison of the gradient-wind level flow (Figs. 6.1:3b and 6.1:4) with the surface pressure pattern (Figs. 6.1:2, parts a and b). Without attention to spatial detail, Fig. 3.3:1 illustrates the convergence and ascending motion resulting from the trade wind flows of both hemispheres meeting in the equatorial trough zone. Furthermore, in a coarse spatial resolution consistent with the maps Figs. 6.1:1 to 6.1:4, Figs. 6.1:5 and 6.1:6 depict the patterns of cloudiness and rainfall. Abundant cloud cover and precipitation are apparent for the equatorial trough zone at large.

6.7.2. STRUCTURE OF THE INTERTROPICAL CONVERGENCE ZONE

The coarse spatial resolution of analyses such as presented in Figs. 6.1:1 to 6.1:6 and Section 6.7.1 may leave the impression that the extrema in the meridional profiles of surface pressure and temperature, convergence, cloudiness and rainfall, and the axis of confluence between the airstreams of Northern and Southern hemispheric origin all coincide. Early computer simulations (e.g. Pike, 1972), raise no conflict with such notions. Thus coarse-resolution empirical analyses together with numerical model experiments (see review in Ramage, 1974) seem to have led to the widely held view that the surface pressure trough, the surface wind discontinuity, and the belts of maximum sea surface temperature, convergence, cloudiness and precipitation all coincide in what is commonly referred to as the Intertropical Convergence Zone, ITCZ (see review in Ramage, 1974). To some, the ITCZ is the locus of most frequent travelling disturbances (review in Lockwood, 1974, pp. 89–93). The time is ripe to abandon these outmoded notions. In the following, we will consider the conceptual difficulties inherent in the aforementioned views and then discuss the observational evidence on the structure of the 'Intertropical Convergence Zone'.

It was pointed out in Section 6.7.1 that the near-equatorial low pressure trough over both continents and oceans has a heat low structure and appears above all thermally induced. Maintenance of the trough throughout the season would then require a heating mechanism to keep the surface temperature maximum at the indicated latitude. If indeed the strongest convection and cloudiness is to stay over the surface temperature maximum, as the widely held concept implies, it is difficult to visualize how this can be maintained through solar radiation. We shall return to this issue after a review of the observational evidence.

6.7.2.1. *Atlantic and Pacific Oceans*

Among the earliest systematic observations of the Intertropical Convergence Zone are the aircraft traverses over the Eastern Equatorial Pacific during World War II (Alpert, 1945, 1946a, b, c; Simpson, 1947). Based on observations on the synoptic time scale, Ramage (1974) for the Indian Ocean and Sadler (1975a) for the Atlantic suggest that the convergence-cloudiness-rainfall maximum is separate from the zone of minimum surface pressure and maximum sea surface temperature. Strongest convection well to the South of the surface confluence is also reported by Frank (1983). Sadler (1975a) contends that "the mean monthly maximum cloud zone does not coincide with the tracks of migratory vortices", and that "synoptic-scale circulations are not important in producing the maximum cloud zones".

These synoptic-scale investigations are on the climatic time scale corroborated by a fine-resolution analysis (one degree square areas, 1911–70 mean) of long-term ship observations in the tropical Atlantic, Eastern Pacific and Indian Oceans (Hastenrath and Lamb, 1977a, b, 1978a, b, 1979; Hastenrath 1977a). This fine-resolution analysis reveals features that have escaped the coarse mesh (two or five degree squares) of other charts (Koninklijk Nederlands Meteorologisch Instituut, 1918–31; U.S. Weather Bureau 1938, 1952; Mintz and Dean, 1952; U.S. Navy, 1974, 1976, 1977, 1978, 1979; Atkinson and Sadler, 1970; Romanov and Romanova, 1976), but which appear essential for the understanding of low-latitude circulations. Still pertinent are Flohn's (1957) early evaluations of ship observations in the equatorial Atlantic along the major shipping route from Europe to South America.

The aforementioned analyses of Hastenrath and Lamb (1977a, b, 1978a, b) for the tropical Atlantic and Eastern Pacific are in part summarized in Fig. 6.7.2.1:1 for the height of the Northern summer. A flat surface pressure trough (Fig. 6.7.2.1:1a) broadly coincides with a wide band of highest sea surface temperature (Fig. 6:7.2.1:1e). Embedded in this zone of low pressure and high sea surface temperature is a band of convergence (Fig. 6.7.2.1:1d) and an axis of confluence between the airstreams originating in the two hemispheres (Fig. 6.7.2.1:1b). Directional steadiness of wind (Fig. 6.7.2.1:1c) is high within the core of the aforementioned airstreams, and smallest in the vicinity of the confluence axis. Prominent features not revealed by earlier maps are as follows: a zonally oriented speed maximum extends over the eastern part of both oceans, near the latitude where the cross-equatorial flow recurves from Southeast to Southwest; this broadly separates a band of distinct divergence immediately to the North of the Equator from the convergence belt further poleward; strongest convergence is situated well to the South of the surface confluence axis, and within a region of high directional steadiness of wind. The latitudinal arrangement of near-equatorial circulation, weather, and oceanic features during July—August is illustrated in more detail in the meridional profiles in Figs. 6.7.2.1:2 and 6.7.2.1:3.

In the Eastern Atlantic, Fig. 6.7.2.1:2c, a band of low sea surface temperature is prominent immediately to the South of the Equator; a flat sea surface temperature maximum extends between 5 and 14°N. This approximately underlies a broad trough of low pressure. The resultant wind discontinuity essentially coincides with the axis of minimum pressure. A maximum of both resultant and scalar wind speed is apparent around 4–5°N. A speed minimum is found near the wind discontinuity. Comparison of resultant and scalar wind speed reveals high directional steadiness within most of the clockwise turning cross-equatorial monsoon flow as well as in the upstream portion of the Northeast trades. The vicinity of the wind discontinuity is characterized by low steadiness values. Fig. 6.7.2.1:2, part c, also identifies a band of pronounced divergence between about 1°S and 5°N, with a wider belt of convergence lying immediately to the North. The convergence maximum stays 3–4 degrees of latitude to the South of the wind discontinuity. The maximum of precipitation frequency is displaced slightly to the North of − or downstream from − the convergence maximum. The cloudiness maximum also extends further North than the convergence and rainfall maxima.

For the Eastern Equatorial Pacific, Fig. 6.7.2.1:2, part a, illustrate conditions similar to the Eastern Atlantic, the major differences being as follows: sea surface temperature increases to the northernmost portion of the transect; the maxima of precipitation frequency, and even more so of cloudiness, are displaced to the South of the convergence maximum, which in turn stays well to the South of the wind discontinuity as in the Atlantic. The southward displacement of the largest cloudiness from the convergence maximum may result from the clockwise-turning flow from the Southern hemisphere being overlain by northeasterlies, which would tend to spread cloud tops southward. The clockwise turning cross-equatorial flow over the Eastern Pacific extends only to about 1 km at the Equator (Hastenrath, 1977a). The cross-equatorial flow over the Eastern Atlantic is less shallow (Hastenrath and Lamb, 1977b).

Fig. 6.7.2.1:1 and Fig. 6.7.2.1:2, parts c and a, show zonally oriented tongues of cold water immediately to the South of the Atlantic and Pacific Equator. Surface waters are coldest at the eastern extremity of either ocean, and meridional temperature gradients are strongest across the Equator. The sea surface temperature signature becomes most pronounced at the height of the Northern summer. Various factors seem conducive to the maintenance of these temperature pattern characteristics.

(i) Wind-driven ocean currents advect cold surface waters from off the West coasts of the Southern hemisphere continents.

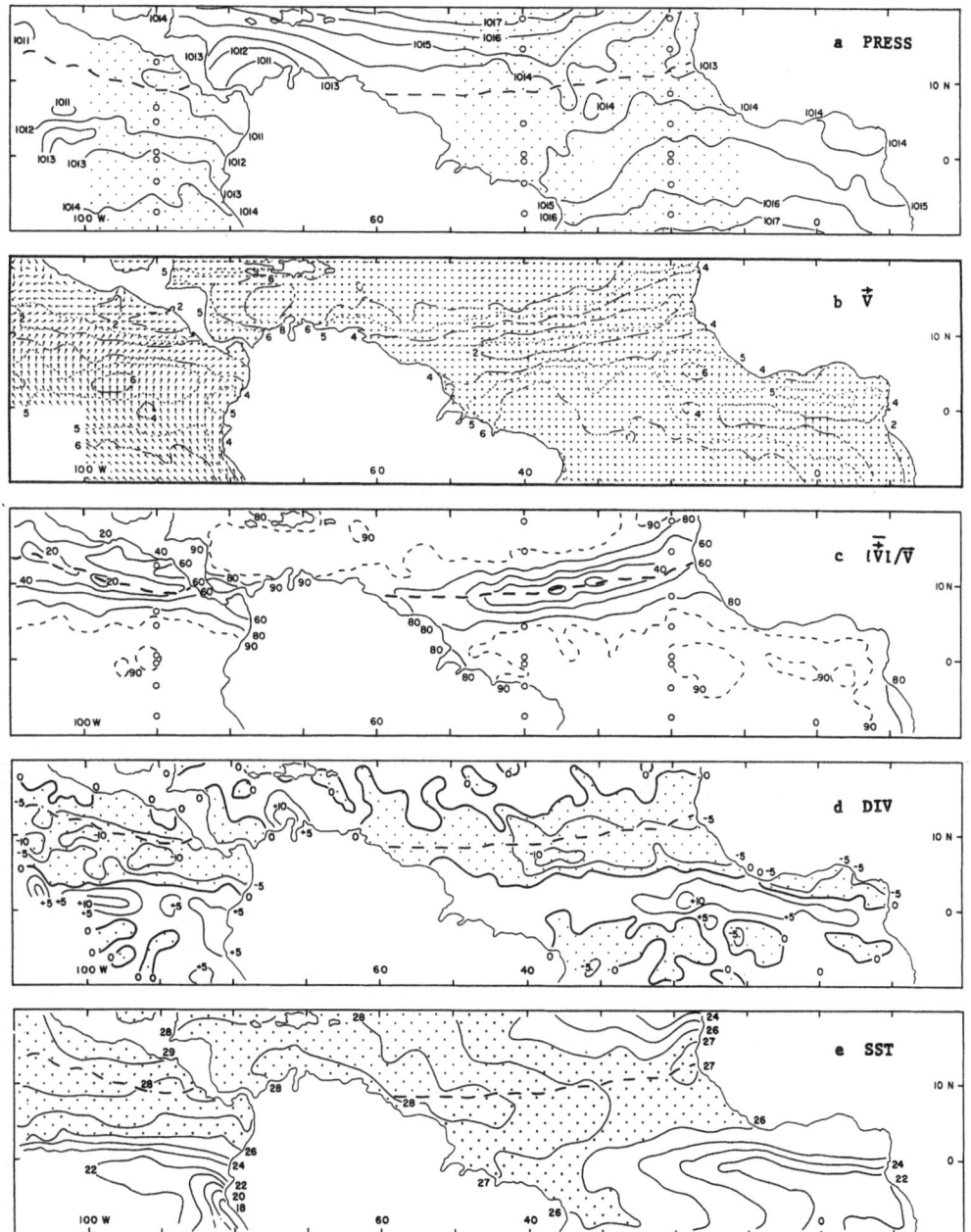

Fig. 6.7.2.1:1. Surface circulation over the Equatorial Atlantic and Eastern Pacific, July–August, 1911–70. (a) pressure PRESS (mb); ocean areas represented in Figs. 6.7.2.1:2 to 6.7.2.1:4 are indicated by dot raster, and latitudes of vector diagrams in Figs. 6.7.3:1 to 6.7.3:3 by open circles; (b) resultant wind direction and speed **V** (m s⁻¹); (c) steadiness of wind $|\mathbf{V}|/\overline{V}$ (percent), latitudes of vector diagrams in Figs. 6.7.3:1 to 6.7.3:3 are marked by open circles; (d) divergence DIV (10^{-6} s⁻¹), with convergence shaded; (e) sea surface temperature SST (°C), with areas above 26°C shaded. Position of resultant confluence axis is shown as heavy broken line. From Hastenrath and Lamb (1978a).

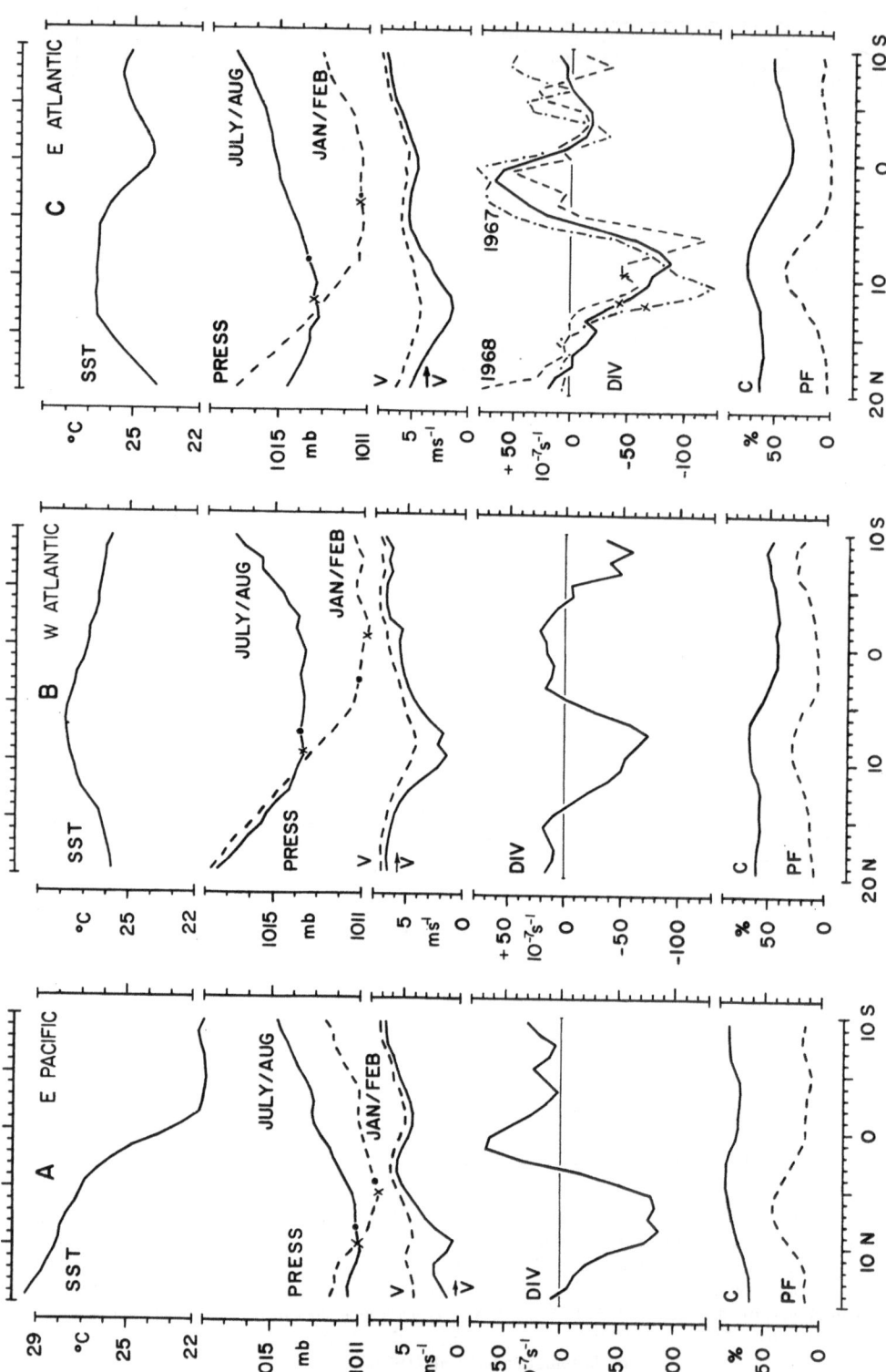

Fig. 6.7.2.1:2. Meridional profiles of selected elements in July–August 1911–70, in three equatorial ocean areas: (a) Eastern Pacific, 80–100°W; (b) Western Atlantic, 30–50°W; (c) Eastern Atlantic 10–30°W. Sea surface temperature, SST; sea level pressure PRESS, with January–February values plotted as broken line, and position of surface confluence axis and maximum convergence indicated by crosses and dots, respectively; resultant V, solid, and scalar wind speed, V, broken line; divergence, DIV, with values for July–August of 1968 (dry and 1967 (wet) entered as broken and dash-dotted lines; total cloudiness C, solid, and precipitation frequency PF, broken line. From Hastenrath and Lamb (1978a).

(ii) The negative vorticity of the surface wind field in the realm of the clockwise turning cross-equatorial air flow, and the associated negative wind stress curl, favor oceanic upwelling in the Southern hemisphere, with particularly large magnitudes in the equatorial zone, in accordance with Eq. (4.1:7); this mechanism is discussed by Yoshida and Mao (1957) and Hantel (1972).

(iii) In the domain of the clockwise turning cross-equatorial surface air flow over the Eastern Equatorial Atlantic and Pacific, the Ekman mass transport according to Eq. (4.1:5) would in the Southern hemisphere be directed away from the Equator and involve upwelling to the South of it, while the recurvature of surface winds from southeasterly to southwesterly near 5°N would favor confluence of Ekman transport and downwelling around 5°N, as discussed in Section 4.3.4.

(iv) It should be noted that westward surface wind stress would according to Eq. (4.1:5) produce meridional Ekman transport directed away from the Equator in either hemisphere and therefore largest upwelling at the Equator, a pattern not consistent with the marked hemispheric asymmetry of the sea surface temperature field.

(v) South to North directed cross-equatorial air flow implies a substantial northward surface wind stress, which according to Eq. (4.1.:5) would be associated with westward/eastward Ekman mass transport in the Southern/Northern hemisphere. A meridionally oriented continental boundary to the East would require upwelling/downwelling to the South/North of the Equator, effects conceivably being most pronounced near the coast and near the Equator; this mechanism is inherent in the model experiments of Philander and Pacanowski (1980, 1981).

Further pertinent are the papers by Schopf and Cane (1983) and Cane and Sarachik (1983). Note that mechanisms (i), (ii), (iii), (v), would contribute in the proper sense and be most effective during the Northern summer. However, candidate (i) would not explain the steep meridional temperature gradient across the Equator; process (ii) would not produce strongest effects in the eastern part of the oceans; while mechanisms (iii) and (v) appear consistent with both of these features in addition to the annual cycle characteristics. Whatever the causes of the aforementioned sea surface temperature pattern, this results in a marked hemispheric asymmetry of the oceanic heat budget (Hastenrath, 1977b).

Turning now to the atmospheric-oceanic conditions of the Western Equatorial Atlantic, refer to Fig. 6.7.2.1:2, part b. The flat extrema of sea surface temperature and pressure essentially coincide, and the kinematic axis — here in the form of an asymptotic confluence (Fig. 6.7.2.1:1) — is embedded in the pressure trough. Wind speed and directional steadiness decrease towards the confluence axis, with no development of a near-equatorial speed maximum as in the two foregoing sea areas. Strongest convergence is also located to the South of the kinematic axis, although the latitudinal separation is smaller than over the eastern part of the oceans. As over the Eastern Atlantic, the maximum of precipitation frequency is found slightly to the North of — that is downstream from — the convergence maximum.

In relation to the climatic mean patterns depicted in Figs. 6.7.2.1:1 and 6.7.2.1:2 aircraft traverses in the Central Pacific during November 1977 to January 1978 (Ramage et al., 1981) are of interest. Ramage et al. (1981) note that turbulent fluxes of sensible and latent heat were not necessarily greatest in the zone of strongest convergence, and conclude that the associated sea surface temperature maximum did not regulate the position or intensity of the convergence zone.

Fig. 6.7.2.1:3 compares the latitudes of the confluence axis and the extrema of sea surface temperature, pressure, convergence, precipitation frequency, and cloudiness, during the course of the year. All of these individual axes undergo a marked annual variation in latitude, but of greater interest are the variations in their collective spatial arrangement.

In the Eastern Atlantic, the sea surface temperature and pressure extrema and confluence axis are in close proximity throughout the year. The convergence maximum is located equatorward of

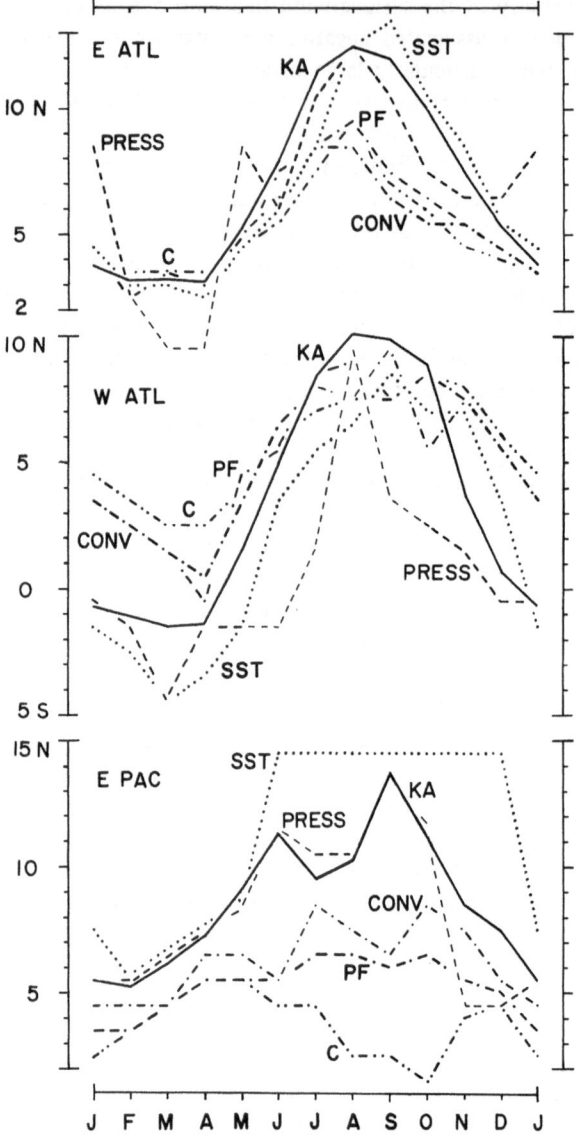

Fig. 6.7.2.1:3. Annual variation in latitude position of resultant confluence axis KA as read from monthly maps, solid; axis of maximum convergence CONV, dash-dotted line; where sufficiently different from convergence maximum, maximum precipitation frequency PF, and maximum total cloudiness C are entered as double dash-dot and double dot-dash lines, respectively; pressure minimum PRESS broken; sea surface temperature maximum SST dotted. Thin signature indicates that maxima are indistinct. Top Eastern Atlantic 10–30°W, middle Western Atlantic 30–50°W, bottom Eastern Pacific 80–100°W. From Hastenrath and Lamb (1978b).

the confluence axis and pressure and sea surface temperature extrema during most of the year, with the latitudinal separation becoming very large in Northern summer. Maxima of precipitation frequency and cloudiness nearly coincide with the convergence maximum during most of the year, but are displaced somewhat to the North in summer and fall.

In the Eastern Pacific, the sea surface temperature maximum stays poleward of the confluence axis, this being most pronounced during the summer half year. The confluence axis and pressure minimum nearly coincide and undergo a double variation in the course of the year. The convergence maximum is found on the equatorward side of the confluence axis, again with a largest latitudinal separation during the summer half-year. The maxima of precipitation frequency and particularly cloudiness tend to stay equatorward of the convergence maximum, with the greatest latitudinal separation occurring during the summer half year.

In the Western Atlantic, the sea surface temperature maximum and pressure minimum tend to stay to the South of the confluence axis throughout most of the year. The convergence maximum, and the maxima of precipitation frequency and cloudiness, are located on the equatorward side of the confluence axis in Northern summer, and on its Northern side in winter.

Figs. 6.7.2.1:1 to 6.7.2.1:3 show that the bands of maximum convergence are largely located in regions of high wind steadiness, namely at values of about 70% over the Eastern Atlantic and 50% in the Western Atlantic, and around 80% over the Eastern Pacific; in the zone of largest cloudiness over the Eastern Pacific directional steadiness of wind is even around 90%. Low steadiness values in the vicinity of the confluence axis are in a region of lesser cloudiness. This detail not revealed by earlier atlas charts is interesting in relation to the aforementioned contention of Sadler (1975a) that the mean monthly maximum cloud zone does not coincide with the tracks of migratory vortices, and that synoptic scale circulations are not important in producing the maximum cloud zone. It is noted, however, that surface observations are analyzed here and the displacement of 'cloud clusters' may reflect conditions in higher layers.

The structure of the Intertropical Convergence Zone is illustrated in the schematic meridional-vertical transects, Fig. 6.7.2.1:4 (parts a to c), for the Equatorial Eastern Atlantic – West African sector (c), the Western Atlantic – Brazil sector (b), and the Eastern Pacific (a). Information on the atmospheric surface layer and the ocean was derived from the aforementioned long-term ship observations. The slope of the large-scale interface between the lower-tropospheric flow originating from the Southern hemisphere and the Northern hemispheric northeasterlies aloft and to the North was drawn from existing radiosonde stations. Regarding the Eastern Atlantic – West African sector the slope in Fig. 6.7.2.1:4, part c, is broadly consistent with values known for the African continent (Leroux, 1970, p. 111, 119). Atmospheric vertical motion was inferred from the lower-tropospheric vergence pattern (Hastenrath and Lamb, 1977a).

Regarding the Eastern Atlantic (Fig. 6.7.2.1:4, part c), the July–August cross-equatorial flow from the Southern hemisphere is markedly divergent from the Equator to near 5°N, in which vicinity it recurves from southeasterly to southwesterly. A band of convergence extends further poleward into the southern fringe of the Northeast trades. However, the maximum convergence occurs about 350 km South of the surface wind discontinuity in a region of high directional steadiness of wind. This suggests that the convergence is primarily associated with the basic flow, rather than migratory disturbances. The maximum of cloudiness may be located somewhat to the North of – or downstream from – the convergence maximum. This may reflect the tendency for a downstream displacement of convective cells in the comparatively deep southwesterly surface flow. The cross-equatorial flow undercuts the Northeast trades, with the latitude of the surface discontinuity and the inclination of the interface increasing toward Northern summer (Leroux, 1970, p. 111, 119).

The cold ocean surface to the South of the Equator is unfavorable for convection. A warmer underlying ocean exists between the Equator and 5°N, but here the marked divergence of the lower-tropospheric flow would mitigate against convection and cloud development. A region of scarce cloudiness near or somewhat to the South of the Equator is commonly apparent in satellite imagery.

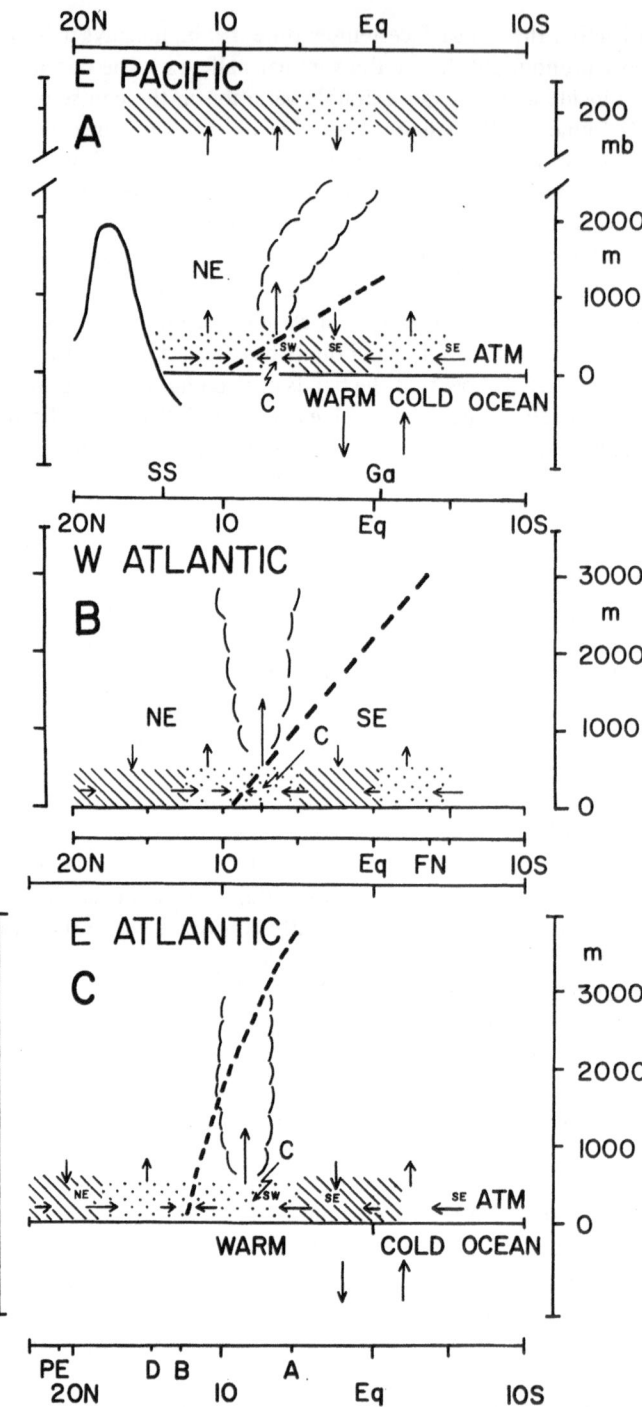

Fig. 6.7.2.1:4. Schematic meridional-vertical transects during July/August. (a) Eastern Pacific, (b) Western Atlantic, (c) Eastern Atlantic. NE, SE, SW are wind directions. Heavy broken line denotes flow discontinuity, hatched area divergence, dot raster convergence, C convergence maximum. Arrows indicate motion in meridional-vertical plane. Station symbols are as follows: PE Port Etienne, D Dakar, B Bamako, A Abidjan; SS San Salvador, Ga Galápagos, FN Fernando de Noronha. Vertical exaggeration is approximately 1:550.

This arrangement of semi-permanent near-equatorial atmospheric and oceanic features differs from the conventional interpretation of the Intertropical Convergence Zone in that the maximum convergence-cloudiness-rainfall stays about 350 km South of the surface confluence axis and pressure minimum and appears essentially associated with the basic flow. The existence of a divergence band to the North of the Equator and a speed maximum at the recurvature latitude near 5°N can only be ascertained through the aforementioned fine-resolution analysis.

Over the Western Atlantic (part (b) of Fig. 6.7.2.1:4), the maximum convergence is located close to the surface confluence axis, but still distinctly to the South of it during the Northern summer when the discontinuity surface is also steepest.

In the Eastern Equatorial Pacific pattern configurations are similar to the Eastern Atlantic (Fig. 6.7.2.1:6, parts a and c). A clockwise turning cross-equatorial airstream of high directional steadiness persists throughout the year. At the height of the Northern summer, this is moderately convergent to the South of the Equator and divergent from there to about 5°N, where it recurves from southeasterly to southwesterly. Maximum convergence is found somewhat further North, but well to the South of the surface wind discontinuity. The surface discontinuity between this current and the Northeast trades is farthest South during Northern winter, moves North until June, then recedes, and reaches its northernmost location in September. The cross-equatorial surface current has a typical vertical extent of only about 1 km near the Equator and the interface between this current and the broad northeasterly flow aloft has a typical slope of 1:1000, with a value as extreme as about 1:1400 at the height of the Northern summer. Highest precipitation frequency is located well to the South of the surface flow discontinuity, associated with the steady and strongly convergent, but very shallow cross-equatorial surface flow (Hastenrath, 1977a). Largest cloudiness is found somewhat equatorward from maximum surface convergence and precipitation frequency, presumably as a consequence of the northeasterly flow aloft.

Characteristically during Northern summer there is a cold ocean surface with upwelling to the South and warm water with downwelling to the North of the Equator, in similarity to the conditions described above for the Eastern Equatorial Atlantic. Both the cold water to the South and the distinct divergence of atmospheric flow immediately to the North of the Equator are not conducive to convection.

For the Eastern Pacific, Sadler's (1975b) streamline-isotach maps allow a qualitative estimate of the vergence pattern at 200 mb. The zonally arranged divergence and convergence bands in the upper troposphere are broadly inverse and thus compensatory to the pattern in the surface layer. This should not be interpreted in terms of closed meridional circulation cells, but merely as reflecting the divergent component of the wind field.

The quasi-permanent circulation features described in Figs. 6.7.2.1:1 to 6.7.2.1:4 seem to extend continuously from the Western Atlantic across Northern South America to the Eastern Pacific (Hastenrath and Guetter, 1978), although documentation for the continent is less detailed than for the oceanic areas that are well covered by long-term ship observations.

Figs. 6.7.2.1:1, 6.7.2.1:2, parts c and a, and 6.7.2.1:4 show the change in flow configuration and structure of the Intertropical Convergence Zone from the eastern to the western side of the Atlantic. Proceeding westward, the confluence axis stays closer to the Equator, the cross-equatorial flow from the Southern hemisphere no longer recurves to the North of the Equator, convergence maximum and confluence axis are less far apart, and the annual march in the latitude position of quasi-permanent circulation features becomes small. A similar longitudinal variation is indicated proceeding from the Eastern Equatorial Pacific, described by Figs. 6.7.2.1:1, 6.7.2.1:2, part a, and 6.7.2.1:3, to the open Central Pacific Ocean.

6.7.2.2. *Indian Ocean*

The surface flow pattern over the Indian Ocean, to be discussed in more detail in Section 6.8, differs from the Atlantic and Pacific in various respects. During November to April flow of Northern hemispheric origin meets the Southern hemisphere Southeast trades in a trough to the South of the Equator, which is broadly embedded within a band of highest sea-surface temperature and lowest pressure. From February to April these features move northward. However, during Northern summer a separate temperature maximum and low pressure trough develop over Southern Asia (Figs. 6.1:1a and 6.1:2a), with surface flow emanating from the South Indian Ocean high, sweeping across the Equator and into the South Asian continent.

The summer monsoon current over the Indian Ocean shows no near-equatorial speed maximum; the areas of highest speed occur over the Arabian Sea and Bay of Bengal. The East African Low Level Jet (Section 6.2.5) has speed maxima on either side of the Equator. In contrast to the Eastern Atlantic and Pacific, a marked zonal component of the pressure gradient is characteristic North of the Equator, with a meridionally oriented ridge near 70°E.

6.7.2.3. *Africa*

The atmospheric circulation over the African continent is to be considered in spatial continuity with conditions over the adjacent Eastern Atlantic, as discussed in Section 6.7.2:1, although data sources for the land and sea areas differ: while a fine-resolution analysis of surface meteorological fields is only possible for the sea areas thanks to the long-term ship observation, the radiosonde network allows a better assessment of the vertical structure over land.

Beyond the overall similarity with the quasi-permanent circulation features over the Eastern Atlantic (Section 6.7.2:1), conditions over the African continent are more extreme in various respects: the heat low characteristics of the equatorial trough are more intense (Figs. 6.1:1 and 6.1:2); the contrast between the properties of air on either side of the flow discontinuity is more pronounced; and the low pressure trough and embedded flow discontinuity undergo a larger latitude displacement in the course of the year, reaching a more poleward position during the Northern summer. These and other features are detailed by the work of British and French meteorologists in part dating back to the colonial era (Hamilton and Archbold, 1945; Walker, 1957; Clackson, 1960; Dettwiller, 1965; Sansom, 1965; Anonymous, 1967; Germain, 1968). With reference to the aforementioned quasi-permanent circulation features, the British school of meteorologists has coined the term 'Intertropical Discontinuity' (ITD) in addition to 'Intertropical Front' (ITF). The French school prefers 'Front Intertropical' (FIT), noting that the discontinuity in the flow field is also associated with marked contrasts in temperature and particularly dew point. Only a selection of this work is reported here.

Fig. 6.7.2.3:1 is a schematic meridional-vertical transect across the discontinuity over West Africa during Northern summer. The slope of the discontinuity is drawn as about 1:100. Various weather zones are distinguished with reference to the surface position of the discontinuity. Well to the North of it the rainless clear-sky desert climate and the northeasterly Harmattan winds prevail (zone A). The vicinity of the surface discontinuity is a region of fair weather with scarce cloudiness (zone B). Several 100 km further to the South, where the layer of moist, cool monsoon airstream is deeper, a broad zone with frequent storms and showers is indicated (Northern part of zone C). Further southward convective cloudiness and storms give way to stratiform cloud decks and more continuous precipitation (southern part of zone C). In this region the discontinuity is no longer marked in flow pattern nor air properties. Finally, some 1000 km to the South of the surface discontinuity, skies are clear and rain scarce (zone D).

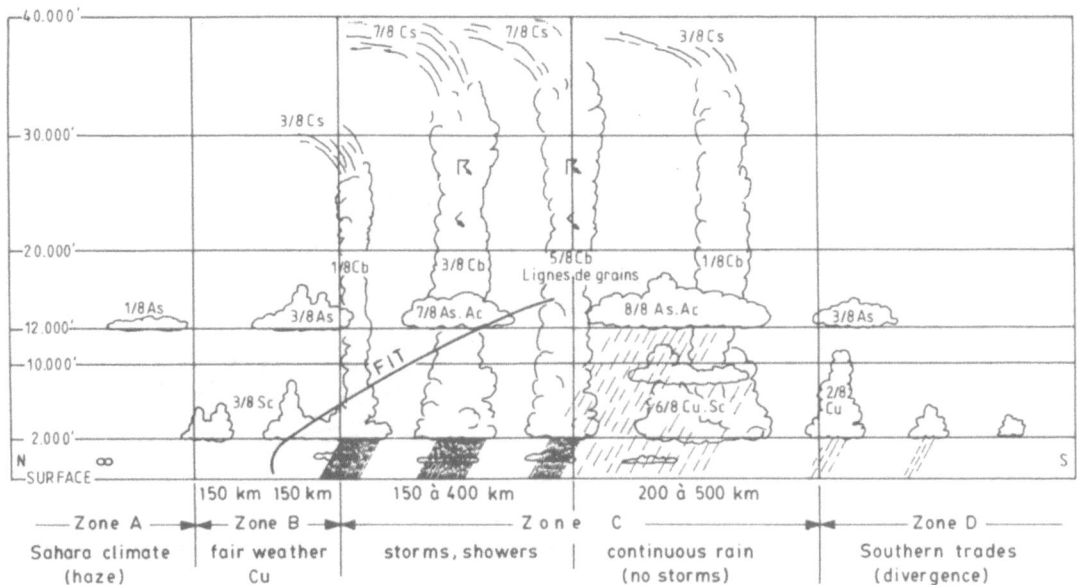

Fig. 6.7.2.3:1. Meridional-vertical transect across the Intertropical Convergence Zone over West Africa along about 0° in Northern summer. From Germain (1968).

The large variation of rainfall with latitude and time of year is illustrated in the topochronopleth diagram Fig. 6.7.2.3:2. For the height of the Northern summer in particular, pronounced latitude gradients are apparent consistent with the schematic meridional-vertical transect Fig. 6.7.2.3:1: rainfall is near to nil to the North of the surface discontinuity, increases to a maximum several 100 km to the South of it, followed by a marked decrease further South towards the Gulf of Guinea coast. A similar topochronopleth diagram for the dew point temperature, Fig. 6.7.2.3:3, shows the steep gradient in atmospheric humidity across the average surface position of the discontinuity.

The time-latitude plot in Fig. 6.7.2.3:4, while a product of the British school of meteorologists in West Africa, is in all major respects consistent with the meridional-vertical transect Fig. 6.7.2.3:1, and the topochronopleth diagrams Figs. 6.7.2.3:2 and 6.7.2.3:3, stemming from French authors. Thus Fig. 6.7.2.3:4 also depicts most abundant rain well to the South of the surface discontinuity, with dry conditions at the Gulf of Guinea coast during Northern summer, and a shift of climatic zones with season. A zonation similar to Walker's (1957) as shown in Fig. 6.7.2.3:4 was earlier proposed by Hamilton and Archbold (1945).

The latitude displacements of the surface discontinuity over West Africa in the course of the year are depicted in the map Fig. 6.7.2.3:5. The range between the extreme seasons is nearly 15 degrees of latitude, with a northernmost location near 20°N in August. In general, the southward shift from Northern summer to winter is faster than the more gradual northward displacement from winter to summer. Concomitant with the seasonal latitude displacements are changes in the vertical structure of the discontinuity, the interface having a gentler slope with the more poleward position during Northern summer, and a steeper slope in winter, as illustrated in Fig. 6.7.2.3:6. This shows a slope of about 1:400 in July and 1:200 in January.

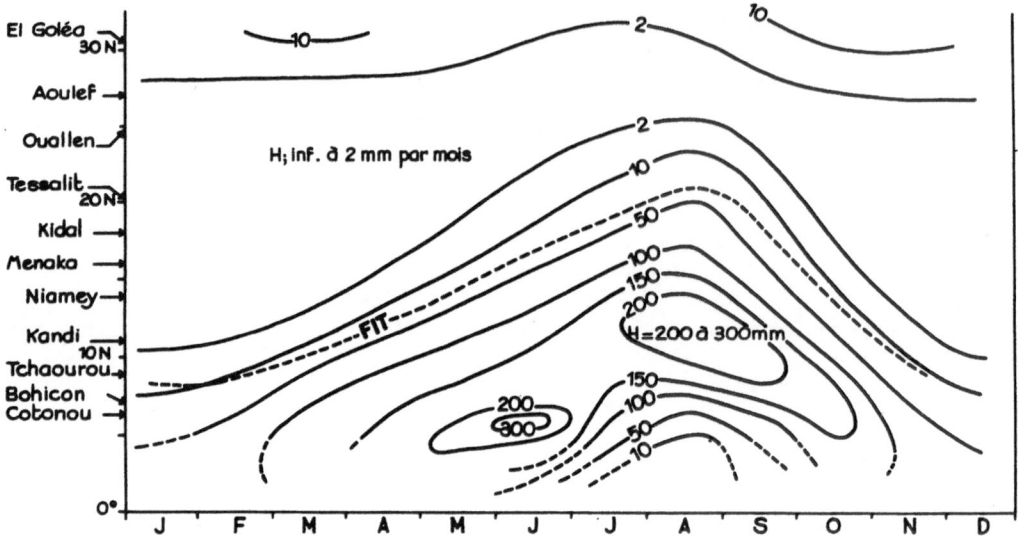

Fig. 6.7.2.3:2. Variation of monthly rainfall totals (in mm) with latitude and time of year along about 0–5°E longitude. Dashed line denotes surface position of discontinuity. From Dettwiller (1965).

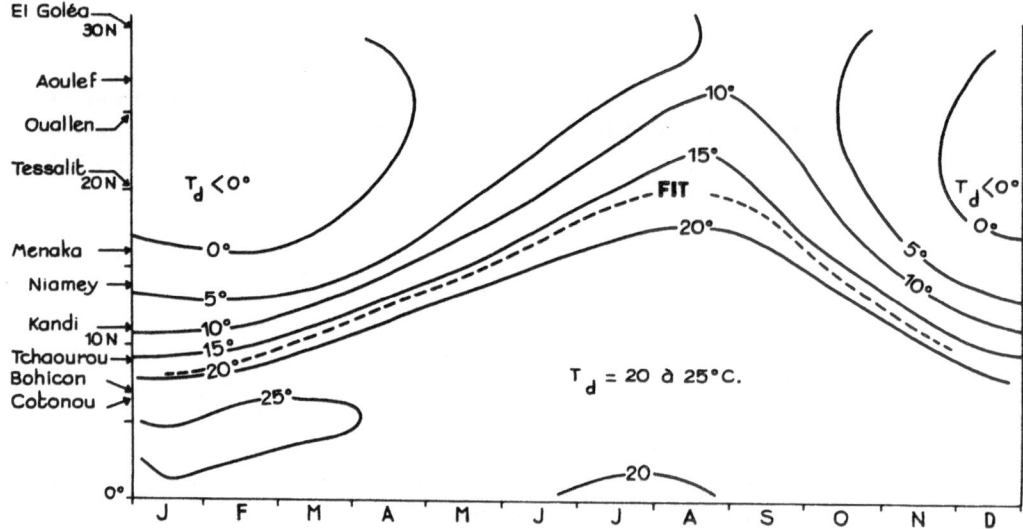

Fig. 6.7.2.3:3. Variation of monthly mean dew point (in °C) with latitude and time of year along about 0–5°E longitude. Dashed line denotes surface position of discontinuity. From Dettwiller (1965).

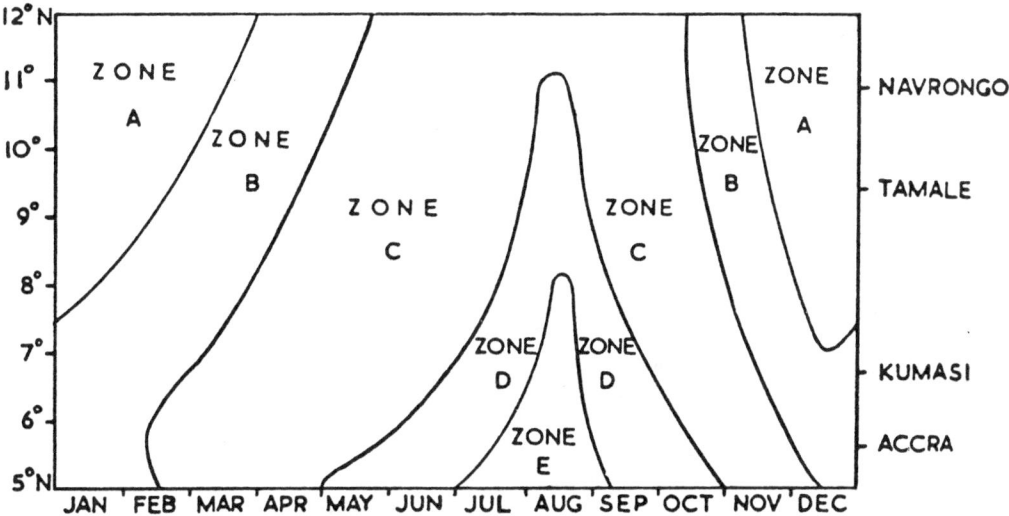

Fig. 6.7.2.3:4. Weather zones associated with the Intertropical Convergence Zone. (A) rainless, clear-sky, hot Northeasterlies; (B) mainly dry, some clouds, Southwesterly at night and Easterly during day; meridional extent about 300 km; (C) disturbance lines, thunderstorms, rain, Southwesterlies, meridional extent about 800 km; (D) rain more continuous and less intense than in Zone C, mostly cloudy to overcast, southwesterly, meridional extent about 300 km; (E) dry, cool, cloudy, southwesterlies, only affecting coastal strip in July–August. From Walker (1957).

Difficulties in comparing analyses of the discontinuity are related to the choice of flow versus thermodynamic criteria and diurnal variations, issues in part discussed in Anonymous (1967, pp. 6–8) and Leroux (1970, pp. 69–78, 118–122). In addition to the more extensive work on West Africa, analyses of the surface discontinuity over the Sudan have been presented by Bhalotra (1963), Osman and Hastenrath (1969), and Hastenrath et al. (1979).

Various interpretation attempts aside, the major merit of the early British and French schools lies in their broad consensus regarding the observational evidence on the quasi-permanent circulation features over West Africa. The spatial continuity to the adjacent Eastern Atlantic suggests common circulation mechanisms, major characteristics being as follows: (i) the surface discontinuity is a region of clear skies over the African continent as over the adjacent Atlantic; (ii) the most abundant cloudiness and rainfall over the continent is found well to the South of the surface flow discontinuity just as the convergence band over the ocean; (iii) further to the South, cloudiness and rainfall over land are less, in similarity to the near-equatorial band of divergence over the ocean. These corollaries suggest that the band-like arrangement of climatic zones in West Africa may at least in part result from kinematic causalities similar to those operative over the adjacent Atlantic Ocean.

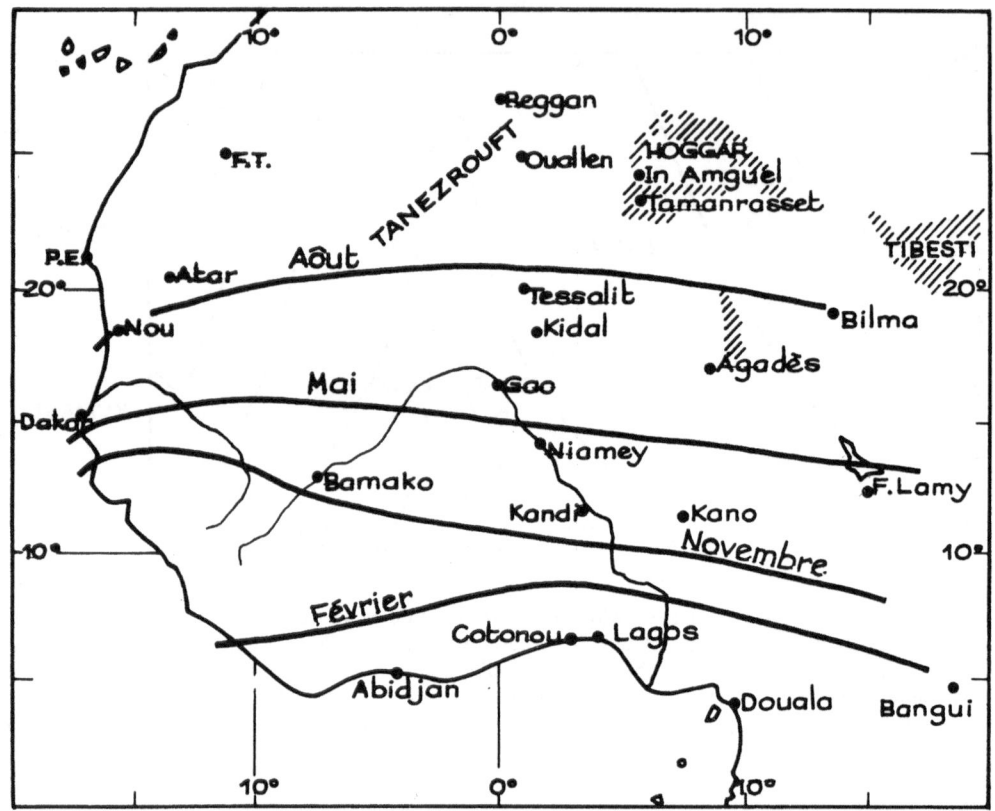

Fig. 6.7.2.3:5. Mean surface position of discontinuity over West Africa in February, May, August, and November. From Dettwiller (1965).

6.7.2.4. *On Atmospheric and Oceanic Controls*

At the beginning of this Section 6.7.2 the contention was made that the equatorial low pressure
trough is largely thermally induced, and that a surface temperature maximum could not well
be maintained by solar radiation under a maximum of cloud cover. However, the analysis of
long-term observations discussed in Sections 6.7.2:1 and 6.7.2:3 shows that, especially in the
longitudes and seasons of a strongly developed and far poleward located Intertropical Convergence
Zone, the convergence-cloudiness-rainfall maximum stays well apart from the belts of highest
surface temperature and lowest pressure and the confluence in the surface wind field. The spread
between convergence maximum and confluence axis during Northern summer amounts to hundreds
of km over the Eastern Equatorial Pacific and Atlantic, whereas it is of the order of thousand km
over the African continent. Thus, a heat-low type mechanism of maintaining surface temperature
maximum and pressure minimum is allowed to operate under the comparatively clear sky, the
convergence-cloudiness-rainfall maximum being located well to the South. In the oceanic regions,
this arrangement of maintaining a surface temperature maximum may further be aided through
cooperation by the upper ocean. Thus, in the Atlantic a particularly shallow oceanic mixed layer is
found in the vicinity of the wind confluence (Lamb, 1983; Hastenrath and Merle, 1986), presumably in response to the surface stress pattern. Radiative heat input at the surface is then distributed

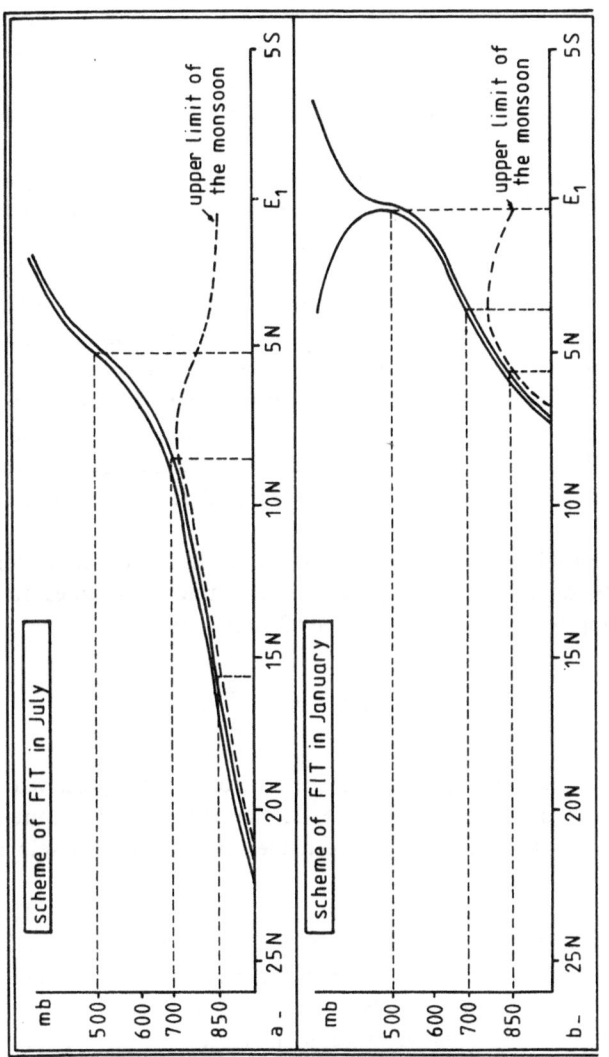

Fig. 6.7.2.3:6. Schematic meridional-vertical transect across discontinuity in January and July. From Leroux (1970).

only over a comparatively thin oceanic layer, thus favoring the maintenance of a warm sea surface.

Another controversy raised at the beginning of this Section 6.7.2 should be reconsidered. The view that the Intertropical Convergence Zone is "the locus of cloud clusters associated with a train of westward propagating wave disturbances" (Lockwood, 1974, pp. 89—93; see also Holton *et al.*, 1971), contrasts with the aforementioned findings that the maximum surface convergence is located in a region of high directional steadiness of wind and can thus not be ascribed to the cumulative effect of travelling disturbances. The schematic meridional-vertical transects in Fig. 6.7.2.1:4 illustrate that clouds piercing from the surface layer to above the large-scale interface encounter a vastly different wind regime aloft. The behavior of satellite-observed clouds may then in large part reflect conditions in higher layers. Such considerations may allow reconciliation of the aforementioned contrasting views on the role of basic flow versus disturbances in the development of the convergence and cloud bands.

Turning to another issue, a particularly intriguing feature of the cross-equatorial surface flow over the eastern part of the Pacific and Atlantic Oceans is the great directional steadiness of wind combined with the remarkable shallowness of the airstream, traits that appear characteristic of the heat-low nature of the equatorial trough. It is recalled from the above discussion that the cross-equatorial flow from the Southern hemisphere as a rule undercuts the Northern hemisphere Northeast trades. The interface between these two airstreams varies in inclination both with longitude and in the course of the year, the slope being gentlest at the height of the Northern summer when the surface confluence in the Northern hemisphere is farthest away from the Equator and when contrasts in the thermodynamic properties of the airstreams tend to be most pronounced. The slope of 1:1400 documented for the interface over the Eastern Equatorial Pacific during Northern summer seems to be the most extreme reported for any large-scale near-surface discontinuity on Earth.

Accordingly, some considerations appear in order as to the factors controlling the slope of the interface between cross-equatorial flow and the Northeast trades. Margules' formula for the slope of an interface can in approximate form be written (Belinskii, 1961, p. 337)

$$\tan \alpha = \frac{2\Omega \sin \phi}{g} \; T \; \frac{\Delta v}{\Delta T} \approx 4 \times 10^{-3} \sin \phi \; \frac{\Delta v}{\Delta T}, \qquad (6.7.2.1:1)$$

where α is the angle between the interface and the horizontal, Ω angular velocity of the Earth's rotation, g acceleration of gravity, T temperature, ΔT the (virtual) temperature contrast across the interface (in °C), and Δv the difference between the interface-parallel velocity component on either side of the interface (in m s^{-1}).

Only a preliminary and qualitative interpretation of Eq. (6.7.2.1:1) is offered here. Although the smallest tan α is observed at the height of Northern summer when the interface is farthest away from the Equator, this is not due to the latitude position as such, which would in fact have the contrary effect. The contrast in v may be rather larger in Northern summer than winter, and would thus not account for the annual variation in interface slope, but the annual variation of Δv (in m s^{-1}) in relative terms seems small compared to the relative changes of ΔT (in °C). The largest (virtual) temperature contrasts ΔT at the height of the Northern summer (cool cross-equatorial airstream from the Southern hemisphere versus warm Northern hemisphere trades), appears as the major factor for the particularly gentle slope of the interface at that time of year. Similar considerations related to Eq. (6.7.2.1:1) are in order concerning the variation of interface slope with longitude.

6.7.3. DYNAMICS OF CROSS-EQUATORIAL FLOW

A limited and preliminary evaluation of observed climatic mean (July–August 1911–70) cross-equatorial flow over the oceans is presented here, largely following Hastenrath and Lamb (1978a).

The equations of motion for large-scale (horizontal) climatic mean (steady-state) flow can be written in component form as

$$\frac{d\overline{u}}{dt} = \overline{u}\,\frac{\partial \overline{u}}{\partial x} + \overline{v}\,\frac{\partial \overline{u}}{\partial y} + \overline{u'\frac{\partial u'}{\partial x}} + \overline{v'\frac{\partial u'}{\partial y}} = +f\overline{v} - \overline{\alpha}\,\frac{\partial \overline{p}}{\partial x} - \overline{\alpha'\frac{\partial p'}{\partial x}} + \overline{F}_x \qquad (6.7.3{:}1a)$$

$$\frac{d\overline{v}}{dt} = \overline{u}\,\frac{\partial \overline{v}}{\partial x} + \overline{v}\,\frac{\partial \overline{v}}{\partial y} + \overline{u'\frac{\partial v'}{\partial x}} + \overline{v'\frac{\partial v'}{\partial y}} = -f\overline{u} - \overline{\alpha}\,\frac{\partial \overline{p}}{\partial y} - \overline{\alpha'\frac{\partial p'}{\partial y}} + \overline{F}_y \qquad (6.7.3{:}1b)$$

or in vector notation

$$\frac{d\overline{\mathbf{V}}}{dt} = \overline{\mathbf{V}} \cdot \nabla \overline{\mathbf{V}} + \overline{\mathbf{V}' \cdot \nabla \mathbf{V}'} = -f\mathbf{k} \times \overline{\mathbf{V}} - \overline{\alpha}\nabla\overline{p} - \overline{\alpha'\nabla p'} + \overline{\mathbf{F}} \qquad (6.7.3{:}2)$$

Bars denote time-averages, taken here for July–August 1911–70, and primes departures from such averages. Following conventional notation, $\mathbf{V}$, u, v, $\mathbf{F}$, F_x, F_y, are the horizontal wind vector, its eastward and northward components, the frictional force per unit mass, and its components, respectively. α, p, f, and ∇ denote specific volume, pressure, Coriolis parameter and horizontal nabla-operator.

Available observations permit the direct evaluation of the first left-hand and first two right-hand terms of Eq. (6.7.3:2). These terms were evaluated in component form for each one-degree latitude strip of the three areas presented in Figs. 6.7.3:1 to 6.7.3:3, namely the Eastern Atlantic, Eastern Pacific, and Western Atlantic Oceans. Resultant and geostrophic wind vectors were also obtained for each one-degree zone. Derivatives in the zonal direction were computed as finite differences between the means of the eastern and western halves of each one-degree strip. Meridional derivatives relating to a given latitude band were computed as finite differences between average values for the contiguous one-degree strips to the North and South.

The terms in Eq. (6.7.3:2) containing primed quantities could not be evaluated directly. They were assumed to be negligible, which allowed the friction vector $\mathbf{F}$ to be obtained as the residual of Eq. (6.7.3:2). Fig. 6.7.3:1 presents examples of friction vectors obtained using this method, together with the other acceleration vectors and those for the resultant and geostrophic winds, for locations where the directional steadiness of wind exceeds 60%. The vectors for a location with directional steadiness as low as 20% (9.5°N in Eastern Pacific) are included in the insert to part a of Fig. 6.7.3:1 for comparison. In view of the predominant zonal uniformity within each area, the vector diagrams for successive latitude strips essentially represent the conditions along a streamline/trajectory.

Inconsistencies in the sets of acceleration vectors obtained by the above procedure could result from covariance in wind speed fluctuations making the primed terms on the left-hand side of Eqs. (6.7.3:1) and (6.7.3:2) non-zero, and from deficiencies in the temporal and spatial sampling of observations and the analysis procedures. How realistic the residual friction vectors in Fig. 6.7.3:1 are, thus indicates whether the above error sources unduly affected the patterns of forces produced. At locations of low directional steadiness of wind, as exemplified by the diagrams for 9.5°N in part a of Fig. 6.7.3:1, the sets of forces obtained are indeed not reasonable and are not considered further. Meaningful constellations of acceleration vectors were, however, produced for

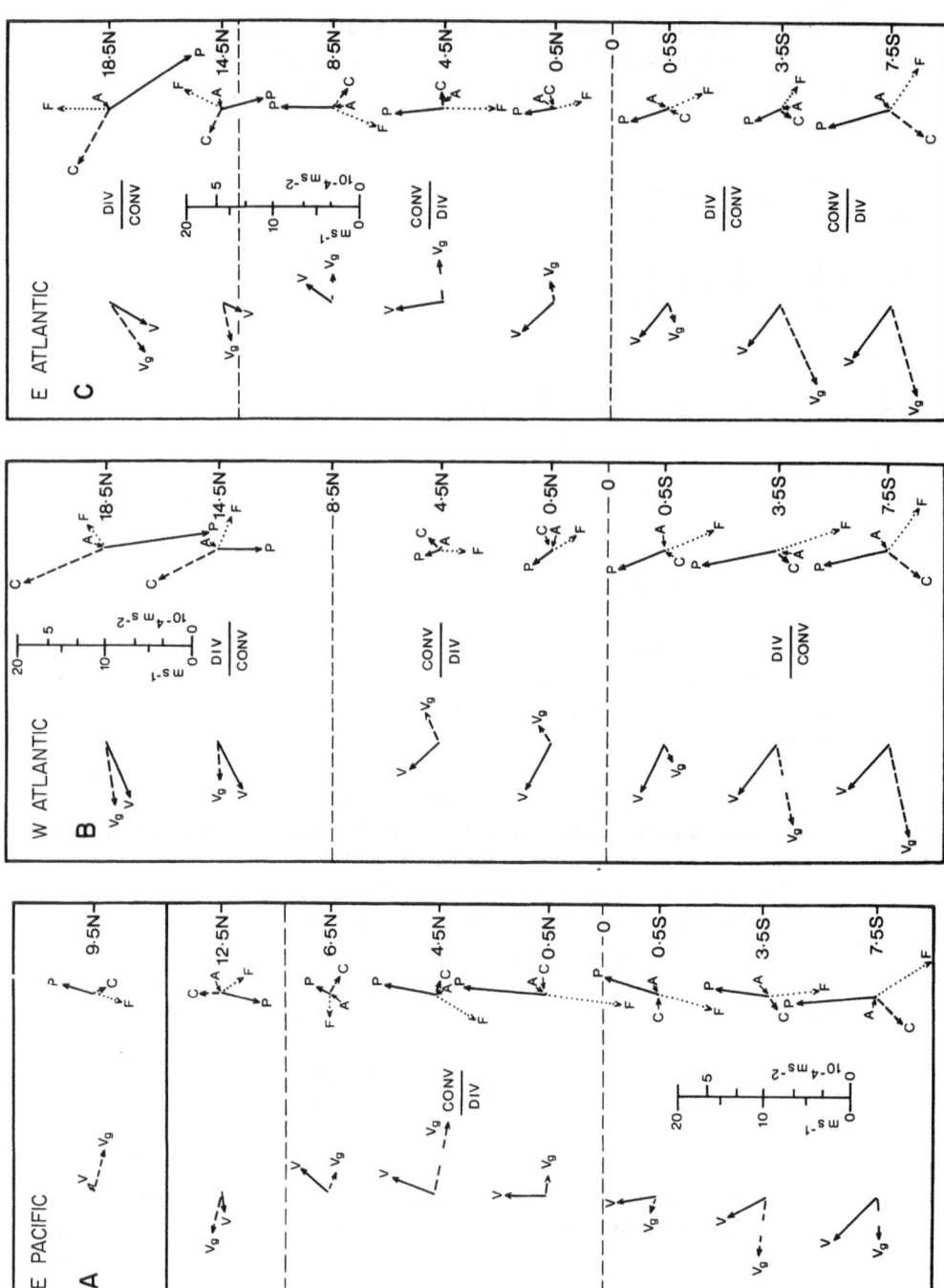

Fig. 6.7.3:1. Dynamics of surface flow, July–August 1911–70, in three equatorial ocean areas: (a) Eastern Pacific, 80–100°W; (b) Western Atlantic, 30–50°W; (c) Eastern Atlantic, 10–30°W. Equator and latitude of resultant confluence axis are schematically entered as broken lines. Left: resultant wind V solid, and geostrophic wind vector V_g broken, see scale of m s^{-1}; very large values of V_g are plotted at one third the regular scale, being denoted by a gap in the middle of the arrow; for 0.5 N and S only the direction of V_g is given. Right: pressure gradient acceleration $P = -\alpha \overline{\nabla p}$ solid, Coriolis acceleration $C = fV \times k$ broken, frictional force per unit mass F dotted, advective acceleration $A = V \cdot \nabla V$, see scale of m s^{-2}. From Hastenrath and Lamb (1978a).

regions of high directional steadiness. As Fig. 6.7.3:1 indicates, the residual friction vectors are in general directed approximately opposite to the resultant wind vector; exact opposition is not even expected from theory. Furthermore, the magnitudes of the friction vectors were checked using two independent comparisons. For an average wind speed at ships' deck level of 7.5 m s^{-1}, Brümmer et al. (1974) found the frictional deceleration of the North Atlantic trades at 10°N during ATEX to be 16.3 $\times$ 10^{-5} m s^{-2}. The frictional coefficient in Guldberg and Mohn's linear formulation $|F| = k|V|$ has been reported to have open ocean values of 1 to 3 $\times$ 10^{-5} s^{-1} (Gordon and Taylor, 1975); Brümmer et al.'s (1974) data would yield a value of 2.3 $\times$ 10^{-5} s^{-1}. The wind speed and residual friction data in the present study yield values of frictional deceleration and k comparable in magnitude to those cited above.

If we assume the primed terms in Eq. (6.7.3:1a) to be negligible, this equation can be used to examine the influence of the pressure field on the recurvature of the Southeast trades. At the latitude where the cross-equatorial flow recurves from southeasterly to southwesterly, Eq. (6.7.3:1a) has the following characteristics: the first left-hand term is zero, and the second left-hand term is positive; the last right-hand term is zero in a first approximation, and the first right-hand term is positive in the Northern and negative in the Southern hemisphere. It follows that recurvature of the mean flow South of the Equator would require the second right-hand term to be positive, implying a pressure decrease towards the East. In contrast, the flow can recurve in the Northern hemisphere with a zero, westward, or very weak eastward directed pressure gradient.

Fig. 6.7.3:1 illustrates the variation of wind and acceleration vectors between significant parts of the cross-equatorial flow over the Eastern part of the equatorial Atlantic and Pacific Oceans. For the Eastern Atlantic, pressure decreases from the Southern hemisphere across the Equator to a minimum at 9–14°N (Fig. 6.7.2.1:2). In the southeasterlies of the Southern hemisphere, frictional energy dissipation not offset by cross-isobaric generation seems to produce a gradual slowing of the current, the direction of which remains essentially unchanged (Figs. 6.7.2.1:1, 6.7.2.1:2, and 6.7.3:1). This is consistent with the modest convergence South of the Equator in Figs. 6.7.2.1:1, 6.7.2.1:2, and 6.7.3:1. Proceeding North from the Equator, the Coriolis force increases and acts in a direction opposite to the Southern hemisphere, individual (or advective) accelerations towards the right of the flow gain importance, and the cross-isobaric angle is enhanced, favoring kinetic energy generation (Fig. 6.7.3:1). As a consequence, the current speeds up and bends to the right, recurving to West of South at about 5°N (Fig. 6.7.2.1:1). Speed increases from the Equator to about this latitude (Figs. 6.7.2.1:1 to 6.7.2.1:2). Romanov and Romanova (1976) noted a similar recurvature latitude, but did not detect the associated speed maximum. The eastward turning of the wind continues with further movement North, reducing the cross-isobaric flow component to the extent that the kinetic energy generation apparently does not offset frictional dissipation (Fig. 6.7.3:1). The current slows down from the recurvature latitude to the flow discontinuity. For the concomitant zonally oriented divergence and convergence bands see Figs. 6.7.2.1:1, 6.7.2.1:2 and 6.7.3:1.

The Northeast trades slow down towards the kinetic discontinuity, presumably also because frictional dissipation is not altogether compensated by generation of kinetic energy through cross-isobaric flow. This leads to convergence on the poleward side of the kinematic discontinuity/ pressure trough (Figs. 6.7.2.1:1, 6.7.2.1:2, and 6.7.3:1). The evaluation of climatic mean conditions over the Eastern Atlantic in Fig. 6.7.3:1, part c, is complemented by Greenhut's (1981) analyses of conditions during GATE in June–September 1974. Greenhut (1981) presents diagrams consistent with Fig. 6.7.3:1, part c, for positions to the North and South of the kinematic discontinuity.

The resultant wind is everywhere weaker than the geostrophic wind; the resultant direction is to the right of the geostrophic wind in the Southern hemisphere and to the left in the Northern hemisphere (Fig. 6.7.3:1). Away from the immediate vicinity of the Equator, the difference between resultant and geostrophic vectors appears due to the substantial frictional deceleration in the Northeast trades and clockwise turning cross-equatorial flow. The ageostrophy close to the Equator is related to the Coriolis acceleration being, by definition, near zero.

Fig. 6.7.3:1, part a, illustrates the latitudinal variation of wind and acceleration vectors in the Eastern Pacific. This reveals flow characteristics and mechanisms comparable to those of the Eastern Atlantic, both for the cross-equatorial current to the South of the kinematic discontinuity/ pressure trough, and the Northeast trades to the North.

Long-term mean monthly charts (Hastenrath and Lamb, 1977a) show, for the Eastern Atlantic and Pacific on a month-to-month basis, a positive relation between the latitude of the surface flow discontinuity, the intensity of the associated convergence belt, and the width and strength of the divergence band immediately to the North of the Equator. The Eastern Atlantic wind and divergence fields for two years with highly contrasting Subsaharan rainy seasons suggest this relationship may also exist on a year-to-year basis. Prominent features of the extreme drought year 1968 as compared to 1967, a wet year in western Subsaharan areas (Lamb, 1978), are (Fig. 6.7.2.1:2) a more equatorward position of the surface wind discontinuity, a southward displacement and weakening of the convergence band, and a weak development of the divergence belt immediately to the North of the Equator.

The foregoing relationship results from the already noted dependence of the divergence/ convergence pattern on the speed maximum, which in turn is dependent on the latitude of the flow discontinuity. If the flow discontinuity and its attendant convergence zone are relatively close to the Equator, then the mechanism offered above to account for the speed maximum would have insufficient space to operate fully. This would restrict the development of the speed maximum, and therefore the divergence belt immediately North of the Equator. The windspeed decrease from the maximum to the discontinuity would also be reduced, which may weaken the convergence band, and place the convergence maximum close to the flow discontinuity rather than further South. In the extreme case of the sixty-year average January and February (Hastenrath and Lamb, 1977a), these processes produce a complete elimination of the speed maximum and concomitant divergence belt, and a coincidence of the convergence maximum and confluence axis (Figs. 6.7.2.1:2 and 6.7.2.1:3). The less extreme comparison of July—August 1967 and 1968 made above, revealed variations in the development of features rather than their elimination (Fig. 6.7.2.1:2).

In remarkable contrast to the Eastern Atlantic and Pacific patterns, the July—August monsoon flow over the Indian Ocean East of 70°E recurves to Southwesterly prior to the Equator being crossed, and does not possess a near-equatorial speed maximum (Koninklijk Nederlands Meteorologisch Instituut, 1952; Hastenrath and Lamb, 1979). The strong eastward directed pressure gradient over the Indian Ocean is fundamental to these flow differences; using Eq. (6.7.3:1a), this was found to be a necessary condition for recurvature South of the Equator. An increase in speed North of the recurvature into the Arabian Sea and Bay of Bengal results from mechanisms similar to those in the Eastern Atlantic and Pacific. However, the continued eastward turning of the wind over the Northern Indian Ocean East of about 70°E does not involve decreased kinetic energy generation, because strong cross-isobaric flow is ensured by the zonally asymmetric pressure configuration. The orientation of isobars over the Western Arabian Sea is unfavorable for cross-isobaric flow, but a speed maximum does occur, abeit well away from the Equator.

Gordon and Taylor (1975) computed trajectories and divergence for the July surface flow over the Indian Ocean, using meridional pressure profiles and winds at an initial latitude given by Koninklijk Nederlands Meteorologisch Instituut (1952). Zonal pressure gradients, however, were neglected. Their trajectories do not recurve South of the Equator, presumably as a consequence of the assumed zonal symmetry of isobars. The associated divergence pattern bears some resemblance to the one observed (Hastenrath and Lamb, 1978b). Over the Western Arabian Sea, where surface speed maxima on either side of the Equator are well developed, the atlas maps (Hastenrath and Lamb, 1979) show convergence bands around 4°S and North of 18°N, the former in similarity with conditions over the Eastern Atlantic and Eastern Pacific.

The flow dynamics in the realm of the asymptotic trade wind confluence of the Western Atlantic are illustrated in Fig. 6.7.3:1, part b. Pressure decreases from the Southern hemisphere across the Equator to a minimum near 8°N (Figs. 6.7.2.1:1 and 6.7.2.1:4), as characterized the other areas considered in Figs. 6.7.2.1:2, 6.7.2.1:3, and 6.7.3:1, part a. The relation between the resultant and geostrophic wind vectors in Fig. 6.7.3:3 is similar to that described for Figs. 6.7.3:1, parts c and a. In similarity to the Eastern Atlantic and Pacific, the Northeast trades slow down as the confluence axis is approached, due to frictional dissipation of kinetic energy not being replaced by cross-isobaric generation. The Southeast trades to the South of the Equator also behave like those of the aforementioned other areas, in that speed decreases downstream conceivably because frictional dissipation exceeds the kinetic energy generation by cross-isobaric flow. The Brazil coast may be a contributing factor here (Fig. 6.7.2.1:1). This speed pattern is consistent with the convergence South of abut 4°S in Figs. 6.7.2.1:1 and 6.7.2.1:4. A band of moderate divergence straddles the Equator in the Western Atlantic (Figs. 6.7.2.1:1 and 6.7.2.1:4), whereas in the Eastern Atlantic and Pacific most of a comparable zone of divergence lies immediately North of the Equator (Figs. 6.7.2.1:1 to 6.7.2.1;3). Figs. 6.7.2.1:1 and 6.7.2.1:4 show little, if any, evidence of the Western Atlantic divergence belt coinciding with a downstream increase in wind speed, unlike the aforementioned other areas. Instead, it seems to result from the flow in the Western Atlantic sector being difficult (Fig. 6.7.2.1:1), which may partly stem from the current in the eastern half of the sector bending to the right with movement from 5°S to 5°N. A reduction in the Coriolis acceleration to the left prior to the Equator, and its subsequent increase to the right, would seem to be involved. The pressure gradient is very slack from the divergence band to the confluence axis, which curtails the generation of kinetic energy, and is manifest in terms of a downstream speed decrease and convergence (Figs. 6.7.2.1:1 and 6.7.2.1:4). The importance of the meridional pressure gradients on either side of the confluence axis for the kinetic energy imbalance and convergence pattern is underlined by the pressure profiles in Fig. 6.7.2.1:4: the convergence maximum switches from the Southern (summer) to Northern (winter) side of the confluence axis, as the Northern meridional pressure gradient becomes the steepest.

For further theoretical treatment of low-level cross-equatorial flow refer to Young (1983) and Stout and Young (1983). The observational data available here limited the empirical study of dynamics to the basic flow. This is unsatisfactory in regions of low wind steadiness, where the perturbation component of the flow is relatively important. However, the zonally oriented convergence and divergence bands within the cross-equatorial flow from the Southern Hemisphere are characterized by very high directional steadiness of wind. The origin of this prominent vergence pattern is at least qualitatively understood in terms of the basic flow. The development of the zones of convergence/divergence is related to the poleward shift of the kinematic discontinuity, both in terms of the annual march and the behavior of individual extreme years.

6.7.4. EQUATORIAL DRY ZONE

As the equatorial trough zone at large is commonly thought of as a region of copious rainfall, the existence of a dry zone along the Equator across much of the Central Pacific is particularly intriguing, all the more as this contrasts with a zonally oriented band of strong convergence and abundant precipitation only a few 100 km to the North. Fig. 6.7.4:1 illustrates the steep meridional climatic gradients in the Equatorial Central Pacific. Annual rainfall increases from less than 500 mm in the vicinity of the Equator to more than 4000 mm in the band of strong convergence centered at about 5°N, and to about 2500 mm in a band to the South of the Equator.

In contrast to the energetics of the equatorial trough zone at large, the atmospheric column in the Equatorial Dry Zone features subsidence, lower-layer export of water vapor and import of geopotential energy and sensible heat (Hastenrath, 1971; see reference list of this paper for further work on the Equatorial Pacific Dry Zone), resulting in a net import of heat within the atmosphere!

In addition to satellite imagery, the cloudiness and rainfall patterns in the Equatorial Central Pacific are well established from numerous island stations. Such platforms are absent for the Equatorial Atlantic, where ship observations (Fig. 6.7.2.1:1) also indicate a belt of strong divergence and scarce cloudiness immediately to the North of the Equator.

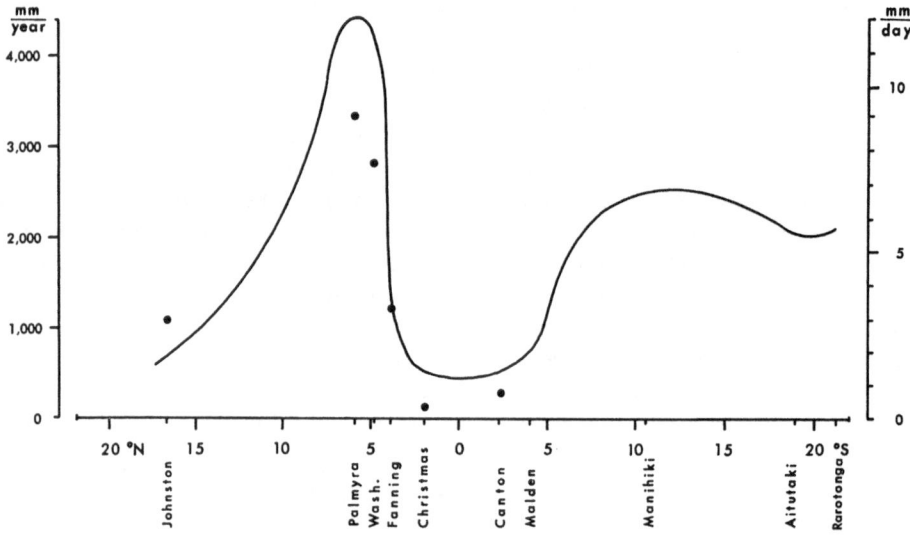

Fig. 6.7.4:1. Annual rainfall totals in the Equatorial Central Pacific in a meridional transect along approximately 160°W, solid line. Average daily rainfall values during the Line Islands Experiment, Spring 1967, are entered as dots. From Hastenrath (1971).

6.8. Monsoons

Annual reversals of lower-tropospheric flow patterns affect a large portion of the global tropics. Various of the quasi-permanent circulation systems discussed in other sections of this chapter are also intrinsic features of the World's monsoon regions.

6.8.1. DEFINITION AND GLOBAL PERSPECTIVE

The term 'monsoon' is traced to an Arabic root meaning 'season'. Das (1986, Chapter 1) presents an extensive account of early awareness of the monsoon phenomenon in the Indian Ocean sector (see also Fein and Stephens, 1986). Since early times the concept has apparently implied a complete reversal of wind regimes in the course of the year, rather than just other more general symtoms of seasonality. In an effort to do justice to the early and accepted meaning, Ramage (1971, pp. 1—6) proposes four simultaneous criteria to delineate the monsoon areas:

(i) the prevailing wind direction shifts by at least 120 degrees between January and July;

(ii) the average frequency of prevailing wind directions in January and July exceeds 40%;

(iii) the mean resultant wind in at least one of the months exceeds 3 m s^{-1};

(iv) fewer than one cyclone-anticyclone alternation occurs every two years in either month in a 5 degree latitude-longitude rectangle.

From these criteria Ramage (1971) delineates the monsoon region of the World in a rectangle bounded by latitudes 35°N and 25°S and longitudes 30°N and 170°E (Fig. 6.8.1:1). The three wind criteria (i) to (iii), which seem most decisive, are met in a zonally oriented band across Subsaharan Africa, in the Northern and Equatorial Indian Ocean and surrounding land areas, and the westernmost Pacific. Much of East Asia, which also satisfies the three wind criteria (i) to (iii), is excluded by the cyclone/anticyclone criterion (iv). Ramage's (1971) delineation of the World's monsoon region is adopted in Rao's (1976, p. 1) and Das' (1986, Chapter 1) authoritative treatises, so it certainly suffices here.

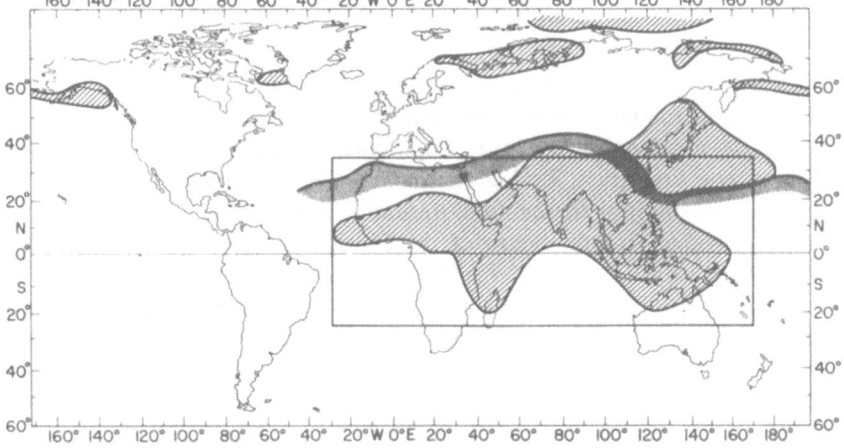

Fig. 6.8.1:1. Delineation of the World's monsoon region. Hatched areas meet simultaneously the wind criteria (i) to (iii), while heavy line marks the Northern limit of the region within the Northern hemisphere where the cyclone/anticyclone criterion (iv) is satisfied. Rectangle encloses the monsoon region. From Ramage (1971).

Over the past decades, various symposium and review volumes have dealt with monsoon problems (Basu *et al.*, eds., 1960; India Meteorological Department, 1965; Krishnamurti, ed., 1977;
Lighthill and Pearce, eds., 1981; Fein and Stephens, eds., 1986). For a global perspective of the
monsoon phenomenon, Fig. 6.8.1:1 along with Figs. 6.1:1 through 6.1:6 are pertinent. The
Northern hemisphere summer monsoon is by far more vigorous than its winter counterpart, except
over East Asia. This circumstance, along with the extent of the monsoon region (Fig. 6.8.1:1),
and the hemispheric asymmetry of land-sea distribution in low latitudes especially in the Africa
— Asia sector, all point to the role of hemispheric asymmetries in surface heating for the monsoon
development. This suggestion is further detailed by Figs. 6.1:1 to 6.1:6. Thus Fig. 6.1:1, part
b, illustrates for July the far northward excursion of the circumglobal band of highest surface
temperature precisely over the Afro — Asian land masses. As discussed in Section 6.1 and consistent
with heat low mechanisms, the circumglobal trough of low pressure (Fig. 6.1:2, part b) attains
its northernmost location together with the surface temperature maximum (Fig. 6.1:1, part b) in
the same longitude domain. The surface pressure gradients are then established for the enormous
lower-tropospheric inflow of air from the Southern hemisphere (Fig. 6.1:4), and convergence,
abundant cloudiness (Fig. 6.1:5), and rainfall (6.1:6).

Conversely, during Northern winter, the Afro-Asian land masses are particularly prone to surface
cooling and comparatively low surface temperatures, which in turn through cold high processes
would favor high surface pressure, consistent with Fig. 6.1:1, part a, and Fig. 6.1:2, part a, while
effects in the opposite sense would concurrently obtain in the summertime Southern hemisphere.
The surface pressure pattern is thus established for an outflow of air from the Northern into the
Southern hemisphere (Fig. 6.1:3), and a more southward position of the band of maximum
cloudiness (Fig. 6.1:5) and rainfall. It is here suggested that the atmosphere over the Northern
hemisphere plays the active role not only for the Northern summer but also for the winter monsoon; the surface heating of the summertime Australian continent would be of secondary and
more regional importance.

The surface pressure maps of the two extreme months, Fig. 6.1:2 are complemented by Fig.
6.8.1:2. This depicts the annual variation in the latitude and longitude position of a low pressure
center in the area of Indonesia and the Indian Ocean, in a plot similar to Fig. 6.3:1 for the subtropical highs. Fig. 6.8.1:2 shows the low center shifting within the greater Australasian region
from about October to April, while in the remainder of the year low pressure is prominent over
Southern Asia and the adjacent Indian Ocean, rather than Australasia.

Being part of the alternation between Northern hemisphere winter and summer conditions, the
manifestations of the monsoons in the lower troposphere are associated with seasonal reversals in
the upper-tropospheric circulation. During Northern hemisphere winter, westerlies prevail in the
upper troposphere over most of the monsoon region (Fig. 6.1.3a), and the Subtropical Westerly
Jet (Section 6.2.2) is found to the South of the Central Asian mountain massifs. By contrast
during Northern summer (Fig. 6.1:4a), the Tropical Easterly Jet (Section 6.2.3) occupies the
upper troposphere to the South of the Himalayas and throughout the monsoon region, from
Southeast Asia to the Atlantic. The development of the Tropical Easterly Jet as a thermal wind
phenomenon resulting from the summertime heating of land surfaces in the subtropics (Section
6.2.3) is intrinsically related to the heat-low induced Northern summer monsoon phenomenon in
the lower troposphere. The gradual change from westerlies to easterlies in the upper troposphere
over Southern Asia accompanies the evolution of the Northern summer lower-tropospheric monsoon flow, while the appearance of upper-tropospheric westerlies from October onward marks the
rather more definite end of the Northern summer monsoon. That the Tibetan Plateau and the
Central Asian mountain massifs in general must play an important role in the functioning of the

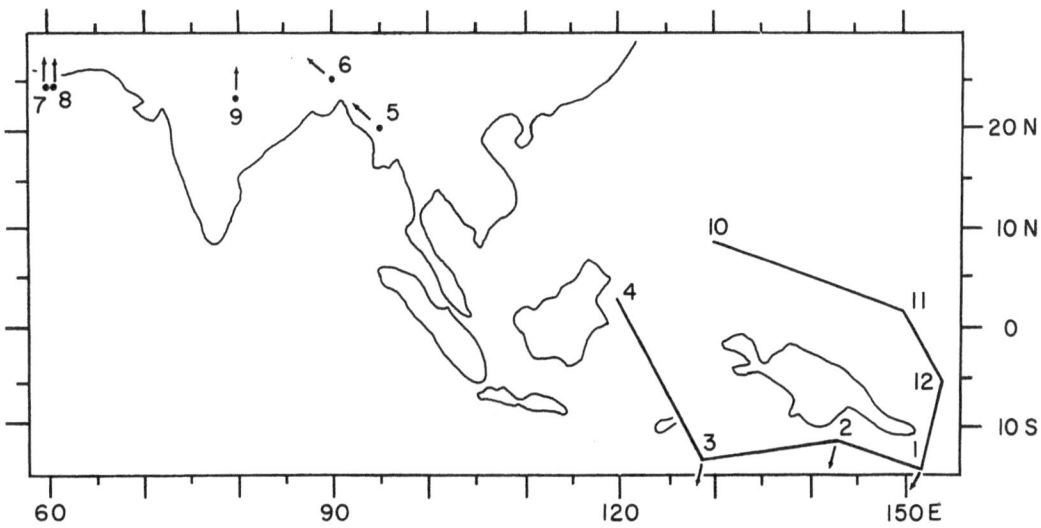

Fig. 6.8.1:2. Annual variation in the latitude and longitude position of low pressure center in the area of Indonesia and the Indian Ocean. The numbers 1 through 12 denote the calendar months. Arrows mean that the lowest pressure is found over land in indicated direction from plotted coordinates. Source: U.S. Navy (1976, 1977, 1979).

monsoons is a contention deserving interest as further observational evidence and analyses from mainland China become accessible.

Finally, as noted by Rao (1976, p. 1) with reference to the global circulation, the Northern winter monsoon is an integral portion of the Northern hemisphere Hadley circulation (Section 3.3), while the summer Southwest monsoon over the Northern Indian Ocean must be regarded as a protrusion of the Southern hemisphere Hadley cell into the Northern hemisphere.

This global overview may serve as a frame of reference for the following discussion of regional monsoon characteristics.

6.8.2. AFRICA

The monsoons in the Subsaharan zone of Africa (Fig. 6.8.1:1) are directly related to the seasonal behavior of the quasi-permanent circulation features discussed in Section 6.7.

During Northern winter, the surface pressure trough (Fig. 6.1:2, part a) and the confluence zone (Fig. 6.1:3, part b) and flow discontinuity (Figs. 6.7.2.1:1, 6.7.2.1:6, 6.7.2.3:1) embedded in it stay closest to the Equator. Accordingly, the Harmattan winds, a stream of dry desert air to be regarded as part of the trade wind system (Section 6.4), sweep far southward. Rains are then confined to the South of the flow discontinuity, with year-round precipitation in a belt extending from the Gulf of Guinea coast eastward into the interior of the continent.

Progressing towards Northern summer, the surface pressure trough and the flow discontinuity shift northward, so that the cross-equatorial moist and cool monsoon airstream penetrates ever deeper into the continent (Fig. 6.1:4, part b, Fig. 6.7.2.3:5). Zonally oriented bands with differing weather characteristics, as described in Section 6.7.2 and illustrated in Figs. 6.7.2.3:1 and 6.7.2.3:4, likewise migrate northward. The reverse progression takes place from the height of the Northern summer to winter. A recent evaluation of satellite and aircraft observations in August 1979 by Desbois *et al.* (1984) documents flow conditions consistent with earlier studies.

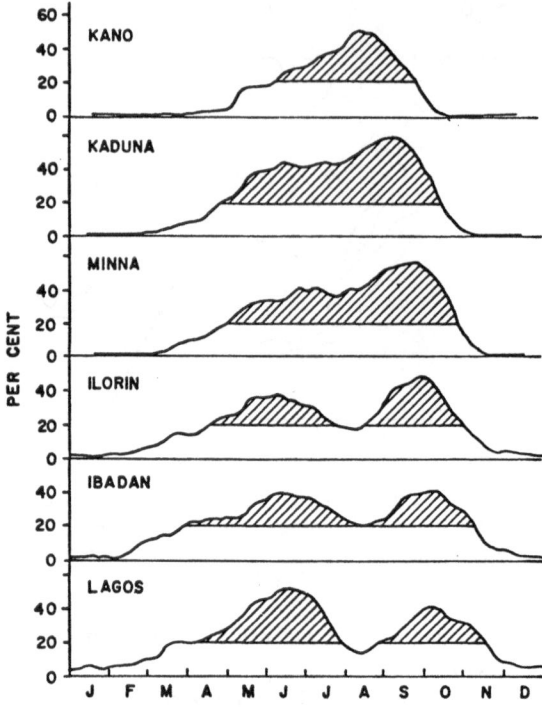

Fig. 6.8.2:1. Annual march of percentage occurrence of rain days (> 0.04 mm) at stations along a meridional profile from about 6 to 12°N in Nigeria. From Griffiths (1972).

In accordance with the annual migration of quasi-permanent circulation systems, the regions nearest to the Equator experience rainfall all year round, the semi-arid desert fringe only at the height of the Northern summer, while double-peaked precipitation regimes with various timings are characteristic for the intervening latitude zones. Such a meridional progression is illustrated in Fig. 6.8.2:1 for a sequence of stations ranging from about 6 to 12°N in Nigeria. Such a seasonal progression of rainfall in Subsaharan Africa is closely related to the varying location of the surface flow discontinuity and the atmospheric humidity of the air to either side of it (Osman and Hastenrath, 1969). For a detailed description of annual rainfall regimes in Subsaharan Africa refer to Leroux (1983, vol. 1, pp. 509–529). Proceeding from the Sudan further eastward, all but the southernmost part of the Arabian Peninsula appears to stay under the dominance of dry hot desert air even at the height of the Northern summer (Hastenrath *et al.*, 1979).

As mentioned in Section 6.7, precipitation in the southern portion of the southward sloping interface, or the northern portion of the rain zone, is predominantly of convective nature and associated with travelling disturbances. These 'West African Disturbance Lines' are discussed in Chapter 7.

6.8.3. INDIAN OCEAN SECTOR

Nowhere else on the globe is the annual reversal of wind and rainfall regimes as spectacular as in the realm of the Indian Ocean and surrounding land areas. On the densely populated Indian subcontinent in particular, the monsoon rains are of vital social and economic consequence. As explained by Rao (1976, p. i), Das (1984, pp. 1–4; 1986, Chapter 1), and in Fein and Stephens (1986), the Indian monsoon has been the subject of study for centuries. Only a limited discussion of this extensive work can be attempted here; among the major sources being the monograph of Ramage (1971), the exhaustive reviews by Rao (1976) and Das (1986), as well as various recent atmospheric and marine atlases (Atkinson and Sadler, 1970; Ramage and Raman, 1972; Ramage et al., 1972; Wyrtki, 1971; Sadler, 1975b; Hastenrath and Lamb, 1979; Young et al., 1980). In the following the annual changes of circulation will be reviewed first, as a basis for the subsequent discussion of regional rainfall regimes.

The upper-tropospheric circulation is characterized by the prevalence of the Subtropical Westerly Jet (Section 6.2.2) over the Northern hemispheric portion of the monsoon area during Northern winter, alternating with the Tropical Easterly Jet (Section 6.2.3) during summer. Reference suffices to the discussion in Section 6.2, and to Figs. 6.1:3a, 6.1:4a, 6.2.2:1 to 6.2.2:3, 6.2.3:1 in particular. However, a more extensive discussion is needed here concerning the sequence of annual changes in the lower-tropospheric circulation. This is documented in detail in the series of monthly maps in a recent atlas (Hastenrath and Lamb, 1979). Only portions of the surface wind charts are reproduced as Figs. 4.4:1 to 4.4:3.

During Northern winter, especially the months from November to March, northeasterlies emanating from the area of surface high pressure over the comparatively cold continent of Southern Asia sweep much of the Northern Indian Ocean including the Arabian Sea, the Bay of Bengal, and the South China Sea (Fig. 4.4.1). These airstreams recurve to northwesterly still to the North of the Equator and meet the Southern hemisphere Southeast trades in a broad trough zone located to the South of the Equator. As discussed in Section 6.7.2.2 this is broadly embedded in a zone of high sea temperature and low surface pressure. This zone also features marked convergence (Hastenrath and Lamb, 1979, vol. 1, charts 40–41, 30–32).

Conditions over the continent are of particular interest in relation to the Northern winter monsoon in the Indian Ocean sector. Earlier work on the general circulation over Eastern Asia is summarized in a series of review articles by Staff Members of the Academia Sinica (1957, 1958a, b). In a more recent series of papers, Murakami (1981a, b, c, d) discusses the Asiatic winter monsoon circulation, with emphasis on the orographic influence of the Tibetan Plateau. Luo and Yanai (1983, 1984) studied the circulation and heat sources over the Tibetan Plateau during the early summer of the FGGE year 1979. Mower et al. (1984) present an analysis of the winter monsoon circulation over Southeast Asia during December 1978. Early studies of the cool season weather conditions over Southeastern Asia are due to Ramage (1952, 1954, 1955). Theoretical studies by Lim and Chang (1981) and Chang and Lau (1982) are directed to the interactions between the mid-latitudes and the tropics during the Asian winter monsoon. The role of cold surges in the winter monsoon circulation over Southeastern Asia is discussed by Murakami (1979), Chang et al. (1979, 1983) Chang and Lau (1980), and Lau and Li (1984) in particular (ref. Section 7.9).

Following is a review of the sequence of changes leading from the height of the Northern winter to the fully developed summer monsoon. Progressing from February to July, the domain of warmest surface waters displaces northward, with similar changes being apparent in the surface pressure pattern (Hastenrath and Lamb, 1979, vol. 1, charts 51–56, 3–8). The trough zone between the Northeast monsoon flow and the Southern hemisphere Southeast trades also shifts northward from February to April, but still stays to the South of the Equator (Hastenrath and Lamb, 1979, vol. 1, charts 15–17). In the transition between the monsoon seasons, and in fact only during April and October, an anticyclonic vortex each is manifest over the Arabian Sea and Bay of Bengal (Hastenrath and Lamb, 1979, vol. 1, charts 17 and 23). There is some indication for a northward displacement of the domain of convergence (Hastenrath and Lamb, 1979, vol. 1, chart 31–33).

In April (Hastenrath and Lamb, 1979, vol. 1, charts 53, 50), the highest sea surface temperature is found in a broad Southwest to Northeast oriented band extending from the equatorial region of the Arabian Sea sector to the Bay of Bengal; this being also a domain of low surface pressure although the lowest values are now found along the South Asia coast and over the continent. Very pronounced changes in the temperature, pressure, and flow fields take place from April to May (Hastenrath and Lamb, 1979, vol. 1, charts 53–54, 5–6, 17–18), marking the changeover from the Northern winter to summer monsoon circulation. Thus, the domain of warmest surface waters moves to the Central Arabian Sea and the Bay of Bengal; a surface pressure pattern establishes itself characterized by a monotonic meridional gradient from the South Indian Ocean high across the Equator into the South Asian continent. The near-equatorial trough zone in the flow field is eliminated, the Southern hemisphere Southeast trades recurve and the continuation of this airstream crosses the Equator and extends into Southern Asia as Southwest monsoon. The independent initiation of heat low processes over the South Asian continent, rather than a gradual northward migration of near-equatorial troughs in the pressure and flow fields, appear instrumental in the establishment of the Northern summer monsoon.

The circulation of the Northern summer monsoon is characterized by highest surface temperatures over the South Asian continent and the adjacent sea areas; surface pressure decreasing monotonically from the South Indian Ocean high across the Equator into South Asia, albeit with some eastward component of the pressure gradient; and a continuation of the Southern hemisphere Southeast trades recurving to the South of the Equator and extending into South Asia as Southwest monsoon, with three separate branches discernible over the Arabian Sea, the Bay of Bengal, and the South China Sea (Hastenrath and Lamb, 1979; vol. 1, charts 55–58, 7–10, 19–22). Speed maxima are apparent over the latter three sea areas and in the core of the Southern hemisphere trades, while the enclosed equatorial zone, where flow recurves, is a zone of weak resultant wind. Of the three aforementioned branches, the airstream over the Western Indian Ocean is by far the most intense, and directional steadiness of wind is high even in the region of flow recurvature and resultant wind minimum near the Equator. Embedded in this lower-tropospheric airstream is the East African Low Level Jet (Section 6.2.5), which shares with the surface wind pattern the speed maxima at the North tip of Madagascar and over the Arabian Sea. Particularly prominent is the decrease of speed from the aforementioned maxima of Arabian Sea, Bay of Bengal, and South China Sea downstream towards the South Asian continent. This is further reflected in marked convergence (Hastenrath and Lamb, 1979, vol. 1, charts 35–38).

The changeover from the Northern summer to the winter monsoon (Hastenrath and Lamb, 1979, vol. 1, charts 58–61, 10–13, 22–25) is heralded by the seasonal cooling of the land surfaces of South Asia and the adjacent sea areas. The domain of the highest surface temperature and lowest pressure gradually returns to lower latitudes. Airstreams originating over South Asia penetrate to the equatorial region. In the course of this transition, anticyclonic vortices are found again over the Arabian Sea and Bay of Bengal in October, in similarity to the April conditions. From October onward, a trough in the flow field establishes itself again to the South of the Equator. This is where the Southern hemisphere trades and the Northeast monsoon meet, as discussed above for the height of the Northern winter monsoon season. For a case study of the onset of the Australian monsoon in December 1978 refer to Davidson *et al.* (1983).

The monsoon circulations are of consequence for the regional climates, as reflected primarily in the rainfall regime. Regarding the annual march of precipitation activity, topographic and other local factors make for a puzzling variety, but three major types stand out: over most of Indonesia and parts of Southeast Asia, rainfall tends to be most abundant during the Northern winter monsoon; much of South Asia and especially India depend for rainfall mainly on the Northern summer monsoon; finally, in large parts of Equatorial Eastern Africa the two major rainy seasons are timed in the transition between the Northern summer and winter monsoons. These three regional regimes are in turn considered in the following.

The greater Indonesian region offers a vast variety of subclimates and annual rainfall regimes, presumably due in large part to distortions of the monsoons by the mountainous islands of the 'maritime continent', as discussed by Braak (1921–1929, part 1, chapter 6, pp. 151–221) and Ramage (1971, pp. 135–141). However, on the whole rainfall activity tends to be concentrated from around November to February. This is the season with a quasi-permanent low pressure center (Fig. 6.8.1:2), and a trough zone in the flow field (Fig. 4.4:1a) developed in the area. This is also a time of year with abundant convergence (Hastenrath and Lamb, 1979, vol. 1, charts 39–41, 30–31). Moreover, conditions would be particularly conducive for rainfall at the coasts and mountain ranges exposed to the Northeast monsoon flow from the South China Sea.

Most of Southern Asia receives the bulk of its annual rainfall with the Southwest monsoon at the height of the Northern summer. The role of the Southwest monsoon circulation in the rainfall regime of India has been most extensively studied, and this research is effectively presented in Rao's (1976) monograph on which the following review is largely based. This also contains a comprehensive bibliography. Patnaik *et al.* (1977) suggest that the summer and the winter monsoon systems are essentially independent of each other, although they find sizeable positive correlations between the summer and the following winter monsoon rainfall in the far North and in portions of South India.

Rao (1976, pp. 62–85) emphasizes that summertime convection and rainfall is greatly favored by the thermodynamic properties of the moist and comparatively cool air mass that penetrates the continent with the Southwest monsoon current, undercutting the dry and hot continental air. While it was shown in Section 6.8.1 that thermally induced low surface pressure over Southern Asia is instrumental in the operation of the Northern hemisphere summer monsoon in the Indian Ocean sector at large, Rao (1976, pp. 86–106) elucidates the *in-situ* role of the monsoon trough for the monsoon weather and climate of North India. This surface pressure trough with heat low structure extends from the desert regions of Iran and Pakistan across North India further eastward. As a further corollary to the discussion of the equatorial trough and the structure of the Intertropical Convergence Zone in Sections 6.7.1, and 6.7.2, Rao (1976, pp. 90–91) notes that the largest rainfall over North India is also found to the South of the mean position of the monsoon trough line. Accordingly, the considerations in Section 6.7.2.5 as to the maintenance of the surface temperature maximum may likewise be pertinent.

Krishnamurti *et al.* (1981) consider the evolution of an 'onset vortex' over the Arabian Sea as a characteristic circulation feature heralding the onset of the Southwest monsoon. The vortex is described as forming on the cyclonic shear side of the low-level jet (see Section 6.2.5), and extending over a deep layer of the troposphere.

Das (1984, p. 10) does not mention an onset vortex and points out that the monsoon onset is usually not a spectacular event, but a gradual process beginning with a period of transition when humidity rises and winds change. Light rains follow. In the Northeastern part of India violent thunderstorms are common in the pre-monsoon months of April and May; it being at times difficult to decide when these 'Nor'westers' (see Section 7.3) end and the monsoon rains proper begin. The difficulties in defining the onset of the Southwest monsoon over India are further discussed by Subbaramayya and Bhanu Kumar (1978).

Just as the rainfall totals vary greatly over the Indian subcontinent, so do the timings of monsoon onset and retreat. According to Rao (1976, p. 35), the India Meteorological Department fixes the onset and withdrawal dates with reference to the rather sharp change of pentade rainfall totals and changes in the circulation.

Das (1984, p. 10) describes the following working rules used to define the onset of the monsoon over Kerala: (i) starting with 10 May, if at least 5 out of 10 stations in Kerala report 24 hour rainfall totals of ≥ 1 mm for two consecutive days, an onset is declared on the second day; (ii) if ≥ 3 out of 7 stations in Kerala report no rainfall for the next 3 days, indications are given for a recession of the monsoon; (iii) after the monsoon has advanced North of 13°N even a temporary recession is a rare event. Das further mentions that there are similar working rules for other parts of India.

Fig. 6.8.3:1a shows the earliest onset in late May to early June over the Bay of Bengal, Burma, and southernmost India, and a progression northward and westward, with a delay to mid July in the arid Northwest region. The retreat (Fig. 6.8.3:1b) shows broadly the reverse pattern, with a timing as early as the beginning of September in the Northwest, a delay to October over Northern and Central India and the Northern Bay of Bengal, and dates as late as December and January in the South. It must be noted that Fig. 6.8.3:1 depicts average onset and retreat dates and that the progression is not ncessarily gradual in individual years. Just as there is a considerable interannual variability in rainfall totals, the onset dates range over several weeks. For the extreme South of the Indian peninsula Das (1986, Chapter 3) gives a standard deviation of 8 days. These and other vagaries of the monsoon are considered in Chapter 8.

Fieux and Stommel (1977) define monsoon onset from the change of wind direction along shipping lanes in the Arabian Sea, and distinguish four types, namely single, multiple, gradual, and indeterminate onset. Subbaramayya et al. (1984) distinguish three phases in the advance of the monsoon over India: the first affecting the southern peninsula, the second reaching the northern peninsula and Central India, and the third penetrating to the central parts of North India. These phases and the intermediate stagnations of the 'northern limit of the monsoon' are schematically mapped in Fig. 6.8.3:2. From an analysis for the FGGE year 1979, Pearce and Mohanty (1984) conclude that the onset consists of two phases, namely a moisture buildup and a subsequent intensification of winds over the Arabian Sea and increased latent heat release.

As described by Rao (1976, pp. 107–185) much of the rainfall during the Southwest monsoon is produced by monsoon depressions. These low pressure centers typically form over the Northern Bay of Bengal, move towards the Westnorthwest at least as far as Central India before decaying, and are associated with abundant precipitation in the Southwest quadrant. Some of these disturbances may develop into 'Cyclones' (Hurricanes; ref. Section 7.2).

In the course of the Southwest monsoon, there are periods when the monsoon trough shifts northward to the foot of the Himalayas, and rains decrease over much of India except along the slopes of the Himalayas and parts of Northeast India, and the Southern Peninsula. This synoptic situation is called 'break in monsoon' (Ramanadham et al., 1973; Rao, 1976, pp. 186–200; Das, 1986, Chapter 3). Fig. 6.8.3:3 shows the typical departure pattern in the pressure field, with largest positive anomalies in a zonally oriented band across North-Central India contrasting with negative departures in the extreme North and South. The typical rainfall departure pattern is mapped in Fig. 6.8.3:4: precipitation appears drastically reduced over most

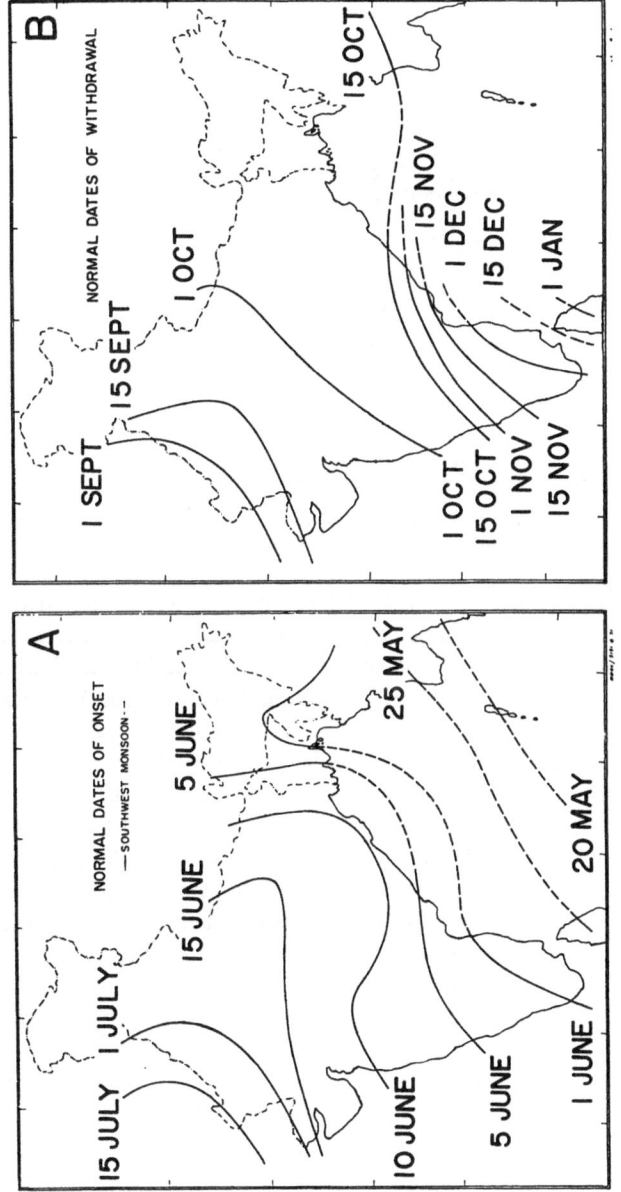

Fig. 6.8.3:1. Normal dates of (a) onset, and (b) withdrawal of Southwest monsoon over India. From Rao (1976) (copyright by the Government of India).

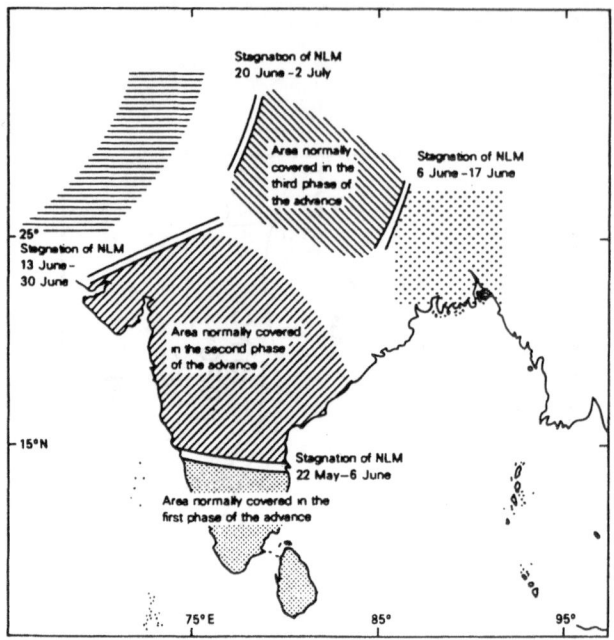

Fig. 6.8.3:2. Areas and periods of different phases in the advance of the Southwest monsoon. NLM means 'northern limit of the monsoon'. From Subbaramayya *et al.* (1984).

of India, but enhanced in the extreme North and South. It is here suggested that the 'break in monsoon' may be associated with a circulation cell essentially in a meridional-vertical plane and involving both the Indian lowlands and the adjacent Tibetan Plateau. This contention is in part substantiated in Section 8.5. Pant (1983) also considered the possible role of the Tibetan Plateau for the 'breaks in monsoon'. While Indian meteorologists have carefully documented the circulation diagnostics of the 'break' over India (see Rao, 1976, pp. 186–200), the lack of concurrent documentation for the Central Asian mountain massifs has been a traditional shortcoming. The 'breaks' are most frequent in July and August, and they typically last from a few days to three weeks. Yanusari (1979), and Sikka and Gadgil (1980) found evidence for a northward shift of a belt of maximum cloudiness, and Krishnamurti and Subrahmanyam (1982) noted a meridional propagation of lower-tropospheric trough and ridge lines from the equatorial region towards the Himalayas. Krishnamurti and Bhalme (1976), Krishnamurti and Ardanuy (1980), and Yasunari (1981) have related oscillations around 10–20 and 40 days in various tropospheric parameters to the occurrence of 'breaks in monsoon'. The mechanisms of 'active' – 'break' sequences were further considered by Pant (1983), Webster (1983), and Cadet (1983).

Annual rainfall totals exceed 2000 mm at much of the Arabian Sea coast where the Western Ghats are exposed to the Southwest monsoon flow. Amounts decrease eastward across the peninsula dropping to less than 300 mm in some areas. The desertic regions to the Northwest receive less than 100 mm and a band of reduced rainfall with 1000 mm or less extends from there eastward across much of North India. Consistent with earlier considerations by Ramage (1971, pp. 194–195), Rao (1976, p. 9) notes that this dry axis paradoxically lies close to the monsoon trough, but explains that the area to the South and towards the Bay of Bengal benefits from the track of monsoon depressions, while he considers the influence of the Himalayas to be responsible for the precipitation increase northward.

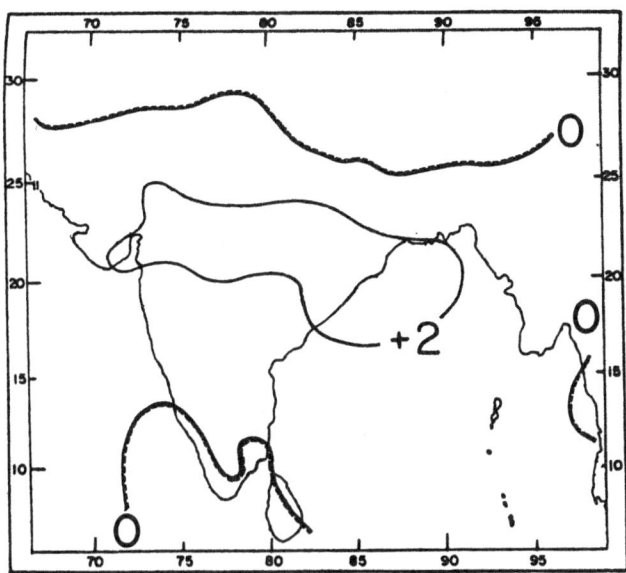

Fig. 6.8.3:3. Mean pressure departure, 03 GMT, in mb, during 'break in monsoon'. From Rao (1976) (copyright by the Government of India).

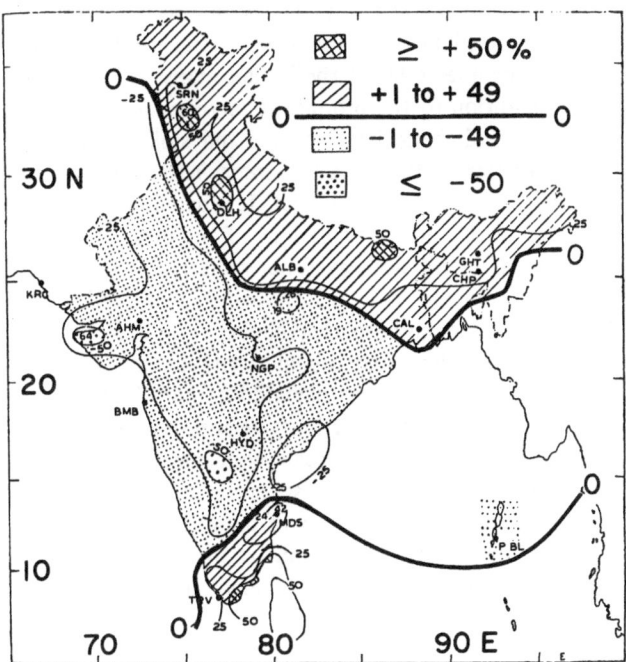

Fig. 6.8.3:4. Percentage rainfall departure in peak phase of 'break in monsoon'. Symbols are as follows: cross-hatching excess of +50% or more; simple hatching excess of +1 to 49%; thin dot raster deficit of −1 to −49%; thick dot raster deficit of −50% and beyond. From Rao (1976) (copyright by the Government of India).

Indonesia and Southern Asia receive the bulk of their annual rainfall at the height of the two opposing monsoons, and at least the major general circulation causalities appear straightforward. By contrast, in Equatorial East Africa the two rainy seasons peaks fall into the transition between the two monsoon circulations; the March–May peak affecting larger areas than the September–December rainy season which is limited to certain regions within East Africa. These two peaks are also referred to as the 'long rains' and 'short rains', respectively. During limited periods in the transition between the two monsoon circulation regimes, airstreams from either hemisphere clash in the equatorial zone (Figs. 4.4:1 to 4.4:3). The convergence is presumably a major factor for the annual march of rainfall, but circulation-rainfall causalities appear more intricate than in the regions where the rainy season coincides with the height of the monsoon circulation.

Kiangi *et al.*'s (1981) evaluation of pibal observations of April and May 1966–75 shows the airstream originating in the Southern hemisphere being undercut by the Northeast monsoon flow in the region of Equatorial East Africa. Conversely in October–November in East Africa, southeasterlies wedge under the northeasterly current from the Northern hemisphere, along a confluence zone then again extending across Equatorial East Africa (Kiangi, personal communication, 1984).

6.8.4. ON THE HEAT AND MOISTURE BUDGET OF THE INDIAN MONSOONS

As described in Sections 6.4.1 and 6.4.3, the monsoons in the Indian Ocean sector entail differential heating and cooling of land and sea surfaces, large-scale transports of atmospheric moisture, and abundant precipitation and latent heat release. Accordingly, atmosphere, hydrosphere, and lithosphere appear all essential for the energetics of the monsoon system. Das (1962), Anjaneyulu (1969, 1971), Rao and Rajamani (1972), and others studied the distribution of *atmospheric* heat sources and sinks and the energetics of the Indian monsoon. Das (1986, Chapter 3) in particular points out that in addition to the prominent meridional pattern, there are marked zonal contrasts in the distribution of diabatic heat sources and sinks in the *atmosphere*. The present review focuses on the energetics of the coupled *atmosphere-ocean* system, and the prominent meridional pattern of the heat and moisture budget. It is based largely on the work of Hastenrath and Lamb (1979, 1980) which includes calculations of the oceanic heat budget from long-term ship observations, evaluation of satellite-observed net radiation at the top of the atmosphere, and estimates of the atmospheric moisture budget. Reference is made to Section 5.1 for the theory of atmospheric and oceanic heat budget and use of symbols.

Fig. 6.8.4:1 illustrates the latitudinal and annual variation of residually determined net oceanic heat gain $(Q_t + Q_v)$ for 30–120°E. The Equatorial and Northern Indian Ocean water body features a sizeable heat gain, except for the more northerly region during Northern winter, while a heat loss prevails in the Southern tropical Indian Ocean, concentrated in Southern winter. Consistent with Fig. 6.8.4:1, the ocean and overlying atmosphere is in Table 6.8.4:1 to 6.8.4:3 considered separately for the domains to the North and South of the Equator, respectively.

During Northern winter, the radiative heat input at the top of the atmosphere and precipitational heating of the atmospheric volume are very small for the Northern hemispheric domain, and much larger to the South of the Equator (Table 6.8.4:1). Oceanic heat export/import is small for both hemispheres at this time of year (Table 6.8.4:2). In contrast, the atmospheric volume over the Northern Indian Ocean imports geopotential energy and sensible heat during this season, while exporting somewhat more latent heat (Table 6.8.4:2). In view of the increase of $(gz + c_pT)$ and decrease of Lq from the lower to the upper troposphere this combination of $(gz + c_pT)$

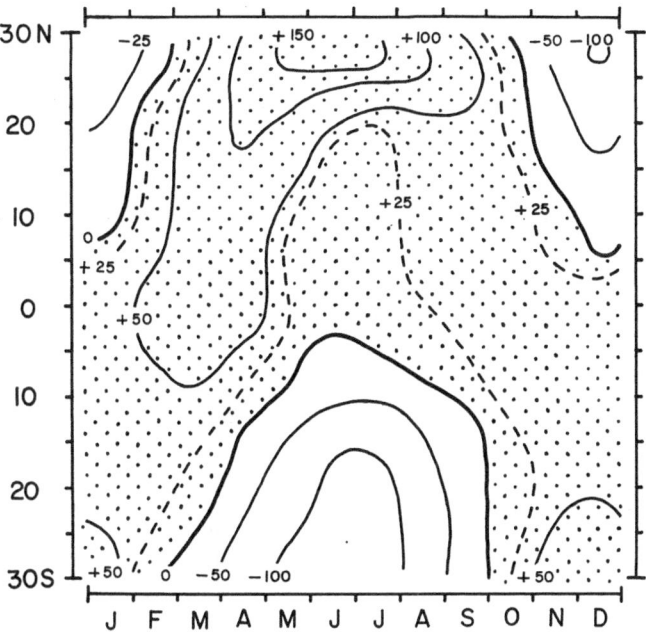

Fig. 6.8.4:1. Latitudinal and annual variation of residually determined net oceanic heat gain ($Q_t + Q_v$) in the Indian Ocean, 30–120°E, in W m^{-2}. Dot raster covers positive areas. From Hastenrath and Lamb (1980).

import and Lq export is indicative of atmospheric subsidence. Approximately inverse conditions simultaneously prevail South of the Equator (Table 6.8.4:2).

During the Northern summer monsoon, the net radiative heat input at the top of the atmosphere is very small for the domain 0–30°S (Table 6.8.4:1). The substantial southward directed oceanic heat transport characteristic of this season imports heat to the Southern tropical Indian Ocean (Table 6.8.4:2). This, together with the seasonal depletion of the local hydrospheric heat content, supports sensible and latent heat losses from the Southern tropical Indian Ocean which greatly exceed the net radiative heat supply (Table 6.8.4:1). The atmospheric volume over the Southern hemisphere waters exports a large amount of latent heat throughout the Northern summer half-year (Table 6.8.4:2). At the height of this semester's monsoon, the same atmospheric volume also imports geopotential energy and sensible heat (Table 6.8.4:2). Such a pattern is consistent with atmospheric subsidence.

A large proportion of the moisture evaporated from the Southern tropical Indian Ocean during Northern summer is carried northward across the Equator in the lower-layer monsoon flow. There is minimal increase in the northward water vapor flux from the Equator to the coasts of Southern Asia (Tables 6.8.4:2 and 6.8.4:3). As a result, the Southern tropical Indian Ocean appears to provide the major moisture supply for the release of precipitation over Southern Asia during the Southwest monsoon. The concurrent export of ($gz + c_pT$) from the Northern hemispheric domain (Table 6.8.4:1) can be inferred to largely occur in the high troposphere, as is characteristic of large-scale ascending motion in the atmosphere.

TABLE 6.8.4:1

'Hemispheric' integrals of atmospheric and oceanic heat and moisture budget components for the *Indian Ocean and peripheral seas* (30–120°E) during the Northern winter and summer monsoon seasons. SWLW↑↓$_{top}$, net radiation at top of atmosphere during December–January–February and June–July–August; SWLW↑↓$_{sfc}$, net radiation at ocean surface in January and July; *LP*, precipitational heating during December–January–February and June–July–August; Q_e and Q_s, latent and sensible heat fluxes from ocean to atmosphere in January and July; $(Q_v + Q_t)_o$, heat storage plus export within the ocean in January and July; Q_{to}, heat storage within the ocean during December–February and June–August, calculated from the annual march of sea surface temperature. Units in 10^{13} W.

	SWLW↑↓$_{top}$	SWLW↑↓$_{sfc}$	*LP*	Q_e	Q_s	$(Q_v + Q_t)_o$	Q_{to}
N winter monsoon season							
0–30°N	+ 23	+163	78	+164	+ 5	− 6	0
0–30°S	+286	+396	261	+272	+ 7	+117	+130
N summer monsoon season							
0–30°N	+148	+192	+124	+169	− 3	+ 26	− 90
0–30°S	− 6	+187	+148	+347	+23	−183	−160

TABLE 6.8.4:2

'Hemispheric' integrals of atmospheric and oceanic heat and moisture budget components for the *Indian Ocean and peripheral seas* (30–120°E) computed from Table 6.8.4:1 and Eqs. (5.1:1), (5.1:4), and (5.1:5). $Q_{vo} = (Q_t + Q_v)_o = Q_{to}$ computed from the annual march of sea surface temperature; Q_{va} = SWLW↑↓$_{top}$ − $(Q_v + Q_t)_o$, Eq. (5.1:1). div $(gz + c_pT)$ = SWLW↑↓$_{top}$ − SWLW↑↓$_{sfc}$ + Q_s + LP, Eq. (5.1:4). div $(Lq)_s = Q_e − LP$, Eq. (5.1:5); div $(Lq)_{aer}$ obtained by aerological method. Units in 10^{13} W.

	Q_{vo}	Q_{va}	div $(gz + c_pT)$	$(div Lq)_s$	$(div Lq)_{aer}$
N winter monsoon season					
0–30N	− 6	+ 29	− 57	+ 86	− 50
0–30S	− 13	+169	+158	+ 11	+ 0
N summer monsoon season					
0–30N	+116	+122	+ 77	+ 45	+100
0–30S	− 23	+177	− 22	+199	+180

TABLE 6.8.4:3

Vertically integrated latent heat transports across vertical transects. Positive values directed northward/landward. Units in 10^{13} W.

	January	July
West of 120°E		
A. N. hemisphere coasts	−125	+300
B. Equator	− 95	+345
C. 30°S	+ 15	+ 30
West of 80°E		
D. N. hemisphere coasts	− 55	+170
E. Equator	+ 10	+170
F. 30°S	− 85	+ 15
60−80°E		
G. Asia coast	+ 55	+125

By way of comparison with Hastenrath and Lamb's (1980) study, it is noted that there has been considerable controversy over the importance of cross-equatorial water vapor transport for the moisture budget of Southern Asia. Pisharoty's (1965) calculations suggest that more than half of the moisture crossing the western coastline of India at the height of the Southwest summer monsoon is supplied by evaporation from the Arabian Sea, with the remainder, representing 80 × 10^{13} W of latent heat, originating in the Southern hemisphere and crossing the Equator between 42 and 75°E. Ghosh et al.'s (1978) and Murakami et al.'s (1984) investigations lends support to Pisharoty's (1965) study. In contrast, the cross-equatorial water vapor transport computed for this season by Saha (1974) and Saha and Bavadekar (1973) is about double Pisharoty's (1965) value. As a result, Saha (1974) and Saha and Bavadekar (1973) propose that 60−80% of the moisture crossing the West coast of India during the summer monsoon emanate from the Southern hemisphere. More recently, attention has focused on the important role of the East African Low Level Jet for the cross-equatorial moisture transport (Hart et al., 1978; Howland and Sikdar, 1983; Van de Boogaard and Rao, 1984). Howland and Sikdar (1983) specifically confirm the results of Hastenrath and Lamb (1980). From an analysis for the summer of 1975, Cadet and Reverdin (1981a) also conclude that about 70% of the water vapor crossing the West coast of India comes from the Southern hemisphere. Cadet and Reverdin (1981b) further note that their evaluation of cross-equatorial water vapor transport yields the same magnitude as that obtained by Saha (1974). Moreover, Peixóto and Oort (1983) presented a global analysis of the atmospheric water vapor budget and transports based on a ten-year data bank of upper -air observations. Their chart for

Northern summer (June–July–August; Peixóto and Oort, 1983, fig. 13c) also indicates the overwhelming importance of the tropical South Indian Ocean (10–30°S) as water vapor source region and the subordinate role of the Arabian Sea. Clearly, Hastenrath and Lamb's (1980) and Peixóto and Oort's (1983) studies support the contention of Saha (1974) and Saha and Bavadakar (1973). Das (1984, p. 20) suggests that interannual variability may be a factor for the differences between the various studies.

The mean annual values and patterns of heat budget parameters for the Indian Ocean region are strongly dictated by their Northern summer monsoon counterparts. For the year as a whole, the oceanic heat import within the Indian Ocean between 20–30°S is comparable in magnitude to the net radiative heat input at the top of the atmosphere. In the realm of the Northern and Equatorial Indian Ocean, in contrast, the atmosphere and hydrosphere cooperate in the heat export to other parts of the globe.

6.8.5. NUMERICAL MODELLING OF THE INDIAN SUMMER MONSOON

The review of empirical analyses in Sections 6.8.1 and 6.8.3 points to the apparent role of various factors in the development of the Indian summer Southwest monsoon. Computer simulation is an obvious choice to test some of these suggestions.

A series of early experiments (Murakami et al., 1970; Godbole, 1973; Hahn and Manabe, 1975) focused on the role of the mountains in the South Asian monsoon circulation. It is found that the mountain topography is essential to produce realistic circulation patterns in the model. In particular, Hahn and Manabe (1975) conclude that the presence of mountains is instrumental in maintaining the South Asian low pressure system, whereas in the simulation without mountain topography the continental low forms far to the North and East. Various shortcomings of these numerical experiments have been discussed by Hahn and Manabe (1975, 1976) and Sadler and Ramage (1976). A more recent computer simulation (Druyan, 1982a, b) further emphasizes the role of the Himalayas in the Indian summer monsoon circulation.

The effect of sea surface temperature anomalies in the Arabian Sea on the monsoon rainfall over India is the subject of Shukla's (1975) computer simulation and subsequent discussions (Sikka and Raghavan, 1976; Shukla, 1976). The model (Shukla, 1975) produces a seemingly plausible positive relation. However, inasmuch as the sea surface temperature anomaly is prescribed and kept fixed, little insight is gained into the conceivably complex feedback processes in the real coupled system. The effect of prescribed and fixed sea surface temperature anomalies in the Arabian Sea on India summer monsoon rainfall was further simulated by Washington et al. (1977) and Druyan (1982a, b), but both studies conclude that the resultant precipitation departures are not statistically significant. More recent simulations with an interactive Arabian Sea water body by Druyan et al. (1983) contain no ocean dynamics.

In various later numerical simulations largely reviewed in Section 6.2.5 (Washington and Daggupaty, 1975; Washington, 1976; Krishnamurti et al., 1976; Anderson, 1976; Hart, 1977; Bannon, 1979a, b; Krishnamurti and Wong, 1979; Krishnamurti et al., 1983) emphasis shifts to the East African Low Level Jet as the 'backbone' of the Southwest monsoon circulation. In addition, Gilchrist (1977) and Krishnamurti and Ramanathan (1982) studied the sensitivity of the monsoon to differential heating through computer simulation.

6.9. Zonal Circulations

In a seminal paper, Bjerknes (1969) proposed the existence of a circulation cell in a zonal-vertical plane along the Pacific Equator. His original cross-sections reproduced here as Fig. 6.9.1 entail a warmer surface and tropospheric column in the West than in the East, westward pressure gradient and easterlies in the lower troposphere and eastward slope of isobaric surfaces and westerlies in the upper troposphere, subsidence over the Eastern Pacific, and ascending motion in the greater Indonesian region. The eastward sea surface temperature gradient is considered the cause of the thermally direct circulation entered in Fig. 6.9:1. Bjerknes named this circulation cell 'Walker Circulation', since he recognized it to be an important part of Walker's (Walker, 1923–24, 1928; Walker and Bliss, 1930, 1932, 1937) 'Southern Oscillation'. A simple circulation as shown in Fig. 6.9:1 would result with zonal vertical walls enclosing the equatorial air. Without such walls, this equatorial circulation enters into exchange with adjacent parts of the atmosphere to the North and South.

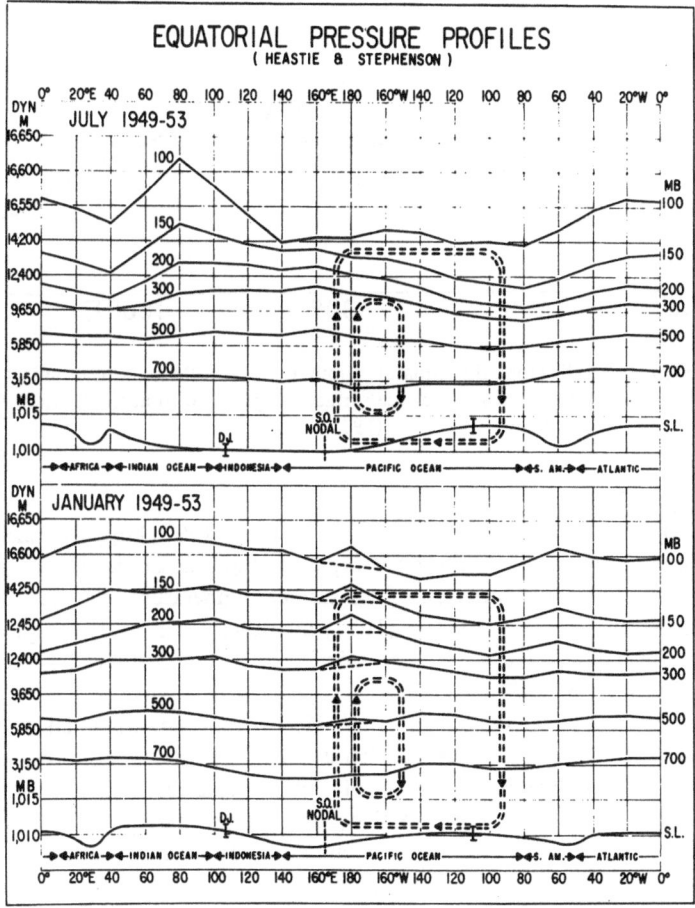

Fig. 6.9:1. Profile of height (dynamic meters) of standard isobaric surfaces along the Equator in January and July. Double broken lines denote 'Walker Circulation' over the Pacific. From Bjerknes (1969) (*Monthly Weather Review*, American Meteorological Society).

ZONAL ("WALKER") CIRCULATION ALONG EQUATOR

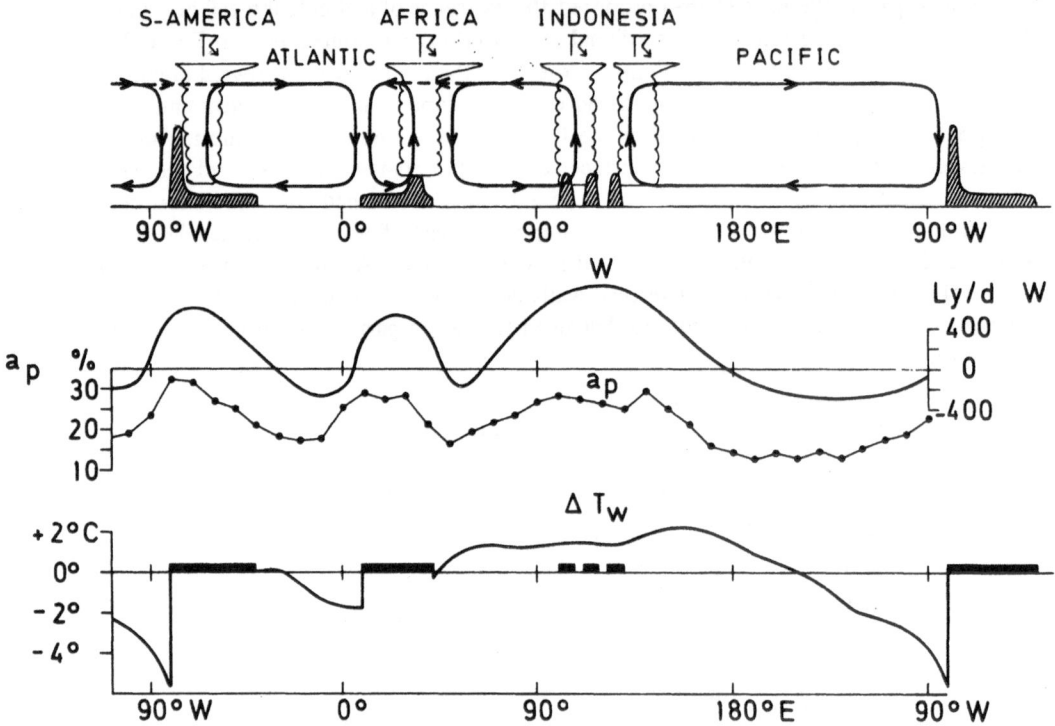

Fig. 6.9:2. Zonal circulation cells along the Equator. Symbols are as follows: w heat budget of an atmospheric column; a_p planetary albedo; ΔT_w departure of sea surface temperature from the latitude mean. From Flohn (1971).

Stimulated by Bjerknes' (1969) ideas for the Pacific, Flohn (1971) hypothesized a circumglobal scheme with various circulation cells along the Equator (Fig. 6.9:2). His diagram plausibly relates the circulation cells to the zonal variation of sea surface temperature, planetary albedo, and heat budget of the atmospheric column. Flohn's (1971) hypothesis calls for four equatorial zonal cells around the globe: the Pacific, Atlantic, Africa (Zaïre), and Indian Ocean cells. Essentially the same scheme is repeated in Wyrtki (1979).

Krishnamurti (1971) and Krishnamurti et al. (1973) inferred the zonal circulations in low latitudes from analyses of the upper-tropospheric wind field over the tropics during Northern winter and summer. Their schematic diagram of zonal circulation cells is shown in Fig. 6.9:3. Results are averaged over 15°S – 15°N for Northern winter and over 0–30°N for summer, in consideration of the active regions of zonal circulations. The diagram differs greatly from both Bjerknes's (1969) and Flohn's (1971) schemes, Figs. 6.9:1 and 6.9:2, as well as from later views to be discussed further below. Particularly noteworthy are the double cells over the Pacific and the complicated configuration in the Americas – Atlantic – Africa sector.

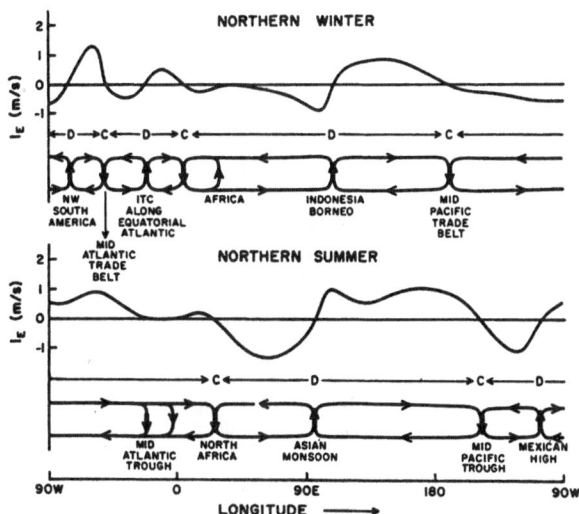

Fig. 6.9:3. Intensity of East-West circulation I_E at 200 mb as a function of longitude, and schematic diagram of zonal circulation cells on mass continuity. C and D indicate regions of upper-tropospheric convergence and divergence, respectively. From Krishnamurti *et al.* (1973) (Krishnamurti, T. N., Kanamitsu, M., Koss, J. W., Lee, J. D., *Journal of Atmospheric Sciences*, American Meteorological Society).

Yet another picture is offered by Newell (1979), as reproduced in Fig. 6.9:4. The scheme resembles that of Flohn (1971) except for Western South America. Newell *et al.*'s (1972, 1974, vol. 2, p. 151) analysis of actual wind observations clearly documents the presence of the Pacific, Atlantic, and Indian Ocean cells during both Northern winter and summer. A cell over Western South America as sketched in Fig. 6.9:4 appears better developed during Northern winter, while the African (Zaïre) cell is at best very weakly indicated in winter, but absent in summer when the Indian Ocean cell is shown to extend to the Atlantic. Newell *et al.*'s (1972, 1974, vol. 2, p. 151) analysis for Northern winter does not confirm the extent of a cell over Western South America and the adjacent Atlantic, and the complicated cell structure in the Americas – Atlantic – Africa sector as depicted in Fig. 6.9:3.

In addition to the wind analyses by Krishnamurti (1971), Krishnamurti *et al.* (1973), and Newell *et al.* (1972, 1974, vol. 2, p. 151), Krueger and Winston (1974, 1975, 1979) draw on satellite observations of cloudiness to substantiate zonal circulation cells along the Equator. Apart from the gross features of the apparently best developed cells centered on the Pacific, Atlantic and Indian Oceans, considerable controversy remains. Particular caution is in order for the inference of interannual variations of zonal circulation cells from indirect evidence other than wind and pressure observations.

In a numerical model experiment, Chervin and Druyan (1984) produced a total of six zonal circulation cells, namely all five contained in Newell's (1979) scheme (present Fig. 6.9:4.), plus an "Indian" cell around 30–60 °E. Of these, the Pacific, South America, Atlantic and African cells are considered major, while two cells in the Indian Ocean sector (Indian and maritime continent) are regarded as minor with respect to strength and coherence of patterns.

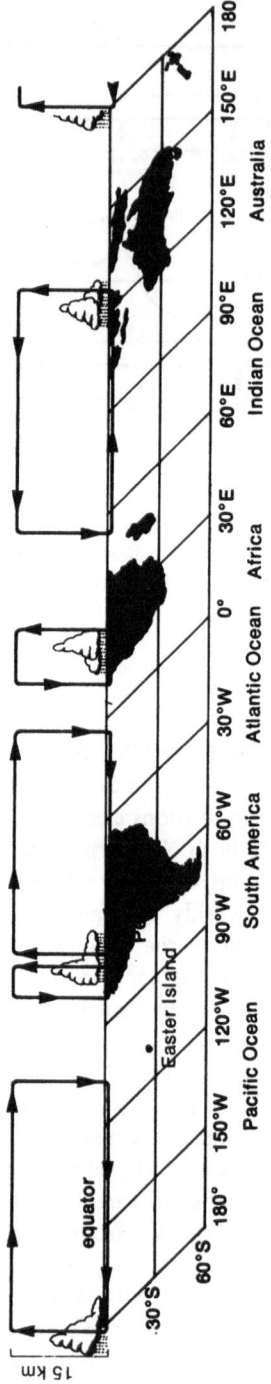

Fig. 6.9:4. Zonal circulation cells along the Equator. From Newell (1979).

6.10. Upper-Tropospheric Anticyclones

A remarkable feature of the large-scale circulation are the thermally induced anticyclones in the upper troposphere over certain low-latitude continental regions. The annual variation in their latitude and longitude positions is illustrated in Figs. 6.10:1 to 6.10:3.

Flohn (1965, 1968) demonstrated that latent and sensible heating related to thunderstorms over the Tibetan Plateau during Northern summer is primarily responsible for the buildup of an upper-level high over the Central Asian mountain massifs. As shown by Fig. 6.10:1, the anticyclonic center appears over Southeast Asia in May, and then moves northwestward reaching the Tibetan Plateau around the height of the Northern summer. From about September the anticyclonic cell migrates again southeastward towards Indonesia, and loses definition after about October.

The annual cycle behavior of this upper-tropospheric anticyclone is interesting in relation to the quasi-permanent surface low in the area of Indonesia and Southern Asia illustrated in Fig. 6.8.1:2. This sits in the greater Indonesia region during the Northern winter half-year (October–April), then also migrates northwestward to Northern summer (May–September) locations over Southern Asia, and thereafter shifts again southeastward towards Indonesia. However, data limitations particularly in the construction of Fig. 6.8.1:2 do not warrant more detailed comparisons beyond these gross annual cycle characteristics.

Conditions somewhat similar to the region just considered are found over Southern Africa during Southern summer. During this season, a surface heat low develops in the interior of Southern Africa, which is topped by a relative high in the upper troposphere (Serviço Meteorológico Nacional, Portugal, 1965, p. 106; Taljaard et al., 1969, p. 133, 21, 30; Hastenrath, 1974). The associated upper-tropospheric anticyclone also undergoes variations in position as illustrated in Fig. 6.10:2. The anticyclone appears in September at a low-latitude location, then migrates to its most poleward position in Southern summer (November–March); thereafter it migrates again equatorward and vanishes around May.

Of greater proportions is the upper-tropospheric anticyclone over South America (Fig. 6.10:3). This appears over the Amazon basin around September, then migrates towards the Peruvian-Bolivian Altiplano, reaching its strongest development in Southern summer or in December–March (Gutman and Schwerdtfeger, 1965; Virji, 1981; Chu and Hastenrath, 1982). At this time of year, the latent and sensible heating associated with thunderstorms (Gutman and Schwerdtfeger, 1965) over the elevated heating surface of the Altiplano may be particularly effective, in addition to the inflation of atmospheric columns over the tropical lowlands. Thereafter the anticyclone shifts equatorward and vanishes around April–May.

The behavior of the three upper-tropospheric anticyclones depicted in Figs. 6.10:1 to 6.10:3 agrees in major respects. All appear first in the equatorial region, then migrate to reach a most poleward position and strongest development at the height of the respective summer season, and thereafter shift again towards lower latitudes where they lose definition. The vigorous latent and sensible heating over extensive high mountain surfaces is instrumental in the particularly strong development of the upper-tropospheric anticyclones over Southern Asia and South America in summer.

CHAPTER 6

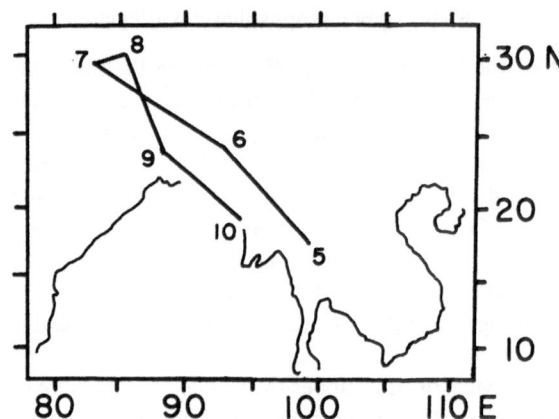

Fig. 6.10:1. Annual variation in the latitude and longitude position of upper-tropospheric (200 mb) anticyclonic center in the area of Southern Asia and Indonesia. The numbers 1 through 12 denote the calendar months. Source: Sadler (1975b).

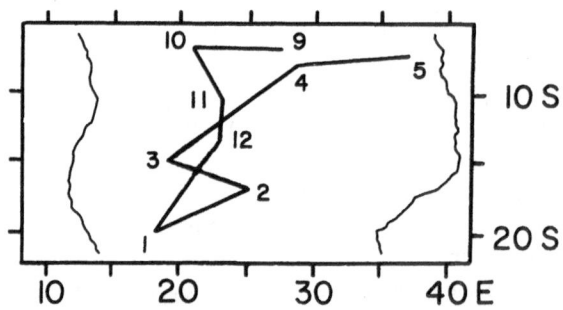

Fig. 6.10:2. Annual variation in the latitude and longitude position of upper-tropospheric (200 mb) anticyclonic center in the area of Southern Africa. The numbers 1 through 12 denote the calendar months. Source: Sadler (1975b).

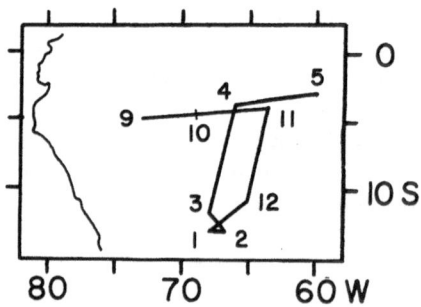

Fig. 6.10:3. Annual variation in the latitude and longitude position of upper-tropospheric (200 mb) anticyclonic center in the area of South America. The numbers 1 through 12 denote the calendar months. Source: Sadler (1975b).

6.11. Wind Regimes of the Equatorial Stratosphere

The eruption of Krakatoa in the Indonesian Archipelago in August 1883 injected large masses of volcanic ash high into the atmosphere and led to the detection of strong and persistent easterlies around 30 km (10 mb) over the equatorial zone, which became known as 'Krakatoa Easterlies'. In 1909 Berson found strong westerlies at 20 km (50 mb) over Lake Victoria in East Africa, and these were promptly named 'Berson Westerlies' (reviews in Reed *et al.*, 1961; Newell *et al.*, 1972, 1974, vol. 2, p. 204). From these discoveries arose the notion of the equatorial strato-spheric circulation as comprising the 'Krakatoa Easterlies' situated above the 'Berson Westerlies', a picture still presented in the review by Palmer (1954). It took half a century to dissipate this misconception.

At a conference of the American Meteorological Society in 1959, McCreary (1959, 1961) presented early analyses of the stratospheric wind regimes over the equatorial Pacific. Shortly thereafter, both Reed *et al.* (1961) and Veryard and Ebdon (1961) described an approximately 26-month alternation between easterlies and westerlies in the equatorial stratosphere. The ampli-tude of the oscillation is as large as 20 m s^{-1} between 40 and 10 mb over the Equator, with a decrease towards adjacent layers and higher latitudes. Easterly and westerly maxima propagate downward at a rate of about 1 km per month, so that the middle stratosphere and tropopause regions are out of phase (reviews in Reed, 1965; Palmén and Newton, 1969, p. 82; Riehl, 1979, pp. 31–34). The downward propagation of alternating easterly and westerly wind regimes is illustrated in Fig. 6.11:1. The pattern is remarkably consistent throughout the equatorial zone, and similar time-altitude plots for Canton Island in the Pacific as well as for various complete latitude circles are found in Reed *et al.* (1961), Reed (1965), Newell *et al.* (1972, 1974, vol. 2, pp. 206–210). The time series plots of zonal wind at 30 mb and 50 mb over Balboa, Panama, Fig. 6.11:2, further illustrate the impressive regularity of the quasi-biennial oscillation throughout decades.

The causes of the quasi-biennial oscillation in equatorial stratospheric winds are not yet understood. Variations with periodicity of somewhat larger than two years are found in many tropospheric and surface parameters (reviews in Brier, 1978, and Trenberth, 1980). As a rule these are small but are revealed by appropriate statistical analysis. It appears tempting to relate these to the pronounced signal in the equatorial stratosphere. However, on the basis of sophisticated spectral techniques Trenberth (1980) finds that the quasi-biennial oscillation in equatorial strato-spheric winds is not coherent with tropospheric variations of similar period. This cautions against high expectations to exploit the regular wind variations in the equatorial stratosphere for the prediction of lower-tropospheric processes.

6.12. Synthesis

The domain of tropical circulation is meaningfully delineated by the subtropical high pressure belts and anticyclonic axes of either hemisphere. From there the lower-tropospheric trade winds emanate to meet within a band of highest surface temperature and low pressure trough near the Equator. The seasonality of heat low processes is primarily responsible for the annual latitude migration of the equatorial trough.

In the low latitudes various jet stream systems are developed at different times of the year. The upper-tropospheric Subtropical Westerly Jet is a feature of the respective winter hemisphere and owes its existence to the convergence of the poleward transport of absolute angular momentum

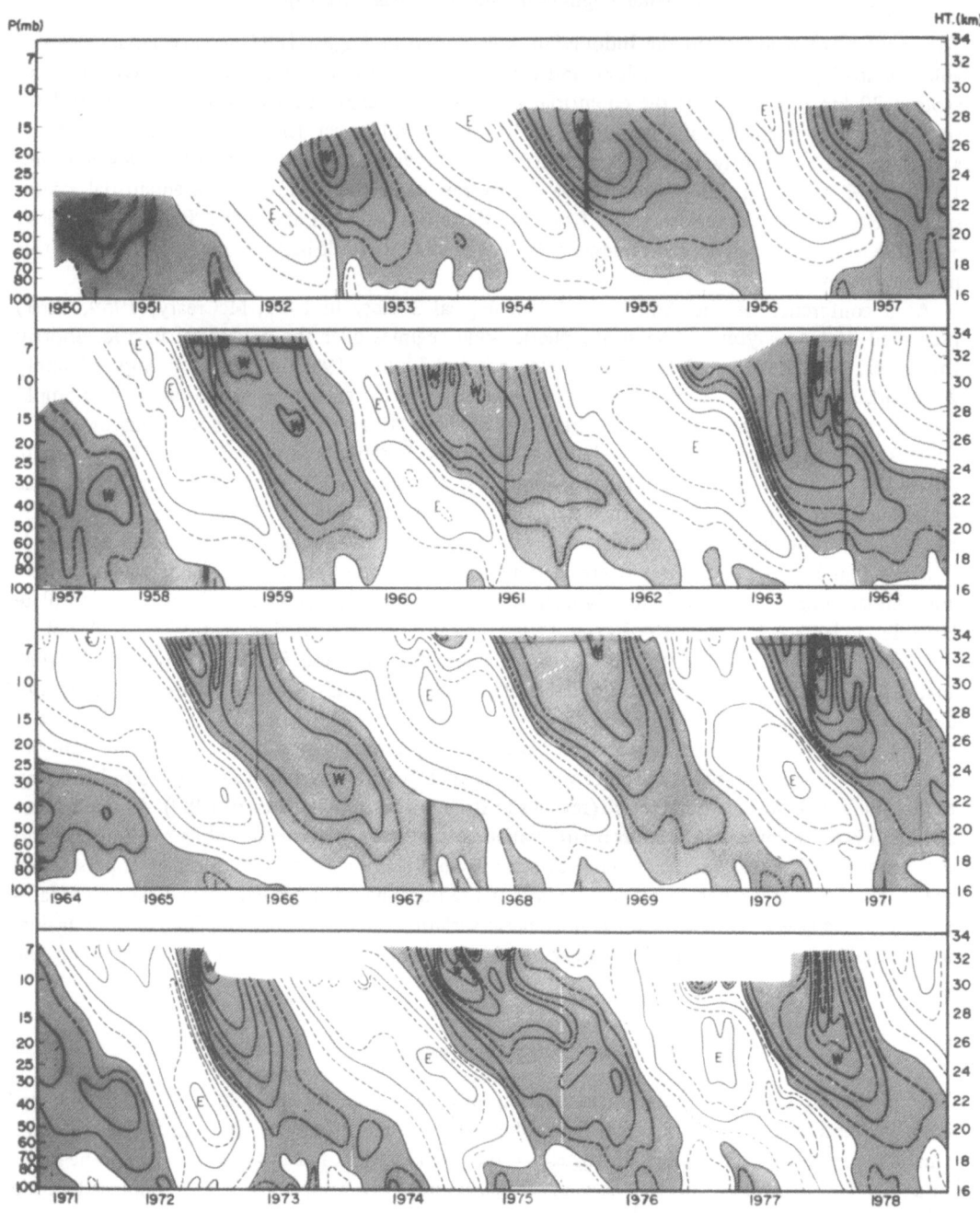

Fig. 6.11:1. Time-height section of the zonal wind near 9°N with the 15 year average of the monthly means subtracted to remove annual and semiannual cycles. Solid isotachs are placed at intervals of 10 m s^{-1}. Shaded areas indicate Westerlies. Data are from the Canal Zone through June 1970, and from Kwajalein thereafter. From Coy (1979, 1980) (*Journal of Atmospheric Sciences*, American Meteorological Society).

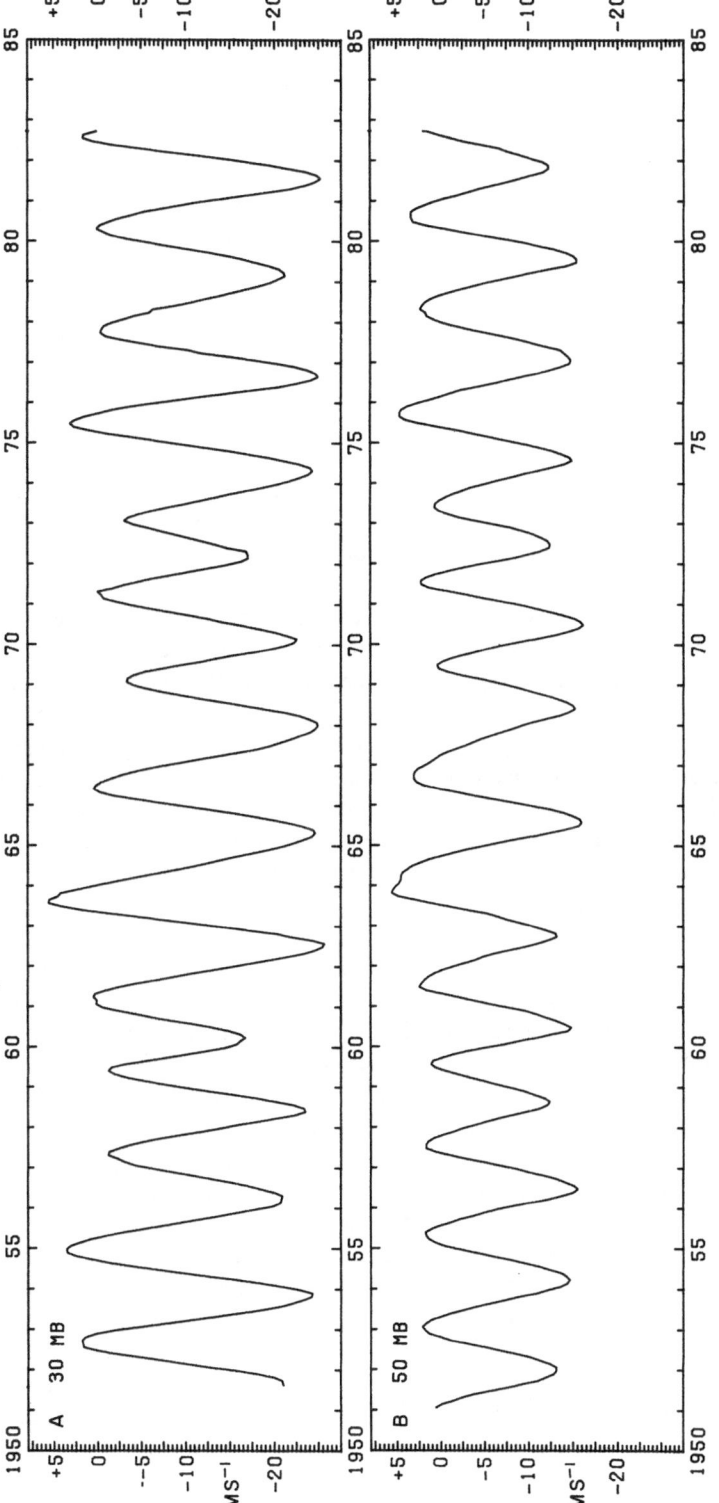

Fig. 6.11:2. Twelve months running means of (a) 30 mb and (b) 50 mb zonal wind over Balboa, Panama. Annual tick marks refer to values of January through December. Westerly positive, in m s⁻¹. Data courtesy of J. Angell, NOAA.

in the upper and poleward portion of the Hadley cells, and to other factors. The likewise upper-tropospheric Tropical Easterly Jet is confined to the height of the summer and is related to the thermal wind pattern associated with the then strongly heated subtropics and the cooler equatorial atmosphere. The best developed Tropical Easterly Jet system is found in July and August extending from Southeast Asia over the Indian Ocean and Africa to the Atlantic. Through cross-circulations in the entrance and exit regions it exerts a control on the surface climate. Corrollaries in other parts of the tropics are weak. The West African Mid-tropospheric Jet is also limited to Northern summer and related to the thermal wind pattern, in this instance associated with the hot desert air to the North and the cool monsoon air to the South of the Intertropical Discontinuity. The East African Low Level Jet appears in Northern summer at about 1 km extending with clockwise curvature from the Southern Indian Ocean across the Equator to the Arabian Sea. As 'backbone' of the Southwest monsoon circulation, it has received much attention both in the MONEX field experiment and in numerical modelling studies.

The subtropical highs are the source of the trades and function as major centers of action for the tropical circulation. They are located farthest away from the Equator during the respective summer, but in both hemispheres they assume a westernmost position in Northern summer. The trades represent the lower-tropospheric portion of the Hadley cells. They pick up moisture (and to a lesser extent sensible heat) from the tropical oceans, accumulate it below the trade inversion, and carry it into the equatorial trough zone, where rainfall and latent heat release are concentrated. The trades thus serve an important role in global energetics. The structure of the lower troposphere over vast expanses of the tropical oceans is characterized by the presence of a spatially continuous extremely stable layer, the trade inversion. This is lowest and best developed in the eastern equatorward sector of the subtropical highs, and rises and weakens both equatorward and towards the central and western part of the Atlantic and Pacific. Large-scale subsidence is the major factor in the origin and maintenance of the trade inversion, with advection and cooling from the surface contributing in limited regions. Mid-tropospheric inversions are common over both Southern and Northern hemispheric Africa and the adjacent oceans. Such a large-scale mid-tropospheric inversion is considered instrumental for the concentration of dust transported from the Sahara to the adjacent Atlantic.

The equatorial trough zone and associated quasi-permanent circulation features are in their development and annual latitude migration controlled by heat low mechanisms. Embedded within broad and coincident bands of high surface temperature and low pressure are an axis of confluence between airstreams of Northern and Southern hemispheric origin and a belt of maximum convergence-cloudiness-rainfall. The latter is typically located well away from the confluence axis, the separation being particularly large over Northern hemispheric Africa and the adjacent Atlantic. Thus insolation in the latitude of highest surface temperature and lowest pressure is ensured by comparatively scarce cloudiness. In the season and longitudes of strong development and far northerly location of the flow discontinuity, the moist cross-equatorial flow from the Southern hemisphere undercuts in a wedge fashion the Northeast trades. The dynamics of the cross-equatorial flow over the Eastern Atlantic and Pacific are dominated by the latitude variation of the Coriolis parameter. Among the resulting characteristics are a recurvature of flow from southeasterly to southwesterly at about 5°N, a speed maximum at that latitude, a band of divergence between Equator and recurvature, and a band of convergence poleward from it. A marked zonal component of the pressure gradient over the Indian Ocean during the Southwest monsoon entails characteristics of the cross-equatorial flow distinctly different from the Eastern Atlantic and Pacific. In addition to the aforementioned band of divergence to the North of the Atlantic Equator, an Equatorial Dry Zone extends across much of the Central Equatorial Pacific. Meridional climatic gradients from the Equatorial Dry Zone to the zonally oriented band of intense convergence-cloudiness-rainfall are among the steepest found on Earth in the absence of topographic effects.

The monsoon area of the World is delineated primarily in terms of the complete annual reversal of wind regimes, thus encompassing the Indian Ocean sector and much of tropical Africa. Over Subsaharan West Africa during Northern summer, a deep moist airstream from the Southern hemisphere replaces and undercuts the dry Northeast trades originating from the Sahara. In the Indian Ocean sector during Northern winter, weak dry winds sweep from Southern Asia across the Equator into the Southern hemisphere. Of far greater proportions is the Northern summer Southwest monsoon. The establishment of a heat-low induced monsoon trough over South Asia is instrumental in its development. In the South Indian Ocean Southeast trades recurve, cross the Equator, and continue into the South Asian continent as Southwest monsoon. Monsoon depressions and 'breaks in monsoon' are among the more important synoptic situations. On the Indian subcontinent and adjacent regions, the bulk of the annual precipitation falls during the Southwest monsoon. By contrast, the greater Indonesian region and parts of Southeast Asia receive much of their annual rainfall during the Northern winter monsoon, with Northeasterlies blowing over the South China Sea. More complicated regimes are found in Equatorial East Africa, where the precipitation peaks are timed in the monsoon transition seasons. The bulk of the water vapor brought to condensation over South Asia during the Northern summer Southwest monsoon stems from South of the Equator. Atmosphere, hydrosphere, and lithosphere are all essential for the energetics of the Southwest monsoon. Numerical simulation of the Indian monsoon has focused on the role of the mountains in the monsoon circulation, the effect of sea surface temperature anomalies on monsoon rainfall, and the dynamics of the East African Low Level Jet.

Zonal circulations along the Equator are caused by East-West differences in surface and tropospheric temperatures in surface and tropospheric temperatures. The Pacific 'Walker Circulation', and cells centered over the Atlantic and Indian Oceans are best developed, while controversy remains concerning some cells hypothesized over the continents and adjacent oceans.

The circulation of the equatorial stratosphere is characterized by an alternation of westerlies and easterlies on the time scale of about 26 month, and a propagation of changeovers from higher to lower layers. Quasi-biennial variations are found in many surface and tropospheric parameters, but a relation to the pronounced quasi-biennial oscillation in equatorial stratospheric winds is not established.

References

Alpert, L., 1945: 'The Intertropical Convergence Zone of the Eastern Pacific region, I'. *Bull. Amer. Meteor. Soc.*, **26**, 426–432.

Alpert, L., 1946a: 'The Intertropical Convergence Zone of the Eastern Pacific region, II'. *Bull. Amer. Meteor. Soc.*, **27**, 15–29.

Alpert, L., 1946b: 'The Intertropical Convergence Zone in the Eastern Pacific region, III'. *Bull. Amer. Meteor. Soc.*, **27**, 62–66.

Alpert, L., 1946c: 'Weather over the tropical Eastern Pacific Ocean, 7 and 8 March 1943'. *Bull. Amer. Meteor. Soc.*, **27**, 384–398.

Anderson, D. L. T., 1976: 'The low-level jet as a western boundary current'. *Mon. Wea. Rev.*, **104**, 907–921.

Anderson, D. L. T., 1984: 'An advective mixed-layer mode with applications to the diurnal cycle of the low-level East African jet'. *Tellus*, **36A**, 278–291.

Anjaneyulu, T. S. S., 1969: 'On the estimates of heat and moisture over the Indian monsoon trough zone'. *Tellus*, **21**, 64–75.

Anjaneyulu, T. S. S., 1971: 'Estimates of kinetic energy over the Indian monsoon trough zone'. *Quart. J. Roy. Meteor. Soc.*, **97**, 103–109.

Anonymous, 1967: 'Notes sur la structure du front intertropical'. ASECNA, Direction de l'Exploitation Météorologique, Notes, Traductions et Informations sélectionées, No. 16, Dakar, 34 pp.

Atkinson, G. D., 1971: *Forecaster's guide to tropical meteorology*. USAF, Air Weather Service, Technical Report No. 240, 341 pp.

Atkinson, G. D., Sadler, J. C., 1970: *Mean cloudiness and gradient-level-wind charts over the tropics*. USAF Air Weather Service, Technical Report No. 215, vol. 1 text, vol. 2 charts.

Augstein, E., 1972: 'Mass and heat budget estimations of the Atlantic SE trade wind flow at the Equator'. *"Meteor" – Forschungsergebnisse, Reihe B*, No. 8, pp. 31–41.

Ball, F. K., 1960: 'Control of inversion height by surface heating'. *Quart. J. Roy. Meteor. Soc.*, 86, 483–494.

Bannon, P. R., 1979a: 'On the dynamics of the East African Jet. I. Simulation of mean conditions for July'. *J. Atmos. Sc.*, 36, 2139–2152.

Bannon, P. R., 1979b: 'On the dynamics of the East African Jet. II. Jet transients'. *J. Atmos. Sci.*, 36, 2153–2168.

Basu, S., Pisharoty, P. R., Ramanathan, K. R., Bose, U. K., 1960: *Symposium on monsoons of the World, New Delhi, February 1958*. Hind Union Press, 271 pp.

Baumgartner, A., Reichel, E., 1975: *The World water balance*. Elsevier, Amsterdam, Oxford, New York, 179 pp.

Belinskii, V. A., 1961: *Dynamic meteorology*. Israel Program for Scientific Translations, Jerusalem, 591 pp.

Bhalotra, Y. P. R., 1963: 'Meteorology of Sudan'. Sudan Meteorological Service, Memoir No. 6, 155 pp.

Bjerknes, J., 1969: 'Atmospheric teleconnections from the Equatorial Pacific'. *Mon. Wea. Rev.*, 97, 163–172.

Boyle, J. S., Chang, C.-P., 1984: 'Monthly and seasonal climatology over the global tropics and subtropics for the decade 1973 to 1983. vol. 1, 200 mb winds'. Naval Postgraduate School, Monterey, California, Technical Report NPS–63–84–006, 172 pp.

Braak, C., 1921–29: *Het climaat van Nederlandsch Indië*. Magnetisch en Meteorologisch Observatorium te Batavia, Verhandelingen, No. 8, parts 1 and 2, 787 and 802 pp.

Brier, G. W., 1978: 'The quasi-biennial oscillation and feedback processes in the atmosphere-ocean-earth system'. *Mon. Wea. Rev.*, 106, 938–946.

Brümmer, B., Augstein, E., Riehl, H., 1974: 'On the low-level wind structure in the Atlantic trade'. *Quart J. Roy. Meteor. Soc.*, 100, 109–121.

Bryson, R. A., Lahey, J. F., 1958: 'The march of the seasons'. Department of Meteorology, University of Wisconsin, Madison. AFCRC–TR–58–223, ASTIA, No. AD–152 500, Final Report Contract AF 19(604)–992, 41 pp.

Burpee, R. W., 1972: 'The origin and structure of Easterly waves in the lower troposphere of North Africa'. *J. Atmos. Sci.*, 29, 77–90.

Cadet, D., 1983: 'The monsoon over the Indian Ocean during summer 1975. Part 2: break and active monsoons'. *Mon. Wea. Rev.*, 111, 95–108.

Cadet, D., Reverdin, G., 1981a: 'Water vapor transport over the Indian Ocean during the summer 1975'. *Tellus*, 33, 476–487.

Cadet, D., Reverdin, G., 1981b: 'The monsoon over the Indian Ocean during summer 1975, Part 1: mean fields'. *Mon. Wea. Rev.*, 109, 148–158.

Cane, M., Sarachik, E., 1983: 'Equatorial oceanography'. *Rev. Geophys. Space Phys.*, 21, 1137–1148.

Carlson, T. N., Prospero, J. M., 1972: 'The large-scale movement of Saharan air outbreaks over the Northern Equatorial Atlantic'. *J. Appl. Meteor.*, 11, 283–297.

Chang, C.-P., Erickson, J. E., Lau, K. M., 1979: 'Northeasterly cold surges and near-equatorial disturbances over the winter MONEX area during December 1974; part I: synoptic aspects'. *Mon. Wea. Rev.*, 107, 812–829.

Chang, C.-P., Lau, K. M. W., 1980: 'Northeasterly cold surges and near-equatorial disturbances over the winter MONEX area during December 1974. Part I: planetary-scale aspects'. *Mon. Wea. Rev.*, 108, 298–312.

Chang, C.-P., Lau, K. M., 1982: 'Short-term planetary-scale interactions over the tropics and midlatitudes during Northern winter, part I: contrasts between active and inactive periods'. *Mon. Wea. Rev.*, 110, 933–946.

Chang, C.-P., Millard, J. E., Chen, G. T. J., 1983: 'Gravitational character of cold surges during winter MONEX'. *Mon. Wea. Rev.*, 111, 293–307.

Chervin, R. M., Druyan, L. M., 1984: 'The influence of ocean surface temperature gradient and continentality on the Walker Circulation. Part 1: prescribed tropical changes.' *Mon. Wea. Rev.*, 112, 1510–1523.

Chu, P.-S., Hastenrath, S., 1982: *Atlas of upper-air circulation over tropical South America*. Department of Meteorology, University of Wisconsin, Madison, 237 pp.

Clackson, J. R., 1960: 'The seasonal movement of boundary of Northern air'. Nigerian Meteorological Service, Technical Note No. 5, 6 pp.

Coy, L., 1979: 'An unusually large westerly amplitude of the quasi-biennial oscillation'. *J. Atmos. Sci.*, 36, 174–176.

Coy, L., 1980: 'Corrigendum'. *J. Atmos. Sci.*, 37, 912–913.

Crutcher, H. L., Jenne, R. L., Taljaard, J. J., Van Loon, H., 1971: 'Climate of the upper air: Southern Hemisphere. vol. 4 – selected meridional cross sections of temperature, dewpoint, and height'. NOAA, NCAR, DOD, Washington D.C., 60 pp.

d'Almeida, G. A., Schütz, L., 1983: 'Number, mass and volume distributions of mineral aerosol and soils of the Sahara'. *J. Climate Appl. Meteor.*, 22, 233–243.

Das, P. K., 1962: 'Mean vertical motion and non-adiabatic heat sources and sinks over India during the monsoon'. *Tellus*, 14, 212–220.

Das, P. K., 1984: 'The monsoons – a perspective'. Indian National Science Academy, Perspective Report Series 4, New Delhi, 52 pp.

Das, P. K., 1986: 'Monsoons'. Fifth IMO lecture, World Meteorological Organization, in press.

Davidson, N. E., McBride, J. L., McAvaney, B. J., 1983: 'The onset of the Australian monsoon during winter MONEX: synoptic aspects'. *Mon. Wea. Rev.*, 111, 496–516.

Desai, B. N., 1967: 'The summer atmospheric circulation over the Arabian Sea'. *J. Atmos. Sci.*, 24, 216–220.

Desbois, M., Pircher, V., Pinty, B., 1984: 'Elements of the West African monsoon circulation deduced from METEOSAT cloud winds and simultaneous aircraft measurements'. *J. Climate Appl. Meteor.*, 23, 161–165.

Dettwiller, J., 1965: 'Note sur la structure du front intertropical boréal sur le Nord-Ouest de l'Afrique'. *La Météorologie*, 6, 337–347.

Dorman, C. E., 1982: 'Indian Ocean rainfall'. *Tropical Ocean-Atmosphere Newsletter*, 10, 4.

Dorman, C. E., Bourke, R. H., 1979: 'Precipitation over the Pacific Ocean, 30°N to 30°S'. *Mon. Wea. Rev.*, 107, 896–910.

Dorman, C. E., Bourke, R. H., 1981: 'Precipitation over the Atlantic Ocean, 30°S to 70°N'. *Mon. Wea. Rev.*, 109, 554–563.

Druyan, L. M., 1982a: 'Studies of the Indian summer monsoon with a coarse-mesh general circulation model, part 1'. *J. Climatol.*, 2, 127–139.

Druyan, L. M., 1982b: 'Studies of the Indian summer monsoon with a coarse-mesh general circulation model, part 2'. *J. Climatol.*, 2, 347–355.

Druyan, L. M., Miller, J. R., Russell, G. L., 1983: 'Atmospheric general circulation model simulations with an interactive ocean: effects of sea surface temperature anomalies in the Arabian Sea'. *Atmosphere-Ocean*, 21, 94–106.

Émon, J., 1948: 'L'inversion de l'alizé dans l'Océan Indian Sud-Ouest'. *Publications du Service Météorologique de Madagascar*, No. 11, 269 pp. + tables and figures.

Fein, J., S., Stephens, P. L., eds., 1986: *Monsoons*. Wiley Interscience Publishers, New York, London, Sidney, Toronto, in press.

Fieux, M., Stommel, H., 1977: 'Onset of the Southwest monsoon over the Arabian Sea from marine reports of surface winds: structure and variability'. *Mon. Wea. Rev.*, 105, 231–236.

Findlater, J., 1969a: 'A major low-level air current near the Indian Ocean during the Northern summer'. *Quart. J Roy. Meteor. Soc.*, 95, 362–380.

Findlater, J., 1969b: 'Interhemispheric transport of air in the lower troposphere over the Western Indian Ocean'. *Quart. J. Roy. Meteor. Soc.*, 95, 400–403.

Findlater, J., 1971: 'Mean monthly air flow at low levels over the Western Indian Ocean'. *Geophys. Memoirs*, No. 115, H.M.S.O., London, 53 pp.

Findlater, J., 1972: 'Aerial exploration of the low-level cross-equatorial current over Eastern Africa'. *Quart. J. Roy. Meteor. Soc.*, 98, 274–289.

Findlater, J., 1974: 'The low-level cross-equatorial air current of the Western Indian Ocean during the Northern summer'. *Weather*, 29, 411–416.

Findlater, J., 1977a: 'A numerical index to monitor the Afro-Asian monsoon during the Northern summer'. *Meteor. Mag.*, 106, 170–180.

Findlater, J., 1977b: 'Observational aspects of the low-level cross-equatorial jet stream of the Western Indian Ocean'. *Pure Appl. Geophys.*, 115, 1251–1262.

Flohn, H., 1957: 'Studien zur Dynamik der äquatorialen Atmosphäre. I. Horizontale und vertikale Windkomponenten auf dem Atlantik'. *Beitr. Phys. Atmos.*, 30, 18–46.

Flohn, H., 1964: 'Investigations of the tropical easterly jet'. *Bonner Meteorol. Abhandl.*, 4, 1–69.

Flohn, H., 1965: 'Thermal effects of the Tibetan Plateau during the Asian monsoon season'. *Australian Meteor. Mag.*, **49**, 55–58.

Flohn, H., 1968: 'Contributions to a meteorology of the Tibetan highlands'. Department of Atmospheric Science, Colorado State University, Fort Collins, Colo., Atmospheric Science Paper No. 130, 120 pp.

Flohn, H., 1971: 'Tropical circulation patterns'. *Bonner Meteorol. Abhandl.*, **15**, 1–55.

Flohn, H., Hantel, M., Ruprecht, E., 1968: 'Air-mass dynamics or subsidence processes in the Arabian Sea summer monsoon?' *J. Atmos. Sci.*, **25**, 527–529.

Frank, W. M., 1983: The structure and energetics of the East Atlantic Intertropical Convergence Zone. *J. Atmos. Sci.*, **40**, 1916–1929.

Germain, H., 1968: 'Météorologie dynamique et climatologie; application au régime des pluies au Sénégal'. ASECNA, Direction de l'Exploitation Météorologique, Dakar, Senegal, 15 pp.

Ghosh, S. K., Pant, M. C., Dewan, B. N., 1978: 'Influence of the Arabian Sea on the Indian Summer Monsoon'. *Tellus*, **30**, 117–124.

Gilchrist, A., 1977: 'A simulation of the African monsoon by general circulation models'. *Pure Appl. Geophys.*, **115**, 1431–1448.

Godbole, R. V., 1973: 'Numerical simulation of the Indian summer monsoon'. *Indian J. Meteor. Geophys.*, **24**, 1–13.

Godbole, R. V., Shukla, J., 1981: 'Global analysis of January and July sea level pressure'. NASA Technical Memorandum 82097, Goddard Space Flight Center, Greenbelt, Maryland, 52 pp.

Gordon, A. H., Taylor, R. C., 1975: 'Computations of surface layer air parcel trajectories, and weather, in the oceanic tropics'. International Indian Ocean Expedition, Meteor. Monogr. No. 7, University Press of Hawaii, 112 pp.

Greenhut, G. K., 1981: 'Analysis of aircraft measurements of momentum flux in the subcloud layer over the tropical Atlantic Ocean during GATE'. *Boundary Layer Meteorology*, **20**, 75–100.

Griffiths, J., 1972: 'Nigeria'. pp. 167–192 in Griffiths, J., ed., *Climates of Africa*, World Survey of Climatology, vol. 10, Elsevier, Amsterdam, London, New York, 604 pp.

Guedalia, D., Estournel, C., Vehil, R., 1984: 'Effects of Sahel dust layers upon nocturnal cooling of the atmosphere (ECLATS Experiment)'. *J. Climate Appl. Meteor.*, **23**, 644–650.

Gutman, G., Schwerdtfeger, W., 1965: 'The role of latent and sensible heat for the development of a high pressure system over the subtropical Andes in the summer'. *Meteor. Rundschau*, **18**, 1–7.

Gutnick, M., 1958: 'Climatology of the trade-wind inversion in the Caribbean'. *Bull. Amer. Meteor. Soc.*, **39**, 410–420.

Hadley, G., 1735: 'Concerning the cause of the general trade-winds. Philosophical Transactions', **29**, 58–62.

Hahn, D. G., Manabe, S., 1975: 'The role of mountains in the South Asian monsoon circulation'. *J. Atmos. Sci.*, **32**, 1515–1541.

Hahn, D. G., Manabe, S., 1976: 'Reply (to comments by J. C. Sadler and C. S. Ramage)'. *J. Atmos. Sci.*, **33**, 2258–2262.

Halley, E., 1686: 'An historical account of the trade-winds and monsoons observable in the seas between and near the tropics with an attempt to assign the physical cause of said winds'. *Philosophical Transactions*, **26**, 153–168.

Haltiner, G. J., Martin, F. L., 1957: *Dynamical and physical meteorology*. McGraw-Hill, New York, Toronto, London, 470 pp.

Hamilton, R. A., Archbold, J. W., 1945: 'Meteorology of Nigeria and adjacent territory'. *Quart. J. Roy. Meteor. Soc.*, **71**, 231–265.

Hantel, M., 1972: 'Wind stress curl – the forcing function for oceanic motions'. *Studies in Physical Oceanography, tribute to Georg Wüst on his 80th birthday*. Ed. A. C. Gordon, Gordon and Breach, New York and London, **1**, pp. 121–131.

Haraguchi, P. Y., 1968: 'Inversion over the tropical Eastern Pacific'. *Mon. Wea. Rev.*, **96**, 177–185.

Hart, J. E., 1977: 'On the theory of the East Africa low-level jet stream'. *Pure Appl. Geophys.*, **115**, 1263–1282.

Hart, J. E., Rao, G. V., Van de Boogaard, H. M. E., Young, J. A., Findlater, J., 1978: 'Aerial observations of the East African low-level jet stream'. *Mon. Wea. Rev.*, **106**, 1714–1724.

Hastenrath, S., 1966: 'On general circulation and energy budget in the area of the Central American Seas'. *J. Atmos. Sci.*, **23**, 694–711.

Hastenrath, S., 1967a: 'Rainfall distribution and regime in Central America'. *Archiv Meteor. Geophys. Bioklim.*, Ser. B, **15**, 201–241.

Hastenrath, S., 1967b: 'The barchans of the Arequipa region, Southern Peru'. *Zeitschrift für Geomorphologie*, **11**, 300–331.

Hastenrath, S., 1971: 'On meridional circulation and heat budget of the troposphere over the Equatorial Central Pacific'. *Tellus*, **23**, 60–73.

Hastenrath, S., 1974: 'Zur Kenntnis der Windverhältnisse über Angola'. *Bonner Meteor. Abhandl.*, **17**, 353–360.

Hastenrath, S., 1977a: 'On the upper-air circulation over the equatorial Americas'. *Archiv Meteor. Geophys. Bioklim.*, *Ser. A*, **25**, 309–321.

Hastenrath, S., 1977b: 'Hemispheric asymmetry of oceanic heat budget in the Equatorial Atlantic and Eastern Pacific'. *Tellus*, **29**, 523–529.

Hastenrath, S., Guetter, P., 1978: 'A contribution to the surface circulation over South America'. *Archiv Meteor. Geophys. Bioklim.*, *Ser. B*, **26**, 97–103.

Hastenrath, S., Hafez, A., Kaczmarczyk, E. B., 1979: 'A contribution to the dynamic climatology of Arabia'. *Archiv Meteor. Geophys. Bioklim.*, *Ser. B*, **27**, 105–120.

Hastenrath, S., Lamb, P. J., 1977a: *Climatic atlas of the tropical Atlantic and Eastern Pacific Oceans*. University of Wisconsin Press, 112 pp.

Hastenrath, S., Lamb, P. J., 1977b: 'Some aspects of circulation and climate over the Eastern Equatorial Atlantic'. *Mon. Wea. Rev.*, **105**, 1019–1023.

Hastenrath, S., Lamb, P. J., 1978a: 'On the dynamics and climatology of surface flow over the equatorial oceans'. *Tellus*, **30**, 436–448.

Hastenrath, S., Lamb, P. J., 1978b: *Heat budget atlas of the tropical Atlantic and Eastern Pacific Oceans*. University of Wisconsin Press, 103 pp.

Hastenrath, S., Lamb, P. J., 1979: *Climatic atlas of the Indian Ocean*. Part I. *Surface climate and atmospheric circulation*. Part II. *The oceanic heat budget*. University of Wisconsin Press, 116 and 110 pp.

Hastenrath, S., Lamb, P. J., 1980: 'On the heat budget of hydrosphere and atmosphere in the Indian Ocean'. *J. Phys. Oceanogr.*, **10**, 694–708.

Hastenrath, S., Merle, J., 1986: Annual cycle of subsurface thermal structure in the tropical Atlantic Ocean. submitted for publication.

Hergesell, H., 1905a: 'Sur les ascensions de cerf-volants executées sur la Méditerranée et sur l'Océan Atlantique au bord du yacht de S.A.S. le Prince de Monaco en 1904'. *Comptes Rendues de l'Académie des Sciences, Paris, 30 Janvier, 1905*, 331–333.

Hergesell, H., 1905b: 'Ascensions de balloon-sondes executées au-dessus de la mer par S.A.S. le Prince de Monaco au mois d'Avril 1905'. *Comptes Rendues de l'Académie des Sciences, 5 Juin, 1905*, 1569–1571.

Hergesell, H., 1906a: 'Ballon-Aufstiege über dem freien Meere zur Erforschung der Temperatur- und Feuchtigkeitsverhältnisse sowie der Luftströmungen bis zu sehr grossen Höhen der Atmosphäre'. *Beiträge zur Physik der freien Atmosphäre*, **1**, 200–204.

Hergesell, H., 1906b: 'Die Erforschung der freien Atmosphäre über dem Atlantischen Ozean nördlich des Wendekreises des Krebses an Bord der Yacht Seiner Durchlaucht des Fürsten von Monaco im Jahre 1905'. *Beiträge zur Physik der freien Atmosphäre*, **1**, 205–207.

Hergesell, H., 1907: 'Über lokale Windströmungen in der Nähe der Kanarischen Inseln'. *Beiträge zur Physik der freien Atmosphäre*, **2**, 51–54.

Holton, J. R., Wallace, J. M., Young, J. A., 1971: 'On boundary layer dynamics and the ITCZ'. *J. Atmos. Sci.*, **28**, 275–280.

Howland, M. R., Sikdar, D. N., 1983: 'The moisture budget over the northeastern Arabian Sea during premonsoon and monsoon onset, 1979'. *Mon. Wea. Rev.*, **111**, 2255–2268.

India Meteorological Department, 1965: *Proceedings of the symposium on meteorological results of the International Indian Ocean Expedition. 22–26 July 1965, Bombay*. Bombay, 437 pp.

Jaw, J., 1937: 'Zur Thermodynamik der Passat-Grundströmung'. *Veröffentlichungen Meteorol. Institut, Universität Berlin*, **2**, no. 6, 24 pp.

Jenne, R. L., Crutcher, H. L., Van Loon, H., Taljaard, J. J., 1971: 'Climate of the upper air: Southern Hemisphere. vol. 3 – vector mean geostrophic winds'. NCAR – TN/STR–58, NCAR, NOAA, DOD, Washington, D.C., 62 pp.

Katzin, M., Pezzner, H., Koo, B. Y. C., Larson, J. V., Katzin, J. C., 1960: 'The trade-wind inversion as a transoceanic duct'. *Journal of Research, National Bureau of Standards*, **64D**, 247–253.

Kiangi, P. M. R., Kavisha, M. M., Patnaik, J. K., 1981: 'Some aspects of the mean tropospheric motion field in East Africa during the long rainy season'. *Kenya Journal of Science and Technology, A*, **2**, 91–103.

Kinuthia, J. H., Asnani, G. C., 1982: 'A newly found jet in North Kenya (Turkana Channel)'. *Mon. Wea. Rev.*, **110**, 1722–1728.

Koninklijk Nederlands Meteorologisch Instituut, 1918–31: 'Oceanographische en meteorologische waarnemingen in den Atlantischen Oceaan'. Publication No. 110, four table volumes and one atlas volume.

Koninklijk Nederlands Meteorologisch Instituut, 1952: 'Indian Ocean oceanographic and meteorological data'. Publication No. 135, 2nd ed., 2 vols, 48 and 31 pp.

Koteswaram, P., 1958: 'The easterly jet stream in the tropics'. *Tellus*, 10, 43–57.

Kraus, E. B., 1959: 'The evaporation-precipitation cycle of the trades'. *Tellus*, 11, 147–158.

Krishnamurti, T. N., 1961: 'The subtropical jet stream of winter'. *J. Meteor.*, 18, 172–191.

Krishnamurti, T. N., 1971: 'Tropical East-West circulations during the Northern summer'. *J. Atmos. Sci.*, 28, 1342–1347.

Krishnamurti, T. N., ed., 1977: 'Special volume dedicated to Indian monsoon problems'. *Pure Appl. Geophys.*, 115, 1082–1529.

Krishnamurti, T. N., 1979: *Tropical meteorology*. Compendium of meteorology, vol. 2, part 4, editor, A. Wiin-Nielsen, WMO No. 364, World Meteorological Organization, Geneva, 428 pp.

Krishnamurti, T. N., Ardanuy, P., 1980: 'The 10 to 20-day westward propagating mode and "Breaks in the Monsoons"'. *Tellus*, 32, 15–26.

Krishnamurti, T. N., Ardanuy, P., Ramanathan, Y., Pasch, R., 1981: 'On the onset vortex of the summer monsoon'. *Mon. Wea. Rev.*, 109, 344–363.

Krishnamurti, T. N., Bhalme, H. N., 1976: 'Oscillations of a monsoon system. Part 1. observational aspects'. *J. Atmos. Sci.*, 33, 1937–1954.

Krishnamurti, T. N., Kanamitsu, M., Koss, J. W., Lee, J. D., 1973: 'Tropical East-West circulation during the Northern winter'. *J. Atmos. Sci.*, 30, 780–787.

Krishnamurti, T. N., Molinari, J., Pan, H.-L., 1976: 'Numerical simulation of the Somali Jet'. *J. Atmos. Sci.*, 33, 2350–2362.

Krishnamurti, T. N., Ramanathan, Y., 1982: 'Sensitivity of the monsoon onset to differential heating'. *J. Atmos. Sci.*, 39, 1290–1306.

Krishnamurti, T. N., Subrahmanyan, D., 1982: 'The 30–50 day mode at 850 mb during MONEX'. *J. Atmos. Sci.*, 39, 2088–2095.

Krishnamurti, T. N., Wong, V., 1979: 'A planetary boundary-layer model for the Somali Jet'. *J. Atmos. Sci.*, 36, 1895–1907.

Krishnamurti, T. N., Wong, V., Pan, H.-L., Pasch, R., Molinari, J., Ardanuy, P., 1983: 'A three-dimensional planetary boundary layer model for the Somali Jet'. *J. Atmos. Sci.*, 40, 894–908.

Krueger, A. F., Winston, J. S., 1974: 'A comparison of the flow over the tropics during two contrasting circulation regimes'. *J. Atmos. Sci.*, 31, 358–370.

Krueger, A. F., Winston, J. S., 1975: 'Large-scale circulation anomalies over the tropics during 1971–72'. *Mon. Wea. Rev.*, 103, 465–473.

Krueger, A. F., Winston, J. S., 1979: 'Further analysis of recent fluctuations in circulation and cloudiness (rainfall) over the tropics'. pp. 84–93 in: *Proceedings of Fourth Annual Climate Diagnostics Workshop, October 1979, Madison, Wis.*, NOAA, Washington, D.C., 439 pp.

Kuhlbrodt, E., 1928: 'Das Strömungsfeld der Luft über dem tropischen Atlantischen Ozean nach den Höhenwindmessungen der Meteor-Expedition'. *Zeitschrift für Geophysik*, 4, 385–386.

Lamb, P. J., 1978: 'Case studies of Tropical Atlantic surface circulation patterns during recent Subsaharan weather anomalies: 1967 and 1968'. *Mon. Wea. Rev.*, 106, 482–491.

Lamb, P. J., 1983: 'On the mixed-layer climatology of the North and tropical Atlantic'. *Tellus*, 35A, 198–212.

Laskowski, H., 1937: 'Das Feld der vertikalen Strömungskomponente im Passatgebiet'. *Veröffentlichungen Meteorol. Institut, Universität Berlin*, 2, no. 1, 19 pp.

Lau, K.-M., Li, M.-T., 1984: 'The monsoon of East Asia and its global associations – a survey'. *Bull. Amer. Meteor. Soc.*, 65, 114–125.

Lee, I.-Y., 1983: 'Simulation of transport and removal processes of the Saharan dust'. *J. Climate Appl. Meteor.*, 22, 632–639.

Leroux, M., 1970: *La dynamique des précipitations en Afrique Occidentale*. Notes de la Direction de l'Exploitation Météorologique, No. 39, ASECNA, Dakar, 281 pp.

Leroux, M., 1983: *Le climat de l'Afrique tropicale*. Champion, Paris, 2 vols., 633 pp. and 24 pp. plus 250 maps.

Lettau, H. H., and Lettau, K., 1969: 'Bulk transport of sand by the barchans of the Pampa de La Joya in Southern Peru'. *Zeitschrift für Geomorphologie*, 13, 182–195.

Lettau, H. H., and Lettau, K., eds., 1978: *Exploring the World's driest climate*. University of Wisconsin, Madison, IES Report No. 101, 264 pps.

Lighthill, J., Pearce, R. P., eds., 1981: *Monsoon dynamics*. Cambridge University Press, 735 pp.

Lim, H., Chang, C. P., 1981: 'A theory for midlatitude forcing of tropical motions during winter monsoons'. *J. Atmos. Sci.*, **38**, 2377–2392.

Lockwood, J. G., 1974: *World climatology: an environment approach*, Arnold, London, 330 pp.

Luo, H., Yanai, M., 1983: 'The large-scale circulation and heat sources over the Tibetan Plateau and surrounding areas during the early summer of 1979. Part I: precipitation and kinematic analyses'. *Mon. Wea. Rev.*, **111**, 922–944.

Luo, H., Yanai, M., 1984: 'The large-scale circulation and heat sources over the Tibetan Plateau and surrounding areas during the early summer of 1979. Part II: heat and moisture budgets'. *Mon. Wea. Rev.*, **112**, 966–989.

McCreary, F. E., 1959: 'Stratospheric winds over the tropical Pacific Ocean'. p. 370, abstract, in: Program of the conference on stratospheric meteorology, 179th National Meeting of the Anerumian Meteorological Society, August–September 1959 at Minneapolis. *Bull. Amer. Meteor. Soc.*, **40**, 366–375.

McCreary, F. E., 1961: 'Variations of the zonal wind in the equatorial stratosphere'. JFMC TF 20, Joint Task Force. Seven Meteorological Center, Pearl Harbor, Honolulu.

Merle, J., 1978: *Atlas hydrologique saisonnier de l'Océan Atlantique Intertropical*. Trav. Doc. ORSTOM, No. 82, 184 pp., 153 cartes.

Mintz, Y., Dean, G., 1952: 'The observed mean field of motion in the atmosphere'. Geophys. Res. Papers No. 17, Air Force Cambridge Research Center, Cambridge, Mass., 55 pp.

Mishra, S. K., Tandon, M. K., 1983: 'A combined barotropic-baroclinic instability study of the upper tropospheric Tropical Easterly Jet'. *J. Atmos. Sci.*, **40**, 2708–2723.

Morales, C., ed., 1979: *Saharan dust, mobilization, transport, deposition*. SCOPE report No. 14, Wiley and Sons, Chichester, New York, Brisbane, Toronto, 297 pp.

Mower, R. N., Chu, J.-H., Martin, D. W., Auvine, B., 1984: 'Mean state of the troposphere over South-east Asia and the East Indies, December 1978'. *Quart. J. Roy. Meteor. Soc.*, **110**, 1023–1033.

Murakami, T., 1979: 'Winter monsoon surges over East and Southeast Asia'. *J. Meteor. Soc. Japan*, **57**, 133–158.

Murakami, T., 1981a: 'Orographic influence of the Tibetan Plateau on the Asiatic winter monsoon circulation. Part I. Large-scale aspects'. *J. Meteor. Soc. Japan*, **59**, 40–65.

Murakami, T., 1981b: 'Orographic influence of the Tibetan Plateau on the Asiatic winter monsoon circulation. Part II. Diurnal variations'. *J. Meteor. Soc. Japan*, **59**, 66–84.

Murakami, T., 1981c: 'Orographic influence of the Tibetan Plateau on the Asiatic winter monsoon circulation. Part III. Short-period oscillations'. *J. Meteor. Soc. Japan*, **59**, 173–200.

Murakami, T., 1981d: 'Orographic influence of the Tibetan Plateau on the Asiatic winter monsoon circulation. Part IV. Long-period oscilllations'. *J. Meteor. Soc. Japan*, **59**, 201–219.

Murakami, T., Godbole, R. V., Kelkar, R. R., 1970: 'Numerical simulation of the monsoon along 80°E'. *Proceedings Conference Summer Monsoon of Southeast Asia*. Navy Weather Research Facility, Norfolk, Virginia, pp. 39–51.

Murakami, T., Nakasawa, T., He, J. H., 1984: 'On the 40–50 days oscillations during the 1979 Northern hemisphere summer part 2: heat and moisture budget'. *J. Meteor. Soc. Japan*, **62**, 469–484.

National Research Council, U.S. Committee for GARP, 1977: 'Plan for U.S. participation in the Monsoon Experiment (MONEX)'. National Academy of Sciences, Washington, D.C., 126 pp.

Neiburger, M., 1960: 'The relation of air mass structure to the field of motion over the eastern North Pacific Ocean in summer'. *Tellus*, **12**, 31–40.

Neiburger, M., Johnson, D. S., Chien, C. W., 1961: 'Studies of the structure of the atmosphere over the Eastern Pacific Ocean. I: the inversion over the Eastern North Pacific Ocean'. *University of California Publications in Meteorology*, **1**, no. 1, University of California Press, 94 pp.

Newell, R. E., 1979: 'Climate and the ocean'. *Amer. Sci.*, **67**, 405–416.

Newell, R. E., Kidson, J. W., Vincent, D. G., Boer, G. J., 1972, 1974: *The general circulation of the tropical atmosphere and interactions with extratropical latitude*. MIT Press, 2 vols., 258 and 371 pp.

Osman, O. E., Hastenrath, S., 1969: 'On the synoptic climatology of summer rainfall over Central Sudan'. *Archiv Meteor. Geophys. Bioklim., Ser. B.*, **17**, 297–324.

Palmén, E., Newton, C. W., 1969: *Atmospheric circulation systems*. Academic Press, New York and London, 603 pp.

Palmer, C. E., 1954: 'The general circulation between 200 mb and 10 mb over the equatorial Pacific'. *Weather*, **9**, 341–349.

Pant, P. S., 1983: 'A physical basis for changes in the phases of the summer monsoon over India'. *Mon. Wea. Rev.*, 111, 487–495.

Patnaik, J. K., Rao, R. R., Ramanadham, R., 1977: 'Some characteristics of Indian monsoon rains'. *Indian Geogr. J.*, 52, 23–30.

Pearce, R. P., Mohanty, U. C., 1984: 'Onsets of the Asian summer monsoon 1979–82'. *J. Atmos. Sci.*, 41, 1620–1639.

Peixóto, J. P., Oort, A. H., 1983: 'The atmospheric branch of the hydrological cycle and climate'. pp. 5–65, in Street-Perrott, A., Beran, M., Ratcliffe, R., eds.: *Variations in the global water budget.* Reidel, Dordrecht, Boston, Lancaster, 518 pp.

Philander, S. G. H., Pacanowski, R. C., 1980: The generation of equatorial currents. *J. Geophys. Res.*, 85, 1123–1136.

Philander, S. G. H., Pacanowski, R. C., 1981: The oceanic response to cross-equatorial winds (with application to coastal upwelling in low latitudes). *Tellus*, 33, 201–210.

Piazzi-Smyth, C., 1858: 'Astronomical experiment on the peak of Teneriffe'. *Philosophical Transactions, Royal Society London*, 48, 465–533.

Pike, A. C., 1972: 'Response of a tropical atmosphere and ocean model to seasonally variable forcing'. *Mon. Wea. Rev.*, 100, 424–433.

Pisharoty, P. R., 1965: 'Evaporation from the Arabian Sea and the Indian Southwest monsoon.. pp. 43–54 in: *Proceedings of Symposium on Meteorological Results of the International Indian Ocean Expedition, Bombay, India*, Meteorological Department–INCOR–WMO–UNESCO, 437 pp.

Prospero, J. M., Carlson, T. N., 1972: 'Vertical and areal distribution of Saharan dust over the Western Equatorial North Atlantic'. *J. Geophys. Res.*, 77, 5255–5265.

Ramage, C. S., 1952: 'Relationship of general circulation to normal weather over Southern Asia and the Western Pacific during the cool season'. *J. Meteor.*, 9, 403–408.

Ramage, C. S., 1954: 'Non-frontal crachin and the cool season clouds of the China Seas'. *Bull. Amer. Meteor. Soc.*, 35, 404–411.

Ramage, C. S., 1955: 'The cool-season tropical disturbances of Southeast Asia'. *J. Meteor.*, 12, 252–262.

Ramage, C. S., 1966: 'The summer atmospheric circulation over the Arabian Sea'. *J. Atmos. Sci.*, 23, 144–150.

Ramage, C. S., 1971: *Monsoon meteorology*. Academic Press, New York and London, 296 pp.

Ramage, C. S., 1974: 'Structure of an oceanic near-equatorial trough deduced from research aircraft traverses'. *Mon. Wea. Rev.*, 102, 754–759.

Ramage, C. S., Khalsa, S. J. S., Meisner, B. N., 1981: 'The central Pacific near-equatorial convergence zone'. *J. Geophys. Res.*, 86, No. C7, 6580–6598.

Ramage, C. S., Miller, F. R. Jefferies, C., 1972: *Meteorological atlas of the International Indian Ocean Expedition.* vol. 1. *The surface climate of 1963 and 1964.* National Science Foundation, U.S. Government Printing Office, 144 charts.

Ramage, C. S., Raman, C. V. R., 1972: *Meteorological atlas of the International Indian Ocean Expedition.* vol. 2. *Upper air.* National Science Foundation, U.S. Government Printing Office, 121 charts.

Ramanadham, R., Rao, P. V., Patnaik, J. K., 1973: 'Break in the Indian summer monsoon'. *Pure Appl. Geophys.*, 104, 635–647.

Rao, G. V., Van de Boogaard, H. M. E., Bolhofer, W. C., 1978: 'Further calculations of sea level air trajectories over the equatorial Indian Ocean'. *Mon. Wea. Rev.*, 106, 1465–1475.

Rao, G. V., Schaub, W. R., Puetz, J., 1981: 'Evaporation and precipitation over the Arabian Sea during several monsoon seasons'. *Mon. Wea. Rev.*, 109, 364–370.

Rao, K. V., Rajamani, S., 1972: 'Study of heat sources and sinks and the generation of available potential energy in the Indian region during the SW-monsoon season'. *Mon. Wea. Rev.*, 100, 383–388.

Rao, Y. P., 1976: *Southwest monsoon*. India Meteorological Department, Meteorological Monograph, Synoptic Meteorology No. 1/1976, Delhi, 367 pp.

Rauh, W., 1952: 'Vegetationsstudien im Hohen Atlas und dessen Vorland'. *Sitzungsberichte der Heidelberger Akademie der Wissenschaften, Math.-Naturwiss. Klasse*, year 1952, No. 1, 118 pp.

Reed, R. J., 1965: 'The present status of the 26 month oscillation'. *Bull. Amer. Meteor. Soc.*, 46, 374–387.

Reed, R. J., Campbell, W. J., Rasmusson, L. A., Rogers, D. G., 1961: 'Evidence of a downward-propagating annual wind reversal in the equatorial stratosphere'. *J. Geophys. Res.*, 66, 813–818.

Reiter, E. R., 1963: *Jet stream meteorology*. University of Chicago Press, Chicago, London, 515 pp.

Riehl, H., 1962: 'Jet streams of the atmosphere'. Department of Atmospheric Science, Colorado State University, Fort Collins, Colorado, Technical Report No. 32, 117 pp.

Riehl, H., 1979: *Climate and weather in the tropics*. Academic Press, London, New York, San Francisco, 611 pp.

Riehl, H., Malkus, J. S., 1957: 'On the heat balance and maintenance of circulation in the trades'. *Quart. J. Roy. Meteor. Soc.,* 83, 21–29.

Riehl, H., Malkus, J. S., 1958: 'On the heat balance of the equatorial trough zone'. *Geophysica,* 6, 503–538.

Riehl, H., Yeh, T. C., Malkus, J. S., La Seur, N. E., 1951: 'The Northeast trades of the Pacific Ocean'. *Quart. J. Roy. Meteor. Soc.,* 77, 598–626.

Robinson, M. K., 1976: *Atlas of North Pacific Ocean monthly mean temperatures and mean salinities of the surface layer*. NAVOCEANO, Washington, D.C., 193 pp.

Robinson, M. K., Bauer, R. A., Schroeder, E. H., 1979: *Atlas of North Atlantic – Indian Ocean monthly mean temperatures and mean salinities of the surface layer*. Naval Oceanographic Office, NOO RP–18, Naval Oceanographic Office, NSTL Station, Bay St. Louis, Mississippi, 234 pp.

Romanov, Yu. A., Romanova, N. A., 1976: 'Some results of the surface wind and atmospheric pressure fields analysis near the Equator'. *Tellus,* 28, 524–532.

Rouse, H., Dodu, J., 1955: 'Diffusion turbulente à travers une discontinuité de densité'. *La Houille Blanche,* 10, 522–532.

Sadler, J. C., 1975a: 'The monsoon circulation and cloudiness over the GATE area'. *Mon. Wea. Rev.,* 103, 369–387.

Sadler, J. C., 1975b: 'The upper tropospheric circulation over the global tropics'. Department of Meteorology, University of Hawaii, UHMET–75–05, 35 pp.

Sadler, J. C., Ramage, C. S., 1976: 'Comments on "The role of the mountains in the South Asian monsoon circulation"'. *J. Atmos. Sci.,* 33, 2255–2258.

Saha, K., 1974: 'Some aspects of the Arabian Sea summer monsoon'. *Tellus,* 26, 464–476.

Saha, K., Bavadekar, S. N., 1973: 'Water vapor budget and precipitation over the Arabian Sea during Northern summer'. *Quart. J. Roy. Meteor. Soc.,* 99, 273–278.

Sansom, H. W., 1965: 'The structure and behavior of the ITCZ'. pp. 91–108 in: *WMO Technical Note,* No. 69, Geneva, 310 pp.

Scherhag, R., and collaborators, 1969: 'Klimatologische Karten der Nordhemisphäre'. Institut für Meteorologie und Geophysik der Freien Universität Berlin, *Meteorologische Abhandlungen,* 100, no. 1, 223 pp.

Schnapauff, W., 1937: Untersuchungen über die Kalmenzone des Atlantischen Ozeans. *Veröffentlichungen, Meteorol. Institut, Universität Berlin,* 2, no. 4, 35 pp.

Schopf, P. S., Cane, M. A., 1983: 'On equatorial dynamics, mixed layer physics and sea surface temperature'. *J. Phys. Oceanogr.,* 13, 517–535.

Serviço Meteorológico Nacional, Portugal, 1965: *Climatología dinâmica da Africa meridional*. Lisboa, 207 pp.

Shukla, J., 1975: 'Effect of Arabian Sea surface temperature anomaly on Indian summer monsoon: a numerical experiment with the GFDL model'. *J. Atmos. Sci.,* 32, 503–511.

Shukla, J., 1976: 'Reply (to comments by D. R. Sikka and K. Raghavan)'. *J. Atmos. Sci.,* 33, 2253–2255.

Sikka, D. R., Gadgil, S., 1980: 'On the maximum cloud zone and the ITCZ over India longitude during the Southwest monsoon'. *Mon. Wea. Rev.,* 108, 1840–1853.

Sikka, D. R., Raghavan, K., 1976: 'Comments on "Effects of Arabian Sea – surface temperature anomaly on Indian summer monsoon: a numerical experiment with the GFDL model."' *J. Atmos. Sci.,* 33, 2252–2253.

Simpson, R. H., 1947: 'Synoptic aspects of the Intertropical Convergence Zone near Central and South America'. *Bull. Amer. Meteor. Soc.,* 28, 335–346.

Staff Members, Academia Sinica, Peking, 1957: 'On the general circulation over Eastern Asia. I'. *Tellus,* 9, 432–446.

Staff Members, Academia Sinica, Peking, 1958a: 'On the general circulation over Eastern Asia. II'. *Tellus,* 10, 58–75.

Staff Members, Academia Sinica, Peking, 1958b: 'On the general circulation over Eastern Asia. III'. *Tellus,* 10, 299–312.

Stearns, C. R., 1969: 'Surface heat budget of the Pampa de La Joya, Peru'. *Mon. Wea. Rev.,* 97, 860–866.

Stout, J. E., Young, J. A., 1983: 'Low-level monsoon dynamics derived from satellite winds'. *Mon. Wea. Rev.,* 111, 774–798.

Subbaramayya, I., Babu, S. V., Rao, S. S., 1984: 'Onset of the summer monsoon over India and its variability'. *Meteor. Mag.,* 113, 127–135.

Subbaramayya, I., Bhanu Kumar, O. S. R. U., 1978: 'The onset and the Northern limit of the south-west monsoon over India'. *Meteor. Mag.,* 107, 37–48.

Sverdrup, H. U., 1917: 'Der Nordatlantische Passat'. *Veröffentlichungen Geophys. Institut, Universität Leipzig*, **2**, no. 1, 96 pp.

Taljaard, J. J., 1953: 'The mean circulation in the lower troposphere over Southern Africa'. *S. Afr. Geogr. J.*, **35**, 33–45.

Taljaard, J. J., 1955: 'Stable stratification in the atmosphere over Southern Africa'. *Notos*, **4**, 217–230.

Taljaard, J. J., Van Loon, H. L., Jenne, R. L., 1969: 'Climate of the upper air, part I, Southern hemisphere. vol. 1, temperature, dewpoint, and height at selected pressure levels'. NAVAIR 50–1c–55, NCAR, ESIA, DOD, Washington, D.C., 134 pp.

Teisserenc de Bort, L., Rotch, L., 1905: 'Sur les preuves directes de l'existence du contrealizé'. *Comptes Rendues de l'Académie des Sciences, 9 Octobre 1905*, 605–612.

Teisserenc de Bort, L., Rotch, L., 1909: 'Étude de l'atmosphère par sondages aériens. Atlantique moyan et intertropicale'. *Annales du Bureau Central Météorologique de France, 1905*, 1, Mémoires, Paris.

Thompson, A. H., Neiburger, M., 1953: 'The radiational temperature changes over the Eastern North Pacific Ocean in July 1949'. *J. Meteor.*, **10**, 167–174.

Trenberth, K. E., 1980: 'Atmospheric quasi-biennial oscillations'. *Mon. Wea. Rev.*, **108**, 1370–1377.

U.S. Navy, Naval Oceanography and Meteorology, 1977: *U.S. Navy Marine Climatic Atlas of the World*. vol. 2, *North Pacific Ocean*, NAVAIR 50–1C–529, U.S. Government Printing Office, Washington, D.C., 20402; 388 pp.

U.S. Navy, Naval Oceanography and Meteorology, 1978: *U.S. Navy Marine Climatic Atlas of the World*. vol. 4, *South Atlantic Ocean*. NAVAIR 50–1C–531, U.S. Government Printing Office, Washington, D.C., 20402; 325 pp.

U.S. Navy, Naval Oceanography Command, 1979: *U.S. Navy Marine Climatic Atlas of the World*, vol. 5, *South Pacific Ocean*. NAVAIR 50–1C–532, U.S. Government Printing Office, Washington, D.C., 20402; 350 pp.

U.S. Navy, Naval Weather Service Command, 1974: *U.S. Navy Marine Climatic Atlas of the World*. vol. 1, *North Atlantic Ocean*. NAVAIR 50–1C–528, U.S. Government Printing Office, Washington, D.C., 20402; 371 pp.

U.S. Navy, Naval Weather Service Command, 1976: *U.S. Navy Marine Climatic Atlas of the World*. vol. 3, *Indian Ocean*. NAVAIR 50–1C–530, U.S. Government Printing Office, Washington, D.C., 20402; 348 pp.

U.S. Weather Bureau, 1938: *Atlas of climatic charts of the oceans*. U.S. Government Printing Office, Washington, D.C., 65 pp.

U.S. Weather Bureau, 1952: 'Normal weather charts for the Northern hemisphere: sea-level pressure, 700 mb height and temperature, thickness (700–1000 mb), 500 mb height'. Technical Paper No. 21, Washington, D.C., 74 pp.

Van de Boogaard, H. M. E., Rao, G. V., 1984: 'Mesoscale structure of the low-level flow near the equatorial East African coast'. *Mon. Wea. Rev.*, **112**, 91–107.

Van Loon, H., Taljaard, J. J., Jenne, R. L., Crutcher, H. L., 1971: 'Climate of the upper air: Southern Hemisphere. vol. 2 – zonal geostrophic winds'. NCAR – TN/STR–57, NCAR, NOAA, DOD, Washington, D.C., 39 pp.

Veryard, R. G., Ebdon, R. A., 1961: 'Fluctuations in tropical stratospheric winds'. *Meteorol. Mag.*, **90**, 125–143.

Virji, H., 1981: 'A preliminary study of summertime tropospheric circulation patterns over South America estimated from cloud winds'. *Mon. Wea. Rev.*, **109**, 599–610.

von Ficker, H., 1936a: 'Die Passatinversion'. *Veröffentlichungen Meteorol. Institut, Universität Berlin*, **1**, 4, 33 pp.

von Ficker, H., 1936b: 'Bemerkungen über den Wärmeumsatz innerhalb der Passatzirkulation'. *Sitzungsberichte der Preussischen Akademie der Wissenschaften Phys. Math. Klasse, 1936*, **11**, 102–114.

Walker, G. T., 1923–24: 'World Weather, I and II'. *Memoirs India Meteor. Dept.*, **24**, 75–131, 275–332.

Walker, G. T., 1928: 'World Weather III'. *Memoirs Roy. Meteor. Soc.*, **2**, 97–106.

Walker, G. T., Bliss, E. W., 1930, 1932, 1937: 'World Weather IV, V, VI'. *Memoirs Roy. Meteor. Soc.*, **3**, 81–95; **4**, 53–84, 119–139.

Walker, H. O., 1957: 'Weather and climate of Ghana'. Ghana Meteorological Department, Department Note No. 5, 42 pp.

Washington, W. M., 1976: 'Numerical simulation of the Asian-African winter monsoon'. *Mon. Wea. Rev.*, **104**, 1023–1028.

Washington, W. M., Chervin, R. M., Rao, G. V., 1977: 'Effects of a variety of Indian Ocean surface temperature patterns on the summer monsoon circulation: experiments with the NCAR general circulation model'. *Pure Appl. Geophys.*, **115**, 1335–1356.

Washington, W. M., Daggupaty, S. M., 1975: 'Numerical simulation with the NCAR global circulation model of the mean conditions during the Asian-African summer monsoon'. *Mon. Wea. Rev.,* 103, 105–114.

Weare, B. C., Strub, P. D., Samuel, M. O., 1980: 'Marine climatic atlas of the tropical Pacific Ocean'. *University of California, Davis, Contributions in Atmospheric Science,* No. 20, 147 pp.

Webster, P. J., 1983: 'Mechanisms of monsoon low-frequency variability: Surface hydrological effects'. *J. Atmos. Sci.,* 40, 2110–2124.

Wyrtki, K., 1966: 'Seasonal variation of heat exchange and surface temperature in the North Pacific Ocean'. Hawaii Institute of Geophysics, Rep. HIG–66–3, University of Hawaii, Honolulu.

Wyrtki, K., 1971: *Oceanographic atlas of the International Indian Ocean Expedition.* National Science Foundation, Washington, D.C., 531 pp.

Wyrtki, K., 1979: 'El Niño'. *La Recherche,* 10, 1212–1220.

Yanusari, T., 1979: 'Cloudiness fluctuations associated with the Northern hemisphere summer monsoon'. *J. Meteor. Soc. Japan,* 57, 227–242.

Yanusari, T., 1981: 'Structure of an Indian summer monsoon system with around 40-day period'. *J. Meteor. Soc. Japan,* 59, 336–354.

Yoshida, K., Mao, H.-L., 1957: 'A theory of upwelling of large horizontal extent'. *J. Mar. Res.,* 16, 40–54.

Young, J. A., 1983: 'Dynamics of cross-equatorial flow'. pp. 270–277, in *Proceedings of First International Conference on Southern Hemisphere Meteorology, Saõ José dos Campos, Brazil, 31 July – 6 August 1983,* Amer. Meteor. Soc., Boston, Mass., 380 pp.

Young, J. A., Virji, H., Wylie, D. P., Lo, C., 1980: *Summer monsoon windsets from geostationary satellite data.* Space Science and Engineering Center and Department of Meteorology, University of Wisconsin, Madison, Wisconsin, 127 pp.

CLIMATOLOGY OF WEATHER SYSTEMS

Embedded within the quasi-permanent large-scale circulation components discussed in Chapter 6 are transient disturbances in which intense weather tends to be concentrated. The purpose of the present chapter is to review the large-scale environmental setting, and the regional and seasonal characteristics of prominent synoptic weather systems. Atmospheric disturbances are individuals rather than identical replicas of each other. Synoptic models are meant to recognize the traits common to a family of individuals, to bring some order into the diversity, and to serve in the practical tasks of weather analysis and forecasting. Where diversity prevails over commonality the design of a synoptic model may be difficult and not helpful. For the midlatitudes, the Norwegian cyclone model continues to provide a successful frame of reference, work throughout the better part of this century notwithstanding. For the tropics, the development of synoptic models (see reviews by Forsdyke, 1960; La Seur, 1964; Johnson, 1964; Riehl, 1954, pp. 210–234, 281–357, 1979, pp. 315–496; Ramage, 1971, pp. 38–85, 91–100) begins only in the 1940's, except for the Tropical Cyclone. Moreover, synthesis is rendered difficult by the great diversity of tropical weather phenomena. Nevertheless, models have been proposed for various weather systems in low latitudes. Of these concepts some, such as the Tropical Storm (Section 7.2), are widely accepted; others as the Subtropical Cyclone (Section 7.7) and the Central American Temporales (Section 7.8) have provoked no major objections; while Waves in the Easterlies (Section 7.5) and Squall Lines (Section 7.4) are at the focus of ongoing research.

Up-to-date discussions of weather systems are found in Ramage (1971), Carlson and Lee (1978), Riehl (1979), Krishnamurti (1979), and Anthes (1982). While these texts collectively offer a useful account of the subject matter, none of them is comprehensive. Thus, Ramage (1971) discusses Squall Lines, Monsoon Depressions, Subtropical Cyclones, and cold surges; Carlson and Lee (1978) Tropical Storms, Waves in the Easterlies, and Squall Lines; Riehl (1979) Tropical Storms and Waves in the Easterlies; Anthes (1982) Tropical Storms; and Krishnamurti (1979) Tropical Storms, Waves in the Easterlies, Squall lines, Monsoon Depressions, and Subtropical Cyclones in the Indian Ocean sector, which he refers to as 'Mid-tropospheric cyclones of the Southwest monsoon'. The Dust Storms of the Sudan and the Temporales of Pacific Central America are not treated in these texts.

In the consideration of the various weather systems in the following sections, discussion of structure, dynamics and energetics is kept to a minimum, and reference is made to the pertinent literature. Instead, emphasis is on problems of synoptic climatology, such as the factors of large-scale circulation conducive to the origin and maintenance of the various weather systems, their spatial distribution and seasonality, and finally their impact on the regional climate and environment, and their role in interannual climate variability.

7.1. Clouds and Convection

Convection, clouds, and precipitation are active components of tropical weather systems, and are therefore pertinent to much of the topics addressed in this chapter. Two accomplishments, above all, in the course of the past 15 years have greatly advanced the understanding of tropical convection and cloudiness: the development of operational satellite imagery and the Global Atmospheric Research Program's Atlantic Tropical Experiment (GATE) in 1974. The highlights of this progress are summarized in two recent review papers (Houze and Betts, 1981; Houze and Hobbs, 1982, pp. 287–303), on which the present section is primarily based. For the theory and modelling of tropical convection through the late 1970's reference is made to Riehl (1979, pp. 123–201) and Krishnamurti (1979, pp. 89–125).

Clouds in the tropics occur in a range of sizes, extending from small isolated cumuli to large cloud ensembles. The latter represent 'meso-scale convective complexes', for which the term 'cloud clusters' was coined from early satellite-derived cloudiness compilations (Houze and Hobbs, 1982, p. 287). Although the smaller, isolated cumuli and cumulonimbi greatly outnumber cloud clusters, the latter are larger and thus dominate the total cloudiness and rainfall of the tropics. The cloud clusters stand out in satellite pictures by their extensive cirrus shields, with dimensions of hundreds of km. Cloud clusters have typical lifetimes of hours to a day, but may form part of much longer-lived weather systems. Thus they may evolve into Tropical Storms (Section 7.2), and be associated with Tropical Squall Lines (Section 7.4).

Houze and Hobbs (1982, pp. 289–296) distinguish between 'squall' and 'nonsquall' clusters. The former are associated with Tropical Squall Lines (Section 7.4) and are noted in geosynchronous satellite imagery by rapid propagation, explosive growth, high brightness, and distinct convex leading edge. Refer to Section 7.4 for the circulation and cloud structure of Tropical Squall Lines. Nonsquall clusters move slower, lack the distinctive oval cirrus shield and sharp arc-shaped leading edge, and are the predominant type of cloud cluster in the tropics. Beneath the large cirrus shield apparent on satellite imagery there may be various areas of active cumulonimbi and rain, with dimensions of hundreds of km. Within the cloud clusters Houze and Betts (1981) recognize both convective-scale updrafts and downdrafts and vertical motion patterns on the meso-scale.

The predominantly convective nature of tropical cloudiness is a major factor for the notorious space and time variability of rainfall in the tropics. However, stratiform cloudiness, in part fuelled by convection, is also a characteristic feature of certain tropical weather systems and extensive areas in low latitudes. The spatial patterns of cloudiness and rainfall are in the following sections considered as integral parts of particular tropical weather systems.

7.2. Tropical Storms

Tropical storms are the most dangerous and destructive weather systems on Earth. It is then not surprising that they have been more extensively studied than any other tropical synoptic entity. While the structure and functioning of the mature storm is now well known, its genesis is still enigmatic. Various publications (Riehl, 1954, pp. 281–317; 1979, pp. 394–458; Gray, 1968, 1979; Atkinson, 1971, pp. 8–1 to 8–30; Carlson and Lee, 1978, pp. 276–357; Krishnamurti, 1979, pp. 186–223; Anthes, 1982) provide extensive and in part up to date surveys of the structure, dynamics, energetics, and climatology of Tropical Storms. In this section, the structure of the mature system is discussed first, then the origin and evolution are considered, and finally a climatology of Tropical Storms is presented.

Early stages of evolution are termed 'Tropical Disturbance' and 'Tropical Depression'; the beginning of the 'Tropical Storm' stage is defined by closed isobars and highest sustained surface

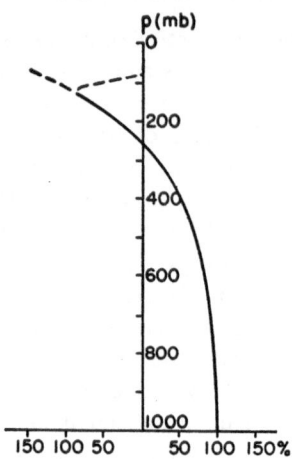

Fig. 7.2:1. Vertical profile of mean pressure difference between inside and outside of hurricanes, expressed in percent of the surface difference. From Riehl (1979).

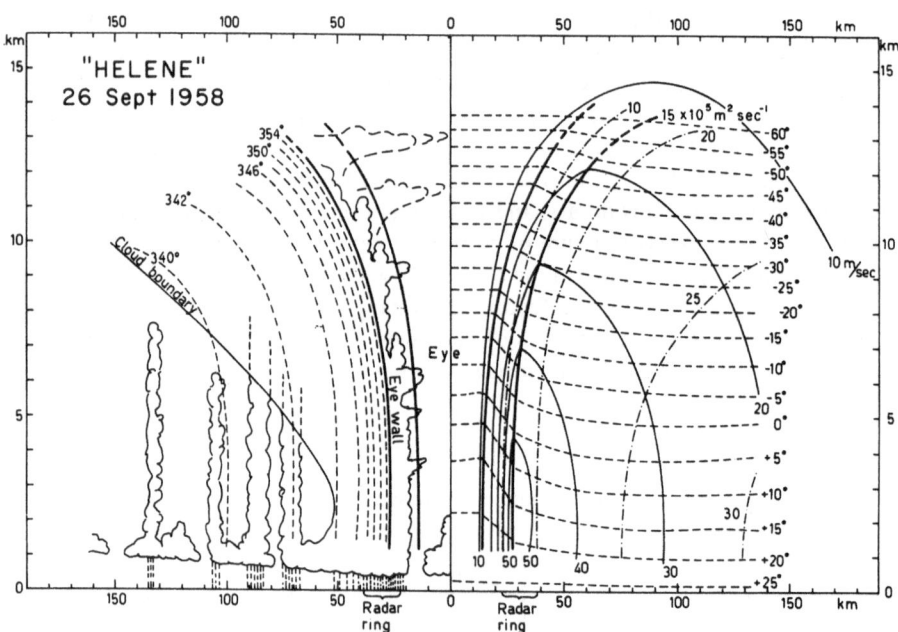

Fig. 7.2:2. Mean structure of a mature hurricane. To the right are given the boundaries of the eye-transition zone, temperature at 5°C interval, tangential with velocity at 10 m s^{-1} interval, and dash-dotted lines showing the angular momentum (for intervals of 5×10^5 m^2 s^{-1}). To the left are given the same eye-wall boundaries, clouds and rain, and lines of constant equivalent potential temperature (2°K interval). From Palmén and Newton (1969).

wind speeds of at least 34 knots (17 m s^{-1}). Tropical Cyclones with highest sustained surface winds of 64 knots (32 m s^{-1}) or more are called 'Hurricanes' in the Americas, 'Typhoons' in the Pacific West of the date line, and 'Cyclones' in the Indian Ocean (Atkinson, 1971, pp. 8–1 and 2; Carlson and Lee, 1978, pp. 276–278; Riehl, 1979, pp. 396–398).

The structure of the mature Hurricane is illustrated in Figs. 7.2:1 to 7.2:5. The pressure gradient force is directed inward in the lower and mid-troposphere, but outward in the upper troposphere (Fig. 7.2:1). From a surface pressure around 1010 mb in the undisturbed environment outside the storm values decrease to a central pressure 50–100 mb lower; Anthes (1982, p. 7, 27) reports as lowest minimum pressure ever recorded a value of 870 mb.

Fig. 7.2:2 is a radial transect across a mature Hurricane, showing the outward sloping eye wall and the distribution of various pertinent elements. Temperature at the surface varies little from the periphery to the center of the hurricane. However, inasmuch as the air flowing inward along the surface trajectory penetrates into a region of substantially lower pressure, potential temperature must increase. This is the result of sensible heat pickup from the ocean surface. In the higher layers, temperature increases from the periphery inward to the rain area and even more to the eye region. Accordingly, isotherms of potential temperature bulge downward in the central portion of the Hurricane. Wind velocity increases inward to a ring of maximum some 30 km from the center, with a decrease to small velocities in the eye. The isopleths of equivalent potential temperature in Fig. 7.2:2 also show an increase inward. There is an inward decrease of absolute angular momentum about the Hurricane axis (Palmén and Newton, 1969, p. 481)

$$M_r = v_\theta r + \frac{f}{2} r^2, \tag{7.2:1}$$

where r is the radial distance from the center, f Coriolis parameter, and v_θ tangential velocity. Cloudiness and rainfall are most abundant in a 30–50 km wide ring outward from the eye, and tend to be concentrated in cyclonically arranged spiral bands. Complementing Fig. 7.2:2 is the very detailed Typhoon model Fig. 7.2:3.

Expanding on Figs. 7.2:1 and 7.2:2, the flow pattern of the axially symmetric Hurricane is illustrated in Figs. 7.2:4 and 7.2:5. In the lower troposphere there is radial inflow and in the upper troposphere outflow (Fig. 7.2:4), in both instances directed down the pressure gradient (Fig. 7.2:1). Tangential velocity (Fig. 7.2:5) is in the cyclonic sense inward of a bell-shaped boundary that stays some 1300 km from the axis at the surface, but ever closer to the axis in the upper layers. Outward from this boundary tangential velocity is in the opposite sense. Together with Fig. 7.2:1 the transects in Figs. 7.2:4 and 7.2:5 illustrate for the lower layers a cyclonic inflow following the inward directed pressure gradient force from great distances to the center, and for the upper layers an anticyclonic outflow originating over the central portion of the Hurricane and directed outward in accordance with the pressure gradient force.

The airstream spiraling inward in the surface layer picks up moisture and to a lesser extent sensible heat from the ocean surface. Local evaporation accounts for a substantial portion of the rainfall in the outer portions of the Hurricane, while in the ring of rain area the bulk of the precipitation results from low level moisture convergence (Gray, 1979, p. 208). Fig. 7.2:6 illustrates the inward increase of rainfall to amounts around 100 mm day^{-1} in the rain area. The latent heat release in the rain area is considered instrumental for the heating of the mid-tropospheric layers and the energetics of the storm. The pickup of sensible heat from the ocean along the surface trajectory is much smaller, but hydrostatically helps to maintain the inward directed pressure gradient, and is highly significant in the narrow annulus of extreme pressure gradient and contributes substantially to the eye wall buoyancy.

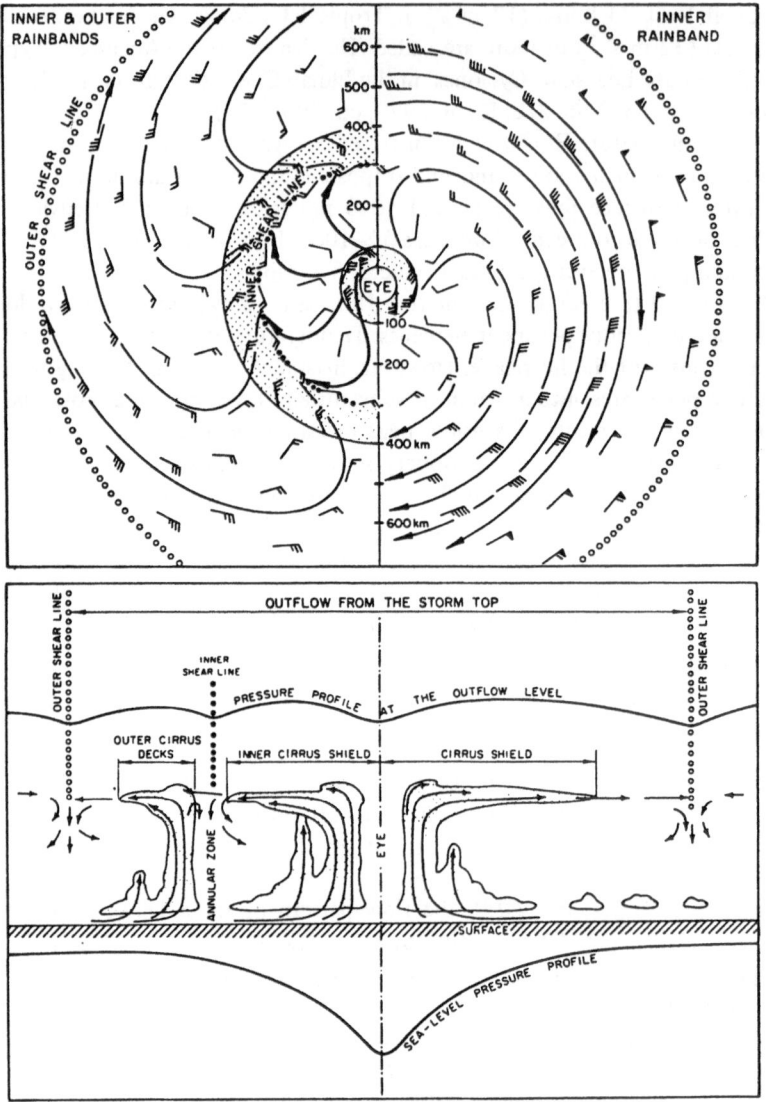

Fig. 7.2:3. Model of typhoons with inner rain band only (right half) and with inner and outer rain bands (left half). Top: upper-tropospheric flow pattern; bottom: vertical cross-section. From Fujita *et al.* (1967) (Fujita, T., Izawa, T., Watanabe, K., Imai, I., *Journal of Applied Meteorology*, American Meteorological Society).

The moist adiabatic ascent of air with surface properties makes for a warmer mid-troposphere in the rain area as compared to the undisturbed environment outside the hurricane. Functional as this is for the inflow of air, it produces temperature and density differences much too small to hydrostatically account for the very low surface pressure in the central region of the Hurricane. This requires even higher temperatures through a deep layer of the troposphere, as produced by the adiabatic descent in the eye. The outward sloping eye wall (Fig. 7.2:2) implies that a certain portion of the tropospheric column is made up of comparatively warm and less dense 'eye' air,

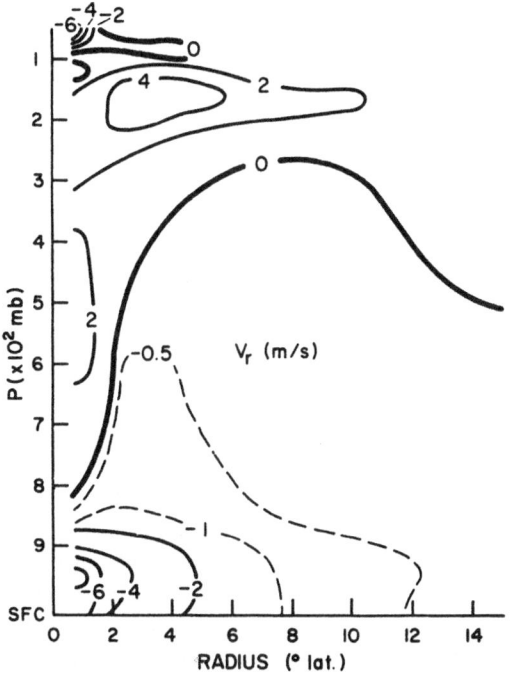

Fig. 7.2:4. Two-dimensional cross-section of radial winds v_r for the mean hurricane. From Gray (1979).

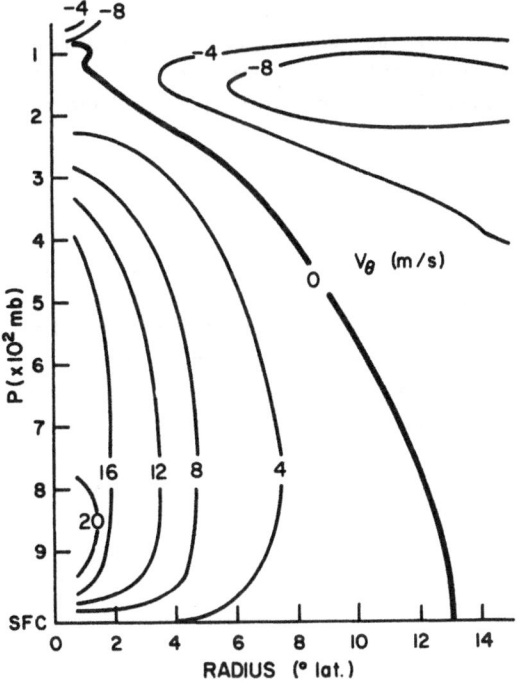

Fig. 7.2:5. Two-dimensional cross-section of tangential winds v_θ for the mean hurricane. From Gray (1979).

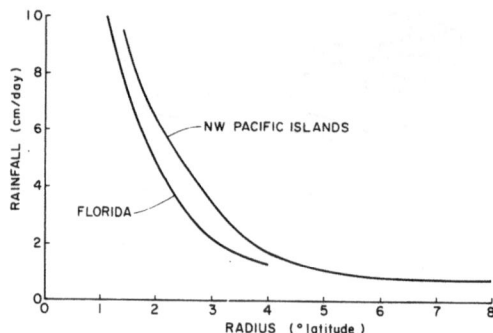

Fig. 7.2:6. Observed precipitation around tropical cyclones from Florida and from Northwest Pacific Islands. From Gray (1979).

thus contributing to the inward directed surface pressure gradient. In fact, well before upper-air observations became available, Haurwitz (1935) already concluded from hydrostatic reasoning and assuming a vanishing pressure gradient at high level, that the eye must be funnel-shaped.

As stated again by Anthes (1982, pp. 32–36), there is no accepted theory to explain the formation and maintenance of the Hurricane eye. A tendency for conservation of absolute angular momentum of the inward spiraling air is generally recognized as conducive to eye formation (Palmén and Newton, 1969, pp. 481–491; Carlson and Lee, 1978, pp. 329–300; Anthes, 1982, pp. 32–36), although frictional effects in the surface layer actually make for an inward decrease of absolute angular momentum, as is also illustrated in Fig. 7.2:2. This is explained in the next two paragraphs.

Refer to Eq. (7.2:1). Although absolute angular momentum is not strictly conserved (Fig. 7.2:2), as the air flows towards the center, v_θ and even more so v_θ^2/r increase. Disregarding friction, the radial equation of motion can be written (Carlson and Lee, 1978, p. 312)

$$\frac{dv_r}{dt} = \frac{v_\theta^2}{r} + fv_\theta - g\frac{\partial Z}{\partial r}, \tag{7.2:2}$$

where g is acceleration of gravity, Z height of an isobaric surface, v_r radial velocity component, and other symbols are as explained for Eq. (7.2:1).

Proceeding inward, the first right-hand term rapidly increases and eventually exceeds the other right-hand terms. Thus the left-hand term becomes positive, providing for outward acceleration of a fluid element. Inward from this critical radius flow is supergradient, and the outward acceleration entails lower layer divergence and compensating subsidence in the eye. Within the critical radius then flow is directed outward and beyond this radius inward, resulting in convergence and upward motion. As the inward directed pressure gradient force decreases with height, the outward directed radial acceleration (left-hand term in Eq. (7.2:2)) increases, so that the rising parcel is thrust outward, which in turn entails a widening of the eye wall with height.

In addition to the angular momentum implications just discussed, Eliassen (1971) and Anthes (1982, pp. 32–36) point out that for a circular vortex in solid rotation, Ekman pumping becomes inefficient near the axis of rotation. Accordingly, the maximum upward motion at the top of the boundary layer would occur at some distance outward from the center, so that boundary layer processes would be further conducive to the development of an eye.

Regarding the energetics, the Hurricane is an open system. As indicated above, the large latent heat release in the rain area is primarily maintained by lower-tropospheric moisture input from outside the Hurricane. At the same time, potential energy is exported through upper-tropospheric outflow. Only some 3% of the latent heat released is converted to kinetic energy (Riehl, 1979, p. 446). Within a 600 km radius, import of kinetic energy is unimportant, and kinetic energy generation approximately balances frictional dissipation. The latent and sensible heat pickup from the ocean surface is essential for the maintenance of the Hurricane. Numerical simulations (Ooyama, 1969) confirm that this factor, rather than just increased surface frictional dissipation, is crucial for the decay of Hurricanes at landfall, as well as with passage over cold ocean areas.

Despite the adverse surface conditions which the Hurricane encounters at landfall, its upper-tropospheric circulation may remain remarkably intact — a factor favorable for the regeneration of a storm system. An example of the remarkable resilience and longevity of a Tropical Storm system occurred during 27 October to 8 November 1961 in the Central American — Caribbean region. The following account is based on analyses of the Servicio Meteorológico Nacional of El Salvador (1961), although Dunn et al. (1962) offer a partly differing view. The Caribbean Hurricane 'Hattie' struck Belize in the morning of 31 October 1961. After causing considerable havoc with a death toll of about 300 and damage estimated at 60 million U.S. Dollars (Dunn et al., 1962), the storm moved inland with a bearing of about Westsouthwest, winds abated, but torrential rains accompanied the storm trajectory through the highlands of Guatemala. At elevations between 3000 and 4000 m, these virtually amputated the lower 300 mb or so of the storm column. Consistent with the storm trajectory over land, on 1 November 1961 a possibly pre-existing surface disturbance over the Pacific waters some 30 km to the South of Tapachula, Mexico, developed, reaching Tropical Storm intensity and thus earning the Pacific name 'Simone'. This moved northwestward, crossing the Pacific coast of Mexico in the morning of 2 November 1961 to the West of Salina Cruz. On the passage into the Mexican Meseta, the system suffered a similar fate as previously in Guatemala, the entire lower-tropospheric circulation being blocked out by the highland terrain. In accordance with the broadly eastward trajectory, a surface disturbance in the Gulf of Campeche part of the Gulf of Mexico, developed to Tropical Storm intensity on 4 November 1961, thus earning the Caribbean alias 'Inga'!

From consideration of the physical processes and climatological evidence, Gray (1979, pp. 167–172) proposes six parameters essential for Hurricane genesis: (i) sea surface temperature and depth of warm water; (ii) vertical gradient of equivalent potential temperature; (iii) middle troposphere relative humidity; (iv) low-level relative vorticity; (v) Coriolis parameter; (vi) vertical shear of horizontal wind. Of these, parameters (i) to (iii) are thermodynamic, while (iv) to (vi) are dynamic.

Regarding the thermodynamic parameters, (i) describes the ocean thermal energy, a plausible factor inasmuch as the ocean is believed to provide a major energy source for the Hurricane. The vertical gradient of equivalent potential temperature (ii) is a measure of convective instability, and an environment of high relative humidity (iii) through the implications of entrainment processes is conducive to deep cumulus convection. It is important to note, however, that all three parameters (i) to (iii) are intimately related to the surface temperature of the tropical ocean.

Turning to the dynamic parameters, intense convection generates a convergent low-level wind field, and this inflow produces cyclonic spin-up in proportion to the existing environmental vorticity field (iv and v). This can be appreciated from the approximate form of the vorticity equation

$$\frac{d(\zeta + f)}{dt} = -(\zeta + f)\nabla_p \cdot \mathbf{V} \tag{7.2:3}$$

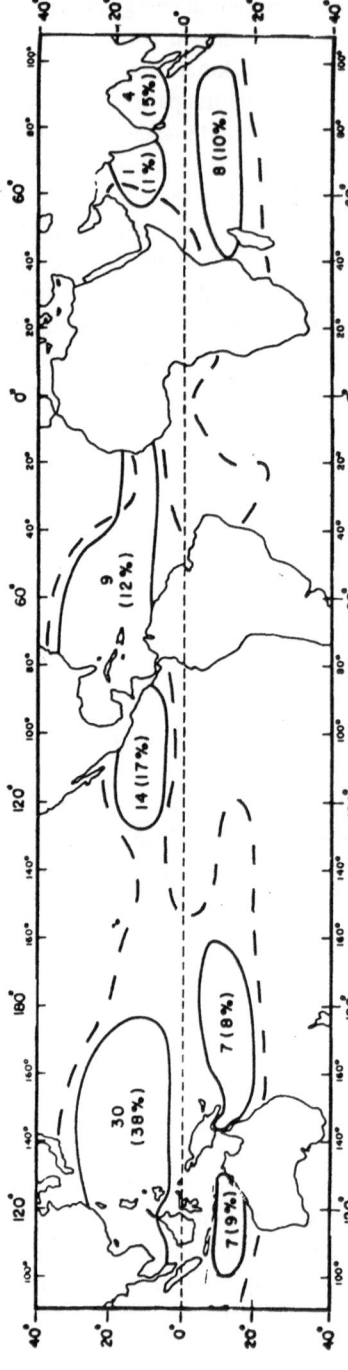

Fig. 7.2:7. Regions of Tropical Storm development: average annual number (and percent of global total). Thin broken lines represent 26.5°C isotherm for August in the Northern and January in the Southern hemisphere. From Gray (1968) (*Monthly Weather Review*, American Meteorological Society).

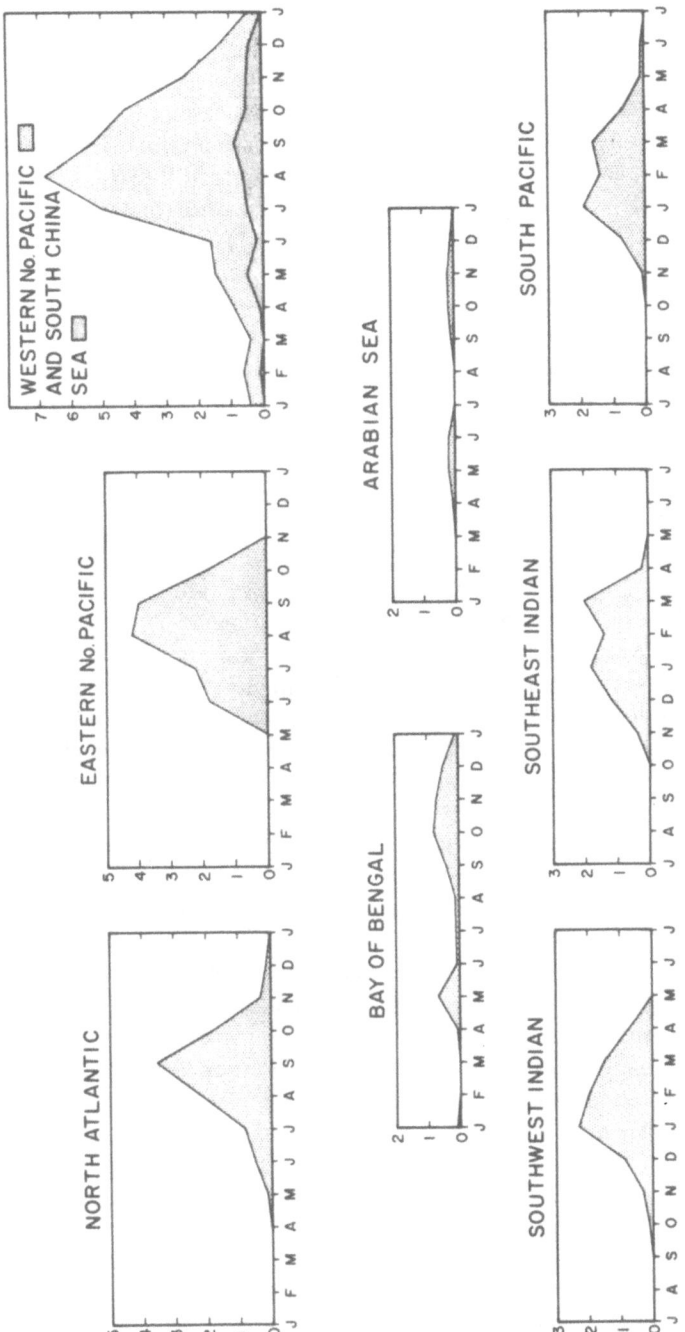

Fig. 7.2:8. Average monthly frequency of Tropical Storms in the eight regions identified in Fig. 6.7:7. From Atkinson (1971).

Note that sign and magnitude of the absolute vorticity are dominated by the Coriolis parameter. Vertical wind shear (vi) determines the extent to which an incipient cloud cluster is ventilated. With small shear and thus weak ventilation, properties imported laterally become concentrated, and enthalpy and moisture increase can occur. This favors a gradual surface pressure drop and cyclone formation. By contrast, if the flow at any level with respect to the incipient disturbance is large, then mass ventilation through the disturbance is too fast to allow concentration and accumulation of heat and is thus not leading to a drop in surface pressure. The positive feedbacks related to parameters (iv) to (vi) are the basis for the concept of CISK (Conditional Instability of the Second Kind) invoked to explain the growth of the large-scale tropical cyclone disturbance (Anthes, 1982, pp. 51–55; Carlson and Lee, 1978, pp. 332–341).

Summarizing the environmental setting requisite for Tropical Storm formation, the following three factors are found essential: sea surface temperature above 26.5°C, latitude beyond 5°, and small vertical wind shear. The global distribution and annual march of Tropical Storm frequency largely conform to these three simultaneous criteria, as is detailed in the following.

Fig. 7.2:7 identifies the eight major regions of tropical storm formation, namely North Atlantic, Eastern North Pacific, Western North Pacific and South China Sea, Bay of Bengal, Arabian Sea, Southwest Indian Ocean, Southeast Indian Ocean, and South Pacific. Excluded are the land surfaces, and oceanic areas with summertime sea surface temperature below 26.5°C, as well as the equatorial belt, 5°N – 5°S. The remaining details of spatial distribution and annual march can be largely understood from the vertical shear conditions.

The annual march of Tropical Storm frequency in the eight aforementioned ocean areas is illustrated in Fig. 7.2:8. In most regions, the annual variation of Tropical Storm occurrence broadly parallels that of sea surface temperature, with largest values towards the end of the respective summer half-year. A remarkable exception are the Arabian Sea and Bay of Bengal, where sea surface temperature during July and August is high, except for a secondary minimum in the Western Arabian Sea (Hastenrath and Lamb, 1979, part 1, charts 50–61), while Tropical Storm frequency shows a marked minimum between the double maxima around May and October–November.

The vertical wind shear is mapped in Fig. 7.2:9 for four cardinal months of the year. During January and April, when sea surface temperature in the Southern hemisphere is most conducive to Tropical Storm formation, vertical shear is small in the three Southern hemispheric genesis regions, namely the Southwest Indian and Southeast Indian Ocean, and the Western South Pacific. Consistent with the coincidence of favorable sea temperature and shear conditions, Fig. 7.2:8 shows large Tropical Storm frequencies in these regions at this time of year. During August when the Northern hemisphere oceans are warm, vertical shear is smaller over the Western North Atlantic and Caribbean, the Eastern North Pacific, and the Western North Pacific, where Tropical Storms abound in this season. By contrast, shear is large in the Central North Pacific during this and other seasons. This may explain the absence of Tropical Storms year-round, despite the adequate sea temperature conditions during the Northern summer. Particularly noteworthy are the large shears over the Arabian Sea and Bay of Bengal in Northern summer, when the sea surface is warm, but Tropical Storm frequency is at a minimum (Fig. 7.2:8). Thus shear appears as a major controlling factor for the annual march of Tropical Storm frequency in the Northern Indian Ocean, although the aforementioned secondary minimum of sea surface temperature in the Western Arabian Sea and the position of the monsoon trough over land to the North (Section 6.1, Fig. 6.1:2) are also interesting in this context (Ramage, 1974).

Gray (1968) concludes that most Tropical Disturbances and Storms form equatorward of 20 degrees latitude on the poleward side of the equatorial trough, where vertical wind shear is a

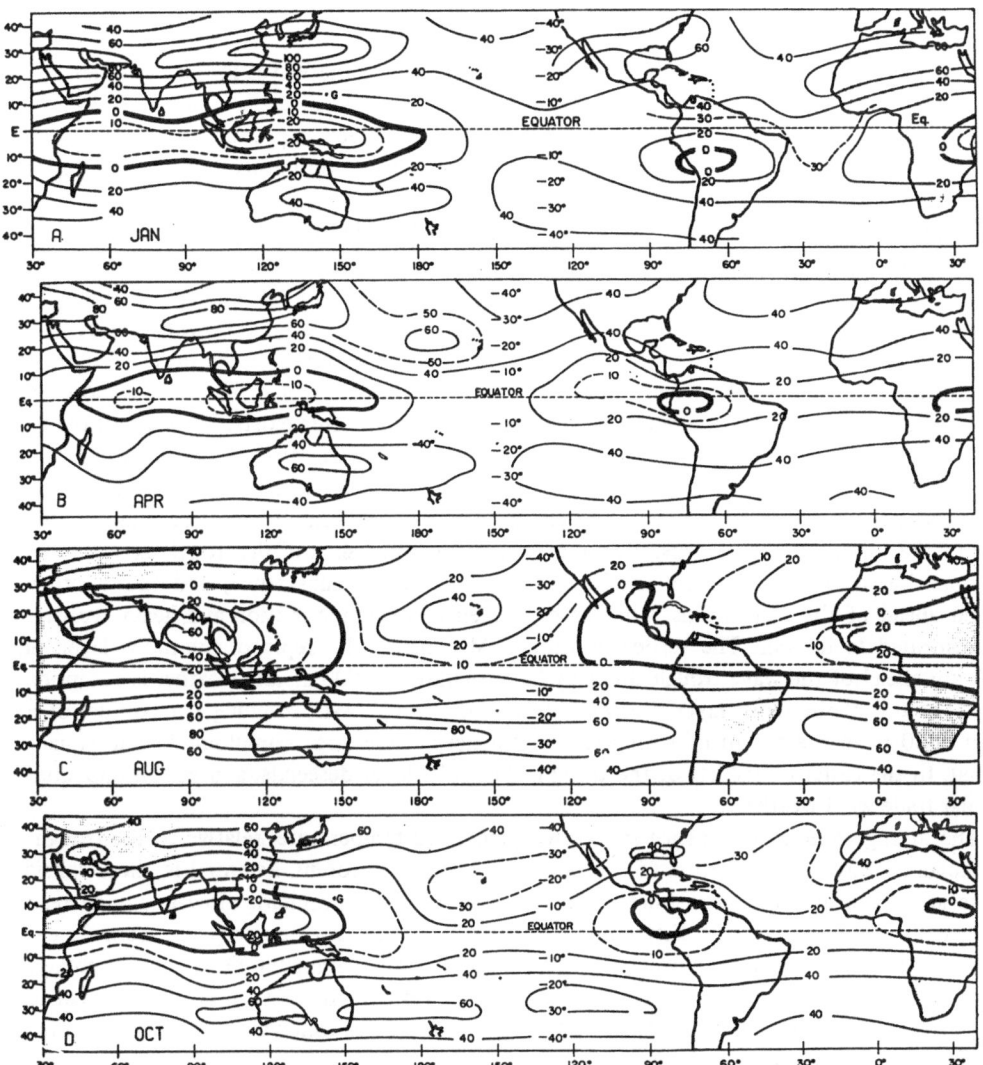

Fig. 7.2:9. Average zonal wind shear between 850 and 200 mb for January, April, August and October: Positive values indicate that the zonal wind at 200 mb is stronger from the West or weaker from the East than the zonal wind at 850 mb. Units are in knots. From Gray (1968) (*Monthly Weather Review*, American Meteorological Society).

minimum and the sea surface is warm enough. Storm development poleward of 20 degrees latitude is found only over the Northwest Atlantic and Northwest Pacific, but these make up only a small percentage of the global annual average of about 80 Tropical Storms per year.

Tropical Storms pose a most severe threat to ship and air traffic at sea, as well as to human settlements along the low-latitude coasts (Simpson and Riehl, 1981, pp. 17–24, 271–353). Efforts at deliberate weather modification (Simpson and Riehl, 1981, pp. 340–353; Anthes, 1982, pp. 119–130) are thus understandable. At the same time the beneficial aspects of rainfall in Tropical Storms must be recognized (Anthes 1982, pp. 6–8), in that Hurricanes play an important role in the hydrological balance and agricultural potential of tropical coastal regions.

7.3. Waves in the Easterlies

The most common disturbances of the trade wind regime are the Waves in the Easterlies. While a vast body of literature over the past three decades bears out the diversity of such systems (Riehl, 1979, p. 342) both within and between various regions, it seems appropriate to begin with the model proposed by Riehl (1954, pp. 210–223) that was based on observations in the Caribbean area. This is in part illustrated in Fig. 7.3:1.

In this model for the Caribbean (Riehl, 1954, pp. 210–223), the wave axis extends in a North-northeast to Southsouthwest orientation from the subtropical high towards the equatorial trough zone. The wave length measured from ridge to ridge is about 1700 km (15 degrees of longitude). Displacement of the system is towards the Westnorthwest, broadly following the basic flow, and at a rate of about 6 m s^{-1}. This would correspond to a period of the order of three days. Ridges/troughs in the wind field are ridges/troughs in the pressure field. The wave is most intense in the middle troposphere, where a weak cyclonic circulation often exists. Throughout the troposphere, the lowest temperature is located to the rear of the surface trough. Cooling in the zone of most active weather results from decreased insolation due to cloud cover, evaporation of falling rain, and downdrafts in large convective clouds. The tropospheric temperature minimum to the rear of the surface trough hydrostatically entails an eastward slant of the trough axis with height, or opposite to the travel direction. Lower-layer divergence, subsidence, and fair weather are found ahead of the trough axis, whereas convergence, ascending motion and heavy weather are concentrated to its rear. The vertical motion field controls the depth of the moist layer. Some 300 km ahead of the wave trough this may be as shallow as 1500 m, and weather is exceptionally fine. The top of the moist layer rises rapidly near the trough axis, reaching a maximum of more than 6500 m in the region of strongest convergence, where large cumulonimbi and squalls are found. In the eastern outskirts of the trough, the moist layer descends, and conditions return to regular trade wind weather.

The arrangement of divergence and convergence, and temperature, relative to the trough axis is explained through application of the potential vorticity theorem. Essential in these considerations are a displacement rate slower than the basic lower-tropospheric flow and a decrease of wind with height (Fig. 7.3:2). Elements of air flowing in the lower troposphere from the eastern outskirts through the wave move poleward (larger Coriolis parameter f) as they approach the trough axis (larger curvature and hence larger positive relative vorticity ζ). This means that to the rear of the trough axis absolute vorticity $\zeta_a = (f + \zeta)$ increases following the motion. In order to satisfy the potential vorticity theorem

$$\frac{\Delta \zeta_a}{\Delta p} = \text{const} \tag{7.3:1}$$

the column thickness Δp must increase accordingly, which corresponds to a stretching of the air column or convergence. Conversely, continuing now from the trough axis towards the forward outskirts of the wave, the air element flows both equatorward (smaller f) and away from the region of largest curvature (smaller positive relative vorticity ζ). Thus absolute vorticity ζ_a decreases following the motion, which according to Eq. (7.3:1) would require a decrease of the column thickness Δp, or shrinking of the air column and divergence. In the upper layers, where the basic flow is rather slower than the wave displacement, the inverse vergence pattern would result, which would allow for mass compensation in the atmospheric columns ahead and to the rear of the trough axis, and vertical circulation. Note that an increase of the basic flow with height (Fig. 7.3:2) and faster wave displacement rate may entail the opposite vergence pattern, namely

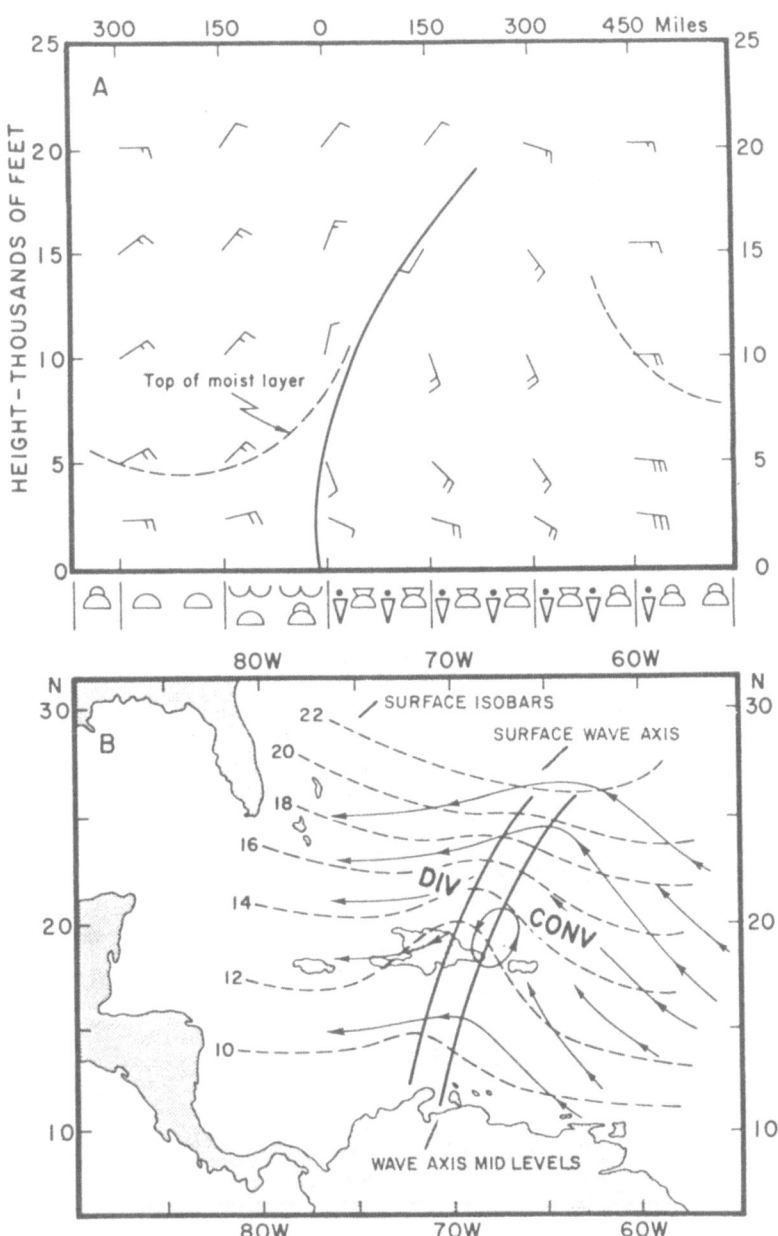

Fig. 7.3:1. Model of a Wave in the Easterlies. (a) Zonal-vertical transect showing eastward sloping wave axis, top of moist layer, and winds; (b) map depicting surface isobars (dashed lines), distribution of divergence (DIV) and convergence (CONV) at surface, wave axis at surface and aloft (heavy solid lines), and streamlines at 10 000– 15 000 feet (thin solid lines). From Carlson and Lee (1978).

lower-tropospheric convergence ahead and divergence to the rear of the trough axis.

Investigations into Waves in the Easterlies were greatly stimulated by the advent of satellite monitoring and intensive field programs over the tropical North Atlantic and West Africa (i.e. GATE in 1974). From an evaluation of satellite cloud images for the Caribbean, Merritt (1964)

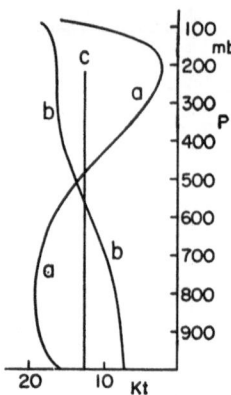

Fig. 7.3:2. Variation of basic current with height relative to wave speed (c) if bad weather is concentrated (a) East and (b) West of trough line. From Riehl (1954).

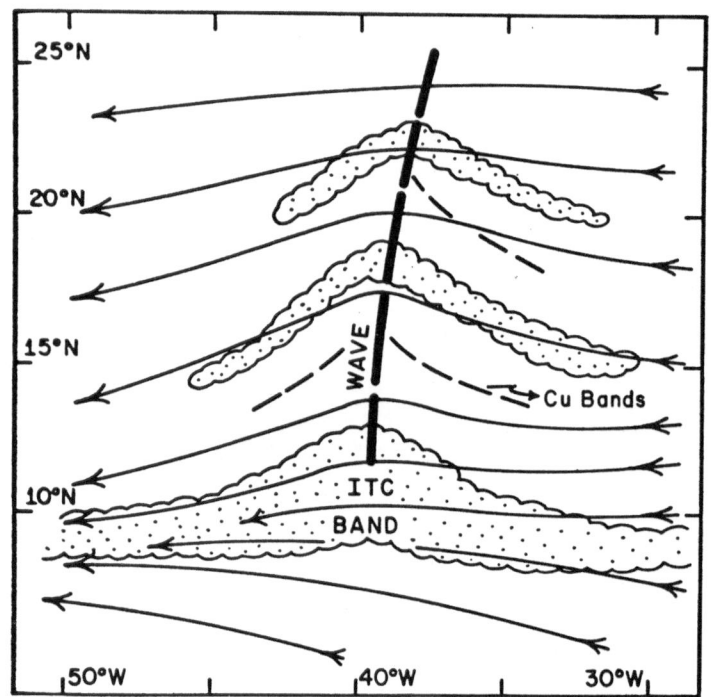

Fig. 7.3:3. A schematic showing the relationship between the lower-tropospheric flow and the 'Inverted V' cloud pattern. From Frank (1969) (*Monthly Weather Review*, American Meteorological Society).

ventured a reappraisal of Easterly Wave models. He concluded that Riehl's (1954, pp. 210–223) model of Waves in the Easterlies conformed with one of five patterns commonly found. In this connection it is noteworthy that before the dawn of the satellite era and also working in the Caribbean – Central American region, Portig (1960) proposed a model of an 'instability wave', which resembles Riehl's (1954, pp. 210–223) model in most respects except for a much faster displacement rate.

Conjectures as to Eastern Atlantic and West African source regions of Caribbean Easterly Waves were systematically pursued in the early years of satellite sensing and as part of the GATE effort (Simpson *et al.*, 1968, 1969; Frank, 1969, 1970; Frank and Johnson, 1969; Carlson, 1969a, b; Burpee, 1972, 1974, 1975; Leary, 1984; Payne and McGarry, 1977; Reed *et al.*, 1977; Norquist *et al.*, 1977; Thompson *et al.*, 1979). The latter three studies relied largely on the methodology developed previously by Reed and Recker (1971) in the Western Pacific. Satellite sensing, compositing, and power spectrum analyses have proven useful tools in the analyses of tropical waves. The numerical modelling and theoretical work by Pedgley and Krishnamurti (1976), Rennick (1976), Simmons (1977), Mass (1979), Estoque *et al.* (1983), and Adejokun and Krishnamurti (1983), complements the observational studies.

Collectively, these investigations shed much light on the behavior of Waves in the Easterlies. They are characteristic of the Northern summer half-year. Waves begin to form mostly East of 15°E on the African continent, but intensify over the land West of this longitude (Carlson, 1969a, b; Burpee, 1972; Reed *et al.*, 1977). The wavelength and period are somewhat longer over land than over sea, typical values being about 2500 km and 3.5 days, respectively, corresponding to a wave speed of about 8 m s^{-1} (Reed *et al.*, 1977). Broadly consistent with Reed *et al.*'s (1977) compositing study, Carlson (1969a, b) finds from synoptic maps a wavelength of 2000 km and period of 3.2 days, and Burpee (1972, 1974) from spectral analysis 4000 km and 3–5 days. Yanai *et al.* (1968) also used spectral analysis for the study of tropospheric waves over the tropical Pacific. During June to October, a wave thus crosses the African West coast about every 3–4 days. Over the ocean, waves are tracked by satellite imagery, the 'Inverted V' cloud pattern (Fig. 7.3:3) being a telltale feature (Frank, 1969). Passing from the African continent onto the cool Eastern Atlantic, waves generally decay (Carlson, 1969a, b), but remnants mostly survive to the Western Atlantic and Caribbean where they regenerate. In fact, African Waves account for about half of the Tropical Cyclones in the Atlantic (Frank, 1970; Burpee, 1972). As discussed in Section 7.2, these form mostly over the Western Atlantic. A rare case of hurricane formation off the West African coast is documented by Erickson (1963). The majority of synoptic scale systems from Africa propagate beyond the Caribbean and the Central American Isthmus into the Eastern Pacific (Frank, 1970), where some intensify into Tropical Storms (Carlson and Lee, 1978, p. 232). Many Typhoons in the Western Pacific are also believed to develop from Easterly Waves, although more work is needed on the relationship of Easterly Waves in the Western and Eastern Pacific (Carlson and Lee, 1978, p. 232).

The compositing approach of Reed *et al.* (1977) serves to enhance the features common to most systems. Individual situations may be considerably more complex. For example, Fortune (1980) refers to a vorticity center accompanying a Squall Line propagating through an Easterly Wave. Krishnamurti (1979, p. 279) even suggests that Squall Lines may be an integral part of African Waves.

The composite structure of African Waves is illustrated in Figs. 7.3:4 to 7.3:6. Following the methodology of Reed and Recker (1971), Reed *et al.* (1977) composited the wave according to 8 phases, category 4 being the trough and 8 the ridge at 700 mb. The latitude is counted with respect to the disturbance center at 700 mb. Fig. 7.3:4 shows a best development in the mid-troposphere (700 mb) and in the equatorward portion of the system a tilt of the trough axis from Southwest to Northeast. A similar configuration was noted by Reed and Recker (1971) for the Pacific. The West African Mid-tropospheric Jet (Section 6.2.4) is seen at about 16°N (Δ latitude = +5). Figs. 7.3:5 and 7.3:6 illustrate the distribution of vertical motion and weather in the various portions of the wave. Clearly, upward motion, lower-tropospheric convergence, upper-tropospheric

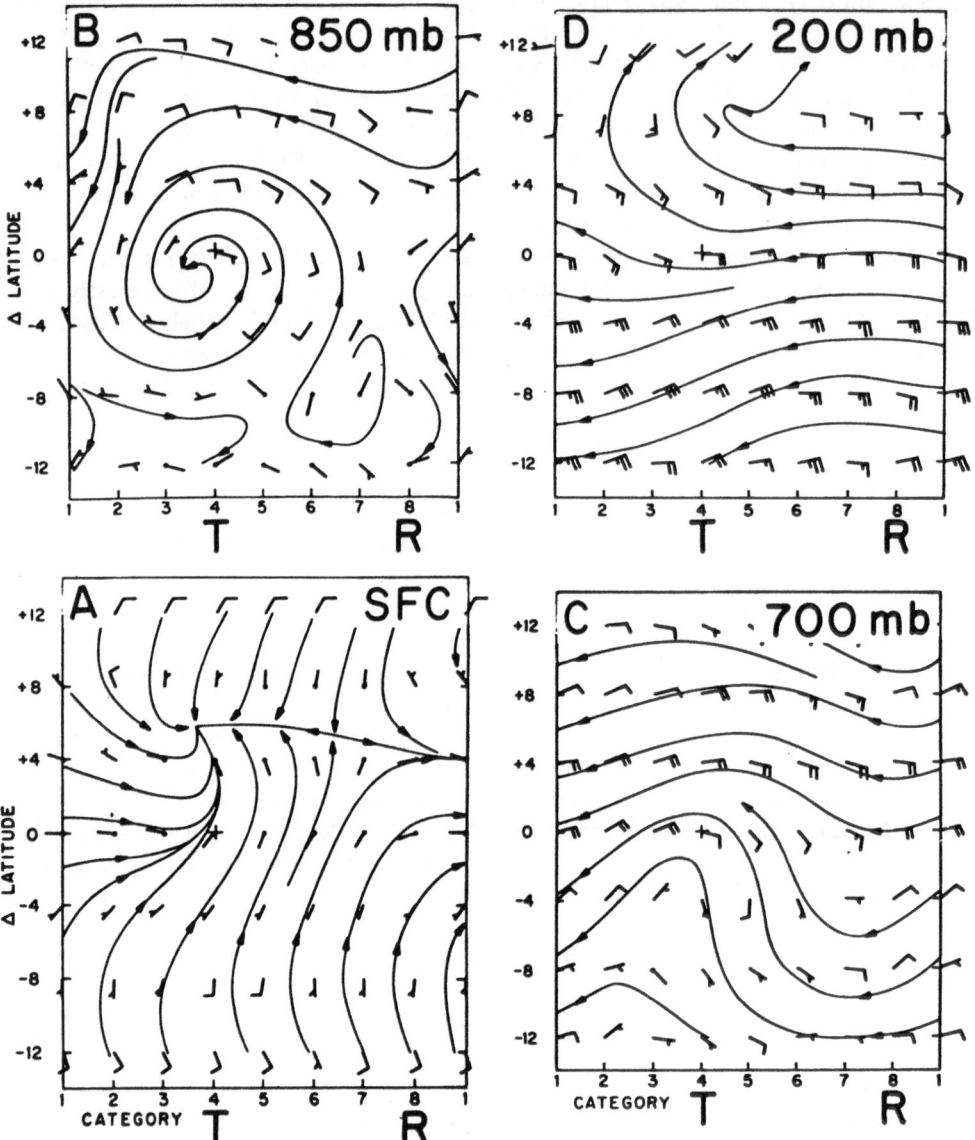

Fig. 7.3:4. Streamlines and wind barbs of total flow field for composite African Wave. Category separation is approximately 3 degrees longitude, T denotes trough, R ridge, and cross disturbance center at 700 mb. Full wind barb corresponds to 5, half barb to 2.5, and no barb to 1 m s^{-1} (a) surface, (b) 850 mb, (c) 700 mb, (d) 200 mb. From Reed *et al.* (1977) (Reed, R. J., Nordquist, D. C., Recker, E. E., *Monthly Weather Review*, American Meteorological Society).

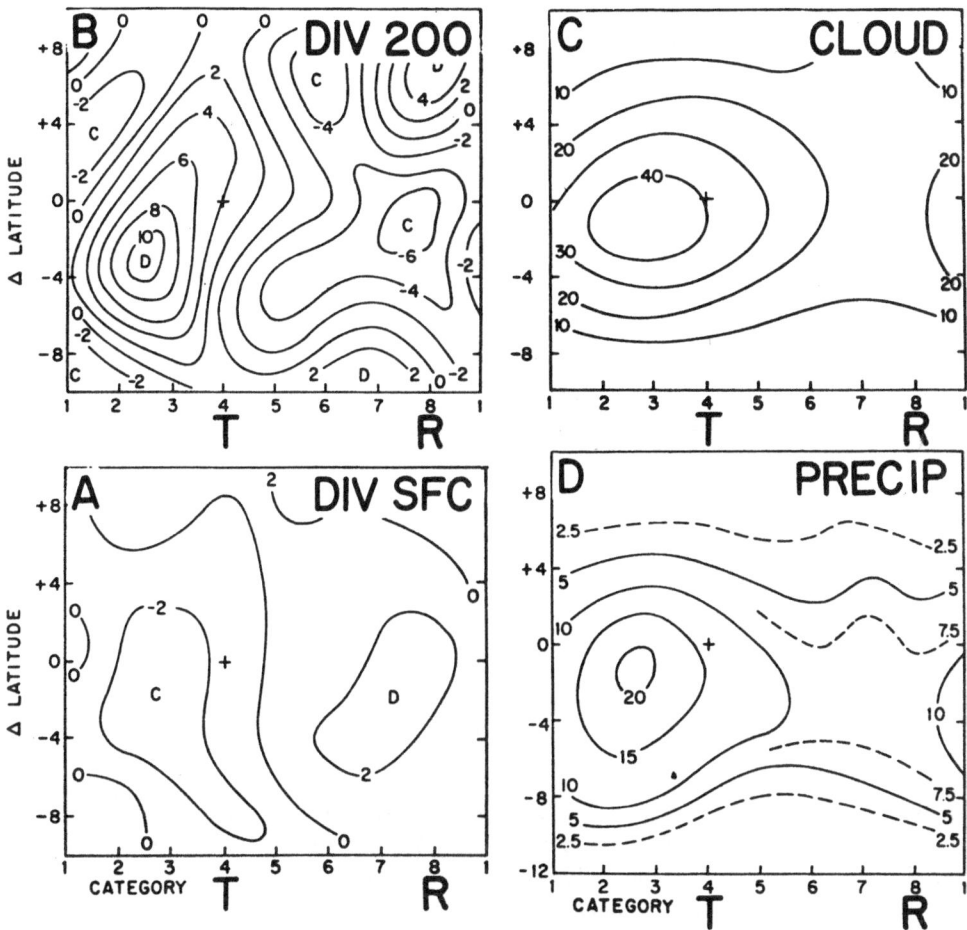

Fig. 7.3:5. Weather structure of composite African Wave. (a) Surface, divergence deviation from zonal mean in 10^{-6}; (b) same but for 200 mb level; (c) percentage of convective cloud cover; (d) average rainfall rate, in mm day^{-1}. Symbols as for Fig. 7.3:4. From Reed *et al*. (1977) (Reed, R. J., Nordquist, D. C., Recker, E. E., *Monthly Weather Review*, American Meteorological Society).

divergence, cloudiness, and rainfall, are concentrated ahead of the wave trough. Reed *et al*.'s (1977) results are consistent with the studies of Carlson (1969a, b) and Burpee (1975).

This pattern contrasts with Riehl's (1954, pp. 216–223) model of the Wave in the Easterlies over the Caribbean, where the wind profile (a) in Fig. 7.3:2 tends to prevail. Over West Africa and the adjacent Eastern Atlantic in Northern summer, easterlies increase with height, and the zonal wind in the lower troposphere is rather slower than the wave displacement, as approximated by profile (b) in Fig. 7.3:2. The location of bad weather ahead, or to the West, of the wave trough thus does not appear inconsistent with Riehl's (1954, pp. 210–220) considerations. Adding to the diversity, Reed and Recker (1971) applied the same compositing technique to the Marshall Islands region of the West Central Pacific, and found lower-layer convergence, ascending motion, and weather concentrated near the wave axis. Referring to Riehl's (1954) theoretical reasoning they note that the vertical wind profile varies markedly across the network.

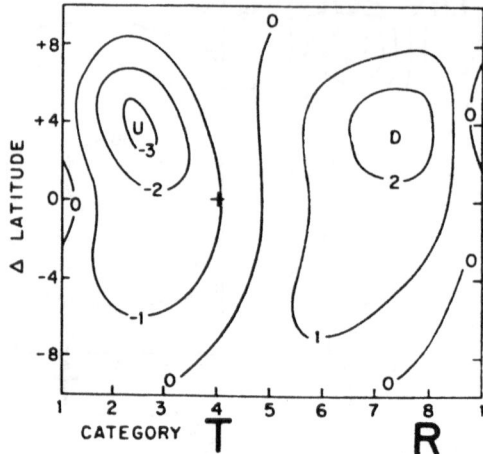

Fig. 7.3:6. Vertical motion deviation from zonal mean at 850 mb for composite wave, in mb hr^{-1}. Symbols as for Figs. 7.2.4 and 7.2.5. From Reed *et al.* (1977) (Reed, R. J., Nordquist, D. C., Recker, E. E., *Monthly Weather Review*, American Meteorological Society).

The West African Mid-tropospheric Jet (Section 6.2.4) is considered instrumental in the origin of the African Waves (Burpee, 1972; Norquist *et al.*, 1977; Carlson and Lee, 1978, pp. 254–273). Fig. 7.3:7 illustrates for the 700 mb region the asymmetry of the wave and the latitude profiles of zonal wind and Coriolis parameter, and shows that the maximum disturbance strength is located some two degrees of latitude equatorward of the jet axis. A necessary criterion for barotropic instability is (Carlson and Lee, 1978, p. 260)

$$\frac{\partial}{\partial y}\left(f - \frac{\partial u}{\partial y}\right) = 0 \qquad (7.3:2)$$

It is seen from Fig. 7.3:7 that this condition is fulfilled somewhere to the South of the jet axis. Note that the criterion is not met in the absence of a well-developed jet such as East of about 15°E and West of 20°W (Fig. 6.2.4:2). For a comprehensive discussion of necessary and sufficient conditions for instability refer to Das (1986, Chapter 4). The Southwest to Northeast tilt of the trough apparent in Fig. 7.3:4, part c, as well as in the left-hand portion of Fig. 7.3:7, is conducive to a flux of relative easterly momentum away from the jet axis, which appears related to removal of zonal kinetic energy from the jet and conversion to wave energy (Carlson and Lee, 1978, p. 256). In addition, Norquist *et al.* (1977) quantitatively evaluated the role of baroclinic processes. These are largest below the level of the jet, where both the meridional temperature gradient and

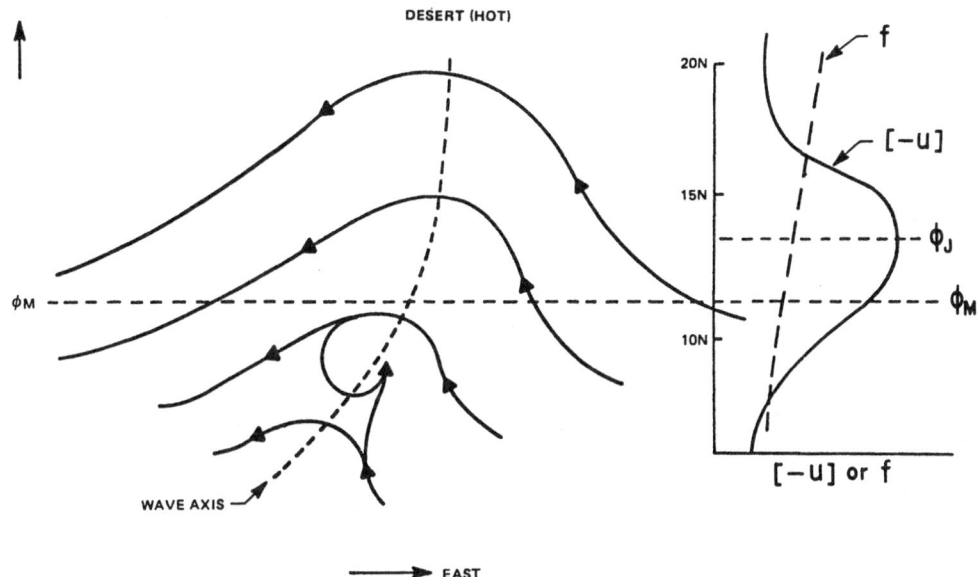

Fig. 7.3:7. African Wave around 700 mb. Left-hand portion shows plan view streamline pattern and tilted wave axis. Right-hand portion illustrates latitudinal profile of Easterly wind component $(-u)$ and Coriolis parameter f. ϕ_M is latitude of maximum disturbance strength, and ϕ_j latitude at West African Mid-tropospheric Jet. From Carlson and Lee (1978).

the transient eddy transport of sensible heat are southward, as is appreciated from the left-hand part of Fig. 7.3:7. In summary Norquist *et al.* (1977) propose that the waves originate to the South of the West African Mid-tropospheric Jet as a consequence of a joint baroclinic-barotropic instability, while they concede that their conclusions concerning the possible role of convection in energizing the waves are somewhat speculative.

These results appear pertinent to the regional distribution and annual march of the Waves in the Easterlies. It is in the Northern summer that a southward directed lower-tropospheric temperature gradient and an easterly jet in the 700 mb region develop over Subsaharan Africa, and consistent with this waves form preferentially during this season, in the mid-troposphere, and between longitudes of about 15°E to 20°W. Disintegration of waves over the Eastern Atlantic appears related not only to the thermodynamic effect of the cool ocean surface but also the fading out of an easterly jet in the mid-troposphere. Regarding the Western Atlantic and mid-Pacific, Palmén and Newton (1969, p. 447) suggest that waves are favored by the deep easterly current found westward of the subtropical highs in summer. Asnani (1982) mentions Waves in the Easterlies over the Southwest Indian Ocean during the Southern summer. Finally, Riehl (1979, p. 342) suggests that wave development requires a deep easterly current, that the asymmetry of the general circulation entails polar westerlies closer to the Equator in the Southern than in the Northern hemisphere, and that therefore waves should be primarily a Northern hemisphere phenomenon.

7.4. Squall Lines

Systems akin to midlatitude Squall Lines (Newton, 1950a, 1963, 1967) are among the most severe storms of the tropics. Following some earlier descriptions of weather phenomena at the West coast of Africa (Regula, 1936, 1943; Piersig, 1936, 1944), operational work in the British colonial weather services led to the discovery of the West African Disturbance Lines (Hamilton and Archbold, 1945; Eldridge, 1957). Further observational studies were contributed by a French school of tropical meteorologists (Anonymous, 1956; Jeandidier and Rainteau, 1957; Tschirhart, 1958, 1959; Bernet, 1966; Dhonneur, 1957, 1970; Voiron, 1964), who coined the term 'Lignes de Grains' for these systems. Bhalotra's (1963, pp. 81–88) compilations complement the picture for the Sudan. Gomes Teixeira (1970) mentions related phenomena from Angola. The advent of satellite imagery opened the perspective for the large-scale setting, and a series of field experiments, including the Line Islands Experiment (LIE) in 1962, the Barbados Oceanographic and Meteorological Experiment (BOMEX) in 1969 and the preparatory work in the preceding year 1968, the Venezuelan International Meteorological and Hydrological Experiment (VIMHEX) in 1972, and the GARP Atlantic Tropical Experiment (GATE) in 1974, furthermore provided an unprecedented detail of surface and upper-air observations (Zipser, 1969, 1977; Obasi, 1974; Aspliden *et al.*, 1976; Betts *et al.*, 1976; Payne and McGarry, 1977; Miller and Betts, 1977; Houze, 1977; Fortune, 1980; Gamache and Houze, 1982; Leroux, 1983, pp. 369–414; Barnes and Sieckman, 1984). Beyond the African continent, these observational studies comprise the tropical North Atlantic, the Caribbean and Northern South America, and the Equatorial Central Pacific. Earlier empirical analyses of Squall Lines are also available for India (Desai, 1950; Newton, 1950b; Das *et al.*, 1957; De, 1963). Largely drawing on the works of Newton (1967), Newton and Newton (1959), and Tschirhart (1958, 1959), Ramage (1971, pp. 91–100) discussed the Squall Line systems of the monsoon regions.

Major characteristics common to Tropical Squall Lines are emerging from this host of observational studies. The system is hundreds of kilometers long and consists of a line of active thunderstorms. While individual cumulonimbi mostly have lifetimes of an hour or less and new convective elements continually replace dying cells, the system as a whole may last hours to days. Afternoon convection is conducive to the development of the Squall Lines, but the system as a whole behaves differently from the sum of component thunderstorm cells. In accordance with the role of convection, Squall Lines form with some preference but not exclusively over land. They displace faster than the environmental wind, with speeds of 10–20 m s^{-1}.

The structure of a Tropical Squall Line is schematically depicted in Fig. 7.4:1. Warm moist air enters the base of the cloud at its leading edge, rises in the convective updraft and condensation takes place. An extensive cloud anvil forms to the rear of the convective tower. Precipitation falls both from the main cloud column and the anvil. Evaporation into comparatively dry mid-tropospheric air leads to cooling and downdrafts, concentrated in the region of intensive convection but extending on the mesoscale to the rear of the Squall Line. It should be noted here that contrary to the generally accepted concept of evaporative cooling as cause of both the convective and mesoscale downdrafts, Miller and Betts (1977) consider two different mechanisms, namely one driven by precipitation within the cumulonimbus cell, and another dynamically driven mesoscale descent over the spreading cold outflow. This stable layer related to the downdrafts and extending along the surface is also indicated in Fig. 7.4:1. Moreover, radar identifies a layer to the rear of the mature convective element where ice precipitation melts (Fig. 7.4:1).

The downward gushing cold air causes a 'pseudo cold front' (Newton, 1950a) or 'gust front' (Houze, 1977; Gamache and Houze, 1982) at the leading edge, undercuts the warm moist air ahead, and thus fosters more convection and new cumuliform cloud cells ahead of the Squall Line.

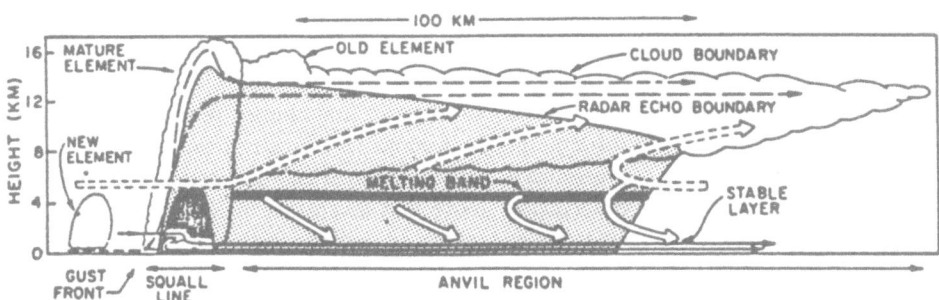

Fig. 7.4:1. Schematic cross-section through Squall Line system. Associated with the mature squall-line elements, dashed streamlines show convective-scale updraft, solid streamlines show downdraft circulation. Associated with the trailing anvil, wide solid arrows show mesoscale downdraft circulation, wide dashed arrows show mesoscale updraft circulation. Dark shading shows strong radar echo in the melting band and in the heavy precipitation zone of the mature squall-line element. Dot raster shows weaker radar echoes. Scalloped line indicates visible cloud boundary. From Gamache and Houze (1982) (*Monthly Weather Review*, American Meteorological Society).

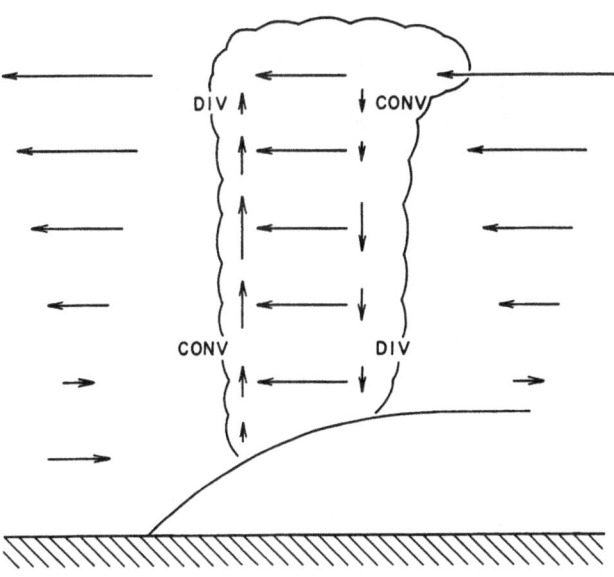

Fig. 7.4:2. Schematic vertical transect across main convective element of Squall Line. Arrows denote horizontal wind velocity and updrafts/downdrafts within cloud, with arbitrary length scale. Regions of horizontal divergence and convergence resulting from vertical exchange of momentum are indicated at cloud boundaries. Source: Newton (1950a).

This then entails a propagation of the convective region. The propagation rate would typically exceed the speed of the large-scale lower-tropospheric flow and be affected by the momentum of mid-tropospheric air brought down in the downdrafts.

As pointed out by Newton (1950a) and as illustrated in Fig. 7.4:2, the upward transfer of lower-tropospheric air possessing comparatively small horizontal momentum near the leading edge, along with the downward transfer of large momentum air to the rear of the convective cell, helps to maintain at the leading edge convergence in the surface layer and divergence aloft, and similarly to the rear of active convection convergence above and divergence below. The boundary layer characteristics of Tropical Squall Lines are the subject of ongoing research (Nicholls and Johnson, 1984).

Various factors conducive to Squall Line development become apparent from this account: (i) warm moist air in the surface layer fuels the convection; (ii) a mid-tropospheric dry layer allows cooling through evaporation of precipitation from the main convective tower and the anvil; (iii) vertical shear of horizontal wind is instrumental in the maintenance of a suitable divergence/ convergence pattern (Fig. 7.4:2) and the propagation of the Squall Line; (iv) land surfaces and especially mountain topography may enhance convection; (v) finally, the large-scale flow and vergence pattern may conceivably be relevant for convection. These factors appear essential in appreciating the regional distribution and annual march of Tropical Squall Lines.

Concerning the Disturbance Lines of Subsaharan Africa, the circulation during Northern summer is characterized by the moist Southwest monsoon airstream undercutting the dry north-easterly Harmattan flow aloft (Section 6.7.2.3) – a situation conforming to criteria (i) to (iii). This is underlined by the recent compilations of Aspliden and Adefolalu (1976). Ramage (1971, p. 100) notes that over Central and West Africa there is little turning of wind with height. Propagation is against the westerly flow component in the surface layer, and mostly faster than the easterly wind component in the mid-troposphere (Aspliden *et al.*, 1976; Fortune, 1980). Land surface properties (iv) also appear favorable. Aspliden *et al.* (1976) found for the GATE period (1974) that the majority of Squall Lines generated and decayed over land, that over the ocean more than three times as many decayed than formed, and that generation took place preferentially at certain longitudes with mountainous terrain, thus confirming various earlier studies mentioned above, in particular Hamilton and Archbold (1945) and Eldridge (1957). As to item (v), Payne and McGarry (1977) and Fortune (1980) call attention to the relation of Squall Lines and Waves. Moreover, Fortune (1980) considers the possible momentum implications of the West African Mid-tropospheric Jet (Section 6.2.4). In a recent study of Squall Lines over Nigeria, Bolton (1984) also proposes that a strong jet around 650 mb is essential for their development, and further notes that the foreward edge of the system tends to move faster than the maximum jet speed. According to Hamilton and Archbold (1945), Disturbance Lines typically extend meridionally some 200– 800 km to the South of the surface confluence axis (Section 6.7.2.3). For well-developed systems propagation is usually westward, that is against the surface flow, while varying directions are observed.

The annual march of frequency of West African Disturbance Lines shows double maxima around the beginning and end of the Southern summer half-year in the South, and a single peak at the height of the summer in the North (Hamilton and Archbold, 1945; Eldridge, 1957). Hamilton and Archbold (1945) report during 1939–41 and for various regions of Nigeria an annual average frequency of about 42; Eldridge (1957) records 283 for 1955 in Ghana; and Aspliden *et al.* (1976) count 176 cases during 63 days of GATE (1974) over West Africa and the adjacent Atlantic, based on satellite imagery rather than surface station observations. Aside from differences in observation practices, part of this apparent spatial and interannual variability may be real (ref. Section 7.10, and Table 7.10:3). The heavy weather and gusty winds associated with West African Disturbance Lines pose hazards to ship and air traffic and various other human activities. This connotation of hazard appears attached to the formerly used synonym 'Tornado' (Regula, 1936, 1943; Piersig, 1936, 1944; Voiron, 1964; Okulaja, 1970). However, the beneficial effects of West African Disturbance Lines are readily appreciated when it is realized that they account for the larger part of annual rainfall.

Over the adjacent Atlantic, Squall Lines are apparently much less frequent than over the African continent (Aspliden *et al.*, 1976). Various of the factors in the large-scale setting consiered above — in particular the superposition of dry over moist air and vertical shear — are much less favorable than over Africa. However, a series of observational studies from the tropical North Atlantic (Payne and McGarry, 1977; Houze, 1977), the Caribbean and Northern South America (Betts *et al.*, 1976; Miller and Betts, 1977; Zipser, 1977), and the Equatorial Central Pacific (Zipser, 1969), indicate that Squall Lines also occur in the oceanic tropics.

The behavior of Squall Lines over Northern India, where they are known as 'Nor'westers', has been described by Desai (1950), Newton (1950b), Das *et al.* (1957), De (1963), and Ramage (1971, p. 96). They develop during spring and fall, when a lower-layer moist southerly airstream undercuts a dry westerly flow aloft. The systems commonly approach from the Northwest quadrant, consistent with the upper-air flow.

7.5. Dust Storms of the Sudan

Bhalotra (1963, pp. 56—72) has described the Dust Storms or Haboobs of the Sudan. Essential are — apart from surface conditions — strong and sustained winds, capable of raising large amounts of dry opaque particles, causing a serious reduction in visibility (< 1000 m). Bhalotra distinguishes three kinds of synoptic situations that are conducive to Dust Storms: (i) instability type associated with thunderstorm activity; (ii) pressure gradient type related to the steepening of the large-scale meridional pressure gradient; (iii) pressure gradient type, associated with cold fronts and strong southward pressure gradient. The former two (i and ii) are found mainly during the advancing Northern summer monsoon or early monsoon period, while the third (iii) occurs with the passage of Mediterranean depressions from February to May. Dust Storms of type (i) tend to be short-lived and localized, with a width of 10—80 km. Types (ii) and (iii) are widespread and blow for long periods.

In Central Sudan, about 20 Dust Storms are counted per year (period 1953—57), more than 70% of which are of types (i) and (ii) and occur between May and August. The development of Dust Storms appears hampered by seasonal vegetation cover, and their occurrence is reduced in large irrigated areas. Similarly the extension of irrigation schemes is believed to be a factor in the observed downward trend in the frequency of Dust Storms.

7.6. Monsoon Depressions

This weather system of the Indian summer Southwest monsoon is described in Ramage (1971, pp. 43—47), Godbole (1975), Rao (1976, pp. 107—185), Krishnamurti *et al.* (1975, 1976), Krishnamurti (1979, pp. 250—255), Nitta and Masuda (1981), and Saha and Chang (1983).

According to Rao (1976, p. 107), Monsoon Depressions account for most of the monsoon rains. They are described as low pressure areas with two or three closed isobars at 2 mb intervals, covering an area of about five degrees square. They mostly form in the northern Bay of Bengal, and move westnorthwestward at least to the central parts of India before weakening or filling up, and have the most widespread rains in their Southwest quadrant. These low pressure systems are called depressions when surface winds over sea are up to 35 knots, and cyclonic storms with higher wind speeds.

Statistics for the 80-year period 1891–1970 (Rao, 1976, p. 107) document an average frequency of formation of 4–5 Monsoon Depressions over the Bay of Bengal for the June-September monsoon season, with a gradual increase from June to September. Only about 1 and 0–1 systems, respectively, formed over land and over the Arabian Sea, for the average over the 1891–1970 period and the June–September season. Over land formation is most frequent in July–August, and over the Arabian Sea in June, in contrast to the Bay of Bengal. The typical lifetime of systems is of the order of 3–5 days.

Rao (1976, pp. 115–116, 120–126, 128–131, 133–134, 136–139, 141–157, 159–181) presents extensive surface and upper-air analyses, and satellite and radar imagery for various synoptic case studies, but deplores (Rao, 1976, p. 114) that the existing upper-air network does not provide sufficient detail about the structure of the Monsoon Depressions at upper levels. In a summary description of the Monsoon Depressions, Krishnamurti (1979, p. 250, 255) states that the disturbance is most intense around 700 mb, with a closed circulation apparent from the surface to 400 mb. Central surface pressure is around 990 mb. The system is embedded in the monsoon trough (Section 6.8.3). The depression has a cold core below 700 mb, the southerly airstreams possess high relative humidity, and maximum speeds exceed 20 m s^{-1}. Strongest upward motion, and most abundant cloudiness and rainfall (up to 20 mm day^{-1}) are concentrated to the West. Krishnamurti (1979, p. 255) admits that little is known about the formative stages of these systems. Statistics of storms and depressions in the Bay of Bengal and the Arabian Sea during 1877–1960 have been published by the India Meteorological Department (1964), and are also summarized by Rao (1976, pp. 107–113) for the period 1891–1970.

7.7. Subtropical Cyclones

The concept of the Subtropical Cyclone was coined by Simpson (1952), and further elaborated and publicized by Ramage (1962; 1971, pp. 47–72). The Subtropical Cyclones of the Eastern North Pacific during Northern winter to spring, known in the Hawaiian Islands as 'Kona Storms' (Simpson, 1952), originate from cutoff lows in the upper-level subtropical westerlies. The heart of the system is in the mid-troposphere (400–600 mb), with pressure gradients, winds and convergence, all increasing from the periphery inward. Strongest mid-tropospheric convergence is compensated by divergence in the upper troposphere and to a lesser extent in the lower layers. Ramage (1962; 1971, pp. 47–72) proposes a conceptual model illustrated in Fig. 7.7:1 including an upward and downward branch. Within about 500 km from the center, strong upward motion entails condensation and deep precipitating clouds. Beyond about 500 km from the center there is downward motion with dry adiabatic warming extending to the subsidence inversion. The relatively warm-cored upper branch is energy-producing, while kinetic energy dissipation within the weak surface circulation is small, thus favoring the longevity of the system. Development of an eye, in analogy to the hurricane, is common. Simpson (1952) mentions the development of Subtropical Cyclones of the Kona type in the North Atlantic to the Southwest of the Azores.

A kind of Subtropical Cyclone of different origin has been studied by Miller and Keshavamurthy (1968) and Ramage (1971, pp. 51–72) over the Northern Indian Ocean during Northern summer. Reference is also made to the discussion in Rao (1976, pp. 249–269). A composite model is shown in Fig. 7.7:2. Convergence is strongest in the mid-troposphere (600–500 mb) near the

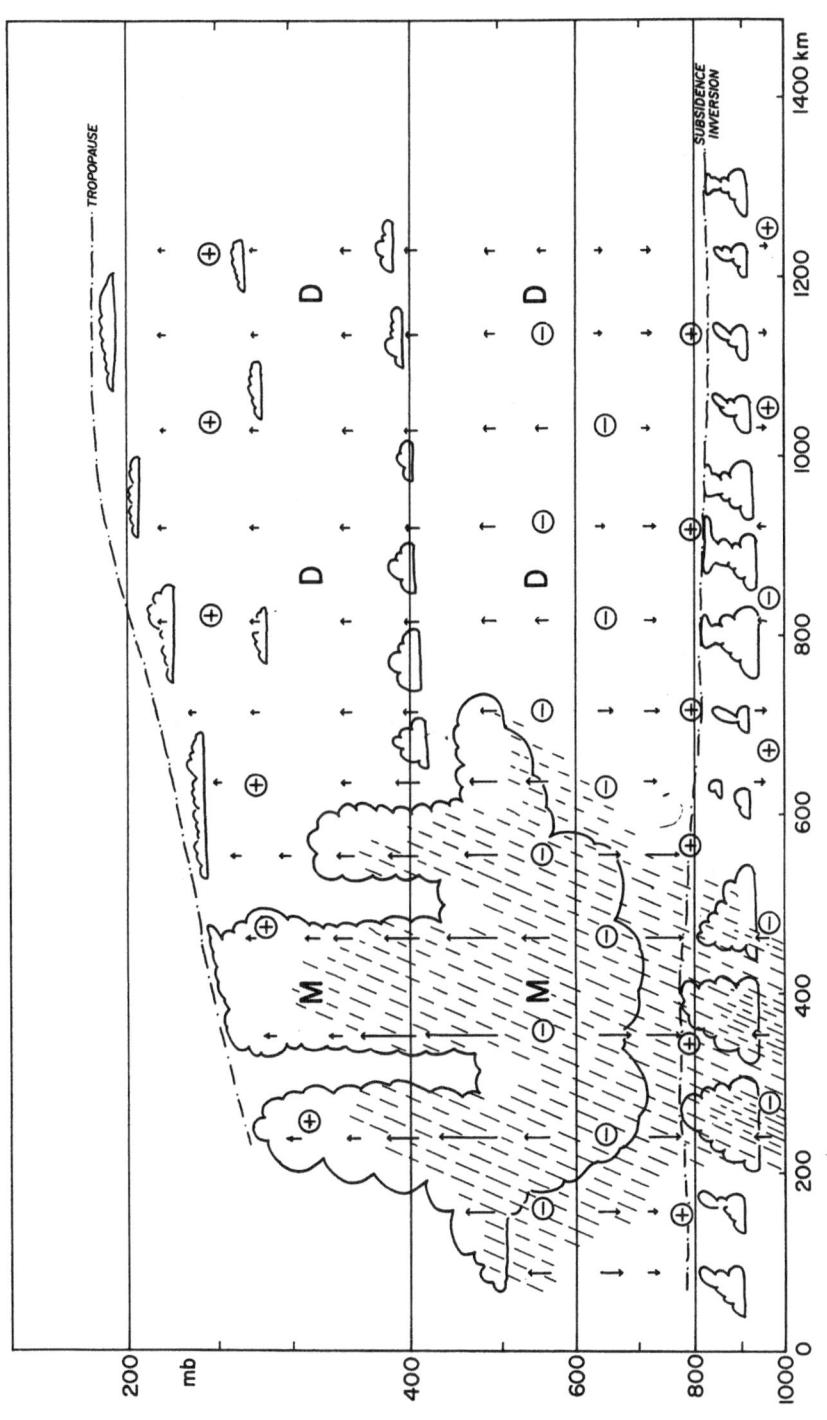

Fig. 7.7:1. Schematic radial cross-section of a Subtropical Cyclone. Severe constriction of the horizontal scale permits only gross features of the vertical motion and cloud systems to be shown. Divergence is indicated by plus signs, convergence by minus signs. Regions of vertically moving air undergoing dry adiabatic temperature changes are denoted by D and regions undergoing moist adiabatic temperature changes by M. From Ramage (1971).

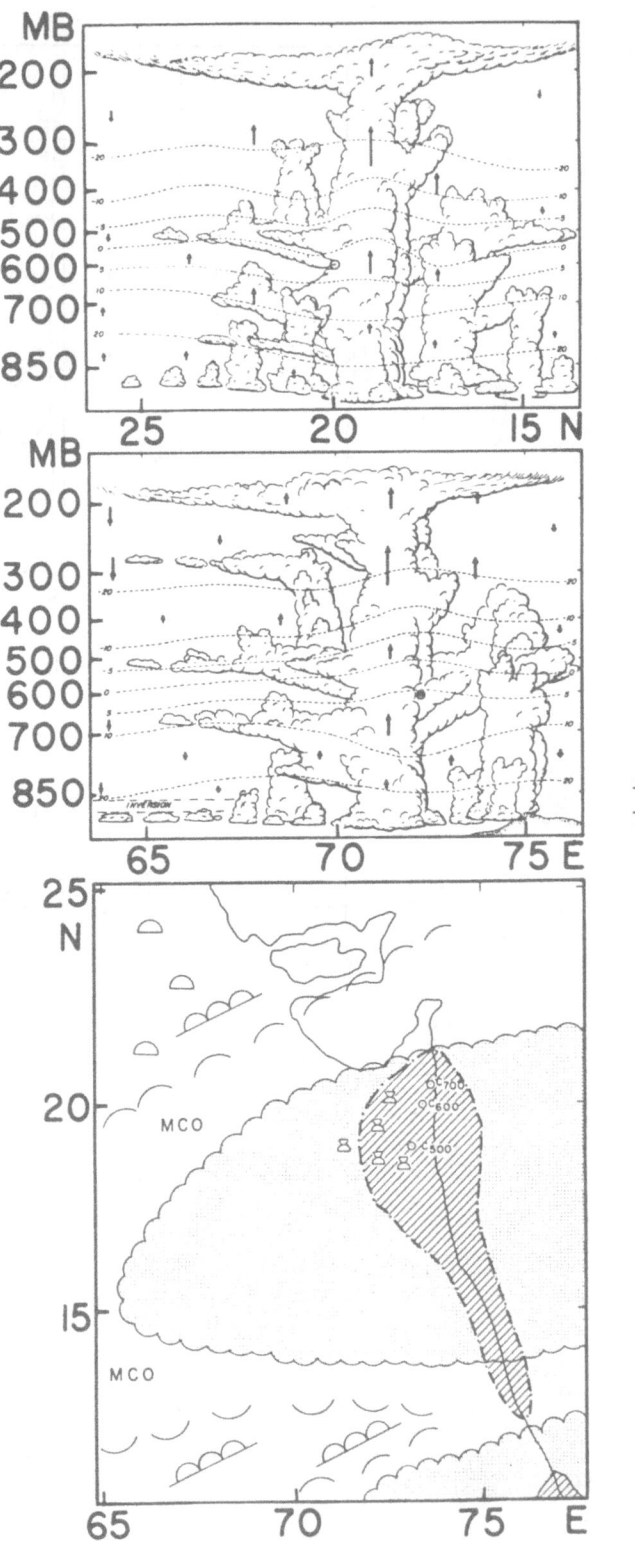

center of the cyclone, compensated by upper-tropospheric divergence. Ascending motion extends from the surface to the upper troposphere and from the center outward to about 400 km. Beyond the limits of the cyclonic circulation, subsidence prevails which is manifested in inversions and hazy conditions. Referring to the system documented by Miller and Keshavamurthy (1968) but without mentioning Ramage (1971), Krishnamurti (1979, p. 255) uses the term 'Mid-tropospheric cyclones of the Southwest monsoons'. He states that one or two of these disturbances form during each month of the Southwest monsoon, and that they occur primarily over the northwestern part of the Arabian Sea and occasionally over the northern part of the Bay of Bengal and over Southern Indochina. Ramage (1971, pp. 60–65) hypothesizes that energy exported from the quasi-stationary heat low over the South Asian continent is instrumental in the origin of the Subtropical Cyclone, calls attention to the apparent role of Subtropical Cyclone development for the onset of monsoon rains, and proposes a symbiotic relationship between continental heat low and Subtropical Cyclones. In contrast to the winter Subtropical Cyclones of the Northeastern Pacific, those in the monsoon area do dissipate, a major cause apparently being the injection of drier air.

Fig. 7.7:2. (on p. 236) Composite model of Subtropical Cyclone over North Indian Ocean, centered at 19°N, 72°E, and 600 mb, showing distribution of clouds, vertical motion, and temperature. (A) The plan view chart in the bottom panel maps the composite weather distribution. Hatching indicates rainfall exceeding 4 cm day^{-1} and stippling broken to overcast middle and high cloud levels. Cumulonimbus symbols denote greatest vertical cloud development, and broken areas extent of broken to overcast low clouds and scattered high clouds. (B) The zonal-vertical cross-section at 19°N in the middle panel schematically shows clouds on a greatly exaggerated vertical scale. Dashed lines are isotherms and solid arrows vertical velocity vectors; longest and shortest arrows denoting 40 and 5 cm^{-1} respectively. Circled dot represents the 600 mb composite center of the cyclone. (C) The meridional-vertical cross-section in the top panel uses the same symbols as the middle panel (B). From Miller and Keshavamurthy, 1968 (reprinted by permission from 'Structure of an Arabian Sea summer monsoon system', by Forrest R. Miller and R. N. Keshavamurthy; copyright 1968 by East-West Center Press).

7.8. Temporales of Pacific Central America

The weather on the Central American Isthmus is mostly dominated by the easterly airflow from the Caribbean and the disturbances imbedded therein. On rare occasions, weather systems originating over the open Pacific Ocean haunt the Pacific coastal regions. In the pre-satellite era these 'Temporales' were difficult to forecast, because of the chronic lack of surface observations away from the coast over the Pacific. Staff members of the National Meteorological Service of El Salvador have spearheaded the exploration of the Temporal (Portig, 1958b, 1960; Pallmann, 1968; Reyes, 1967, pp. 25–40; Pallmann and Murino, 1967).

Pallmann (1968) in particular analyzed both satellite radiation data and conventional surface and upper-air observations and developed a model of the structure, dynamics, and energetics of the Temporal, as illustrated in Figs. 7.8:1 and 7.8:2. The thermodynamic, cloud and weather characteristics are sketched in Fig. 7.8:1. Convective cells consisting mainly of cumulus congestus pump moisture into an extended altostratus deck, thus maintaining continuous rain from that layer. Sometimes thunder indicates the presence of isolated cumulonimbus cells, which act as feeders for both the altostratus and cirrostratus layers. The hydrostatic structure, dynamics and energetics of the Temporal are schematically shown in Fig. 7.8:2. In contrast to the Hurricane, the Temporal is characterized by an all-layer depression without significant deepening with altitude. Accordingly, a quasi-barotropic stratification is proposed, energy being supplied mainly from the upper-layer shear flow. Upper- and lower-tropospheric mechanisms are coupled by downward transfer of cyclonic vorticity from the shear layer. This downward vorticity transfer overpowers the weak anticyclonic vorticity in the upper part of the lower-tropospheric convective system. Three mechanisms cooperate in the maintenance of the extended altostratus cloud layer and the continuous rainfall: large-scale lower-tropospheric convergence, convection, and upslope motion of moist air at coastal mountain ranges of Central America.

In the large-scale circulation setting, Pallmann (1968) recognizes the following factors as conducive to development of the Temporal disturbance: high pressure over the Greater Antilles, a trough over the Eastern tropical Pacific, and cross-equatorial flow resulting in an activation of the Intertropical Convergence Zone. Temporales are in fact regarded as disturbances originating in the Intertropical Convergence Zone (Section 6.7.2) over the Eastern Tropical Pacific.

Table 7.8:1 shows the occurrence of Temporales during 1952–83. While the rainy season in much of Central America extends over the Northern summer half-year, Table 7.8:1 reveals that Temporales occur with preference in June and September–October. This double-peak pattern in the frequency of Temporales is consistent with the annual march in the latitude of the Intertropical Convergence Zone (Section 6.7.2). Alpert (1945) already recognized that the Intertropical Convergence Zone over the Eastern tropical Pacific migrates northward from February–March to June, recedes closer to the Equator in July–August, and then reaches its northernmost location in September. The seasonality of the surface circulation over the Eastern tropical Pacific is more fully documented in Hastenrath and Lamb (1977). The average frequency of Temporales is about one per year, but there are considerable interannual variations.

Most characteristic of the Temporal is abundant rainfall, while winds are weak. Over land Temporales may last a few days. Rainfall is nearly continuous, reaching daily totals of more than 300 mm, and more than 700 mm in the course of the Temporal. Reyes (1967, p. 40) even reports a record 24 h total of 426 mm. These are, however, record quantities not obtained in the entire area during all events. By comparison, rainfall from showers in other weather situations rarely yields more than 200 mm per day, although the bulk of this may fall within 90 min. The disasters caused by the Temporal are landslides, erosion of roads, interruption of telephone lines, and

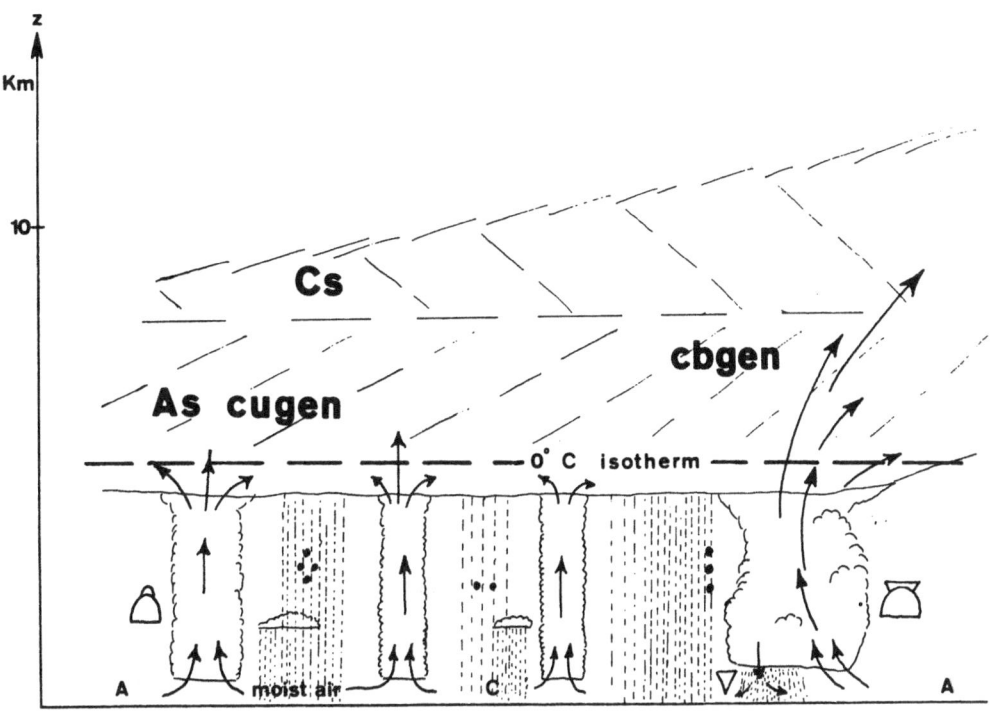

Fig. 7.8:1. Zonal-vertical transect across Temporal. At C Temporal is closest to section. Convective cells mainly of Cu congestus pump moisture into extended As layer providing feedback to maintain continuous rain from As. Isolated Cb cells act as feeder for both As and Cs. From Pallmann (1968).

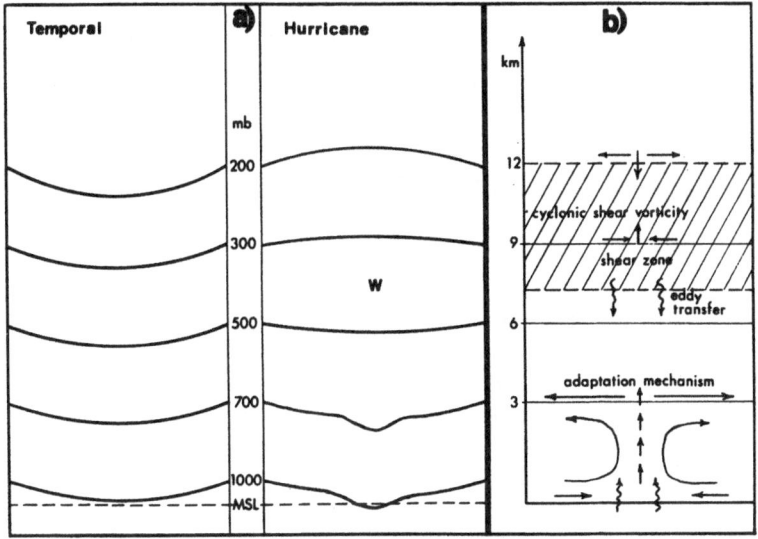

Fig. 7.8:2. Model vertical structure of Temporal. (a) the Temporal is not a warm-core system; (b) upper-tropospheric shear zone superimposed on lower-tropospheric low pressure system. From Pallmann (1968).

TABLE 7.8:1

Temporales in El Salvador, number of occurrences during 1952–83. (sources: Servicio Meteorológico Nacional, El Salvador, 1983; and H. Lessmann, pers. comm., 1984).

(a) annual march (monthly totals for 32 year period)

J	F	M	A	M	J	J	A	S	O	N	D	Year
0	0	0	0	2	4	2	3	13	9	1	0	34

(b) interannual variability (annual totals)

	0	1	2	3	4	5	6	7	8	9
1950's	–	–	1	3	3	7	3	2	1	0
1960's	2	2	0	1	0	0	0	0	1	2
1970's	1	0	0	1	1	0	0	1	1	0
1980's	0	0	1	0						

flooding of rivers, often associated with loss of human lives. It is then easy to overlook the possible positive consequences of the Temporales. The Pacific coastal regions of Central America have annual rainfall totals of the order of 1000 mm. In this context it appears that the Temporales are an important component in the regional hydrological budget.

In closing it must be mentioned that the term 'Temporal' is also used in a less specific sense in various regions of Central America. The notions of extended and persistent stratiform cloudiness and continuous rainfall seem essential. Such situations are not uncommon on the windward side of mountain ranges in Guatemala and Costa Rica in the Northern winter. These various namesakes seem in origin and structure different from the Temporales of the Pacific side of Central America, but have not been systematically studied.

7.9. Cold Surges

Throughout this chapter discussion concentrated on genuinely tropical weather systems. However, during the respective winter seasons, disturbances of midlatitude origin may penetrate deep into the tropics. Cold surges or equatorward intrusions of midlatitude air are in fact common occurrences in various regions of the global tropics. Characteristic of these events is the build-up of a cold air reservoir in the midlatitudes, the establishment of a steep meridional pressure gradient in the lower troposphere, and a subsequent abrupt cold-air outbreak. The present section reviews evidence from various tropical regions.

The North American continent with its predominantly meridionally oriented mountain ranges provides in winter a suitable setting for the generation of a vast pool of cold air and its outpour into the Central American – Caribbean region to the South. These episodes of cold air penetrations into the tropics and strong winds are known in the region as 'Nortes' or 'Northers'. Extensive operational experience on the 'Nortes' has been gathered by the National Meteorological Service of El Salvador, but little of this is now commonly accessible (Portig, 1958a; Hastenrath, 1962; Guzmán López, 1967, pp. 69–71). Modification of the cold continental air over the warm surface waters of the Gulf of Mexico and the Caribbean Sea is rapid, so that these airstreams can bring abundant cloudiness and rainfall to the windward side of the large mountain chains. Subsidence effects and clear skies, but likewise cold conditions and strong winds are characteristic of the leeward areas. In particular, funnel effects make for a concentration of winds in the large topographic depressions across the Mexican – Central American land bridge. Typically, increase of wind speed and shift in direction are observed several hours prior to the arrival of the cold air. Overall, the 'Nortes' appear concentrated in the lower 1000 m. Aged midlatitude cold fronts can retain considerable identity (Portig, 1958a). Lowest temperature and killing frosts in coffee plantations typically occur in nights after the 'Nortes' have abated (Guzmán López, 1967).

   Wintertime cold air invasions and midlatitude frontal passages also haunt the Southern hemispheric portion of the tropical Americas (Kousky, 1979; Girardi, 1983). For the Southern winter Girardi (1983) describes 'polar outbreaks' and cold Southern hemisphere air penetrating into the Amazon basin, with maintenance of remarkable midlatitude frontal characteristics. Kousky (1979) finds that Southern hemisphere cold fronts, or their remains, enter the Northeast of Brazil throughout the year and suggests a modulating influence on tropical weather developments.

   Rather less explored seem the wintertime cold air invasions into tropical Africa. Concerning Northern hemispheric Africa, such midlatitude influences reportedly account for some of the Dust Storms in the Sudan (Section 7.5), and affect dust loads and visibility in Subsaharan West Africa (Leroux, 1983, vol. 1, pp. 315–319). Invasions of polar air are responsible for the wintertime 'heug' rains of the Sahara (Leroux, 1983, vol. 1, pp. 319–334). The tropical part of Southern Africa is also known to be reached by extratropical influences during the winter (Thompson, 1965, pp. 8–9; Leroux, 1983, vol. 1, pp. 334–349).

   By comparison with the other tropical regions, considerable attention has been given recently to the winter-monsoon surges and cool-season disturbances of Southeast Asia (Ramage, 1954, 1955; Murakami, 1979; Chang et al., 1979, 1983; Chang and Lau, 1980; Chu and Park, 1984). Murakami (1979) recognizes the wintertime cold surface high over Siberia and North China as a source region of cold air outbursts, and considers the mountain barrier of the Himalayas as a factor for cold surges to rarely reach West of about 90°E. He notes that surface cold surges from the Siberian high tend to be preceded by extratropical cyclone development over the East China Sea, where air mass modification is favored by the abundant latent and sensible heat transfer from the warm Kuroshio Current. The penetration of cold surges into the equatorial region is seen as part of a regional scale mean meridional (Hadley type) circulation, with rising motion in the domain of heaviest rainfall in the Malaysia-Indonesia area and subsidence over Siberia. Murakami (1979) finds pulsations in the lower-tropospheric surge activity over East and Southeast Asia of the order of several days. Chang et al. (1979) underline the role of cold air surges emanating from the Siberian high in enhancing convection in the equatorial belt and thus contributing to the winter monsoon heat engine. In the development of the cold surge they distinguish two major phases: a widespread freshening of the northeasterly monsoon flow over the open South China Sea, followed by the southward incursion of cold air concentrated along the Vietnam coast, which progresses at a distinctly slower pace than the aforementioned airstream over the South China Sea at large. The freshening of surface winds prior to the arrival of the cold air as described by Chang et al. (1979) finds a remarkable parallel in the Central American – Caribbean region, where however the particulars of flow conditions are different.

7.10. Interannual Variability

Individual weather systems as described in Sections 7.2 to 7.9 coin the character of the rainy season in tropical regions. As part of the interannual variability of circulation and climate (Chapter 8), the occurrence and intensity of weather systems undergoes variations from year to year.

Best documented is the varying frequency of Tropical Storms. For pertinent data sources refer to the bibliographies in Milton (1974), Hastenrath (1976), Hastenrath and Wendland (1979), Shapiro (1982a, b), Hastenrath and Rosen (1983), and Gray (1983). As emphasized in Section 7.2, Tropical Storm formation is strongly controlled by certain factors in the large-scale atmospheric and oceanic environment. Of primary concern as dynamic and thermodynamic controls are vertical wind shear and sea surface temperature. Attention is also to be paid to surface pressure, because this is among the more commonly available elements, and because it is related to the wind field and thus both vertical shear and wind stress and sea surface temperature.

Figs. 7.10:1 to 7.10:3 show the secular variation of storms over the tropical North Atlantic and Eastern Pacific along with plots of pertinent elements of the large-scale atmospheric-oceanic environment. Most conspicuous are the broadly inverse variations of sea level pressure and sea surface temperature in the Atlantic (Fig. 7.10:2). This and other relationships apparent in Figs. 7.10:1 to 7.10:3 are quantitatively summarized in Table 7.10:1. Vertical wind shear is strongly negatively correlated with Tropical Cyclone frequency and positively with sea level pressure, which in turn, for the Atlantic, has a strong negative correlation with sea surface temperature. Except for Tropical Cyclone frequency and sea level pressure in the Pacific, correlations with sea level pressure and sea surface temperature are weak. Finally, the Tropical Cyclones of the Eastern Pacific are negatively correlated with the occurrence of Temporales.

On the time scale of decades, Milton (1974) found broadly parallel variations of Tropical Cyclone frequency in the North Atlantic and North Indian Ocean, as well as some similarity of variations in the latter region and the South Indian Ocean and Australian sector. Such sympathetic behavior is, however, not indicated on the interannual time scale (Gray, 1979, Table 4). The interannual variation of Tropical Cyclone frequency in the Northern and Southern hemispheres seems to compensate in part (Gray, 1979, Table 1), in such a way that during 1958–77 the global annual total ranged from 67 to 97, with an average of 79.

The annual frequency of West African Wave disturbances during the years 1968 through 1979 has been compiled in a series of notes in the Monthly Weather Review (Frank, 1970, 1971, 1972, 1973, 1975, 1976; Frank and Clark, 1977, 1978, 1979, 1980; Frank and Hebert, 1974; Simpson et al., 1969). Results are summarized in Table 7.10:2. This shows little variation from year to year and no relation to the rainfall conditions in Subsaharan West Africa.

The available information on the annual frequency of West African Squall Lines is summarized in Table 7.10:3. The studies of Hamilton and Archbold (1945) and Eldridge (1957), while from

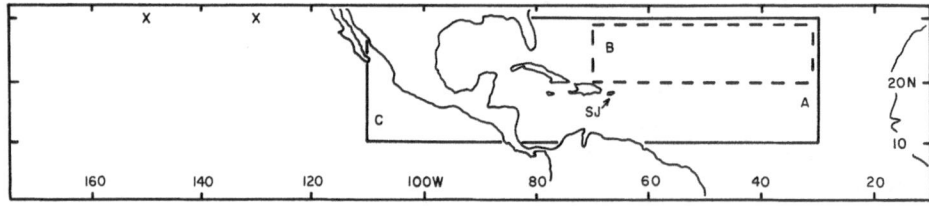

Fig. 7.10:1. Orientation map. Areas A and C have time series of SST, B of sea level pressure, compiled from ship observations. SJ signifies rawin station San Juan, Puerto Rico. Crosses indicate longitudes for which sea level pressure was obtained from Historical Weather Maps. From Hastenrath and Wendland (1979).

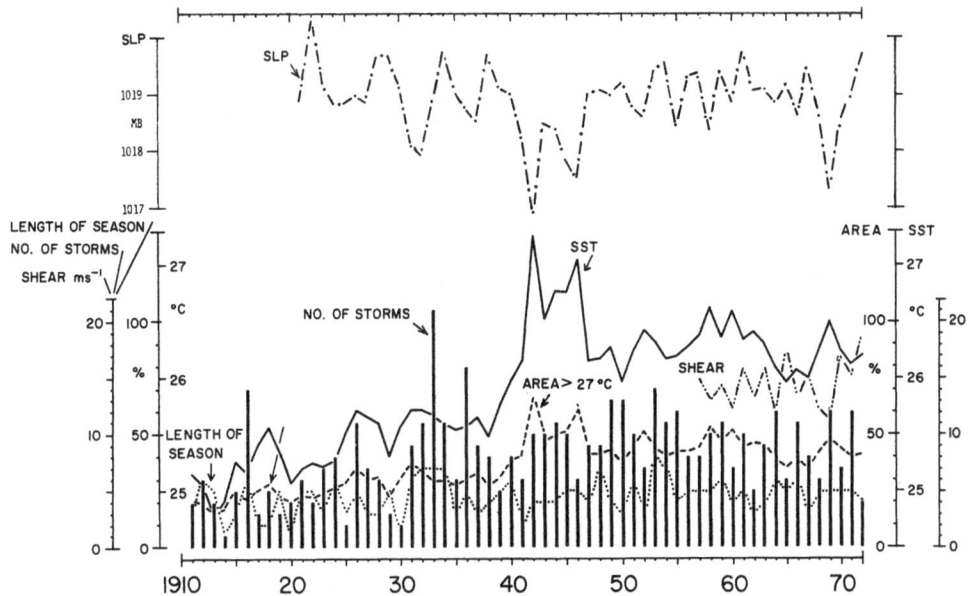

Fig. 7.10:2. North Atlantic. Annual frequency of Tropical Cyclones, solid vertical columns; length of storm season dotted, sea-level pressure dash-dotted, sea surface temperature solid, and percentage area above 27°C, broken lines; May−October mean vertical shear of zonal wind component 200−850 mb at San Juan, Puerto Rico dash-double-dotted line. From Hastenrath and Wendland (1979).

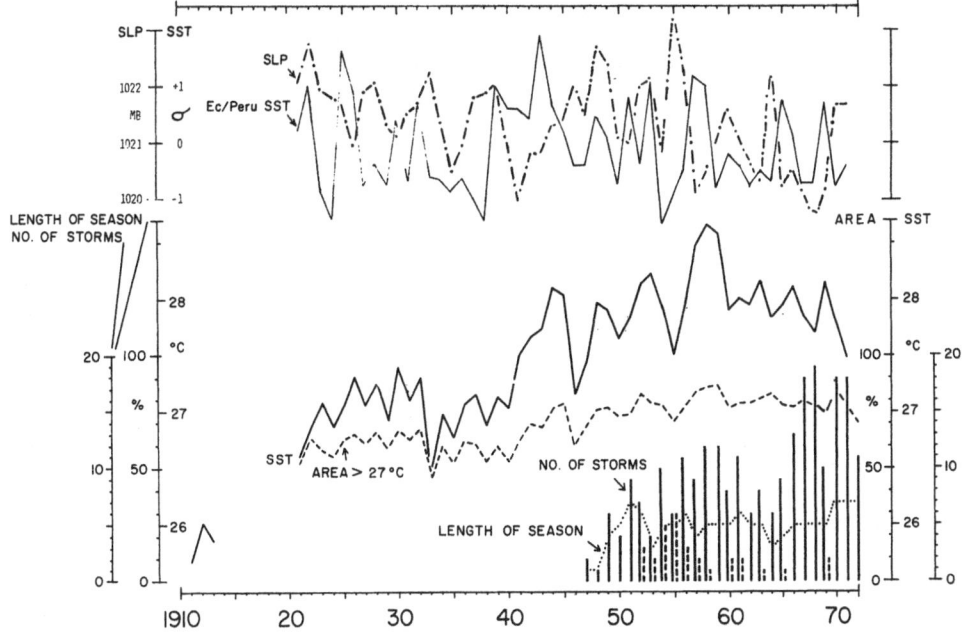

Fig. 7.10:3. Eastern Pacific. Annual frequency of Tropical Cyclones, solid vertical columns; and Temporales, broken vertical columns; index of sea surface temperature off Ecuador-Peru coast, Ec/Peru SST, thin solid line (departures from mean in terms of standard deviation, $\sigma = 0.8°C$); other symbols as in Fig. 7.10:2. From Hastenrath and Wendland (1979).

CHAPTER 7

TABLE 7.10:1

Correlation coefficients, in hundredths, between sea level pressure (SLP), sea surface temperature (SST), index of sea surface temperature conditions off the Ecuador-Peru coast (Ec/P SST), Tropical Cyclone frequency (TS), Temporales (Te), and vertical wind shear over San Juan, Puerto Rico (SHEAR), as plotted in Figs. 7.10:2 and 7.10:3. One, two and three asterisks denote significance at the 5, 1, and 0.1 percent levels, respectively. Quenouille's (1952, p. 168) method was used to account for the reduction of effective number of degrees of freedom due to persistence. Period is generally 1921–71, but correlations with Atlantic SHEAR are limited to 1957–72, correlations with Pacific TS to 1947–71, and correlations with Pacific Te to 1952–69.

	Pacific	Atlantic
SST – SLP	−16	−55***
SHEAR – SLP	−	+67***
SHEAR – TS	−	−49
TS – SST	+ 5	+24
TS – SLP	−50*	−22
Te – Ec/P SST	−24	−
Te – SST	−30	−
Te – TS	−43	−

TABLE 7.10:2

Annual frequency of West African Waves (May–October), at longitude of Dakar (sources: Frank, 1970, 1971, 1972, 1973, 1975, 1976; Frank and Clark, 1977, 1978, 1979, 1980; Frank and Hebert, 1974; Simpson et al., 1969), and rainfall index for Subsaharan West Africa (source: Lamb, 1982), in 10^{-2} σ.

Year	No. of waves	Rainfall index
1968	46	− 78
69	45	+ 6
70	33	− 72
71	47	− 88
72	54	−139
73	53	− 91
74	47	− 60
75	51	− 27
76	56	− 62
77	64	−129
78	54	+ 2
79	47	− 46

different regions, are both based on the surface station network. A larger apparent frequency can be expected in the study of Aspliden et al. (1976) because they used satellite imagery and analyzed all of West Africa. With these qualifications, Table 7.10:3 suggests a distinctly lower frequency of Squall Lines during dry years in Subsaharan West Africa. It appears plausible that with fewer rain-bearing weather systems there is less rainfall. However, as discussed in Section 7.4,

TABLE 7.10:3

Annual frequency of West African Squall Lines.

Reference	Region	Observation	Year(s)	Number per year	Character of season in Sahel
Hamilton and Archbold (1945)	Nigeria	surface stations	1939–41	42	DRY
Eldridge (1957)	Ghana	surface stations	1955	283	WET
Aspliden et al. (1976)	West Africa	satellite	1974 (63 days of GATE)	176	(DRY)

TABLE 7.10:4

Annual frequency of cyclonic disturbances (Monsoon Depressions and cyclonic storms) over the Bay of Bengal and the Arabian Sea. Source: Rao (1976).

	0	1	2	3	4	5	6	7	8	9
1890's	–	10	9	11	12	12	11	13	12	15
1900's	10	11	11	10	14	11	10	10	14	8
1910's	8	11	7	10	9	7	6	12	7	10
1920's	8	8	12	9	10	10	10	14	10	13
1930's	9	7	8	12	10	8	11	11	4	12
1940's	9	12	9	10	12	8	12	9	9	7
1950's	11	10	7	7	9	6	8	4	5	11
1960's	6	11	6	6	9	6	8	7	8	8
1970's	7	–	–	–	–	–	–	–	–	–

the formation of Squall Lines is favored by distinct features in the large-scale circulation, which themselves are subject to interannual variations.

The interannual variability of cyclonic disturbances (Monsoon Depressions and cyclonic storms) over the Bay of Bengal and the Arabian Sea during 1891–1970, as compiled by the India Meteorological Department (Rao, 1976, p. 108) is shown in Table 7.10:4. Rao (1976, pp. 107–108) calls attention to the apparent decrease in frequency since the beginning of the record but notes that comparatively little trend is apparent in recent decades. Rao (1976, p. 107) also mentions some studies suggesting various periodicities. At any rate, Table 7.10:4 documents a considerable variation in disturbance frequency from year to year.

Concerning the Temporales of Pacific Central America, Table 7.8:1 indicates a substantial interannual and long-term variability. Table 7.10:1 shows a positive correlation with sea level pressure and a negative correlation with Tropical Cyclone frequency over the Eastern North Pacific, as well as a weak negative coupling with the sea surface temperature conditions off the Ecuador-Peru coast.

Little information can be found in the current literature on the interannual variability of the Dust Storms of the Sudan (Section 7.5), and the Subtropical Cyclones of the North Pacific and North Indian Oceans (Section 7.7).

7.11. Synthesis

The objectives of this chapter have been to outline the structure, large-scale environmental setting, regional distribution, and seasonal characteristics of certain tropical weather systems.

Convection, cloudiness, and rainfall are instrumental for the functioning of tropical weather systems. The understanding of convection has been aided by satellite imagery and tropical field experiments. 'Mesoscale convective complexes' or 'cloud clusters' are recognized on satellite imagery from their extensive cirrus shields, and dominate the total cloudiness and rainfall in the tropics. While having lifetimes of only up to a day, they may evolve into Tropical Storms and be associated with Tropical Squall Lines, although 'nonsquall' clusters prevail.

Tropical cyclones with sustained surface winds of 32 m s^{-1} or more are called 'Hurricanes' in the Americas, 'Typhoons' in the Pacific West of the dateline, and 'Cyclones' in the Indian Ocean. Water vapor carried in from the outskirts is brought to condensation in the rain area, and thus diabatic heating accounts for a relatively warm mid troposphere and contributes to the inward directed surface pressure gradient. However, the very low central pressures of often 900 mb and less are only possible with adiabatic descent in the eye. A major factor in the formation of the eye is the approximate conservation of absolute angular momentum of the inward spiralling flow, which leads to high speeds at small radial distances from the center, and centrifugal acceleration exceeding the inward directed pressure gradient. Energetically, the Hurricane is an open system. It imports latent heat with the surface layer inflow, and exports geopotential energy and sensible heat aloft. Only a few percent of the latent heat release are converted to kinetic energy within the Hurricane circulation. Latent and sensible heat transfer from the ocean surface rather than merely increased frictional dissipation, are crucial for the decay of Hurricanes at landfall.

Essential for Hurricane genesis are (a) ocean surface temperature above $26.5°C$, (b) latitude beyond 5 degrees from the Equator, and (c) small vertical wind shear. Factor (a) affects the thermodynamic processes, (b) relates to cyclonic spin-up associated with the inflow, and (c) is pertinent to the concentration of latent heat release by convection in a limited area. The major regions of storm formation are the Western Atlantic and Caribbean, the Eastern Pacific, Western Pacific and South China Sea, Western South Pacific, Eastern South Indian Ocean, Arabian Sea and Bay of Bengal. This spatial distribution as well as the annual march of storm formation are consistent with the aforementioned criteria. Thus, the cold surface waters exclude much of the Southern hemisphere, the small Coriolis parameter eliminates the equatorial belt, and vertical shear accounts for the scarceness of cyclones over the Arabian Sea and Bay of Bengal at the height of the Northern summer, when the sea surface there is warm. Aside from their destructive effects, Hurricanes are beneficial for the hydrologic balance of tropical coastal regions.

Waves in the Easterlies are the most common disturbances of the trade wind regime. The distribution of vertical motion and weather within the wave and the displacement rate appears related to the vertical wind shear. For a regime with upward decrease of the basic easterly flow, Riehl's (1954) model calls for the heavy weather to be located to the rear of the trough axis. Structures differing from this pattern have been described for the African continent and the Pacific, where other vertical wind profiles are found. Wave length is of the order of 1700 to more than 2500 km, the period 3–5 days, and the displacement rate 6–9 m s^{-1}. Waves form with preference between about 15°E and 20°W to the South of the West African Mid-tropospheric Jet, as a consequence of a joint baroclinic-barotropic instability. In Northern summer, a wave crosses the West coast of Africa about every 3–4 days. Passing onto the cold Eastern Atlantic the waves generally decay, but remnants survive to the Western Atlantic and the Caribbean, where some regenerate and account for about half of the Atlantic Tropical Cyclones.

Squall Lines are among the more severe storms of the tropics. The system is typically hundreds of km in meridional extent and consists of a line of active thunderstorms. Displacement rate is faster than the environmental wind, with speeds of 10–20 m s^{-1}. Essential to the structure of a Tropical Squall Line is warm moist air entering the cloud base at the leading edge and rising in the convective updraft. Precipitation falls from both the main cloud column and a trailing anvil. Evaporation into comparatively dry mid-tropospheric air leads to cooling and downdrafts concentrated in the region of intense convection. The downward gushing cold air causes a "gust front" at the leading edge, undercuts the warm moist air ahead, thus causing more convection and a propagation of the Squall Line. Vertical shear and the vertical transfer of horizontal momentum are conducive to lower-tropospheric convergence and upper-layer divergence on the leading side of the system. Among the environmental factors favoring Squall Line formation are (a) warm moist surface air, (b) dry mid-tropospheric layer, (c) vertical shear of horizontal wind. These conditions are satisfied over Subsaharan Africa in Northern summer, and under the circumstances Squall Lines are particularly common. However, Tropical Squall Lines have also been described from the Western Atlantic, the Equatorial Central Pacific, and Northern India. Over Subsaharan West Africa they account for the bulk of annual rainfall.

Dust Storms over the Sudan arise in diverse synoptic situations with strong surface winds. Seasonal changes in surface conditions as well as artificial irrigation affect the frequency of Dust Storms.

Monsoon Depressions are most intense in the mid-troposphere, and possess a cold core in the lower layers. Ascending motion, cloudiness, and rainfall are most pronounced in the West.

Subtropical Cyclones are characterized by largest pressure gradients, strongest winds, and greatest convergence in the mid-troposphere. They are energy-producing and relatively long-lived because of modest frictional dissipation at the surface. Subtropical Cyclones over the Eastern North Pacific originate from cutoff lows in the subtropical Westerlies. Systems over the Northern Indian Ocean appear to maintain a symbiotic relationship with the summertime heat low over the South Asian continent.

The Temporales of Pacific Central America originate as disturbances in the Intertropical Convergence Zone over the Eastern Tropical Pacific especially during May–June and September–October. They are all-layer depressions without significant deepening with altitude. Individual convective cells feed an extensive altostratus deck, from which falls a continuous rain of moderate intensity. Isolated cumulonimbus cells act as feeders for both the altostratus and cirrostratus decks. Convergence of the large-scale tropospheric flow, convection, and mountain topography contribute about equally to the vertical motion and maintenance of an extended altostratus layer. Precipitation from Temporales amounts to records of more than 400 mm in 24 h, and more than 700 mm in three days. Temporales are feared for the hazards of landslides, and flooding, but they are an essential component of the regional water balance on the Pacific side of Central America.

Wintertime cold air intrusions from midlatitudes affect various regions of the global tropics, namely the Central American – Caribbean region, eastern South America, both Northern and Southern hemispheric Africa, and Southeast Asia. While being of extratropical origin, they interact with tropical weather systems and play a distinct role in the regional climate.

Documentation on interannual variability exists only for some of the tropical weather systems, namely for Tropical Storms, West African Wave disturbances, and Squall Lines, and the Temporales of Pacific Central America. Such information is of interest in relation to the interannual variability of large-scale circulation and climate.

In comparison with the higher latitudes, there is a vast diversity of weather systems in the tropics. Some models such as the Tropical Cyclone are widely accepted; others as the Subtropical

Cyclone and the Central American Temporales are of interest in limited regions; and much attention is being paid to Waves in the Easterlies and Tropical Squall Lines. Various of these weather systems through the associated rainfall play an essential role in the regional climate. Their inter-annual variability is an integral part of circulation and climate anomalies in the tropics.

References

Adejokun, J. A., Krishnamurti, T. N., 1983: 'Further numerical experiments on tropical waves'. *Tellus*, **35**A, 398–416.

Alpert, L., 1945: 'The Intertropical Convergence Zone of the Eastern Pacific region. I'. *Bull. Amer. Meteor. Soc.*, **26**, 426–432.

Anonymous, 1956: *Aperçus sur la climatologie de l'A.E.F.* Monographies de la Météorologie Nationale, No. 1, Paris, 21 pp.

Anthes, R. A., 1982: *Tropical cyclones, their evolution, structure, and effects*. Meteorological Monographs, vol. 19, no. 41, Amer. Meteor. Soc., Boston, 208 pp.

Asnani, G. C., 1982: 'The climate of Africa, including feasibility study of climate alert system'. pp. 107–129 in: *Proceedings of Technical Conference on Climate – Africa, Arusha, Tanzania, Jan. 1982*, WMO – No. 596, Geneva, Switzerland, 535 pp.

Aspliden, C. I., Adefolalu, D., 1976: 'The mean troposphere of West Africa'. *J. Appl. Meteor.*, **15**, 705–716.

Aspliden, C. I., Tourre, Y., Sabine, J. B., 1976: 'Some climatological aspects of West African disturbance lines during GATE', *Mon. Wea. Rev.*, **104**, 1029–1035.

Atkinson, G. D., 1971: *Forecaster's guide to tropical meteorology*. USAF Air Weather Service, Technical Report No. 240, 364 pp.

Barnes, G. M., Sieckman, K., 1984: 'The environment of fast- and slow-moving tropical mesoscale convective cloud lines'. *Mon. Wea. Rev.*, **112**, 1782–1794.

Bernet, G., 1966: 'Recherche d'un mode de formation des lignes de grains en Afrique Centrale'. Publications de la Direction de l'Exploitation Météorologique, No. 5, ASECNA, Dekar, 16 pp.

Betts, A. K., Grover, R. W., Moncrieff, M. W., 1976: 'Structure and motion of tropical squall-lines over Venezuela'. *Quart. J. Roy. Meteor. Soc.*, **102**, 395–404.

Bhalotra, Y. P. R., 1963: 'Meteorology of Sudan'. Sudan Meteorological Service, Memoir No. 6, Khartoum, 155 pp.

Bolton, D., 1984: 'Generation and propagation of African squall lines'. *Quart. J. Roy. Meteor. Soc.*, **110**, 695–721.

Burpee, R. W., 1972: 'The origin and structure of Easterly Waves in the lower troposphere of North Africa'. *J. Atmos. Sci.*, **29**, 77–90.

Burpee, R. W., 1974: 'Characteristics of North African Easterly Waves during the summers of 1968 and 1969'. *J. Atmos. Sci.*, **31**, 1556–1570.

Burpee, R. W., 1975: 'Some features of synoptic-scale waves based on a compositing analysis of GATE data'. *Mon. Wea. Rev.*, **103**, 921–925.

Carlson, T. N., 1969a: 'Synoptic histories of three African disturbances that developed into Atlantic hurricanes'. *Mon. Wea. Rev.*, **97**, 256–276.

Carlson, T. N., 1969b: 'Some remarks on African disturbances and their progress over the tropical Atlantic'. *Mon. Wea. Rev.*, **97**, 716–726.

Carlson, T. N., Lee, J. D., 1978: *Tropical meteorology*. Pennsylvania State University, Independent Study by Correspondence, University Park, Pennsylvania, 387 pp.

Chang, C.-P., Erickson, J. E., Lau, K. M., 1979: 'Northeasterly cold surges and near-equatorial disturbances over the winter MONEX area during December 1974, part 1: synoptic aspects'. *Mon. Wea. Rev.*, **107**, 812–829.

Chang, C.-P., Lau, K. M., 1980: 'Northeasterly cold surges and near-equatorial disturbances over the winter MONEX area during December 1974, part 2: planetary-scale aspects'. *Mon. Wea. Rev.*, **108**, 298–312.

Chang, C.-P., Millard, J. E., Chen, G. T. J., 1983: 'Gravitational character of cold surges during winter MONEX'. *Mon. Wea. Rev.*, **111**, 293–307.

Chu, P. S., Park, S. U., 1984: 'Regional circulation characteristics associated with a cold surge event over East Asia during winter MONEX'. *Mon. Wea. Rev.*, **112**, 955–965.

Das, P. K., 1986: *Monsoons*. Fifth IMO lecture, World Meteorological Organization, in press.

Das, P. M., De, A. C., Gangopadhyaya, M., 1957: 'Movements of two Nor'westers of West Bengal: a radar study'. *Indian J. Meteorol. Geophys.*, 8, 399–406.

De, A. C., 1963: 'Movement of pre-monsoon squall lines over gangetic West Bengal as observed by radar at Dum Dum airport'. *Indian J. Meteorol. Geophys.*, 14, 32–45.

Desai, B. N., 1950: 'Mechanisms of nor'westers in Bengal'. *Indian J. Meteorol. Geophys.*, 1, 74–76.

Dhonneur, G., 1970: 'Essai de synthèse sur les théories des lignes de grains en Afrique Centrale et Occidentale'. Notes de la Direction de l'Exploitation Météorologique, No. 38, ASECNA, Dakar, 55 pp.

Dhonneur, M., 1957: 'Étude climatologique des routes aériennes aboutissant à Fort Lamy'. Monographies de la Météorologie Nationale, No. 2, Paris, 49 pp.

Dunn, G. E. and staff, 1962: 'The hurricane season of 1961'. *Mon. Wea. Rev.*, 90, 107–119.

Eldridge, B. H., 1957: 'A synoptic study of West African disturbance lines'. *Quart. J. Roy. Meteor. Society*, 83, 303–314.

Eliassen, A., 1971: 'On the Ekman layer in a circular vortex'. *J. Meteor. Soc. Japan*, 49, special issue, 784–789.

Erickson, C. O., 1963: 'An incipient hurricane near the West African coast'. *Mon. Wea. Rev.*, 91, 61–68.

Estoque, M. A., Shukla, J., Jiing, J. G., 1983: 'African wave disturbances in a general circulation model'. *Tellus*, 35A, 287–295.

Forsdyke, A. G., 1960: 'Synoptic models of the tropics'. pp. 14–23 in: Bargman, D. J., ed., *Tropical meteorology in Africa*, Proceedings of WMO – Munitalp Foundation Symposium, Munitalp Foundation, Nairobi, 446.

Fortune, M., 1980: 'Properties of African squall lines inferred from time-lapse satellite imagery'. *Mon. Wea. Rev.*, 108, 153–168.

Frank, N. L., 1969: 'The "inverted V" cloud pattern – an Easterly wave?' *Mon. Wea. Rev.*, 97, 130–140.

Frank, N. L., 1970: 'Atlantic tropical systems of 1969'. *Mon. Wea. Rev.*, 98, 307–314.

Frank, N. L., 1971: 'Atlantic tropical systems of 1970'. *Mon. Wea. Rev.*, 99, 281–285.

Frank, N. L., 1972: 'Atlantic tropical systems of 1971'. *Mon. Wea. Rev.*, 100, 268–275.

Frank, N. L., 1973: 'Atlantic tropical systems of 1972'. *Mon. Wea. Rev.*, 101, 334–338.

Frank, N. L., 1975: 'Atlantic tropical systems of 1974'. *Mon. Wea. Rev.*, 103, 294–300.

Frank, N. L., 1976: 'Atlantic tropical systems of 1975'. *Mon. Wea. Rev.*, 104, 446–474.

Frank, N. L., Clark, G., 1977: 'Atlantic tropical systems of 1976'. *Mon. Wea. Rev.*, 105, 676–683.

Frank, N. L., Clark, G., 1978: 'Atlantic tropical systems of 1977'. *Mon. Wea. Rev.*, 106, 559–565.

Frank, N. L., Clark, G., 1979: 'Atlantic tropical systems of 1978'. *Mon. Wea. Rev.*, 107, 1035–1041.

Frank, N. L., Clark, G., 1980: 'Atlantic tropical systems of 1979'. *Mon. Wea. Rev.*, 108, 966–972.

Frank, N. L., Hebert, P. J., 1974: 'Atlantic tropical systems of 1973'. *Mon. Wea. Rev.*, 102, 290–295.

Frank, N. L., Johnson, H. M., 1969: 'Vortical cloud systems over the tropical Atlantic during the 1967 hurricane season'. *Mon. Wea. Rev.*, 97, pp. 124–129.

Fujita, T., Izawa, T., Watanabe, K., Imai, I., 1967: 'A model of typhoon accompanied by inner and outer rainbands'. *J. Appl. Meteor.*, 6, 3–19.

Gamache, J. F., Houze, R. A., 1982: 'Mesoscale air motions associated with a tropical squall line'. *Mon. Wea. Rev.*, 110, 118–135.

Girardi, C., 1983: 'O poço dos Andes'. Centro Técnico Aerospacial, Relatorio Técnico ECA 01/81, Saõ José dos Campos, S.P., 25 pp.

Godbole, R. V., 1975: 'The composite structure of the monsoon depression'. Florida State University, Department of Meteorology, NSF grant ATM 75–18945, Report No. 75–9, 31 pp.

Gomes Teixeira, L., 1970: 'Alguns aspectos das linhas de instabilidade que atingem Angola'. Serviço Meteorológico de Angola, SMA 178, MEM 71, Luanda, 6 pp.

Guzmán López, G. T., 1967: 'Über das Wetter und Klima in Mittelamerika, insbesondere in El Salvador'. *Institut für Meteorologie und Geophysik der Freien Universität Berlin, Meteorologische Abhandlungen*, 71, no. 3, 136 pp.

Gray, W. M., 1968: 'Global view of the origin of tropical disturbances and storms'. *Mon. Wea. Rev.*, 96, 669–700.

Gray, W. M., 1979: 'Hurricanes, their formation, structure, and likely role in the tropical circulation'. pp. 155–218 in: Shaw, O. B., ed., *Meteorology over the tropical oceans*, Conference, August 1978, Royal Meteorological Society, Bracknell, 278 pp.

Gray, W. M., 1983: 'Atlantic seasonal hurricane frequency'. Part 1. El Niño and 30 mb QBO influences. Part 2. forecasting its variability. Department of Atmospheric Science, Colorado State University, Fort Collins Colo., Atmospheric Science Paper No. 370, 48 pp.

Hamilton, R. A., Archbold, J. W., 1945: 'Meteorology of Nigeria and adjacent territory'. *Quart. J. Roy. Meteor. Soc.*, **71**, 230–265.

Hastenrath, S., 1962: 'Resultados de los vuelos de reconocimiento del tiempo sobre El Salvador'. Servicio Meteorológico Nacional, El Salvador, 15 pp.

Hastenrath, S., 1976: 'Variations in low-latitude circulation and extreme climatic events in the tropical Americas'. *J. Atmos. Sci.*, **33**, 202–215.

Hastenrath, S., Lamb, P. J., 1977: *Climatic atlas of the tropical Atlantic and Eastern Pacific Oceans*. University of Wisconsin Press, 112 pp.

Hastenrath, S., Lamb, P. J., 1979: *Climatic atlas of the Indian Ocean*. Part 1. *Surface climate and atmospheric circulation*. Part 2. *The oceanic heat budget*. University of Wisconsin Press, 116 and 110 pp.

Hastenrath, S., Rosen, A., 1983: 'Patterns of Indian monsoon rainfall anomalies'. *Tellus*, **35A**, 324–331.

Hastenrath, S., Wendland, W. M., 1979: 'On the secular variation of storms in the tropical North Atlantic and Eastern Pacific'. *Tellus*, **31**, 28–38.

Haurwitz, B., 1935: 'The height of tropical cyclones and the eye of the storm'. *Mon. Wea. Rev.*, **63**, 45–49.

Houze, R. A., 1977: 'Structure and dynamics of a tropical squall-line system'. *Mon. Wea. Rev.*, **105**, 1540–1567.

Houze, R. A. Jr., Betts, A. K., 1981: 'Convection in GATE'. *Rev. Geophys. Space Phys.*, **19**, 541–576.

Houze, R. A. Jr., Hobbs, P. V., 1982: 'Organization and structure of precipitating cloud systems'. *Advances in Geophysics*, **24**, 225–315, Academic Press.

India Meteorological Department, 1964: *Tracks of storms and depressions in the Bay of Bengal and the Arabian Sea, 1877–1960*. New Delhi.

Jeandidier, G., Rainteau, P., 1957: 'Prévision du temps sur de bassin du Congo'. Monographies de la Météorologie Nationale, No. 9, Paris, 13 pp.

Johnson, D. H., 1964: 'Weather systems of West and Central Africa'. pp. 339–346 in: Hutchings, J. W., ed., *Proceedings of the Symposium on Tropical Meteorology*, New Zealand Meteorological Service, Wellington, 737 pp.

Kousky, V. E., 1979: 'Frontal influences on Northeast Brazil'. *Mon. Wea. Rev.*, **107**, 1140–1153.

Krishnamurti, T. N., 1979: *Tropical meteorology*. Compendium of meteorology, vol. 2, part 4, editor A. Wiin-Nielsen, WMO No. 364, World Meteorological Organization, Geneva, 428 pp.

Krishnamurti, T. N., Kanamitsu, M., Godbole, R., Chang, C. B., Carr, F., Chow, J. H., 1975: 'Study of a monsoon depression. I. Synoptic structure.' *J. Meteor. Soc. Japan*, **53**, 227–239.

Krishnamurti, T. N., Kanamitsu, M., Godbole, R., Chang, C. B., Carr, F., Chow, J. H., 1976: 'Study of a monsoon depression, II. Dynamical structure.' *J. Meteor. Soc. Japan*, **54**, 208–225.

Lamb, P. J., 1982: 'Persistence of Subsaharan drought'. *Nature*, **299**, 46–48.

La Seur, N. E., 1964: 'Synoptic models in the tropics'. pp. 319–328 in: Hutchings, J. W., ed: *Proceedings of the Symposium on Tropical Meteorology Nov. 1963, Rotorura, New Zealand*, New Zealand Meteorological Service, Wellington, 737 pp.

Leary, C. A., 1984: 'Precipitation structure of the cloud clusters in a tropical easterly wave'. *Mon. Wea. Rev.*, **112**, 313–325.

Leroux, M., 1983: *Le climat de l'Afrique tropicale*. Champion, Paris, 2 vols., 633 pp. and 24 pp. plus 250 maps.

Mass, C., 1979: 'A linear primitive equation model of African Wave disturbances'. *J. Atmos. Sci.*, **36**, 2075–2092.

Merritt, N. S., 1964: 'Easterly waves and perturbations, a reappraisal'. *J. Appl. Meteor.*, **3**, 367–382.

Miller, F. R., Keshavamurthy, R. N., 1968: *Structure of an Arabian Sea summer monsoon system*. International Indian Ocean Expedition Meteorological Monographs, No. 1, University Press of Hawaii, Honolulu, 94 pp.

Miller, M. J., Betts, A. K., 1977: 'Traveling convective storms over Venezuela'. *Mon. Wea. Rev.*, **105**, 833–848.

Milton, D., 1974: 'Some observations of global trends in tropical cyclone frequencies'. *Weather*, **29**, 267–270.

Murakami, T., 1979: 'Winter monsoonal surges over East and Southeast Asia'. *J. Meteor. Soc. Japan*, **57**, 133–158.

Newton, C. W., 1950a: 'Structure and mechanisms of the prefrontal squall line'. *J. Meteor.*, **7**, 210–222.

Newton, C. W., 1950b: 'Note on the mechanisms of nor-westers of Bengal'. *Indian J. Meteorol. Geophys.*, **2**, 48–50.

Newton, C. W., 1963: *Dynamics of severe convective storms*. Meteorol. Monographs, vol. 5, no. 27, pp. 33–58, Amer. Meteor. Soc., Boston.

Newton, C. W., 1967: 'Severe convective storms'. *Advances in Geophysics*, **12**, 257–308.

Newton, C. W., Newton, H. R., 1959: 'Dynamical interaction between large convective clouds and environment with vertical shear'. *J. Meteor.*, **16**, 483–496.

Nicholls, M. E., Johnson, R. H., 1984: 'A model of a tropical squall line boundary layer wake'. *J. Atmos. Sci.*, **41**, 2774–2792.

Nitta, T., Masuda, K., 1981: 'Observational study of a monsoon depression developed over the Bay of Bengal during summer MONEX.' *J. Meteor. Soc. Japan*, **59**, 672–682.

Norquist, D. C., Recker, E. E., Reed, R. J., 1977: 'The energetics of African wave disturbances as observed during Phase III of GATE'. *Mon. Wea. Rev.*, **105**, 334–342.

Obasi, G. O. P., 1974: 'The environmental structure of the atmosphere near West African Disturbance Lines'. pp. 62–66, Preprints Symposium Tropical Meteorology Part II, Nairobi, Amer. Meteor. Soc., Boston.

Okulaja, P. O., 1970: 'Synoptic flow perturbations over West Africa'. *Tellus*, **6**, 663–668.

Ooyama, K., 1969: 'Numerical simulation of the life cycle of tropical cyclones'. *J. Atmos. Sci.*, **26**, 3–40.

Pallmann, A., 1968: 'The synoptics, dynamics, and energetics of the temporal using satellite radiation data'. Department of Geophysics, Saint Louis University, NESS-ESSA Grant WBG-89, second year's reports 1 and 2, 30 and 97 pp.

Pallmann, A., Murino, C. J., 1967: 'The synoptics, dynamics, and energetics of the temporal, using satellite radiation data'. Department of Geophysics, Saint Louis University, NESS-ESSA Grant WBG-89, First Year's Final Report, 90 pp.

Palmén, E., Newton, C. W., 1969: *Atmospheric circulation systems*. Academic Press, New York and London, 603 pp.

Payne, S. W., McGarry, M. M., 1977: 'The relationship of satellite inferred convective active to Easterly waves over West Africa and the adjacent ocean during Phase III of GATE'. *Mon. Wea. Rev.*, **105**, 413–420.

Pedgley, D. E., Krishnamurti, T. N., 1976: 'Structure and behavior of a monsoon cyclone over West Africa'. *Mon. Wea. Rev.*, **104**, 149–167.

Piersig, W., 1936: 'Schwankungen von Luftdruck und Luftbewegung sowie ein Beitrag zum Wettergeschehen im Passatgebiet des östlichen Nordatlantischen Ozeans'. *Archiv der Deutschen Seewarte*, **54**, No. 6, 41 pp. (+ 4 plates).

Piersig, W., 1944: 'The cyclonic disturbances of the sub-tropical North Atlantic'. *Bull. Amer. Meteor. Soc.*, **25**, pp. 2–17.

Portig, W., 1958a: 'Frontdurchgang in Mittelamerika'. *Meteor. Rundschau*, **11**, 112–116.

Portig, W., 1958b: 'Der Temporal von Ende Oktober 1957'. *Meteor. Rundschau*, **11**, 150–156.

Portig, W., 1960: 'Beiträge zur Meteorologie Mittelamerikas (insbesondere El Salvadors)'. Deutscher Wetterdienst, Seewetteramt, Einzelveröffentlichungen, No. 28, 24 pp.

Quenouille, M. H., 1952: *Associated measurements*. Butterworths Scientific Publications, London, 242 pp.

Ramage, C. S., 1954: 'Non-frontal crachin and the cool season clouds of the China Seas'. *Bull. Amer. Meteor. Soc.*, **35**, 404–411.

Ramage, C. S., 1955: 'The cool-season tropical disturbances of Southeast Asia'. *J. Meteor.*, **12**, 252–262.

Ramage, C. S., 1962: 'The subtropical cyclone'. *J. Geophys. Res.*, **67**, 1401–1411.

Ramage, C. S., 1971: *Monsoon meteorology*. Academic Press, New York and London, 296 pp.

Ramage, C. S., 1974: 'Monsoonal influences on the annual variation of tropical cyclone frequency development over the Indian and Pacific Oceans'. *Mon. Wea. Rev.*, **102**, 745–753.

Rao, Y. P., 1976: *Southwest monsoon*. India Meteorological Department Meteorological Monograph, Synoptic Meteorology, No. 1/1976, Delhi, 367 pp.

Reed, R. J., Norquist, D. C., Recker, E. E., 1977: 'The structure and properties of African wave disturbances as observed during phase III of GATE'. *Mon. Wea. Rev.*, **105**, 317–333.

Reed, R. J., Recker, E. E., 1971: 'Structure and properties of synoptic-scale wave disturbances in the Equatorial Western Pacific'. *J. Atmos. Sci.*, **28**, 1117–1133.

Regula, H., 1936: 'Druckschwankungen und Tornados an der Westküste von Afrika'. *Annalen der Hydrographie und Maritimen Meteorologie*, **64**, 107–111.

Regula, H., 1943: 'Pressure changes and "tornadoes" (squalls) on the West coast of Africa'. *Bull. Amer. Meteor. Soc.*, **24**, pp. 311–317.

Rennick, M. A., 1976: 'The generation of African waves'. *J. Atmos. Sci.*, **33**, 1955–1969.

Reyes, L. R., 1967: 'Modelos sinópticos utilizados en la Sección de Pronósticos del Servicio Meteorológico Nacional'. Servicio Meteorológico Nacional de El Salvador, Publicación Ténica No. 9, San Salvador, 44 pp.

Riehl, H., 1954: *Tropical meteorology*. McGraw-Hill, New York, Toronto, London, 392 pp.

Riehl, H., 1979: *Climate and weather in the tropics*. Academic Press, London, New York, San Francisco, 611 pp.

Saha, K., Chang, C.-P., 1983: 'The baroclinic processes of monsoon depressions. *Mon. Wea. Rev.*, **111**, 1506–1514.

Servicio Meteorológico Nacional, El Salvador, 1961: Cartas del tiempo, 30 October to 9 November, 18–19 December 1961. San Salvador.

Servicio Meteorológico Nacional, El Salvador, 1983: *Almanaque Salvadoreño*. San Salvador, 90 pp.

Shapiro, L. J., 1982a: 'Hurricane climatic fluctuations. Part 1: patterns and cycles'. *Mon. Wea. Rev.*, 110, 1007–1013.

Shapiro, L. J., 1982b: 'Hurricane climatic fluctuations. Part 2. relation to large-scale circulation'. *Mon. Wea. Rev.*, 110, 1014–1023.

Simmons, A. J., 1977: 'A note on the instability of the African easterly jet'. *J. Atmos. Sci.*, 34, 1670–1674.

Simpson, R. H., 1952: 'Evolution of the Kona Storm, a subtropical cyclone'. *J. Meteor.*, 9, 24–35.

Simpson, R. H., Frank, N., Shideler, D., Johnson, H. M., 1968: 'Atlantic tropical disturbances, 1967'. *Mon. Wea. Rev.*, 96, 251–259.

Simpson, R. H., Frank, N., Shideler, D., Johnson, H. M., 1969: 'Atlantic tropical disturbances of 1968'. *Mon. Wea. Rev.*, 97, 240–255.

Simpson, R. H., Riehl, H., 1981: *The hurricane and its impact*. Louisiana State University Press, Baton Rouge and London, 398 pp.

Thompson, B. W., 1965: *The climate of Africa*. Oxford University Press, Nairobi, New York, 132 pp.

Thompson, R. M., Payne, S. W., Recker, E. E., Reed, R. J., 1979: 'Structure and properties of synoptic-scale wave disturbances in the Intertropical Convergence Zone of the Eastern Atlantic'. *J. Atmos. Sci.*, 36, 53–72.

Tschirhart, G., 1958: 'Les conditions aérologiques à l'avant des lignes de grains en Afrique Équatoriale'. Monographies de la Météorologie Nationale, No. 11, Paris, 28 pp.

Tschirhart, G., 1959: 'Les perturbations atmosphériques intéressant l'A.E.F.'. Monographies de la Météorologie Nationale, No. 13, Paris, 32 pp.

Voiron, H., 1964: 'Quelques aspects de la météorologie dynamique en Afrique Occidentale'. *La Météorologie*, 215–231.

Yanai, M., Maruyama, T. Hayashi, Y., 1968: 'Power spectra of large-scale disturbances over the tropical Pacific'. *J. Meteor. Soc., Japan*, 46, 308–323.

Zipser, E. J., 1969: 'The role of organized unsaturated convective downdrafts in the structure and rapid decay of an equatorial disturbance'. *J. Appl. Meteor.* 8, 799–814.

Zipser, E. J., 1977: 'Mesoscale and convective-scale downdrafts as distinct components of squall-line structure'. *Mon. Wea. Rev.*, 105, 1568–1589.

CHAPTER 8

INTERANNUAL VARIABILITY OF THE ATMOSPHERE-OCEAN SYSTEM

The functioning of the tropical atmosphere in the course of the average annual cycle and the role of weather disturbances in these processes are discussed in Chapters 6 and 7, while Chapter 4 provides an orientation on the characteristic average mode of operation of the low-latitude oceans. This background is required for the understanding of year-to-year variations which are to be explored in the present chapter. The mechanisms of interannual variability of the combined atmosphere-hydrosphere system are manifest in changes of the atmospheric and oceanic circulation and extreme events – especially rainfall anomalies – in the regional climates. Regional climate anomalies appear in part related to circulation changes on the scale of the global tropics.

In the following sections, the Southern Oscillation is discussed first (Section 8.1), as this reflects conditions throughout the low latitudes and even the extra-tropical caps. The El Niño phenomenon of the greater Pacific basin (Section 8.2) is now known to be an integral part of the Southern Oscillation. The circulation mechanisms operative in the climate anomalies of key tropical land areas have been elucidated in various low-latitude analyses, including a series of circum-Atlantic studies (Sections 8.6 to 8.10). Typical time scales of climate variability are identified in Section 8.11, while in Section 8.12 the various results are placed into context.

8.1. Surface Patterns of the Southern Oscillation

Large-scale long-term surface pressure seesaws between key regions of the tropics are increasingly recognized as being associated with circulation changes and regional climate anomalies. Research on this problem complex can be traced over nearly a century to Hildebrandsson's (1897–1914) classical study on the 'centers of action', in which an inverse pressure variation between South-eastern Australia and Southern South America is already noted (Hildebrandsson, 1897). Lockyer and Lockyer (1902a, b, 1904) and Lockyer (1906) substantiated low-frequency pressure seesaws spanning large part of the globe. These early publications must be regarded as essential background to the landmark studies of Walker (1909, 1923, 1924, 1928a) and Walker and Bliss (1930, 1932) who described the surface pressure seesaws in relation to rainfall and temperature fluctuations.

Walker (1924, pp. 317–332; 1928, pp. 99–106) and Walker and Bliss (1932, pp. 54–75; 1937, pp. 119–131) recognized three major long-term surface pressure seesaws: (i) the North Atlantic Oscillation (NAO), (ii) the North Pacific Oscillation (NPO), both referred to as 'Northern Oscillations', and (iii) the Southern Oscillation. The North Atlantic Oscillation represents inverse pressure variations at the Azores high and the Icelandic low, is associated with temperature anomalies in Europe consistent with the zonal wind pattern over the North Atlantic, and appears most pronounced in winter (ref. Van Loon and Rogers, 1978; Rogers and Van Loon, 1979; Meehl and Van Loon, 1979). Similarly, the North Pacific Oscillation consists of a surface pressure seesaw between the North Pacific high and the Aleutian low, and is likewise associated with variations in the zonal wind field and temperature conditions in downstream North America. The Southern

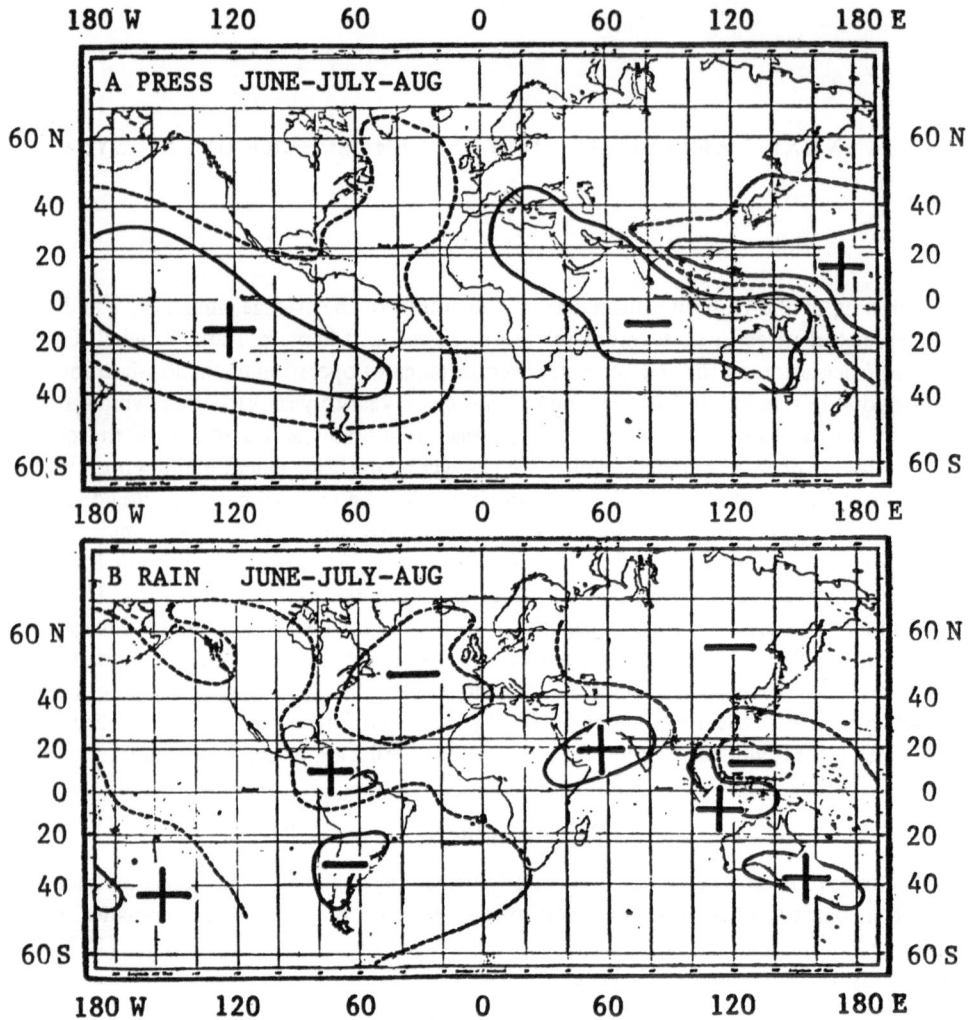

Fig. 8.1:1. Global maps of the Southern Oscillation (SO) during June–July–August. Shown are correlation coefficients between an index of SO and contemporary (a) pressure (top), and (b) rainfall (bottom). Lines of ± 0.5 are solid, and zero line is broken. Source: Walker and Bliss (1932).

Oscillation is defined by Walker and Bliss (1932, p. 60), in expansion of Walker's (1924, p. 323) earlier description, as follows: "In general terms, when pressure is high in the Pacific Ocean, it tends to be low in the Indian Ocean from Africa to Australia; these conditions are associated with low temperatures in both these areas, and rainfall varies in the opposite direction to pressure. Conditions are related differently in winter and summer, and it is therefore necessary to examine separately the seasons December to February and June to August." Walker and Bliss (1932, p. 66) state further: "The connection between SO and the NAO is negligible, but the SO has a considerable negative influence on the NPO six months in advance as well as simultaneously."

Of these three systems it is the Southern Oscillation which appears primarily relevant for the interannual variability of circulation and climate in the tropics. Walker and Bliss' (1932) original global maps of Southern Oscillation patterns are reproduced here as Figs. 8.1:1a and b and

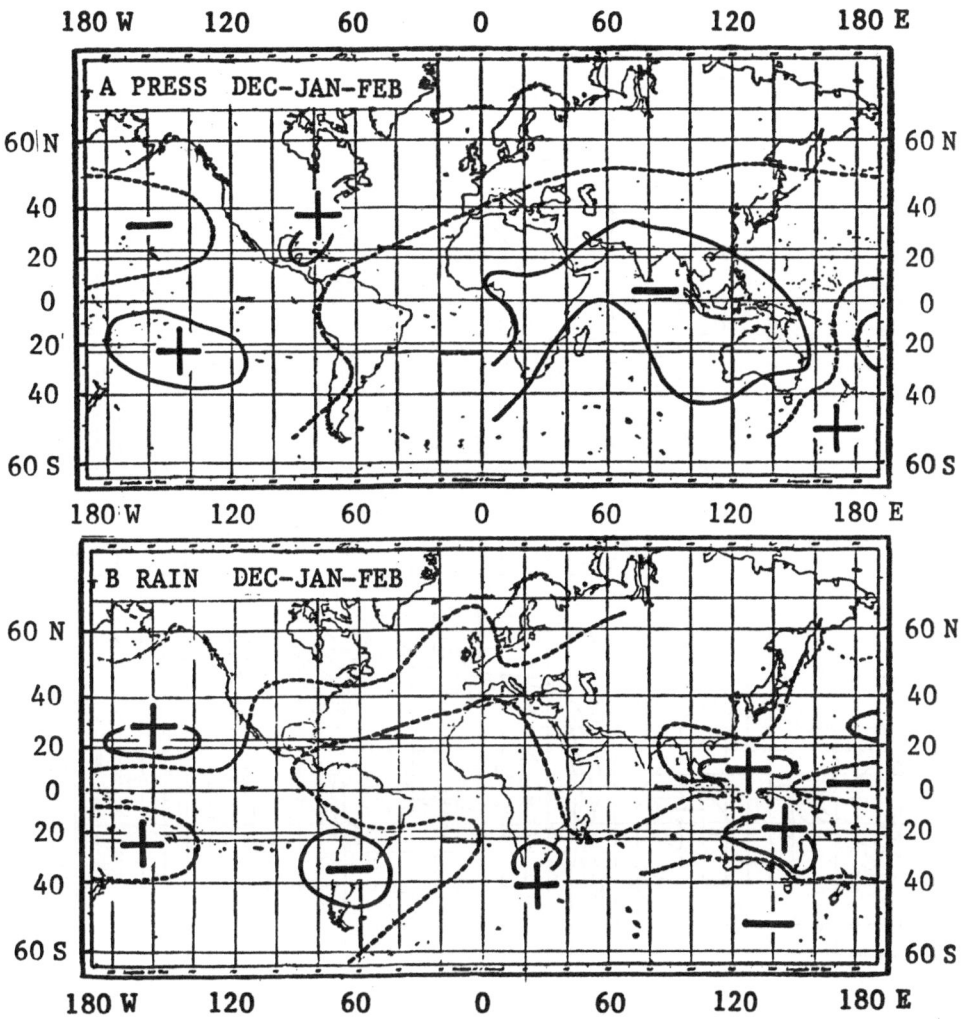

Fig 8.1:2. Global maps of the Southern Oscillation (SO) during December–January–February. Shown are correlation coefficients between an index of SO and contemporary (a) pressure (top), and (b) rainfall (bottom). Lines of ± 0.5 are solid, and zero line is broken. Source: Walker and Bliss (1932).

8.1:2a and b. Some explanation is needed for the use of these maps. Walker and Bliss (1932) first calculated an index of the Southern Oscillation as follows: (Santiago pressure) + (Honolulu pressure) + (India rain) + (Nile flood) + 0.7 (Manila pressure) − 0.7 (Darwin pressure) − 0.7 (Chile rain). Then correlation coefficients were calculated of this index with pressure, temperature, and rainfall at a network of stations. These results are mapped in Figs. 8.1:1 and 8.1:2. The aforementioned formula already indicates some elementary characteristics of the Southern Oscillation, such as the inverse pressure relationship between the Pacific and the Australasian region, and the positive coupling of India rainfall with the Southern Oscillation. Figs. 8.1:1a and 8.1:2a illustrate the prominent pressure patterns, which can be characterized by two main dipoles in the Eastern South Pacific and the Australasian region, respectively, and a nodal line in the Atlantic – South American sector, albeit with details differing between the two extreme seasons. Figs. 8.1:1b and

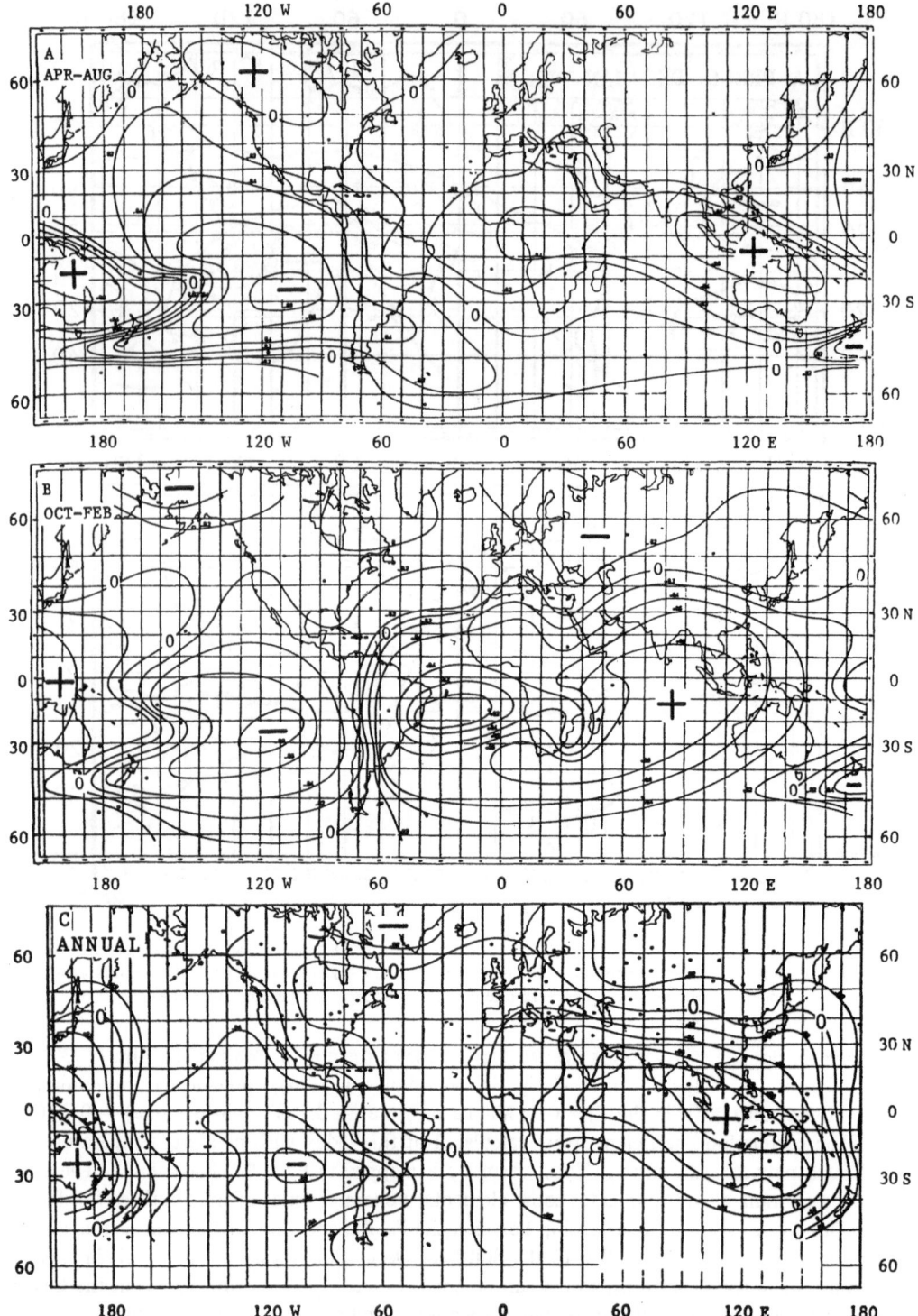

Fig. 8.1:3. Maps of simultaneous correlation of air pressure anomalies with Djakarta air pressure anomalies. Isoline spacing is 0.2. (a) April–August (top), (b) October–February (middle), (c) annual (bottom). Source: Berlage (1957, 1966).

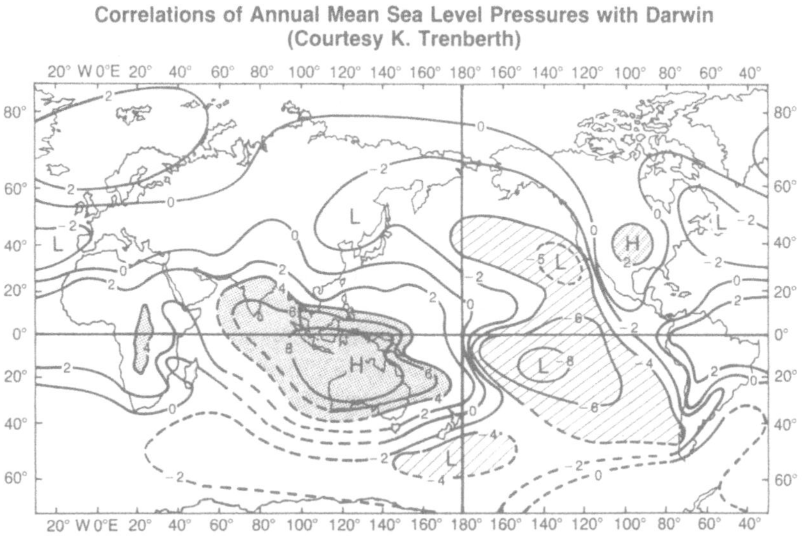

Correlations of Annual Mean Sea Level Pressures with Darwin
(Courtesy K. Trenberth)

Fig. 8.1:4. Isocorrelation map of annual mean sea level pressure with Darwin. Isoline spacing is 0.2. Courtesy of K. Trenberth (1984).

8.1:2b show rather noisier map patterns for rainfall, but a polarity between the Eastern South Pacific versus the Australasia — Indian Ocean sector is generally apparent at least in the June—July—August map (Fig. 8.1:1b).

Walker and Bliss pursued their studies of the Southern Oscillation with a keen practical motivation for seasonal forecasting (or 'foreshadowing' as they later called it; Walker and Bliss, 1930), as will be discussed in Chapter 9. However, for decades thereafter little attention was given to the Southern Oscillation problem, except for the works of Berlage (1957, 1966), Troup (1965), and Bjerknes (1966, 1969).

Research into the Southern Oscillation was resumed by Berlage (1957, 1966), Berlage and de Boer (1959, 1960) and Schove and Berlage (1965), an essential asset being the start of regular pressure observations at Easter Island in the Eastern South Pacific. In particular, Berlage (1957, 1966) produced isocorrelate charts of pressure with respect to the pressure variations at Djakarta, Indonesia. In addition to an annual map (Berlage, 1957, p. 59), he constructed charts for two opposing times of the year, namely April—August and October—February (Berlage, 1966, pp. 14—15). These three maps are here reproduced as Figs. 8.1:3a to c. These maps confirm the major pattern characteristics identified by Walker and Bliss (1932), as illustrated in Figs. 8.1:1a and 8.1:2a, in particular the inverse relation between major dipoles in the Eastern South Pacific and the greater Australasian region. However, as Walker and Bliss (1932) before, Berlage (1966) also finds substantial pattern differences between the two opposing times of year. Drawing on a more extensive data base, Trenberth (personal communication, 1984), also produced maps of pressure correlations with Darwin, North Australia. His annual map is reproduced as Fig. 8.1:4. The major similarity with Figs. 8.1:1a, 8.1:2a and 8.1:3a to c consists in the overwhelming polarity between the Eastern Pacific and the greater Australasian region.

Berlage (1957, pp. 70—74; 1966, pp. 109—119) also expanded on Walker and Bliss' (1932) earlier work concerning the rainfall variations associated with the Southern Oscillation. More importantly, Berlage (1966, pp. 21—23, 106—116) called attention, for the first time, to the tendency of El Niño events (Section 8.2) to coincide with the phase of the Southern Oscillation characterized by relatively low pressure over the Eastern South Pacific, and high pressure in the greater Australasian region. Berlage (1957, pp. 9–27; 1966, p. 9, 27–29, 36–43) searched extensively for periodicities in the operation of the Southern Oscillation, without definite success except for concluding that 1–5 years represented a typical time scale.

A series of later studies (Troup, 1965; Quinn, 1974; Quinn and Burt, 1970, 1972; Quinn et al., 1978; Wright, 1975, 1977, 1985; Kidson, 1975; Trenberth, 1976; Julian and Chervin, 1978; Rasmusson and Carpenter, 1982; Newell et al., 1982; Philander, 1983; Kousky et al., 1984) confirmed the persistent and markedly aperiodic nature of the Southern Oscillation. While some preference is indicated for the time scale of 3–6 years (Trenberth, 1976; Julian and Chervin, 1978; Rasmusson and Carpenter, 1982; Philander, 1983), the Southern Oscillation seems to vary over the range of 2–10 years (Trenberth, 1976; Philander, 1983). In addition, Quinn (1976) and Trenberth (1976) call attention to spatial variations of the Southern Oscillation. These may in part be reflected in the pattern differences between the maps, Figs. 8.1:1 to 8.1:4, although the differing analysis modes and reference stations may account for a major part of the apparent diversity.

Figs. 8.1:1 to 8.1:4 all illustrate the global nature of long-term pressure variations associated with the Southern Oscillation. Van Loon and Madden (1981), Van Loon and Rogers (1981), and Rogers (1984), in particular, documented the associations with the extratropical caps of the two hemispheres. They find that with low pressure over the tropical South Pacific, pressure tends to be high over the extratropical South Pacific and East Antarctica and low in the West wind belt of the Indian and South Atlantic Oceans. Moreover, in this phase of the Southern Oscillation meridional pressure gradients appear steepened in the low, and flattened in the high latitudes of both hemispheres.

However, these and other analyses confirm as major dipoles of the Southern Oscillation the greater Indonesian – Australasian region and the Eastern South Pacific. Recognizing the remarkable spatial coherence, regularity, and persistence of the long-term pressure variations, various authors (Quinn and Burt, 1970, 1972; Quinn, 1974; Trenberth, 1976; Wright, 1975, 1977, 1984; McBride and Nicholls, 1983) devised simple indices to describe the phases of the Southern Oscillation, observations at the aforementioned dipoles being as a rule the most important information.

While the coincidence of warm water anomalies off the South American West coast with a certain phase of the Southern Oscillation had already been noted by Berlage (1966, pp. 21–23, 106–116), Bjerknes (1966, 1969) developed a conceptual model relating the zonal pressure seesaw between the extremities of the great Pacific basin to the 'Walker Circulation' (Section 6.9) and sea surface temperature anomalies in the equatorial Pacific. His hypothesis provided a major stimulus for investigations into the mechanisms of the El Niño phenomenon to be discussed in the following Section 8.2.

At this point it should be realized that the Southern Oscillation involves both atmosphere and ocean. While the wind stress forcing of the upper ocean has been studied extensively by theoretical oceanographers, the causes for the pressure seesaw in the lower atmosphere – which is instrumental in the alteration of the zonal wind field – has remained unexplained. After the review of pertinent evidence in Sections 8.2 and 8.3, this issue will be considered in Section 8.4. While the causes of the Southern Oscillation are not well understood, its climatic consequences are more obvious. Inasmuch as the Southern Oscillation entails variations in the large-scale pressure pattern and vertical motion field, it can be expected to play a role for climate anomalies in many tropical regions.

8.2. El Niño

The Pacific littoral from Ecuador through Peru to Northern Chile is among the most extensive coastal deserts on Earth. Natural vegetation and agriculture is essentially confined to the beds of small rivers fed from the Andes. Elsewhere, certain modest plants of particular design make use of dew, as no rain falls for many years (Schweigger, 1959, pp. 247–254; Koepcke, 1961, pp. 217–232). The wealth of this region is derived from the marine resources. The upwelling waters of the Humboldt Current system are rich in nutrients and plankton and thus allow an abundant fish population (Schweigger, 1959, pp. 254–363). This coastal ocean was in fact renowned as one of the most copious fishery regions of the World. Guano birds also feed on the fish (Schweigger, 1959, pp. 399–415); their excrement deposited on the small islands off the coast represents an important mineral resource which is commercially exploited for fertilizer and other purposes (Schweigger, 1959, pp. 415–442; Koepcke, 1961, pp. 212–214).

In rare years, the delicately tuned ecology of this coastal environment is severely disturbed (refer to Section 10.5). While in the average annual cycle sea surface temperature rises to a maximum around March and April, the annual march is greatly enhanced in certain years, a phenomenon known as 'El Niño.' The term refers to the Christ Child, and reflects the notion that the anomalous warming tends to begin around Christmas time. Schweigger (1959, pp. 98–127; 1961, pp. 3–4) and Schütte (1968, pp. 48–50) also mention a southward flowing warm "El Niño Current" common around this time of year. Rasmusson and Carpenter (1982) point out that more recently the term 'El Niño' has been used to encompass the large-scale features of the warming event, that is the upwelling area along both the Equator and the South American coast. Fig. 8.2:1 illustrates the marked interannual variability of sea surface temperature along the Ecuador/Peru coast. From the 19th century, Schweigger (1959, pp. 104–121) and Murphy (1926) mention the catastrophic 1891 El Niño, and Murphy (1926) further reports heavy rainfall indicative of El Niño for 1884 and 1878. Based on a variety of sources, Rasmusson (1984) lists the onset years of El Niño type events since the early 18th century as follows: 1726, 28, 63, 70, 91; 1803, 04, 14, 17, 19, 21, 24,

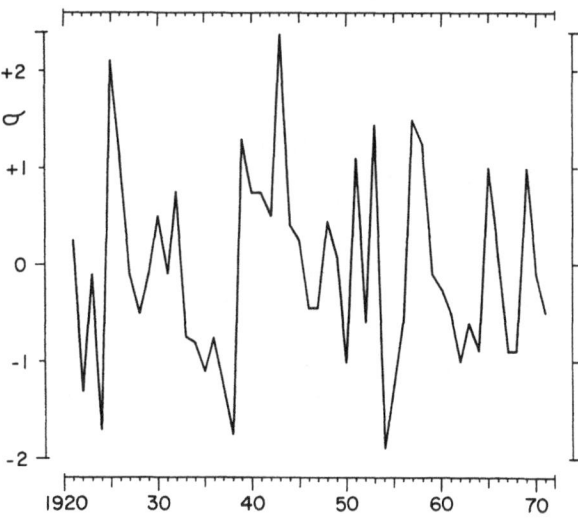

Fig. 8.2:1. Normalized departures of calendar-year mean sea surface temperature along the Ecuador/Peru coast ($\sigma = 0.8°C$). From Covey and Hastenrath (1978).

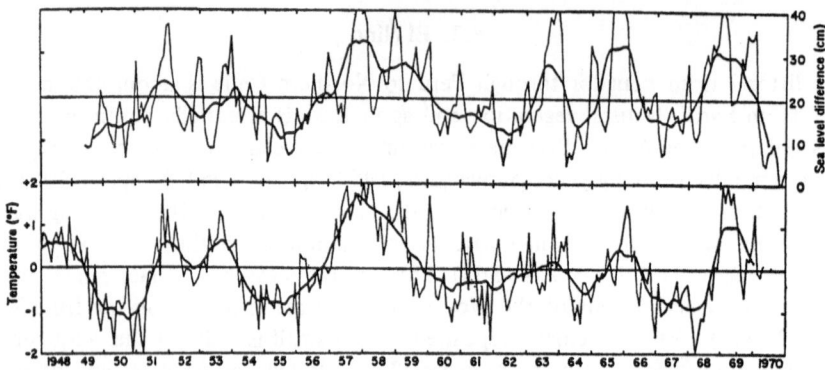

Fig. 8.2:2. Time series (1948 to 1970) of (top) sea level difference across the Pacific Equatorial Countercurrent and (bottom) the sea surface temperature anomaly off Central America. Thin curves give the monthly mean values, heavy curves the 12-month running mean. From Wyrtki, 1973 (*Science, 180,* 66–68, Copyright by American Association for the Advancement of Science).

28, 29, 32, 37, 44, 45, 46, 50, 52, 54, 55, 57, 62, 64, 66, 68, 71, 73, 75, 77, 78, 80, 84, 85, 87, 88, 89, 91, 96, 99; 1900, 02, 05, 11, 12, 14, 17, 19, 23, 25, 26, 29, 30, 32, 39, 40, 41, 43, 44, 46, 48, 51, 53, 57, 58, 63, 65, 69, 72, 73, 75, 76, 82, 83.

The anomalous warming of the sea surface during El Niño appears related to changes in nutrient and plankton content of the upper ocean, which causes a mass death of fish. Continuing in the food chain, marine birds die in great numbers, adding to the ecological catastrophe. Refer to Barber and Chavez (1983), Schreiber and Schreiber (1984), and Feldman *et al.* (1984) for recent reviews of the biological consequences of El Niño. El Niño involves both hydrosphere and atmosphere (see Schweigger, 1959, 1961, for detailed discussion). Concomitant with the warming of the upper ocean, the extended stratiform cloud deck, which provides at best negligible precipitation amounts, breaks up and gives way to convective cloudiness and torrential rains. Although the coastal desert lacks precipitation for years, rains of this abundance and intensity are no blessing but a further component of the natural disaster: it causes erosion of the loose surface material, destruction of roads, communication lines, houses, and untold other damage.

Considering the many disastrous consequences of El Niño, it is not surprising that there has been a long-standing interest in the atmospheric and oceanic causes of this phenomenon. Over decades, causalities have been sought in the region itself, that is in the domain of the South American West coast and the adjacent easternmost South Pacific (see, for example, Schweigger, 1961; Schütte, 1968; Schell, 1965; Lettau, 1967, pp. 67–68), but with remarkably little success. Schweigger (1961) noted the tendency for seasonal sea surface temperature anomalies in the Peruvian waters to coincide with departures of the same sign off California. This finding went unnoticed for well over a decade, not only because of the remote medium of publication, but also because its implications were not appreciated. As mentioned in Section 8.1, Berlage (1966, pp. 21–23, 106–116), and Bjerknes (1966, 1969) first realized the coincidence of El Niño events with a certain phase of the Southern Oscillation, and Bjerknes (1969) advanced a preliminary conceptual model of a possible causal relationship. A further milestone was Wyrtki's (1973) discovery of changes in the Pacific Equatorial Countercurrent related to El Niño events off South America. In addition, Namias (1973) pointed out the similarity of variations in the Pacific Equatorial Countercurrent and the subtropical westerlies over the Pacific. Fig. 8.2:2 illustrates the remarkably parallel long-term variations of Equatorial Countercurrent velocity and of sea surface temperature off the coast of the Americas. The works of Bjerknes (1969) and Wyrtki (1973) in

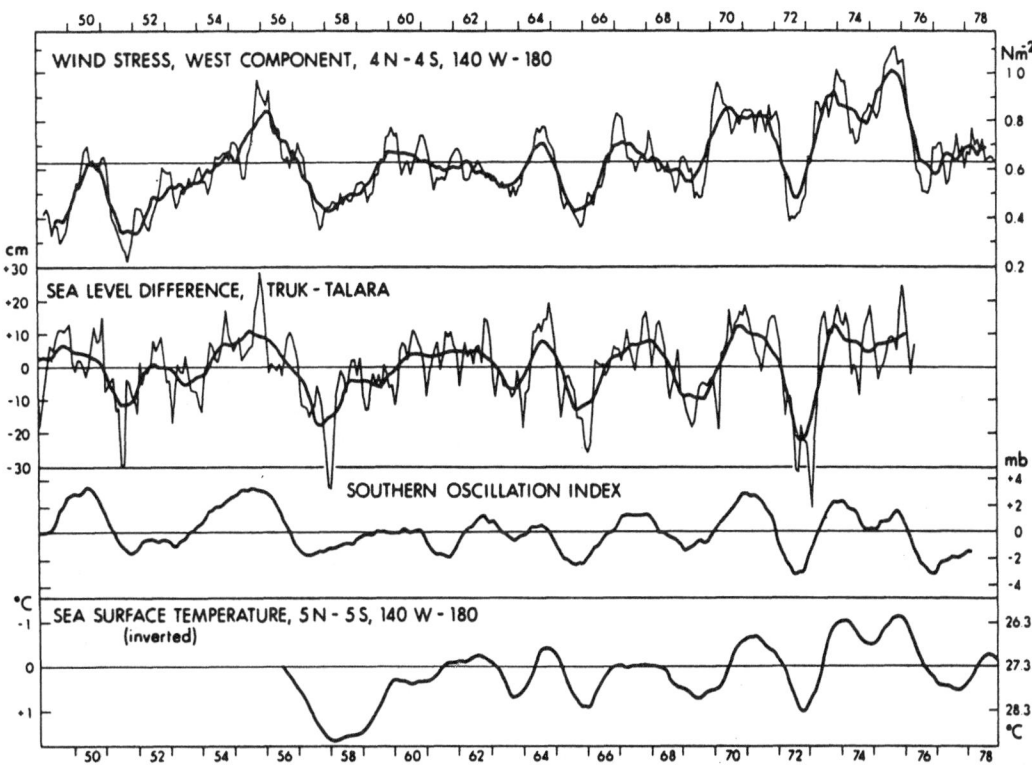

Fig. 8.2:3. Time series of the wind stress in the central equatorial Pacific, of the sea level difference between Truk Island and Talara, Peru, of the Southern Oscillation (index is atmospheric pressure difference between Easter Island and Darwin, Australia, relative to a mean of 10.3 mb), and of sea surface temperature in the central Pacific from 1949 through 1978. The heavy curves are 12-month running means, the thin curve gives 5 month running means for the wind stress and monthly means of the sea level difference. From Wyrtki, 1982 (*Marine Technology Society Journal*).

particular opened the perspective to the relation of the El Niño phenomenon off Ecuador and Peru to processes spanning the great Pacific basin as a whole.

In a seminal paper Wyrtki (1975b) outlined the basin-wide response of the equatorial Pacific ocean circulation to wind-stress forcing and suggested equatorial Kelvin waves as a mode of fast transmission of signals from the western to the eastern extremity of the Pacific. An essential foundation for this study was laid by a series of his earlier observational analyses (Wyrtki, 1974a, b, 1975a). Wyrtki's (1975a, b) empirical work gave the impulse to the pioneering theoretical studies of Hurlburt *et al.* (1976), and McCreary (1976) and various later numerical modelling efforts (Busalacchi and O'Brien, 1980; O'Brien *et al.*, 1981; Busalacchi, 1982; Busalacchi *et al.*, 1983a, b; Schopf and Harrison, 1983; Inoue and O'Brien, 1984; Cane, 1984; Cane and Zebiak, 1985), as well as numerous observational studies of wind and sea surface temperature fields in the equatorial Pacific (Barnett, 1977, 1981, 1983, 1984a; Wyrtki and Eldin, 1982; Weare, 1982, 1983, 1984; Luther and Harrison, 1983).

For a review of equatorial Pacific Ocean circulation refer to Section 4.3. Fig. 4.3.1:4. summarizes much of the pertinent background. The atmospheric pressure pattern is dominated by the quasi-permanent subtropical high pressure cells over the Eastern North and South Pacific and low pressure in the equatorial zone of the greater Indonesian – Australasian region. Consistent with this pressure distribution, easterlies sweep the entire expanse of the equatorial Pacific. This easterly wind stress piles up waters in the Western Pacific, thus causing an eastward slope of the free ocean surface and of constant pressure topographies with a maximum effect in the thermocline region. A direct result is an eastward flowing undercurrent centered in the thermocline along the Equator, that is directed opposite to the westward flowing South Equatorial Current at the surface. As explained in Section 4.3, this atmospheric-oceanic circulation complex is associated with upwelling and cold surface waters at the Equator. There is a distinct annual cycle in the intensity of wind and ocean current systems, the Equatorial Undercurrent being strongest in March to May and weakest during September to December (Fig. 4.3.1:5).

This background is essential for the consideration of El Niño events, inasmuch as the El Niño and its antithesis appear to a large extent as reductions and enhancements, respectively, of the average annual cycle. The notion of an enhanced annual cycle of the trades during the negative phase of the Southern Oscillation was stated again recently by Van Loon (1984). El Niño events are associated with the negative phase of the Southern Oscillation (Section 8.1), characterized by a weakening of the Eastern South Pacific high and a filling of the low in the greater Indonesian — Australasian region. This altered pressure distribution is accompanied by weakened trade winds over the equatorial Pacific. The diminished easterly wind stress can no longer balance the marked eastward slope of the free ocean surface along the Equator (Figs. 4.3:6 and 4.3:7). As a result, the waters piled up in the Western equatorial Pacific in part slosh back eastward, the slope of the free ocean surface decreases, as do the subsurface zonal pressure gradients, and the Equatorial Undercurrent weakens and may surface or fade out altogether. The decrease of the trade wind easterlies is also conducive to weakened upwelling along the Equator (ref. Section 4.3). Fig. 8.2:3 illustrates concomitant variations of Southern Oscillation Index, zonal wind stress, zonal slope of free ocean surface, and sea surface temperature in the Central Equatorial Pacific. More importantly the relaxation of easterly wind stress is thought to excite eastward travelling Kelvin waves, which are manifest in changes of thermocline depth and sea surface temperature. These waves propagate fast in terms of the climatic time scale and can transmit signals from the western to the eastern extremity of the equatorial Pacific within 2–3 months (ref. Section 4.3). Hurlburt *et al.* (1976) and McCreary (1976) concluded on theoretical grounds that upon impinging on the (South American) coast, the equatorial Kelvin wave would split and coastal Kelvin waves would propagate poleward into either hemisphere. Interestingly, Hurlburt *et al.* (1976) and McCreary (1976) were throughout the theoretical investigation unaware of Schweigger's (1961) much earlier empirical work indicating an interhemispheric symmetry, with sea surface temperature anomalies off the coasts of Peru and California being broadly in phase.

Despite considerable commonality, each El Niño event is an individual. For example, much attention has been paid to the unusual evolution of the 1982/83 El Niño in particular (reviews in Philander, 1983; Cane, 1983; Gill and Rasmusson, 1983; Rasmusson and Wallace, 1983; Rasmusson, 1984; Namias and Cayan, 1984; Meyers and Donguy, 1984, Toole and Borges, 1984; Tang and Weisberg, 1984; Wyrtki, 1985; Sardeshmukh and Hoskins, 1985; Inoue *et al.*, 1985; Busalacchi and Cane, 1985; Anderson and McCreary, 1985; Kok and Opsteegh, 1985). Donguy and Dessier (1983) call attention to El Niño-like events in the western Pacific. Conceivably, different patterns of wind stress forcing may lead to diverse oceanic responses. Busalacchi and O'Brien (1981), Busalacchi (1982), and Busalacchi *et al.* (1983a, b) investigated this problem complex through a combination of numerical modelling experiments and empirical analysis of Pacific ship observations during 1961–78. The model allows for wind stress forcing of the ocean but contains no thermodynamic or radiation processes. Their results suggest that the response of the upper hydrosphere differs depending on the longitude distribution of easterly wind stress relaxation over the equatorial Pacific. Common to all of the El Niño events studied was a weakening of the trades West of the dateline, late in the preceding year. However, during January–April of 1972 and 1976, the easterlies decreased considerably over the central equatorial Pacific. This is considered essential for the different upper ocean response, as compared to the 1963, 1965 and 1969 El Niño events. Busalacchi *et al.* (1983a) concluded that location and timing of wind changes are instrumental in the diversity of El Niño events, and that internal Kelvin and Rossby waves play important roles in the sea surface temperature variability throughout the tropical Pacific.

While investigations into remote forcing and response mechanisms first concentrated on the Pacific, similar causalities seem to be operative in the tropical Atlantic. A key experience was Houghton's (1976) finding that the seasonal upwelling in the Gulf of Guinea is not well explained by the local wind stress. In fact, a series of theoretical and empirical studies (Moore *et al.*, 1978; Adamec and O'Brien, 1978; O'Brien *et al.*, 1978; Hisard, 1980; Servain *et al.*, 1982; Picaut, 1983; Hirst and Hastenrath, 1983a; Busalacchi and Picaut, 1983; McCreary *et al.*, 1984) indicates that easterly wind-stress forcing over the Western Equatorial Atlantic may induce sea surface temperature variations at the eastern extremity of the ocean. Sea surface temperature in turn through its influence on moisture content and stability of the boundary layer flow, is a relevant factor for rainfall anomalies. Accordingly remote forcing – response relationships as discussed above more extensively for the Pacific, may represent important climate anomaly mechanisms.

Although controversy continues on the details, a broad consensus is emerging in the sense that relaxation of easterly wind stress represents the major forcing for the basin-wide response of the tropical Pacific Ocean manifest in the sea surface temperature anomalies of El Niño.

By comparison, the atmospheric part of the El Niño Southern Oscillation (ENSO) phenomenon appears rather less well explored. In a landmark study based on the judicious compositing of surface wind and sea surface temperature fields during six individual El Niño events, Rasmusson and Carpenter (1982) constructed the canonical evolution of El Niño. Figs. 8.2:4 to 8.2:6 summarize their main results. Although the timing of El Niño varies somewhat, the anomalous warming tends to be remarkably well locked to the average annual cycle. Thus the compositing by seasons of three calendar months (Figs. 4.2:4 and 4.2:5) seems appropriate.

During August–September–October (ASO, not shown here) and November–December–January (NDJ) preceding the El Niño peak (Figs. 8.2:4a, 8.2:5a), most of the Eastern equatorial Pacific is anomalously cold, while positive sea surface temperature anomalies cover the Southern tropical South Pacific and the Western Equatorial Pacific in November–December–January (Fig. 8.2:4a), with maxima off the Peru coast and in the equatorial region near the dateline. Most prominent in the wind field during November–December–January (NDJ) preceding the El Niño peak (Fig. 8.2:5a) is anomalous northwesterly flow over most of the tropical South Pacific, tantamount to weakened Southeast trades; this effect being most pronounced over the Eastern and the West Central South Pacific.

During the March–April–May (MAM) peak phase of the El Niño year (Figs. 8.2:4b, 8.2:5b), most of the Eastern tropical Pacific is anomalously warm, with largest departures off the coast of Peru and in a band extending from there into the equatorial region of the Eastern Pacific. Separate but much weaker maxima persist in the central equatorial Pacific. Most conspicous in the wind field (Fig. 8.2:5b) are the weakened trades over the Central Equatorial Pacific.

During the August–September–October following the peak phase of El Niño (Fig. 8.2:4c, 8.2:5c), pronounced positive sea surface temperature anomalies prevail in the entire equatorial Pacific from the Americas to well West of the dateline. The largest departures are no longer found off the Peru coast, but the warm water anomalies have spread to the open ocean in a band to the South or near the East Pacific Equator. Separate maxima continue to persist in the Central

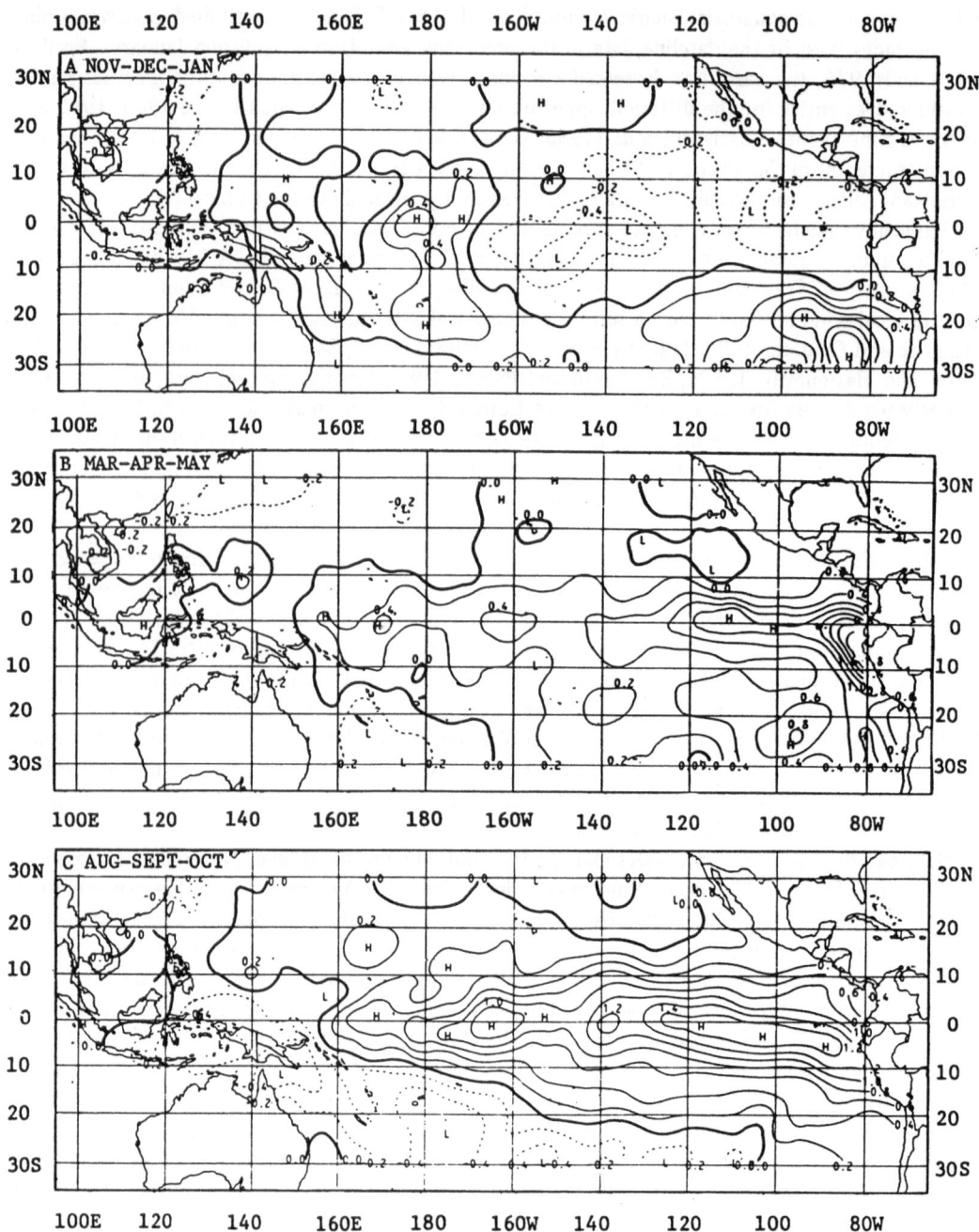

Fig. 8.2:4. Sea surface temperature anomaly (°C) pattern of the canonical El Niño. (a) November–December–January preceding, (b) March–April–May peak, (c) August–September–October after El Niño peak. Source: Rasmusson and Carpenter, 1982 (*Monthly Weather Review*, American Meteorological Society).

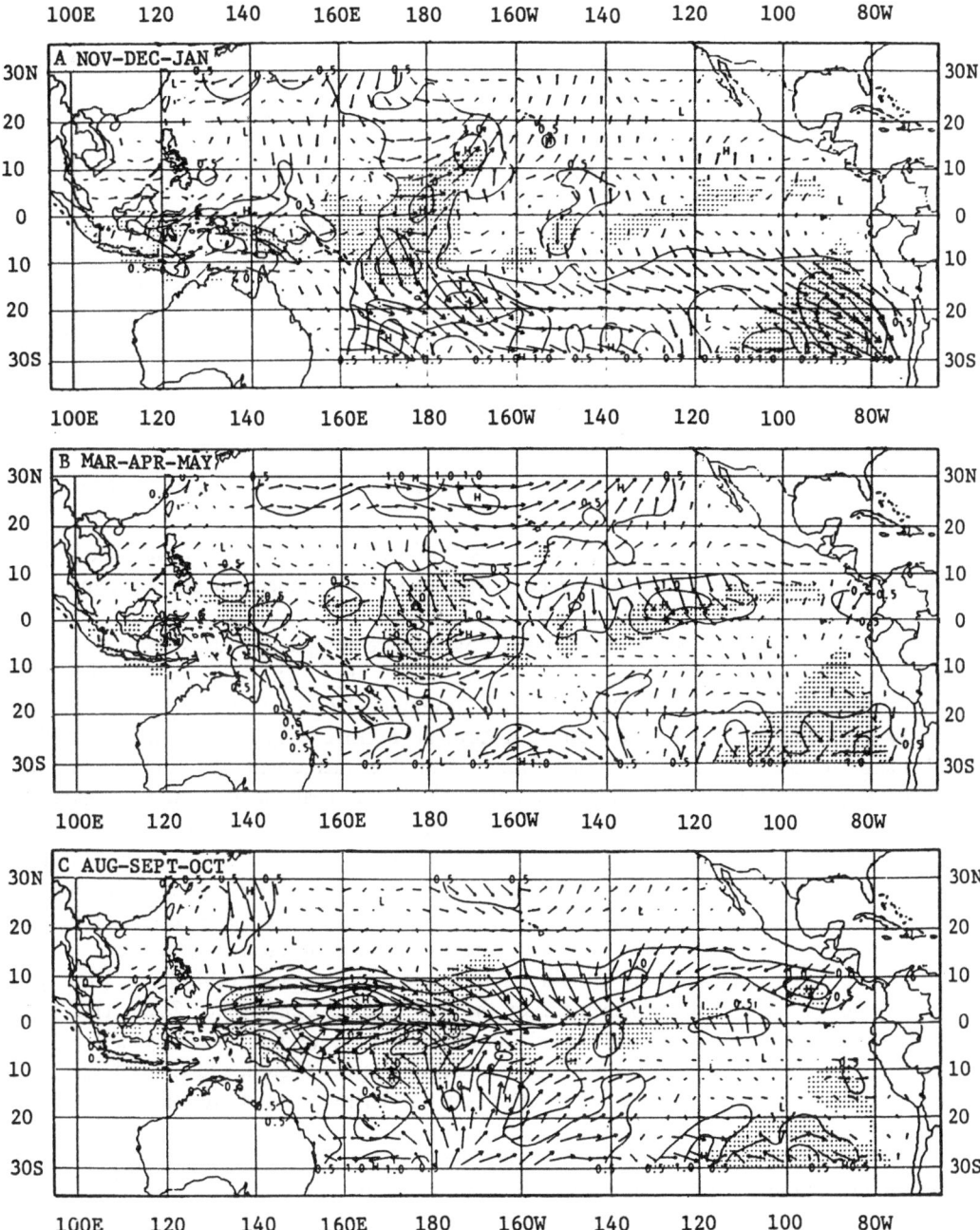

Fig. 8.2:5. Wind anomaly patterns (m s^{-1}) of the canonical El Niño. (a) November–December–January preceding, (b) March–April–May peak, (c) August–September–October after El Niño peak. Dot raster denotes areas with sparse data. Source: Rasmusson and Carpenter, 1982 (*Monthly Weather Review*, American Meteorology Society).

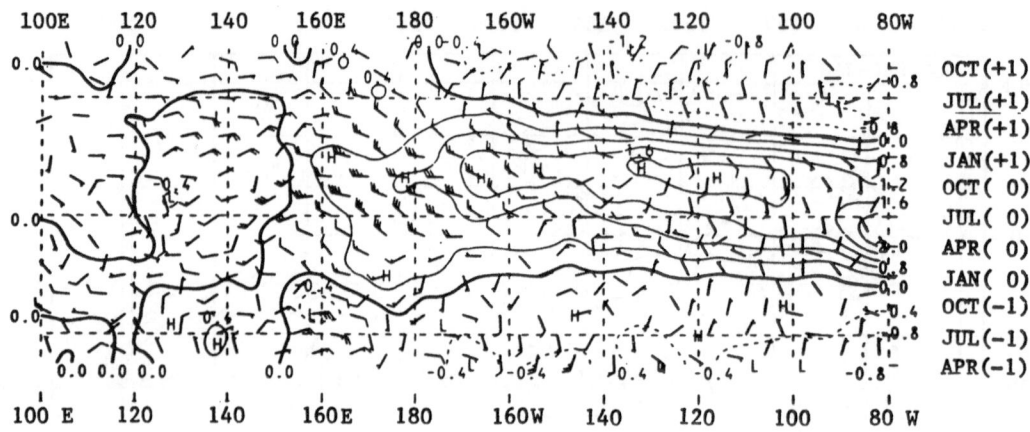

Fig. 8.2:6. Time section of composite sea surface temperature (°C) and wind (m s^{-1}) anomalies along the Equator (West of 95°W) and axis of maximum SST anomalies. (95°W to Peru coast) during evolution of the canonical El Niño. From Rasmusson and Carpenter, 1982 (*Monthly Weather Review*, American Meteorological Society).

Equatorial Pacific. The wind field (Fig. 8.2:5c) is characterized by drastically weakened trades over the Central and Western Equatorial Pacific. The sea surface temperature and wind anomalies linger on into subsequent months (not shown here).

The map sequences Figs. 8.2:4a to c and 8.2:5a to c are complemented by the time section of composite sea surface temperature and wind anomalies in Fig. 8.2:6. The area of warm water anomaly which appears near the dateline during November–December preceding El Niño subsequently spreads both eastward and westward. Also, the positive temperature anomalies off Peru and Ecuador spread westward along the Equator and then merge with the anomaly near the dateline. In the meantime positive departures intensify further near the coast of South America and extend westward. In the Western Equatorial Pacific sea surface temperature anomalies are much smaller then, and broadly out-of-phase with those observed in the Eastern Equatorial Pacific. The time section Fig. 8.2:6 also shows the tendency for departure westerlies especially over the Central and Western Equatorial Pacific. Kraus and Hanson (1983) propose air-sea interaction processes as propagator mechanisms of equatorial ocean surface temperature anomalies.

As further developments during El Niño, Rasmusson and Carpenter (1982) and Kousky *et al.* (1984) note a southward displacement of the Intertropical Convergence Zone (ITCZ) and a northeastward shift of the South Pacific Convergence Zone (SPCZ), resulting in a reduced Equatorial Dry Zone (Section 6.7.4) and enhanced rainfall in the Eastern and Central Equatorial Pacific and below normal precipitation over Indonesia. These patterns are supported by the analyses of Lau and Chan (1983a, b).

Figs. 8.2:4b and c demonstrate that during El Niño Equatorial Pacific sea surface temperature anomalies span about a third of the Earth's circumference. Bjerknes (1966) already hypothesized that the sudden introduction of such a large anomalous heat source for the atmosphere near the Equator cannot remain without consequence for the mid-latitude circulation. Wallace and Gutzler (1981) and Horel and Wallace (1981) studied the extra-tropical circulation features associated with El Niño. For the Pacific – North American area Horel and Wallace (1981) find pattern characteristics in part documented in Namias' (1969) earlier work. Fig. 8.2:7 shows their schematic illustration of hypothesized global pattern. A feature which attracted attention in particular is a wave train of geopotential height anomalies extending from the Equatorial Central Pacific along a great

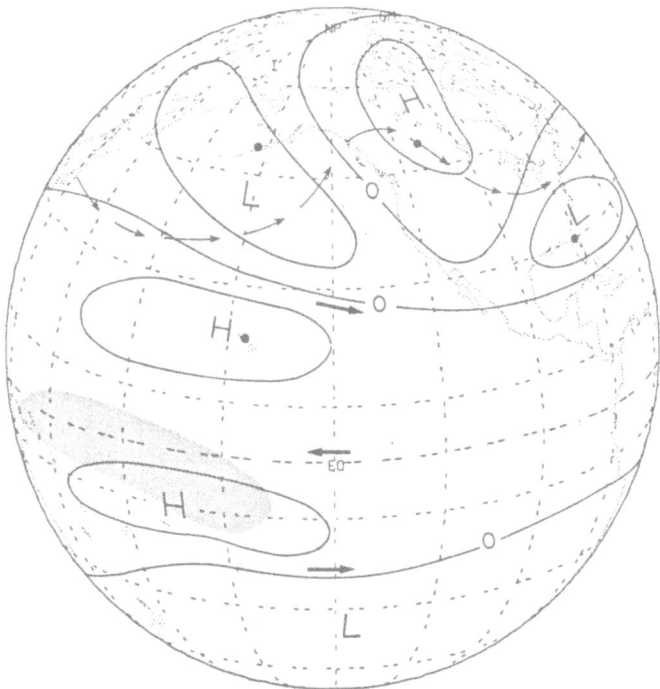

Fig. 8.2:7. Schematic illustration of the hypothesized global pattern of middle and upper-tropospheric geo-potential height anomalies (solid lines) in Northern winter during an equatorial Pacific warm episode. Arrows in darker type reflect the strengthening of the subtropical jets in both hemispheres along with stronger easterlies near the Equator, and arrows in lighter type depict a mid-tropospheric streamline as distorted by the anomaly pattern, with pronounced 'troughing' over the central Pacific and 'ridging' over western Canada. Shading indicates regions of enhanced cirriform cloudiness and rainfall. From Horel and Wallace, 1981 (*Monthly Weather Review*, American Meteorological Society).

circle route to North America. Horel and Wallace (1981) emphasize the resemblance of this pattern with results of various theoretical studies (Egger, 1977; Opsteegh and Van den Dool, 1980; Hoskins and Karoly, 1981, Webster, 1981). It is noteworthy that Nobre and Moura (1984) recently found a similar standing wave train in the upper-air circulation over the Atlantic sector associated with droughts in Northeast Brazil (Section 8.6).

The aforementioned studies suggest that an equatorial sea surface temperature anomaly, once existing, could have important consequences for the midlatitude circulation. However, the origin and maintenance of such an oceanic temperature anomaly remain as the more central problems. Recognizing the seeming importance of ocean-atmosphere interaction for the functioning of the Southern Oscillation and El Niño, McCreary (1983) and McCreary and Anderson (1984) include both atmosphere and hydrosphere in their numerical modelling studies, but so far the model atmosphere consists merely of two wind states which are switched on and off as specified by the ocean state.

It appears that a realistic concept of the El Niño Southern Oscillation phenomenon must include not only atmosphere and ocean, but more particularly also upper-air processes. Empirical studies seem essential for a realistic formulation of models. In the following Section 8.3 climate anomalies in the greater Indonesian region are discussed, as this is recognized as a major dipole of the Southern Oscillation. On that basis, upper-air patterns of the El Niño Southern Oscillation phenomenon are then considered in Section 8.4.

8.3. Rainfall Anomalies in Indonesia

The greater Indonesia region merits particular interest in the functioning of the tropical climate system because of its role as a dipole of the Southern Oscillation. Refer to Section 6.8.3 for a discussion of background circulation and climate. Although Indonesia receives precipitation all year round and encompasses a puzzling variety of rainfall regimes (Braak, 1921−29, Part 1, Chapter 6, pp. 151−221), the rainy season peak is centered around December−January, or the height of the Northern winter, when the quasi-permanent low-pressure center is best developed (Fig. 6.1:2) and lower-tropospheric convergence is pronounced.

Fig. 8.3:1 illustrates the interannual variability of Java rainfall. The index represents the 'all station average normalized departure' of annual (seasonalized for July−June) precipitation totals for a collective of 40 long-term stations on the island of Java. Also plotted for comparison is Wright's (1975, 1977) Southern Oscillation index. The two indices vary in a remarkably parallel fashion, such that abundant rainfall in Java tends to coincide with the positive phase (pressure anomalously high over the Eastern South Pacific and low over Indonesia) of the Southern Oscillation, while in the negative phase rainfall is deficient in Indonesia. This behavior was already known to Walker and Bliss (1932; present Figs. 8.1:1 and 8.1:2), and absorbed the attention of

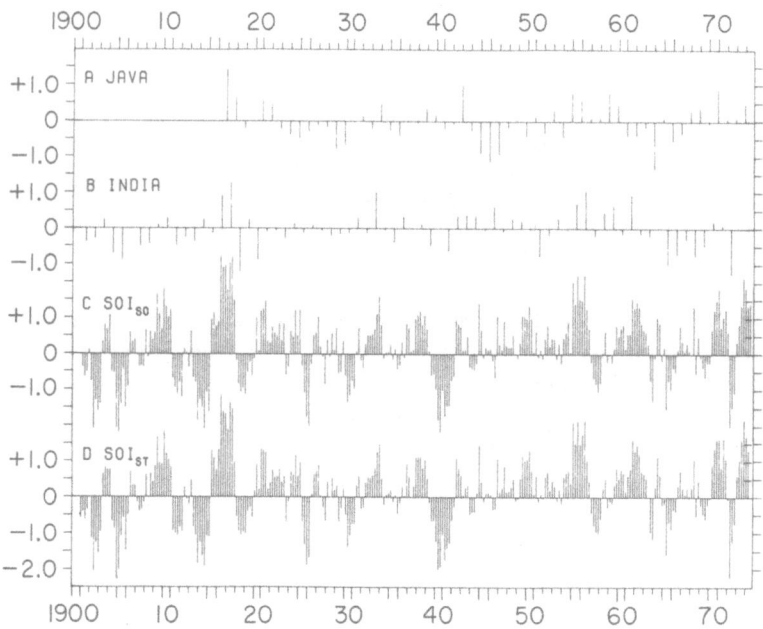

Fig. 8.3:1. Time series plots of (a) Java rainfall index ('all-station average normalized departure' of July−June rainfall totals of 40 stations; expressed in σ), (b) all-India rainfall index (arithmetic mean of normalized rainfall departure in the 31 subdivisions; expressed in σ), (c) Southern Oscillation (SO) index of Wright (1975, 1977) for the 'SO seasons' (February−March−April, May−June−July, August−September−October, and November−December−January), and (d) the same for the 'standard seasons' (December−January−February, March−April−May, June−July−August, September−October−November). Note that all indices are plotted according to the respective times of year (i.e. Java rainfall index centered on December−January, all-India rainfall index July−August, and Southern Oscillation index on March, June, September, December ('SO seasons'), and on January, April, July, October ('standard seasons')).

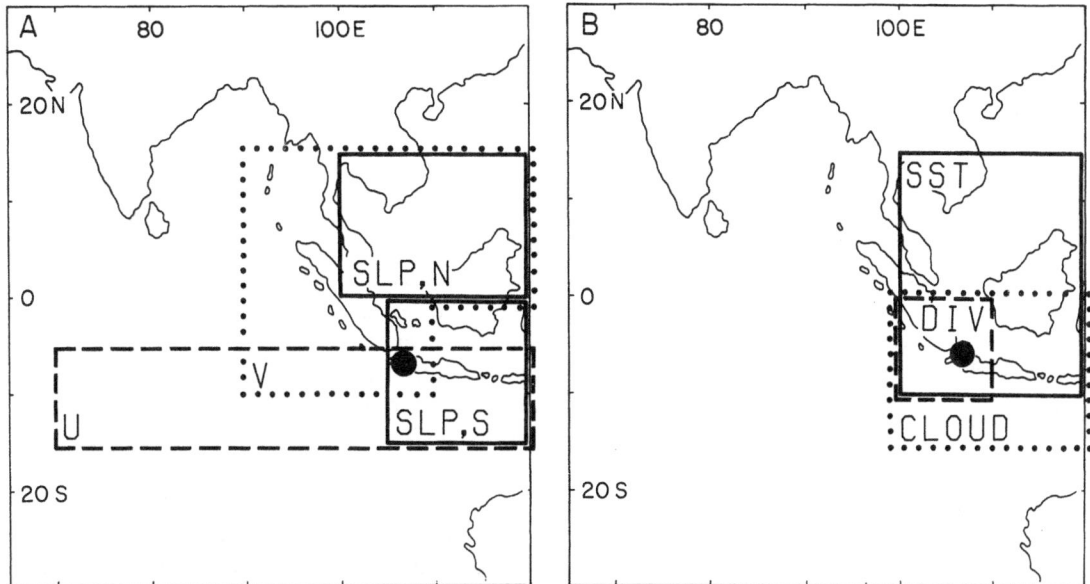

Fig. 8.3:2. Orientation maps for Figs. 8.3:3 and Table 8.3:1. Shown are areas for which time series of indicated elements were compiled as follows. Map A: U zonal wind component in Southern Indian Ocean, 5–15°S, 70–120°E; V meridional wind component over Australasia, 0–15°N, 90–120°E and 0–10°S, 90–110°E; SLP, N sea level pressure to the North of the Equator, 0–15°N, 100–120°E; SLP, S sea level pressure to the South of the Equator, 0–15°S, 105–120°E; Map B: SST sea surface temperature in Southeast Asia, 10°S – 15°N, 100–120°E; DIV surface wind divergence over Indonesia, 0–10°S, 100–110°E; CLOUD cloudiness over Indonesia, 0–15°S, 100–120°E. Dot in maps A and B denotes Djakarta, for which sea level pressure and rainfall series are used. From Hackert and Hastenrath (1986).

generations of Dutch meteorologists (Braak, 1919; Berlage, 1927, 1934, 1957, 1966; Schregardus, 1938; de Boer, 1947; de Boer and Euwe, 1949a, b; Euwe, 1949a, b; Schmidt – ten Hoopen and Schmidt, 1951; Schmidt and van der Vecht, 1952; Reesinck, 1952). More recently the relation between Indonesia rainfall and the Southern Oscillation was mentioned by Quinn *et al.* (1978) and Rasmusson and Carpenter (1982).

In the light of these pronounced teleconnections it is desirable to substantiate the regional circulation characteristics associated with rainfall anomalies in Indonesia. Circulation conditions in the greater Indonesian region are of primary interest, and anomalies are related to the annual cycle of the large-scale atmospheric and oceanic fields. The most sensitive areas in the various fields have been identified from correlation and stratification map analyses of the Indian Ocean domain with respect to Java rainfall (Hackert and Hastenrath, 1986), which are not reproduced here. The most indicative of these elements and areas are shown in Fig. 8.3:2. The pertinent circulation departures associated with Indonesia rainfall anomalies are illustrated in Fig. 8.3:3 and Table 8.3:1. The graphical display, Fig. 8.3:3, provides a visual impression of time series relationships, while Table 8.3:1 substantiates correlations quantitatively.

Fig. 8.3:3 (parts a, b, c) bears out the remarkable spatial coherence of pressure variations throughout the greater Indonesian region, noted before by Braak (1919, p. 4, 7). It is furthermore apparent that the amplitude of pressure variations increases southward, or in fact towards the 'center of action', the nucleus of the climatic mean low pressure domain in the North Australian area (Fig. 6.1:2). This means that pressure variations are typically associated with changes of the meridional pressure gradient over Indonesia. Negative pressure departures would correspond to a steeper southward pressure gradient, or departure westerly geostrophic wind to the South of the Equator. Fig. 8.3:3 (parts e and c) actually demonstrates the tendency for inverse pressure and zonal (westerly) wind variations. A steepened southward pressure gradient is also conducive to enhanced northerly ageostrophic wind, and Fig. 8.3:3 (parts d, e, c), shows that the meridional (southerly) wind component over Australasia tends to vary broadly parallel to pressure, but inverse to the zonal (westerly) wind component.

Consider the implications of the aforementioned pressure gradient and wind variations for the lower-tropospheric divergence. Thus, steeper meridional pressure gradient and departure northerly wind component over the eastern equatorial Indian Ocean would be conducive to enhanced convergence over Java. Fig. 8.3:3 (parts g, c, d, e) indeed shows that divergence in the Java area tends to vary broadly parallel with pressure, and thus parallel to the meridional (southerly) wind departure, but inverse to the zonal (westerly) wind departure curves. Plausibly, rainfall and cloudiness variations run broadly parallel to those of convergence, and thus inverse to pressure (Fig. 8.3:3, parts h, i, g, c). Finally, note that sea surface temperature lags behind pressure by about half a year, and accordingly also behind zonal wind (Fig. 8.3:3, parts f, c, e), whereas temperature extrema follow the opposite extrema of cloudiness (Fig. 8.3:3, parts f and i). These results should be appreciated in context with the review by Lau and Li (1984) who also call attention to the tendency for buildup of a cold anticyclone over the East Asian continental regions and the development of a baroclinic zone between the cold polar air mass to the North and the warm tropical air mass to the South. The resultant cold surges (Section 7.9) within the Northern winter monsoon are regarded as important synoptic-scale agents contributing to the convective activity over the greater Indonesian region, within the frame of the seasonal circulation patterns.

Table 8.3:1, parts A and B, not only substantiates the relationships displayed in Fig. 3.3:3 quantitatively, but also details conditions in the rainy (November–April) and drier (May–October) halves of the year. The larger correlations are of the same sign in both half-years. Sea surface temperature shows a prevailingly negative correlation with rainfall.

Table 8.3:1, part B, bears out some marked lag relationships. Thus, high pressure – which is broadly concurrent with departure easterlies, weak convergence, cloudiness, and rainfall in May–October – tends to be followed by high sea surface temperature in November–April, while November–April pressure has little coupling with the subsequent semester's sea surface temperature. This characteristic was already noted by Braak (1919, p. 24–26) and Berlage (1957, p. 40–44), for regions somewhat different from those represented in Figs. 8.3:2 and 8.3:3 and Table 8.3:1. Sea surface temperature shows strong persistence from May–October semester. Furthermore, sea surface temperature anomalies in any semester are followed by inverse SLP departures in the subsequent semester, presumably due to hydrostatic effects.

Complementing the Southern Oscillation teleconnection illustrated in Fig. 8.3:1, the Table 8.3:1 and Fig. 8.3:3 substantiate the regional circulation characteristics associated with Indonesia rainfall anomalies. In relation to the turnover between opposing phases of the Southern Oscillation, negative feedback mechanisms at the Indonesia dipole deserve attention, in particular processes capable to produce alternations between cold and warm water regimes. Two candidates are offered by Fig. 8.3:3 and Table 8.3:1. A first mechanism involves the radiative effects of clouds. Thus cloudiness, which becomes particularly abundant during that phase of the Southern Oscillation characterized by low Indonesia pressure, would curtail surface net radiation and thus be conducive to cooling the top layer of the ocean; the opposite effect would result in epochs of high pressure and scarce cloudiness. This mechanism has been suggested by Hastenrath and Wu (1982). Another independent mechanism, first proposed by Braak (1919, p. 7) early in this century, entails wind variations associated with pressure changes, and their mechanical and

TABLE 8.3:1

Correlation coefficient in hundredths between semester and yearly values of elements mostly identified in Fig. 8.3:2. SLP, sea level pressure at Djakarta; U zonal wind component over Southern Equatorial Indian Ocean, 5–15°S, 70–120°E; V meridional wind component over Australasia, 0–15°N, 90–120°E, and 0–10°S, 90–110°E; DIV surface wind divergence over Indonesia, 0–10°S, 100–110°E; CLOUD cloudiness over Indonesia, 0–15°S, 100–120°E; RAIN, monthly rainfall totals at Djakarta; JRI annual (July–June) rainfall index compiled for a collective of 40 stations on the Island of Java (Fig. 8.3:1); SST sea surface temperature in Southeast Asia, 10°S – 15°N, 100–120°E; Part A of the table shows concurrent, and part B lag correlations. One, two, and three asterisks denote significance at the 10, 5, and 1% level, respectively. Quenouille's (1952, pp. 168–170) method was used to account for the reduction of the effective number of degrees of freedom due to persistence.
Source: Hackert and Hastenrath (1986).

A. concurrent correlations

	SLP		RAIN			JRI
	Nov–Apr	May–Oct	Nov–Apr	May–Oct	July–June	July–June
SLP						
Nov–Apr			−06			−10
May–Oct				−43***	+03	−14
U						
Nov–Apr	−22*		+33**		+42***	+55***
May–Oct		−35***		+31**	+10	+21
V						
Nov–Apr	+35***		−31**		−29***	−33**
May–Oct		−11		−28**	−00	−15
DIV						
Nov–Apr	+36**		−08		−22	+03
May–Oct		+10		−19	−06	−04
RAIN						
Nov–Apr	−06				+86***	+32***
May–Oct		−43***			−29**	+18
CLOUD						
Nov–Apr	−37***		+12		+14	+20
May–Oct		−34***		+34***	+04	+10
SST						
Nov–Apr	+10		−24*		−40***	−41***
May–Oct		−25*		+15	−14	−14

Part B. lag correlations

	SLP leads SST		SST leads SLP		SST leads SST	
	Nov–Apr	May–Oct	Nov–Apr	May–Oct	Nov–Apr	May–Oct
SST						
Nov–Apr		+34***		−27**		+71***
May–Oct	−07		−30**		+52***	

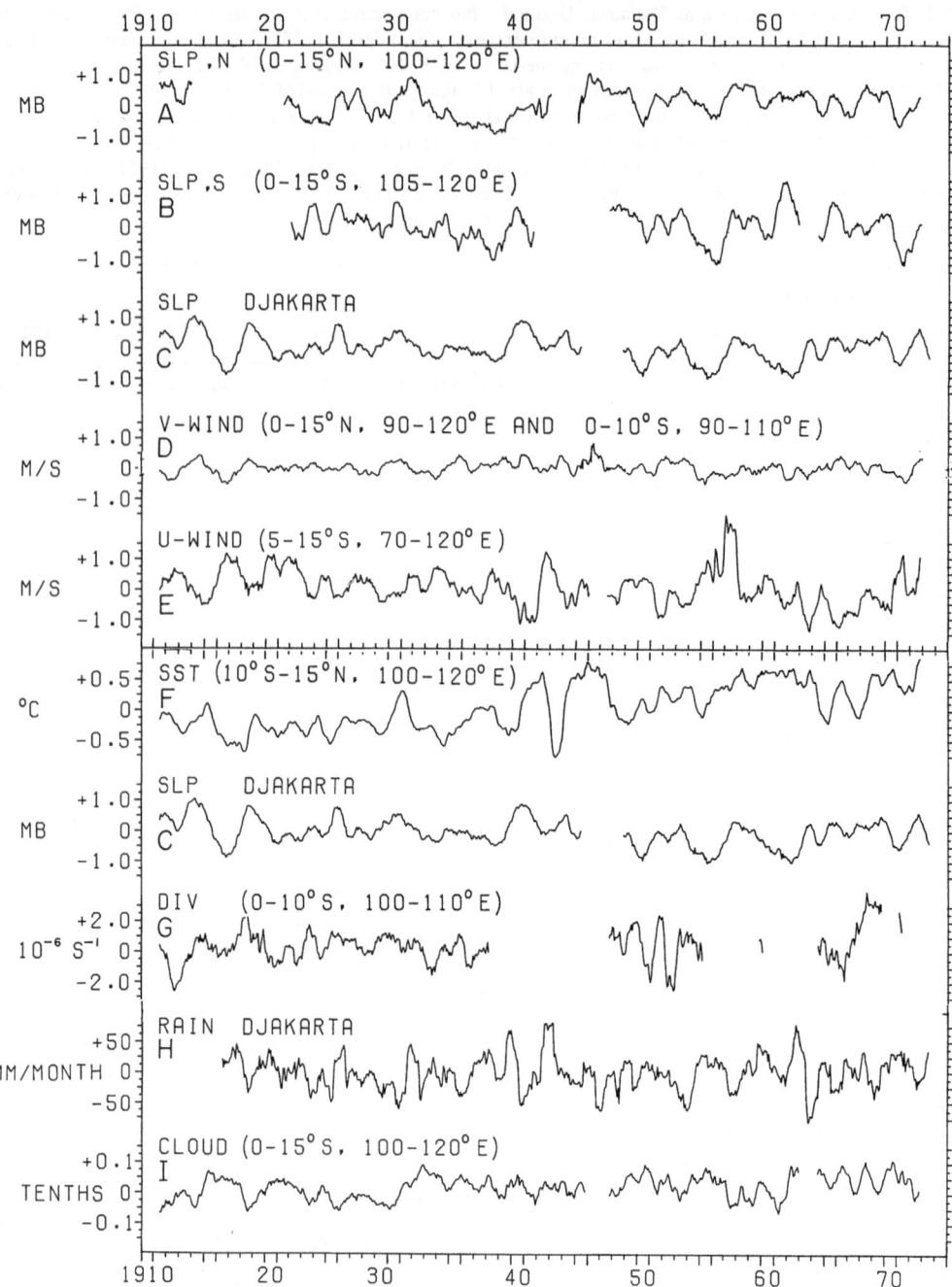

evaporative forcing on the upper ocean. As indicated by Fig. 8.3:3 and Table 8.3:1, anomalously low pressure in the Indonesia region entails during the rainy season semester (November–April) enhanced westerly flow in the Equatorial Zone and an increased northerly wind component over Southeast Asia, which combined with the basic flow (Figs. 4.4:1, part a; 4.4:3, parts d–f; 4.4:2, parts a–c; Hastenrath and Lamb, 1979, vol. 1, charts 24–25, 14–17) results in higher wind speed. By contrast, during the drier half of the year (May–October) when the direction of the basic flow is broadly reversed (Figs. 4.4:1, part b; 4.4:2, parts a–f; 4.4:3, parts a–c; Hastenrath and Lamb, 1979, vol. 1, charts 18–23), the aforementioned pressure and wind departures would be reflected in weaker wind. Braak (1919, p. 7, 24–26) already hypothesized that enhanced wind stress, through stirring and evaporation, serves to cool the surface waters, so that anomalously low/ high pressure would lead to strong/weak wind and cold/warm surface waters in November–April, but reverse effects in May–October. The possible role of such seasonally varying atmosphere-ocean couplings for the functioning of the Southern Oscillation is considered in Section 8.4.

[On p. 272]

Fig. 8.3:3. Twelve-month running mean plots of anomalies of time series identified in the maps Figs. 8.3:2a and b. (a) SLP, N sea level pressure, in mb, $0-15°$N, $100-120°$E; (b) SLP, S sea level pressure, in mb, $0-15°$S, $105-120°$E; (c) SLP DJAKARTA sea level pressure at Djakarta, in mb; (d) V meridional wind component over Australasia, in m s^{-1}, $0-15°$N, $90-120°$E, and $0-10°$S, $90-110°$E; (e) U zonal wind component in Southern equatorial Indian Ocean, in m s^{-1}, $5-15°$S, $70-120°$E; (f) SST sea surface temperature in Southeast Asia, in °C, $10°$S $- 15°$N, $100-120°$E; (c) SLP DJAKARTA, sea level pressure at Djakarta, in mb. (g) DIV surface wind divergence over Indonesia, in 10^{-6} s^{-1}, $0-10°$S, $100-110°$E; (h) RAIN DJAKARTA rainfall at Djakarta, in mm per month; (i) CLOUD cloudiness over Indonesia, $0-15°$S, $100-120°$E; Note that the Djakarta pressure series (c) is repeated for convenient reference. From Hackert and Hastenrath (1986).

8.4. Upper-Air Patterns of ENSO

The functioning of the El Niño Southern Oscillation phenomenon involves both atmosphere and ocean, as is apparent from the review of surface patterns in Sections 8.1 to 8.3. However, inasmuch as the surface pressure seesaw of the Southern Oscillation reflects a redistribution of mass in the atmospheric column, upper-air processes must be instrumental in its operation. These are considered here, in part based on Hastenrath and Wu's (1982) work which included the analysis of 37 long-term upper-air stations throughout the tropics, numerous surface land stations, and ship observations from selected areas. Fig. 8.4:1 shows the location for all the series, but only a few of these are referred to here.

The time series plots in Fig. 8.4:2 display the familiar lower-tropospheric pattern of Southern Oscillation pressure variations. Thus, stations in the greater Indonesian – Australasian region (Fig. 8.4:2a, stations Singapore, Darwin, Djakarta) show surface pressure variations broadly parallel to sea surface temperature in the Equatorial Central Pacific, but inverse to pressure stations in the Eastern to Central South Pacific (Fig. 8.4:2, c; Tahiti, Antofagasta, Easter Island). Transitional patterns are found in intermediate regions (Fig. 8.4:2b).

In conspicuous contrast to the seesaw pattern in the lower layers, topography variations in the upper troposphere run approximately synchronous throughout the tropics (Fig. 8.4:3a, b, c). In fact, the 200 mb height variations in the greater Indonesian – Australasian region (Fig. 8.4:3a, stations Singapore, Darwin, Cocos Island) are broadly in phase with those over the Eastern to Central South Pacific (Fig. 8.4:3c, stations Tahiti, Antofagasta, Nandi), with intermediate regions (Fig. 8.4:3b, stations Truk, Johnston Island, Abidjan), as well as with sea surface temperature in the Central Equatorial Pacific (Fig. 8.4:3a). Relative maxima appear in the years 1963, 66, 69, 70, 73, 75, 77, and minima in the years 1962, 64, 65, 71, 74, 76. However, there is a striking contrast in the amplitude of height variations, being large over the Eastern to Central South Pacific and much smaller over the Indonesian – Australasian region.

For comparison with the time series plots of Fig. 8.4:3, the Figs. 8.4:4 and 8.4:5 map the spatial pattern of standard deviation of 200 mb height variations and of zonal wind components throughout the tropics. The latter two maps show similar distributions, thus supporting the internal consistency between the height and the independent wind observations. Fig. 8.4:4 shows that the standard deviation of 200 mb height anomalies is relatively large in the open Eastern Pacific, large parts of Africa, and India, but small in the greater Indonesian – Australasian region, the Central Equatorial and Western Pacific, the Pacific coast of South America and the Southern

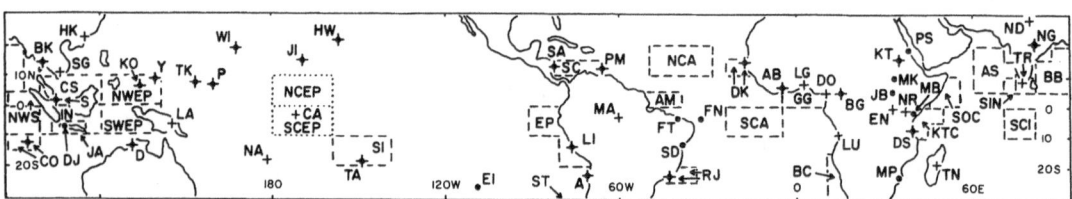

Fig. 8.4:1. Location map of stations/areas referred to in Figs. 8.4:2 to 8.4:9. Broken lines bound areas with sea surface temperatures and dotted lines areas with both sea surface temperature and surface winds. Dots denote stations with sea level or station pressure, and crosses upper-air stations. Dots-crosses indicate stations with both sea level or station pressure and upper-air data. The code for the series specifically referred to here is as follows. Areas: SCEP = South Central Equatorial Pacific; EP = Ecuador/Peru; Stations with both surface and upper-air data: A = Antofagasta, AB = Abidjan, CO = Cocos Island, D = Darwin, JI = Johnston Island, S = Singapore, TA = Tahiti, TK = Truk; Stations with surface data only: DJ = Djakarta, EI = Easter Island; Stations with upper-air data only: ND = Nandi. Source: Hastenrath and Wu (1982).

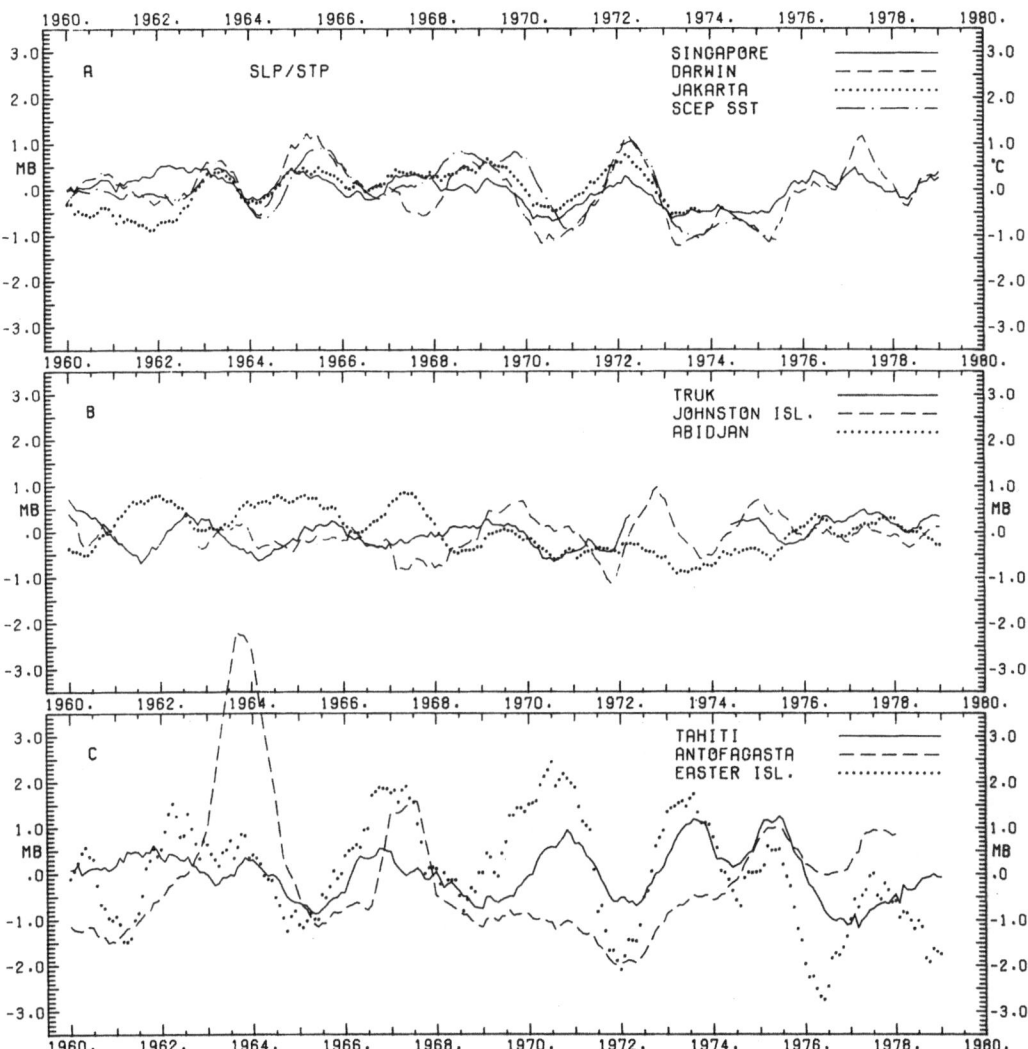

Fig. 8.4:2. Twelve-month running mean plots of SLP/STP (sea level pressure, station pressure) anomalies (mb) with reference to SST (sea surface temperature, °C) in the South-Central Equatorial Pacific (SCEP). (a) greater Indonesian – Australasian region and SCEP, (b) transitional, (c) Eastern to Central South Pacific. From Hastenrath and Wu (1982).

Caribbean, and Equatorial East Africa. Of particular interest in relation to the time series plots of Fig. 8.4:3 is the fact that the Eastern to Central South Pacific is a region of large height variations, while variability is considerably smaller over Indonesia and the adjacent Western Equatorial Pacific.

Fig. 8.4:6 emphasizes the main results of Figs. 8.4:2 and 8.4:3. Surface pressure in the greater Indonesian region varies inversely to the Eastern to Central South Pacific, and broadly parallel with sea surface temperature along the Ecuador/Peru coast. 200 mb variations are more nearly in phase throughout the tropics, but are particularly large over the Eastern to Central South Pacific.

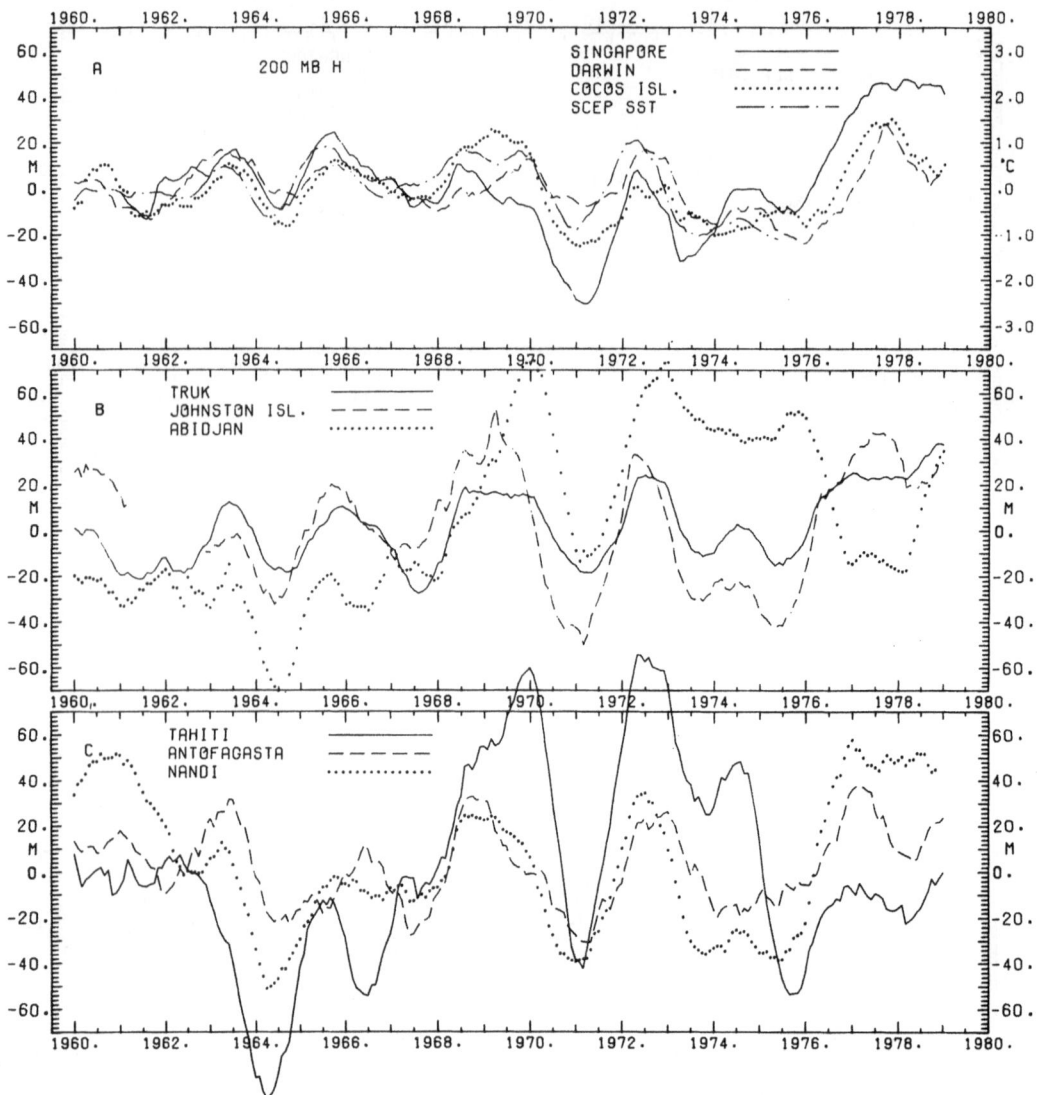

Fig. 8.4:3. Twelve-month running mean plots of 200 mb height anomalies (m) with reference to SST (°C) in the South-Central Equatorial Pacific (SCEP). (a) greater Indonesian − Australasian region and SCEP, (b) transitional, (c) Eastern to Central South Pacific. From Hastenrath and Wu (1982).

From numerous time series plots such as shown in Figs. 8.4:2, 8.4:3, and 8.4:6 relative phases of sea surface temperature, pressure, geopotential height, and wind variations were determined, with reference to the running mean plot of sea surfaces temperature in the South Central Equatorial Pacific (SCEP; Figs. 8.4:1 to 8.4:3). These relative phases are displayed in Fig. 8.4:7 for sea surface temperature, surface pressure, and 200/850 mb thickness, which behaves essentially as the 200 mb geopotential height. Fig. 8.4:7 a exhibits the phase opposition in surface pressure between the greater Indonesian − Australasian region and the Eastern to Central South Pacific familiar for the Southern Oscillation, the former area being broadly in phase with sea surface temperature in the Central Equatorial Pacific. Noteworthy in Fig. 8.4:7b are the moderate phase

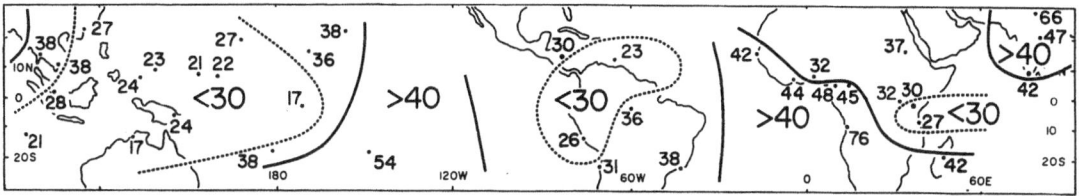

Fig. 8.4:4. Spatial distribution of standard deviation of 200 mb height anomalies, in m. Solid and dotted lines bound areas with values > 40 and < 30 m, respectively. From Hastenrath and Wu (1982).

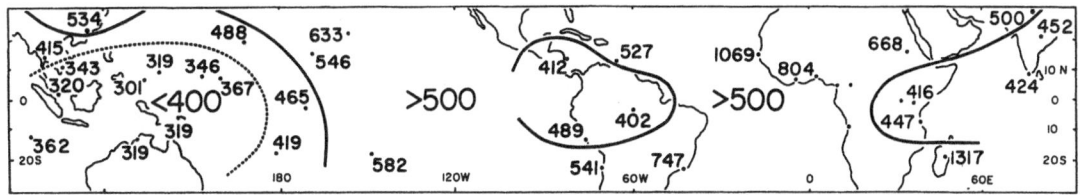

Fig. 8.4:5. Spatial distribution of standard deviation of 200 mb zonal wind anomalies in 10^{-2} m s^{-1}. Solid and dotted lines bound areas with values > 500 and < 400 × 10^{-2} m s^{-1}. From Hastenrath and Wu (1982).

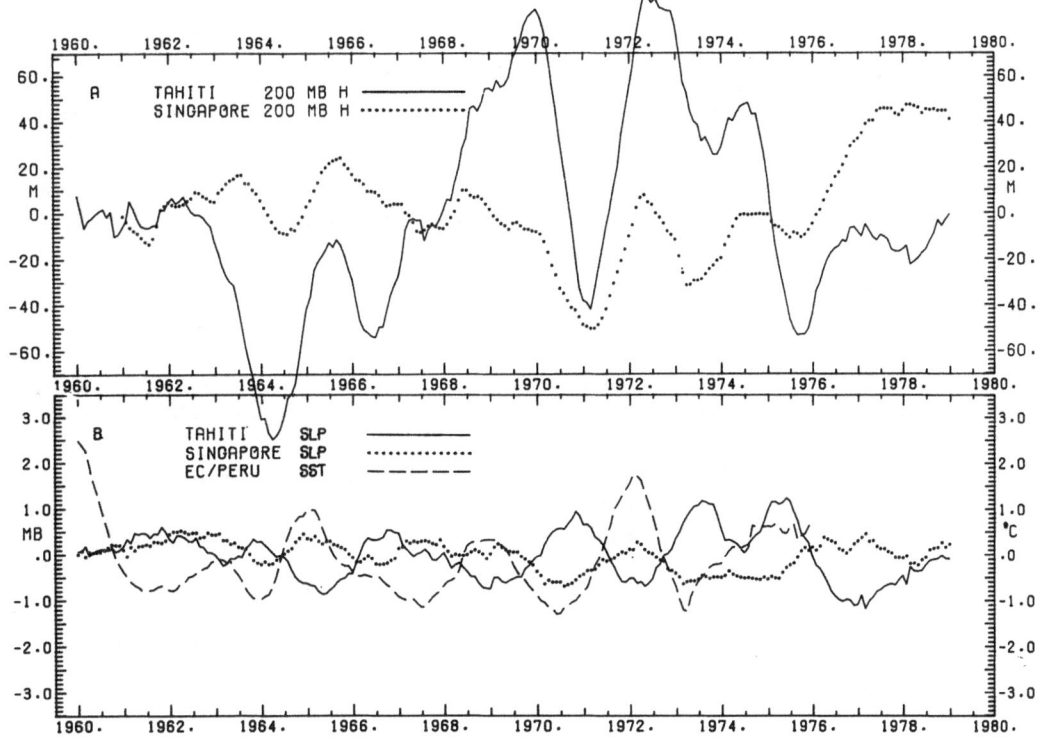

Fig. 8.4:6. Twelve-month running mean plots of (a) 200 mb height variations over Singapore and Tahiti, (b) surface pressure variations at Singapore and Tahiti, and sea surface temperature variations off the Ecuador/Peru coast (Ec/Peru).

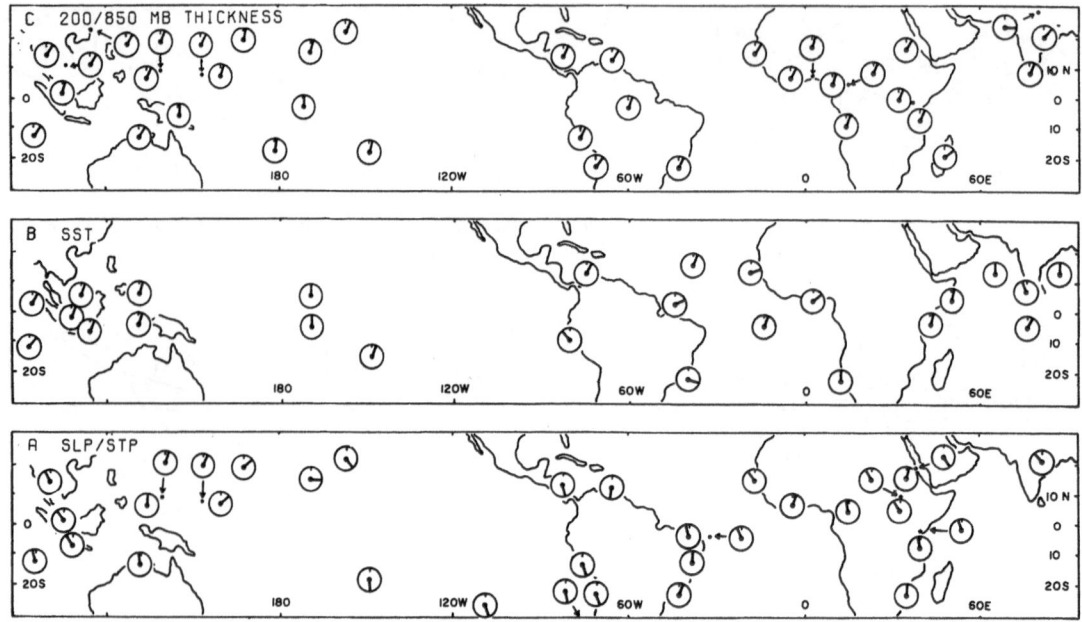

Fig. 8.4:7. Spatial distribution of the relative phase of (a) surface pressure (SLP/STP), (b) sea surface temperature (SST), (c) 200/850 mb thickness. The phases are represented by the dial at each station/area. 0, 3, 6, and 9 o'clock indicate, respectively, 0, 90, 180, and 270 degrees phase lag behind SST variations in the South Central Equatorial Pacific (360 degrees correspond to order 2–4 years; ref. Figs. 8.4:2, 8.4:3, 8.4:6). Dots without dial denote stations that were not used. Dials without arm indicate stations where phase relationship is not well defined. From Hastenrath and Wu (1982).

difference from the Eastern to the Western Pacific, and the circumstance that sea surface temperature variations in the Indian and the Atlantic Oceans tend to lag somewhat behind the Pacific. The 200/850 mb thickness, Fig. 8.4:7c, which essentially represents the mean temperature of the tropospheric column, is broadly in phase or lags somewhat behind sea surface temperature, Fig. 8.4:7b, differences in timing being small throughout the global tropics.

As discussed in Section 8.3, the pressure seesaw between the eastern and western extremities of the Pacific, through the associated zonal wind stress variations in the equatorial zone, exerts a major control on the sea surface temperature regime of the tropical Pacific. Figs. 8.4:2, 8.4:3, 8.4:6, and 8.4:7 indicate that, in turn, the sea surface temperature is a major factor for the temperature of the overlying atmospheric column. The tendency for sympathetic variations of tropospheric mean temperature throughout the global tropics has been noted by Newell and Weare (1976a, b), Newell (1980), Angell (1981), and Navato et al. (1981). Refer also to the more recent study by Angell and Korshover (1984). The largest 200/850 mb thickness and 200 mb absolute topography are found somewhat after the highest sea surface temperature in the Central Pacific (Fig. 8.4:7). Of importance further below is the circumstance that the largest upper-tropospheric topography variations occur over the Eastern to Central Pacific, where sea surface temperature variatins are also particularly large.

Lower- and upper-tropospheric topography variations such as illustrated in Figs. 8.4:2, 8.4:3, 8.4:6, and 8.4:7, are associated with systematic departures in the wind patterns. Fig. 8.4:8 shows the lower- and upper-tropospheric wind departures associated with warm events in the Equatorial Pacific, while Fig. 8.4:9 depicts conditions during cold events.

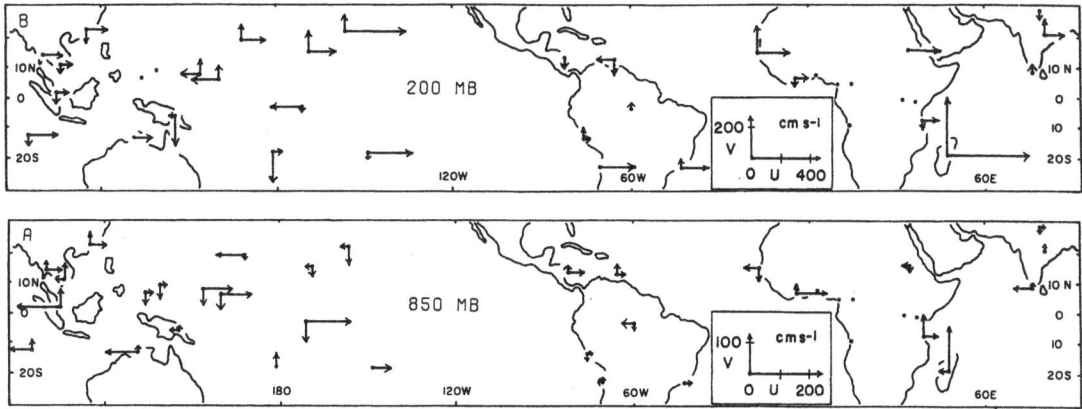

Fig. 8.4:8. Average wind departures at (a) 850 mb, (b) 200 mb, concurrent with maximum of sea surface temperature (SST) in the South-Central Equatorial Pacific. The average departures are constructed from the following four running mean values. May 1963 to April 1964, August 1965 to July 1966, March 1969 to February 1970, and May 1972 to April 1973. Arrows at each station denote zonal and meridional departure wind components. Length of arrows is scaled as indicated by orientation diagram in lower right-hand portion of map. From Hastenrath and Wu (1982).

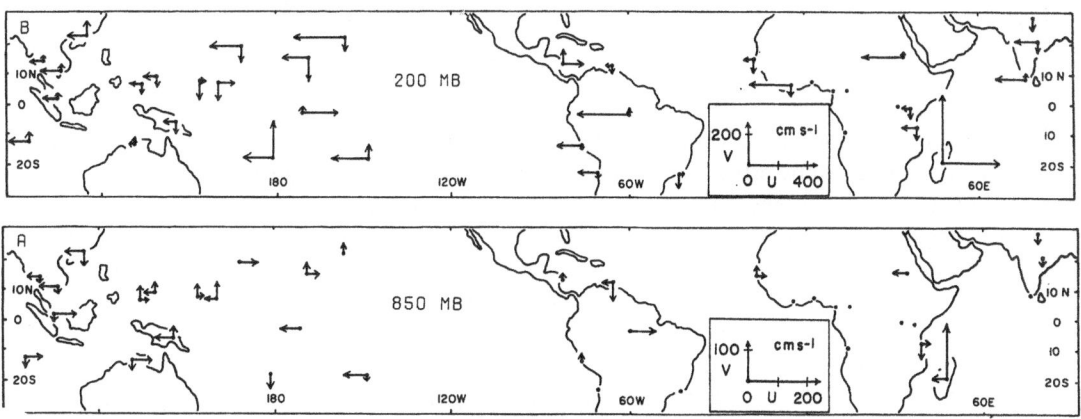

Fig. 8.4:9. Average wind departures at (a) 850 mb, (b) 200 mb, concurrent with minimum sea surface temperature (SST) in the South Central Equatorial Pacific. The average departures are constructed from the following four running mean values: May 1964 to April 1965, October 1966 to September 1967, January 1971 to December 1971, and September 1973 to August 1974. Symbols are as in Fig. 8.4:8. From Hastenrath and Wu (1982).

During the Pacific warm events, the lower-tropospheric (Fig. 8.4:8a) departure zonal wind component is largely westerly extending from the Western Pacific eastward to the Southern Caribbean, and strongest easterly from the greater Indonesian — Australasian region westward to West Africa and the Amazon basin. The lower-tropospheric departure meridional wind is strong equatorward in much of the tropics, except for Southeast Asia, South India, and parts of South America. The upper troposphere (Fig. 8.4:8b) has strong westerly departure throughout most of the tropics, except in the Western and Equatorial Central Pacific, and in the Southern Caribbean. Westerly wind departure is largest in the open Eastern Pacific and Southeast Africa. Enhanced

Subtropical Westerly Jets (Section 6.2.2) associated with El Niño have also been noted by Kousky *et al.* (1984), and have been reproduced in numerical model experiments (Keshavamurty, 1982, 1983). Upper-tropospheric meridional wind departure is strong poleward except in Southeast Asia, and in part of South America and Africa. During the Pacific cold events (Figs. 8.4:9a, b), characteristics of wind departure are broadly opposite to those shown in Fig. 8.4:8 for the warm events. Thus Figs. 8.4:8 and 8.4:9 are complementary. When the Pacific sea surface is warm, lower-tropospheric zonal flow appears divergent in the Indonesian – Australasian region, concomitant with weaker easterlies over the Pacific and opposite departures over the Indian Ocean; upper-tropospheric westerly flow prevails throughout most of the tropics; equatorward wind departures are found in the lower, and poleward wind departures in the upper troposphere. Flow patterns are essentially opposite when the Pacific sea surface is cold.

The upper-tropospheric wind departure patterns during warm and cold events depicted in Figs. 8.4:8b and 8.4:9b are consistent with the results of Chiu and Lo (1979), who investigated the statistical linkage between sea surface temperature variations at the Peru coast and the 250 mb zonal wind component over the Central Pacific. Chiu and Lo (1979, p. 21) in fact found a positive coupling with the zonal (westerly) wind in the outer tropics of either hemisphere, but a negative linkage with the equatorial zone.

The wind departure patterns displayed in Figs. 8.4:8 and 8.4:9 should be considered in relation to the pressure/height variations (Figs. 8.4:2, 8.4:3, 8.4:5 to 8.4:7). Discussion is here in terms of the episodes of warm and cold surface waters in the Equatorial Pacific, since these exemplify opposing extreme modes of operation. Characteristic of the warm episodes is anomalously low surface pressure over the Eastern South Pacific and positive pressure departures in the Indonesian – Australasian region, and lower-tropospheric zonal wind departures directed away from the positive pressure anomaly in the Indonesian – Australasian region. Meridional wind departures in the lower troposphere are predominantly directed equatorward, or towards low surface pressure related to the hydrostatic effect of the warm ocean surface. The prevailing westerly wind departures in the upper troposphere are consistent with the largest warming of the equatorial ocean and related increase in tropospheric thickness. Contrasting with the westerly anomalies in much of the tropics are the departure easterlies in the upper troposphere over the equatorial central Pacific, where the greater warming to the East as compared to the West entails a diminished eastward slope of upper-tropospheric topographies.

Some exceptions to these major patterns are as follows. The poleward departures in the lower troposphere over Southeast Asia may be related to the anomalously high pressure/height in the Indonesian – Australasian region; and the poleward departures along the Pacific coast of South America to the South of the Equator seem consistent with the negative pressure/height anomalies in the Eastern South Pacific. This pressure/height anomaly pattern may also be the cause for the equatorward wind departures in the Pacific leading the Equatorial Pacific sea surface temperature. Consistent with this phase difference between wind and sea surface temperature departures, Weare (1984) noted that in the evolution of El Niño changes of atmospheric moisture convergence over the Eastern Pacific precede those of sea surface temperature. Analogous to the above considerations, the broadly opposite departures of zonal and meridional wind components appear consistent with the inverse pressure and sea surface temperature anomaly field.

Figs. 8.4:8 and 8.4:9 allow qualitative inference on the departure vergence pattern, as indicated above. In the lower troposphere during warm episodes the zonal wind components appears divergent in the Indonesian – Australasian region and convergent over the Eastern Pacific, while the meridional wind component results in convergence for the low latitudes as a whole. This is consistent with Reiter's (1979) findings for the Pacific. The upper-tropospheric pattern is qualitatively consistent with the lower-tropospheric conditions in terms of mass continuity, inasmuch as the zonal component is convergent over the Indonesian – Australasian region, while meridional flow in low latitudes is predominantly divergent. The departure vergence pattern during cold episodes is approximately inverse, and again qualitatively consistent with mass continuity.

During warm episodes the tropospheric thickness is anomalously large throughout the tropics, but with remarkable regional contrasts (Figs. 8.4:4 and 8.4:5): values are largest over the Central Pacific, the Atlantic and India, and smallest in the Western Pacific, in the Indonesian – Australasian region, and the Indian Ocean. Consistent with such an upper-tropospheric contour configuration are certain features in the zonal wind departure pattern; while westerly wind departures prevail in the low latitudes as a whole, easterly departures over the Central and Western Pacific (Figs. 8.4:8b) appear related to the anomalous westward directed pressure/height gradient borne out by Fig. 8.4:4. The latter feature is confirmed by Arkin's (1984) analysis of a different data set. On the whole, the observed wind departures appear broadly consistent with the pressure/height anomalies. The topography variations in the upper troposphere can in turn be traced to sea surface temperature anomalies as primary factor, via both latent and sensible heating mechanisms.

This analysis of surface and upper air observations provides a preliminary description of long-term large-scale pressure oscillations in the tropics, and may serve as a basis for hypotheses on the mechanisms of atmospheric mass redistributions.

Fig. 8.4:10 is a scheme of the two major modes of operation of the equatorial Pacific atmosphere-ocean system. During the positive phase of the Southern Oscillation (Fig. 8.4:10, part a), surface pressure is particularly high over the Eastern and low over the Western Pacific, the trade wind easterlies are strong, and the sea surface is cold, especially in the Eastern Equatorial Pacific. Concomitant with the ocean temperature pattern, the atmosphere is cold, possesses small thickness over the Eastern Pacific, but is warm and has greater relative topographies to the West. Accordingly, upper-tropospheric topographies slope strongly from West to East, with correspondingly vigorous westerly flow aloft. Pronounced subsidence over the Eastern Pacific and ascending motion at the western dipole of the Southern Oscillation complete the picture of a vigorous Walker Circulation (Section 6.9). Conversely, during the negative phase of the Southern Oscillation (Fig. 8.4:10, part b), the lower-tropospheric westward pressure gradient is slack, the trade wind easterlies over the equatorial Pacific are weak, and the sea surface is warm, notably in the Eastern Pacific. In

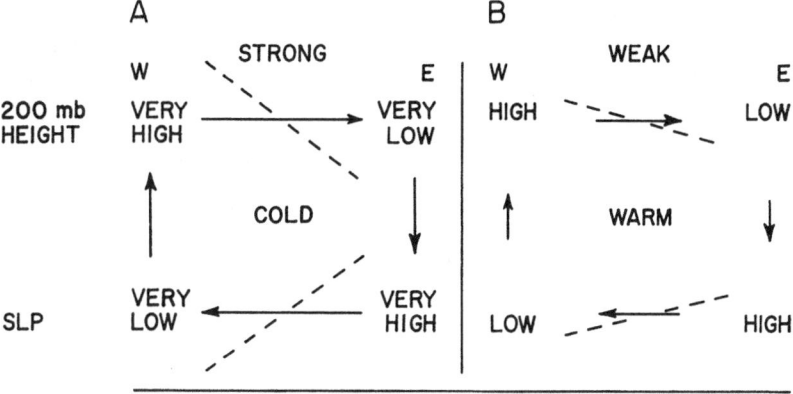

Fig. 8.4:10. Scheme of two modes of operation of the equatorial Pacific atmosphere-ocean system (a) positive phase of Southern Oscillation, strong Walker Circulation, (b) negative phase of the Southern Oscillation, weak Walker Circulation.

response to the ocean temperature pattern, the atmospheric column is particularly warm over the Eastern Pacific, so that the upper-tropospheric topographies are higher than during the positive phase of the Southern Oscillation; the temperature and inflation effect being less to the West. The weaker upper-tropospheric zonal pressure gradient entails weaker westerly flow aloft. Subsidence over the Eastern Pacific and diminished ascending motion at the western dipole of the Southern Oscillation are the further features of a weak Walker Circulation. Essential in the consideration of ocean temperature effects on the atmospheric thickness pattern is the observed fact emphasized further above, that sea surface temperature and upper-tropospheric topography variations are most pronounced over the Eastern to Central Pacific, and much less in the Indonesian − Australasian area.

Either of the two modes of operation of the Pacific atmosphere-ocean system sketched in Fig. 8.4:10 is conceivable and may persist over a long time span. A crucial and unexplained issue is the way in which the turnover from one to the other regime takes place. A negative feedback mechanism appears essential in this context. Consider for example the scheme of the strong Walker Circulation, Fig. 8.4:10, part a. A mechanism intrinsic to this mode, that would cool the ocean surface at the western dipole, would have as a thermodynamic consequence the cooling of the overlying atmospheric column, and a 'slumping' of upper-tropospheric topographies. As a result, the eastward pressure gradient is reduced, as is the westerly flow in the upper troposphere across the Pacific. Up to this instant the mass content of atmospheric columns, and hence the surface pressure remain unaltered, so that the easterlies in the lower troposphere are not yet affected. Consequently, the altered zonal topography and flow pattern aloft lead to a mass gain over the western and a loss over the eastern dipole. This is reflected in a reduced westward pressure gradient and hence weaker easterlies at the surface, which in turn furnish the atmospheric forcing to which the ocean responds in the El Niño warming (Section 8.2). The scene is now set for further thermodynamic effects of ocean and atmosphere and the evolution of a weak Walker Circulation. A similar negative feedback mechanism in reverse can be envisaged for the turnover from the weak to the strong Walker regime. This would involve an effect intrinsic to the strong Walker mode, which would lead to a warming of the sea surface at the western dipole of the Southern Oscillation.

Deliberately, only the simple example of a closed Pacific domain was considered here, and the triggering mechanism for the turnover between regimes was visualized at the western dipole of the Southern Oscillation. At this stage, the nature of a possible negative feedback mechanism deserves attention (Section 8.3). Hastenrath and Wu (1982) considered the possible effect of abundant cloudiness over the western dipole, characteristic of the positive phase of the Southern Oscillation. Through a reduction of surface net radiation this would be conducive to decreasing the sea surface temperature, which in turn would thermodynamically affect the overlying atmospheric column. Another mechanism would be related to enhanced surface wind speed during the positive phase of the Southern Oscillation. This would, through mixing and evaporation, cause a cooling of the ocean surface, with subsequent links in the causality chain as sketched above. Without reference to a negative feedback mechanism responsible for the turnover between the two Walker regimes, and without analysis of wind observations, Nicholls (1978, 1981) suggested that cooling of surface waters during the positive phase of the Southern Oscillation might be the result of enhanced surface westerlies concomitant with strong meridional pressure gradient. A recent analysis of long-term ship observations (Section 8.3) confirms that westerlies to the South of the Equator and northerlies over Southeast Asia during the Java rainy season peak are enhanced concomitant with anomalously low pressure in the Indonesian region, and wind speed and sea surface temperature are indeed inversely related. It is historically interesting that the relation of pressure, wind, and sea surface temperature anomalies in the Indonesian region was already known to Braak (1919, pp. 24–26), but his works did not receive the attention they deserve.

While a series of general circulation model simulations with prescribed and fixed sea surface temperature patterns, but directed at the Southern Oscillation and Walker Circulation (Rowntree, 1972; Julian and Chervin, 1978; Chervin and Druyan, 1984; Stone and Chervin, 1984) have produced interesting results, the design limitations of such experiments with an inert ocean should be noted. In fact, the empirical evidence reviewed above indicates that negative feedback mechanisms are instrumental for the turnover between the opposing modes of the Southern Oscillation, and that sea-air interaction and upper-air processes are essential in its functioning.

8.5. Vagaries of the Indian Monsoon

India's agriculture and economy depend critically on the rains of the summer Southwest monsoon. It is therefore understandable that the great famines of 1877 and 1899 (Shukla, 1986) gave the impetus for the extension of a network of precipitation gauges spanning all of the country. In fact, India possesses one of the densest and most continuous surface climatological station networks in the tropics. The India Meteorological Department compiles rainfall observations for 31 subdivisions as illustrated in Fig. 8.5:1. These series form the basis for numerous studies of long-term, interannual, and shorter term variability of India rainfall (Subbaramayya, 1968; Koteswaram and Alvi, 1969; Bhargava and Bansal, 1969; Jagannathan and Bhalme, 1973; Jagannathan and Parthasarathy, 1973; Parthasarathy and Dhar, 1974; Parthasarathy and Mooley, 1978; Raghavan et al., 1978; Bhalme and Mooley, 1980; Bedi and Bindra, 1980; Mooley et al., 1981; Hastenrath and Rosen, 1983; Dhar and Rakhecha, 1983; Bhalme et al., 1983; Mooley and Parthasarathy, 1983, 1984a, b; Cadet and Diehl, 1984; Raman and Maliekal, 1985; Shukla, 1986). Most relevant are the rainfall totals of the Northern summer half-year.

The interannual variability of rainfall is not uniform throughout a subcontinent as large as India. The maps Fig. 8.5:2 show the spatial correlation of annual rainfall between the various subdivisions identified in Fig. 8.5:1. While positive correlation is generally largest between neighboring subdivisions, distinct spatial patterns not merely related to distance are apparent. Thus the Northwest has only weak positive or even negative correlation with the Eastern to Southern parts of the country; North Central India exhibits only weak correlation with both the Southern tip of the peninsula and the Northeast; and central India, or a substantial portion of the subcontinent, shows considerable correlation even with the peripheral subdivisions. Accordingly, an all-India index of monsoon rainfall appears meaningful for general orientation. Such an index is included in Fig. 8.3:1. This is the arithmetic mean of normalized rainfall departures in the 31 subdivisions. The plot illustrates the considerable interannual variability of rainfall, and indicates some correlation with both Java rainfall and the Southern Oscillation indices of Wright (1975, 1977).

The background circulation and climate of the Indian Ocean sector is discussed in Section 6.8:3, which also includes a review of characteristic synoptic events in the course of the Southwest summer monsoon. Specific synoptic situations are associated with peculiar spatial distributions of rainfall anomalies. The intuitive synoptic experience is supported by stratification, correlation, and principal component analysis (Hastenrath and Rosen, 1983). Thus, Bay of Bengal cyclonic storms favor rainfall in the northeastern to central part of the country (centered on subdivisions 5–7 and 18; Fig. 8.5:1), while 'breaks in monsoon' affect negatively a broad zone across North Central India. The time series of annual number of break monsoon days is negatively correlated with both the annual frequency of cyclonic storms and the all-India rainfall index. All three of these series are correlated with a specific principal component pattern of monsoon rainfall anomalies.

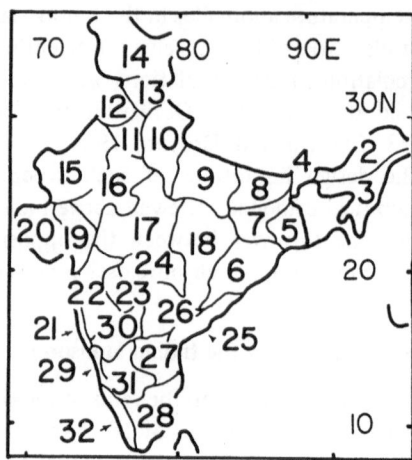

Fig. 8.5:1. Map showing rainfall subdivisions of India. Identification numbers are as used by the India Meteorological Department. From Hastenrath and Rosen (1983).

In particular, the all-India rainfall index (Fig. 8.3:1) possesses a correlation of +0.97 with the first
principal component pattern of monsoon rainfall anomalies, which is characterized by largest
positive factor loadings over the central to western part of the country, and which explains 29%
of the total variance (Hastenrath and Rosen, 1983).

Long-term variations of India monsoon rainfall have been the subject of various investigations
(Koteswaram and Alvi, 1969, 1970; Bhargava and Bansal, 1969; Jagannathan and Bhalme, 1973;
Jagannathan and Parthasarthy, 1973; Parthasarathy and Dhar, 1974; Parthasarathy and Mooley,
1978; Bhalme and Mooley, 1980; Mooley et al., 1981; Bhalme et al., 1983). Upward trends in
earlier decades are reversed in the 1960's and 1970's. The often used 1931—60 reference period
stands out as an era of relatively copious rainfall. Rainfall variability concentrated at somewhat
over two years was found in various studies (Koteswaram and Alvi, 1969, 1970; Bhargava and
Bansal, 1969; Jagannathan and Bhalme, 1973; Jagannathan and Parthasarathy, 1973; Parthasarathy
and Dhar, 1974; Partharathy and Mooley, 1978).

A negative relation between Southwest monsoon rainfall and snowfall of the preceding winter
was suggested as long as a century ago by Blanford (1884). Based on satellite measurements,
Hahn and Shukla (1976) and Dickson (1984) find a similar relationship between Eurasian winter
snow cover and Indian monsoon rainfall. Moreover, Dickson (1984) notes a correlation between
Eurasian and Himalayan region snow cover extent, as derived from satellite observations, and
between both of these quantities and the subsequent Indian monsoon rainfall. Concerning a
possible relationship between Eurasian spring snow cover and an advance period of the summer
monsoon, results of satellite data evaluations seem inconclusive (Dey and Bhanu Kumar, 1982,
1984; Ropelewski et al., 1984). Refer also to Walsh's (1984) recent review of snow cover and
atmospheric variability.

The strong relation between India monsoon rainfall and the Southern Oscillation as illustrated
by Fig. 8.3:1 was already at the focus of Walker's (1923, 1924, 1928a) and Walker and Bliss'
(1930, 1932, 1937) early statistical investigations (Section 8.1), and has aroused curiosity again
in recent years (Tsuchiya, 1970; Alexander et al., 1974; Pant and Parthasarathy, 1981; Barnett,
1983, 1984b, c; Shukla and Paolino, 1983; Bhalme et al., 1983; Rasmusson and Carpenter, 1983;
Mooley and Parthasarathy, 1983, 1984a; Parthasarathy and Pant, 1984). While such teleconnection

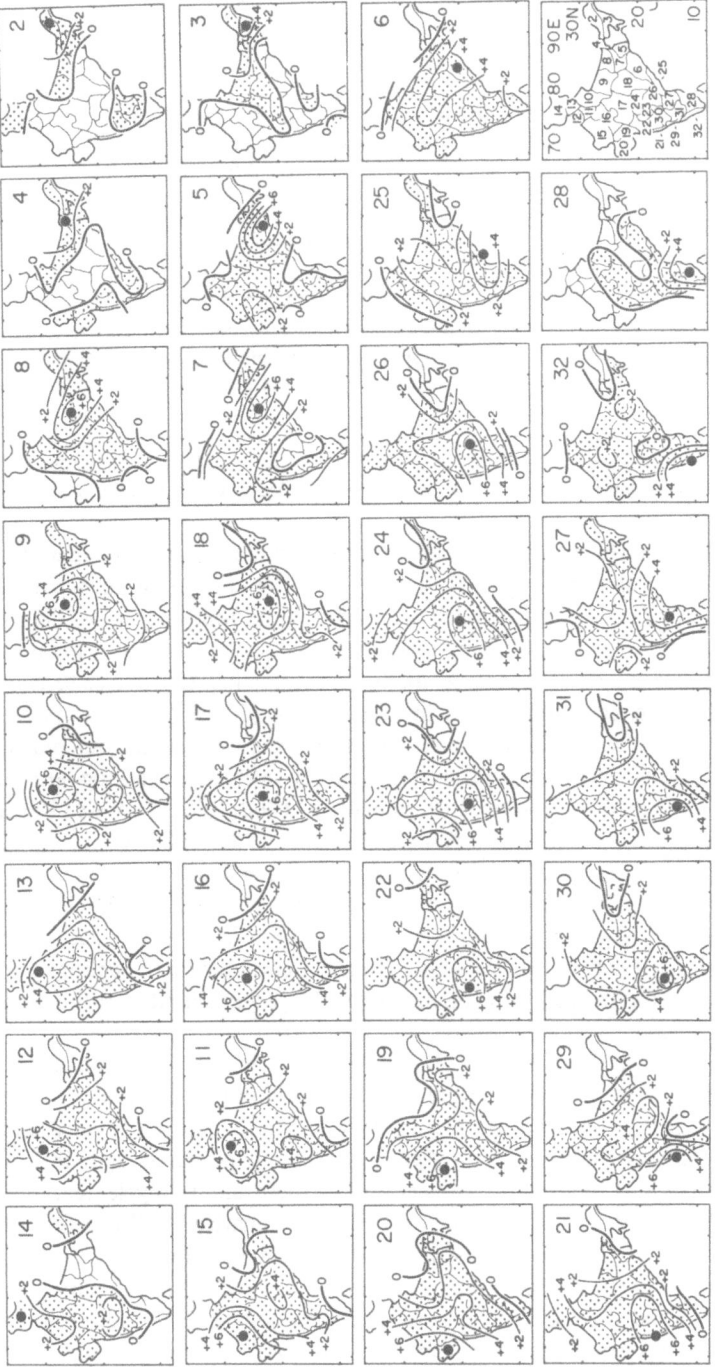

Fig. 8.5:2. Maps of correlation between annual rainfall in subdivisions of India. Isocorrelates are in tenths, zero line is heavy, and dot raster indicates positive areas. Subdivision orientation map is in lower right-hand corner. From Hastenrath and Rosen (1983).

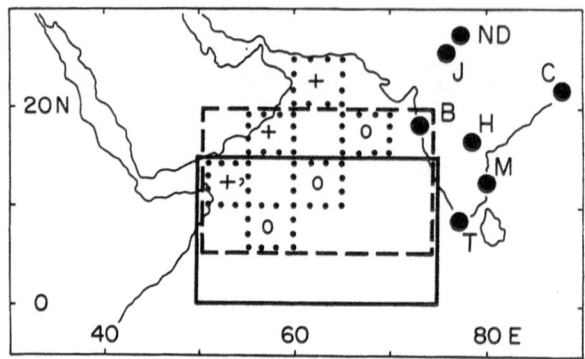

Fig. 8.5:3. Orientation map for Tables 8.5:1 to 8.5:4. Shown are areas for which time series of indicated elements were compiled. Solid line refers to total wind speed (W) and meridional wind component (v); broken line denotes sea surface temperature (SST), sea level pressure (SLP), and cloudiness (C); and index of position of surface jet (SJI) is indicated by dotted lines. SJI is computed as difference in wind speed of area A (denoted by crosses) to the Northwest, minus the wind speed in area B (denoted by circles), to the Southeast, of the climatic mean position of the jet axis. Dots indicate locations of surface stations in India: Bombay B, Jaipur J, Trivandrum T, New Delhi ND, Hyderabad H, Madras, M, Calcutta C. Source: Wu (1984, 1985).

patterns are statistically established, the general circulation causes are not well understood. In fact, Normand (1953) suggests that the Indian monsoon should be considered an 'active' rather than a 'passive' feature in world weather, inasmuch as it tends to be followed rather than preceded by variations elsewhere on the globe. Normand (1953) also recognized the unfortunate implications of this state of affairs for efforts of long-range monsoon forecasting. However, a series of diagnostic and prognostic studies, in part summarized in Jagannathan (1960) and Rao (1964), have shed some light on the general circulation settings typical of years with anomalous monsoon rainfall. While most of the aforementioned studies are based on conventional long-term surface land stations, and some investigations (reviews in Jagannathan, 1960; Rao, 1964; also Ramaswamy and Pareek, 1978; Bhalme and Mooley, 1980; Kung and Sharif, 1980, 1982; Verma, 1982; Thapliyal, 1982) use short upper-air records, Wu (1984, 1985), Cadet and Diehl (1984) and Cadet (1985) added a further dimension by analyzing also the long-term ship observations in the tropical Indian Ocean with reference to the Indian monsoon and the Southern Oscillation. These diagnostic investigations along with the review of background circulation and climate in Sections 6.8.1 and 6.8.3 provide the foundation for the following discussion of the general circulation functioning of India monsoon rainfall anomalies.

The evolution of a heat low over the South Asian continent is an important annual cycle precursor of the Southwest monsoon (Section 6.8.1), favoring as it does a strong meridional pressure gradient from the South Indian Ocean high to South Asia, and hence a strong lower-tropospheric cross-equatorial flow, which ferries large amounts of water vapor into the Northern hemisphere (Section 6.8.3). Plausibly consistent with these annual cycle features appear the circulation characteristics of monsoon rainfall anomalies summarized in Tables 8.5:1 to 8.5:4 and Fig. 8.5:3. Thus pre-season temperature and pressure over India are strongly negatively correlated, and temperature is positively and pressure negatively correlated with the subsequent monsoon rainfall (Tables 8.5:2 and 8.5:1). Temperature over land and the neighboring Arabian Sea are spatially coherent, as is surface pressure (Table 8.5:2). Pressure over land and wind speed over the Arabian Sea are coupled inversely (Table 8.5:2). All these pre-season characteristics support the key role of continental heat low evolution for the monsoon rainfall. It appears tempting to

TABLE 8.5:1

Correlation coefficients, in hundredths, between all-India rainfall index INR and monthly and seasonally averaged sea level pressure SLP, surface wind speed W, sea surface temperature SST, cloudiness C in the Arabian Sea; upper-air ridge position UAR (Fig. 9.1:3), onset date at Trivandrum ONS (Fig. 8.5.5), and index of surface jet axis over Arabian Sea SJI. See Fig. 8.5:3 for locations of time series. One and two asterisks indicate significance at the 5 and 1% levels. In estimating the significance of correlation coefficients, Quenouille's (1952, p. 168) method was used to account for the reduction of effective degrees of freedom due to persistence. MAM, JJA, and SON denote, respectively, the seasons of March–April–May, June–July–August, and September–October–November. Source: Wu (1984, 1985).

	J	F	M	A	M	J	J	A	S	O	N	D	MAM	JJA	SON
INR–SLP	− 2	+ 1	+ 2	−12	−38**	−17	− 5	−41**	−44**	−16	− 3	−15	−24	−35**	−31*
INR–W	−21	− 0	+ 8	− 7	+35**	+12	+ 7	+12	+44**	+15	−11	− 6	+24	+16	+23
INR–SST	+21	+ 2	+12	+15	+ 8	+ 1	+ 3	−13	−27	−44	−26	− 9	+13	− 3	−35*
INR–C	+ 3	+15	− 2	+ 5	+49**	+17	+29*	+53**	+33**	+11	−10	−24	+33**	+50**	+15
INR–UAR				+62**											
INR–SJI			− 3	+ 4	−31	+10	+22	+ 0	−16						
ONS–UAR				+20											
ONS–SJI			− 8	−34**	+11	− 8	+ 7	− 2	+13						

TABLE 8.5:2

Correlation coefficients, in hundredths, between all-India rainfall index INR (Fig. 8.3:1), arithmetic means of surface maximum temperature T_{max}, and station pressure STP, at seven India stations (Bombay, Jaipur, Trivandrum, New Delhi, Hyderabad, Madras, and Calcutta), and sea level pressure SLP, as well as sea surface temperature SST and wind speed W in the Arabian Sea. See Fig. 8.5:3 for locations of time series. One and two asterisks indicate, respectively, significance at the 5 and 1 percent levels. In estimating the significance of correlation coefficients, Quenouille's (1952, p. 168) method was used to account for the reduction of effective degrees of freedom due to persistence. MAM, JJA, and SON denote, respectively, seasons of March–April–May, June–July–August, and September–October–November. Source: Wu (1984, 1985).

	M	A	M	J	J	A	S	MAM	JJA	SON
T_{max}–STP	−45*	−51*	−48**	−19	+18	+ 8	+29*	−57**	−12	− 3
T_{max}–SLP	+23*	− 4	+15	+ 3	+10	+27*	+35**	+ 9	+15	+ 2
T_{max}–SST	+51**	+54**	+30**	+51**	+38*	+18	+35**	+36**	+28*	+40**
T_{max}–INR	+17	+15	−24	−26*	−59**	−65**	−56**	+ 4	−66**	−61**
STP–INR	− 6	− 9	−12	−26	− 9	−19	−37**	−11	−20	−33**
STP–W	+ 8	+ 3	−40**	+ 7	−18	−25	−42**	− 3	− 6	−28
SLP–W	+25	+ 1	−42**	− 8	− 7	−35*	−24	+ 5	− 9	−12
STP–SLP	+48**	+64**	+56**	+36**	+54**	+56**	+52**	+56**	+44**	+48**

TABLE 8.5:3

Correlation coefficients, in hundredths, between monthly sea surface temperature SST, surface wind speed W, and cloudiness C in the Arabian Sea from May through September. See Fig. 8.5:3 for locations of time series. One and two asterisks indicate, respectively, significance at the 5 and 1% levels. In estimating the significance of correlation coefficients Quenouille's (1952, p. 168) method was used to account for the reduction of effective degrees of freedom due to persistence. Source: Wu (1984, 1985).

		W					C				
		M	J	J	A	S	M	J	J	A	S
	M	+ 4	+ 5	+24	+37**	+15	+ 4	+23	+31**	+24	+25*
	J	+12	+ 2	+19	+39**	+ 3	− 3	+39**	+25	+22	+ 3
SST	J	+11	+ 0	+16	+43**	+ 2	+13	+12	+27	+13	+15
	A	+14	+ 8	+ 8	+22	+ 3	+13	+10	+11	+ 9	+13
	S	+ 9	+ 2	+ 8	+13	− 9	+ 6	+ 3	+16	+ 2	+ 2

TABLE 8.5:4

Correlation coefficients, in hundredths, of monthly index of positions of surface jet SJI with surface wind speed W and sea surface temperature SST in the Arabian Sea from May through September. SJI was constructed by determining the difference in average wind speed between two areas, on the Northwest and the Southeast side of the climate mean position of the jet axis, respectively. See Fig. 8.5:3 for locations. One and two asterisks indicate, respectively, significance at the 5 and 1% levels. In estimating the significance of correlation coefficients, Quenouille's (1952, p. 168) method was used to account for the reduction of effective degrees of freedom due to persistence. Source: Wu (1984, 1985).

		W					SST				
		M	J	J	A	S	M	J	J	A	S
SJI	M	−29	−11	+15	−11	−11	+16	+ 7	+ 1	− 2	+ 9
	J	+10	− 9	+19	− 4	+22	−14	−23	−28*	−23	−24
	J	−23	− 9	−16	+12	−16	−19	−22	+ 1	−14	−28
	A	− 8	+ 4	+ 5	−37**	+ 3	−37**	−27	−34*	−31*	−22
	S	−27	− 4	−39**	−21	−70**	−34**	−23	−15	−18	.11

interpret the positive correlation between pre-season sea surface temperature in the Arabian Sea and the subsequent monsoon rainfall as reflecting a thermodynamic enhancement of moisture content and instability of the boundary layer flow. This would conform with Shukla's (1975) numerical model experiment (ref. Section 6.8.5). It should be realized, however, that Arabian Sea surface temperature is itself controlled by a complex of processes (Colón, 1964; Düing and Leetmaa, 1980; Hastenrath and Lamb, 1980; Bruce, 1983; Ramesh Babu and Sastry, 1984). Moreover, the correlation with monsoon rainfall fades out and changes sign at the height of the monsoon.

In fact, the evolution of relationships from the pre-season (March–April–May) through the peak (July–August) to the post-monsoon season (September) is illuminating (Tables 8.5:1 to 8.5:4). From the pre-monsoon to the post-monsoon season, pressure maintains a negative correlation with wind speed over the Arabian Sea (Table 8.5:2), as well as with monsoon rainfall (Tables 8.5:1 and 8.5:2). Likewise, the correlation between the latter two parameters remains positive throughout (Tables 8.5:1 and 8.5:2). By contrast, the correlation between temperature over land and monsoon rainfall reverses sign after the monsoon onset (Table 8.5:2), conceivably due to the rainfall and subsequent evaporation. In accord with this, the correlation between temperature and pressure also reverses (Table 8.5:2).

Interestingly, the correlation between sea surface temperature in the Arabian Sea and monsoon rainfall also fades out from the pre-monsoon season to the onset and then reverses sign (Table 8.5:1). The explanation given for the land areas (rainfall and subsequent evaporation) cannot

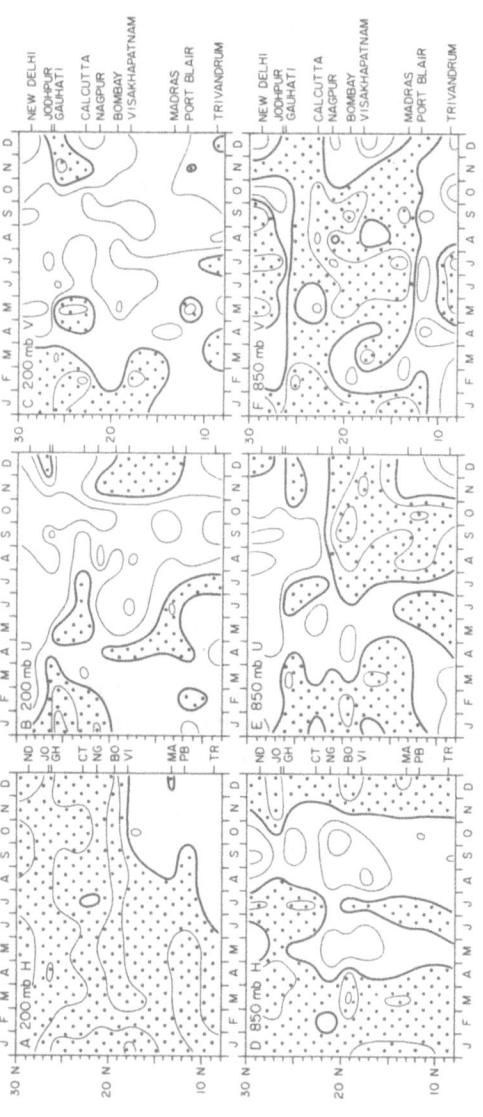

Fig. 8.5:4. Topochronopleth diagrams of correlation between India monsoon rainfall (INR) and (a) geopotential height H, (b) zonal wind component U, (c) meridional wind component V, at 200 mb; and (d), (e), (f) same elements at 850 mb. Stations are plotted according to latitude along ordinate scale, while calendar months are given on the abscissa. Zero line is heavy, thin lines are isocorrelates of ± 0.3, and dot raster denotes positive areas. From Wu (1984, 1985).

apply. Therefore, various other hypotheses were tested (Wu, 1984, 1985). A negative cloudiness-radiation feedback is ruled out by the consistently positive correlation between cloudiness and sea surface temperature (Table 8.5:3). Wind-stress enhanced mixing and evaporation cannot be a major factor for surface cooling, because wind speed and sea surface temperature continue to be positively correlated from the pre-monsoon through the post-monsoon season (Table 8.5:3). A third hypothesis involves ocean dynamics. The surface wind field to the right (Southeast) of the jet axis (Fig. 8.5:3) entails oceanic Ekman mass transport directed towards the right and decreasing away from the jet axis. The resultant downwelling seems to be a cause for the observed deepening of the mixed layer in the Central Arabian Sea in the course of the monsoon season (Wyrtki, 1971, charts 29–32; Robinson et al., 1979, charts 56, 70, 84, 98, 112, 126), and surface cooling. In order to test this hypothesis, an index of the surface wind jet (SJI) was computed as indicated in Fig. 8.5:3. This index is negatively correlated with wind speed over the Arabian Sea (Table 8.5:4), as a more northwesterly position of the jet entails weaker winds in the area to the Southeast, namely the Central Arabian Sea. The surface jet index shows no consistent correlations with monsoon rainfall (Table 8.5:1). However, it is negatively correlated with sea surface temperature in the Arabian Sea (Table 8.5:4), consistent with the hypothesis to be tested. Thus, details of the wind stress pattern, through the dynamics of the upper ocean, appear to be a major factor for the seasonal reversal of the correlation between Arabian Sea temperature and monsoon rainfall. While the summer cooling of the Arabian Sea has attracted the attention of various researchers (Düing and Leetmaa, 1980; Bruce, 1983; Ramesh Babu and Sastry, 1984), the role of this seasonal hydrospheric evolution in the functioning of the Southwest monsoon remains to be explored.

The upper-air circulation also experiences a drastic changeover from the winter to the summer monsoon (Section 6.8.1). The replacement of upper-tropospheric westerlies to the South of the Himalayas by easterlies appears as the foremost annual cycle precursor of the summer Southwest monsoon. Concomitantly, southerly wind components in the upper troposphere give way to northerlies. The circulation characteristics of monsoon rainfall anomalies depicted in Figs. 8.5:4a to f also appear plausibly consistent with the aforementioned annual cycle features. Thus, Figs. 8.5:4d to f, referring to the lower troposphere, reflect the evolution, from the pre-season onward, of particularly low pressure and strong southwesterly wind over India during copious rainfall years. The upper-tropospheric patterns (Figs. 8.5:4a to c) exhibit a similarly plausible relation to the annual cycle. Abundant monsoon rainfall is associated with high 200 mb topography (Fig. 8.5:4a), seemingly reflecting a warm troposphere resulting from copious latent heat release. Good monsoon years are further characterized by enhanced easterlies (Fig. 8.5:4b) and northerly flow component (Fig. 8.5:4c) in the upper troposphere. The latter feature is consistent with the notion of a strengthened return flow in the upper-tropospheric part of the mean meridional circulation cell in the Indian Ocean sector. The content of Fig. 8.5:4 is consistent with various upper-air studies mentioned above and recognition of the mid-tropospheric ridge position as monsoon precursor. Thus, Table 8.5:1 shows a strong correlation between the latitude position of the 500 mb ridge in April and the rainfall of the following Southwest monsoon.

A recent study by Dhar and Rakhecha (1983) is interesting in regard to the relation between interannual variability and annual cycle suggested by the above consideration of circulation characteristics. Dhar and Rakhecha (1983) note that the Northeast monsoon (October–December) is of considerable economic importance for Tamil Nadu in South India, as it provides nearly half of the annual rainfall total in this region. In their efforts at climate prediction they find that a deficient Southwest monsoon tends to be followed by abundant Northeast monsoon rainfall. The topic of climate prediction for India is discussed in Section 9.1.

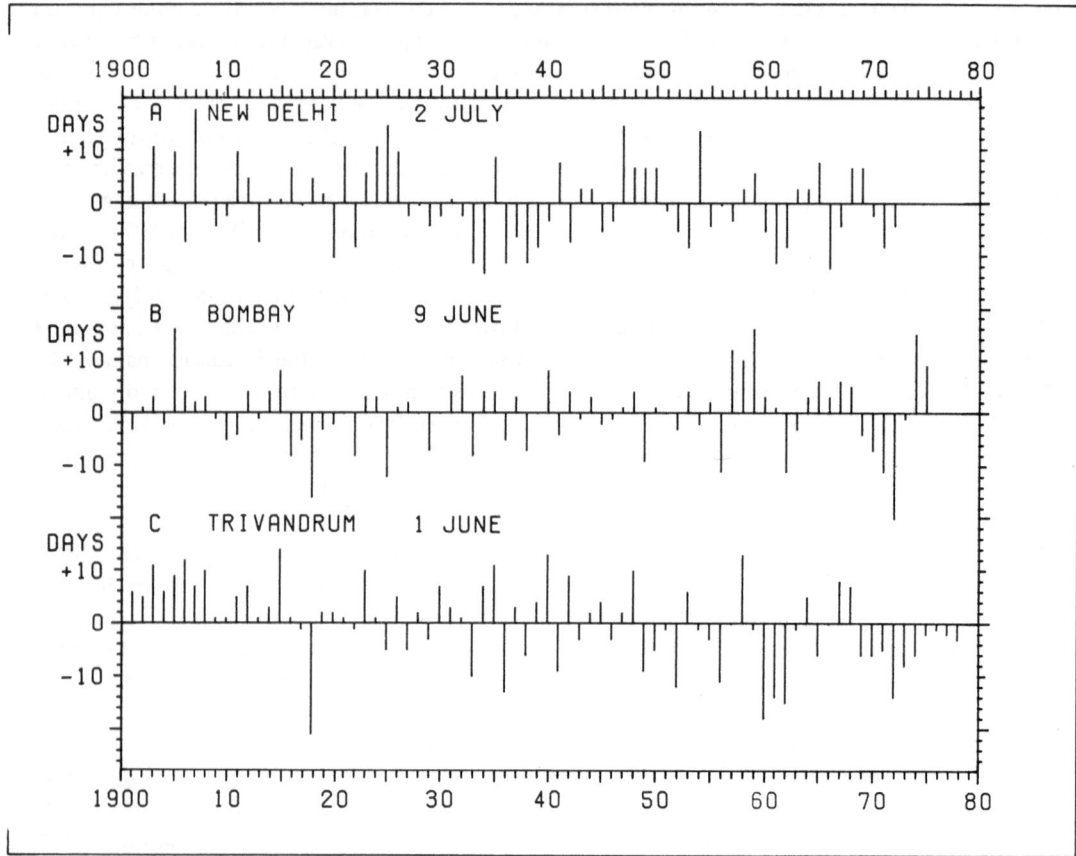

Fig. 8.5:5. Time series plots of monsoon onset dates over Trivandrum, Bombay, and New Delhi: Departures, in days, are shown from the average onset dates of 1 June, 9 June, and 12 July. The correlation coefficient between the Trivandrum and Bombay series is +0.63, significant at the one percent level. New Delhi is correlated with Bombay and Trivandrum only at +0.08 and +0.17, respectively. Source: India Meteorological Department, unpublished data.

As discussed in Section 6.8.3, the date of monsoon onset may vary over several weeks at a given location, aside from the common delay northward and westward (Fig. 6.8.3:1a). Fig. 8.5:5 depicts the interannual variability of onset dates over Trivandrum, Bombay, and New Delhi. For Trivandrum and Bombay the anomalies of onset date are similar, whereas New Delhi shows only weak correlation with the other two stations. Among the annual cycle precursors of monsoon onset are the northward progression of the surface wind axis over the Arabian Sea and the northward shift of the upper-air ridge between tropical easterlies and temperate latitude westerlies (Table 8.5:1). The latter circulation parameter is also correlated with the monsoon rainfall (Table 8.5:1). However, a late onset may be compensated by copious rainfall later on. Accordingly, onset date and total monsoon rainfall are essentially uncorrelated.

8.6. The Sêcas of Northeast Brazil

The Sêcas or droughts of Northeast Brazil are among the most disastrous climatic hazards of the tropics. Throughout recorded history, the droughts have haunted the region referred to as 'Nordeste', which actually constitutes the easternmost corner of Brazil (Figs. 8.6:1 and 8.6:2), and have caused famine and mass exodus. In his monumental work 'Os sertoês' from the latter part of the 19th century, Euclides da Cunha (1979) vividly describes the human and societal impact of the Sêcas, and reports numerous proverbs and farmers' rules, and popular efforts at foretelling the quality of the next rainy season. Extensive bibliographies pertaining to the Sêcas of the Nordeste are found in Markham (1967), Conselho Nacional de Pesquisas (1980), Chu and Hastenrath (1981), Moura and Shukla (1981), Aldaz (1983), Hastenrath and Heller (1977), and Hastenrath *et al.* (1984). The latter two papers provide the major basis for the present section. For the background circulation and climate, Atkinson and Sadler (1970), Sadler (1975), Ratisbona (1976), Hastenrath and Lamb (1977a), and Chu and Hastenrath (1982), are further useful sources. Reference is also made to Figs. 6.1:1 to 6.1:6, 6.7.2.1:1 to 6.7.2.1:4, 6.7.3:1.

The general circulation in the Brazil — tropical Atlantic sector is dominated by a near-equatorial trough of low pressure sandwiched between the subtropical highs of the two hemispheres (Figs. 6.1:2, 6.7.2.1:1a, 6.7.2.1:2b). In this trough the Northeast trades and cross-equatorial flow from the Southern hemisphere meet (Figs. 6.1:3b, 6.1:4b, 6.7.2.1:1b, 6.7.2.1:2b). The trough contains a zone of convergence and maximum cloudiness and rainfall (Figs. 6.1:5, 6.1:6, 6.7.2:1:1d, 6.7.2.1:2b). The surface temperature is highest in the tropical North Atlantic, while the South equatorial waters are cold. The aforementioned complex of quasi-permanent circulation features attains a farthest poleward position at the height of the Northern summer, and a southernmost location in March to April when interhemispheric sea surface temperature gradients are weakest· (Hastenrath and Lamb, 1977a, charts 4, 5, 16, 17, 32, 33, 52, 53).

March/April is also the time of the narrowly concentrated rainy season in Northern Northeast Brazil (Figs. 8.6:2 and 8.6:3). At least three factors in the surface circulation are conducive to rainfall at this time of year: (i) the convergence band over the Atlantic is closest to the Nordeste; (ii) the surface waters of the Equatorial South Atlantic upstream from the Nordeste are least cold, which serves to enhance the moisture content and instability of the boundary layer flow; (iii) the perennial contrast between warm North equatorial waters and cold waters to the South of the

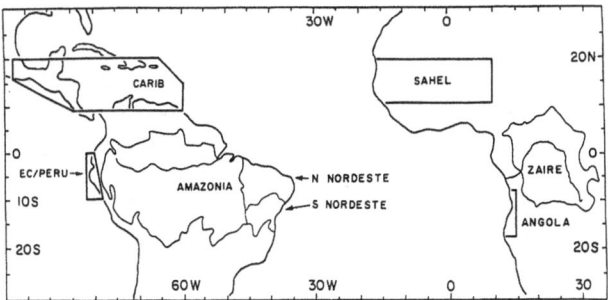

Fig. 8.6:1. Orientation map. Key tropical land areas in the Americas and Africa: Central American — Caribbean region (CARIB), Ecuador-Peru coast (EC/PERU), Northern and Southern Northeast Brazil (N NORDESTE, S NORDESTE), Amazonia (AMAZON), Subsaharan Africa (SAHEL), Zaïre (Congo) basin (ZAIRE), Angola coast (ANGOLA). Source: Hastenrath (1984).

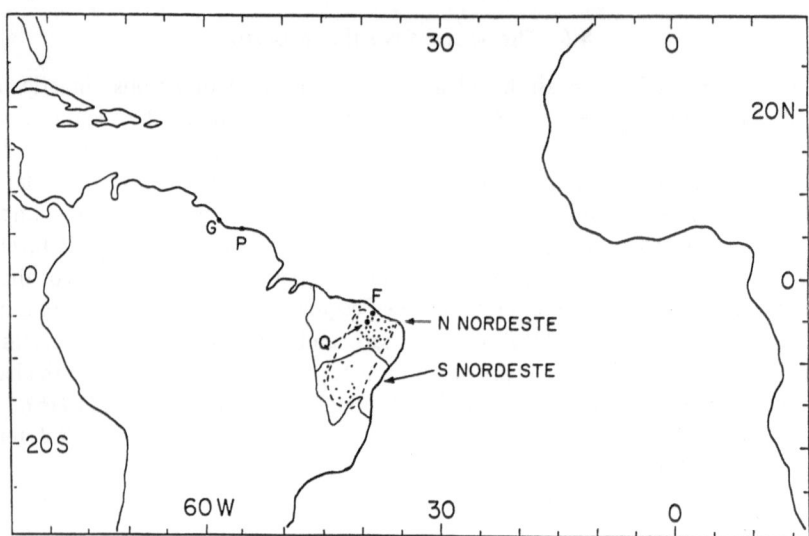

Fig. 8.6:2. Orientation map. Thin solid lines delineate the Northern and Southern Nordeste, and broken line the region of Nordeste with annual rainfall below 800 mm. Dots denote rainfall stations, with letters identifying Fortaleza (F), Quixeramobim (Q), Georgetown (G), and Paramaribo (P). Source: Hastenrath *et al.* (1984).

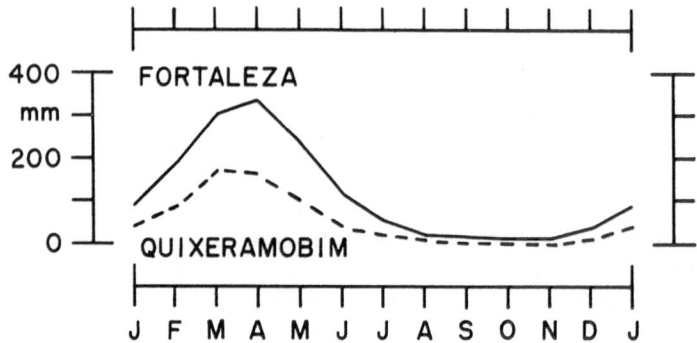

Fig. 8.6:3. Annual march of rainfall at Fortaleza and Quixeramobim (ref. Fig. 8.6:2) in Northern Northeast Brazil.

Equator is weakest. Moura and Shukla (1981) hypothesized that the juxtaposition of warm surface waters to the North and cold waters to the South would induce a meridional circulation cell in the atmosphere, with subsidence over the Nordeste. This process is expected to be weakest in the season of weakest meridional sea surface temperature gradients.

After March, the equatorial trough and associated convergence band migrate northward, the South equatorial waters cool, and cross-equatorial sea surface temperature gradients steepen — and the Northern Nordeste rainy season is over. While the general circulation background of rainfall in Northern Northeast Brazil has only recently been conclusively documented by analyses of long-term ship observations in the tropical Atlantic (Hastenrath and Lamb, 1977a), the peasants of the Nordeste have been intuitively aware of this state of affairs for more than a century. In fact, a local proverb notes that, if rains have not started by Saint Joseph's day (19 March) they will not come (Euclides da Cunha, 1979, pp. 92–93).

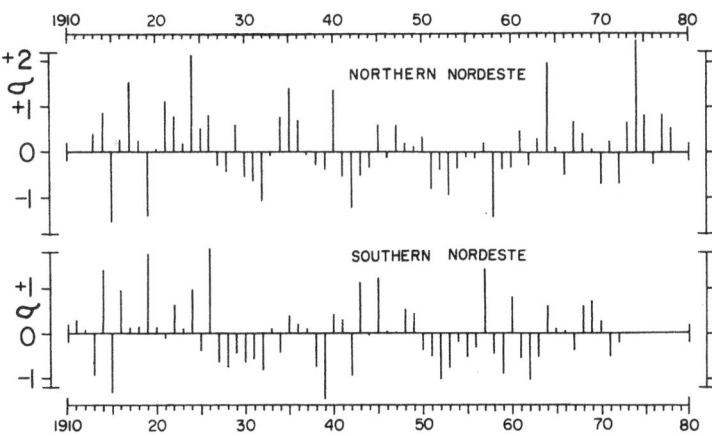

Fig. 8.6:4. Time series plots of rainfall indices for the Northern (October–September) and Southern (July–June) Nordeste. For both series values are ascribed to the later of the two calendar years. From Hastenrath *et al.* (1984).

The annual march of rainfall shows a gradual transition from the Northern to the Southern Nordeste, where the maximum is found around November/December. The Southern Nordeste in particular is affected by frontal influences from the Southern hemisphere (Sampiao Ferraz, 1925, 1929; Ratisbona, 1976; Kousky, 1979; Chu, 1983). The rainfall mechanisms at the Eastern Nordeste coast include the land-sea breeze system and westward propagating cloud clusters (Ramos, 1975; Yamazaki and Rao, 1977; Kousky, 1980). The role of the Intertropical Convergence Zone (Section 6.7.2.1) in Northeast Brazil rainfall has further been studied by Valgas (1982), Oliveira (1982), and Rao *et al.* (1984). The atmospheric moisture budget during contrasting extreme years has been analyzed by da Silva Marques *et al.* (1983) and Rao and da Silva Marques (1984). Of particular interest here is the Central Northern Nordeste, for which the average annual precipitation is low, the peak of the rainy season is narrowly confined to March/April, and both the average annual cycle and the interannual variability of rainfall are conspicuously related to the large-scale circulation.

The interannual variability of rainfall in the Northern and Southern Nordeste (Fig. 8.6:2) is illustrated in Fig. 8.6:4. For the Northern Nordeste, 32 stations are used that have a correlation of at least +0.7 with Quixeramobim in central Ceará state (Fig. 8.6:2). The index for the Southern Nordeste (Chu, 1983) is based on 14 stations. The rainfall indices were compiled as described before for Java (Section 8.3) and India (Section 8.5). As described in Hastenrath and Heller (1977), long-term rainfall series are complemented by seasonalized (October–September) annual river discharge during 1913–64 at gauging sites on the Jaguaribe and Banbuiu rivers in Ceará state. These discharge series are highly correlated with the rainfall index. Fig. 8.6.4 indicates similarity but also considerable differences between the Northern and Southern Nordeste. Both series exhibit a long-term downward trend from the 1920's to the 1950's. The infamous 1958 drought year stands out for the Northern Nordeste in particular. Based on Fig. 8.6:4 and river discharge information, extreme years in Northern Northeast Brazil were identified as follows: 19*15*, *19*, 30, 31, 32, 36, *42*, 51, 53, *58*, as DRY; and 19*17*, 21, 22, *24*, 26, 34, *35*, 40, *64*, 67, as WET; italicized years being very extreme. For all ten DRY years the rainfall index is −0.5 σ or beyond; for the ten WET years it is greater than or equal to +0.7 σ. For the Southern Nordeste, Chu (1983) lists the following extreme years; 19*13* (i.e. July 1912 to June 1913), *15*, 28, 32, *39*, 42, 52, 53, 59, 62, as DRY, and 19*14*, 16, *19*, 24, *26*, 43, 45, 57, 60, 69, as WET; italicized years being very extreme. For all ten of the DRY years the rainfall index is less than −0.8 σ, and for the ten WET years it is greater than +0.7 σ.

For composites of the extreme DRY and WET years in the Northern Nordeste, as identified from Fig. 8.6:4, maps of departures from the 60-year mean of the pertinent atmospheric and oceanic fields were constructed (Hastenrath and Heller, 1977). Similarly, Chu (1983) produced stratification maps for the Southern Nordeste, but only the results for the Northern Nordeste will be detailed here. For the various fields, composite departure maps for the DRY years are approximately inverse to those for the WET years. Therefore, only the maps of departures for the composite of WET years minus the departures for the composite of DRY years for the March/April rainy season peak are reproduced here as Fig. 8.6:5.

The WET minus DRY maps, Fig. 8.6:5, can be thought of as showing the circulation departures typical of abundant rainfall years. The following description of pertinent features is largely repeated from Hastenrath (1984). The near-equatorial trough of low pressure is displaced southward (Fig. 8.6:5a), as are the confluence between the Northern and Southern hemisphere trade wind air-streams (Figs. 8.6:5b and c), and the associated band of maximum convergence and cloudiness (Figs. 8.6:5e and f); surface waters of the equatorial and South Atlantic are anomalously warm, while negative sea surface temperature departures prevail in much of the tropical North Atlantic (Fig. 8.6:5d). During drought years in Northern Northeast Brazil, the departure patterns of the various atmospheric and oceanic fields are approximately inverse to those in abundant rainfall years. The *t*-test shows statistical significance in the areas of positive/negative pressure departures in the North/South Atlantic (Fig. 8.6:5a), in the zonal wind field over the equatorial region (Fig. 8.6:5b), in the meridional wind field over the equatorial zone and the adjacent North Atlantic (Fig. 8.6:5c), in the area of positive cloud departure over the equatorial Atlantic adjacent to Northeast Brazil (Fig. 8.6:5f), and in the core regions of negative/positive sea surface temperature departures in the low-latitude North/South Atlantic (Fig. 8.6:5d). Concerning the pressure map (Fig. 8.6:5a), the large areas of statistical significance in the domain of positive departures in the North Atlantic are of particular interest. In the region of negative departures in the South Atlantic, where data coverage is less satisfactory, the area with statistical significance is more limited.

The departure characteristics of the pressure, wind, and convergence fields are internally consistent and symptomatic of an anomalous latitude position of the near-equatorial trough. The band of anomalous convergence (Fig. 8.6:5e) extending from Brazil to equatorial Africa is matched by a corresponding zone of enhanced cloudiness (Fig. 8.6:5f). The substantive features of the sea surface temperature field identified in Hastenrath and Heller (1977) have been duplicated by Markham and McLain (1977), Moura and Shukla (1981), and Lough (1986), and early results concerning the wind field have also been duplicated later (review in Hastenrath, 1983). However, the origin of the sea surface temperature departures is not yet understood.

The role of the anomalous latitude positions of near-equatorial low pressure trough, confluence axis, and convergence and cloudiness bands, for Nordeste rainfall anomalies, as documented by ship observations in the tropical Atlantic (Fig. 8.6:5), is further supported by rainfall observations on the South American continent. An anomalous March/April rainy season peak is heralded by precipitation departures in the same sense during the preceding month in the Northern Nordeste itself. Conversely, rainfall anomalies in the Guyanas (stations Paramaribo and Georgetown, Fig. 8.6:2) are opposite to those in the Nordeste (Hastenrath and Heller, 1977; Hastenrath *et al.*, 1984). This is consistent with the notion that an anomalously far poleward position of the near-equatorial trough is conducive to abundant rainfall in Northern South America and drought in

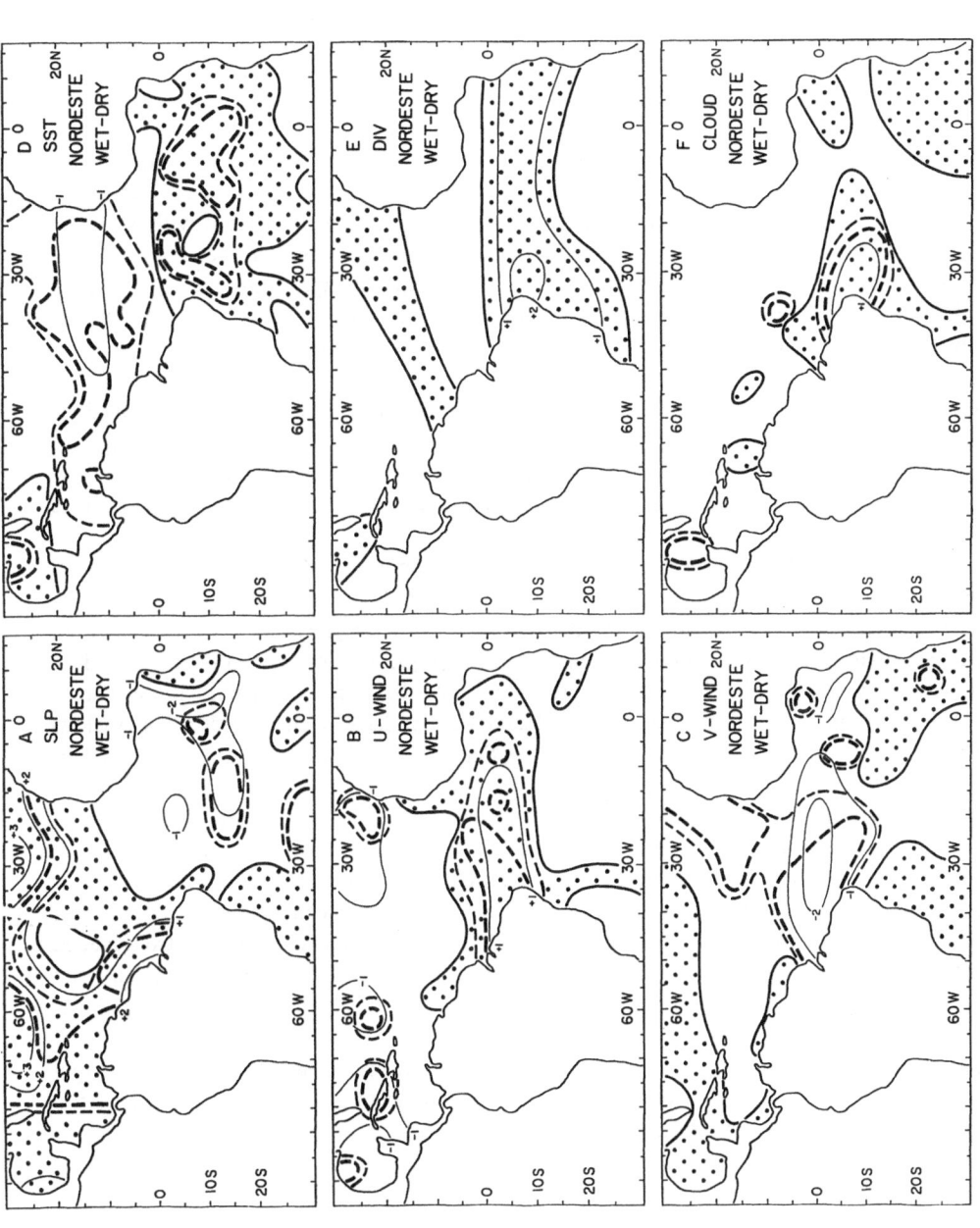

Fig. 8.6:5. Maps of March/April circulation departures associated with rainfall anomalies in the Northern Nordeste of Brazil. Charts represent the difference of departures for the composite of WET years (1917, 21, 22, 24, 26, 34, 35, 40, 64, 67) minus the departure for the composite of DRY years (1915, 19, 30, 32, 36, 42, 51, 53, 58). (a) sea level pressure SLP in mb; (b) zonal (u), and (c) meridional (v) components of wind in m s^{-1}; (d) sea surface temperature SST in °C; (e) divergence DIV in 10^{-6} s^{-1}; (f) total cloudiness CLOUD in tenths. Dot raster denotes positive areas, except for convergence in chart (e). Areas significant at the 5 and 10 percent levels are enclosed, respectively, by heavy and thin broken lines; being excepted from testing. From Hastenrath (1984).

Brazil's Nordeste. The gradual evolution of circulation anomaly patterns in the half-year preceding the rainy season peak offers the prospect for climate prediction (Section 9.7).

The aforementioned stratification and correlation studies are complemented by principal component analysis (Hastenrath, 1978). While the former two methods yield the circulation departure patterns associated with Nordeste rainfall anomalies, the latter technique identifies the most common spatial patterns. In an experiment of principal component analysis involving pressure and sea surface temperature (Hastenrath, 1978) two modes stand out. The first principal component explaining 34% of the variance, shows positive factor loadings for pressure over the Pacific, and negative loadings for pressure over the Atlantic, with largest values in the South Atlantic, while sea surface temperature has negative factor loadings in both the tropical Pacific and Atlantic. This first principal component possesses a significant positive correlation with Nordeste rainfall, and can be considered as a Southern Oscillation mode, involving inverse pressure variations over the Eastern South Pacific and the South Atlantic. The second principal component, explaining 15% of the variance, shows positive factor loadings for pressure over the South, but negative loadings over the North Atlantic, while sea surface temperature has positive factor loadings in a broad band across the tropical North Atlantic and negative loadings in the South Atlantic. Thus the pattern configurations are similar to the ones obtained with respect to Nordeste rainfall (Figs. 8.6:5a and d). Indeed, this second principal component possesses a highly significant negative correlation with Nordeste rainfall. Thus the principal component experiment complements the stratification and correlation studies, in showing that certain departure modes related to Nordeste rainfall anomalies are common.

The circulation over South America and the adjacent tropical Atlantic (Fig. 8.6:5), as discussed above, is of most immediate importance for the functioning of Nordeste rainfall anomalies. However, inasmuch as the pressure distribution over the Atlantic (Fig. 8.6:4a) — with which are associated particular flow and convergence configurations (Figs. 8.6:5b, c, e, f) — are related to long-term large-scale mass redistributions (Sections 8.1 and 8.4), couplings with more distant parts of the globe are also of interest. Such teleconnections have been considered in various earlier studies (Walker, 1928b; Sampaio Ferraz, 1929; Serra, 1956; Namias, 1972; Caviedes, 1973; Kousky and Moura, 1981).

In particular, Namias (1972) pointed out that increased cyclonic activity in the Newfoundland area is associated with abundant rainfall in the Nordeste. The cause for this statistical linkage seems to involve the 'North Atlantic Oscillation' (Section 8.1), consisting of a pressure seesaw between the subpolar and subtropical latitudes of the Atlantic. A recent analysis of the latter phenomenon is available in Van Loon and Rogers (1978). As explained above, a strengthened North Atlantic high is conducive to a more southerly position of quasi-permanent circulation features over the tropical Atlantic and hence abundant rainfall in the Northern Nordeste. A similar pressure seesaw is conceivable for the South Atlantic.

A relation of Nordeste rainfall to the intensity and position of the summertime upper-tropospheric anticyclone over the South American Altiplano and adjacent Southern Amazonia (Section 6.10) is suggested by Kousky and Moura (1981).

The general circulation causes for coupling between the Sêcas of the Nordeste and the Pacific El Niño phenomenon (Section 8.2) have been discussed by Caviedes (1973), Hastenrath and Heller (1977), Hastenrath (1978), and Covey and Hastenrath (1978). The 1983 drought in Northeast Brazil has been discussed by Rao *et al.* (1986). Not only El Niño, but also the Sêcas, appear to be related to the Southern Oscillation. Note that the pressure over the South Atlantic is immediately relevant for Northern Nordeste rainfall. It is then not surprising that various departure characteristics of the Southern Oscillation (Section 8.1) in remote regions may coincide with rainfall anomalies in the Nordeste. The correlation of Nordeste rainfall with that of the preceding Indian monsoon (Hastenrath *et al.*, 1984) appears as yet another component of the Southern Oscillation pattern.

The departure patterns of the surface circulation associated with Northeast Brazil rainfall anomalies, as discussed above, appear as enhancements and reductions of the average annual cycle. Recent analyzes (Nobre, 1984; Nobre and Moura, 1984) show that the Nordeste rainfall anomalies are also accompanied by organized departure patterns in the upper-air circulation. Nobre (1984) and Nobre and Moura (1984) find a standing wave train in the flow departures over the Atlantic sector, with departures of the same sign from the upper to the lower troposphere. In particular, drought years in Northeast Brazil feature an anomalous cyclonic circulation cell in the lower troposphere over the tropical North Atlantic. The latter feature appears consistent with surface circulation departures described above as characteristic of the Sêcas of the Nordeste. Nobre and Moura's (1984) study thus sheds new light on the relation between the anomalies in the surface climate of the Nordeste region and the circulation of the global tropics.

Considerable interest has been directed to the typical time scales of interannual climate variability in Northeast Brazil. Various papers (Markham, 1974; Jones and Kearns, 1976; Kousky and Chu, 1978; Hastenrath and Kaczmarczyk, 1981; Chu, 1984; de Mesquita and Morettin, 1984) have presented evidence to the effect that rainfall variability in the Northern Nordeste is concentrated with some preference in the frequency bands of somewhat over two years, at around 14 years, and around 28 years. Moreover, the studies of Hastenrath and Kaczmarczyk (1981) and Chu (1984) suggest that the circulation and climate anomaly mechanisms identified by the diagnostic analyses

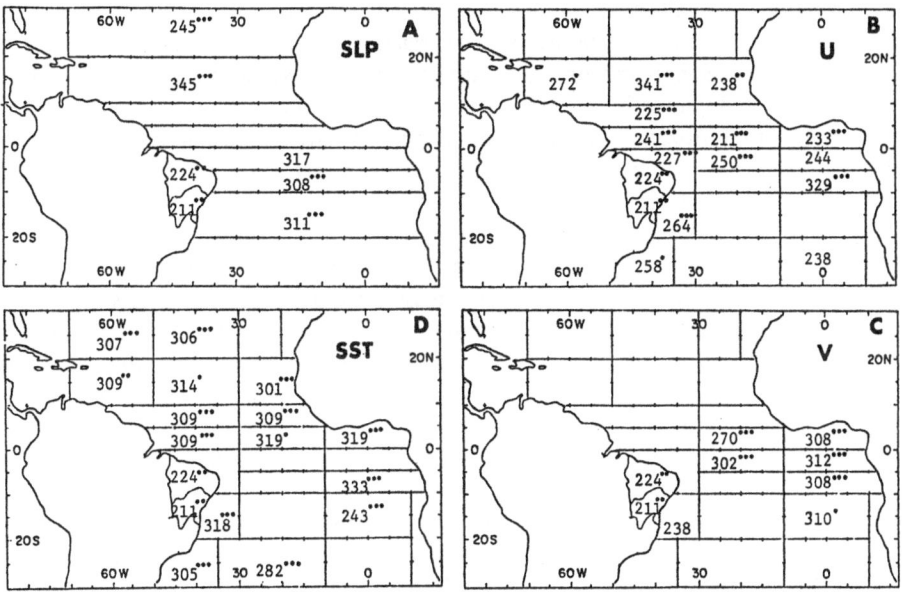

Fig. 8.6:6. Phase relationships in the frequency band of 12.7–14.9 years between Northern and Southern Nordeste rainfall index series (Fig. 8.6:4) and departures of (a) sea level pressure SLP, (b) zonal wind component u, (c) meridional wind component v, (d) sea surface temperature SST. The timing of maxima is indicated by numbers in tenths of years (i.e. 211 signifies 1921.1). One, two, and three dots denote, respectively, ten, five, and one percent significance levels. From Chu (1984).

summarized above are with some preference operative at the time scales of 2+ and about 14 and 28 years. For the 12.7–14.9 year frequency band in particular, phase relationships are mapped in Fig. 8.6:6. In this frequency band, rainfall in the Northern Nordeste varies inversely to pressure on the equatorward side of the South Atlantic high, but approximately in phase with pressure over the North Atlantic (Fig. 8.6:6a). Westerly wind departures are broadly in phase, while southerly wind departures are out of phase with Northern Nordeste rainfall (Fig. 8.6:6b, and c). Finally, sea surface temperature in a broad zone of the tropical North Atlantic varies inversely to Northern Nordeste rainfall (Fig. 8.6:6d). In context, these patterns indicate that the effect of near-equatorial trough position on Nordeste rainfall, as discussed above, is particularly prominent at this time scale. Furthermore, preferred time scales for rainfall variability in the Nordeste can be understood as resulting from just such time preferences in the variation of large-scale circulation patterns in the Brazil – tropical Atlantic sector.

Throughout the long history of the Nordeste, droughts have disrupted the economy and societal patterns of the area, and have remained a central National issue to the present. Continuing a proud tradition of coping with the environmental problems in one of its earliest colonized regions (Carvalho, 1973; Gonçalves de Souza, 1979), the Government of Brazil (Conselho Nacional de Pesquisas, 1980) is taking an active interest in the possibility of forecasting the Sêcas of the Nordeste. The basis and potential of climate prediction for the Nordeste are discussed in Section 9.7, while the human impact of these natural disasters is considered in context in Section 10.7.

8.7. Rainfall Variations in the Central American – Caribbean Region

As elsewhere in the tropics, climatic variations in the Central American – Caribbean region (Fig. 8.6:1) are primarily reflected in the rainfall activity. These have received little publicity, as compared to the Pacific El Niño phenomenon, the vagaries of the Indian monsoon, the Sêcas of Northeast Brazil, or the Subsaharan droughts, discussed in other sections of this chapter. However, in this area runs of predominantly wet years have commonly alternated with prevailingly dry ones, throughout the 20th century. Such variations in rainfall have important consequences for the large-scale water budget, natural vegetation, and land use (Hastenrath, 1970). The major sources for this section include Hastenrath (1966, 1967, 1976).

The Caribbean and Central America contain a great variety of rainfall regimes, due to stress-differential induced divergence effects along variously oriented coastlines, orography, and other factors (Chapter 2). However, setting aside such meso-scale patterns, the area as a whole experiences its rainy season during the Northern summer half-year, when the North Atlantic high is displaced northward, the downstream portion of the trades is convergent, the trade inversion (Section 6.5) high and weak (Hastenrath, 1966, 1967), the ocean is warm, and atmospheric moisture abundant. By contrast, February/March is the driest time of the year in most of the area, coincident with a tendency for lower-tropospheric divergence and subsidence, strong trade inversion, cold ocean, and reduced atmospheric humidity.

The interannual variability of rainfall in the Central American – Caribbean area is illustrated in the time series plot Fig. 8.7:1. This index is the all-station average of normalized departure of calendar-year annual rainfall totals (Hastenrath, 1976) compiled from a collective of 48 stations as described in Section 8.3. Fig. 8.7:1 shows appreciable variations from year to year, but also over longer intervals. For example, the 1930's and 1950's stand out as particularly wet, while the 1940's and 1960's were considerably drier. The long-term variations run broadly inverse to the U.S. Central Great Plains, which in fact experienced their disastrous 'dust bowl' years in the mid 1930's. This hydrometeorological index based on rainfall stations is corroborated by records of

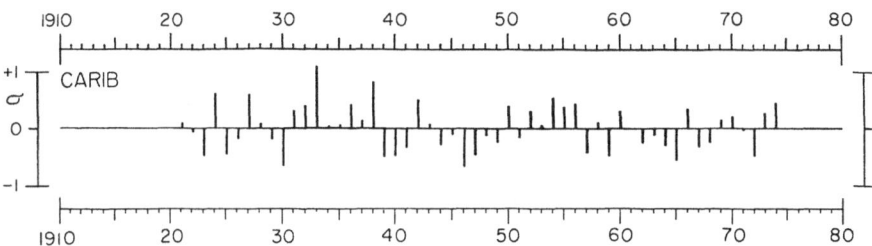

Fig. 8.7:1. Time series plot of hydrometeorological index (in σ) for the Central American – Caribbean area (all-station average of normalized departure of calendar-year rainfall totals of network of 48 stations). Source: Hastenrath (1976).

large rivers in Northern South America. From Fig. 8.7:1 extreme rainfall years in the Central American – Caribbean area identified as follows: 1925, *39, 40, 46*, 47, 57, 59, *65*, 72 as DRY; and 19*24, 27*, 32, *33*, 36, *38, 54*, 56, 60, 66, as WET; italicized years being very extreme. For all ten of the DRY years the index is −0.45 σ or beyond, for the WET years it is greater or equal to +0.30 σ.

For these collectives of DRY and WET years, composite maps of departures from the sixty-year mean were constructed for indicative atmospheric and oceanic fields and the July/August middle of the rainy season (Hastenrath, 1976, 1984). In general, the DRY and WET composites show broadly opposite departure configurations, so that the pertinent anomaly patterns appear enhanced in the WET minus DRY difference maps reproduced in Fig. 8.7:2. These can be understood in terms of abundant rainy season conditions. Thus, an abundant rainy season is associated with anomalously low pressure on the equatorward side of the North Atlantic high (Fig. 8.7:2a), a poleward displacement of the North Atlantic trades (Fig. 8.7:2b), anomalously warm North Atlantic waters in a band between about 10–20°N extending from the African coast all the way to the Americas (Fig. 8.7:2d), and enhanced convergence (Fig. 8.7:2e) and cloudiness (Fig. 8.7:2f) over the Caribbean and the adjacent low-latitude Atlantic. For a deficient rainy season a sign convention broadly inverse to abundant rainfall years would apply. The t-test shows statistical significance at the chosen bands in the core areas of the prominent positive and negative zonal wind departures in the North Atlantic (Fig. 8.7:2b), in the band of positive sea surface temperature departures across the North Atlantic (Fig. 8.7:2d), and less markedly in the realm of negative sea level pressure departures in the North Atlantic (Fig. 8.7:2a).

The essential departure features of the pressure, wind, convergence, and cloudiness fields appear internally consistent and related to decreased pressure on the equatorward side of the North Atlantic subtropical high. The sea surface temperature departures are prominent, but their origin is not yet understood. The interannual variations of circulation and rainfall in the Central American – Caribbean area appear in part as enhancements and reductions of the annual cycle. The gradual evolution of circulation anomalies in the course of the half-year preceding the rainy season is of interest for climate forecasting (Chapter 9).

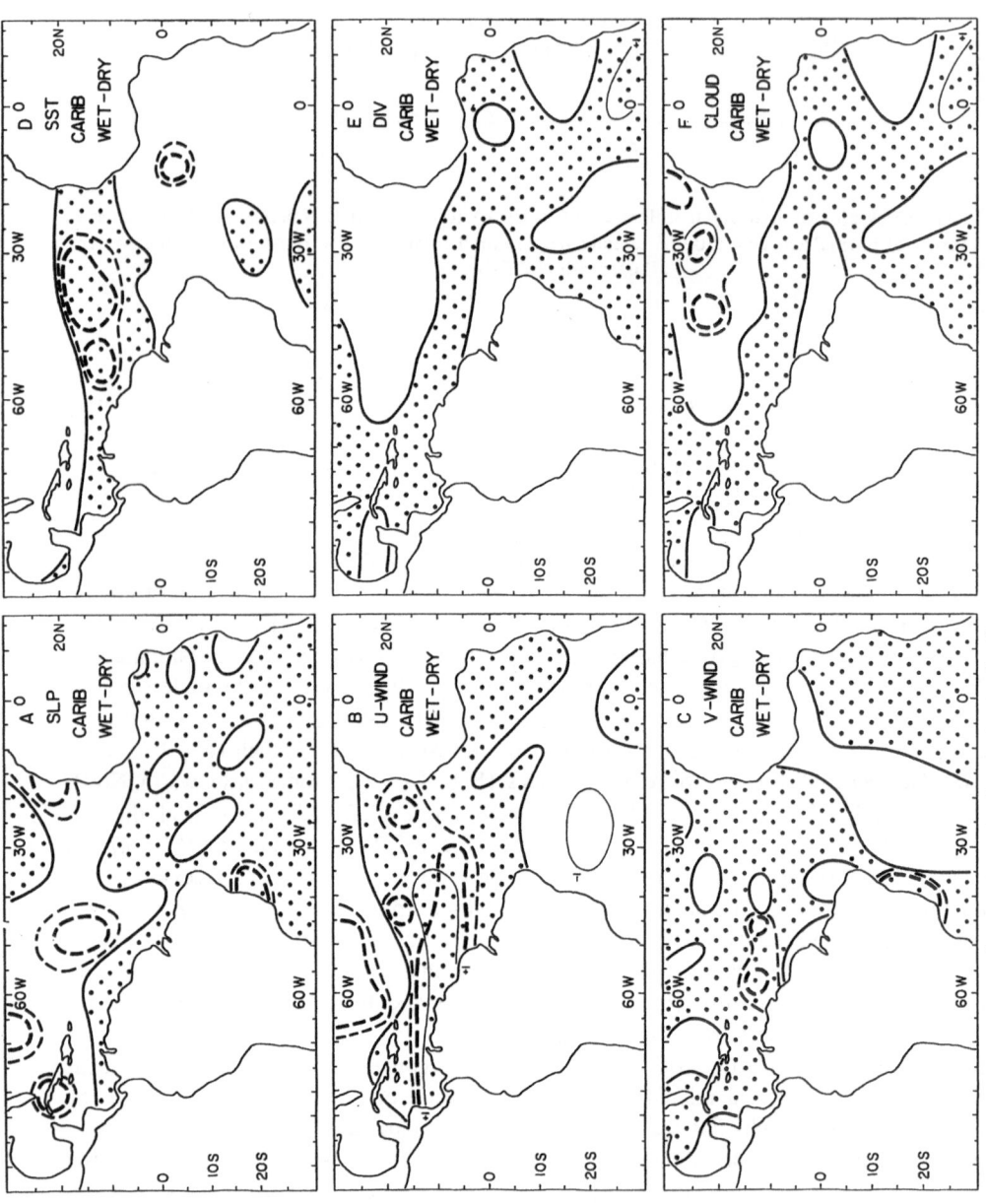

Figs. 8.7:2. Maps of July/August circulation departures associated with rainfall anomalies in the Central American – Caribbean region. Charts represent the difference of departures for the composite of WET years (1924, 27, 32, 33, 36, 54, 56, 60, 66) minus the departures for the composite of DRY years (1925, 30, 39, 40, 46, 47, 57, 59, 65, 72). (a) sea level pressure SLP in mb; (b) zonal (u) and (c) meridional (v) components of wind in m s^{-1}; (d) sea surface temperature SST in °C; (e) divergence in 10^{-6} s^{-1}; (f) total cloudiness in tenths. Dot raster denotes positive areas, except for convergence in chart (e). Areas significant at the 5 and 10% levels are enclosed, respectively, by heavy and thin broken lines; (e) being excepted from testing. From Hastenrath (1984).

8.8. Drought and Flood Regimes in Subsaharan Africa

The semi-arid Sahel zone of West Africa (Fig. 8.6:1) has experienced sequences of extremely dry years repeatedly in this century (Grove, 1972, 1973; Davy, 1974; Landsberg, 1975a, b; Hastenrath, 1976), but none of these has provoked as wide publicity as the drought episode which began in the late 1960's. The general circulation setting of Subsaharan Africa is discussed in Sections 6.7.2.3 and 6.8.2. Reference is also made to Figs. 6.1:1 to 6.1:6, 6.2.4:1, 6.2.4:2, 6.7.2.1:1, 6.7.2.1:2c, 7.7.2.1:3 to 6.7.2.1:6. Rainfall in Subsaharan West Africa is limited to the Northern summer, when the near-equatorial trough of low pressure and associated confluence axis and convergence band are relatively far poleward, and the moist cross-equatorial monsoon flow penetrates deep into the interior of the continent (Lamb, 1978a, b, 1983a; Hastenrath and Lamb, 1977a, b; 1978b). Near the Gulf of Guinea to the South, the annual rainfall regime exhibits two peaks, early and late in the summer half-year (Fig. 6.8.2:1), related to the poleward and equatorward migrations of the near-equatorial trough. In the North, a single rainfall maximum around July–August is found, and as one proceeds closer to the fringe of the desert this becomes of ever shorter duration. Accordingly, the circulation conditions at the height of the Northern summer are of particular interest.

The interannual variability of hydrometeorological conditions in Subsaharan Africa is illustrated in the time series plots in Fig. 8.8:1. These include a rainfall index compiled as the 'all-station average of normalized departure' of April–October precipitation totals of 14–20 stations West of 9°E (Lamb, 1983a). Refer also to Lamb (1985) for an updated index series for 1941–84. As long-term rainfall stations are rare, this index only begins in 1941. Also plotted are the water level records of the Senegal and Niger rivers and the annual change in water level of Lake Chad, all expressed in standard deviation. These also represent the hydrometeorological conditions of large catchments. Although rainfall anomalies vary in both space and time, the time series plots in Fig. 8.8:1 bear out considerable spatial coherence. Collectively, these plots show the tendency for runs of deficient and abundant rainfall years. Moreover, they indicate extreme drought at the beginning of the century, in the 1910's, and around 1940, favorable hydrometeorological conditions in the 1950's to early 1960's, and then drought beginning in the late 1960's and persisting to the present (mid 1980's).

A series of papers by Lamb (1978a, b, 1979, 1980, 1982a, b, 1983a, b) serve as major basis for the present section. The episode of droughts beginning in the late 1960's and affecting, in varying degree (Motha et al., 1980), the Subsaharan zone of Africa from the Atlantic Ocean to the Red Sea, received unprecedented publicity and aroused curiosity into the general circulation causes of this natural diaster.

Hypotheses included cooling of the Northern hemisphere extratropical cap, causing an expansion of the circumpolar vortex and an equatorward displacement of the subtropical high pressure belt, which would in turn affect the tropical circulation systems controlling Sahel rainfall (Bryson, 1973; Winstanley, 1973a, b). Angell and Korshover (1974) also suggest that a southward displacement of the Azores high since the 1940's may be related to the Subsaharan drought. Following the works of Smagorinsky (1963), Flohn (1964, 1965), and Henning (1967), Bryson (1973) discusses the implications of a 'Z criterion' which specifies the latitude of subtropical anticyclones in terms of meridional temperature gradient and lapse rate. By more remote inference the latitude of the subtropical anticyclones is related to that of the Intertropical Discontinuity (Section 6.7.2). With reference to the 'Z criterion' Bryson (1973) suggests that man-made increase of carbon dioxide and air pollution in the atmosphere of the Northern hemisphere extratropical cap may be responsible for the Sahel drought. These arguments were reconsidered by Beer et al. (1977) and Greenhut (1977). Greenhut (1977) concludes that increased carbon dioxide and particulate

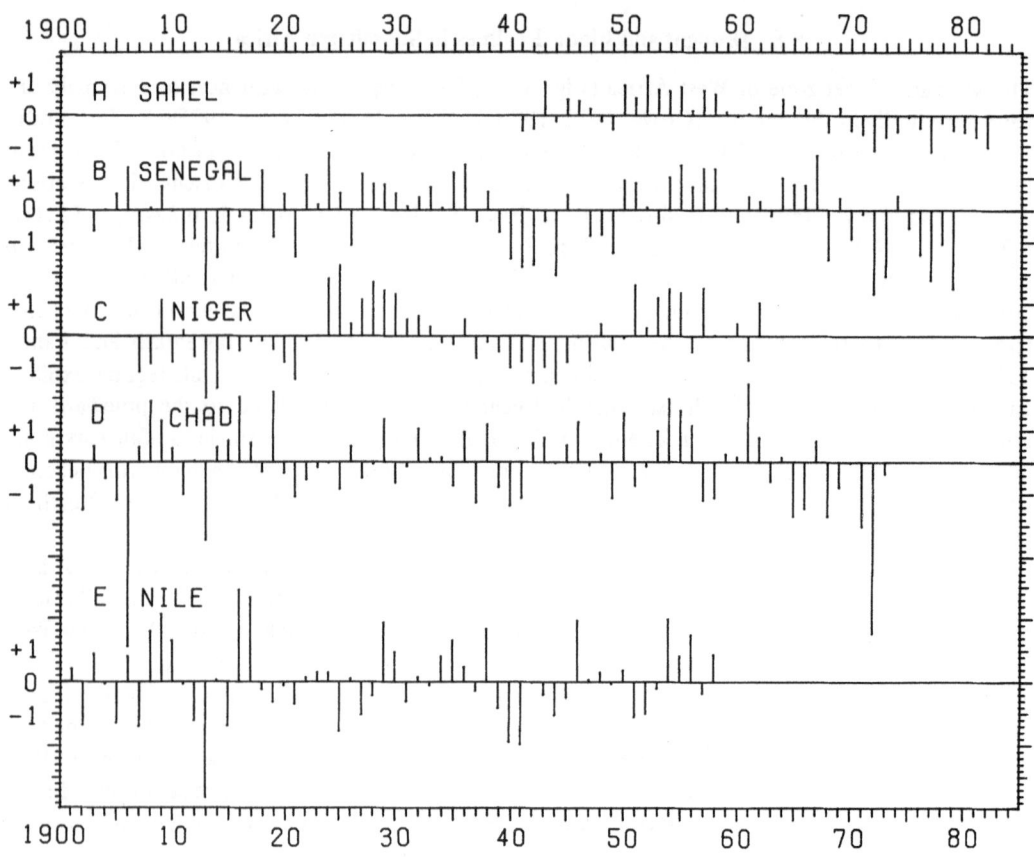

Fig. 8.8:1. Time series plots of hydrometeorological indices for Subsaharan Africa. (a) SAHEL, all-station average of normalized departures of April–October rainfall for 14–20 stations West of 9°E (Source: Lamb, 1983b; base period 1941–82). (b) SENEGAL, normalized variations of river discharge at Dagana (base period 1903–64, σ = 156 m^3 s^{-1}; (c) NIGER, normalized variations of river discharge at Koulikoro (base period 1907–57, σ = 368 m^3 s^{-1}); (d) CHAD, normalized variations of annual changes in lake level at Bol (base period 1921–70, σ = 29 cm year^{-1}); (e) NILE, normalized variations of Nile discharge at Aswan (base period 1921–50, σ = 960 m^3 s^{-1}).

matter would have an effect opposite to the one proposed by Bryson (1973). Moreover, these suggested relationships are not supported by various empirical studies (Miles and Follard, 1974; Tanaka *et al.*, 1975; Bunting *et al.*, 1976; Nicholson, 1979, 1980a, b, 1981). Kraus (1977a, b) considered the energy demands of the wintertime Southern hemisphere as a factor for the latitude position of the Intertropical Convergence Zone (Section 6.7.2).

Apart from the aforementioned speculations on the role of the Northern hemisphere extratropics, Charney (1975) and Charney *et al.* (1977) advanced an hypothesis whereby Subsaharan drought may be initiated or reinforced largely from within the region itself. Charney's hypothesis of a 'biogeophysical feedback mechanism' (ref. Section 10.2) considers that deterioration of the vegetation cover increases both the reflection of shortwave radiation and the longwave radiation emitted by the earth surface. This results in a reduction of surface net allwave radiation and hence of the heat transfer to the overlying atmosphere. The heat budget of the atmospheric column then implies decreased ascending motion or enhanced subsidence, which would be associated with reduced cloudiness and precipitation. The decreased rainfall would further hamper the plant cover, and so on. Numerical modelling experiments (Charney *et al.*, 1975, 1976, 1977) suggest this hypothesis. Other theoretical studies on albedo feedback effects include Otterman (1974), Ellsaesser *et al.* (1976), Berkofsky (1977), and Sud and Fennessy (1982). As candidate causes for vegetation depletion, Lamb (1979) mentions man-induced overgrazing, some natural biological disruption of the regional ecosystem, and sand transport from the Sahara associated with changes in the wind field. Reservations on the effectiveness of the biogeophysical feedback mechanism have been raised by Ripley (1976a, b).

The above review of hypotheses for extratropical forcing and *in situ* causes of Sahel drought demonstrates the need for comprehensive empirical analyses of the circulation features in the West African — tropical Atlantic sector characteristic of Subsaharan rainfall anomalies. A useful step in this direction are Kidson's (1977) August maps of 850 and 200 mb flow over Africa during the wet years 1959 and 1961, and the dry years 1972 and 1973. For the dry years, Kidson (1977) finds a virtual disappearance of the 850 mb trough near 8°N (ref. Fig. 6.1:4b) and weakening of the Tropical Easterly Jet (Section 6.2.3) above it. The latter feature is confirmed by Kanamitsu and Krishnamurti (1978) for the drought year 1972. Newell and Kidson's (1984) results suggest a stronger West African Mid-tropospheric Jet (Section 6.2.4) during the period of dry years 1970–73 as compared to the composite of wet years 1958–62. Lamb (1978a, b) analyzed the surface circulation in the tropical Atlantic for composites of WET (1943, 50, 52, 54, 57) and DRY (1942, 49, 70, 71, 72) years, as well as for individual extreme years (1967 wet; 1968 dry), giving particular attention to the seasonal evolution of departure patterns. These are complemented (Hastenrath, 1984) by analyses for partly different composites (1943, 50, 52, 54, 55, 57 for WET, and 1941, 42, 44, 68, 70, 71, 72 for DRY), but limited to the July/August peak of the Subsaharan rainy season. Results from the latter study are presented in Fig. 8.8:2.

Fig. 8.8:2 shows departure patterns of the WET minus those of the DRY composite, a plausible procedure since these patterns tend to be broadly inverse. The prominent patterns thus appear enhanced and can be interpreted as reflecting the conditions characteristic of an abundant rainy season. The essential features are as follows. The near-equatorial trough of low pressure (Fig. 8.8:2a), the associated confluence between the Northern and Southern hemispheric trade wind air streams (Figs. 8.8:2b, c), and the band of maximum convergence (Fig. 8.8:2e), are displaced northward, and the surface waters of much of the tropical North Atlantic are comparatively warm (Fig. 8.8:2d). The *t*-test shows statistical significance in the realm of positive sea surface temperature departure (Fig. 8.8:2d) and in the negative pressure area (Fig. 8.8:2a) in the North Atlantic.

The departures in the pressure and wind fields are internally consistent in indicating a northward displacement of quasi-permanent circulation features over the tropical Atlantic. SST departures are prominent, but their origin is not yet understood. The substantive departure features first pointed out by Lamb (1978a, b) were later duplicated by other analysts (review in Lamb, 1982b; Bah, 1985; Lough, 1986). The statistical test supports the prominent characteristics of the SST maps (Fig. 8.8:2d), but only to a limited extent the other internally consistent fields. In brief, the generation circulation mechanisms of interannual rainfall variability illustrated by Fig. 8.8:2 appear as described above for the annual cycle.

The map of divergence, Fig. 8.8:2e, is complemented by the meridional profiles Fig. 8.8:3 (Hastenrath and Lamb, 1978b; Hastenrath, 1984). During the Northern summer rainy season, the convergence band assumes a far poleward position and a zone of distinct divergence is found between the Equator and about 5°N, while in the winter dry season the convergence band is

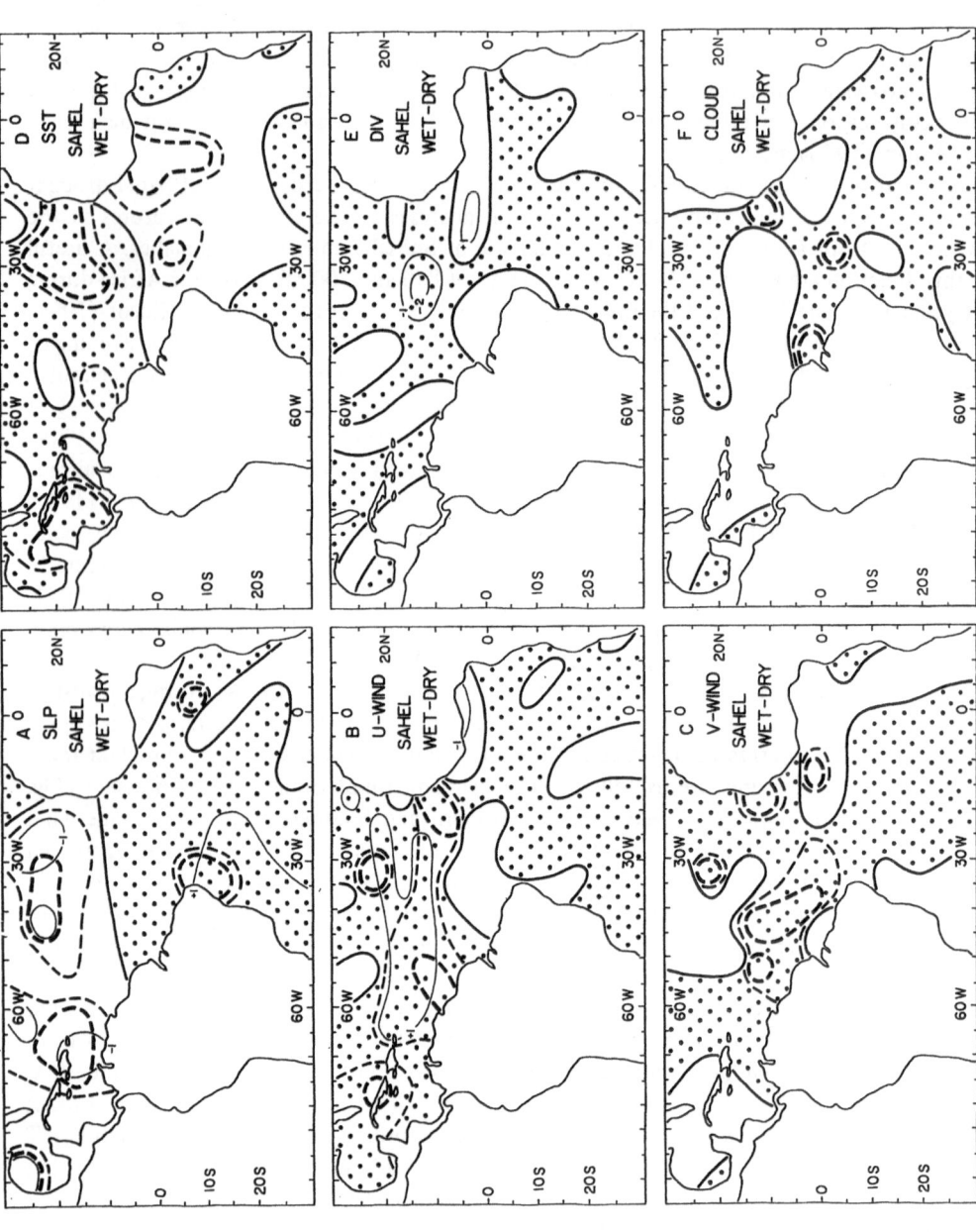

Fig. 8.8:2. Maps of July/August circulation departures associated with rainfall anomalies in Subsaharan Africa. Charts represent the difference of departures for the composite of WET years (1943, 50, 52, 54, 55, 57) minus the departures for the composite of DRY years (1941, 42, 44, 68, 70, 71, 72). (a) sea level pressure SLP in mb; (b) zonal (u) and (c) meridional (v) components of wind, in m s^{-1}. (d) sea surface temperature SST in °C; (e) divergence DIV in 10^{-6} s^{-1}; (f) total cloudiness CLOUD in tenths. Dot raster denotes positive areas, except for convergence in chart e. Areas significant at the 5 and 10 percent levels are enclosed, respectively, by heavy and thin broken lines; (e) being excepted from testing. From Hastenrath (1984).

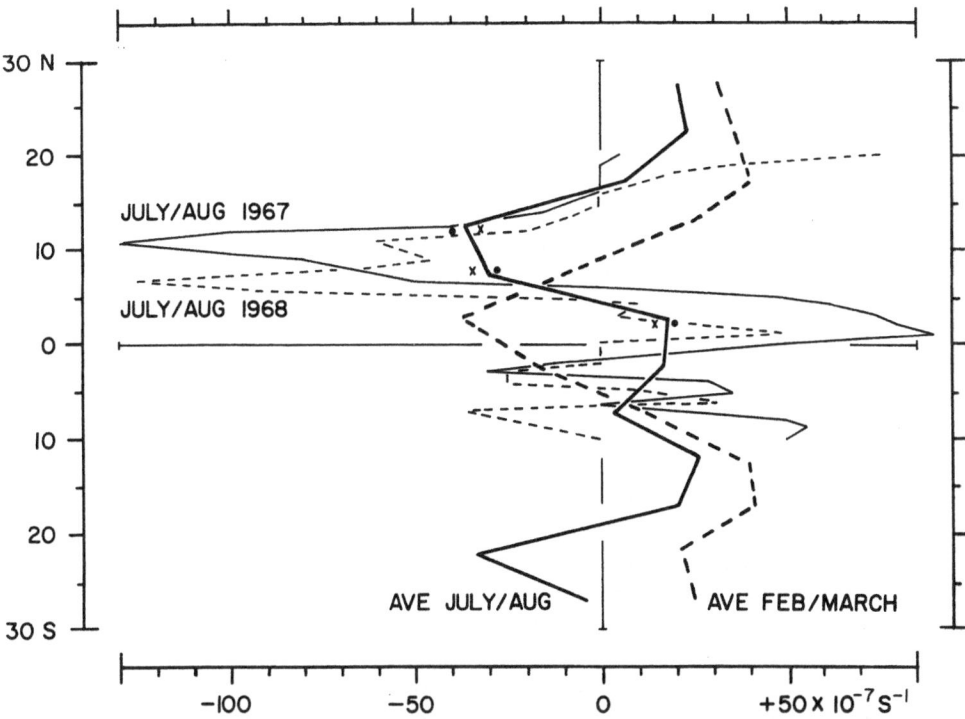

Fig. 8.8:3. Meridional profiles of divergence (10^{-7} s^{-1}) in the strip 10–30°W. The 60 year means of July/August and February/March are shown as heavy solid and broken lines, respectively. Thin solid lines refer to the abundant rainfall year 1967, and thin broken lines to the extreme drought year 1968. In the latitude bands 0–5, 5–10, and 10–15°N, the values for the composites of 6 WET years and of 7 DRY years are denoted by dots and crosses, respectively. From Hastenrath (1984), and Hastenrath and Lamb (1978b).

displaced southward under elimination of the near-equatorial divergence belt. In similarity to these annual cycle characteristics, the profile for the extreme drought year 1968 is characterized by a comparatively far equatorward position of the convergence band, while the moderately wet year 1967 features a poleward location of the convergence band along with a well-developed near-equatorial divergence zone. The composites of DRY and WET years depart from the 60 year mean July/August profile in the same sense as just described for the individual extreme years 1968 (dry) and 1967 (wet), but with much smaller amplitude. The features for the latter two profiles are more conspicuous in part because they are drawn with a one degree latitude resolution, whereas values for five degree strips are plotted for the DRY and WET composites and the 60 year average. Divergence is excluded from statistical significance testing for reasons explained above.

Complementing Fig. 8.8:2, the meridional profiles in Fig. 8.8:3 illustrate that the divergence departures during dry/wet years appear as reductions/enhancements of the average annual cycle.

Lamb's (1978a, b) analyses reveal further important details of large-scale circulation anomalies. During DRY, the near-equatorial low pressure trough, confluence axis, and the sea surface temperature maximum were all displaced 200–300 km South of their 60-year mean location. At the same time, the North Atlantic high extended further equatorward than average although its center was some 150 km further poleward, that is displaced in a sense opposite to trough and confluence axis. However, during the extreme drought year 1968 the center of the North Atlantic high was located up to 500 km further South than average. These findings are consistent with the results of Tanaka et al. (1975). Comparing the extreme drought year 1968 with the wetter 1967, locations some 300–500 km further South are found for band of maximum sea surface temperature, near-equatorial trough of low pressure, confluence axis, and zones of maximum convergence, cloudiness, and rainfall frequency. Also, all of these features were found distinctly further South during 1968 than in the DRY composite.

The evolution of anomaly patterns is of particular interest. Lamb (1978a, b) found that many of the departures characteristic of the atmospheric and oceanic fields as documented for the height of the Northern summer evolve gradually in the course of the preceding half-year. These features of the DRY composite include the enhanced trades during the preceding January–June, the southward displacement of the confluence axis during April–June, positive sea surface temperature anomalies South of 10°N in January–June, as well as a more poleward position of the center of the North Atlantic high during most of January–June. However, various details of anomaly patterns in the preceding half-year differ from conditions at the height of the rainy season: there are positive sea surface temperature departures also between 10–20°N, the zone of maximum sea surface temperature appears expanded both northward and southward, and divergence, rainfall, and cloudiness fields in the equatorial Atlantic differ little from average. Moreover, during April–June of the DRY composite the near-equatorial low pressure trough appears somewhat poleward, rather than equatorward, of its mean position. Compared to the DRY composite, most departures of 1968 evolved earlier in the year.

For the WET composite Lamb (1978b) finds less pronounced anomaly patterns than for the DRY counterpart, the more important departures being as follows. The center of the North Atlantic high is located some 100–150 km further poleward, and its southern flank extends less equatorward than average, the confluence axis appears some 100–150 km further poleward, and the trades are rather weaker. For the individual wet year 1967, anomalies are more pronounced than for the WET composite. Little pre-season evolution of anomaly patterns is apparent for the WET composite.

These analyses show that the droughts and floods of Subsaharan Africa are associated with distinct but small anomalies in the large-scale circulation. In fact, displacements of the surface confluence axis of a few 100 km can be ascertained only with a fine resolution analysis. These small meridional displacements of the quasi-permanent circulation components should, however, be seen in perspective with the steep annual rainfall gradients which characterize the Subsaharan zone of Africa. Values of order 180 mm a^{-1} per degree latitude are, for example, suggested by Ilesanmi (1971) and Greenhut (1977). A meridional displacement of the surface confluence axis of, say 200 km – which would escape any coarse-mesh analysis – would thus translate into the drastic rainfall anomaly of about 400 mm a^{-1}. Motha et al. (1980) substantiate the impact of displacements of order 200 km for the annual water budget of Subsaharan Africa. Clearly, interannual variations in the location of the Intertropical Discontinuity (Section 6.7.2) cannot be appreciated from the sole evaluation of precipitation data.

A tendency has been noted (Hookey, 1970; Tanaka *et al.*, 1975; Lamb, 1978a; Motha *et al.*, 1980) for inverse precipitation anomalies in the Sahel and more southerly regions closer to the Gulf of Guinea. Such contrasts appear consistent with the meridional divergence profiles in Fig. 8.8:3, but this pattern does not seem to occur exclusively. Apart from numerous years with inverse precipitation anomalies in the Sahel and the subhumid zone to the South, Motha *et al.* (1980) also identified years with a widespread reduction of rainfall throughout the entire region. Motha *et al.* (1980) further document both some diversity of spatial anomaly patterns and their large zonal extent.

An important consideration for the water budget of anomalous rainfall years is the water vapor transport by the summer Southwest monsoon (Section 6.7.2.3, 6.8.2). Using all existing radiosonde observations, Lamb (1983a) evaluated the water vapor flux pattern during July, August, and September of the drought years 1968, 1971, 1972, and of the nondrought years 1967, 1969, 1975. The main results of this investigation are as follows. Inflow of unusually dry surface air from the tropical Atlantic is not an important factor for Subsaharan drought. In fact, the specific humidity of the boundary layer flow across the African coast tends to be higher than average during the drought years. However, the extreme 1972 drought year featured a particularly shallow Southwest monsoon flow across the coast, while thicker monsoon layers were characteristic of the less severe 1968 and 1971 dry years, although these features changed further inland. At Dakar, in a sensitive location, the advective water vapor flux during deficient rainy seasons tended to be North of West, whereas westerly to southwesterly directions prevailed in the more nearly normal rainfall years. Other thermodynamic factors of possible importance for anomalous rainy seasons have been considered by Adedokun (1981, 1982).

At the beginning of this section and with reference to Fig. 8.8:1 it was pointed out that drought regimes in the Sahel occurred repeatedly during this century, and that they showed a tendency to persist for a series of years. While the biogeophysical feedback mechanism (Charney, 1975) may be instrumental in ensuring such a persistence, it is noteworthy that Courel *et al.* (1984) document a substantial decrease of albedo in the Sahel zone from 1973 to 1979, even though concurrently the rainfall stayed low (Fig. 8.8:1). In fact, Lamb (1982a, 1983b) called attention to the continuation of the drought regime which began in the late 1960's, and stated the intention to continue the monitoring of annual rainfall in the Sahel. Without referring to Lamb (1982a), Nicholson (1983) later also mentioned the continuation of drought conditions. In addition to Fig. 8.8:1, the long-term nature and persistence of anomalous rainfall regimes in the Sahel are noted in various papers (Wood and Lovett, 1974; Landsberg, 1975a, b; Palutikof *et al.*, 1981; Faure and Gac, 1981a, b). The positive correlation with rainfall variations in the Central American — Caribbean region results primarily from variations at the long time scales (Hastenrath, 1976; Hastenrath and Kaczmarczyk, 1981). There appears to be a considerable temptation to extrapolate quasi-periodic variations in the past record for prognostic application (Winstanley, 1973a; Wood and Lovett, 1974; Faure and Gac, 1981a, b). Prospects are also appearing (Rasool, 1982) for seasonal forecasts based on an understanding of atmosphere-ocean processes. The possibility of climate prediction for the Sahel will be considered in Section 9.5, and the human impact of the natural disasters in Sections 10.2 and 10.8.

8.9. Climate Anomalies at the Angola Coast

Stretching as it does along the eastern extremity of the tropical Atlantic, the Angola littoral (Fig. 8.6:1) appears an obvious corollary to the El Niño (Section 8.2) region of the South American West coast. A tendency for abundant rainfall years to coincide with anomalously warm coastal waters has in fact long been noted locally. Moreover, various recent studies (Moore *et al.*, 1978; Adamec and O'Brien, 1978; O'Brien *et al.*, 1978; Hisard, 1980; Servain *et al.*, 1982; Picaut, 1983; Busalacchi and Picaut, 1983; McCreary *et al.*, 1984) focused on remote forcing and response relationships in the Atlantic that would be akin to mechanisms operative during the Pacific El Niño phenomenon (Section 8.2). The present section is primarily based on Hirst and Hastenrath (1983a) and Hastenrath (1984). For the background circulation and climate, useful references include Jackson (1964), Thompson (1965), Atkinson and Sadler (1970), Hastenrath and Lamb (1977a, 1978a).

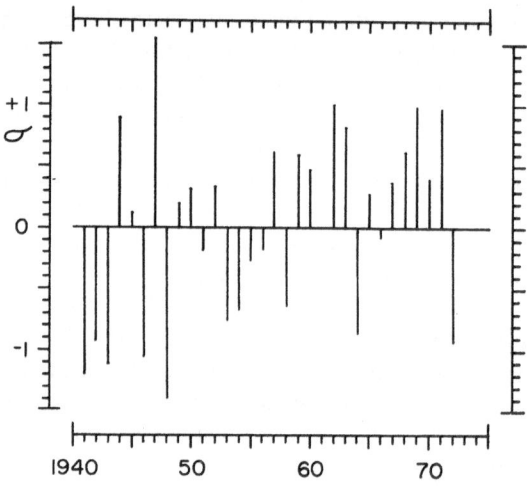

Fig. 8.9:1. Time series plot of Angola coast rainfall index, in σ. From Hirst and Hastenrath (1983a).

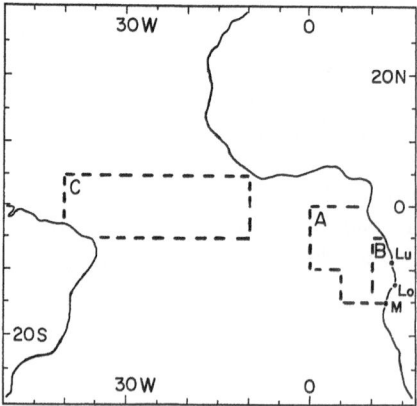

Fig. 8.9:2. Orientation map. Atlantic areas A, B, C, are outlined by broken lines. Luanda, L, Lobito Lo, and Mocâmedes M, are coastal rainfall stations. Source: Hirst and Hastenrath (1983a).

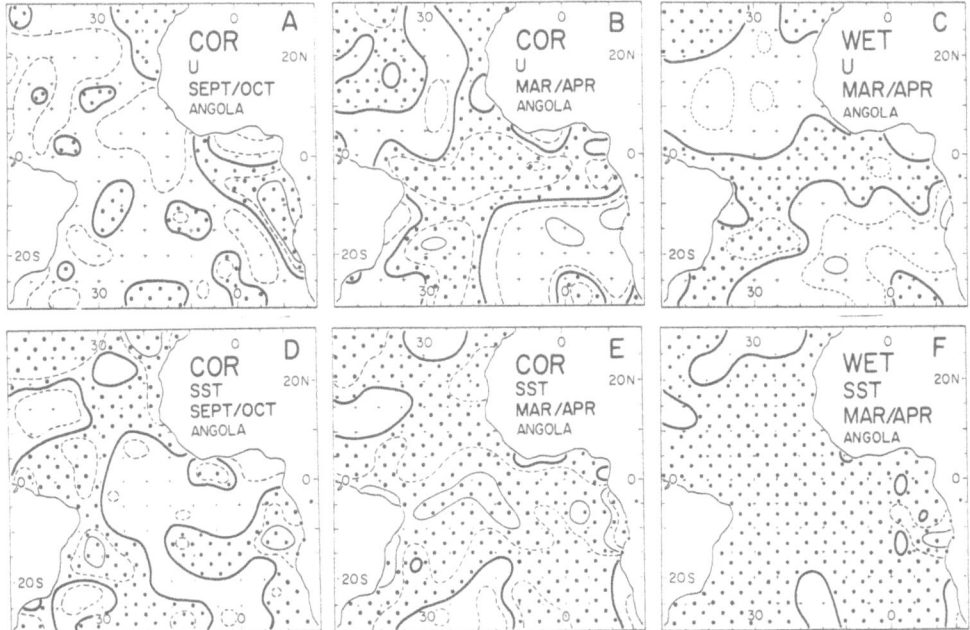

Fig. 8.9:3. Maps of relation between (i) zonal wind speed (panels A, B, C) and sea surface temperature (panels D, E, F) in the tropical Atlantic and (ii) Angola rainfall. Panels A and D are isocorrelate maps of fields during September/October preceding the March/April rainy season; panels B and E show the concurrent March/April correlation; and panels C and F depict departure patterns for the composite of WET years (1957, 62, 63, 69, 71). Zero lines are heavy, dot raster denotes positive areas. Isocorrelates are at intervals of 0.2. For the stratification maps C and F, isolines are in m s^{-1} and °C, respectively. Source: Hirst and Hastenrath (1983a).

The rainy season on the Angola coast is narrowly concentrated in March/April, when the surface waters of the Eastern South Atlantic are warmest, thus favoring the instability and moisture content of the boundary-layer flow into the continent. The interannual climate variability at the Angola coast is illustrated by the time series plot of a rainfall index in Fig. 8.9:1. This index is the three-station average of normalized departure of annual (July–June) rainfall compiled as described in Section 8.3. Within the 1948–72 interval, during which ship observations are more plentiful than in the earlier years of the plot, the most extreme years are identified from Fig. 8.9:1 as follows: 1948 (i.e. July 1947 to June 1948), 53, 54, 58, 64, 72, as DRY; and 1957, 62, 63, 69, 71 as WET.

Hirst and Hastenrath (1983a) find that wind-stress relaxation over the remote western equatorial Atlantic provides a major forcing for sea surface temperature variations off Angola (area A in Fig. 8.9:2), which in turn affect the moisture content and instability of the boundary layer flow and hence the coastal rainfall. This mechanism is considered part of the 'normal' annual cycle, but appears extended or reduced in years with anomalous sea surface temperature and rainfall.

The maps Fig. 8.9:3 illustrate the relationships between zonal wind and sea surface temperature patterns in the tropical Atlantic and Angola rainfall. In the September/October preceding the March/April rainy season (Figs. 8.9:3a and d), enhanced easterlies over much of the central and western equatorial Atlantic presage abundant rainfall to come (Fig. 8.9:3a), while the isocorrelate pattern for sea surface temperature (Fig. 8.9:3d) is not distinct. In the concurrent March/April maps (Fig. 8.9:3b and e), patterns appear reversed: weak easterlies in the equatorial Atlantic

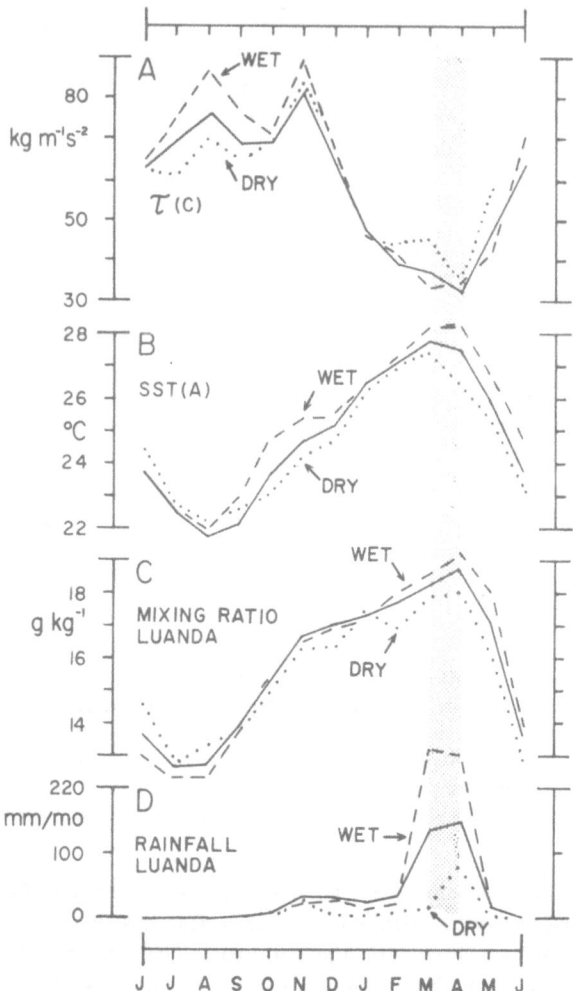

Fig. 8.9:4. Annual march of (A) westward wind stress in area (B) SST in area A, (C) surface mixing ratio at Luanda, (D) rainfall at Luanda (ref. Fig. 8.9:2). Solid, dashed and dotted lines refer to the 1948–72 average, and the WET and DRY components, respectively. From Hirst and Hastenrath (1983a).

(Fig. 8.9:3b) are concurrent with abundant rainfall and warm surface waters in much of the tropical Atlantic including the area off Angola (Fig. 8.9:3e). The stratification maps for the March/April rainy season (Figs. 8.9:3c and f) corroborate the content of the correlation maps (Figs. 8.9:3b and e).

Figs. 8.9:3a versus 8.9:3b and c already identify seasonal wind stress relaxation over the western equatorial Atlantic as a characteristic of abundant Angola rainy seasons. Fig. 8.9:4 further illustrates the pertinent relationships. In panel A of Fig. 8.9:4 the annual march of equatorial wind stress (area C) appears amplified during the WET and reduced for the DRY years. Likewise, sea surface temperature anomalies off Angola (part B of Fig. 8.9:4) during the wet and DRY composites appear as enhancements and suppressions of the average annual cycle. Similar differences are apparent for the curves of atmospheric humidity and rainfall at the coast (parts C and D of Fig. 8.9:4). Thus Fig. 8.9:4 supports the notion of remote wind stress forcing of sea surface

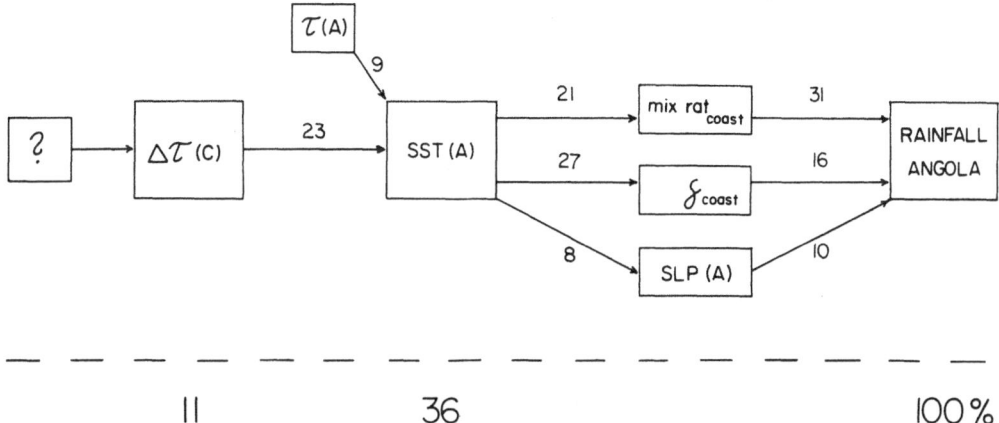

Fig. 8.9:5. Causality chain of atmospheric-oceanic anomalies. $\Delta\tau$ (C) is SON−FMA difference of westward wind stress in area C (Fig. 8.9:2), τ (A) is total wind stress in area A (Fig. 8.9:2), SST (A) is sea surface temperature in area A, mix rat$_{coast}$ the mixing ratio at Luanda, γ_{coast} the surface minus 700 mb temperature difference at Luanda and SLP (A) is sea level pressure in area A. Numbers next to the arrows signify percentage of variance in the downstream element explained by the upstream element. Numbers at bottom of graph are percentage of variance of Angola coast rainfall explained by element in box above. From Hirst and Hastenrath (1983a).

temperature off the Angola coast, and the effect of the latter on the moisture and instability of the boundary layer flow, and hence coastal rainfall.

Fig. 8.9:5 illustrates a causality chain. Whatever the causes of anomalous seasonal wind stress relaxation in the western equatorial Atlantic, this makes for 23 percent of the sea surface temperature variance off the Angola coast, while the local wind stress forcing is less important. The sea surface temperature variations over the eastern South Atlantic in turn, presumably through moisture content and instability, accounts statistically for 36 percent of the rainfall variability along the Angola coast. Thus the remote wind stress forcing appears as an important contributor for the interannual variability of Angola climate, although the major part of the variance remains to be explained from other factors.

8.10. Hydrometeorological Anomalies in the Zaïre (Congo) Basin

Investigations into climate diagnostics of the equatorial rainforest regions have been hampered by the deficient stations network. However, long-term observations of water level near the mouth of the Zaïre river are a valuable indicator of the hydrometeorological conditions over this equatorial river basin as a whole. The present section is based primarily on Hirst and Hastenrath (1983b). For the background circulation and climate reference is also made to Jackson (1964), Thompson (1965), Flohn (1960), Taljaard et al. (1969), Atkinson and Sadler (1970), Bultot and Griffiths (1972), Torrance (1972), Newton (1972), Hastenrath and Lamb (1977, 1978a), and Trewartha (1981, pp. 125−133).

The Zaïre catchment (Fig. 8.6:1) is throughout the year affected by a rain zone associated with the equatorial trough system. In January, this zone lies over the southern extremity of the catchment and little rain falls North of the Equator. Thereafter, the rain zone migrates northward, lying across the center of the Zaïre catchment in April and North of it by July, when rains are scarce South of the Equator. On its southward sweep, the rain zone passes again over the center of the Zaïre catchment in November. As a result, the annual march of rainfall over the Zaïre basin exhibits two peaks around March/April and October/November, the latter being the more pronounced. Accordingly, two semesters or rainy seasons are distinguished: February to July and August to January. The annual march of the Zaïre river level at Kinshasa lags the catchment average rainfall by 1–2 months.

The moist westerly flow from the eastern South Atlantic is the main moisture source for precipitation in the Zaïre catchment throughout the year. Therefore, anomalous climatic conditions over the eastern equatorial and South Atlantic can be expected to contribute to extreme rainfall conditions over the Zaïre catchment. Furthermore, a variety of processes originating over equatorial Africa, as well as frontal influences from higher latitudes of the Southern hemisphere, may affect rainfall in the Zaïre basin.

Rainfall over the Zaïre catchment in the first half of the year is in fact positively related to sea surface temperature over the eastern tropical Atlantic, and to atmospheric humidity and instability at the coast. However, lapse rate is also strongly controlled by upper-air conditions, in that cold air aloft may enhance instability and thus rainfall. As discussed in Section 8.9, sea surface temperature conditions off Angola are remotely forced by wind stress over the western equatorial Atlantic. This is not the case in the second semester. Equatorial southwesterlies from the Atlantic into the interior of the Zaïre catchment are strong/weak during second semesters of high/low water level. Rainfall over the Zaïre basin also appears modulated by a zonal circulation along the Atlantic Equator.

8.11. Time Scales of Climate Variability

The reviews in the preceding sections of this chapter exemplify that extreme events in the regional rainfall conditions and the large-scale circulation are not exceptions, but an intrinsic part of tropical climate. Much attention has been given to the search for distinct and statistically significant periodicities (for example, Wood and Lovett, 1974; Jones and Kearns, 1976; Schickedanz and Bowen, 1977; Jagannathan and Bhalme, 1973; Faure and Gac, 1981a, b; Palutikof et al., 1981; and others), in part motivated by the desire for long-range prediction through spectral extrapolation. While few climate variations may be strictly periodic, preferred time scales of variability stand out, and are mentioned in various of the preceding sections.

A preference for variability at somewhat more than 2 years is apparent in numerous climatic elements and areas (Trenberth, 1980), for example India rainfall (Section 8.5; Bhargava and Bansal, 1969; Koteswaram and Alvi, 1969, 1970; Hastenrath and Rosen, 1983), the circulation over the equatorial Atlantic and the precipitation in Northeast Brazil (Section 8.6; Hastenrath and Kaczmarczyk, 1981; Kousky and Chu, 1978; Chu, 1984), the frequency of Atlantic hurricanes and the position of the North Atlantic high (Hastenrath and Wendland, 1979; Shapiro, 1982), and rainfall in various parts of Africa (Tyson et al., 1975; Ogallo, 1979). Brier (1978) has demonstrated in very general terms how quasi-biennial variations can arise from a combination of positive and negative feedbacks reversing between seasons. Wu (1984) substantiated such a possibility for the domain of India and Tibet. Concurrently, quite independent quasi-biennial feedback systems may be operative in different regions of the tropics. For the Brazil – tropical Atlantic sector it was possible to show (Hastenrath and Kaczmarczyk, 1981; Chu, 1984) that the surface circulation and rainfall vary plausibly in unison at the quasi-biennial time scale.

Spectral peaks around 3–5 years have been reported for the rainfall records of various tropical to subtropical regions (Tyson et al., 1975; Rodhe and Virji, 1976; Kousky and Chu, 1978; Ogallo, 1979), without reference to corresponding variations in the large-scale circulation.

The Southern Oscillation (ref. Section 8.1; Philander, 1983) is manifest in the pressure, wind, and sea surface temperature fields, and rainfall, of many tropical regions, is markedly aperiodic, and operates over the range of about 2 to 10 years.

In the rainfall of Northern Northeast Brazil and in the circulation over the adjacent equatorial Atlantic (Section 8.6) a marked preference for variability around 13–14 years has been found (Markham, 1974; Jones and Kearns, 1976, Hastenrath and Kaczmarczyk, 1981; Chu, 1984), phase relationships between the rainfall and circulation variations being plausibly consistent. A concentration of spectral power in rainfall is further apparent around double that period, namely around 27–28 years (Markham, 1974).

Variability concentrated at the scale of 2 to 3 decades is characteristic for various regions in the outer tropics to subtropics around the Atlantic. This is the case for the Central American – Caribbean area (Hastenrath, 1970, 1976; Hastenrath and Kaczmarczyk, 1981) and Subsaharan Africa (Hastenrath and Kaczmarczyk, 1981; Faure and Gac, 1981a, b) in the Northern hemisphere. A similar long time scale is indicated for the variability of rainfall over Southern Africa and the upper-air circulation in this sector of the Southern hemisphere (Tyson and Dyer, 1978; Tyson, 1981, 1984, Miron and Tyson, 1984), as well as for the rainfall variability over subtropical South America (Compagnucci and Vargas, 1983). Although spectral peaks in these regions are not identical, the persistence of drought and flood regimes indicated in these long time scales is remarkable. A 'biogeophysical feedback mechanism' involving surface albedo variations (Section 8.8), as proposed by Charney (1975), appears a plausible candidate for ensuring the longevity of departure regimes in the semi-arid outer tropics to subtropics.

It is noteworthy that various preferred time scales of interannual variability coexist in the tropical Atlantic and surrounding land areas: a quasi-biennial oscillation and variations around 13–14 years in the equatorial Atlantic and Northeast Brazil, and variations on the scale of 2–3 decades in the outer tropics to subtropics of the Americas and Africa and both hemispheres. All of these 'circum-Atlantic' characteristics differ from the El Niño Southern Oscillation time scale of 2–10 years. The causes for these prominent and diverse preferences are not yet understood.

8.12. Synthesis

Marked interannual variations of the atmosphere-ocean system are an essential characteristic of tropical climate. Instrumental in various regional climate anomalies is the surface pressure seesaw of the Southern Oscillation, on a time scale of 2–10 years, and with dipoles over the Eastern South Pacific and the greater Indonesian – Australasian region, but spanning the global tropics. Although these large-scale long-term pressure variations have been studied since the latter part of the last century, their causal relation to the El Niño phenomenon on the West coast of South America has been recognized only since the 1970's.

During the positive phase of the Southern Oscillation, both the Eastern South Pacific high and the Indonesia low are strongly developed, vigorous easterly trade winds sweep the vast expanses of the equatorial Pacific piling up waters at its western extremity. Accordingly, the eastward slope of the free ocean surface, the westward deepening of the oceanic mixed layer, and the Equatorial Undercurrent are pronounced, and waters are particularly cold off the South American West coast and in an extended zone immediately to the South of the Equator stretching from the coast of the Americas far into the open Pacific. Within the atmosphere, the Walker circulation along the Pacific Equator is strong, featuring not only vigorous surface easterlies but also an enhanced westerly return flow aloft, as well as pronounced convection and ascending motion over the Indonesian dipole and marked subsidence over the East Pacific dipole.

During the negative phase of the Southern Oscillation, the atmosphere-ocean system in the Pacific operates in a remarkably different mode. Both the Eastern South Pacific high and the Indonesia low are weak, the slackened zonal pressure gradient entails weaker surface easterly winds in the equatorial zone, and accordingly the zonal slopes of the free ocean surface, of constant pressure topographies at depth, and of the thermocline, diminish, the Equatorial Undercurrent slows down, and may surface or vanish altogether. The relaxation of surface wind stress incites equatorial Kelvin waves which travel to the eastern extremity of the Pacific within 2–3 months, where they are manifest in a warming of surface waters, with a maximum around the March/April peak in the annual march. The warm ocean and torrential rains at the otherwise desertic coast of Peru and Ecuador trigger an ecological catastrophe including the mass death of fish and Guano birds, floods, destruction of roads and houses, and loss of human life. Within the atmosphere, the Walker circulation along the Pacific Equator is weak, as manifest in the slackened trade winds and westerly return flow aloft, as well as reduced convection, rainfall, and ascending motion over the Indonesia dipole and lesser subsidence over the Eastern Pacific.

Rainfall anomalies in Indonesia, which has its rainy season peak around December/January, are thus closely tied into the Southern Oscillation. During its positive phase, surface pressure in the area is low, lower-tropospheric inflow and convergence pronounced, convection and ascending motion vigorous, and rainfall abundant. During the negative phase of the Southern Oscillation, the opposite departure characteristics of circulation and rainfall conditions prevail in the greater Indonesian – Australasian region.

Variations of the upper-air circulation are an essential part of the El Niño Southern Oscillation (ENSO) phenomenon. During the positive or cold phase, the troposphere is anomalously cold throughout the global tropics, so that upper-tropospheric topographies are low particularly in the lower latitudes, thus entailing departure easterlies and equatorward flow aloft. By contrast, during the negative or warm phase, the tropical troposphere is warm, upper-tropospheric topographies are inflated especially in the tropics, and departure westerlies and poleward flow prevail in the upper troposphere. These variations appear broadly in phase throughout the tropical belt, but the oceanic warming and cooling, and consequently the temperature variations in the overlying atmosphere, as well as the upper-tropospheric height variations, are most pronounced in the Eastern to Central Pacific. This allows for zonal flow departures in the upper troposphere over the equatorial zone of the Western Pacific to run essentially inverse to most of the remainder of the tropics. The aforementioned zonal flow variations in the upper troposphere over the Western Equatorial Pacific are an integral part of modulations in the Walker cell.

For the alternation between the El Niño and counter El Niño modes of operation not only ocean-atmosphere coupling but also upper-air processes appear essential, inasmuch as surface pressure variations reflect changes of mass in the atmospheric column. This in fact calls attention, in the first place, to altered inflow and outflow aloft, which requires changes in upper-tropospheric topography and therefore tropospheric mean temperature, with sea surface temperature change being a more remote link in the causality chain. In this context, negative feedback mechanisms involving the upper ocean merit particular attention.

India Southwest summer monsoon rainfall is associated with characteristic circulation departure patterns from the pre-monsoon through the post-monsoon seasons. Abundant rainfall is heralded by strong heat low development over the continent, strong surface wind and a warm Arabian Sea preceding the monsoon onset. The negative correlation with pressure and the positive coupling with wind persist to the post-monsoon season. Enhanced upper-tropospheric easterlies throughout the summer half-year are further indicative of a good monsoon year. Abundant rainfall years tend to coincide with the positive or cold phase of the Southern Oscillation.

A series of circum-Atlantic studies undertaken at the University of Wisconsin since the early 1970's, as summarized in Sections 8.6 to 8.10, has demonstrated the usefulness of long-term ship observations for investigating the mechanisms of tropical climate anomalies. Historical ship files have since become a standard data source in analyses of interannual climate variability in low latitudes.

The droughts of Northern Northeast Brazil, which has its rainy season mainly concentrated in March/April, are characterized by an anomalously far poleward position of the near-equatorial trough and embedded confluence axis and convergence band, positive sea surface temperature departures in the tropical North Atlantic, and anomalously cold waters in the South equatorial Atlantic. Surface circulation features conducive to drought include the distant position of the convergence band, the cold South equatorial waters, and resulting effects on moisture and instability of the boundary layer flow, and enhanced meridional temperature contrasts across the Equator which drive a thermally direct meridional circulation cell in the atmosphere, featuring subsidence over the Nordeste. Departure patterns in the large-scale atmospheric and ocean fields, are approximately inverse in the dry and wet years, and evolve during the half-year preceding the Nordeste rainy season, thus offering the prospect of climate prediction.

Rainfall variations during the Northern summer half-year in the Central American – Caribbean region are related to the position of the North Atlantic high. During drought years, the pressure on the equatorward side of the subtropical high is enhanced, the trades appear stronger albeit in a band further South, and most of the North Atlantic waters are anomalously cold. The approximately inverse departure configurations are characteristic of abundant rainfall years.

Droughts in Subsaharan Africa which experiences its rainfall at the height of the Northern summer are associated with anomalously far equatorward position of quasi-permanent circulation features, in particular the surface confluence axis and convergence band, and negative sea surface temperature anomalies are found in a broad band across the tropical North Atlantic. Abundant rainfall years in the Sahel show the approximately inverse departure characteristics.

Rainfall and sea surface temperature at the Angola coast shows antecedents of remote forcing by seasonal wind stress relaxation over the Western Equatorial Atlantic. Strong easterly wind stress relaxation in that remote region during the preceding months is associated with warm coastal waters and abundant rainfall around the March/April maxima in the annual cycle. Abundant/deficient rainy seasons appear as enhancements and reductions of the 'normal' annual cycle. A strong affinity between interannual variability and annual cycle mechanisms is also indicated in various other tropical regions.

The Zaïre (Congo) basin experiences two rainy season peaks, during the first and second semesters of the calendar year. Rainfall anomalies during the first semester appear related to mechanisms mentioned above for the Angola coast. The second semester rainfall is also related to sea surface temperature conditions, conceivably through moisture and instability of the boundary layer flow from the Atlantic into the interior of the continent.

Interannual variability appears concentrated in various preferred time scales. A quasi-biennial oscillation is apparent in India rainfall, in the circulation over the equatorial Atlantic and rainfall over Northeast Brazil, as well as in other elements and areas. Northeast Brazil rainfall and the equatorial Atlantic circulation further exhibit spectral power around 13–14 years. By contrast, interannual variability in various regions of the outer tropics to subtropics of the Americas and Africa (Central American – Caribbean area, Subsaharan Africa, Southern Africa, subtropical South America), is concentrated at the time scale of 2 to 3 decades. Positive feedback mechanisms may play a role in the persistence of anomaly regimes in these semi-arid to semi-humid regions. The coexistence of various preferred time scales of interannual variability in the Atlantic and surrounding continents is remarkable.

References

Adamec, D., O'Brien, J. J., 1978: 'The seasonal upwelling in the Gulf of Guinea due to remote forcing'. *J. Phys. Oceanogr.*, **8**, 1050–1060.

Adedokun, J. A., 1981: 'Potential instability and precipitation occurrence within an inter-tropical discontinuity environment'. *Archiv Meteor. Geophys. Bioklim., Ser. A*, **30**, 69–86.

Adedokun, J. A., 1982: 'On an instability index relevant to precipitation forecasting in West Africa'. *Archiv Meteor. Geophys. Bioklim., Ser. A*, **31**, 221–230.

Aldaz, L., 1983: 'Aplicación del concepto de interacción de escalas temporales al fenómeno de las sequías extremas en el Nordeste del Brazil'. Doctoral Dissertation, Universidad Complutense, Madrid, 329 pp.

Alexander, B., Keshavamurty, R. N., Mikhopadhyay, R. K., Bhosale, S. G., 1974: 'Pacific equatorial pressure gradient and monsoon rainfall'. *Nature*, **252**, 463–464.

Anderson, D. L. T., McCreary, J. P., 1985: 'Slowly propagating disturbances in a coupled ocean-atmosphere model.' *J. Atmos. Sci.*, **42**, 615–629.

Angell, J. K., 1981: 'Comparison of variations in atmospheric quantities with sea surface temperature variations in the Equatorial Eastern Pacific'. *Mon. Wea. Rev.*, **109**, 230–243.

Angell, J. K., Korshover, J., 1974: 'Quasi-biennial and long-term fluctuations in the centers of action'. *Mon. Wea. Rev.*, **102**, 669–678.

Angell, J. K., Korshover, J., 1984: 'Some long-term relations between equatorial sea-surface temperature, the four centers of action and 700 mb flow'. *J. Climate Appl. Meteor.*, **23**, 1326–1332.

Arkin, P. A., 1984: 'An examination of the Southern Oscillation in the upper troposphere tropical and subtropical wind field'. Ph.D. Dissertation, University of Maryland, 240 pp.

Atkinson, G. D., Sadler, J. C., 1970: 'Mean cloudiness and gradient-level wind charts over the tropics'. Air Weather Service Technical Report No. 215, 2 vols.

Bah, A., 1985: 'About possible linkage between sea surface temperature in the Gulf of Guinea and precipitation in the Sahel'. Unpublished manuscript.

Barber, R. T., Chavez, F. P., 1983: 'Biological consequences of El Niño'. *Science*, **222**, 1203–1210.

Barnett, T. P., 1977: 'An attempt to verify some theories of El Niño'. *J. Phys. Oceanogr.*, **7**, 633–647.

Barnett, T. P., 1981: 'Statistical relations between ocean-atmosphere fluctuations in the tropical Pacific'. *J. Phys. Oceanogr.*, **11**, 1043–1058.

Barnett, T. P., 1983: 'Interaction of the monsoon and Pacific trade wind system at interannual time scales. part 1: The equatorial zone'. *Mon. Wea. Rev.*, **111**, 756–773.

Barnett, T. P., 1984a: 'Prediction of El Niño of 1982–83'. *Mon. Wea. Rev.*, **112**, 1403–1407.

Barnett, T. P., 1984b: 'Interaction of the monsoon and Pacific trade wind system at interannual time scales, part 2: the tropical band'. *Mon. Wea. Rev.*, **112**, 2380–2387.

Barnett, T. P., 1984c: 'Interaction of the monsoon and Pacific trade wind system at interannual time scales, part 3: a partial anatomy of the Southern Oscillation'. *Mon. Wea. Rev.*, **112**, 2388–2400.

Bedi, H. S., Bindra, M. M. S., 1980: 'Principal components of monsoon rainfall'. *Tellus*, **32**, 296–298.

Beer, T., Greenhut, G. K., Tandoh, S. E., 1977: 'Relations between the Z criterion for the subtropical high, Hadley cell parameters and the rainfall in northern Ghana'. *Mon. Wea. Rev.*, **105**, 849–855.

Berkofsky, L., 1977: 'The relation between surface albedo and vertical velocity in a desert'. *Beitr. Phys. Atmos.*, **50**, 312–320.

Berlage, H. P., 1927: 'East-monsoon forecasting in Java'. Koninklijk Magnetisch en Meteorologisch Observatorium te Batavia, Indonesia, Verhandelingen No. 20, 42 pp.

Berlage, H. P., 1934: 'Further research into the possibility of long range forecasting in Netherlands – India'. Koninklijk Magnetisch en Meteorologisch Observatorium te Batavia, Indonesia, Verhandelingen no. 26, 31 pp.

Berlage, H. P., 1957: 'Fluctuations of the general circulation of more than one year, their nature and prognostic value'. Koninklijk Nederlands Meteorologisch Instituut, Mededelingen en Verhandelingen No. 69, 152 pp.

Berlage, H. P., 1966: 'The Southern Oscillation and World weather'. Koninklijk Nederlands Meteorologisch Instituut, Mededelingen en Verhandelingen No. 88, 152 pp.

Berlage, H. P., de Boer, H. J., 1959: 'On the extension of the Southern Oscillation throughout the World during the period 1 July 1949 to 1 July 1957'. *Geofisica pura e applicata*, **44**, 286–295.

Berlage, H. P., de Boer, H. J., 1960: 'On the Southern Oscillation, its way of operation and how it affects pressure patterns in the higher latitudes'. *Geofisica pura e applicata*, **46**, 329–351.

Bhalme, H. N., Mooley, D. A., 1980: 'Large-scale droughts/floods and monsoon circulation'. *Mon. Wea. Rev.*, **108**, 1197–1211.

Bhalme, H. N., Mooley, D. A., Jadhav, S. K., 1983: 'Fluctuations in the drought/flood area over India and relationships with the Southern Oscillation'. *Mon. Wea. Rev.*, 111, 86–94.

Bhargava, B. N., Bansal, R. K., 1969: 'A quasi-biennial oscillation in precipitation at some Indian stations'. *Indian J. Meteorol. Geophys.*, 20, 127–128.

Bjerknes, J., 1966: 'A possible response of the atmospheric Hadley circulation to equatorial anomalies of ocean temperature'. *Tellus*, 18, 820–829.

Bjerknes, J., 1969: 'Atmospheric teleconnections from the equatorial Pacific'. *Mon. Wea. Rev.*, 97, 163–172.

Blanford, H. F., 1884: 'On the connexion of the Himalaya snowfall with dry winds and seasons of drought in India'. *Proc. Roy. Soc. London*, 37, 3–22.

Braak, C., 1919: 'Atmospheric variations of short and long duration in the Malay Archipelago and neighboring regions, and the possibility to forecast them'. Koninklijk Magnetisch en Meteorologisch Observatorium te Batavia, Indonesia, Verhandelingen No. 5, 57 pp.

Braak, C., 1921–29: 'Het climaat van Nederlandsch Indië'. Magnetisch en Meteorologisch Observatorium te Batavia, Verhandelingen, No. 8, parts 1 and 2, 787 and 802 pp.

Brier, G. W., 1978: 'The quasi-biennial oscillation and feedback processes in the atmosphere-ocean-earth system'. *Mon. Wea. Rev.*, 106, 938–946.

Bruce, J., 1983: 'The wind field in the Western Indian Ocean and the related ocean circulation'. *Mon. Wea. Rev.*, 111, 1442–1452.

Bryson, R. A., 1973: 'Drought in the Sahelia: who or what is to blame?' *The Ecologist*, 3, 366–371.

Bultot, F., Griffiths, J. F., 1972: 'The equatorial wet zone'. Chapter 8, pp. 259–312, World survey of climatology, vol. 10: *Climates of Africa*. Elsevier, New York, 604 pp.

Bunting, A. H., Dennett, M. D., Elston, J., Milford, J. R., 1976: 'Rainfall trends in the West African Sahel'. *Quart. J. Roy. Meteor. Soc.*, 102, 59–64.

Busalacchi, A. J., 1982: 'Linear wind-driven variability of the tropical Pacific and Atlantic Oceans'. Department of Oceanography, Florida State University, Tallahassee, Fla., Technical Report NSF Grants ATM 7920485 and OCE 8119052, 138 pp.

Busalacchi, A. J., Cane, M. A., 1985: 'Hindcasts of sea level variations during the 1982–83 El Niño.' *J. Phys. Oceanogr.*, 15, 213–221.

Busalacchi, A. J., O'Brien, J. J., 1980: 'The seasonal variability in a model of the tropical Pacific'. *J. Phys. Oceanogr.*, 10, 1929–1951.

Busalacchi, A. J., O'Brien, J. J., 1981: 'Interannual variability of the Equatorial Pacific in the 1960's'. *J. Geophys. Res.*, 86, 10901–10907.

Busalacchi, A. J., Picaut, J., 1983: 'Seasonal variability from a model of the tropical Atlantic Ocean'. *J. Phys. Oceanogr.*, 13, 1564–1588.

Busalacchi, A. J., Takeuchi, K., O'Brien, J. J., 1983a: 'Interannual variability of the equatorial Pacific – revisited'. *J. Geophys. Res.*, 88, 7551–7562.

Busalacchi, A. J., Takeuchi, K., O'Brien, J. J., 1983b: 'On the interannual wind-driven response of the tropical Pacific Ocean'. pp. 155–195, in Nihoul, J. C. J. (ed.), *Hydrodynamics of the equatorial ocean*, Elsevier, Amsterdam.

Cadet, D. L., 1985: 'The Southern Oscillation over the Indian Ocean'. *J. Climatol.*, 5, 189–212.

Cadet, D. L., Diehl, B. L., 1984: 'Interannual variability of surface fields over the Indian Ocean during recent decades'. *Mon. Wea. Rev.*, 112, 1921–1935.

Cane, M. A., 1983: 'Oceanographic events during El Niño'. *Science*, 222, 1189–1195.

Cane, M. A., 1984: 'Modeling sea level during El Niño'. *J. Phys. Oceanogr.*, 14, 1864–1874.

Cane, M. A., Zebiak, S. E., 1985: 'A theory for El Niño and the Southern Oscillation.' *Science*, 228, 1085–1087.

Carvalho, O., ed., 1973: 'Plano integrado para o combate preventivo aos efeitos das Sêcas do Nordeste'. Ministerio do Interior Serie Desenvolvimento Regional, 1, 237 pp.

Caviedes, C. N., 1973: 'Sêcas and El Niño: two simultaneous climatological hazards in South America'. *Proc. Assoc. Amer. Geogr.*, 5, 44–49.

Charney, J. G., 1975: 'Dynamics of deserts and drought in the Sahel'. *Quart. J. Roy. Meteor. Soc.*, 101, 193–202.

Charney, J. G., Quirk, W. J., Chow, S. H., Kornfield, J., 1977: 'A comparative study of the effects of albedo change on drought in semi-arid regions'. *J. Atmos. Sci.*, 34, 1366–1385.

Charney, J. G., Stone, P. H., Quirk, W. J., 1975: 'Drought in the Sahara: a biogeophysical feedback mechanism'. *Science*, 187, 434–435.

Charney, J. G., Stone, P. H., Quirk, W. J., 1976: 'Drought in the Sahara: insufficient biogeophysical feedback? Reply'. *Science*, 191, 100–102.

Chervin, R. M., Druyan, L. M., 1984: 'The influence of ocean surface temperature gradient and continentality on the Walker Circulation, part 1: prescribed tropical changes'. *Mon. Wea. Rev.*, 112, 1510–1523.

Chiu, W.-C., Lo, A., 1979: 'A preliminary study of the possible statistical relationship between the tropical Pacific sea surface temperature and the atmospheric circulation'. *Mon. Wea. Rev.*, 107, 18–25.

Chu, P.-S., 1983: 'Diagnostic studies of rainfall anomalies in Northeast Brazil'. *Mon. Wea. Rev.*, 111, 1655–1664.

Chu, P.-S., 1984: 'Time and space variability of rainfall and surface circulation in the Northeast Brazil – tropical Atlantic sector'. *J. Meteor. Soc. Japan*, 62, 363–370.

Chu, P.-S., Hastenrath, S., 1981: 'Diagnostic studies of Brazil climate: preliminary results'. Department of Meteorology, University of Wisconsin, Project Report NOAA Grant NA79AA-D-00131, 56 pp.

Chu, P.-S., Hastenrath, S., 1982: 'Atlas of upper-air circulation over tropical South America'. Department of Meteorology, University of Wisconsin, Madison, Wisconsin, 237 pp.

Colón, J. A., 1964: 'On interactions between the Southwest monsoon current and the sea surface over the Arabian Sea'. *Indian J. Met. Geophys.*, 15, 183–200.

Compagnucci, R. H., Vargas, W. M., 1983: 'Análisis espectral de las series de precipitatión estival'. pp. 150–153 in: Preprints of First International Conference on Southern Hemisphere Meteorology, 31 July – 6 August 1983, São José dos Campos, Brazil. Amer. Meteor. Soc., 380 pp.

Conselho Nacional de Pesquisas, Brazil, 1980: 'Workshop on drought forecasting for Northeast Brazil'. Instituto Nacional de Pesquisas Espaciais, São José dos Campos, São Paulo, Brazil, 45 pp.

Courel, M. F., Kandel, R. S., Rasool, S. I., 1984: 'Surface albedo and the Sahel drought'. *Nature*, 307, 528–538.

Covey, D. L., Hastenrath, S., 1978: 'The Pacific El Niño phenomenon and the Atlantic circulation'. *Mon. Wea. Rev.*, 106, 1280–1287.

da Cunha, E., 1979: *Os sertões*. 28th edition, Franscisco Alves, Rio de Janeiro, 416 pp.

da Silva Marques, V., Rao, V. B., Molión, L. C. B., 1983: 'Inter-annual and seasonal variations in the structure and energetics of the atmosphere over Northeast Brazil'. *Tellus*, 35A, 136–148.

Davy, E. G., 1974: 'Drought in West Africa'. *WMO Bulletin*, 23, 18–23.

de Boer, H. J., 1947: 'On forecasting the beginning and the end of the dry monsoon in Java and Madura'. Koninklijk Magnetisch en Meteorologisch Observatorium te Batavia, Indonesia, Verhandelingen No. 32, 20 pp.

de Boer, H. J., Euwe, W., 1949a: 'On long-periodical temperature variations'. Koninklijk Magnetisch en Meteorologisch Observatorium te Batavia, Indonesia, Verhandelingen No. 35, 16 pp.

de Boer, H. J., Euwe, W., 1949b: 'Forecasting rainfall in the period July–August–September for parts of Celebes and South Borneo'. Koninklijk Magnetisch en Meteorologisch Observatorium te Batavia, Indonesia, Verhandelingen No. 36, 14 pp.

de Mesquita, A. R., Morettin, P. A., 1984: 'Interannual variations of precipitation at Fortaleza, Ceará, Brazil'. *Tropical Ocean-Atmosphere Newsletter*, 27, 9–10.

Dey, B., Bhanu Kumar, O. S. R. U., 1982: 'An apparent relationship between Eurasian snow cover and the advance period of Indian Summer monsoon'. *J. Appl. Meteor.*, 21, 1929–1932.

Dey, B., Bhanu Kumar, O. S. R. U., 1984: 'Reply (to comments by Ropelewski, C. F., Robock, A., Matson, M.)'. *J. Climate Appl. Meteor.*, 23, 343–344.

Dhar, O. N., Rakhecha, P. R., 1983: 'Foreshadowing Northeast monsoon rainfall over Tamil Nadu, India'. *Mon. Wea. Rev.*, 111, 109–112.

Dickson, R. R., 1984: 'Eurasian snow cover versus Indian monsoon rainfall – an extension of the Hahn-Shukla results'. *J. Climate Appl. Meteor.*, 23, 171–173.

Donguy, J. R., Dessier, A., 1983: 'El Niño-like events in the tropical Pacific'. *Mon. Wea. Rev.*, 111, 2136–2139.

Düing, W., Leetmaa, A., 1980: 'Arabian Sea cooling: a preliminary heat budget'. *J. Phys. Oceanogr.*, 10, 307–312.

Egger, J., 1977: 'On the linear theory of the atmospheric response to sea surface temperature anomalies'. *J. Atmos. Sci.*, 34, 603–614.

Ellsaesser, H. W., MacCracken, M. C., Potter, G. L., Luther, F. M., 1976: 'An additional model test of positive feedback from high desert albedo'. *Quart. J. Roy. Meteor. Soc.*, 102, 543–544.

Euwe, W., 1949a: 'Forecasting rainfall in the periods December–January–February and April–May–June for parts of Celebes and South Borneo'. Koninklijk Magnetisch en Meteorologisch Observatorium te Batavia, Indonesia, Verhandelingen, No. 38, 23 pp.

Euwe, W., 1949b: 'Forecasting rainfall for the period December–January–February for Java and Madoera'. Koninklijk Magnetisch en Meteorologisch Observatorium te Batavia, Indonesia, Verhandelingen No. 39, 16 pp.

Faure, H., Gac, J. Y., 1981a: 'Will the Sahelian drought end in 1985?' *Nature*, 291, 475–478.

Faure, H., Gac, J. Y., 1981b: 'Senegal river runoff, reply'. *Nature*, 293, 414.

Feldman, G., Clark, D., Halpern, D., 1984: 'Satellite color observations of the phytoplankton distribution in the eastern equatorial Pacific during the 1982–83 El Niño'. *Science*, 226, 1069–1071.

Flohn, H., 1964: 'Grundfragen der Paläoklimatologie im Lichte einer theoretischen Klimatologie'. *Geologische Rundschau*, 54, 504–515.

Flohn, H., 1965: 'Probleme der theoretischen Klimatologie'. *Naturwissenschaftliche Rundschau*, 18, 385–392.

Gill, A. E., Rasmusson, E. M., 1983: 'The 1982–83 climate anomaly in the equatorial Pacific'. *Nature*, 306, 229–234.

Gonçalves de Souza, J., 1979: *O Nordeste Brasileiro*. Banco do Nordeste do Brazil, Fortaleza, 409 pp.

Greenhut, G. K., 1977: 'A new criterion for locating the subtropical high in West Africa'. *J. Appl. Meteor.*, 16, 729–734.

Grove, A. T., 1972: 'Climatic change in Africa in the last 20,000 years'. *Les Problèmes de Dévelopment du Sahara Septentrional*, vol. 2, Alger.

Grove, A. T., 1983: 'A note on the remarkably low rainfall of the Sudan in 1913'. *Savanna*, 2, 133–138.

Hackert, E. C., Hastenrath, S., 1986: 'On mechanisms of Indonesia rainfall anomalies', submitted for publication.

Hahn, D. G., Shukla, J., 1976: 'An apparent relationship between Eurasian snow cover and Indian monsoon rainfall'. *J. Atmos. Sci.*, 33, 2461–2462.

Hastenrath, S., 1966: 'On general circulation and energy budget in the area of the Central American Seas'. *J. Atmos. Sci.*, 23, 604–711.

Hastenrath, S., 1967: 'Rainfall distribution and regime in Central America'. *Archiv Meteor. Geophys. Bioklim., Ser. B*, 15, 201–241.

Hastenrath, S., 1970: 'Lake-level changes and recent climatic fluctuations in Central America'. Preprints Symposium Tropical Meteorology, Honolulu, Hawaii, Amer. Meteor. Soc., L IX 1–4.

Hastenrath, S., 1976: 'Variations in low-latitude circulation and extreme climatic events in the tropical Ameicas'. *J. Atmos. Sci.*, 33, 202–215.

Hastenrath, S., 1978: 'On modes of tropical circulation and climate anomalies'. *J. Atmos. Sci.*, 35, 2222–2231.

Hastenrath, S., 1983: 'Comments on "correlation between the tropical Atlantic trade winds and precipitation in Northeastern Brazil"'. *J. Climatol.*, 3, 207–209.

Hastenrath, S., 1984: 'Interannual variability and annual cycle: mechanisms of circulation and climate in the tropical Atlantic sector'. *Mon. Wea. Rev.*, 112, 1097–1107.

Hastenrath, S., Heller, L., 1977: 'Dynamics of climate hazards in Northeast Brazil'. *Quart. J. Roy. Meteor. Soc.*, 103, 77–92.

Hastenrath, S., Kaczmarczyk, E. B., 1981: 'On spectra and coherence of tropical climate anomalies'. *Tellus*, 33, 453–462.

Hastenrath, S., Lamb, P. J., 1977a: 'Climatic atlas of the tropical Atlantic and eastern Pacific Oceans'. The University of Wisconsin Press, Madison, Wisconsin, 113 pp.

Hastenrath, S., Lamb, P. J., 1977b: 'Some aspects of circulation and climate over the eastern equatorial Atlantic', *Mon. Wea. Rev.*, 106, 1280–1287.

Hastenrath, S., Lamb, P. J., 1978a: 'Heat budget atlas of the tropical Atlantic and Eastern Pacific Oceans'. The University of Wisconsin Press, Madison, Wis., 103 pp.

Hastenrath, S., Lamb, P. J., 1978b: 'On the dynamics and climatology of surface flow over the equatorial oceans'. *Tellus*, 30, 436–448.

Hastenrath, S., Lamb, P. J., 1980: 'On the heat budget of hydrosphere and atmosphere in the Indian Ocean'. *J. Phys. Oceanogr.*, 10, 694–708.

Hastenrath, S., Rosen, A., 1983: 'Patterns of India monsoon rainfall anomalies'. *Tellus*, 35A, 324–331.

Hastenrath, S., Wendland, W., 1979: 'On the secular variation of storms in the tropical North Atlantic and Eastern Pacific'. *Tellus*, 31, 28–38.

Hastenrath, S., Wu, M.-C., 1982: 'Oscillations of upper-air circulation and anomalies in the surface climate of the tropics'. *Archiv Meteor. Geophys. Bioklim., Ser. B*, 31, 1–37.

Hastenrath, S., Wu, M.-C., Chu, P.-S., 1984: 'Towards the monitoring and prediction of Northeast Brazil droughts'. *Quart. J. Roy. Meteor. Soc.*, 110, 411–425.

Henning, D., 1967: 'Zur Interpretation eines Zirkulationskriteriums'. *Archiv Meteor. Geophys. Bioklim., Ser. A*, 16, 126–136.

Hildebrandsson, H., 1897, 1899, 1909, 1910, 1914: 'Les centres d'action de l'atmosphère, I–V'. *Kongliga Svenska Vetenskaps-Akademiens Handlingar*, vol. 29, no. 3, 36 pp.; vol. 32, no. 4, 22 pp.; vol. 45, no. 2, 11 pp.; vol. 45, no. 11, 22 pp.; vol. 51, No. 8, 16 pp.

Hirst, A. C., Hastenrath, S., 1983a: 'Atmosphere-ocean mechanism of climate anomalies in the Angola – tropical Atlantic sector'. *J. Phys. Oceanogr.,* **13**, 1146–1157.

Hirst, A. C., Hastenrath, S., 1983b: 'Diagnostics of hydrometeorological anomalies in the Zaïre (Congo) basin'. *Quart. J. Roy. Meteor. Soc.,* **109**, 879–890.

Hisard, P., 1980: 'Observation de réponses du type "El Niño" dans l'Atlantique tropical oriental – Golfe de Guinée'. *Ocean. Acta,* **3**, 69–78.

Hookey, P., 1970: 'Revenge of the gods'. *Weather,* **25**, 425–428.

Horel, J. D., Wallace, J. M., 1981: 'Planetary scale atmospheric phenomena associated with the Southern Oscillation'. *Mon. Wea. Rev.,* **109**, 813–823.

Hoskins, B. J., Karoly, D., 1981: 'The steady linear response of a spherical atmosphere to thermal and orographic forcing'. *J. Atmos. Sci.,* **38**, 1179–1198.

Houghton, R. W., 1976: 'Circulation and hydrographic structure over the Ghana Shelf during the 1974 upwelling'. *J. Phys. Oceanogr.,* **6**, 909–924.

Hurlburt, H., Kindle, J., O'Brien, J. J., 1976: 'A numerical simulation of the onset of El Niño'. *J. Phys. Oceanogr.,* **6**, 621–631.

Ilesanmi, O. O., 1971: 'An empirical formulation of an ITD rainfall model for the tropics: A case study in Nigeria'. *J. Appl. Meteor.,* **10**, 882–891.

Inoue, M., O'Brien, J. J., 1984: 'A forecasting model for the onset of a major El Niño'. *Mon. Wea. Rev.,* **112**, 2326–2337.

Inoue, M., O'Brien, J. J., White, W. B., Pazan, S. E., 1985: 'Interannual variability in the tropical Pacific prior to the onset of the 1982/83 ENSO event'. *J. Geophys. Res.,* submitted.

Jackson, S. P., 1964: *Climatic atlas of Africa.* Government Printer, Pretoria, South Africa, 117 pp.

Jagannathan, P., 1960: 'Seasonal forecasting in India, a review'. Meteorological Office, Poona, 79 pp.

Jagannathan, P., Bhalme, H. N., 1973: 'Changes in the pattern of distribution of Southwest monsoon rainfall over India associated with sunspots'. *Mon. Wea. Rev.,* **101**, 691–700.

Jagannathan, P., Parthasarathy, B., 1973: 'Trends and periodicities of rainfall over India'. *Mon. Wea. Rev.,* **101**, 371–375.

Jones, R. H., Kearns, J. P., 1976: 'Fortaleza, Ceará, Brazil, rainfall'. *J. Appl. Meteor.,* **15**, 307–308.

Julian, P. R., Chervin, R. M., 1978: 'A study of the Southern Oscillation and Walker circulation phenomenon'. *Mon. Wea. Rev.,* **106**, 1433–1451.

Kanamitsu, M., Krishnamurti, T. N., 1978: 'Northern summer tropical circulations during drought and normal rainfall months'. *Mon. Wea. Rev.,* **106**, 331–347.

Keshavamurthy, R. N., 1982: 'Response of the atmosphere to sea surface temperature anomalies over the equatorial Pacific and the teleconnections of the Southern Oscillation'. *J. Atmos. Sci.,* **39**, 1241–1259.

Keshavamurthy, R. N., 1983: 'Southern Oscillation: further studies with a GFDL general circulation model'. *Mon. Wea. Rev.,* **111**, 1988–1997.

Kidson, J. W., 1975: 'Tropical eigenvector analysis and Southern Oscillation'. *Mon. Wea. Rev.,* **103**, 187–196.

Kidson, J. W., 1977: 'African rainfall and its relation to the upper air circulation'. *Quart. J. Roy. Meteor. Soc.,* **103**, 441–456.

Koepcke, H. W., 1961: 'Synökologische Studien an der Westseite der peruanischen Anden'. *Bonner Geographische Abhandlungen,* No. 29, 320 pp.

Kok, C. J., Öpsteegh, J. D., 1985: Possible causes of anomalies in seasonal mean circulation patterns during the 1982–83 El Niño event. *J. Atmos. Sci.,* **42**, 677–694.

Koteswaram, P., Alvi, S. M. A., 1969: 'Secular trends and periodicities in rainfall at West coast stations in India'. *Current Science,* **38**, 229–231.

Koteswaram, P., Alvi, S. M. A., 1970: 'Secular trends and variations in rainfall of Indian regions'. *Időjárás,* **74**, 176–183.

Kousky, V. E., 1979: 'Frontal influences on Northeast Brazil'. *Mon. Wea. Rev.,* **107**, 1140–1153.

Kousky, V. E., 1980: 'Diurnal rainfall variations in Northeast Brazil'. *Mon. Wea. Rev.,* **108**, 488–498.

Kousky, V. E., Chu, P.-S., 1978: 'Fluctuations in annual rainfall for Northeast Brazil'. *J. Meteor. Soc. Japan,* **57**, 457–465.

Kousky, V. E., Kagano, M. T., Cavalcanti, I. F. A., 1984: 'The Southern Oscillation: oceanic-atmospheric circulation changes and related rainfall anomalies'. *Tellus,* **36A**, 490–504.

Kousky, V. E., Moura, A. D., 1981: 'Previsão de precipitacaõ no Nordeste do Brazil: o aspecto dinâmico'. *IV Simpósio Brasileiro de Hidrología e Recursos Hídricos, Nov. 1981,* Fortaleza.

Kraus, E. B., 1977a: 'Subtropical droughts and cross-equatorial energy transports'. *Mon. Wea. Rev.,* **105**, 1009–1018.

Kraus, E. B., 1977b: 'The seasonal excursion of the intertropical convergence zone'. *Mon. Wea. Rev.*, 105, 1052–1055.

Kraus, E. B., Hanson, H. P., 1983: 'Air-sea interaction as a propagator of equatorial ocean surface temperature anomalies'. *J. Phys. Oceanogr.*, 13, 130–138.

Kung, E. C., Sharif, T. A., 1980: 'Regression forecasting of the onset of the Indian summer monsoon with antecedent upper air conditions'. *J. Appl. Meteor.*, 19, 370–380.

Kung, E. C., Sharif, T. A., 1982: 'Long-range forecasting of the Indian summer monsoon onset and rainfall with upper air parameters and sea surface temperature'. *J. Meteor. Soc. Japan*, 60, 672–681.

Lamb, P. J., 1978a: 'Case studies of tropical Atlantic surface circulation pattern during recent Sub-Saharan weather anomalies: 1967 and 1968'. *Mon. Wea. Rev.*, 106, 482–491.

Lamb, P. J., 1978b: 'Large-scale tropical Atlantic circulation patterns associated with Subsarahan weather anomalies'. *Tellus*, 30, 240–251.

Lamb, P. J., 1979: 'Some perspectives on climate and climate dynamics'. *Progress Phys. Geogr.*, 3, 215–235.

Lamb, P. J., 1980: 'Sahelian drought'. *New Zealand J. Geogr.*, 68, 12–16.

Lamb, P. J., 1982a: 'Persistence of Subsaharan drought'. *Nature*, 299, 46–48.

Lamb, P. J., 1982b: 'Comments on "West African rainfall variations and tropical Atlantic sea surface temperatures"'. *Climate Monitor*, 11, 46–49.

Lamb, P. J., 1983a: 'West African water vapor variations between recent contrasting Subsaharan rainy seasons'. *Tellus*, 35A, 198–212.

Lamb, P. J., 1983b: 'Sub-Saharan rainfall update for 1982: continued drought'. *J. Climatol.*, 3, 419–422.

Lamb, P. J., 1985: 'Rainfall in Subsaharan West Africa during 1941–83'. *Zeitschrift für Gletscherkunde und Glazialgeologie*, 21, 131–139.

Landsberg, H. E., 1975a: 'Sahel drought: change of climate or part of climate'. *Archiv Meteor. Geophys. Bioklim., Ser. B*, 23, 193–200.

Landsberg, H. E., 1975b: 'Drought, a recurrent element of climate'. Drought, WMO Special Environmental Report No. 5 (WMO-No. 403), 41–90.

Lau, K. M., Chan, P. H., 1983a: 'Short-term climate variability and atmospheric teleconnection from satellite-observed outgoing longwave radiation: part I: simultaneous relationships'. *J. Atmos. Sci.*, 40, 2735–2750.

Lau, K. M., Chan, P. H., 1983a: 'Short-term climate variability and atmospheric teleconnection from satellite-observed outgoing longwave radiation, part II: lagged correlations'. *J. Atmos. Sci.*, 40, 2751–2767.

Lau, K. M., Li, M. T., 1984: 'The monsoon of East Asia and its global associations – a survey'. *Bull. Amer. Meteor. Soc.*, 65, 114–125.

Lettau, H. H., 1967: 'Small to large-scale features of boundary layer structure over mountain slopes', pp. 1–72 in: Reiter, E. R., Rasmusson, J. L., ed., *Proceedings of the Symposium on Mountain Meteorology, June 1967, Fort Collins, Colo.*, Department of Atmospheric Science, Colorado State University, Atmospheric Science Paper No. 122, 221 pp.

Lockyer, N., Lockyer, W. J. S., 1902a: 'On some phenomena which suggest a short period of solar and meteorological changes'. *Proc. Roy. Soc. London*, 70, 500–504.

Lockyer, N., Lockyer, W. J. S., 1902b: 'On the similarity of the short-period pressure variation over large areas'. *Proc. Roy. Soc. London*, 71, 134–136.

Lockyer, N., Lockyer, W. J. S., 1904: 'The behavior of the short-period atmospheric pressure variation over the earth's surfaces'. *Proc. Roy. Soc. London*, 73, 457–470.

Lockyer, W. J. S., 1906: 'Barometric variations of long duration over large areas'. *Proc. Roy. Soc. A*, 78, 43–60.

Lough, J. M., 1986: 'Tropical Atlantic sea surface temperatures and rainfall variations in Subsaharan Africa and Northeast Brazil'. *Mon. Wea. Rev.*, submitted.

Luther, D. S., Harrison, D. E., 1983: 'Zonal winds in the central equatorial Pacific and El Niño'. *Science*, 222, 327–330.

Markham, C. G., 1967: 'Climatological aspects of drought in Northeastern Brazil'. Ph.D. Dissertation, Department of Geography, University of California, Berkeley, 352 pp.

Markham, C. G., 1974: 'Apparent periodicities in rainfall at Fortaleza, Ceará, Brazil'. *J. Appl. Meteor.*, 13, 176–179.

Markham, C. G., McLain, D. R., 1977: 'Sea surface temperature related to rain in Ceará, Northeastern Brazil'. *Nature*, 265, 320–323.

McBride, J. L., Nicholls, N., 1983: 'Seasonal relationships between Australian rainfall and the Southern Oscillation'. *Mon. Wea. Rev.*, 111, 1998–2004.

McCreary, J. P., 1976: 'Eastern tropical ocean response to changing wind systems: with application to El Niño'. *J. Phys. Oceanogr.*, 6, 632–645.

McCreary, J. P., 1983: 'A model of tropical ocean-atmosphere interaction'. *Mon. Wea. Rev.*, 111, 370–387.

McCreary, J. P., Anderson, D. L. T., 1984: 'A simple model of El Niño and the Southern Oscillation'. *Mon. Wea. Rev.*, 112, 934–946.

McCreary, J. P., Picaut, J., Moore, D. W., 1984: 'Effects of remote annual forcing in the eastern tropical Atlantic Ocean'. *J. Mar. Res.*, 42, 45–81.

Meehl, G. A., Van Loon, H., 1979: 'The see-saw in winter temperatures between Greenland and Northern Europe, Part 3: Teleconnections with lower latitudes'. *Mon. Wea. Rev.*, 107, 1095–1106.

Merle, J., 1980: 'Variabilité thermique annuelle et interannuelle de l'Océan Atlantique équatorial Est. L'hypothèse d'un "El Niño" Atlantique'. *Ocean. Acta,* 3, 209–220.

Meyers, G., Donguy, J. R., 1984: 'The North Equatorial Countercurrent and heat storage in the Western Pacific Ocean during 1982–83'. *Nature,* 312, 258–260.

Miles, M. K. and Follard, C. K., 1974: 'Changes in the latitude of the climatic zones of the Northern hemisphere'. *Nature,* 252, 616.

Miron, O., Tyson, P. D., 1984: 'Wet and dry conditions and pressure anomaly fields over South Africa and the adjacent oceans, 1963–79'. *Mon. Wea. Rev.*, 112, 2127–2132.

Mooley, D. A., Parthasarathy, B., 1983: 'Variability of the Indian summer monsoon and tropical circulation features'. *Mon. Wea. Rev.*, 111, 967–978.

Mooley, D. A., Parthasarathy, B., 1984a: 'Indian summer monsoon and the East Equatorial Pacific sea surface temperature'. *Atmosphere-Ocean,* 22, 23–35.

Mooley, D. A., Parthasarathy, B., 1984b: 'Fluctuations in all-India summer monsoon rainfall during 1871–1978'. *Climatic Change,* 6, 286–301.

Mooley, D. A., Parthasarathy, B., Sontakke, N. A., Munot, A. A., 1981: 'Annual rain-water over India, its variability and impact on the seasons'. *J. Climatol.*, 1, 167–186.

Moore, D. W., Hisard, P., McCreary, J. P., Merle, J., O'Brien, J. J., Picaut, J., Verstraate, J. M., Wunsch, C., 1978: 'Equatorial adjustment in the Eastern Atlantic'. *Geophys. Research Letters,* 5, 637–640.

Motha, R. P., Leduc, S. K., Steyaert, L. T., Sakamoto, C. M., and Strommen, N. D., 1980: 'Precipitation patterns in West Africa'. *Mon. Wea. Rev.*, 109, 1567–1578.

Moura, A. D., Shukla, J., 1981: 'On the dynamics of droughts in Northeast Brazil: Observations, theory, and numerical experiments with a general circulation model'. *J. Atmos. Sci.*, 38, 2653–2675.

Murphy, R. C., 1926: 'Oceanic and climate phenomena along the West coast of South America during 1925'. *Geogr. Rev.*, 16, 26–54.

Namias, J., 1969: 'Seasonal interactions between the North Pacific Ocean and the atmosphere during the 1960's'. *Mon. Wea. Rev.*, 97, 173–192.

Namias, J., 1972: 'Influence of Northern hemisphere general circulation on drought in Northeast Brazil'. *Tellus,* 24, 336–343.

Namias, J., 1973: 'Response of the Equatorial Countercurrent to the subtropical atmosphere'. *Science,* 181, 1244–1245.

Namias, J., Cayan, D. R., 1984: 'El Niño: the implications for forecasting'. *Oceanus,* 27, no. 2, 41–47.

Navato, A' R., Newell, R. E., Hsiung, J. C., Billing, C. B., 1981: 'Tropospheric mean temperature and its relationship to the oceans and atmospheric aerosols'. *Mon. Wea. Rev.*, 109, 405–416.

Newell, R. E., 1980: 'Climate and the ocean'. *Amer. Scientist,* 67, 405–416.

Newell, R. E., Kidson, J. W., 1984: 'African mean wind changes between Sahelian wet and dry periods'. *J. Climatol.*, 4, 27–33.

Newell, R. E., Selkirk, R., Ebisuzaki, W., 1982: 'The Southern Oscillation: sea surface temperature and wind relationships in a 100-year data set'. *J. Climatol.*, 2, 357–373.

Newell, R. E., Weare, B. C., 1976a: 'Factors governing tropospheric mean temperature'. *Science,* 194, 1413–1414.

Newell, R. E., Weare, B. C., 1976b: 'Ocean temperature and large-scale atmospheric variations'. *Nature,* 262, 40–41.

Newton, C. W., ed., 1972: 'Meteorology of the Southern hemisphere'. *Amer. Meteor. Soc., Monographs,* 13, No. 35, 263 pp.

Nicholls, N., 1978: 'Air-sea interaction and the quasi-biennial oscillation'. *Mon. Wea. Rev.*, 106, 1505–1508.

Nicholls, N., 1981: 'Air-sea interaction and the possibility of long-range weather prediction in the Indonesian Archipelago'. *Mon. Wea. Rev.*, 109, 2435–2443.

Nicholson, S. E., 1979: 'Revised rainfall series for the West African subtropics'. *Mon. Wea. Rev.*, 107, 620–623.

Nicholson, S. E., 1980a: 'The nature of rainfall fluctuations in subtropical West Africa'. *Mon. Wea. Rev.*, 108, 473–487.

Nicholson, S. E., 1980b: *Saharan climates in historic times. The Sahara and the Nile*, M. A. J. Williams and H. Faure, eds., A. A. Balkema, Rotterdam, 400 pp.

Nicholson, S. E., 1981: 'Rainfall and atmospheric circulation during drought and wetter periods in West Africa'. *Mon. Wea. Rev.,* 109, 2191–2208.

Nicholson, S. E., 1983: 'Sub-Saharan rainfall in the years 1976–80: evidence of continued drought'. *Mon. Wea. Rev.,* 111, 1646–1654.

Nobre, P., 1984: 'Fontes de calor nos trópicos e escoamentos anómalos de larga escala associadas com anomalías de precipitação no Nordeste do Brasil'. Instituto Nacional de Pesquisas Espaciais, Publicação No. INPE–3211–TDL/175, Saõ José dos Campos, S.P., 90 pp.

Nobre, P., Moura, A. D., 1984: 'Large scale tropical heat sources and global atmosphere energy propagation associated with droughts in Northeast Brazil, pp. 83–86 in: Extended abstracts of papers presented at the second WMO symposium on meteorological aspects of tropical droughts. September 1984. World Meteorological Organization, TMP Report Series No. 15, Geneva, Switzerland, 134 pp.

Normand, C. W. B., 1953: 'Monsoon seasonal forecasting'. *Quart. J. Roy. Meteor. Soc.,* 79, 463–473.

O'Brien, J. J., Adamec, D., Moore, D. W., 1978: 'A simple model of upwelling in the Gulf of Guinea'. *Geophys. Research Letters,* 5, 641–644.

O'Brien, J. J., Busalacchi, A. J., Kindle, J., 1981: 'Ocean models of El Niño', pp. 159–212 in: Glantz, M. H., Thompson, J. D. (eds.), *Resource management and environmental uncertainty: lessons from coastal upwelling fisheries*. Wiley, Somerset, N.J., 491 pp.

Ogallo, L., 1979: 'Rainfall variability in Africa'. *Mon. Wea. Rev.,* 107, 1133–1139.

Oliveira, L. L. de, 1982: 'Zonas de convêrgencia no Atlantico Sul e suas influêneias no regime de precipitação no Nordeste do Brazil'. Instituto Nacional de Pesquisas Espaciais, Publicão No. INPE–2307–TDL/074, 103 pp.

Opsteegh, J. D., Van den Dool, H. M., 1980: 'Seasonal differences in the stationary response of a linearized primitive equation model: prospects for long-range forecasting'. *J. Atmos. Sci.,* 37, 2169–2185.

Otterman, J., 1974: 'Baring high-albedo soils by overgrazing: A hypothesized desertification mechanism'. *Science,* 186, 531–553.

Palutikof, J. P., Lough, J. M., Farmer, G., 1981: 'Senegal River runoff'. *Nature,* 293, 414.

Pant, G. B., Parthasarathy, B., 1981: 'Some aspects of an association between the Southern Oscillation and Indian summer monsoon'. *Archiv Meteor. Geophys. Bioklim., Ser. B,* 29, 245–252.

Parthasarathy, B., Dhar, D. N., 1974: 'Secular variations of regional rainfall over India'. *Quart, J. Roy. Meteor. Soc.,* 100, 245–257.

Parthasarathy, B., Mooley, D. A., 1978: 'Some features of a long homogeneous series of Indian summer monsoon rainfall'. *Mon. Wea. Rev.,* 106, 771–781.

Parthasarathy, B., Pant, G. B., 1984: 'The spatial and temporal relationships between the Indian summer monsoon rainfall and the Southern Oscillation'. *Tellus,* 36A, 269–277.

Philander, S. G. H., 1983: 'El Niño Southern Oscillation phenomena'. *Nature,* 302, 295–301.

Picaut, J., 1983: 'Propagation of the seasonal upwelling in the Eastern Equatorial Atlantic'. *J. Phys. Oceanogr.,* 13, 18–37.

Quenouille, M. H., 1952: *Associated measurements*. Butterworths Scientific Publications, London, 242 pp.

Ouinn, W. H., 1974: 'Monitoring and predicting El Niño invasions'. *J. Appl. Meteor.,* 13, 825–830.

Quinn, W. H., 1976: 'An improved approach for following and predicting Equatorial Pacific changes and El Niño'. *Proceedings of the NOAA Climate Diagnostics Workshop, 4–5 Nov., 1976*. Washington, D.C., article 21, pp. 1–15.

Quinn, W. H., Burt, W. V., 1970: 'Prediction of abnormally heavy precipitation over the Equatorial Pacific Dry Zone'. *J. Appl. Meteor.,* 9, 20–28.

Quinn, W. H., Burt, W. V., 1972: 'Use of the Southern Oscillation in weather prediction'. *J. Appl. Meteor.,* 11, 616–628.

Quinn, W. H., Zopf, D. O., Short, K. S., Kuo Yang, R. T. W., 1978: 'Historical trends and statistics of the Southern Oscillation, El Niño, and Indonesian droughts'. *Fishery Bulletin,* 76, 663–678.

Raghavan, K., Puranik, P. V., Mujumdar, V. R., Ismail, P. M. M., Paul, D. K., 1978: 'Interaction between the Arabian Sea and the Indian monsoon'. *Mon. Wea. Rev.,* 106, 719–724.

Raman, C. R. V., Maliekal, J. A., 1985: 'A "northern oscillation" relating northern hemispheric pressure anomalies and the Indian summer monsoon'. *Nature,* 314, 430–432.

Ramaswamy, C., Pareek, R. S., 1978: 'The Southwest monsoon over India and its teleconnections with the middle and upper tropospheric flow patterns over the Southern hemisphere'. *Tellus,* 30, 126–135.

Ramesh Babu, V., Sastry, J. S., 1984: 'Summer cooling in the East Central Arabian Sea – a process of dynamic response to the Southwest monsoon'. *Mausam,* 35, 17–26.

Ramos, R. P. L., 1975: 'Precipitation characteristics in the Northeast Brazil dry region'. *J. Geophys. Res.,* **80,** 1665–1678.

Rao, K. N., 1964: 'Seasonal forecasting for India'. WMO Technical Note no. 66, WMO no. 162, TP 79, Geneva, pp. 17–30.

Rao, V. B., da Silva Marques, V., 1984: 'Water vapor characteristics over Northeast Brazil during two contrasting years'. *J. Climate Appl. Meteor.,* **23,** 440–444.

Rao, V. B., da Silva Marques, V., Bonatti, J. P., 1984: 'On the possibility of barotropic instability over Northeast Brazil'. *Tellus,* **36A,** 207–210.

Rao, V. B., Satyarmurty, P., de Brito, J. I. B., 1986: 'On the 1983 drought in Northeast Brasil'. *J. Climatol.,* **6,** in press.

Rasmusson, E. M., 1984: 'El Niño: The ocean/atmosphere connection'. *Oceanus,* **27,** 5–12.

Rasmusson, E. M., Carpenter, T. H., 1982: 'Variations in tropical sea surface temperature and surface wind fields associated with the Southern Oscillations/El Niño'. *Mon. Wea. Rev.,* **110,** 354–384.

Rasmusson, E. M., Carpenter, T. H., 1983: 'The relationship between Eastern Equatorial Pacific sea surface temperatures and rainfall over India and Sri Lanka'. *Mon. Wea. Rev.,* **111,** 517–528.

Rasmusson, E. M., Wallace, J. M., 1983: 'Meteorological aspects of the El Niño/Southern Oscillation'. *Science,* **222,** 1195–1202.

Rasool, S. I., 1982: 'Are Sahelian droughts predictable?' *Nature,* **297,** 19–20.

Ratisbona, L. R., 1976: 'The climate of Brazil', in *World Survey of Climatology,* vol. 12, pp. 219–269. Elsevier, New York, 532 pp.

Reesinck, J. J. M., 1952: 'Some remarks on monsoon forecasting for Java'. Kementerian Perhubungan Djawatan Meteorologi dan Geofisiki, Djakarta, Indonesia, Verhandelingen No. 44, 22 pp.

Reiter, E. R., 1979: 'Trade-wind variability, Southern Oscillation, and quasi-biennial oscillation'. *Archiv Meteor. Geophys. Bioklim., Ser. A,* **28,** 113–126.

Ripley, E. A., 1976a: 'Drought in the Sahara: insufficient biogeophysical feedback?' *Science,* **191,** 100.

Ripley, E. A., 1976b: 'Comments on the paper "Dynamics of deserts and drought in the Sahel" by J. G. Charney'. *Quart. J. Roy. Meteor. Soc.,* **102,** 466–467.

Robinson, M., Bauer, R. A., Schroeder, E., 1979: 'Atlas of North Atlantic – Indian Ocean mean temperatures and mean salinities of the surface layer'. Naval Oceanographic Office, NOO RP–18, NSTL Station, Bay St. Louis, Missouri, 39522, 234 pp.

Rodhe, H., Virji, H., 1976: 'Trends and periodicities in East African rainfall data'. *Mon. Wea. Rev.,* **104,** 307–315.

Rogers, J. C., 1984: 'The association between the North Atlantic Oscillation and the Southern Oscillation in the Northern Hemisphere'. *Mon. Wea. Rev.,* **12,** 1999–2005.

Rogers, J. C., Van Loon, H., 1979: 'The see-saw in winter temperature between Greenland and Northern Europe, Part 2: some oceanic and atmospheric effects in middle and high latitudes'. *Mon. Wea. Rev.,* **107,** 509–519.

Ropelewski, C. F., Robock, A., Matson, M., 1984: 'Comments on "An apparent relationship between Eurasian spring snow cover and the advance period of the Indian Summer monsoon"'. *J. Climate Appl. Meteor.,* **23,** 341–342.

Rowntree, P. R., 1972: 'The influence of tropical East Pacific Ocean temperature on the atmosphere'. *Quart. J. Roy. Meteor. Soc.,* **98,** 290–321.

Sadler, J. C., 1975: 'The upper-tropospheric circulation over the global tropics'. Department of Meteorology, University of Hawaii, Honolulu, Hawaii, UHMET–75–05, 35 pp.

Sampaio Ferraz, J. de, 1925: 'Causes provaveis das secas do Nordeste Brasileiro'. Ministerio de Agricultura, Directoría de Meteorologia, Rio de Janeiro, 12 pp.

Sampaio Ferraz, J. de, 1929: 'Sir Gilbert Walker's formula for Ceará droughts: suggestions for its physical explanation'. *Meteor. Mag.,* **64,** 81–84.

Sardeshmukh, P. D., Hoskins, B. J., 1985: 'Vorticity balances in the tropics during the 1982–83 El Niño – Southern Oscillation event'. *Quart. J. Roy. Meteor. Soc.,* **111,** 261–278.

Schell, I. I., 1965: 'The origin and possible prediction of the fluctuations in the Peru Current and upwelling'. *J. Geophys. Res.,* **70,** 5529–5540.

Schickedanz, P. T., Bowen, E. G., 1977: 'The computation of climatological power spectra'. *J. Appl. Meteor.,* **16,** 359–367.

Schmidt, F. H., van der Vecht, J., 1952: 'East monsoon fluctuations in Java and Madoera during the period 1880–1940'. Kementerian Perhubungan Djawatan Meteorologi dan Geofisiki, Djakarta, Indonesia, Verhandelingen No. 43, 36 pp.

Schmidt – ten Hoopen, K. J., Schmidt, F. H., 1951: 'On climatic variations in Indonesia'. Kementerian Perhubungan Djawatan Meteorologi dan Geofisiki, Djakarta, Indonesia, Verhandelingen, No. 41, 43 pp.

Schopf, P. S., Harrison, D. E., 1983: 'On equatorial waves and El Niño, I: influence of initial state on wave-induced currents and warming'. *J. Phys. Oceanogr.*, 13, 936–948.

Schove, D. J., Berlage, H. P., 1965: 'Pressure anomalies in the Indian Ocean area 1796–1960'. *Geofisica pura e applicata*, 61, 219–251.

Schregardus, M. W. F., 1938: 'Sea-surface temperatures on some steamer routes in Netherlands India, third series 1933–1937'. Koninklijk Magnetisch en Meteorologisch Observatorium te Batavia, Verhandelingen. No. 28, 35 pp.

Schreiber, R. W., Schreiber, E. A., 1984: 'Central Pacific seabirds and the Southern Oscillation: 1982 to 1983'. *Science*, 225, 713–716.

Schütte, K., 1968: 'Untersuchungen zur Meteorologie und Klimatologie des El Niño – Phänomens in Ecuador and Nordperu'. *Bonner Meteorologische Abhandlungen*, no. 9, 152 pp.

Schweigger, E., 1959: *Die Westküste Südamerikas im Bereich des Perustroms*. Keysersche Verlagsbuchhandlung, Heidelberg und München, 513 pp., plus maps and photos.

Schweigger, E., 1961: 'Anomalías térmicas en el Océano Pacífico Oriental y su pronóstico'. *Boletín de la Sociedad Geográfica de Lima*, 78, 3–50.

Serra, A., 1956: 'As Secas do Nordeste'. *Boletim Geográfico*, 14, 269–270.

Servain, J., Picaut, J., Merle, J., 1982: 'Evidence of remote forcing in the equatorial Atlantic Ocean'. *J. Phys. Oceanogr.*, 12, 457–463.

Shapiro, L. J., 1982: 'Hurricane climatic fluctuations, part 1: patterns and cycles'. *Mon. Wea. Rev.*, 110, 1007–1013.

Shukla, J., 1975: 'Effect of Arabian sea-surface temperature anomaly on Indian summer monsoon: a numerical experiment with the GFDL Model'. *J. Atmos. Sci.*, 32, 503–511.

Shukla, J., 1986: 'Interannual variability of monsoons', in Fein, J., Stephens, P., ed., Monsoons, Wiley Interscience Publishers, New York, London, Sidney, Toronto, in press.

Shukla, J., Paolino, D. A., 1983: 'The Southern Oscillation and long-range forecasting of the summer monsoon rainfall over India'. *Mon. Wea. Rev.*, 111, 1830–1837.

Smagorinsky, J., 1963: 'General circulation experiments with the primitive equations'. *Mon. Wea. Rev.*, 91, 99–164.

Stone, P. H., Chervin, R. M., 1984: 'The influence of ocean surface temperature gradient and continentality on the Walker Circulation, part 2: prescribed global changes'. *Mon. Wea. Rev.*, 112, 1524–1534.

Subbaramayya, I., 1968: 'The interrelations of monsoon rainfall in different subdivisions of India'. *J. Meteor. Soc. Japan*, 46, 77–84.

Sud, Y. C., Fennessy, M., 1982: 'A study of the influence of surface albedo on July circulation in semi-arid regions'. *J. Climatol.*, 2, 105–125.

Taljaard, J. J., van Loon, H., Crutcher, H. L., Jenne, R. L., 1969: 'Climate of the upper air: Southern Hemisphere, vol. 1, temperatures, dew points and heights at selected pressure levels'. NAVAIR 50–1C–55, Chief Naval Operations, Washington, D.C., 135 pp.

Tanaka, M., Weare, B. C., Navato, A. R., Newell, R. E., 1975: 'Recent African rainfall patterns'. *Nature*, 255, 201–203.

Tang, T. Y., Weisberg, R. H., 1984: 'On the equatorial Pacific response to the 1982/83 El Niño – Southern Oscillation event'. *J. Mar. Res.*, 42, 809–829.

Thapliyal, V., 1982: 'Stochastic dynamic model for long-range prediction of monsoon rainfall in peninsular India'. *Mausam*, 33, 399–404.

Thompson, B. W., 1965: *The climate of Africa*. Oxford University Press, Nairobi, New York, 132 pp.

Toole, J. M., Borges, M. D., 1984: 'Observations of horizontal velocities and vertical displacements in the Equatorial Pacific Ocean associated with the early stages of the 1982/83 El Niño'. *J. Phys. Oceanogr.*, 14, 948–959.

Torrance, J. D., 1972: Malawi, Rhodesia and Zambia, ch. 13, pp. 409–460, World survey of climatology, vol. 10: Climates of Africa. Elsevier, New York, 604 pp.

Trenberth, K. E., 1976: 'Spatial and temporal variations of the Southern Oscillation'. *Quart. J. Roy. Meteor. Soc.*, 102, 639–653.

Trenberth, K. E., 1980: 'Atmospheric quasi-biennial oscillations'. *Mon. Wea. Rev.*, 108, 1370–1377.

Trewartha, G. T., 1981: *The Earth's problem climates*. 2nd edition, University of Wisconsin Press, 371 pp.

Troup, A. J., 1965: 'The Southern Oscillation'. *Quart. J. Roy. Meteor. Soc.*, 91, 490–506.

Tsuchiya, I., 1970: 'Year-to-year variations over the India Equatorial Pacific region and of low and middle latitude circulations in the Southern hemisphere'. *Papers in Meteorology and Geophysics, Japan,* **21**, no. 2, pp. 73–87.

Tyson, P. D., 1981: 'Atmospheric circulation variations and the occurrence of extended wet and dry spells over Southern Africa'. *J. Climatol.,* **1**, 115–130.

Tyson, P. D., 1984: 'The atmospheric modulation of extended wet and dry spells over South Africa, 1958–1978.' *J. Climatol.,* **4**, 621–635.

Tyson, P. D., Dyer, T. G. J., 1978: 'The predicted above-normal rainfall in the seventies and the likelihood of droughts in the eighties in South Africa'. *South African J. of Science,* **74**, 372–377.

Tyson, P. D., Dyer, T. G. J., Mametse, M. N., 1975: 'Secular changes in South African rainfall'. *Quart. J. Roy. Meteor. Soc.,* **102**, 817–833.

Valgas, P. R. L., 1982: 'Um estudo climatológico da zona de convergencia intertropical (ZCIT) e sua influencia sobre o Nordeste do Brazil'. *Anais Hidrográficos,* **39**, 175–215.

Van Loon, H., 1984: 'The Southern Oscillation, Part III: associations with the trades and with the trough in the westerlies of the South Pacific Ocean'. *Mon. Wea. Rev.,* **112**, 947–954.

Van Loon, H., Madden, R. A., 1981: 'The Southern Oscillation, Part I: global associations with pressure and temperature in Northern winter'. *Mon. Wea. Rev.,* **109**, 1150–1162.

Van Loon, H., Rogers, J. C., 1978: 'The see-saw in winter temperatures between Greenland and Northern Europe, Part I: General description'. *Mon. Wea. Rev.,* **106**, 296–310.

Van Loon, H., Rogers, J. C., 1981: 'The Southern Oscillation, Part II: associations with changes in the middle troposphere in the Northern winter'. *Mon. Wea. Rev.,* **109**, 1163–1168.

Verma, R. K., 1982: 'Long-range prediction of monsoon activity: a synoptic-diagnostic study'. *Mausam,* **33**, 35–44.

Walker, G. T., 1909: 'Correlation in seasonal variation of climate'. Mem. Indian Meteor. Dept. **20**, part 6, pp. 117–124.

Walker, G. T., 1923: 'Correlation in seasonal variations of weather VIII'. *Mem. India Meteor. Dept.,* **24**, part 4, p. 75–131.

Walker, G. T., 1924: 'Correlation in seasonal variations of weather IX'. *Mem. India Meteor. Dept.,* **24**, part 9, pp. 275–332.

Walker, G. T., 1928a: 'World Weather III'. *Mem. Roy. Meteor. Soc.,* **2**, no. 17, 97–106.

Walker, G. T., 1928b: 'Ceará (Brazil) famines and the general air movement'. *Beitr. Phys. d. freien Atm.,* **14**, 88–93.

Walker, G. T., Bliss, E. W., 1930: 'World weather IV'. *Mem. Roy. Meteor. Soc.,* **3**, no. 24, pp. 81–95.

Walker, G. T., Bliss, E. W., 1932: 'World weather V'. *Mem. Roy. Meteor. Soc.,* **4**, no. 36, pp. 53–84.

Walker, G. T., Bliss, E. W., 1937: 'World weather VI'. *Mem. Roy. Meteor. Soc.,* **4**, no. 39, pp. 119–139.

Wallace, J. M., Gutzler, D. S., 1981: 'Teleconnections in the geopotential height field during the Northern hemisphere winter'. *Mon. Wea. Rev.,* **109**, 784–812.

Walsh, J., 1984: 'Snow cover and atmospheric variability'. *Amer. Scientist,* **72**, 50–56.

Weare, B. C., 1982: 'El Niño and tropical Pacific Ocean surface temperatures'. *J. Phys. Oceanogr.,* **12**, 17–27.

Weare, B. C., 1983: 'Interannual variation in net heating at the surface of the tropical Pacific Ocean'. *J. Phys. Oceanogr.,* **13**, 873–885.

Weare, B. C., 1984: 'Interannual moisture variations near the surface of the tropical Pacific Ocean'. *Quart J. Roy. Meteor. Soc.,* **110**, 489–504.

Webster, P. J., 1981: 'Mechanisms determining the atmospheric response to sea surface temperature anomalies'. *J. Atmos. Sci.,* **38**, 554–571.

Winstanley, D., 1973a: 'Recent rainfall trends in Africa, the Middle East and India'. *Nature,* **243**, 464–465.

Winstanley, D., 1973b: 'Rainfall patterns and the general atmospheric circulation'. *Nature,* **245**, 190–194.

Wood, C. A., Lovett, R. R., 1974: 'Rainfall, drought and the solar cycle'. *Nature,* **251**, 594–596.

Wright, P. B., 1975: 'An index of the Southern Oscillation'. Climatic Research Unit, University of East Anglia, Research Publication No. 4, 22 pp.

Wright, P. B., 1977: 'The Southern Oscillation – patterns and mechanisms of the teleconnections and persistence'. Hawaii Institute of Geophysics, HIG–77–13, 107 pp.

Wright, P. B., 1984: 'Relationships between indices of the Southern Oscillation'. *Mon. Wea. Rev.,* **112**, 1913–1919.

Wright, P. B., 1985: 'The Southern Oscillation: an ocean-atmosphere feedback system?' *Bull. Amer. Meteor. Soc.,* **66**, 398–412.

Wu, M.-C., 1984: 'On the interannual variability of the Indian monsoon and the Southern Oscillation'. Ph.D. Dissertation, Department of Meteorology, University of Wisconsin, Madison, 110 pp.

Wu, M.-C., 1985: 'Studies of the Indian monsoon and the Southern Oscillation', unpublished manuscript.

Wyrtki, K., 1971: *Oceanographic atlas of the International Indian Ocean Expedition*. National Science Foundation, Washington, D.C., 531 pp.

Wyrtki, K., 1973: 'Teleconnections in the Equatorial Pacific'. *Science,* 180, 66–68.

Wyrtki, K., 1974a: 'Sea level and the seasonal fluctuations of the equatorial currents in the western Pacific Ocean'. *J. Phys. Oceanogr.,* 4, 91–103.

Wyrtki, K., 1974b: 'Equatorial currents in the Pacific 1950 to 1970 and their relations to the trade winds'. *J. Phys. Oceanogr.,* 4, 372–380.

Wyrtki, K., 1975a: 'Fluctuations of the dynamic topography in the Pacific Ocean'. *J. Phys. Oceanogr.,* 5, 450–459.

Wyrtki, K., 1975b: 'The dynamic response of the Equatorial Pacific Ocean to atmospheric forcing'. *J. Phys. Oceanogr.,* 5, 572–582.

Wyrtki, K., 1982: 'The Southern Oscillation, ocean-atmosphere interaction and El Niño'. *Marine Technology Society Journal,* 16, 3–10.

Wyrtki, K., 1985: 'Sea level fluctuations in the Pacific during the 1982–83 El Niño'. *Geophys. Res. Lett.,* 12, 125–128.

Wyrtki, K., Eldin, G., 1982: 'Equatorial upwelling events in the Central Pacific'. *J. Phys. Oceanogr.,* 12, 984–988.

Yamazaki, Y., Rao, V. B., 1977: 'Tropical cloudiness over the South Atlantic Ocean'. *J. Meteor. Soc. Japan,* 55, 205–207.

CHAPTER 9

CLIMATE PREDICTION

Climatic disasters of great magnitude are common in many tropical regions (Chapter 8), and they severely affect a range of human activities (Chapter 10). It is therefore not surprising that endeavors to forecast them well in advance extend over a century, and that both the World Climate Program (World Meteorological Organization, 1980, p. 42) and the U.S. National Climate Program (National Climate Program Office, NOAA, 1980, pp. 23–24) identify climate prediction as a major objective. Approaches can be grouped into five major categories: (i) the extrapolation of presumed periodicities, such as the sunspot cycle or other empirically determined time series characteristics; (ii) the assessment of statistical relationships between rainfall and pressure and other data at distant locations; (iii) the relation between rainfall in the pre-season and at the height of the rainy season; (iv) comprehensive diagnostic studies of climate and circulation anomalies combined with statistical methods; and anticipated for the future (v) numerical modelling.

Livezey and Jamison (1977), Nicholls (1980), and Preisendorfer and Mobley (1982) have reviewed the limited success of long-range forecasting in the mid-latitudes. Charney and Shukla (1981) suggest that climatic variability in the tropics should be more predictable because it is in large part due to slowly varying anomalies of the lower boundary of the atmosphere. However, Nicholls' (1980) comprehensive review is still unable to quote climate prediction work proper for the tropics. Major breakthroughs were achieved in the early 1980's.

Some explanation seems in order concerning the meaning of 'climate prediction'. 'Prediction' should refer to a statement about an event from information on antecedent conditions. 'Climate prediction' is a statement on the quality of a (rainy) season or year as a whole, based on conditions more than a month preceding. Contrariwise, where a statement is made from concurrent conditions, terms such as 'specify', 'determine', 'deduce', or 'infer', rather than 'predict' should be used.

More particularly, 'climate prediction' in a real-time operational mode will make statements about future events, and will therefore require the input information on a timely basis. This involves logistic and organizational tasks separate from the 'science' accomplishment of method development.

Crucial to climate prediction is the verification of forecast performance. However, information pertinent to this issue is not available for most works. The survey of climate prediction research for the tropics in the following sections is presented primarily in a regional perspective.

9.1. Indian Monsoon

It seems appropriate to begin with the studies for India, as these include the earliest climate prediction efforts in the tropics. Jagannathan (1960, p. 5) mentions from the pre-instrumental period the almanacs of astrologers and varied notions on weather mechanisms passed on in folk-lores. Progressing to the regular instrumental records, the modern era of climate prediction in India dawned about a century ago. After the great 1877 famine, H. F. Blanford, the then 'Meteorological

Reporter', was called upon to make tentative monsoon forecasts. Referring to that occasion, both Normand (1953) and Jagannathan (1960, p. 5) quote from a Government Famine Commission report: "So far as it may become possible, with the advance of knowledge, to form a forecast in the future, such aids should be made use of, though with due caution". The human impact of droughts in India is discussed in Section 10.6.

Blanford issued monsoon forecasts from 1882 to 1885 (Normand, 1953; Jagannathan, 1960, p. 5), calling attention in particular to the negative relation between Southwest monsoon rainfall and winter snowfall in the Himalayas (Blanford, 1884). Then it was decided to issue monsoon forecasts annually as a matter of routine, but they were held confidential in 1902–1906, during World War II, and later (Jagannathan, 1960, p. 5). Blanford was succeeded by Sir John Eliot, who used the subjective methods of analogs and parallel curves (Normand, 1953), and under whose direction, according to Sir Charles Normand (1953), "the monsoon forecasts became more and more ambitious, though based on slender foundation". Some papers by Dallas (1900, 1902) stem from this era. Sir Gilbert Walker followed Sir John Eliot. In his pioneering work on the Southern Oscillation, Sir Gilbert Walker (ref. Section 8.1) was motivated by the desire to predict India monsoon rainfall anomalies.

Walker introduced the use of spatial correlation with precursors considered indicative from various parts of the globe, apparently in part drawing on earlier work by Blanford (Normand, 1953) and the Lockyers (Lockyer and Lockyer, 1902a, b, 1904; Lockyer, 1906). In a series of papers pre-dating his Southern Oscillation work discussed in Section 8.1 (Walker, 1923, 1924b, 1928a; Walker and Bliss, 1930, 1932, 1937), he dealt with the foundations of the correlation method (Walker, 1909, 1914), with the relation of sunspots to rainfall, temperature, and pressure (Walker, 1915a, b, c), and with the development of regression formulae for monsoon forecasting (Walker, 1910, 1924a).

The formula for India monsoon rainfall published in 1910 (Walker, 1910) has as predictors snowfall accumulation to the North of India, Mauritius pressure, South American (Argentina and Chile) pressure, and Zanzibar rainfall. In 1924 (Walker, 1924a), the Peninsula and Northwest India are treated separately. The predictors for Peninsula rain are South American pressure, Zanzibar rain, and Java rain. The formula for Northwest India rain contains in addition to the former two predictors snowfall to the North of India and Ceylon (Sri Lanka) rain. Apart from rainfall for the entire monsoon period, formulae are also presented for August-September rain in India, with partly different predictors. Not content with the tall job of forecasting India monsoon rainfall, Walker also published forecast formulae for the Nile floods, and Australia rainfall (Walker, 1910; Walker and Bliss, 1930), for Canada winter temperature and South Africa rainfall (Walker and Bliss, 1930), as well as for Ceará (Northeast Brazil) rainfall (Walker, 1928b).

Progressively updated review papers on seasonal forecasting of India monsoon rainfall have been presented by Banerji (1950), Normand (1953), Rao and Ramamoorthy (1960), Jagannathan (1960), and Rao (1965). More recent but short references to climate prediction for India are found in Rao (1976, p. 354–360), Das (1984, 1986) and Thapliyal (1981, 1982). Banerji (1950) states that Walker finally used 28 factors in six forecast formulae differentiated regionally and seasonally, and that at the time of Banerji's writing the India Meteorological Department still used part of the aforementioned factors in five such forecasting formulae, none of which then contained upper-air information. Rao and Ramamoorthy (1960) report that at the time the India Meteorological Department issued three forecasts, two for the summer monsoon period and one for the winter. In addition to various surface elements familiar since Walker's work, upper-air wind at the 2, 4, and 6 km levels over India are mentioned as predictors. In his approximately contemporaneous but more comprehensive review, Jagannathan (1960, pp. 14–31) lists the various elements/areas chosen for prediction, including upper-air winds between 2 and 8 km over India apparently used to a limited extent since the 1920's. He details the varying correlations over the decades between the various surface and upper-air parameters and India monsoon rainfall, and concludes (Jagannathan, 1960, p. 25) that none of the factors show any consistent relationship and that correlations wax and wane over the decades. He further suggests (Jagannathan, 1960, pp. 30–31) that forecast failures can in part be attributed to long-term changes of spatial correlations. Jagannathan (1960, p. 20) and Rao (1965, p. 22) indicate that at the time primarily five factors were used to predict Southwest monsoon rainfall anomalies. These still include four of Walker's favorites, namely South American pressure, South Rhodesia (Zimbabwe) rain, Java rain, and equatorial pressure, while snow accumulation in the Himalaya was omitted. New predictors include the aforementioned upper-air winds over India, and for the prediction of August–September rainfall also Indian river discharge and Punjab temperature range. The incorporation of upper-air information is exemplified in Jagannathan and Khandekar's (1962) study.

Banerjee *et al.* (1978) present a regression formula for predicting the ratio of the number of sub-divisions with normal or above normal rainfall in June–September to the total number of subdivisions (31). The latitude position of the 500 mb ridge along 75°E in April serves as sole predictor, and the performance tested on 16 years of independent data appears remarkable.

On the part of the India Meteorological Department (Jagannathan, 1960, p. 26; Rao, 1965, p. 4; Das, 1984, p. 33; 1986, Chapter 8), the desired standard of performance for forecasting monsoon rainfall has been variously expressed as being an 80% chance of success in a three category scheme. In appraising forecast power, it is essential to distinguish between dependent and independent data sets, but it is not clear whether this distinction has always been made.

Endeavors to forecast the onset of the Southwest monsoon are more recent. Reddy (1977) proposes as predictor the May 50 mb zonal wind component over Singapore, with westerlies presaging an early and easterlies a late onset data. Das (1984, p. 32; 1986, Chapter 8) mentions a prediction scheme for monsoon onset used by the India Meteorological Department, based on various upper-tropospheric wind observations over India.

Kung and Sharif (1980, 1982) developed regression methods for forecasting both the India Southwest monsoon rainfall and the onset date in Kerala, South India, based on April upper-air patterns in the India — Australia region and sea surface temperature around India in the pre-season. For the earlier of the two papers they had available the 700 and 100 mb geopotential height, temperature, and geostrophic wind at grid points over the greater India region. The simple correlations between April upper-air parameters over Kerala and monsoon onset data can be described as follows: an early onset date is heralded by low 700 mb height, but high 100 mb topography and temperature, strong southerly and westerly wind components at 700 mb, but strong easterlies at 100 mb. The highest simple correlation is obtained for kinetic energy at 700 mb. As overall the simple correlations are small, multi-regression is the preferred alternative. Of the aforementioned upper-air parameters 8 are retained in the regression model. Results are summarized in a diagram containing three time series plots for the period 1958–78, namely (i) the observed onset dates, (ii) the onset dates obtained from the regression model for the entire interval 1958–78, while (iii) shows the dates calculated from regression equations based, successively, on all years except the particular year plotted. Curves (i) and (ii) agree closely but (iii) differs more appreciably. This presentation is prompted by the shortness of the upper-air record.

In the second paper, Kung and Sharif (1982) examine the dependence of monsoon onset and rainfall on upper-air parameters over India and Australia, sea surface temperature around India, and Eurasian snow cover. In the regression models 5 parameters are retrieved for monsoon onset and 6 for rainfall. Accumulation of winter snow cover over Eurasia is approximated from monthly precipitation data at selected stations. In contrast to the first paper (Kung and Sharif, 1980), parameters now come from a large portion of the Indian Ocean sector rather than a single location. Nonetheless various of the parameters retained are plausibly consistent with the earlier study. Thus, an early monsoon onset is heralded by high 100 mb topography over Pakistan, large kinetic energy at 700 mb over North India, strong southerly 700 mb wind component over Australia, but furthermore by warm Arabian Sea waters. Abundant monsoon rainfall is preceded by strong 100 mb easterly and southerly wind components over the Persian Gulf, warm 200 mb conditions over New Delhi, warm ocean waters at the South tip of India, and other factors from diverse regions.

Kung and Sharif's (1982) results for the monsoon onset are illustrated in Fig. 9.1:1. This shows for all years during 1958–77 (i) the observed onset dates, (ii) the dates calculated from regression equations developed in Kung and Sharif (1982), based successively on all years except the particular year plotted, and (iii) the same as (ii) but based on the equations in Kung and Sharif (1980). Fig. 9.1:2 similarly shows the results for monsoon rainfall. The observed monsoon rainfall amounts are compared with the values calculated from regression equations developed in Kung and Sharif (1982) based successively on all years except the particular year plotted.

Aside from the considerable diagnostic value of Kung and Sharif's (1980, 1982) results, it must be recognized that such specification of onset dates and rainfall totals within the 1958–78 time span provides no fair measure of performance for forecasts proper of conditions beyond this base interval. Ramage (1983) also points out that in the earlier paper of Kung and Sharif (1980) the performance of onset forecasts is inferior to climatology. Kung (1983) calls attention to the considerable improvement of onset forecasting in the second paper (Kung and Sharif, 1982).

The important role of antecedent upper-air conditions for India monsoon rainfall, as recognized in various earlier studies reviewed above, is emphasized in the recent work of Thapliyal (1981, 1982), who chooses the April latitude position of the 500 mb ridge over India as sole predictor. This limitation appears dictated by his use of an ARIMA (auto-regressive integrated moving average) method. He developed a stochastic dynamic model for prediction. Fig. 9.1:3 shows the input time series of subtropical ridge position together with the output series of monsoon rainfall over the Indian Peninsula. More particularly, he applied the model to predict for the independent data set of the years 1977–80, and concluded that its performance is superior to the multivariate methods hitherto used by the India Meteorological Department.

Further pertinent to India monsoon rainfall prediction is the recent work by Campbell et al. (1983), who show that a substantial portion of interannual rainfall variability in North India may be related to periodic soli-lunar tidal variations and that this component of rainfall variability can be predicted a year in advance. While the effect accounts for only about 10% of the total variance, this portion may in part be independent from the variance accessed by other methods. Accordingly, mechanical tidal forcing appears a useful addition to long-range prediction.

The recent papers by Dhar and Rakhecha (1983) and Shukla and Paolino (1983) already referred to in Section 8.5 are diagnostic rather than prognostic in nature. Shukla and Paolino (1983) explore the possibility of predicting the all-India summer monsoon rainfall from an index of the Southern Oscillation, while Dhar and Rakhecha (1983) consider as possible predictand the winter rainfall of Tamil Nadu in South India, and as predictor the precipitation of the preceding summer monsoon. Both papers may add to a useful diagnostic foundation for the development of climate prediction.

More recent work on climate prediction for India by Wu (1984) is summarized below. This effort follows the method used for Northeast Brazil described in Section 9.4.

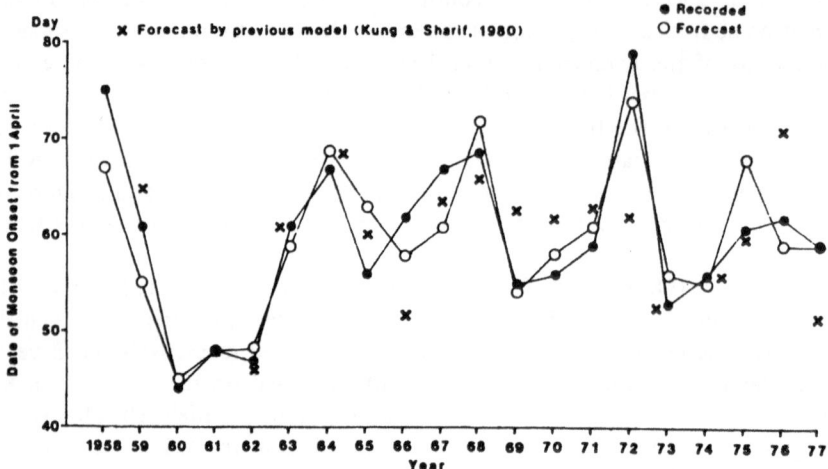

Fig. 9.1:1 Monsoon onset dates on the Kerala coast 1958–77. (i) Dots denote the observed dates; (ii) open circles show dates obtained from Kung and Sharif's (1982) regression equations based, successively, on all years except the particular year plotted; (iii) crosses represent the same as (ii) but for regression equations in Kung and Sharif (1980). From Kung and Sharif (1982).

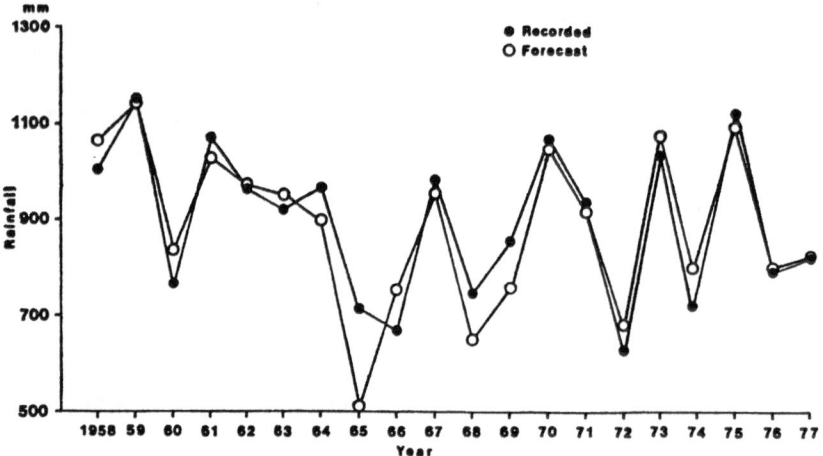

Fig. 9.1:2 Monsoon rainfall in central India (1958–77). (i) Dots denote the observed totals; (ii) open circles show amounts obtained from Kung and Sharif's (1982) regression equations based, successively, on all years except the particular year plotted. From Kung and Sharif (1982).

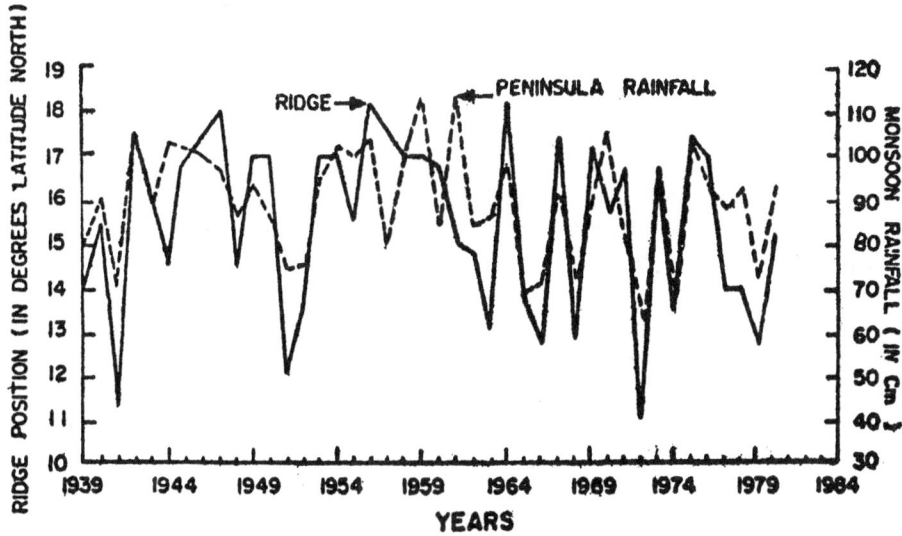

Fig. 9.1:3. Time series plots of latitude position of 500 mb ridge along 75°E over India (input) and of rainfall over Indian peninsula (output). From Thapliyal, 1981 (copyright by the Government of India).

Based on extensive diagnostic studies of the circulation in the Indian Ocean sector (Section 8.5; Tables 8.5:1 to 8.5:4, Fig. 8.5:3) elements/areas most promising as predictors are identified. These include pre-season temperature and pressure over land, the pressure, temperature, wind and cloudiness in the Arabian Sea, a Southern Oscillation index, and various upper-air parameters. These predictor candidates are input to a stepwise multiple regression scheme. The resulting regression model expresses the departure of a rainfall index in terms of a linear combination of departures in selected circulation parameters. This equation is then to be used for prediction on an independent data set.

In Wu's (1984) study, observations of important elements were limited to the period 1951–72. This precluded the possibility to split the entire record into a portion used for the development of the regression model and an independent portion used in prediction experiments, a strategy followed in the Northeast Brazil drought prediction described in Section 9.4. Instead, the afore-mentioned procedure of Kung and Sharif (1980, 1982) was used, whereby values of an annual rainfall index are calculated for individual years from regression equations based on all other years of the series. The coefficient of correlation between the rainfall departures thus calculated and the observed anomalies is +0.92, a performance comparable to that of Kung and Sharif (1982).

9.2. Indonesia Rainfall

The year-to-year variations of rainfall on the densely populated and agriculturally productive islands of Indonesia (ref. Sections 6.8.3, 8.3), very early attracted the attention of Dutch meteorologists. The considerable agricultural and economic implications led to a decades-long effort to predict the vagaries of monsoon rainfall.

Braak (1919) already recognized the essential relationship between long-term pressure, wind, and rainfall variations in the Indonesian region. Pressure variations are remarkably coherent throughout the area but possess largest amplitude over Northern Australia. Consistent with this, during high pressure episodes the West monsoon (November–February) is weaker, and the East monsoon (June–September) is stronger. The inverse wind departures are characteristic of low pressure episodes. Braak (1919, p. 8) further notes that high pressure is associated with deficient rainfall during the East monsoon, although the inverse rainfall departures occur in parts of Java during the West monsoon. Braak (1919, p. 4) further finds a tendency for cloudiness to vary inverse to, but for air temperature to change parallel to pressure. By contrast, sea surface temperature extrema appear to lead the opposite extrema in pressure (Braak, 1919, p. 29). Braak (1919, p. 26) suggests that the enhanced winds of low pressure episodes are, through deeper mixing of the upper ocean, conducive to a cooling of the sea surface, while the opposite effects would prevail during the high pressure episodes. Braak (1919, p. 19) considers that forecasting for the East monsoon rainfall holds the better prospect, and that pressure is to be taken as sole predictor. He then gives (Braak, 1919, pp. 40–41) three rules to anticipate the evolution of pressure departures beyond April.

Braak's (1919) works provide the major basis for Berlage's (1927) paper on East monsoon forecasting for Java. Berlage (1927, pp. 15–27, 38) proposes a prediction of East monsoon rainfall in Java from Batavia (Djakarta) pressure, Singapore air temperature, and an index of wind direction calculated from the West-East rainfall contrast on an exposed mountain. As Braak (1919) before him, he is intrigued by a 3 year periodicity and the sunspot cycle, further proposes a 7 year periodicity, and suggests a relation between the 3 and 7 year cycles. In a sequel paper Berlage (1934) reviews the Southern Oscillation work of Walker and various other studies, reiterates his interest in a 3 year cycle, calls attention to the possible usefulness of tree ring measurements in reconstructing long climate series, comments on the relation of rainfall variations to the Southern Oscillation Indonesia, and expresses reservation on the prospects for monsoon rainfall forecasting.

De Boer (1947) constructed regression equations to forecast at the beginning of September the end of the dry monsoon for 12 separate districts of Java and Madura. In addition he presents regression equations for predictions as early as the beginning of August for 4 districts in Java. The pressure at Kupang, Honolulu, and Mauritius, and the temperature at Hong Kong serve as input information. For the forecasting of the beginning of the dry monsoon at the beginning of March, similar regression equations are presented for 13 districts of Java and Madura, based on pressure at Mauritius, Santiago, and Manila, temperature at Hong Kong, and rainfall at Port Blair. This ambitious formalism, as well as the aforementioned scheme of Berlage (1927), are not supported by verification of performance on an independent data set.

De Boer and Euwe (1949a, b) constructed regression equations to forecast the July–August–September rainfall for parts of Celebes and South Borneo. Input information consists of all or part of the following: pressure at Kupang, Honolulu, Mauritius, Apia, and Hokitika, and temperature at Hong Kong. The same weather factors are chosen as input in Euwe's (1949a, b) attempts to forecast the rainfall during December–January–February and April–May–Jane in parts of Celebes and South Borneo, and in December–January–February on Java and Madura. Again, the method is not tested on an independent data set.

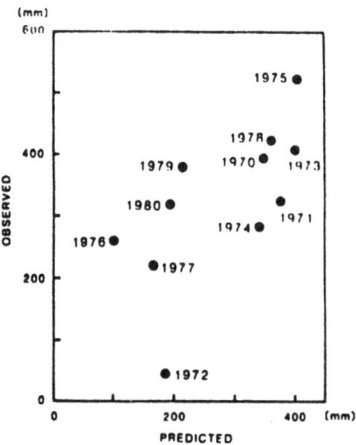

Fig. 9.2:1. Predicted and observed September–November Jakarta rainfall in 1970–80. Predictions made from observed Darwin August pressure using linear regression based on 1951–69 data. Correlation coefficient between predicted and observed is +0.66. From Nicholls (1983).

Schmidt – ten Hoopen and Schmidt (1951) analyzed the secular variations of pressure, temperature, rainfall, and sunshine duration in Indonesia, with particular attention to the 11-year sunspot cycle. They confirm the relation between pressure and rainfall noted earlier by Braak (1919) and Berlage (1927) and consider the latitude position of the confluence between airstreams from the Northern and Southern hemispheres as a factor in Indonesia rainfall. They recommend caution in using the correlation method for seasonal forecasting. Also with reference to seasonal forecasting Schmidt and van der Vecht (1952) call attention to the variation of pressure rainfall correlations over the decades. Reesinck (1952) concludes that forecasting of the beginning of the West monsoon should be based almost exclusively on pressure in the Darwin-Kupang region, and that correlations used to forecast the end of the West monsoon are probably real, but weak and not persistent.

All of the aforementioned schemes of the Dutch school for Indonesia monsoon forecasting lack verification of performance on an independent data set. Accordingly, the practical usefulness of these exercises appears uncertain. However, the various studies are not without interest diagnostically. In hindsight, much of the large-scale long-term relations referred to by the Dutch school of meteorologists are now recognized as symptoms of the Southern Oscillation.

The early Dutch work on Indonesia also provides an essential background for the recent studies of Nicholls (1978, 1979a, b, 1981, 1983). Nicholls (1979a) shows that Darwin January–March temperature departures precede May–October rainfall anomalies of the same sign in Tasmania. Nicholls (1979b) also shows that June–August pressure at Darwin is negatively correlated with tropical cyclone frequency in October–December.

Of greatest interest for Indonesia monsoon forecasting is Nicholls' (1981, 1983) direct follow-up of Braak's (1919) ideas more than 60 years earlier, namely to use Darwin pressure in the first half of the calendar year to predict Java rainfall during the second semester. Most important is Nicholls' (1981, 1983) verification of forecast performance on an independent data set. In particular, Nicholls (1983) constructed a linear single-parameter regression model of Djakarta

September–November rainfall versus Darwin August pressure during 1951–69. He then used
this relationship to predict the Djakarta September–November rainfall during each of the years
1970–80. Results are shown in the scatter plot Fig. 9.2:1. This performance of forecast from
a single predictor for the early West monsoon rainfall is remarkable. Prospects for predicting other
details of Indonesia monsoon rainfall anomalies are less apparent.

Expanding on Nicholls' (1981, 1983) work for Indonesia, Nicholls et al. (1982) and Nicholls
(1984) developed a scheme for predicting the onset of the wet season (around beginning of
November and later) in Northern Australia, again based on the pre-season pressure at Darwin; an
early onset date being heralded by low pressure.

9.3. El Niño

The intimate relation of El Niño events to the Southern Oscillation pressure seesaw (Sections
8.1 to 8.4) suggests that the episodes of warm water and abundant rainfall of the Equatorial
Pacific could be predicted from such straightforeward information as surface pressure observations.
This possibility was explored by Quinn and Burt (1970, 1972) and Quinn (1974, 1976). The
biological and economic impact of El Niño is considered in Section 10.5.

In their first paper, Quinn and Burt (1970) rely solely on the pressure at Darwin. Surface
pressure rising above 1012 mb during May–July is taken as an indication for abnormally heavy
rainfall to occur in the Equatorial Pacific Dry Zone by the latter part of the year. This convention
would have been correct in 76 percent of historical cases. In their second paper (Quinn and Burt,
1972), they use the pressure difference Easter Island minus Darwin, or in fact a simple 'Southern
Oscillation Index'. The decrease of this pressure difference to less than 5 mb is considered critical
for heavy equatorial precipitation to occur. Quinn and Burt (1972) find that this method would
have provided correct forecasts in 88% of cases. Encouraged by these results, Quinn (1974, 1976)
developed the scheme for real-time prediction in an operational mode. In fact, his contribution
was an integral part of the El Niño Watch project in the mid 1970's. However, forecast failures
discredited the approach. Although the pressure seesaw between the western and eastern extremities of the Pacific basin are regarded instrumental in the origin of El Niño, it is realized that
ENSO is a complex phenomenon and can be affected by factors in various parts of the globe.

With a view to El Niño prediction, Barnett (1981) analyzed the wind field over the equatorial
Pacific in relation to sea surface temperature conditions at Talara on the Peru coast, and suggested
that sea surface temperature anomalies could be predicted approximately a year in advance from
knowledge of the wind field alone. More recently, Barnett (1984) applied his statistical prediction
model to the 1982/83 El Niño and concluded that the warming of surface waters off South
America could have been predicted some four months in advance.

Complementing the empirical work of Barnett (1984), Inoue and O'Brien (1984) used the
ocean numerical model of Busalacchi and O'Brien (1980, 1981) in a predictive mode. On this
basis, they concluded that, if the wind changes required to generate a large 'El Niño type' Kelvin
wave (Sections 4.3 and 8.2) have already taken place in the western and/or central Pacific, El Niño
could be predicted one to three months in advance from knowledge of the wind field alone. Inoue
and O'Brien's (1984) paper appears a historically important step towards climate prediction by
numerical modelling.

9.4. The Droughts of Northeast Brazil

The Sêcas of Northeast Brazil represent a climate problem which is unusually well defined in its large-scale circulation setting (Section 8.6) and which has an extraordinary economic and societal impact (Section 10.7). The possibility of predicting them has accordingly long been a central concern. The present section focuses on the Northern Nordeste (Section 8.6), and is based primarily on Hastenrath et al. (1984) and Hastenrath (1984).

In the course of the 20th century, the meteorological community has sporadically attempted to provide seasonal forecasts of this phenomenon (see also Aldaz, 1971, 1983, pp. 6–26; Carvalho, 1973, pp. 193–229). The approaches proposed include: (i) the extrapolation of presumed periodicities such as the sunspot cycle (Sampaio Ferraz, 1950; Markham, 1974; Jones and Kearns, 1976); (ii) the assessment of statistical relationships with pressure and other data at distant locations (Walker, 1928b; Sampaio Ferraz, 1929; Serra, 1956); and (iii) the relation of pre-season rainfall to the precipitation amounts at the peak of the rainy season (Freise, 1938; Markham, 1967; Serra, 1973a, b). Continuing a tradition of coping with the environmental problems of the Nordeste (Carvalho, 1973; Conçalves de Souza, 1979), the Government of Brazil (Conselho Nacional de Pesquisas, 1980) is taking an active interest in the possibility of forecasting the Sêcas.

For purposes of seasonal prediction for the Northern Nordeste, the circulation departure characteristics in the tropical Atlantic sector appear most immediately relevant. However, extreme climatic events in the Nordeste are further associated with circulation anomalies in more remote regions. These have in part been reviewed by Kousky and Moura (1981). As a rule, these teleconnections can be understood in terms of the circulation mechanisms operative in the tropical Atlantic domain, as described above. Thus Namias (1972) pointed out that increased cyclonic activity in the Newfoundland area is associated with abundant rainfall in the Nordeste. There is an inverse coupling of surface pressure between the subpolar and subtropical Atlantic as was shown by Van Loon and Rogers (1978). As explained above, the strengthened North Atlantic high is conducive to a more southerly position of quasi-permanent circulation features over the tropical Atlantic and rainfall in the Northern Nordeste. A similar pressure seesaw is conceivable for the South Atlantic.

Both Northeast Brazil Sêcas and El Niño (Sections 8.2, 9.3) must be viewed as manifestations of the Southern Oscillation; the pressure over the South Atlantic being of immediate relevance for the rainfall in the Northern Nordeste. It is then not surprising that various departure characteristics of the Southern Oscillation in remote regions may coincide with rainfall anomalies in the Nordeste. To the extent that the circulation over the Atlantic is related to processes on a much larger scale, specific predictors from more remote regions may also be pertinent.

As reviewed in Section 8.6, various papers (Markham, 1974; Jones and Kearns, 1976; Kousky and Chu, 1978; Hastenrath and Kaczmarczyk, 1981; Chu, 1984) have presented evidence to the effect that rainfall variability in Northeast Brazil is concentrated with some preference in the frequency bands of somewhat over 2 years, around 14, and around 28 years. Moreover, the studies of Hastenrath and Kaczmarczyk (1981) and Chu (1984) suggest that the circulation and climate anomaly mechanisms identified in Section 5 are with some preference operative at the time scales of 2+ and about 14 and 28 years. It appears tempting to exploit such time series characteristics for long-range rainfall prediction (Sampaio Ferraz, 1950, and others), especially with lead times of several years. Such endeavors hinge on two criteria: (i) the percentage of total rainfall variance explained by the particular frequency band or harmonic; and (ii) the stationarity of the time series.

A limited investigation was made into the possibility of predicting rainfall from past spectral characteristics (Hastenrath et al., 1984). The harmonics corresponding to 26 and 28 months each explain about 2% of the total variance of twelve monthly (October–September) rainfall. The pairs of harmonics corresponding to 13 and 26, or 13.5 and 27 years, explain only about 18% of the total variance of the twelve-monthly (October–September) rainfall. Averaging over 5 year intervals eliminates the higher harmonics, so that the aforementioned pairs of harmonics (14/28 and 13.5/27 years) account for about 40% of the 5 year average annual rainfall. Spectral and harmonic analyses for various portions of the historical record do not yield identical results, so that strict stationarity is not expected for the future. However, sample calculations not presented here suggest the possible usefulness of spectral extrapolation for estimating 5 year average rainfall. Spectral extrapolation yields low rainfall for the mid 1980's.

In the following a method (Hastenrath et al., 1984) is developed for predicting Northern Nordeste rainfall, in the form of March–April and March–September precipitation indices (ref. Section 8.6), using as input information that would exist by certain dates preceding the March/April rainy season peak in the Northern Nordeste, namely through the end of October, November, January, and February. Plausible predictor candidates are identified from the diagnostic studies summarized in Section 8.6, and input to a stepwise multiple regression model (years 1931–40, 46 – Feb 56). The resulting regression equations are then applied in a predictive mode on an independent data set (years 1957–72).

Concerning possible predictors it is recalled from Section 8.6 that droughts in the Northern Nordeste are heralded by an anomalously far northerly position of the equatorial trough and associated symptoms. As the mechanisms of interannual circulation and climate variability in the Brazil – tropical Atlantic sector are to be regarded as modulations of the average annual cycle, departures of pre-season rainfall in Northeast Brazil and Northern South America may precede the precipitation anomalies around the March/April peak of the Northern Nordeste rainy season. Inasmuch as the pressure distribution over the Atlantic – which is instrumental in the Northeast Brazil rainfall anomalies – is related to large-scale mass redistributions of the Southern Oscillation type (Hastenrath and Heller, 1977), various indicators of the Southern Oscillation and parameters from more remote parts of the globe are of interest.

With a view towards identifying suitable predictor candidates, isocorrelate maps were constructed of the October–September Northeast Brazil rainfall index (ref. Section 8.6) with pressure, sea surface temperature, zonal and meridional wind components, and cloudiness over the Atlantic. The regions of correlation exceeding 0.3 in absolute value were identified, and time series compiled as input to a regression model. The screening is intended to safeguard against noise. Other predictor candidates considered include pre-season rainfall in the Northern Nordeste itself, rainfall at Paramaribo and Georgetown in the Guyanas, pressure at Port Stanley in the South Atlantic, temperature in the Southeastern United States, India rainfall, and Wright's (1975, 1977) Southern Oscillation index. Of 17 candidate series, 11 possess a correlation with Northeast Brazil exceeding 0.3 in absolute value, and were input into a stepwise multiple regression scheme with 5% significance margin. This identified as most important predictors pressure, sea surface temperature, and zonal and meridional wind components over the tropical Atlantic. Of further use was the pre-season rainfall in Northern Northeast Brazil itself, and the May–June–July Southern Oscillation index value.

Regression models were constructed for the record 1921–40 plus 1946–56 for all months from October through March for two 'predictands', namely the March–April and the March–September rainfall indices. Details are summarized in Table 9.4:1. This is interesting in relation to the general circulation diagnostics of rainfall anomalies (Section 8.6) which identifies an *ensemble* of departure characteristics. In fact, the various elements list in Table 9.4:1 are in part strongly correlated with each other. Accordingly, a given element may be rejected by the regression model,

TABLE 9.4:1

Stepwise multiple regression model, with significance level α = 5%, period 1921−40 plus 46−56, for regression on March−April and March−September rainfall index. Symbols are as follows: pressure P, sea surface temperature T, zonal U and meridional wind component V; n and s denote location in Northern and Southern hemisphere, respectively. NER denotes pre-season rainfall in Northeast Brazil, and SOI the May−June−July Southern Oscillation Index (Wright, 1975, 1977). The column marked VAR lists the percentage variance explained, and x indicates the predictors retained by the regression scheme.

Predictor	VAR	P_n	T_n	T_s	U	V	NER	SOI
Regression on March−April
OCT	18				x			
NOV	55		x	x	x			
DEC	25	x			x			
JAN	63				x		x	
FEB	56	x	x	x				
MAR	71			x	x	x	x	
Regression on March−September
OCT	34		x		x			
NOV	61	x		x	x			
DEC	48	x				x		
JAN	66				x		x	x
FEB	71	x	x	x	x			

TABLE 9.4:2

Coefficients (in 10^{-3}) of stepwise multiple regression model, with significance level α = 5%, period 1921−40, 46−56, for regression of March−April and March−September rainfall index. Symbols are as follows: pressure P, sea surface temperature T, zonal U and meridional with component V; n and s denote location in Northern and Southern hemisphere, respectively. NER signifies pre-season rainfall in Northeast Brazil, and SOI the May−June−July Southern Oscillation Index (Wright, 1975, 1977).

Predictor	CONST	P_n	T_n	T_s	U	V	NER	SOI
Month
Regression on March−April
NOV	+ 99		−437	+649	+ 343			
JAN	+142				+1125		+497	
Regression on March−September
NOV	+ 82	+461		+615	+ 214			
JAN	+ 13				+1093		+541	+372

Example:
March−April rainfall index = +0.099
    −0.437 × (Nov SST anomaly in °C in Northern hemisphere region)
    +0.649 × (Nov SST anomaly in °C in Southern hemisphere region)
    +0.343 × (Nov zonal wind anomaly in m s^{-1} in equatorial region (ref. Hastenrath *et al.*, 1984)

TABLE 9.4:3

Correlation coefficients in 10^{-2} for (A) regression and (B) forecast periods. One and two asterisks denote significance at the 5, and 1% levels, respectively. Quenouille's (1952, p. 168) method was used to account for the reduction of effective degrees of freedom due to persistence.

	A Regression Period 1921−40, 46−50	B Forecast period 1958−72
March−April rainfall index		
NOV	+74	+21
JAN	+79	+64**
March-September rainfall index		
NOV	+78	+40
JAN	+81	+68**

as its information content is already available from other elements with stronger correlation. It is noted from Table 9.4:1 that cloudiness is not retained as a predictor. Conceivably, cloudiness as such is indicative of circulation departures, but this element is not well observed from aboard ship.

Table 9.4:1 shows for both 'predictands' (March−April and March−September rainfall indices) a general increase of the explained variance as one approaches the peak rainy season; no hypothesis being offered for the remarkably looser relation in December. Thus the explained variance is largest for the months November, January, and February (respectively March). The March−April and the March−September rainfall are specified to a very high degree by the departure ensemble of the general circulation in February and March. However, February leaves no lead time for the practical use of the forecast, while November accounts only for a small percentage of the rainfall variability. Accordingly, the January models merit particular attention.

Table 9.4:2 lists the coefficients of the regression models (period 1921−40 plus 45−46) with significance level $\alpha = 5\%$, for November and January, and parts A of Figs. 9.4:1 and 9.4:2 are scatter diagrams for January of 'regressed' versus 'observed' values of the rainfall indices. Parts A of Figs. 9.4:1 and 9.4:2 illustrate the goodness of fit of the models within the input period (1921−40 plus 45−56). For November, which is less suited for prediction, only correlation coefficients are presented in Table 9.4:3 but no scatter diagrams. Parts A of Figs. 9.4:1 and 9.4:2 and of Table 9.4:3 serve as reference for rainfall index forecasts for years beyond the base period of the regression models, as presented in parts B of Figs. 9.4:1 and 9.4:2 and Table 9.4:3.

The regression models developed from the 1921−40 plus 1946−56 record were used to predict the rainfall anomalies of the 15 years 1958−1972. Parts B of Figs. 9.4:1 and 9.4:2 are scatter plots of the January models predicting the March−April and March−September rainfall indices, respectively. Information such as parts A of Figs. 9.4:1 and 9.4:2 serves as background reference for rainfall index forecasts for years beyond the base period of the regression models, as presented in parts B of Figs. 9.4:1 and 9.4:2.

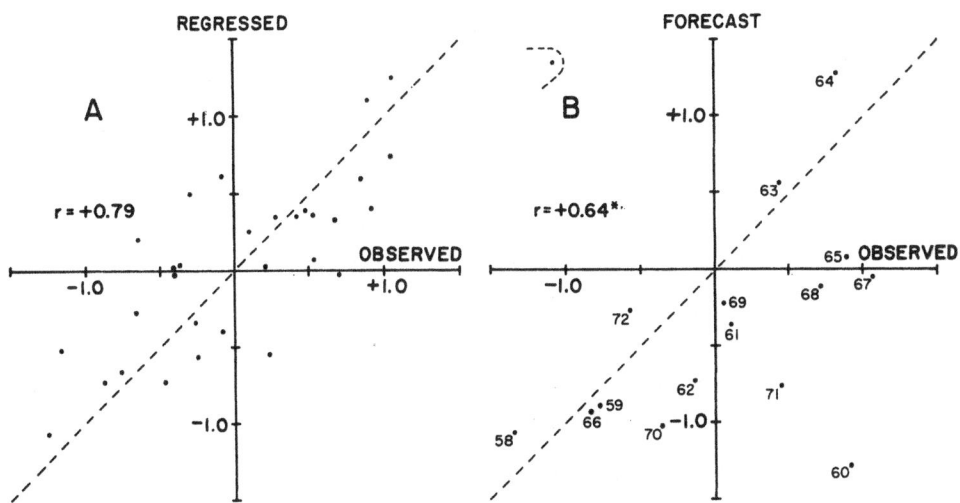

Fig. 9.4:1 Scatter diagrams of March–Arpil Nordeste rainfall index based on observations through January. (A) Regression, period 1921–40, 46–56, correlation coefficient r = +0.79. (B) Forecast, period 1958–72, with numbers indicating the years. r = +0.64*. Broken lines denote 45 degree angle. Asterisk indicates significance of correlation coefficient at the one percent level. Quenouille's (1952, p. 168) method is used to account for the reduction of effective degrees of freedom due to persistence. From Hastenrath *et al.* (1984).

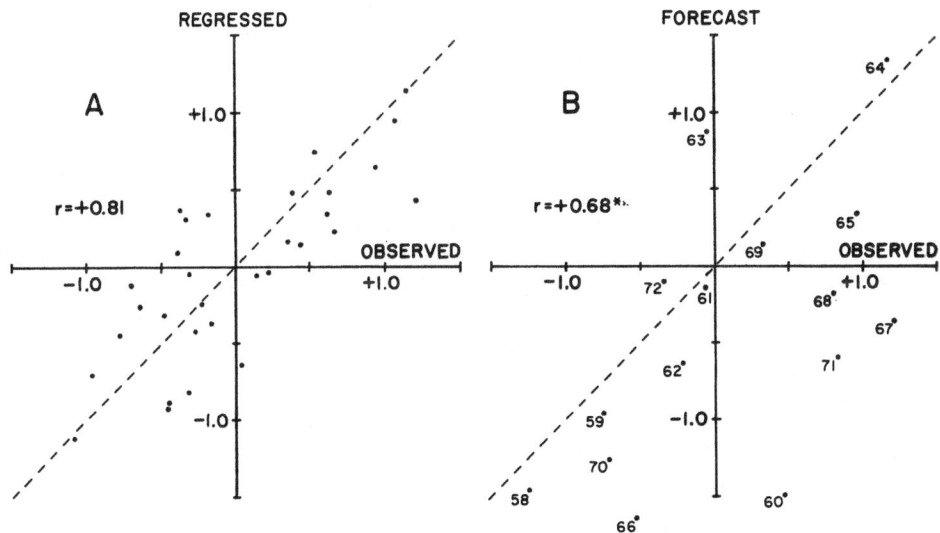

Fig. 9.4:2. Scatter diagrams of March–September Nordeste rainfall index based on observations through January. (A) Regression, r = +0.81. (B) Forecast, r = +0.68* Details as for Fig. 9.4:1. From Hastenrath *et al.* (1984).

As a basis for the appraisal of forecast performance, various forms of evidence are considered (i) The scatter diagram in parts B of Figs. 9.4:1 and 9.4:2 show a positive relation between forecast and observed values, although the scatter is larger than for the regression model plotted in part A. (ii) Correlation coefficients between forecast and observed values such as given in parts B of Figs. 9.4:1 and 9.4:2 are high. (iii) In addition, skill scores (Panofsky and Brier, 1968, pp. 191–208) and stochastic joint probabilities (Preisendorfer and Mobley, 1982) were calculated. These also show appreciable forecast skill. The pertinent information is contained in Figs. 9.4:1 and 9.4:2, but details are not discussed here. In context, the various measures of prediction quality all bear out a remarkable forecast performance. The forecasts are particularly good for the extreme drought year 1958.

Figs. 9.4:1 and 9.4:2 show particularly good forecasts for the extreme drought year 1958, as well as for the dry years 1959 and 1966, and the wet year 1964. By contrast, the discrepancy is worst for the January forecasts of the March–April and March–September rainfall of 1960. This becomes understandable from the peculiar seasonal evolution from late 1959 into the March/April rainy season: the large-scale circulation conditions (as well as pre-season rainfall) remained as characteristic of droughts through February 1960, and then changed unusually abruptly to a setting typical of abundant rainfall.

The study (Hastenrath *et al.*, 1984) shows that the March–April and entire season rainfall is specified to a very high degree by the departure ensemble of the large-scale circulation especially in the tropical Atlantic around the peak rainy season. To a lesser extent, the rainfall anomalies are presaged by the antecedent establishment of circulation anomaly patterns. The severe 1958 drought in particular was highly predictable from this methodology.

Although a wide range of dynamically plausible predictor candidates was examined, it is found that the large-scale atmospheric and oceanic fields in the tropical Atlantic are most essential for the prediction of extreme hydrometeorological regimes in the Northern Nordeste. For operational applications, such parameters are required on a timely basis. Remote sensing by satellite appears an obvious choice. However, for climate diagnostics and prediction, long homogeneous time series of internally consistent parameters are needed, absolute calibration not being of essence. Such a satellite-derived reference climatology, which would have to extend over more than ten years, has not been created and would require a substantial effort by a large organization.

With a view towards practical usefulness it must further be recognized that the Northeast Brazil drought problem poses demands that go beyond meteorological expertise. Questions as the following two arise: Should a meteorological drought prediction become possible, what use would be made of it? What form and what details of prediction are most needed? A similar pragmatic concern has recently been voiced by Lamb (1981) for the climatic forecasting effort in general. Clearly, the wisdom of agriculturists, economists, and other planners is called upon, not only in the eventual application but also in the very design of the prediction effort.

In conclusion, the Northeast Brazil Sêcas represent a climate problem that is unusually well defined in terms of the large-scale circulation, and that has an extraordinary economic and societal impact. Diagnostic research may serve as foundation for monitoring and prediction systems. Development of such systems on an operational basis requires a substantial material commitment and will thus hinge on the realization of priorities in science and society.

9.5. Subsaharan Drought

The widespread publicity given to the Sahel drought which began in the late 1960's (Section 8.8) and its severe social and economic consequences (Section 10.8) was followed by various papers proposing to forecast climatic disasters in this part of the World. Based on the rainfall record at 8 stations spread between the Atlantic and India, Winstanley (1973a, b) ventures a prophecy into the third millenium. Less ambitiously, Winstanley (1974) proposes to forecast June–September rainfall anomalies in Subsaharan West Africa from June rainfall totals, an exercise queried by Bunting *et al.* (1975) who point out that June precipitation is essentially uncorrelated to that for July–September. Wood and Lovett (1974) propose rainfall forecasts for Ethiopia based on the 11-year sunspot cycle. Faure and Gac (1981a, b) extrapolate a 30-year periodicity in the Senegal River runoff to infer a termination of the current drought around 1985.

While the aforementioned papers give little attention to the underlying general circulation conditions, a recent note by Adedokun (1979) seems a step in the right direction. Drawing on the annual cycle latitudinal shifts of the Intertropical Discontinuity (Section 6.7.2.3) and the associated steep meridional precipitation gradients over West Africa (Section 8.8), Adedokun (1979) suggests seasonal rainfall prediction for Subsaharan Africa based on pre-season precipitation at stations in the South.

9.6 Southern Africa

As pointed out in Section 8.11, rainfall in parts of Southern Africa exhibits a preference for variability on the time scale of about two decades. This behavior is noted for the northeastern portion of South Africa in particular (Tyson *et al.*, 1975; Tyson and Dyer, 1978, 1980; Tyson, 1981). Tyson (1981) has related this quasi-periodicity in summer rainfall to variations in the Southern hemisphere upper-air circulation. From an extrapolation of these time series characteristics, Tyson and Dyer (1978, 1980) suggest the likelihood of above normal rainfall in the 1970's and of drought in the 1980's, but they refrain from statements concerning individual years and caution against a possible non-stationarity of time series behavior.

9.7. Kenya Rainfall

Sansom (1955) described a procedure for the prediction of precipitation anomalies in a portion of the Central Kenya highlands, where the rainfall activity is concentrated in March–May and September–December, the so-called 'long rains' and 'short rains', respectively (ref. Section 6.8.3). For the March–May rainfall he proposed a formula using as predictors the December–February values of pressure at Cape Town, South Africa, and Alexandria, Egypt, and of rainfall at Tabora in Tanzania. The formula for the October–December precipitation contains as predictors the July–September values of pressure at Mauritius and of rainfall at Khartoum, Sudan, as well as the March–May rainfall of the same year in Central Kenya. The comparison of forecast versus observed rainfall departures presented for three years appears encouraging, but no information seems available concerning any operational application after 1954.

Since 1984, the Kenya Meteorological Department issues forecasts for the two rainy seasons in the first and second semesters of the year. The method has not been published, but relies on a form of extrapolation of rainfall time series of Kenya stations.

9.8. North Atlantic Hurricanes

Drawing on his extensive investigations into the large-scale environmental factors for Tropical Storm formation (Section 7.2), Gray (1983a, 1984c, d) studied the general circulation diagnostics of interannual variability in the frequency of North Atlantic Hurricanes. He finds a decreased frequency during El Niño (Section 8.2) years, presumably related to the enhanced vertical wind shear resulting from strengthened subtropical upper-tropospheric westerlies. He further notes that Hurricanes tend to be less frequent when 30 mb equatorial winds are easterly or becoming easterly during the Northern summer. Although the regression models have not yet been tested on an independent data set, Gray (1983b, 1984a, b, 1985a, b) applies these diagnostic results to the seasonal forecasting of West Atlantic seasonal Hurricane activity.

9.9. Hong Kong Climate

This section reports on a series of thought-provoking papers on seasonal forecasting for the Far East by G. Bell (1976a, b, 1977). Hong-Kong summer (June–September) rainfall tends to be negatively correlated with the Irkutsk minus Tokyo pressure difference in January. This relationship (Bell, 1976a) is strikingly illustrated in Fig. 9.9:1. This finding is supported by analyses of 700 and 500 mb charts (Bell, 1976b). High/low index Northern hemisphere circulations in January tend to be followed by low/high index circulations in the early summer, which in turn are associated with high/low Hong Kong rainfall. Successful forecasts of Hong Kong summer rainfall based on these results have been issued for many years (Bell, 1976a). There is, however, a considerable secular variation in these relationships (Bell, 1977).

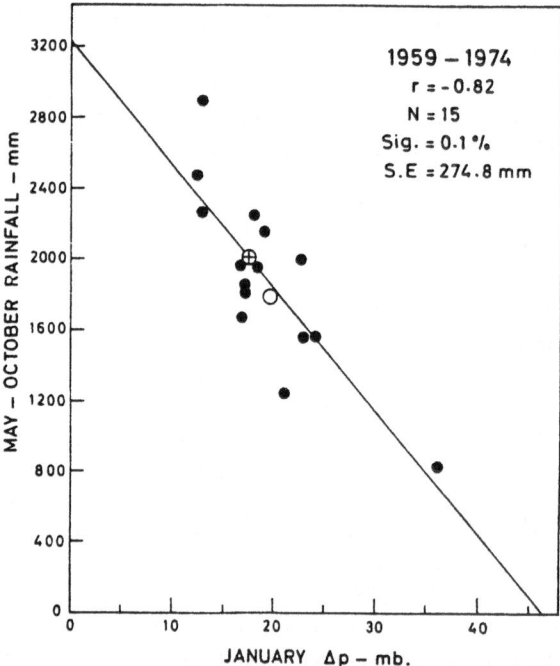

Fig. 9.9:1. Scatter plot of June–September rainfall in Hong Kong versus the January pressure difference Δp Irkutsk minus Tokyo. The open circle refers to the 1974 values not included in the calculation of the correlation coefficient r. The cross corresponds to the 1975 value of Δp. From Bell (1976a).

9.10. Synthesis and Outlook

In large part of the tropics, climate prediction is of far greater practical importance than daily weather forecasting, with the notable exception of such severe weather systems as Tropical Storms and Squall Lines. It is then not surprising that attempts at climate prediction have a long tradition in some densely populated agricultural lands of the tropics, in particular India and Indonesia.

In India the early efforts in the latter part of the last century relied heavily on winter snow accumulation in the Himalaya as a telltale factor for Southwest monsoon rainfall. Later on, other parameters from various parts of the tropics were added, in particular pressure and rainfall. In the early years of the 20th century the seasonal monsoon forecasts developed into ambitious detail not warranted by the limited input information, and fell into disrepute. The subsequent forecasting contributions of Sir Gilbert Walker were closely related to his studies of the Southern Oscillation. His multiple regression schemes and input surface parameters remained the mainstay of the India Meteorological Department's monsoon forecasting activities for decades. New predictors were introduced later on; most significant among these was the inclusion of upper-air parameters over India. Forecast equations and input parameters are continuously being updated and changed. Although predictions are issued every year, verifications of forecast performance over the decades does not seem to be available in published form. A desired performance standard, as stated by the India Meteorological Department, is to be correct in 80% of cases for forecasts formulated in three categories. The most relevant recent studies include Thapliyal (1981) and Kung and Sharif (1980, 1982). Seemingly dictated by the constraints of the sophisticated statistical procedure used, Thapliyal's (1981) scheme is limited to a single predictor, namely the April latitude position of the 500 mb ridge over India. Verification for 4 years of independent data suggests a good performance. Kung and Sharif's (1980, 1982) work is based on a larger number of upper-air and surface parameters and suggests appreciable skill.

In Indonesia Braak (1919) recognized the associations of pressure, wind, and rainfall variations, and suggested pressure as sole predictor for monsoon rainfall anomalies. Prediction formulae developed later on excelled in ambitious detail but their performance was apparently never verified on an independent data set. Only recently did Nicholls (1983) present a convincing verification of performance of rainfall prediction for 11 years of an independent data set, his sole predictor being Darwin pressure — as suggested by Braak (1919) more than 60 years earlier!

The Southern Oscillation pressure seesaw has been used for prediction of El Niño, but repeated forecast failures discredited this attempt.

Various methods have been tried over the decades to predict the droughts of Northeast Brazil. Based on extensive diagnostic analysis and stepwise multiple regression, a climate prediction method was constructed, tested on 15 years of independent data, and found to perform well (Hastenrath et al., 1984). Development of suitable data banks is essential for applications of this method on a real-time operational basis.

For Subsaharan Africa, Kenya, as well as for Southern Africa extrapolation of time series behavior has been proposed as a means to estimate future rainfall anomalies. Diagnostic studies indicate the possibility to foretell such diverse phenomena as summer rainfall in Hong Kong and seasonal frequency of Tropical Storms in the North Atlantic. Bell's (1976a, b, 1977) method for Hong Kong has been used operationally for several years, and Gray (1984a, b) issues and verifies predictions of North Atlantic Tropical Storm frequency on an experimental basis.

In closing this review of endeavors at climate prediction, it seems appropriate to consider the potential of the various approaches. Prediction from extrapolation of time series behavior hinges on two criteria, namely (i) stationarity, and (ii) percentage of variance explained by the quasi-periodic variation. Because of the latter aspect, the approach seems more appropriate for the average of several years rather than for an individual year. Another approach which offers greater prospect combines extensive diagnostic analysis with statistical methods, such as stepwise multiple regression. The choice of good predictors on the basis of sound general circulation diagnostics appears essential. It is here suggested that the sophistication of the statistical method chosen in the prediction exercise should be commensurate with the quality of the diagnostic analysis and of the input observations. The performance of the empirical method combining general circulation diagnostics and statistics should be verified on an independent data set. Various results of this class of approach indicate that climate prediction on an empirical basis is possible for *certain* tropical climate problems. Finally, numerical models have been suggested as a possible tool in climate prediction. While some of the empirical methods reviewed here could already be translated into operation at least for *certain* tropical regions, numerical models may achieve a comparable performance only in the more distant future. At any rate, a sound empirical understanding seems a prerequisite for the design of realistic numerical models. Finally, the meteorological community must give serious consideration to the credibility of climate forecasts: publication of the method and verification of results are imperative.

References

Adedokun, J. A., 1979: 'An empirical method for forecasting the April–August precipitation in parts of West Africa'. pp. 546–557 in: Adefolalu, D. O., ed., 'Pre-WAMEX Symposium on the West African Monsoon'. Publication of the Nigerian Meteorological Services, 600 pp.

Aldaz, L., 1971: 'A partial characterization of the rainfall regime of Brazil'. SUDENE, Publicacaõ Técnica No. 4, DINMET vol. 1., Rio de Janeiro.

Aldaz, L., 1983: 'Aplicación del concepto de interacción de escalas temporales al fenómeno de las sequías extremas en el Nordeste del Brasil'. Doctoral Diss., Universidad Complutense, Madrid, 329 pp.

Banerjee, A. K., Sen, P. N., Raman, C. R. V., 1978: 'On foreshadowing Southwest monsoon rainfall over India with mid-tropospheric circulation anomaly of April'. *Indian J. Meteor. Hydrol. Geophys.*, **29**, 425–431.

Banerji, S. K., 1950: 'Methods of foreshadowing monsoon and winter rainfall in India'. *Indian J. Meteor. Geophys.*, **1**, 4–14.

Barnett, T. P., 1981: 'Statistical relations between ocean-atmosphere fluctuations in the tropical Pacific'. *J. Phys. Oceanogr.*, **11**, 1043–1058.

Barnett, T. P., 1984: 'Prediction of the El Niño of 1982–83'. *Mon. Wea. Rev.*, **112**, 1403–1407.

Bell, G. J., 1976a: 'Seasonal forecasts of Hong Kong summer rainfall'. *Weather*, **31**, 208–212.

Bell, G. J., 1976b: 'Seasonal forecasts and northern hemisphere circulation anomalies'. *Weather*, **31**, 282–292.

Bell, G. J., 1977: 'Changes in sign of the relationship between sunspots and pressure, rainfall and the monsoons'. *Weather*, **32**, 26–32.

Berlage, H. P., 1927: 'East-monsoon forecasting in Java'. Koninklijk Magnetisch en Meteorologisch Observatorium te Batavia, Indonesia, Verhandelingen, No. 20, 42 pp.

Berlage, H. P., 1934: 'Further research into the possibility of long range forecasting in Netherlands-India'. Koninklijk Magnetisch en Meteorologisch Observatorium te Batavia, Indonesia, Verhandelingen No. 26, 31 pp.

Blanford, H. F., 1884: 'On the connexion of the Himalaya snowfall with dry winds and seasons of drought in India'. *Proc. Roy. London*, **37**, 3–22.

Braak, C., 1919: 'Atmospheric variations of short and long duration in the Malay Archipelago and neighboring regions, and the possibility to forecast them'. Koninklijk Magnetisch en Meteorologisch Observatorium te Batavia, Indonesia, Verhandelingen No. 5, 57 pp.

Bunting, A. H., Dennett, M. D., Elston, J., Milford, J. R., 1975: 'Seasonal rainfall forecasting in West Africa'. *Nature*, **253**, 622–623.

Busalacchi, A. J., O'Brien, J. J., 1980: 'The seasonal variability in a model of the tropical Pacific'. *J. Phys. Oceanogr.*, **10**, 1929–1951.

Busalacchi, A. J., O'Brien, J. J., 1981: 'Interannual variability of the equatorial Pacific in the 1960's'. *J. Geophys. Res.*, **86**, 10901–10907.

Campbell, W. H., Blechman, J. B., Bryson, R. A., 1983: 'Long-period tidal forcing of Indian monsoon rainfall: an hypothesis'. *J. Atmos. Sci.*, **22**, 287–296.

Carvalho, O., ed., 1973: 'Plano integrado para o combate preventivo aos efeitos das Sêcas do Nordeste', *Ministerio do Interior, Serie Desenvolvimento Regional*, **1**, 267 pp.

Charney, J., Shukla, J., 1981: 'Predictability of monsoons'. pp. 99–109 in: Lighthill, Sir J., Pearce, R. P., eds., *Monsoon dynamics*, Cambridge University Press, 735 pp.

Chu, P. S., 1984: 'Time and space variability of rainfall and surface circulation in the Northeast Brazil – tropical Atlantic sector'. *J. Meteor. Soc. Japan*, **62**, 363–370.

Conselho Nacional de Pesquisas, Brazil, 1980: 'Workshop on drought forecasting for Northeast Brazil, INPE, February 1980'. São José dos Campos, 71 pp.

Dallas, W. L., 1900: 'A discussion on the failure of the Southwest monsoon rains in 1899'. *Mem. India Meteorol. Dept.*, **12**, 1–30.

Dallas, W. L., 1902: 'Meteorological history of the seven monsoon seasons 1893–1899, in relation to the Indian rainfall'. *Mem. India Meteorol. Dept.*, **12**, 409–486.

Das, P. K., 1984: 'The monsoons – a perspective'. India National Science Academy, Perspective Report Series 4, New Delhi, 52 pp.

Das, P. K., 1986: 'Monsoons'. Fifth IMO lecture, World Meteorological Organization, Geneva, in press.

de Boer, H. J., 1947: 'On forecasting the beginning and the end of the dry monsoon in Java and Madura'. Koninklijk Magnetisch en Meteorologisch Observatorium te Batavia, Indonesia, Verhandelingen No. 32, 20 pp.

de Boer, H. J., Euwe, W., 1949a: 'On long-periodical temperature variations'. Koninklijk Magnetisch en Meteorologisch Observatorium te Batavia, Indonesia, Verhandelingen No. 35, 16 pp.

de Boer, H. J., Euwe, W., 1949b: 'Forecasting rainfall in the period July–August–September for parts of Celebes and South Borneo'. Koninklijk Magnetisch en Meteorologisch Observatorium te Batavia, Indonesia, Verhandelingen No. 36, 14 pp.

Dhar, O. N., Rakhecha, P. R., 1983: 'Foreshadowing Northeast monsoon rainfall over Tamil Nadu, India'. *Mon. Wea. Rev.*, **111**, 109–112.

Euwe, W., 1949a: 'Forecasting rainfall in the periods December–January–February and April–May–June for parts of Celebes and South Borneo'. Koninklijk Magnetisch en Meteorologisch Observatorium te Batavia, Indonesia, Verhandelingen, No. 38, 23 pp.

Euwe, W., 1949b: 'Forecasting rainfall for the period December–January–February for Java and Madoera'. Koninklijk Magnetisch en Meteorologisch Observatorium te Batavia, Indonesia, Verhandelingen No. 39, 16 pp.

Faure, H., Gac, J. Y., 1981a: 'Will the Sahelian drought end in 1985?' *Nature*, **291**, 475–478.

Faure, H., Gac, J. Y., 1981b: 'Senegal river runoff, reply'. *Nature*, **293**, 414.

Freise, F. W., 1938: 'The drought region of Northeastern Brazil'. *Geogr. Rev.*, **28**, 363–378.

Gonçalves de Souza, J., 1970: *O Nordeste Brasileiro*. Banco do Nordesto do Brasil, S. A., Fortaleza, 409 pp.

Gray, W. M., 1983a: 'Atlantic seasonal hurricane frequency. part 1, El Niño and 30 mb QBO influences, part 2. forecasting its variability'. Department of Atmospheric Sciences, Colorado State University, Fort Collins, Colorado, Atmospheric Science Paper No. 370, 48 pp.

Gray, W. M., 1983b: 'Forecasts of West Atlantic seasonal hurricane activity for 1983 and 1984'. Department of Atmospheric Science, Colorado State University, 8 pp.

Gray, W. M., 1984a: 'Forecast of Atlantic seasonal hurricane activity for 1984'. Department of Atmospheric Science, Colorado State University, 34 pp.

Gray, W. M., 1984b: 'Summary of 1984 Atlantic seasonal tropical cyclone activity and verification of author's forecast'. Department of Atmospheric Science, Colorado State University, 27 pp.

Gray, W. M., 1984c: 'Atlantic seasonal hurricane frequency, part 1: El Niño and 30 mb quasi-biennial oscillation influences'. *Mon. Wea. Rev.*, **112**, 1649–1668.

Gray, W. M., 1984d: 'Atlantic seasonal hurricane frequency, part 2: forecasting its variability'. *Mon. Wea. Rev.*, **112**, 1669–1683.

Gray, W. M., 1985a: 'Forecast of Atlantic seasonal hurricane activity for 1985. Department of Atmospheric Science, Colorado State University, Fort Collins, Colo., 34 pp.

Gray, W. M., 1985b: 'Updated (as of 27 July 1985) forecast of Atlantic seasonal hurricane activity for 1985.' Department of Atmospheric Science, Colorado State University, Fort Collins, Colo., 12 pp.

Hastenrath, S., 1984: 'Predictability of Northeast Brazil drought'. *Nature*, **307**, 531–533.

Hastenrath, S., Heller, L., 1977: 'Dynamics of climatic hazards in Northeast Brazil'. *Quart. J. Roy. Meteor. Soc.*, **103**, 77–92.

Hastenrath, S., Kaczmarczyk, E. B., 1981: 'On spectra and coherence of tropical climate anomalies'. *Tellus*, **33**, 453–462.

Hastenrath, S., Wu, M.-C., Chu, P.-S., 1984: 'Towards the monitoring and prediction of Northeast Brazil droughts'. *Quart. J. Roy. Meteor. Soc.*, **110**, 411–425.

Inoue, M., O'Brien, J. J., 1984: 'A forecasting model for the onset of a major El Niño'. *Mon. Wea. Rev.*, **112**, 2326–2337.

Jagannathan, P., 1960: 'Seasonal forecasting in India, a review'. Meteorological Office, Poona, 79 pp.

Jagannathan, P., Khandekar, M. L., 1962: 'Predisposition of the upper air structure in March to May over India to the subsequent monsoon rainfall of the peninsula'. *Indian J. Meteorol. Geophys.*, **13**, 305–316.

Jones, R. H., Kearns, J. P., 1976: 'Fortaleza, Ceará, Brazil, rainfall'. *J. Appl. Meteor.*, **15**, 307–308.

Kousky, V. E., Chu, P.-S., 1978: 'Fluctuations in annual rainfall for Northeast Brazil'. *J. Meteor. Soc. Japan*, **57**, 457–465.

Kousky, V. E., Moura, A. D., 1981: 'Previsão de precipitacão no Nordeste do Brazil: o aspecto dinâmico'. IV Simpósio Brasileiro de Hidrología e Recursos Hídricos, Nov. 1981, Fortaleza.

Kung, E. C., 1983: 'Reply (to comments by C. S. Ramage)'. *J. Climate Appl. Meteor.*, **22**, 1134–1135.

Kung, E. C., Sharif, T. A., 1980: 'Regression forecasting of the onset of the India summer monsoon with antecedent upper air conditions'. *J. Appl. Meteor.*, **19**, 370–380.

Kung, E. C., Sharif, T. A., 1982: 'Long-range forecasting of the Indian summer monsoon onset and rainfall with upper air parameters and sea surface temperature'. *J. Metor. Soc. Japan*, **60**, 672–681.

Lamb, P J., 1981: 'Do we know what we should be trying to forecast – climatically?' *Bull. Amer. Meteor. Soc.*, **62**, 1000–1001.

Livezey, R. E., Jamison, S. W., 1977: 'A skill analysis of Soviet seasonal weather forecasts'. *Mon. Wea. Rev.*, **105**, 1491–1500.

Lockyer, N., Lockyer, W. J. S., 1902a: 'On some phenomena which suggest a short period of solar and meteorological changes'. *Proc. Roy. Soc. London*, **70**, 500–504.

Lockyer, N., Lockyer, W. J. S., 1902b: 'On the similarity of the short-period pressure variation over large areas'. *Proc. Roy. Soc. London*, **71**, 134–136.

Lockyer, N., Lockyer, W. J. S., 1904: 'The behavior of the short-period atmospheric pressure variation over the earth's surface'. *Proc. Roy. Soc. London*, **73**, 457–470.

Lockyer, W. J. S., 1906: 'Barometric variations of long duration over large areas'. *Proc. Roy. Soc. A*, **78**, 43–60.

Markham, C. G., 1967: 'Climatological aspects of drought in Northeastern Brazil'. Ph.D. dissertation, Department of Geography, University of California, Berkeley, 352 pp.

Markham, C. G., 1974: 'Apparent periodicities in rainfall at Fortaleza, Ceará, Brazil'. *J. Appl. Meteor.*, **13**, 176–179.

Namias, J., 1968: 'Long-range weather forecasting – history, current status and outlook'. *Bull. Amer. Met. Soc.*, **49**, 438–470.

Namias, J., 1972: 'Influence of Northern hemisphere general circulation on drought in Northeast Brazil'. *Tellus*, **24**, 336–343.

Namias, J., 1974: 'Suggestions for research leading to long-range precipitation forecasting for the tropics'. Preprints International Tropical Meteorology Meeting, Nairobi, Kenya, Amer. Meteor. Soc., pp. 141–144.

National Climate Program Office, NOAA, 1980: 'National Climate Program'. Washington, D.C., 101 pp.

Nicholls, N., 1978: 'Air-sea interaction and the quasi-biennial oscillation'. *Mon. Wea. Rev.*, **106**, 1505–1507.

Nicholls, N., 1979a: 'A simple air-sea interaction model'. *Quart. J. Roy. Meteor. Soc.*, **105**, 93–105.

Nicholls, N., 1979b: 'A possible method for predicting seasonal tropical cyclone activity in the Australian region'. *Mon. Wea. Rev.*, **107**, 1221–1224.

Nicholls, N., 1980: 'Long-range weather forecasting: value, status, and prospects'. *Rev. Geophys. Space Phys.*, **18**, 771–788.

Nicholls, N., 1981: 'Air-sea interaction and the possibility of long-range weather prediction in the Indonesian Archipelago'. *Mon. Wea. Rev.*, **109**, 2435–2443.

Nicholls, N., 1983: 'Prospects for empirical long-range weather prediction'. pp. 154–166 in: *Proceedings of the WMO/CAS/JB expert study meeting on long-range forecasting, Princeton, December 1983*, WMO, Programme on Weather Prediction Research, Long-Range Forecasting Research Publication Series, Geneva, 238 pp.

Nicholls, N., 1984: 'A system for predicting the onset of the North Australian wet season'. *J. Climatol.*, **4**, 425–435.

Nicholls, N., McBride, J. L., Ormerod, R. J., 1982: 'On predicting the onset of the Australian wet season at Darwin'. *Mon. Wea. Rev.*, 110, 14–17.

Normand, C. W. B., 1953: 'Monsoon seasonal forecasting'. *Quart. J. Roy. Meteor. Soc.*, 79, 463–473.

Panofsky, H. A., Brier, G. W., 1968: 'Some applications of statistics to meteorology'. Pennsylvania State University, 224 pp.

Preisendorfer, R. W., Mobley, C. D., 1982: 'Climate forecast verifications, U.S. mainland, 1974–82'. NOAA Technical Memorandum ERL PMEL-36, Seattle, Washington, 235 pp.

Quenouille, M. H., 1952: *Associated measurements*. Butterworths, London, 242 pp.

Quinn, W. H., 1974: 'Monitoring and predicting El Niño invasions'. *J. Appl. Meteor.*, 13, 825–830.

Quinn, W. H., 1976: 'An improved approach for following and predicting Equatorial Pacific changes and El Niño'. *Proceedings of the NOAA Climate Diagnostics Workshop, 4–5 Nov. 1976, Washington, D.C.*, article 21, 5.

Quinn, W. H., Burt, W. V., 1970: 'Prediction of abnormally heavy precipitation over the equatorial Pacific Dry Zone'. *J. Appl. Meteor.*, 9, 20–28.

Quinn, W. H., Burt, W. V., 1972: 'Use of the Southern Oscillation in weather prediction'. *J. Appl. Meteor.*, 11, 616–628.

Ramage, C. S., 1983: 'Comments on "Regression forecasting of the onset of the Indian summer monsoon with antecedent upper air conditions"'. *J. Climate Appl. Meteor.*, 22, 1132–1134.

Rao, K. N., 1965: 'Seasonal forecasting – India'. WMO Technical Note No. 66, WMO No. 162 TP. 79, Geneva, pp. 17–30.

Rao, K. N., Ramamoorthy, K. S., 1960: 'Seasonal (monsoon) rainfall forecasting in India'. pp. 237–249 in: Basu, S., Pisharothy, P. R., Ramanathan, K. R., Bose, U.K., eds., *Symposium on monsoons of the World, New Delhi, February 1958*. Hindi Union Press, 271 pp.

Rao, Y. P., 1976: *Southwest monsoon*. India Meteorological Department, Meteorological Monograph, Synoptic Meteorology, No. 1/1976, 307 pp.

Reddy, S. J., 1977: 'Forecasting the onset of Southwest monsoon over Kerala'. *Indian J. Meteor. Hydrol. Geophys.*, 28, 113–114.

Reesinck, J. J. M., 1952: 'Some remarks on monsoon forecasting for Java'. Kementerian Perhubungan Djawatan Meteorologi dan Geofisiki, Djakarta, Indonesia, Verhandelingen No. 44, 22 pp.

Sampaio Ferraz, J. de, 1929: 'Sir Gilbert Walker's formula for Ceara droughts: suggestion for its physical explanation'. *Meteor. Mag.*, 64, 81–84.

Sampaio Ferraz, J. de, 1950: 'Iminencia duma "Grande" Seca Nordestina'. *Revista Brasileira de Geografía*, 12, no. 1, pp. 3–15.

Sansom, H. W., 1955: 'Prediction of the seasonal rains of Kenya by means of correlation'. *Weather*, 10, 81–86.

Schmidt – ten Hoopen, K. J., Schmidt, F. H., 1951: 'On climatic variations in Indonesia'. Kementerian Perhubungan Djawatan Meteorologi dan Geofisiki, Djakarta, Indonesia, Verhandelingen, No. 41, 43 pp.

Schmidt, F. H., van der Vecht, J., 1952: 'East monsoon fluctuations in Java and Madoera during the period 1880–1940'. Kementerian Perhubungan Djawatan Meteorologi dan Geofisiki, Djakarta, Indonesia, Verhandelingen No. 43, 36 pp.

Serra, A., 1956: 'As Sêcas do Nordeste'. *Boletim Geográfico*, 14, 269–270.

Serra, A., 1973a: 'Previsão das secas nordestinas – testes estatatísticos'. *Boletim Geográfico*, 32, 78–104.

Serra, A., 1973b: 'Previsão das secas nordestinas'. Banco do Nordeste do Brazil, Fortaleza, 55 pp.

Shukla, J., Paolino, D. A., 1983: 'The Southern Oscillation and long-range forecasting of the summer monsoon rainfall over India'. *Mon. Wea. Rev.*, 111, 1830–1837.

Thapliyal, V., 1981: 'ARIMA model for long-range prediction of monsoon rainfall in peninsular India'. India Meteor. Monograph, Climatology, 12, India Meteorological Department, Poona, India, 22 pp.

Thapliyal, V., 1982: 'Stochastic dynamic model for long-range prediction of monsoon rainfall in peninsular India'. *Mausam*, 33, 399–404.

Tyson, P. D., 1981: 'Atmospheric circulation variations and the occurence of extended wet and dry spells over Southern Africa'. *J. Climatol.*, 1, 115–130.

Tyson, P. D., Dyer, T. G. J., 1978: 'The predicted above-normal rainfall of the seventies and the likelihood of droughts in the eighties in South Africa'. *South African J. Science*, 74, 372–377.

Tyson, P. D., Dyer, T. G. J., 1980: 'The likelihood of droughts in the eighties in South Africa'. *South African J. Science*, 76, 340.

Tyson, P. D., Dyer, T. G. J., Mametse, M. N., 1975: 'Secular changes in South African rainfall: 1880 to 1972'. *Quart. J. Roy. Meteor. Soc.*, 101, 817–833.

Van Loon, H., Rogers, J. C., 1978: 'The seesaw in winter temperatures between Greenland and Northern Europe, part I: General description'. *Mon. Wea. Rev.,* 106, 296–310.

Walker, G. T., 1909: 'Correlation in seasonal variation of climate', *Mem. Indian Meteor. Dept.,* 20, part 6, pp. 117–124.

Walker, G. T., 1910: 'Correlation in seasonal variations of weather II'. *Mem. Indian Meteor. Dept.,* 21, part 2, pp. 22–45.

Walker, G. T., 1914: 'Correlation in seasonal variations of weather III'. *Mem. Indian Meteor. Dept.* 21, part 9, pp. 13–15.

Walker, G. T., 1915a: 'Correlation in seasonal variations of weather IV: sunspots and rainfall'. *Mem. Indian Meteor. Dept.,* 21, part 10, pp. 17–60.

Walker, G. T., 1915b: 'Correlation in seasonal variations of weather, V: *Mem. Indian Meteor. Dept.,* 21, part 11, pp. 61–90.

Walker, G. T., 1915c: 'Correlation in seasonal variations of weather, VI: *Mem. Indian Meteor. Dept.,* 21, part 12, pp. 91–118.

Walker, G. T., 1923: 'Correlation in seasonal variations of weather, VIII'. *Mem. India Meteor. Dept.,* 24, part 4, p. 75.

Walker, G. T., 1924a: 'Correlation in seasonal variations of weather, VII: The local distribution of monsoon rainfall'. *Mem. Indian Meteor. Dept.,* 23, part 2, pp. 23–40.

Walker, G. T., 1924b: 'Correlation in seasonal variations of weather IX'. *Mem. India Meteor. Dept.,* 24, part 9, pp. 275–332.

Walker, G. T., 1928a: 'World Weather III'. *Mem. Roy. Meteor. Soc.,* 2, no. 17, pp. 97–106.

Walker, G. T., 1928b: 'Ceará (Brazil) famines and the general air movement'. *Beitr. Phys. d. freien Atm.,* 14, 88–93.

Walker, G. T., Bliss, E. W., 1930: 'World weather IV'. *Mem. Roy. Meteor. Soc.,* 3, no. 24, pp. 81–95.

Walker, G. T., Bliss, E. W., 1932: 'World weather V'. *Mem. Roy. Meteor. Soc.,* 4, no. 36, pp. 53–84.

Walker, G. T., Bliss, E. W., 1937: 'World weather VI'. *Mem. Roy. Meteor. Soc.,* 4, No. 39, pp. 119–139.

Winstanley, D., 1973a: 'Recent rainfall trends in Africa, the Middle East and India'. *Nature,* 243, 464–465.

Winstanley, D., 1973b: 'Rainfall patterns and the general atmospheric circulation'. *Nature,* 245, 190–194.

Winstanley, D., 1974: 'Seasonal rainfall forecasting in West Africa'. *Nature,* 248, 464.

Wood, C., Lovett, R. R., 1974: 'Rainfall, drought, and the solar cycle'. *Nature,* 231, 594–596.

World Meteorological Organization, 1980: 'Outline plan and basis for the World Climate Programme 1980–1983'. WMO-No. 540, Geneva, 64 pp.

Wright, P. B., 1975: 'An index of the Southern Oscillation'. University of East Anglia, Climatic Research Unit, Research Publication No. 4, 22 pp.

Wright, P. B., 1977: 'The Southern Oscillation – patterns, and mechanisms of the teleconnections and the persistence'. Hawaii Intitute of Geophysics, HIG–77–13, 107 pp.

Wu, M.-C., 1984: 'On the interannual variability of the Indian monsoon and the Southern Oscillation'. Ph.D. Dissertation, Department of Meteorlogy, University of Wisconsin, Madison, 110 pp.

CHAPTER 10

THE HUMAN IMPACT

The fundamental role of climate for the fate of mankind is increasingly being recognized at both the international and national levels (World Meteorological Organization, 1980, pp. 7–10, 28–37; World Meteorological Organization-ICSU, 1975, pp. 1–3; United Nations Environment Programme, 1979, pp. 1–16; National Climate Program Office NOAA, 1980, pp. 1–4, 55–58). There is a two-way interaction between the climatic environment and human activities. The climatic disasters discussed in Chapters 8 and 9 are part of the 'natural' interannual variability of the atmosphere-ocean system, and can have serious consequences for agriculture, fisheries, economy, and society. In turn, humanity has taken it upon itself to modify its environment, albeit in part inadvertently and without rational control. The purpose of this chapter is to raise the awareness for both the impact of climate variations on human activities, and the impact of man on the climatic environment. A comprehensive treatment is precluded not only by the vast scope of this subject, but also because various aspects of this problem complex have so far been insufficiently studied. The selected bibliography to this chapter is meant to provide a preliminary orientation. Sections 10.1 to 10.4 deal mainly with man's impact on the climatic environment and are process-oriented, while Sections 10.5 to 10.8 are primarily concerned with the impact of climate variations on human activities and have regional focus. The practical use of climate prediction is considered in Section 10.9, and a synopsis is presented in Section 10.10.

10.1. Deforestation

Properties of the Earth's surface, such as soil moisture, surface roughness, and albedo, are pertinent to the regional heat and water budget, and hence climate. A variety of human activities can alter these surface properties, wholesale deforestation having particularly drastic effects. According to Sagan et al. (1979) a substantial portion of the tropical rainforests of South America, Africa, and Indonesia has been converted to grassland over the past millenia, and they are currently being cleared at a rate of $5-25 \times 10^4$ km^2 per year. Brown (1984, pp. 74–79) finds that over half of the closed forests cleared annually are in South and Central America, and identifies the conversion of forestland to cropland as the leading direct cause of tropical deforestation. For the Amazon basin in particular, Sanford et al. (1985) conclude that numerous rain-forest fires have occurred since the mid-Holocene, and consider both climatic changes and human activities as possible causes. Whatever the causes, the possible consequences of deforestation for the regional and global climate are manifold.

The possible effects of deforestation on the global carbon dioxide budget have been considered by Bolin (1977), Woodwell et al. (1978), Zimmerman et al. (1982), and others. Bolin (1977) attempts a preliminary global budget of carbon dioxide and suggests that the observed increase of atmospheric carbon dioxide over the past decades is less than the input by fossil fuel combustion. Enhanced growth of land biota due to increased atmospheric carbon dioxide may have a compensating effect. Woodwell et al. (1978) estimate that the global biosphere may represent a carbon dioxide source to the atmosphere more copious than fossil fuel burning. Broecker *et al.* (1979) find no evidence for a decrease of the global biomass comparable to that of fossil fuel combustion. Zimmerman *et al.* (1982) propose termites as a potentially large source of atmospheric carbon dioxide, with largest emissions in tropical areas disturbed by human activities. These studies exemplify the lack of a representative closed global budget of carbon dioxide. It appears that the World Ocean may play an important role. In fact, temperature variations may affect the atmospheric budget not only through the carbon dioxide holding capacity in the top layer of the ocean, but perhaps more substantially through altered marine biological activity. The possible warming effect of increased atmospheric carbon dioxide has received much publicity (references in Woodwell *et al.*, 1978; Kellogg and Schware, 1981; Zimmerman *et al.*, 1982). By contrast, little attention has been given to global warming as a *cause* of an increase in carbon dioxide. In this context it is interesting to compare the rates of atmospheric carbon dioxide and temperature changes from 18 000 to 9000 year B.P. (Neftel *et al.*, 1982; Crowley, 1983), with trends over 1958–81 (Bolin, 1977; Woodwell *et al.*, 1978; Peixoto and Oort, 1984). Over the former interval the atmospheric carbon dioxide content increased by about 100 ppm (Fig. 12.4:1), concomitant with a rise of global temperature of the order of 2–4°C (CLIMAP project members, 1981; Section 12.5; Fig. 12.5:1). The corresponding rate of change of the order of 30 ppm per one °C appears comparable to rates indicated for the 20th century. The present ignorance about the quantitative role of marine biological activity, among other factors, precludes realistic estimates of the global carbon dioxide budget. The potential effects of primary productivity changes during El Niño on atmospheric carbon dioxide has been considered by Chavez *et al.* (1984).

Various modelling studies (Newell, 1971; Lettau and Molion, 1975; Potter *et al.*, 1975; Schneider and Mesirow, 1976, pp. 343–344; Shukla and Mintz, 1982; Henderson-Sellers and Gornitz, 1984) examined the effects of albedo, soil moisture, and evaporation changes, as produced by deforestation, on the global climate. Newell (1971) suggests that reduction of the water cycling between the forest and the air would alter the latent heating of the overlying atmospheric column and thus the pattern of diabatic heat sources in the global atmosphere. Potter *et al.* (1975) deduce from a computer simulation of deforestation effects the following chain of causality: deforestation in the zone 5°N – 5°S → increased surface albedo → reduced surface net shortwave radiation → surface cooling → reduced evaporation and sensible heat flux from surface to atmosphere → reduced convection and rainfall → reduced latent heat release, weakened Hadley circulation, cooling in the middle and upper tropical troposphere → increased tropical lapse rates → increased precipitation in the zones 5–25°N and 5–25°S and decreased equator-pole temperature gradient → reduced poleward transport of heat and moisture from the equatorial zone → global cooling and decreased precipitation in the zones 45–85°N and 40–60°S. However, Potter *et al.* (1975) caution that these inferences should be understood merely as indications of possible change, and that no case can be made for this cascade of events to occur in the real atmosphere. From a computer simulation of the global atmospheric circulation Shukla and Mintz (1982) conclude that the global fields of rainfall, temperature, and motion strongly depend on the land surface evapotranspiration, which in turn is related to the vegetation cover; deforestation of large magnitude and large horizontal extent could affect the global climate, but the response would vary regionally. Henderson-Sellers and Gornitz (1984) suggests that replacement of rainforest by grassland in large portion of the Amazon basin would be of little consequence for the large-scale circulation.

For a narrow strip along the East coast of Northeast Brazil, Roth (1984) calculated the possible climatic effect of deforestation resulting from a change in surface roughness. Considerations are based on the stress-differential divergence mechanism discussed in Section 2.3 (Eq. (2.3:1)). Roth

(1984) concludes that in this particular near-equatorial coastal situation deforestation would lead to a considerable reduction in annual rainfall. It should be recalled (Section 2.3), however, that in this region the major rain-producing mechanism is confined to late night to early morning.

While there is considerable uncertainty on the large-scale circulation consequences of deforestation, some major effects on the climatic environment within the particular rainforest regions are unambiguous. Destruction of the forest canopy eliminates the peculiar humidity, temperature, ventilation, and radiation conditions within the forest ensemble (Bestandsklima). Clearing of the vegetation cover reduces the water holding and buffering capacity of the soil. Accordingly runoff increases, but also soil erosion and sediment load of rivers (Salati et al., 1983; Salati and Vose, 1984). Gentry and López-Parodi (1980, 1982) ascribed the measured flooding of the upper Amazon to recent deforestation, but this contention has been challenged by Nordin and Meade (1982) and Sternberg (1984). Sternberg (1984) also calls attention to the considerable chemical contamination of Amazon waters by increased mining operations in various portions of the catchment. At any rate, as a result of forest clearing evapotranspiration is reduced and Bowen ratio and diurnal temperature range are enhanced. Hydrological consequences to be expected for the large tropical rivers include increased water discharge and sediment freight, and larger annual range in water level. The fast recycling of water through repeated evaporation and precipitation so characteristic of the Amazon basin (Lettau and Molion, 1975; Lettau et al., 1979; Salati et al., 1983), would also be inevitably reduced as a result of deforestation.

Last but not least, the large-scale destruction of the rainforest in various parts of the tropics has biological consequences: extinction of floral and faunal species is irreversible. Biological diversity and a rich gene pool are, however, essential in the breeding for plant cultivation and animal husbandry. Discussions of the biological consequences of deforestation and extensive bibliographies are found in numerous recent publications (Gómez-Pompa et al., 1972; Prance and Elias, 1977; U.S. Department of State, U.S. Agency for International Development, 1978; National Academy of Sciences, 1980a, b; Denevan, 1981; Salati and Vose, 1983; Salati et al., 1983; Myers, 1983; Iltis, 1983). The modern distribution of vegetation types in the Amazon basin, as well as areas now under some form of protection are shown in a recent official map (Ministerio das Minas e Energía, 1983). The climate and vegetation history of the Amazon basin is considered in Section 12.9.

10.2. Land Use and Surface Albedo

Albedo changes related to certain land use practices have received increased attention recently in connection with the concept of 'biogeophysical' feedback processes in semi-arid regions (Charney, 1975; Sections 8.8, 8.11). That the effect of land use on the vegetation cover can be substantial is strikingly illustrated in the satellite and ground photographs published by Otterman (1974, 1977), Wade (1974), and others.

Otterman (1974) reproduces a satellite image (ERTS–1, 22 October 1972) showing high albedo regions of the Sinai and Gaza Strip in contrast to the darker western Negev, and a sharp demarkation line between these two domains. Otterman (1974) explains that this line coincides with the 1949 armistice line between Israel and Egypt and that a fence was erected there in the late 1960's. He states that ground observations indicate denuding of the bright sandy soil on the formerly Egyptian-held side, and ascribes this to overgrazing by goats, camels, and sheep, as well as the picking of shrubs for construction purposes and shallow ploughing in some areas. Otterman (1977) points out that effects such as described above for the Sinai/Negev demarkation line are found in many parts of the World, where across a political boundary different cultivation and settlement patterns exist. He offers two such examples, namely the USSR/Afghanistan and Namibia/Botswana border regions. The reflection of political boundaries in the surface conditions is further documented in images of the United States (GOES–1 geosynchronous satellite, 23 February 1977, 22:00 GMT; photograph with hand-held camera from Apollo 9 spacecraft, 12 March 1969; ERTS satellite, 1 February 1977) published by Kessler *et al.* (1978). They relate differences in surface conditions across the Texas – New Mexico border to different groundwater management and land use practices. In one of these pictures, in particular, they call attention to the beginning of a dust streak just to the East of this boundary. Concerning the effect of grazing practices, Otterman (1977) shows an interesting air photograph of the semi-arid region of the lower Volta in West Africa. This shows dense vegetation cover near the river, large rectangular fields to either side of it, and most importantly bright circular haloes of about 5 km diameter around many of the settlements, indicating the range of grazing by sheep. Similarly, Flohn and Ketata (1971) suggest for Southern Tunisia that the observed degradation of natural vegetation outside the oases is caused by overgrazing, but that vegetation can be restored in protected areas.

Wade (1974) reproduces satellite and ground photographs of 'Ekrafane ranch', a 250 000 acre estate in Niger. These indicate a marked contrast between abundant vegetation cover inside a barbed-wire fence and the surrounding desert. The ranch is divided into five sectors with cattle allowed to graze one sector a year. Wade (1974) notes that the ranch was started only 5 years earlier, or around the beginning of the recent great Subsaharan drought, but that the simple protection made the difference between pasture and desert.

These pictures suggest that land use practices may affect the vegetation cover and hence surface albedo in relatively short time, vegetated areas being characterized by lesser albedo. Norton *et al.* (1979) report an increase of satellite-derived surface albedo in Subsaharan West Africa from 1967 to 1973, but find no relation between albedo and subsequent rainfall. Further important observational evidence on long-term vegetation and albedo changes in Subsaharan West Africa is contained in Courel *et al.* (1984), who present evaluations of satellite sensing for four sample regions of the Sahel over the period 1967–79. For the time span 1967–72 when the Subsaharan drought was intensifying, a moderate increase of surface albedo is indicated. Courel *et al.* (1984) report that the dry season albedo in the Sahel reached a maximum of about 0.30 in 1973, but then decreased to values close to 0.20 in 1979. It should remain open to what extent their satellite-derived estimates of surface albedo may be affected by the abundant dust in the lower atmosphere over the Sahel.

Charney (1975), Charney *et al.* (1975; 1976), Otterman (1974, 1975), and Berkofsky (1976, 1978) suggest that surface albedo has a positive feedback effect on climate. Decreased surface net shortwave radiation would entail a lesser heat input to the overlying atmosphere, which in turn would be conducive to subsidence, and would thus further aggravate drought conditions. Consistent with these notions and assuming a uniform emissivity of 0.9, Otterman (1974) interprets infrared measurements as indicating lower surface temperature over the high albedo areas. This contention is challenged by Jackson and Idso (1975), who point out that emissivity tends to be substantially larger for vegetated surfaces than for desert soil. Idso (1977) and Idso and Deardorff (1978) also raise reservations concerning model experiments of albedo feedback.

Ripley (1976a, b) queries that in their computer simulation Charney (1975) and Charney *et al.* (1975) ignore the effect of evapotranspiration, which would lead to lower surface temperature for vegetated ground because much of the absorbed radiation is used to evaporate water. Charney *et al.* (1977) thereupon conducted further numerical experiments including evaporation. They conclude that for the case of appreciable evaporation increased albedo first acts to reduce net shortwave radiation and therefore the sensible plus latent heat transfer into the atmosphere. As a result convective cloudiness is reduced, which enhances the downward directed shortwave radiation but reduces the downward directed longwave radiation even more, so that the net allwave radiation is decreased. Thus, in these computer simulations with or without evaporation, albedo increase causes a decrease of net allwave radiation and consequently of convective cloudiness and precipitation. More recent computer simulations by Sud and Fennessy (1982, 1984) support the results of Charney (1975) and Charney *et al.* (1977). Gutman *et al.* (1984) and Gutman (1984) conclude that the importance of the biosphere-albedo feedback mechanism may have been over-estimated in previous studies. At any rate, it should be realized that these computer experiments ignore the possible implications of the diurnal cycle. As mentioned in Section 2.4, a late night to early morning rainfall maximum prevails in much of Subsaharan West Africa (Delorme, 1963). Accordingly, the daytime albedo effects on rainfall appear less straightforward than assumed in the model calculations. Recent pertinent work includes d'Almeida and Schütz (1983) and Guedalia *et al.* (1984).

Ellsaesser *et al.* (1976) tested the biogeophysical feedback hypothesis in a further numerical experiment and find agreement with Charney *et al.*'s (1975) earlier results, in confirming a positive feedback through precipitation but a negative feedback in terms of temperature. They suggest that tropical deserts may play an air conditioning role in placing an upper bound on temperature. They further note that Charney *et al.* (1975) consider the mechanism described by Otterman (1974) not to be in accord with Charney's (1975). Finally, Ellsaesser *et al.* (1976) quote Schnell's (1975) suggestion that convective precipitation is reduced by decreased availability of biogenic ice nuclei following denudation and reduced plant litter.

The 'biogeophysical' feedback mechanism discussed above appears interesting as a possible factor for the persistence of drought or wet regimes in semi-arid regions. Otterman (1974) considers a self-regulating biological cycle, such that population explosion of grazing herds during 'the seven fat years' is the cause of 'the seven lean years' and vice versa. In this context it is noteworthy that according to Courel *et al.* (1984) albedo in Subsaharan West Africa decreased after 1973. This may be viewed in the context with decreased stress concomitant with the diminished land use of this area in consequence of the drought. However, such recovery of the vegetation cover would have taken place despite the continuation and in fact deterioration of drought conditions at least into 1984 (Lamb, 1982, 1983, 1985). Satellite documentation of albedo changes in the 1980's would be of particular interest.

10.3. Effects of Dust on Climate

Man-made destruction of the plant cover would, in addition to the effects discussed in Sections 10.1 and 10.2, be conducive to enhanced dust input to the atmosphere. Bryson and Baerreis (1967), Bryson and Murray (1977; pp. 107–114), and Sagan *et al.* (1979) suggest that this could provide a further positive feedback mechanism for desertification: dust would serve to increase the net radiative cooling of the tropospheric column, which in turn favors subsidence and would suppress convection and rainfall. Bryson and Baerreis (1967), Bryson (1976), and Sagan *et al.* (1979) point out that several millenia ago the Indus Valley civilization occupied what is now the Rajasthan dry area of India. Overgrazing is proposed as a cause for the break-up of the vegetation cover, dust input to and enhanced infrared cooling of the atmosphere. However, there appears to be considerable uncertainty (Prospero and Nees, 1977; Sagan *et al.*, 1979) concerning the magnitude of this contribution to self-enforced desertification. Bryson and Baerreis (1967) suggest that, as the desert is man-made, surface stabilization should make it possible to reduce the dust and consequently modify the subsidence and rainfall patterns. By contrast, Allchin *et al.* (1978, pp. 305–330) consider the Thar Desert an ancient phenomenon, rather than the result of a combination of man-induced desertification and climate desiccation in the Holocene.

MacLeod (1976) studied the possible effect of dust on the climate of Subsaharan West Africa. He suggests that extensive bare surfaces represent an effective dust source, and that dust in the atmosphere is conducive to a more southerly position of the Intertropical Convergence Zone (Section 6.7.2.3) and hence deficient rainfall in the Sahel. An extensive, though inconclusive discussion of the effect of Saharan dust on the climate system is found in Morales (1979).

10.4. Intentional Climate Modification

That human activities may have profound albeit inadvertent side effects on the climatic environment is strongly suggested by the evidence reviewed in the preceding Sections 10.1 to 10.3. In the present section the prospects for intentional climate modification are considered. Two major categories can be distinguished: (i) efforts to reverse the inadvertent anthropogenic changes discussed in Sections 10.1 to 10.3; and (ii) deliberate attempts to modify the 'original' environment and climate.

Pertaining to the former category, various suggestions have been made. Deforestation, the climatic consequences of which are considered in Section 10.1, is contrasted by reforestation. The evidence presented in Section 10.2 indicates that just as certain excessive land exploitation practices can lead to destruction of the plant cover, changes in surface albedo and soil moisture, and positive feedbacks with the atmosphere, so land management may aid in the recovery of vegetation, and thus be conducive to the restoration of 'original' surface properties and, more remotely, of climate. Anthes (1984) hypothesizes that planting bands of vegetation with width of the order of 50–100 km in semi-arid regions could favor convective precipitation through a combination of three different mechanisms. While man-made dust is considered a factor for large-scale subsidence and aridity over Northwest India, Bryson and Baerreis (1967) also propose a grandiose scheme of stabilizing the desert surface, in order to reduce dust sources, atmospheric aerosol loads, and ultimately the large-scale subsidence (Section 10.3). If the lack of biogenic precipitation nuclei is a factor in the Subsaharan droughts as Schnell (1975) suggests, then recovery of the plant cover may mitigate the drought conditions.

Deliberate modifications of the 'original' environment and climate have also been variously proposed. With reference to the rainfall in Brazil's Nordeste (Sections 8.6, 9.4, 10.7), Gonçalves de Souza (1979, p. 94) mentions the proposition of massive artificial injection of carbon dust into the atmosphere, with the purpose of enhancing the absorption of solar radiation, thus creating an additional heat source, and fostering convection. This idea, ascribed to W. M. Gray (Gray *et al.*, 1976), is also mentioned by Glantz and Parton (1976) and Glantz (1977b) for Subsaharan West Africa (Sections 8.8, 9.5, 10.8). Glantz and Parton (1976) and Glantz (1977b) describe various other schemes. Dark asphalt coating of land surfaces is thought to stimulate convection and rainfall. Creation of a large lake in the Sahara is intended to stimulate the evaporation-precipitation cycle. Cloud seeding is suggested to enhance rainfall on a large scale.

It is noteworthy that none of these propositions has as yet been translated into practical application. Proposals to deliberately modify the 'original' environment and climate, namely 'category (ii)', would not only be exorbitantly costly, but reservations are also in order concerning the implied mechanisms and possible undesirable and unforeseen side effects. Some of the more modest suggestions to revert the inadvertent effects of human activities and to restore the 'original' environment and climate, namely 'category (i)', merit attention. Finally mankind may be well advised to live with, rather than against its climatic environment.

10.5. Ocean Climate and Fisheries

Fisheries are concentrated in a few coastal cold water regions of the low-latitude oceans (Thompson, 1977), namely the waters off Ecuador, Peru, and Chile, off California, off Northwest and Southwest Africa. Other cold water regions are found off Somalia and Arabia. Less than a thousandth of the world ocean area yields half the World's fish catch. The most productive fishery of the World's ocean is situated within 100 km of the South American West coast. Referring to the underlying great disparity in biological productivity, Thompson (1977) coined the terms 'ocean oases' and 'ocean deserts'. Three trophic levels are distinguished: primary organic production through photosynthesis provides phytoplankton, which feeds zooplankton, which is eaten by fish. Given sufficient light, plants in the sea are limited by the availability of nutrients. In the sunlit or euphotic zone of the upper ocean, nutrients are assimilated by phytoplankton. As this dies and sinks, nutrients are depleted from the euphotic zone. Phytoplankton growth in the euphotic zone can only be maintained by continuous supply of nutrients. It is through upwelling and turbulent mixing that decomposed organic material and nutrients are returned to the euphotic zone; land runoff being a nutrient source near coastal margins. The limited areas of the 'ocean oases' are of primary interest for fisheries. Thompson (1977) points to the threat of 'ocean desertification' as a result of natural and anthropogenic causes.

The interannual variability of the marine environment and its impact on fishery and economy has been most extensively studied for the West coast of South America (Sections 8.2, 9.3). Glantz and Thompson (1981) edited a volume on the complex physical, biological, economic, societal, and political problems of fisheries along the coasts of Peru and Ecuador, and Barber and Chavez (1983) and Chavez et al. (1984) present an updated synopsis of the biological consequences of El Niño. Similar problems of human impact are described for the fishery regions of the Gulf of Guinea (Stretta, 1977; Bakun, 1978), and off Southwest Africa (Cram, 1981).

At the Peruvian coast, the guano industry (Section 8.2) greatly overshadowed anchoveta fishery until World War II, but by 1963 Peru had become the leading fishing nation of the world, harvesting some 15% of the global catch (Murphy, 1981; Thompson, 1981, p. 28). This phenomenal growth is demonstrated in Fig. 10.5:1, which through 1970 shows only a moderate correlation between sea surface temperature and anchoveta catch. Factories along the coast process the anchoveta to fish meal, marketed as a high-value, high protein animal feed (Vondruska, 1981, p. 283; Thompson, 1981, p. 28; Paulik, 1981, p. 56). With the growth of the fishery the bird population declined (Paulik, 1981, p. 57), as is illustrated in Fig. 10.5:2. Paulik (1981, p. 56) explains that the cormorants which produce the highest quality guano have a narrow diet consisting to 95% of anchoveta. A panel of fisheries economists in 1970 recommended the quota of 9.5 million metric tons per year as maximum sustainable yield for the Peru Current anchoveta. However, the capacity of Peruvian fishery was far above that level, and in 1970 a record of 12.5 million metric tons of anchoveta were landed. This performance was followed by the 1972 El Niño and the drastic decline of fishery yields as documented in Fig. 10.5:1. Thompson (1981, p. 30) sees the demise of Peruvian fishery industry as caused by a combination of adverse circumstances. The natural variability of the ocean-atmosphere system is considered responsible for the reduced anchoveta harvest in 1972–73. Two other factors added to the catastrophe: large-scale ocean conditions changed in such a way as to concentrate anchoveta near the coast; at the same time, the fishing fleet had grown too large for controlled exploitation of anchoveta under such circumstances. Recovery of the Peruvian anchoveta fishery appears uncertain. Fig. 10.5:3 illustrates the increased yield of various fish for consumption concomitant with the decline of the anchoveta harvest.

Barber and Chavez (1983) detail the processes leading to the reduction of fish population during El Niño. Cold sea surface, shallow mixed layer, and shallow thermocline, are characteristics

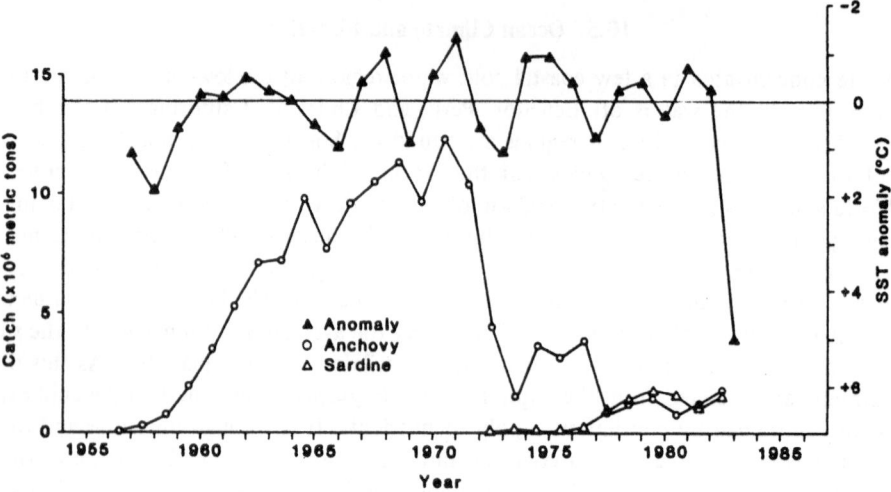

Fig. 10.5:1. Association of annual sea surface temperature anomalies with annual catch of anchovies and sardines off Peru. The anomaly temperature scale is inverted: upward indicates cool, downward warmer, that is El Niño. The annual temperature anomaly is relative to the 26 year mean temperature of 16.9°C at Chicama; it is calculated for the thermal year from July to June spanning half of two calendar years, and the value is plotted in the middle of the thermal year at January. The fish catch for each calendar year is plotted in the middle of the calendar year at July. From Barber and Chavez, 1983 (*Science,* **222**, 1203–1210, copyright by American Association for the Advancement of Science).

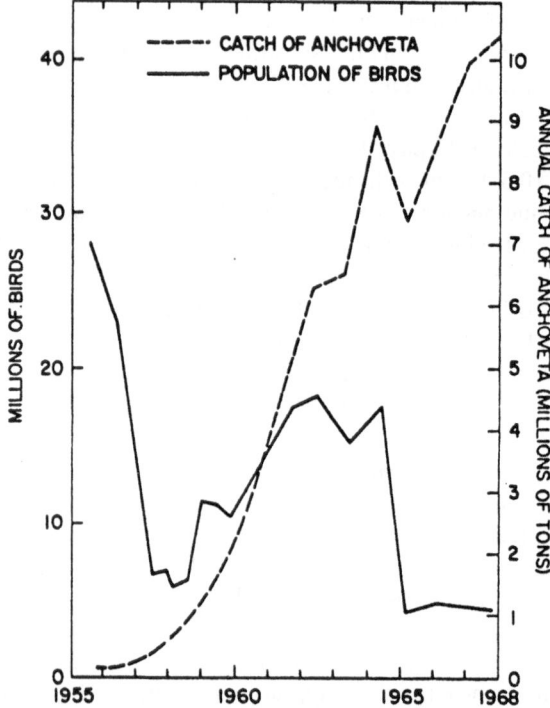

Fig. 10.5:2. Guano bird population census figures and anchoveta catch by year for 1955–68. From Paulik (1981).

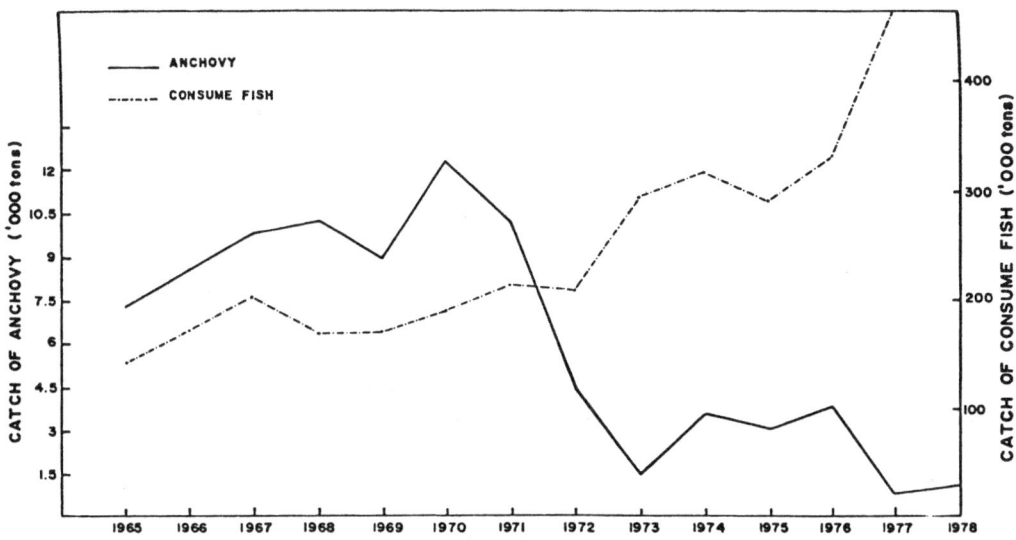

Fig. 10.5:3. Annual catches of anchoveta and fish for consumption along the Peru coast. From Guillén and Calienes, 1981 ('Biological productivity and El Niño', in: Glantz, M. H., Thompson, J. D., eds.: *Resource management and environmental uncertainty; lessons from coastal upwelling fisheries*; copyright 1981, Wiley; reprinted by permission of John Wiley and Sons, Inc.).

of the easternmost tropical Pacific and are all conducive to high productivity at all trophic levels of the ecosystem. El Niño disrupts this high productivity. The theoretical basis of biological productivity during El Niño has two causal aspects, one dealing with inorganic plant nutrients and the other with light for photosynthesis. The main inorganic nutrient reservoir is below the thermocline. Depression of the thermocline away from the surface layer where light for photosynthesis abounds, reduces productivity. The depth of the mixed layer in which the phytoplankton is distributed determines the amount of light that can be captured by it. With a deeper mixed layer phytoplankton spend a greater proportion of time in the dark, and most of the light is absorbed by water molecules rather than phytoplankton. The two parameters referred to here, thermocline and mixed layer depth, experience substantial variations during El Niño. Changes in these parameters rather than upwelling seem to be the biologically important factors. Coastal upwelling may continue during El Niño, but the water entrained is warmer and poorer in nutrients. Schreiber and Schreiber (1984) and Feldman *et al.* (1984) describe the biological consequences of the 1982–83 El Niño in particular.

Cram's (1981) essay on the fisheries off Southwest Africa indicates certain parallels to the human impact problems at the Peru coast. Fishery is geared to the processing for fish meal in large factories, and the fishing fleet possesses a capacity far exceeding the level warranted by the marine resources.

For the Gulf of Guinea, Bakun (1978) examined the five years 1964–68 and found that the colder years tended to be deficient in rainfall but abundant in herring catch. Stretta (1977) reports similar results for the years 1972–75.

10.6. India

The intimate and long relation between Indian civilization and the monsoons has been described by Rao (1976, p. i), Das (1983; 1984, pp. 1–4; 1986, Chapter 1), and others (see also Fein and Stephens, 1986). Detailed quantitative accounts of the human impact are available from the 19th century onward (Ghosh, 1944; Bhatia, 1967; McAlpin, 1979).

Coping with the occasional failures of the summer Southwest monsoon is the central societal issue. Ghosh (1944, pp. 1–2) mentions famine relief measures by the Moghul Emperors in the 17th century. However, for pre-British India McAlpin (1979) finds no systematic government programs to prevent or relieve famine, but reports various measures below the level of the state (villages, castes, lineages, families). Objectives are considered in hierarchic priority: first avoid starvation, second sustain the capacity to carry on agricultural operations, third are all other goals. In western India drought-resistant varieties were available for some crops, and a mix of crops with different rainfall demands were commonly cultivated in the same field, presumably in order to reduce the variance of yield from a plot, as well as in response to subsistence needs, and spread of harvesting jobs. Options available beyond these measures include grain storage and migration, and in areas of rather unequal land distribution the possibility of loans from a 'big man'.

Ghosh (1944, pp. 1–3) mentions famines in 1629–30, 1661, and for Bengal events in 1770, 1783, 1866, 1873–74, 1892, 1897, and 1943. Major famines reported for India are (Bhatia, 1967, pp. 58, 161, 238–270, 309): 1860–61 (Northwest Provinces), 1865–66 (Orissa), 1868–69 (Rajputana and Northwest Provinces), 1873–74 (Bengal and Bihar) 1876–77 (Bombay and Madras), 1878 (Northwest Provinces); 1891–92 (Upper Burma), 1888–89 (Madras); 1896–97 (countrywide), 1899–1900 (countrywide), 1905–1907 (Bombay Deccan), 1907–08 (United Provinces of Agra and Oudh, Central Provinces, Bombay, Madras); 1943 (Bengal). Bhatia (1967) traces the history of famines in India from the middle of the 19th century to Independence. He recognizes major societal transformations associated with the construction of railroads and trade developments after 1860, which led to steady impoverishment of certain classes, thereby creating a class of potential victims of drought and scarcities. This set the scene for a change in nature of famine and its increased frequency (Bhatia, 1967, p. 2). Rice is the staple food in much of the country, is dependent on the summer rains, and reaped in the *kharif* harvest. Winter rains allow the *rabi* crop, mainly wheat. Famines in India are primarily related to the failure of the *kharif* harvest (Bhatia, 1967, pp. 2–3).

Under British administration it was found undesirable for the government to intervene in the daily operation of the economy and instead policies were favored that would prevent famines (McAlpin, 1979), in particular the construction of railroads and irrigation schemes. Bhatia (1967, pp. 2–10) explains that high food prices during drought, rather than the scarcity of food most victimized the poorer sections of the community. Improved transport conditions served to spread local scarcity evenly over the country as a whole, resulting in a general price increase. Grain exports and destructive speculation in foodgrains are considered as further aggravating factors (Ghosh, 1944, pp. 12–13; 27–51; Bhatia, 1967, pp. 2–10). Ghosh (1944, pp. 11–13) reports that under British rule and during times of famine, rice was exported from Bengal, on the grounds of providing food for Indian laborers and troops overseas. Ghosh (1944, pp. 52–55) further describes that for fear of a military invasion in 1943, means of transport were destroyed and rice removed from Bengal, thus greatly aggravating the famine.

In an attempt to guarantee a subsistence level, a set of famine relief policies, a Famine Relief Code, was initiated (McAlpin, 1979). The development of the Indian Famine Codes in the 1880's is discussed by Brennan (1984). These measures included (McAlpin, 1979) employment on 'relief works' of able-bodied persons whose normal agricultural activities were interrupted by the drought;

the 'gratuities relief' or 'village dole' for persons unable to work; loans to private agriculturists to generate employment (*takavi*); and famine 'reconstruction', that is loans for seed and cattle necessary to resume agricultural activities after the famine. These four measures of famine relief shifted gradually from the 1876–77 to the 1918–19 famine, with the relative shares of 'village dole' and *takavi* loans increasing, and 'relief works' declining.

Ghosh (1944, pp. 26–123) explains that widespread poverty and food consumption near subsistence level are the long-term background against which India famines must be seen, and considers the lack of indigenous control of the economy in the colonial era as a major factor. His text and photographic documentation underlines the severe human consequences of the famines.

Bhatia (1967, pp. 112–114) discusses a variety of societal problems related to the famine relief operations. It was difficult to distinguish between 'ordinary poverty' and 'extraordinary poverty' caused by famine. This and the considerable surplus of labor would make employment at ordinary wage rates overly attractive, and thus relief operations unmanageable. Objection was taken to poor houses and free food because of caste prejudices (Ghosh, 1944, p. 11; Bhatia, 1967, pp. 112–114). Bhatia (1967, p. 307) sees the ultimate solution to poverty and famine in economic development, increased agricultural production, and diversification of occupations. Bhatia (1967, p. 361) concludes by identifying food supply as the central national problem for India.

Estimates of the death toll of the famines vary. For the era 1875–1900 McAlpin (1979, p. 146) lists a total of less than 1 million as compared to 15 million reported by Das (1983, p. 124). According to McAlpin (1979, p. 146), the droughts of 1896–97, 1899–1900, and 1900–1901, caused the death of 700 000 people, while later on famines were effectively mitigated by increased agricultural trade and demand for labor, and governmental credit and relief services (McAlpin, 1979), although Ghosh's (1944, pp. 26–123) account to 1944 indicates that these measures were counteracted by a variety of unfortunate factors. The famines of India are among the topics of a recent edited volume (Currey and Hugo, 1984). It is also recalled (Sections 8.5 and 9.1) that the famine of 1876–77 provided an impulse for the extension of the exemplary surface meteorological station network and the early endeavors at climate prediction in India.

10.7. Northeast Brazil

The 'Nordeste' (Sections 8.6, 9.4) is the earliest colonized region of Brazil. The Sêcas or droughts form part of the cultural tradition of the Nordeste and are the theme of various classics in Brazilian literature (da Cunha, 1979; Ramos, 1978; Amado, 1978). A historical perspective of the human impact of the recurrent droughts is given in Carvalho (1973, pp. 95–103, 179–192), Gonçalves de Souza (1979, pp. 73–83, 305–307), Sudene (1981, p. 17–24), Araújo (1982, pp. 27–35), and Cuniff (1970). Subsistence agriculture (of millet beans, maize, and manioc), cattle ranching, and cotton growing prevail in the traditional land use pattern of the semi-arid interior. Since the early days, the impact of the Sêcas was aggravated by the poor road communications. Beyond the limited possibility of food storage, the early and drastic societal response consisted of mass migrations to the coastal towns and cities , but also to more distant parts of Brazil. For most of the peasants the idea may always have been to get back to their land with the return of the rains, but many stayed away forever. Nordestinos contributed to the labor force of the coast, the mines of Minas Gerais, and the industrial areas of Saõ Paulo, and carried foreward the colonization of the Amazon basin, penetrating deep into Peruvian territory (Gonçalves de Souza, 1979, pp. 73–76; Rabelo, 1983, pp. 235–265).

Carvalho (1973, pp. 95–103, 179–192) and Gonçalves de Souza (1979, pp. 73–83, 305–307) offer a detailed account of the social and economic consequences of the Sêcas from the early days of Portuguese colonization to the 20th century. The earliest record of a drought dates to 1583, when some 5000 Indians driven by hunger descended from the interior to join the Whites at the coast. Another drought in 1587 was apparently limited to Pernambuco. In the 17th century the following six droughts are recorded: 1603, 1608, 1614, 1645, 1652 and 1692. With the arrival of the Dutch in 1649 there was some migration into the interior. After 1692, rural populations began to emigrate in search for work in the mines of Minas Gerais. This era also marks the appearance of bandits in the interior. In addition to the subsistence cultivation of manioc for flour, cattle raising became an essential economic activity of the Nordeste by the end of the 17th century.

In the 18th century, the cattle herds increased and the white population penetrated ever deeper into the interior. The repercussions of the prolonged 1721–27 drought was brought to the attention of the King of Portugal D. Joaõ V. Among the administrative responses of the crown was the mandatory cultivation of manioc and fines for those refusing to contribute to the production of flour. An administrative measure of some consequence was established in 1766. The crown decided that the famine refugees, or 'flagelados', should be gathered in settlements of more than 50 households, along the rivers, and adjacent land distributed to them. The towns of Sobral and Russas in Ceará state owe their foundation to this order. A census in 1782 reports a population of 137 000 in the drought-prone region. The 'Pia Sociedade Agricola' created to provide work for the 'flagelados' of the 1790–93 drought was the first famine relief organization.

At the beginning of the 19th century, the population in the drought-prone region is estimated at 726 000. In 1831 the government ordered the drilling of artesian wells to solve the problem of water supply during the droughts. The construction of dams and reservoirs was also proposed by naturalists and scholars travelling through the interior Nordeste. A great drought struck in 1845–46. This was followed by the creation in 1856 of the 'Comissão Científica de Exploração', and various studies directed at the drought problem. Suggestions of reforestation, dams, irrigation, and fishery connected with the dams, as well as road construction date back to the middle of the 19th century. The use of wind power for the pumping of water and irrigation was suggested in the 1880's. Gonçalves de Souza (1979, p. 81), explains that most measures of drought relief implemented in the 20th century had already been proposed in the 19th century, and that the people familiar with the region tended to suggest the simpler and more practical solutions.

It took the great Sêca of 1877–79 to make the drought problems of the Nordeste a matter of national consciousness. By this time the population had grown to 1 754 000, and more than half of them died during this famine. The death toll due to the droughts from 1877 to 1913 is estimated at 2 million.

Carvalho (1973, pp. 95–103) and Gonçalves de Souza (1979, pp. 305–307) find that the interior Nordeste has served as a source of manpower for various other regions of Brazil, the recurrent Sêcas being an essential factor. Migrations to Minas Gerais date to the 17th century. The Amazon basin continued to be the preferred destination of migrants from the Nordeste to the latter part of the 19th and the early part of the 20th century. Later emigration was directed mainly towards the Southeast, and Nordestinos were attracted by the construction of the new Capital Brasilia. While the population of the semi-arid Nordeste increased to about 12 million, only some 200 000 inhabitants were borne elsewhere, whereas nearly 4 million Nordestinos live in other regions of Brazil (Gonçalves de Souza, 1979, pp. 305–306; Carvalho, 1973, p. 95).

In the developments during the 19th century, parallels with India (Section 10.6) are remarkable. The restricted traffic communications traditionally accentuated the effects of droughts in both regions of the World. 1877 and the years around the turn of the century were times of severe drought in both India and the Nordeste, and in both countries they jolted society and government into some action. The construction of roads, railways, dams, and irrigation schemes was undertaken in the Nordeste (Cuniff, 1970, pp. 176–221), while railways and irrigation were advanced under British rule in India. Government relief operations were linked to local land owners and entrepreneurs in the Nordeste (Cuniff, 1970, pp. 176–221), just as the private sector gained importance in the famine relief programs of the India Government.

Among the tangible governmental responses to the Nordeste drought problems during the 20th century is the foundation in 1909 of a specialized agency, the Departamento Nacional de Obras Contra as Secas (DNOCS). The dense network of rain gauge stations in the Nordeste began shortly thereafter. Gonçalves de Souza (1979, pp. 111–191) details the history and functioning of various other institutions concerned with economic and environmental problems of the Nordeste. The great drought of 1958 gave the final impetus for the foundation of the Superintendencia do Desenvolvimento do Nordeste (SUDENE). The tasks and functions of SUDENE range from agronomy and animal husbandry over road and dam constructions, to meteorological and hydrological networks, and traffic and industrial planning (Carvalho, 1973; Gonçalves de Souza, 1979, pp. 193–326; Sudene, 1985).

Carvalho (1973, pp. 182–192, 193–229) emphasizes various major solutions to the drought problems of the Nordeste: dams, reservoirs and irrigation; reforestation; dry farming; composite solution through the adjustment of the physical and social environment to new situations, with agricultural research playing an important role; emergency measures including food supply, medical assistance, and employment of 'flagelados' in various public work projects; and drought prediction. Araújo (1982) presents a comprehensive account of dams in the Northeast of Brazil. Carvalho (1973, p. 184) also calls attention to the unfortunate political and entrepreneurial manipulation of drought-related support characterized by the term 'indústria das secas'.

The complex of measures described above seemingly reduced the impact of droughts on the population of the interior Nordeste. The development of operationally viable climate prediction (Carvalho, 1973, pp. 193–229; Conselho Nacional de Pesquisas, 1980) continues to be recognized as an important practical task.

10.8. Subsaharan Africa

The transition zone between the Sahara desert to the North and the perennially humid equatorial rainforest belt in the South (Sections 8.8, 9.5) is a region where the balance between natural environment and human activities is particularly precarious (Cochemé and Franquin, 1967; Wade, 1974). A series of researchers have studied the patterns of human adaptation to this delicate climatic environment (reviews in Wade, 1974; Wiseberg, 1976; Lovejoy and Baier, 1976; Imperato, 1976, Ware, 1975, 1977; Franke and Chasin, 1980). Moreover, the drought which began in the late 1960's not only spawned a host of papers related to the interannual variability of circulation and climate (Sections 8.8, 9.5), but also prompted investigations into the human impact of this natural disaster (Seaman et al., 1973; Davy, 1974; Wade, 1974; Greene, 1974; Cohn, 1975; Glantz, 1976, 1977a; Breman, 1976; Breman and Cissé, 1977; Ball, 1978; Franke and Chasin, 1980). In the present section, the background pattern of human adjustment to the Subsaharan climatic environment is considered first. This forms the basis for the subsequent discussion of the economic and societal impact of the recent Subsaharan drought.

Grove (1973), Wade (1974), Wiseberg (1976), Lovejoy and Baier (1976), Imperato (1976), Ware (1975, 1977), Ball (1978), Franke and Chasin (1980), and others, have described the traditional patterns of land use, economy, and society in the semi-arid desert fringe of Subsaharan West Africa. The nomadic way of life is recognized as an adaptation to the climatic environment, characterized as it is by the poleward shift of the rain belt towards the height of the Northern hemisphere summer followed by a southward displacement from the summer towards the winter (Sections 6.7.2.3 and 8.8).

As Wade (1974) describes, during the dry season the nomads with their herds are far to the South but just out of range of the tsetse fly. There is a symbiotic arrangement between the nomads and the sedentary farmers in this area. The grazing cattle provide manure for the harvested fields, and the nomads receive millet from the farmers. Lovejoy and Baier (1976) and Imperato (1976) also mention nomad control of servile villages and economic specialization based on ethnic stratification. Imperato (1976) describes sedentary groups practicing both farming and livestock raising, as well as primarily sedentary groups engaging in transhumance. Imperato (1972, 1976) further points out that fishing is a significant activity on the Senegal and Niger and their tributaries, carried out by particular ethnic groups. The environmental history, human use, and health hazards of the Niger and other West African rivers are the subject of a forthcoming edited volume (Grove, 1986). As to the nomadic pattern, with the onset of precipitation, grass begins to grow and the herds move northward, following the northward shifting rain belt. When the desert edge is reached, the return to the South begins. The cattle graze grass that grew up behind them on their northward trek, and they drink standing water left over from the rainy season. At the southern extremity of their travel they find mature grass to carry them to the next growing season.

Tribal rules govern the traditional migration routes and the time spent at the various wells (Wade, 1974; Imperato, 1972, 1976). The timing of herd migrations was carefully calculated. Just as the herders, the sedentary farmers may have made optimal use of the land by allowing long periods of fallow before recropping, and by developing a large number of varieties of millet and sorghum adapted to different growing seasons and situations. In the latter respect, the parallels with India (Section 10.6) are striking. Wade (1974) further suggests that the peoples of the Sahel have been impressively innovative rather than unduly conservative. In his account of the 1913 drought in the Sahel, Grove (1973) describes various indigenous emergency responses, and suggests that the establishment of certain new patterns in long-distance cattle trading was stimulated by this drought.

The historical significance of the desert edge in the development of the adjacent savanna belt to the South has been stressed by Seaman et al. (1973), Wade (1974), Lovejoy and Baier (1976), and others. According to Lovejoy and Baier (1976), the desert edge provided a substantial market for grain, manufactures, and other products imported from the savanna, served as a source of livestock, mineral salts, and transport services, and effected access to North African markets. In this context, Wade (1974) calls attention to the legendary trading empires of Ghana, Mali, and Songhai, which flourished in the Sahel during the middle ages. While these empires declined for complex reasons perhaps not fully understood, a major rupture in the economic and societal structure of the Sahel occurred with the onset of colonialism. This historical development is widely considered to be a major factor for the great economic and societal impact of the recent Subsaharan drought (Wade, 1974; Lovejoy and Baier, 1976; Ball, 1978; Franke and Chasin, 1980).

Introduction of cash crops profoundly affected the traditional system. The division into separate states under French rule hampered the free movement of the nomadic tribes in particular (Wade, 1974). Moreover with the decline of trans-Saharan trade and the introduction of peanuts as agricultural export crop, trade from the savanna began to be directed towards the coast (Lovejoy and Baier, 1976). In Africa at large, increased threat of famines appears in part attributable to government policies and development strategies which emphasize production of cash crops for export at the expense of food for domestic consumption (Food and Agriculture Organization of UN and OAU, 1982, p. 14). The causes of famine and famine relief in Africa are among the topics of a recent edited volume (Currey and Hugo, 1984).

For Subsaharan West Africa in particular Wade (1974) suggests that most harmful of all may have been the deliberate attempts to do good, on the part of the French colonial administration until Independence in 1960, and by various Western countries thereafter. While cattle numbers may have been rising since earlier in the 20th century, the drilling of thousands of wells (Ambroggi, 1966) and application of Western medicine prompted a drastic increase in human and animal population (Wade, 1974). The first few years of Independence were an era of unusually abundant rainfall, and the number of cattle in the Sahel increased from some 18 to 25 million between 1960 and 1971, while the optimum number is estimated at only 15 million (Wade, 1974). Not only were the pastures overtaxed, but the same was done to the arable land. As population had increased and the best plots were reserved for the cultivation of cotton and peanuts, more marginal lands had to be used for subsistence agriculture (Wade, 1974). The strain of intensive agriculture proved fatal for such ecologically fragile zones. As more land was taken under cultivation, the nomads and their herds were squeezed into a narrower and more northerly strip of the desert fringe. Wade (1974) further explains that the boreholes interfered with the traditional tribal agreements on pasture use. As a consequence the deep-rooted perennial grasses began to disappear and were replaced by coarse annual grasses, but due to heavy grazing and trampling these gave way to leguminous plants that dry up quickly and do not protect the soil, so that the scene is set for widespread erosion. Wade (1974) and Eckholm (1977) further suggest that the search for the daily firewood seriously affects the natural vegetation. Sai (1984) considers both the demand for the export of hardwoods and need of household fuel as factors for the progressive deforestation.

It is against this background that the economic and societal impact of the recent drought should be seen which began in the late 1960's. A reasoned estimate of the numbers of deaths from famine in Subsaharan Africa may be precluded by the lack of pertinent information. Various sources (Wade, 1974; Sheets and Morris, 1976; Shear and Stacy, 1976; El-Khawas, 1976; Imperato, 1976) estimate a toll of more than 100 000 deaths and millions of refugees as well as a loss of some 40% of the livestock from the onset of the drought to the early 1970's. The scope and effect of the various international relief programs has been discussed by Wade (1974), Sheets and Morris (1976), El-Khawas (1976), Wiseberg (1976), Shear and Stacy (1976), Imperato (1976), and others. It appears that in some countries, Ethiopia in particular, the human consequences of the drought could in large part have been handled on a merely national basis — given the commitment and determination to do so! Poverty, rather than just drought, is seen as a major cause of famine and hunger (Food and Agriculture Organization of UN and OAU, 1982, p. 3). As Sheets and Morris (1976) explain, the human cost of the drought in Subsaharan West African is not only in the death toll but in the destruction of a way of life for millions of nomads. With their herds and livelihood wiped out, this mass of humanity driven from its economy and culture would be another burden to the already impoverished regions to the South and a potential source of social and political unrest in the future (Sheets and Morris, 1976).

It is noteworthy that the first few years of the recent drought (early 1970's) brought the fall of the three governments of particularly long standing: presidents Hamani Diori of Niger and Tombalbaye of Chad, and Emperor Haile Selassie were toppled in this period. In their analysis of the 1974 coup d'état in Niger, Higgott and Fuglestad (1975) discuss the complex political background of events but also note that as a consequence of the prolonged drought the country was in a general state of economic crisis and low morale. While Hess (1970) draws on optimistic picture of social progress in Ethiopia, Oriana Fallaci's (1974, pp. 295—314) interview with the aging emperor in 1972 reveals a lack of compassion and sensitivity — forebodings of the impending political change. A decade later and under a different political regime, large international famine relief operations and airlifts of food supplies into Ethiopia are again taking place and are receiving the attention of the news media; news reports in 1985 (Anonymous, 1985) indicate that at the same time food commodities are being exported from Ethiopia to international markets. Wiseberg (1976) describes a dismal pattern of mismanagement and corruption in Subsaharan Africa, which can least be tolerated in times of natural crises. Wiseberg (1976) further suggests that the drought situation in Upper Volta played some role in the downfall of the civilian government of Gerard Ouedraogo. In turn, political disturbances are considered a factor for the disruption of food production and distribution, and the evolution of refugee populations of unprecedented proportions, which generate additional pressure on the food resources of host countries (Food and Agricultural Organization of UN and OAU, 1982, p. 14).

The unprecedented media coverage of the human impact of the recent drought during the early 1970's faded out when a few years had more copious rainfall in the mid 1970's (Fig. 8.8:1). While publicity has ceased, the lack of rains continues and has become ever more accentuated in recent years. The socio-economic consequences of this situation are gradually being recognized (Toose, 1984; Walsh, 1984; United Nations, Economic and Social Council, 1984; Colwell, 1984; Schumacher, 1985; Kerr, 1985; Weisburd and Raloff, 1985; Raloff, 1985; Todorov, 1985). A recent report by the United Nations Economic and Social Council (1984, pp. 4–5) points out that the persistence of deficient rainfall in the Sahel has curtailed water supplies for human consumption, animal husbandry, irrigation, fisheries, and hydroelectric power generation, which has also reduced supply of electricity in various Gulf of Guinea countries, thus affecting basic industries and services. The report further mentions that since 1973 Lake Chad has been cut into two parts, with the northern portion drying up every year. Aside from the notoriously short attention span of news media, a gradual adaptation to the persistence of drought may be taking place thus reducing its human impact. Of interest in this context are Courel et al.'s (1984) findings of an albedo decrease in the Sahel in the course of the 1970's (Section 8.10) despite the aggravating rainfall deficiency.

10.9. The Use of Climate Prediction

The concern for climate prediction expressed in both the World Climate Program (World Meteorological Organization, 1980, p. 42) and the U.S. National Climate Program (National Climate Program Office, NOAA, 1980, pp. 23–24) is motivated by the perceived human impact of interannual climate variability. Accordingly, the practical application of climate forecasts is of particular interest.

In the case of the drought prediction for Northeast Brazil (Sections 8.6, 9.4, 10.7) specific propositions have been made concerning practical applications (Conselho Nacional de Pesquisas, 1980; Hastenrath, 1984; Hastenrath et al., 1984). Thus the question arises, should operational drought predictions become a reality, what use would be made of them? What form and what details of prediction are most needed? The wisdom of agriculturists, economists, and other planners is called upon, not only in the eventual application but also in the very design of the prediction effort. Brazil possesses the national resources that would permit it to benefit from climate prediction, as these resources are mobilized.

For Subsaharan West Africa (Sections 8.8, 9.5, 10.8), Glantz (1977c) attempted to ascertain the benefit which a forecast, six months in advance, of the 1973 drought would have yielded. He concludes that, given the national structures of states in the Sahel in which a potential technological capability would be used, the value of a long-range forecast would be limited, but that its value could be much enhanced through removal of numerous social, political, and economic obstacles. Glantz (1981) also considered the societal value of El Niño (Sections 8.2, 9.3, 10.5) forecasts and concludes that their practical usefulness would be limited.

In conclusion it is suggested that climate prediction may have a considerable human impact and socio-economic benefit, but that implementation of this task exceeds meteorological expertise. It should also be considered to what extent speculation in commodities could interfere with the societal value of climate forecasts. Lamb (1981) considers three prerequisites to this end: the human activities most severely affected by climatic fluctuations should be identified; it is essential to assess the flexibility of regional economies to adjust sufficiently to benefit substantially from climate forecasts, should they become reality; and skillful climate forecast schemes themselves need to be developed. Thus, for climate forecasts to be useful it will be essential to concentrate on elements with strong human impact in regions where the economic and social systems are flexible enough to adjust in response to the forecasts. It also appears an important task to develop such a societal flexibility and responsiveness.

10.10. Synthesis

The interannual variability of tropical climate has considerable economic and societal relevance, and conversely a variety of human activities affect the climatic environment. A particularly drastic interference with the natural setting is the deliberate deforestation throughout extensive regions of the tropics. While some immediate *in situ* consequences on soils, water holding capacity, evaporation and runoff are unmistakable, the possible effects of changes in surface albedo and soil moisture on the large-scale circulation are still controversial. Less conspicuous than the wholesale deforestation are certain land use patterns that affect surface albedo, and through 'biogeophysical' feedback processes may be conducive to subsidence and thus reinforcement and persistence of drought conditions. Increased dust input to the atmosphere resulting from overgrazing has been suggested as a factor for enhanced net radiative cooling of the atmospheric column, subsidence, and hence perpetuation of desert conditions. Among various proposals for climate modification, efforts to restore the 'original' environment and climate merit particular attention.

Fish life concentrated in a few coastal regions of the tropical oceans is vulnerable to both the interannual variability of the combined atmosphere-ocean system and excessive fishing. The impact on regional and World economy is appreciable.

The human impact of droughts has been extensively studied for a few tropical land areas. In India, diversification of crops and various land use practices were among the early endeavors to cope with recurrent droughts. Government measures included the construction of railroads and irrigation schemes and the introduction of famine relief projects. These were prompted by the great famines of 1876–77 and 1918–19 in particular.

Throughout its long history Northeast Brazil has repeatedly been plagued by drought catastrophes. Famines are recorded since the early colonial era, and emigration to other parts of Brazil began in the 17th century. Government measures to cope with drought problems and the first famine relief organization date back to the 1700's. In the 19th century the government directed the drilling of artesian wells, as well as the study of possible other measures to cope with the economic and social consequences of drought. The great famine of 1877–79 led to the recognition of the Nordeste Sêcas as a national problem. The great drought of 1958 gave the final impetus for the foundation of a specialized government agency concerned with the development of the Nordeste (SUDENE). A variety of measures, including the combination of dams, irrigation schemes, and roads, as well as diverse emergency schemes reduced the impact of droughts on the population of the interior Nordeste. The development of operational climate prediction remains an important task.

In the semi-arid desert fringe of Subsaharan Africa traditional economic and societal patterns evolved in a delicate balance with the natural environment, with a symbiotic relationship between sedentary agriculturists in the savanna and nomadic herdsmen closer to the desert fringe. The destruction of traditional patterns caused by colonialism and the introduction of cash crops for overseas export is considered a factor in the devastating impact of the recent Subsaharan drought. More importantly, the drilling of numerous wells and the introduction of Western medicine led to a greatly increased human and animal population in the Sahel, which proved utterly vulnerable when the recent drought developed.

Climate prediction is considered to be of great economic and societal relevance. It must, however, be recognized that operational climate forecasts — should they become a reality — can yield substantial benefits only where certain prerequisites are met. Thus it seems essential that wisdom from disciplines other than meteorology is brought to bear not only in the eventual application but also in the very design of the prediction schemes. Forecasts can only be useful where the economic, social, and political systems are flexible enough to respond effectively to natural disasters. Development of such a societal flexibility and responsiveness is an important task.

References

Allchin, B., Goudie, A., Hegde, K., 1978: *The prehistory and paleogeography of the Great Indian Desert.* Academic Press, London, New York, San Francisco, 370 pp.

Amado, J., 1978: *Seara vermelha.* 32nd edition, Editora Record, Saõ Paulo, 335 pp.

Ambroggi, R. P., 1966: 'Water under the Sahara'. *Scientific American, 214,* 21–29.

Anonymous, 1985: Zehntausende wandern aus, um zu sterben. *Der Spiegel,* 21 Jan 85, pp. 93–99.

Anthes, R. A., 1984: 'Enhancement of convective precipitation by mesoscale variations in vegetative covering in semiarid regions'. *J. Climate Appl. Meteor., 23,* 541–554.

Araújo, J. A. de A., 1982: 'Dams in the Northeast of Brazil; DNOCS experience in dams in semi-arid region'. Departamento Nacional de Obras contra as Secas (DNOCS), Fortaleza, 158 pp.

Bakun, A., 1978: 'Guinea Current upwelling'. *Nature, 271,* 147–150.

Ball, N., 1978: 'Drought and dependence in the Sahel'. *International Journal of Health Services,* 8, 271–298.

Barber, R. T., Chavez, F. P., 1983: 'Biological consequences of El Niño'. *Science, 222,* 1203–1210.

Berkofsky, L., 1976: 'The effect of variable surface albedo on the atmospheric circulation in desert regions'. *J. Appl. Meteor., 15,* 1139–1144.

Berkofsky, L., 1978: 'Reply (to comments by S. B. Idso and J. W. Deardorff)'. *J. Appl. Meteor., 17,* 561.

Bhatia, B. M., 1967: *Famines in India.* Asia Publishing House, Bombay, second edition, 389 pp.

Bolin, B., 1977: 'Changes of land biota and their importance for the carbon cycle'. *Science, 196,* 613–615.

Breman, H., 1976: 'Beelzebub als duivelsuitbanner (plattelandsontwikkeling in de Sahel)'. *Wending, 31,* 126–136.

Breman, H., Cissé, A. M., 1977: 'Dynamics of Sahelian pastures in relation to drought and grazing'. *Oecologia, 28,* 301–315.

Brennan, I., 1984: 'The development of the Indian Famine Codes: personalities, politics, and policies', pp. 91–111, in Currey, B., Hugo, G., eds.: *Famine as a geographical phenomenon.* Geojournal Library, Reidel, Dordrecht, Boston, Lancaster, 202 pp.

Broecker, W. S., Takahashi, T., Simpson, H. J., Peng, T. H., 1979: 'Fate of fossil fuel carbon dioxide and the global carbon budget'. *Science, 206,* 409–418.

Brown, L. R., 1984: *State of the World 1984.* Norton, New York, 252 pp.

Bryson, R. A., 1973: 'Drought in Sahelia: who or what is to blame?' *The Ecologist, 3,* 366–371.

Bryson, R. A., 1976: 'The lessons of climatic history'. *The Ecologist,* 6, 205–211.

Bryson, R. A., Baerreis, D. A., 1967: 'Possibilities of major climatic modification and their implications: Northwest India, a case for study'. *Bull. Amer. Meteor. Soc., 48,* 136–142.

Bryson, R. A., Murray, T. J., 1977: *Climates of hunger.* University of Wisconsin Press, 171 pp.

Carvalho, O., ed., 1973: 'Plano integrado para o combate preventivo aos efeitos das Secas do Nordeste'. Ministerio do Interior, Serie Desenvolvimento Regional, No. 1, 267 pp.

Charney, J. G., 1975: 'Dynamics of deserts and droughts in the Sahel'. *Quart. J. Roy. Meteor. Soc., 101,* 193–202.

Charney, J. G., Quirk, W. J., Chow, S. H., Kornfield, J., 1977: 'A comparative study of the effects of albedo change on drought in semi-arid regions'. *J. Atmos. Sci., 34,* 1366–1385.

Charney, J. G., Stone, P. H., Quirk, W. J., 1975: 'Drought in the Sahara: a biogeophysical feedback mechanism'. *Science*, 187, 434–435.

Charney, J. G., Stone, P. H., Quirk, W. J., 1976: 'Drought in the Sahara: insufficient biogeophysical feedback? Reply'. *Science*, 191, 100–102.

Chavez, F. P., Barber, R. T., Kogelschatz, J. E., Thayer, V. G., 1984: 'El Niño and primary productivity: potential effects on atmospheric carbon dioxide and fish production'. Tropical Ocean – Atmosphere Newsletter, No. 28, 1–2.

Cochemé, J., Franquin, P., 1967: 'An agroclimatology survey of a semiarid area in Africa South of the Sahara'. WMO Technical Note No. 86, WMO No. 210. TP. 110, 136 pp.

Cohn, T., 1975: 'The Sahelian drought: problems of land use'. *International Journal*, Canadian Institute of International Affairs, 30, 428–444.

Colwell, A., 1984: 'South of Sahara, the intrusive politics of hunger'. *New York Times*, 3 Dec. 84.

Conselho Nacional de Pesquisas, Brazil, 1980: 'Workshop on drought forecasting for Northeast Brazil'. INPE, Feb. 1980, Saõ José dos Campos, 71 pp.

Courel, M. F., Kandel, R. S., Rasool, S. I., 1984: 'Surface albedo and the Sahel drought'. *Nature*, 307, 528–531.

Cram, D., 1981: 'Hidden elements in the development and implementation of marine resource conservation policy: the case of the South West African/Nambian fisheries', pp. 137–156 in: Glantz, M. H., Thompson, J. D., eds.: *Resource management and environmental uncertainty; lessons from coastal upwelling fisheries*. Wiley, New York, Chichester, Brisbane, Toronto, 491 pp.

Crowley, T. J., 1983: 'The geologic record of climatic change'. *Rev. Geophys. Space Phys.*, 21, 828–877.

Cuniff, R. L., 1970: 'The great drought: Northeast Brazil 1877–1880'. Ph.D. Diss., Modern History, University of Texas at Austin, 347 pp.

Currey, B., Hugo, G., 1984: *Famine as a geographical phenomenon*. Geojournal Library, Reidel, Dordrecht, Boston, Lancaster, 202 pp.

da Cunha, E., 1979: *Os sertoẽs*. 28th ed., Francisco Alves, Rio de Janeiro, 416 pp.

d'Almeida, G. A., Schütz, L., 1983: 'Number, mass, and volume distributions of mineral aerosol and soils in the Sahara'. *J. Climate Appl. Meteor.*, 22, 233–243.

Das, P. K., 1983: 'Droughts and famines in India – a historical perspective'. *Mausam*, 34, 123–130.

Das, P. K., 1984: 'The monsoons – a perspective'. India National Science Academy, Perspective Report Series 4, New Delhi, 52 pp.

Das, P. K., 1986: *Monsoons*. Fifth IMO lecture, World Meteorological Organization, in press.

Davy, E. G., 1974: 'Drought in West Africa'. *WMO Bulletin*, 23, 18–23.

Delorme, G. A., 1963: 'Repartition et durée des précipitations en Afrique Occidentale'. Monographies de la Météorologie Nationale, No. 28, Paris, 27 pp.

Denevan, W. W., 1981: 'Swiddens and cattle versus forest: the imminent demise of the Amazon rain forest reexamined'. Studies in Third World Societies, Publication No. 13, pp. 25–44.

Eckholm, E. P., 1977: 'The other energy crisis', pp. 39–79 in: Glantz, M. H., Thompson, J. D., eds.: *Resource management and environmental uncertainty; lessons from coastal upwelling fisheries*. Wiley, New York, Chichester, Brisbane, Toronto, 491 pp.

El-Khawas, M., 1976: 'A reassessment of international relief programs', pp. 77–100 in: Glantz, M. H., Thompson, J. D., eds.: *Resource management and environmental uncertainty; lessons from coastal upwelling fisheries*. Wiley, New York, Chichester, Brisbane, Toronto, 491 pp.

Ellsaesser, H. W., MacCracken, M. C., Potter, G. L., Luther, F. M., 1976: 'An additional model test of positive feedback from high desert albedo'. *Quart. J. Roy. Meteor. Soc.*, 102, 543–544.

Fallaci, O., 1974: *Intervista con la storia*. Rizzoli, Milano, 391 pp.

Fein, J. F., Stephens, P. L., eds., 1986: *Monsoons*. Wiley Interscience Publishers, New York, London, Sidney, Toronto, in press.

Feldman, G., Clark, D., Halpern, D., 1984: 'Satellite color observations of the phytoplankton distribution in the eastern equatorial Pacific during the 1982–1983 El Niño'. *Science*, 226, 1069–1071.

Flohn, H., Ketata, M., 1971: 'Investigations on the climatic conditions of the advancement of the Tunisian Sahara'. WMO Technical Note No. 116, WMO Publ. No. 279, World Meteorological Organization, Geneva, 30 pp.

Food and Agriculture Organization of the United Nations and Organization of African Unity, 1982: 'Famine in Africa; situation, cause, prevention, control'. FAO, Rome, 36 pp.

Franke, R. W., Chasin, B. H., 1980: *Seeds of famine; ecological destruction and the development dilemma in the West African Sahel*. Allanheld, Osmun, and Co., New York, 266 pp.

Gentry, A. H., López-Parodi, J., 1980: 'Deforestation and increased flooding of the upper Amazon'. *Science,* **210**, 1354–1356.

Gentry, A. H., López-Parodi, J., 1982: 'Deforestation and increased flooding of the upper Amazon (reply to Comments by Nordin, C. F., and Meade, R. H.)'. *Science,* **215**, 427.

Ghosh, K. Ch., 1944: *Famines in Bengal, 1770–1943*. Indian Associated Publishing, Calcutta, 204 pp.

Glantz, M. H., ed., 1976: *The politics of natural disaster: the case of the Sahel drought*. Praeger Publishers, New York, Washington, London, 340 pp.

Glantz, M. H., ed., 1977a: *Desertification: environmental degradation in and around arid lands*. Westview Press, Boulder, Colo., 346 pp.

Glantz, M. H., 1977b: 'Climate and weather modification in and around arid lands in Africa', pp. 307–337 in: Glantz, M. H., ed.: *Desertification: environmental degradation in and around arid lands*. Westview Press, Boulder. Colo., 346 pp.

Glantz, M. H., 1977c: 'The value of a long-range weather forecast for the West African Sahel'. *Bull. Amer. Meteor. Soc.,* **58**, 150–158.

Glantz, M. H., 1981: 'Considerations of the societal value of an El Niño forecast and the 1972–73 El Niño', pp. 449–476 in: Glantz, M. H., Thompson, J. D., eds.: *Resource management and environmental uncertainty; lessons from coastal upwelling fisheries*. Wiley, New York, Chichester, Brisbane, Toronto, 491 pp.

Glantz, M. H., Parton, W., 1976: 'Weather and climate modification and the future of the Sahara', pp. 303–324 in: Glantz, M. H., ed.: *The politics of natural disaster; the case of the Sahel drought*. Praeger Publishers, New York, Washington, London, 340 pp.

Glantz, M. H., Thompson, J. D., eds., 1980: *Resource management and environmental uncertainty. Lessons from coastal upwelling fisheries*. Wiley and Sons, New Jersey, 491 pp.

Gómez-Pompa, A., Vázquez-Yanes, C., Guevara, S., 1972: 'The tropical rainforest: a nonrenewable resource'. *Science,* **177**, 762–765.

Gonçalves de Souza, J., 1979: *O Nordeste Brasileiro*. Banco do Nordeste do Brazil, S.A., Fortaleza, 409 pp.

Gray, W. M., Frank, W. M., Covin, M. L., Stokes, C. A., 1976: 'Weather modification by carbon dust absorption of solar energy'. *J. Appl. Meteor.,* **15**, 355–386.

Greene, M. H., 1974: 'Impact of the Sahelian drought in Mauretania'. *African Environment,* **1**, 11–21. (also: *The Lancet*, 1 June 1974)

Grove, A. T., 1973: 'A note on the remarkably low rainfall of the Sudan in 1913'. *Savanna,* **2**, 133–138.

Grove, A. T., 1974: 'Desertification in the African environment'. *African Affairs,* **73**, 137–151.

Grove, A. T., ed., 1986: *The Niger and its neighbours*. Balkema, Rotterdam, in press.

Guedalia, D., Estournel, C., Vehil, R., 1984: 'Effects of Sahel dust layers upon nocturnal cooling of the atmosphere (ECLATS Experiment)'. *J. Climate Appl. Meteor.,* **23**, 644–650.

Guillén, O. G., Calienes, R. Z., 1981: 'Biological productivity and El Niño', pp. 255–281 in Glantz, M. H., Thompson, J. D., eds.: *Resource management and environmental uncertainty; lessons from coastal upwelling fisheries*. Wiley, New York, Chichester, Brisbane, Toronto, 491 pp.

Gutman, G., 1984: 'Numerical experiments on land surface alterations with a zonal model allowing for interaction between geobotanic state and climate'. *J. Atmos. Sci.,* **41**, 2679–2685.

Gutman, G., Ohring, C., Joseph, J. H., 1984: 'Interaction between the geobotanic state and climate: a suggested approach and a test with a zonal model'. *J. Atmos. Sci.,* **41**, 2663–2678.

Hastenrath, S., 1984: 'Predictability of Northeast Brazil droughts'. *Nature,* **307**, 531–533.

Hastenrath, S., Wu, M.-C., Chu, P.-S., 1984: 'Towards the monitoring and prediction of Northeast Brazil droughts'. *Quart. J. Roy. Meteor. Soc.,* **110**, 411–425.

Henderson-Sellers, A., Gornitz, V., 1984: 'Possible climatic impacts of land cover transformations, with particular emphasis on tropical deforestation'. *Climatic Change,* **6**, 231–257.

Hess, R. L., 1970: *Ethiopia; the modernization of autocracy*. Cornell University Press, Ithaca, New York, 272 pp.

Higgott, R., Fuglestad, F., 1975: 'The 1974 coup d'état in Niger: towards an explanation'. *Journal of Modern African Studies,* **13**, 383–398.

Hookey, P., 1970: 'Revenge of the gods'. *Weather,* **25**, 425–428.

Idso, S. B., 1977: 'A note on some recently proposed mechanisms of genesis of deserts'. *Quart. J. Roy. Meteor. Soc.,* **103**, 360–370.

Idso, S. B., Deardorff, J. W., 1978: 'Comments on "The effect of variable surface albedo on the atmospheric circulation in desert regions"'. *J. Appl. Meteor.,* **17**, 560.

Iltis, H. H., 1983: 'Tropical forests, what will be their fate?' *Environment,* **25**, 55–60.

Imperato, P. J., 1972: 'Nomads of the Niger'. *Natural History,* **81**, pp. 60–69, 78–79.

Imperato, P. J., 1976: 'Health care systems in the Sahel: before and after the drought', pp. 283–302 in: M. H. Glantz, ed.: *The politics of natural disaster; the case of the Sahel drought*. Praeger Publishers, New York, Washington, London, 340 pp.

Jackson, R. D., Idso, S. B., 1975: 'Surface albedo and desertification'. *Science*, 189, 1012–1013.

Kellogg, W. W., Schware, R., 1981: *Climate change and society; consequences of increasing atmospheric carbon dioxide*. Westview Press, Boulder, Colo., 178 pp.

Kerr, R. A., 1985: 'Fifteen years of African drought'. *Science*, 227, 1453–1454.

Kessler, E., Alexander, D. Y., Rarick, J. F., 1978: 'Dust storms from the U.S. high plains in late winter 1977 – search for cause and implications'. *Proc. Okla. Acad. Sci.*, 58, 116–128.

Lamb, P. J., 1981: 'Do we know what we should be trying to predict – climatically?' *Bull. Amer. Meteor. Soc.*, 62, 1000–1001.

Lamb, P. J., 1982: 'Persistence of Subsaharan drought'. *Nature*, 299, 46–48.

Lamb, P. J., 1983: 'Sub-Saharan rainfall update for 1982: continued drought'. *J. Climatol.*, 3, 419–422.

Lamb, P. J., 1985: 'Rainfall in Subsaharan West Africa during 1941–83'. *Zeitschrift für Gletscherkunde und Glazialgeologie*, 21, 131–139.

Lee, I.-Y., 1983: 'Simulation of transport and removal processes of the Saharan dust'. *J. Climate Appl. Meteor.*, 22, 632–639.

Lettau, H. H., Lettau, K., Molion, L. B. C., 1979: 'Amazonia's hydrological cycle and the role of atmospheric recycling in assessing deforestation effects'. *Mon. Wea. Rev.*, 107, 227–228.

Lettau, H. H., Molion, L. B. C., 1975: 'Amazonas water basin supply, storage, and discharge'. *Transactions Amer. Geophys. Union*, 56, 597–598.

Lovejoy, P. E., Baier, S., 1976: 'The desert-side economy of the Central Sudan', pp. 145–175 in: M. H. Glantz, ed.: *The politics of natural disaster; the case of the Sahel drought*. Praeger Publishers, New York, Washington, London, 340 pp.

MacLeod, N. H., 1976: 'Dust in the Sahel: cause of drought?' pp. 215–231, in Glantz, M. H., ed.: *The politics of natural disaster: the case of the Sahel drought*. Praeger Publishers, New York, Washington, London, 340 pp.

McAlpin, M. B., 1979: 'Dearth, famine and risk: the changing impact of crop failures in Western India, 1870–1920'. *J. Econ. History*, 39, 143–157.

Ministerio das Minas e Energía, Brazil, 1983: Amazonia Legal, Projeto RADAMBRASIL, scale 1:2 500 000, two sheets. Rio de Janeiro.

Morales, C., ed., 1979: 'Saharan dust, mobilization, transport, deposition'. SCOPE report No. 14, Wiley and Sons, Chichester, New York, Brisbane, Toronto, 297 pp.

Murphy, R. C., 1981: 'The guano and the anchoveta fishery', pp. 81–106 in: Glantz, M. H., Thompson, J. D., eds.: *Resource management and environmental uncertainty; lessons from coastal upwelling fisheries*. Wiley, New York, Chichester, Brisbane, Toronto, 491 pp.

Myers, N., 1983: *A wealth of wild species*. Westview Press, Boulder, 274 pp.

National Academy of Sciences, National Research Council, 1980a: *Research priorities in tropical biology*. Washington, D.C., 116 pp.

National Academy of Sciences, National Research Council, 1980b: *Conversion of tropical moist forests*. Washington, D.C., 205 pp.

National Climate Program Office, NOAA, 1980: *National Climate Program*. Washington, D.C., 101 pp.

Neftel, A., Oeschger, H., Schwander, J., Stauffer, B., Zumbrunn, R., 1982: 'Ice core sample measurements give atmospheric CO_2 content during the past 40 000 years'. *Nature*, 295, 220–223.

Newell, R. E., 1971: 'The Amazon forest and atmospheric general circulation', pp. 457–459 in Matthew, W. J., Kellogg, W. H., Robinson, C. D., eds.: *Man's impact on climate*. MIT Press, Cambridge, Mass., 594 pp.

Nordin, C. F., Meade, R. H., 1982: 'Deforestation and increased flooding of the upper Amazon: comments'. *Science*, 215, 426–427.

Norton, C. C., Mosher, F. R., Hinton, B., 1979: 'An investigation of surface albedo variations during the recent Sahel drought'. *J. Appl. Meteor.*, 18, 1252–1262.

Otterman, J., 1974: 'Baring high-albedo soils by overgrazing: a hypothesized desertification mechanism'. *Science*, 186, 531–533.

Otterman, J., 1975: 'Surface albedo and desertification, reply'. *Science*, 189, 1013–1014.

Otterman, J., 1977: 'Anthropogenic impact on the albedo of the earth'. *Climatic Change*, 1, 137–155.

Paulik, G. J., 1981: 'Anchovies, birds, and fishermen in the Peru Current', pp. 156–185 in: Glantz, M. H., Thompson, J. D., eds.: *Resource management and environmental uncertainty; lessons from coastal upwelling fisheries*. Wiley, New York, Chichester, Brisbane, Toronto, 491 pp.

Peixoto, J. P., Oort, A. H., 1984: 'Physics of climate'. *Reviews of Modern Physics,* **56**, 365–429.

Potter, G. L., Ellsaesser, H. W., McCracken, M. C., Luther, F. M., 1975: 'Possible climatic impact of tropical deforestation'. *Nature,* **258**, 697–698.

Prance, G. T., Elias, T. S., eds., 1977: *Extinction is forever.* New York Botanical Gardens, New York, 437 pp.

Prospero, J. M., Nees, R. T., 1977: 'Dust concentration in the atmosphere of the equatorial North Atlantic: possible relationship to the Sahelian drought'. *Science,* **196**, 1196–1198.

Rabelo, S., 1983: *Euclides da Cunha.* Third edition, Editora Civilização Brasileira, Rio de Janeiro, 361 pp.

Raloff, J., 1985: 'Africa's famine: the human dimension'. *Science News,* **127**, 299–301.

Ramos, G., 1978: *Vidas secas.* 41st edition, Editora Record, São Paulo, 167 pp.

Rao, Y. P., 1976: *Southwest monsoon.* India Meteorological Department, Meteorological Monograph. Synoptic Meteorology No. 1/1976, Delhi, 367 pp.

Ripley, E. A., 1976a: 'Drought in the Sahara: insufficient biogeophysical feedback?' *Science,* **191**, 100.

Ripley, E. A., 1976b: 'Comments on paper "Dynamics of deserts and drought in the Sahel" by J. G. Charney'. *Quart. J. Roy. Meteor. Soc.,* **102**, 466–467.

Roth, R., 1984: 'The effect of change in surface roughness on the precipitation regime in the coastal area of NE-Brazil'. *Geojournal,* **8**, 205–209.

Sagan, C., Toon, O. B., Pollack, J. B., 1979: 'Anthropogenic albedo changes and the Earth's climate'. *Science,* **206**, 1363–1368.

Sai, F. T., 1984: 'The population factor in Africa's development dilemma'. *Science,* **226**, 801–805.

Salati, E., Lovejoy, T. E., Vose, P. B., 1983: 'Precipitation and water recycling in tropical rain forests with special reference to the Amazon basin'. *The Environmentalist,* **3**, 67–72.

Salati, E., Vose, P. B., 1983: 'Depletion of tropical rain forests'. *Ambio,* **12**, 67–71.

Salati, E., Vose, P. B., 1984: 'Amazon basin: a system in equilibrium'. *Science,* **225**, 129–138.

Sanford, R. L. Jr., Saldarriaga, J., Clark, K. E., Uhl, C., Herrera, R., 1985: 'Amazon rain-forest fires'. *Science,* **227**, 53–55.

Schneider, S. H., Mesirow, L. E., 1976: *The genesis strategy.* Plenum Publishing Co., New York, 419 pp.

Schnell, R. C., 1975: 'The biogenic component of Sahelian eolian dust: a possible drought factor'. *EOS,* **56**, 994–995 (abstract).

Schreiber, R. W., Schreiber, E. A., 1984: 'Central Pacific seabirds and the Southern Oscillation: 1982 to 1983'. *Science,* **225**, 713–716.

Schumacher, E., 1985: 'Drought turns nomads' world upside down'. *New York Times,* 2 March 85.

Seaman, J., Holt, J., Rivers, J., 1973: 'An inquiry into the drought situation in Upper Volta'. *The Lancet,* 6 Oct., 774–778.

Shear, D., Stacy, R., 1976: 'Can the Sahel survive? Prospects for long-term planning and development', pp. 128–144 in: M. H. Glantz, ed.: *The politics of natural disaster; the case of the Sahel drought.* Praeger Publishers, New York, Washington, London, 340 pp.

Sheets, H., Morris, R., 1976: 'Disaster in the desert', pp. 25–76 in: M. H. Glantz, ed.: *The politics of natural disaster; the case of the Sahel drought.* Praeger Publishers, New York, Washington, London, 340 pp.

Shukla, J., Mintz, Y., 1982: 'Influence of land-surface evapotranspiration on the Earth's climate'. *Science,* **215**, 1498–1501.

Sternberg, H. O'R., 1984: 'Human-induced changes in the natural environment of Amazonia: some fluvio-ecological aspects'. Paper presented to the 25th International Geographical Congress, Paris, 27–31 Aug. 1984.

Stretta, J. M., 1977: 'Température de surface et pêche thonière dans la zone frontale du Cap Lopez (Atlantique Tropical Oriental) en Juin et Juillet 1972, 1974 et 1975'. *Cahiers ORSTOM, Sér. Oceanogr.,* **15**, no. 2, pp. 163–180.

Sud, Y. C., Fennessy, M., 1982: 'A study of the influence of surface albedo on July circulation in semi-arid regions using the GLAS GCM'. *J. Climatol.,* **2**, 105–125.

Sud, Y. C., Fennessy, M., 1984: 'Influence of evaporation in semi-arid regions on the July circulation: a numerical study'. *J. Climatol.,* **4**, 383–398.

Sudene, Brazil, 1981: As sêcas do Nordeste; uma abordagem historica de causas e efeitos. Recife, 119 pp.

Sudene, Brazil, 1985: Recursos naturais do Nordeste; investigação e potencial. Recife, 195 pp.

Thompson, J. D., 1977: 'Ocean deserts and ocean oases', pp. 103–139 in: Glantz, M. H., ed.: *Desertification: environmental degradation in and around arid lands.* Westview Press, Boulder, Colo., 346 pp.

Thompson, J. D., 1981: 'Climate, upwelling, and biological productivity: some primary relationships', pp. 13–33 in: Glantz, M. H., Thompson, J. D., eds.: *Resource management and environmental uncertainty; lessons from coastal upwelling fisheries.* Wiley, New York, Chichester, Brisbane, Toronto, 491 pp.

Todorov, A. V., 1985: 'Sahel: the changing rainfall regime and the "normals" used for its assessment'. *J. Climate Appl. Meteor.*, **24**, 97–107.

Toose, S., 1984: 'Call for joint action'. *Nature*, **307**, 497.

United Nations Economic and Social Council, 1984: 'Critical economic situation in Africa'. E/1934/109, 14 June 84, New York, 9 pp.

United Nations Environment Programme, 1979: 'The Environment Programme, report of the Executive Director'. UNEP NA 79–0075, Nairobi, 147 pp.

U.S. Department of State, U.S. Agency for International Development, 1978: 'Proceedings of the U.S. Strategy Conference on Tropical Deforestation'. Washington, D. C., 78 pp.

Vondruska, J., 1981: 'Postwar production, consumption, and prices of fish meal.' pp. 285–316 in: Glantz, M. H., Thompson, J. P., eds: *Resource management and environmental uncertainty; lessons from coastal upwelling fisheries.* Wiley and Sons, New Jersey, 491 pp.

Wade, N., 1974: 'Sahelian drought: no victory for Western aid'. *Science*, **185**, 234–237.

Walsh, J., 1984: 'Sahel will suffer even if rains come'. *Science*, **224**, 467–471.

Ware, H., 1975: 'The Sahelian drought: some thoughts of the future'. U.N. Special Sahelian Office, New York, 47 pp.

Ware, H., 1977: Desertification and population: Sub-Saharan Africa, pp. 165–202 in: Glantz, M. H., ed.: *Desertification: environmental degradation in and around arid lands.* Westview Press, Boulder, Colo., 346 pp.

Weisburd, S., Raloff, J., 1985: 'Climate and Africa: why the land goes dry?' *Science News*, **127**, 282–285.

Wiseberg, L., 1976: 'An international perspective on the African famines', pp. 101–127 in: Glantz, M. H., ed.: *The politics of natural disaster; the case of the Sahel drought.* Praeger Publishers, New York, Washington, London, 340 pp.

Woodwell, G. M., Whittaker, R. H., Reiners, W. A., Likens, G. E., Delwiche, C. C., Botkin, D. B., 1978: 'The biota and the World carbon budget'. *Science*, **199**, 141–146.

World Meteorological Organization, 1980: 'Outline plan and basis for the World Climate Programme 1980–1983'. WMO-No. 540, Geneva, 64 pp.

World Meteorological Organization-ICSU, 1975: *The physical basis of climate and climate modelling.* GARP Publication Series, No. 16, 265 pp.

Zimmerman, R. R., Greenberg, J. P., Wandiga, S. O., Crutzen, P. J., 1982: 'Termites: a potentially large source of atmospheric methane, carbon dioxide, and underwater hydrogen'. *Science*, **218**, 563–565.

CHAPTER 11

TROPICAL GLACIERS AND CLIMATE

Tropical glaciers and ice caps are a particularly climate-sensitive component of the environment. It is then not surprising that international efforts directed at climatic and environmental change (UNESCO, 1970; World Meteorological Organization-ICSU, 1975, pp. 7, 11, 16; International Association of Hydrological Sciences -UNESCO, 1977; Temporary Technical Secretariat for World Glaciers Inventory, UNESCO-UNEP-IASH-ICSI, 1977; United Nations Environment Programme, 1979, pp. 5–17) have identified the study of glacier and climate relationships as a task of high priority.

Glaciers are relevant to the study of climate in various respects. The spatial distribution of ice is, through conditions of heat and ice mass budgets, related to the large-scale and local climatic patterns (Section 11.1). A combination of geomorphic and other evidence allows a preliminary reconstruction of glacier variations in Pleistocene to early Holocene times (Section 11.2). This is an important contribution to the study of past climates (Chapter 12, Section 12.4). For the quantitative inference of more recent climate variations (scale of decades to 1–2 millenia) from glacier observations, two independent methods are being developed for application in the tropics: the reconstruction of the unknown climatic forcing through numerical modelling of the observed terminus response (Section 11.3), and climatic ice core studies (Section 11.4). The purpose of this chapter, based on Hastenrath (1981, 1984) and other sources, is a synopsis of work to date and an introduction to the pertinent literature.

11.1. Spatial Patterns

At present, glaciers still exist in three regions of the tropics (Fig. 11.1:1): in the South American Andes and on the Mexican volcanoes; on Kilimanjaro, in the Ruwenzori, and on Mount Kenya in East Africa; and on Mounts Jaya, Idenburg, and Mandala in the Indonesian part of New Guinea. An abundance of geomorphic evidence attests to a formerly much larger ice extent including on mountains that are presently not glaciated, such as the Altos de Cuchumatanes (Guatemala) and the Cordillera de Talamanca (Costa Rica) on the Central American land bridge; the Aberdares and Mount Elgon in East Africa, and various mountains in Ethiopia (High Semyen, Badda, Cacca, Cilalo, Enguelo, Bale); mountains in both Irian Jaya and Papua New Guinea, as well as in Borneo (Kinabalu); and in Hawaii.

A first orientation of large-scale spatial patterns is obtained from meridional profiles of the elevations of (i) the modern ice equilibrium line (MEL), (ii) the ice equilibrium line during a formerly larger ice extent in the Pleistocene (PEL) whose time is not as a rule known, and (iii) the 0°C mean annual isotherm (Fig. 11.1:2). MEL drops from the tropics towards high latitudes, but is in general higher in the outer tropics than in the equatorial zone. Particularly high MEL is characteristic for the hyper-arid regions of the southern part of the South American Altiplano. PEL follows a pattern similar to MEL, with a vertical separation of the order of 1000 m. The 0°C mean annual isothermal surface drops monotonically from the equatorial belt poleward.

The ice equilibrium line is defined as the region where solid precipitation equals the ablation through melting and/or sublimation, and is accordingly determined by specific combinations of the mass and heat budgets, or climate. Fig. 11.1:2 draws attention to the great diversity of the climatic environment at the ice equilibrium line in the tropics. In the humid inner tropics, the equilibrium line is low, the environment is relatively warm, and ablation is effected predominantly through the energetically inexpensive melting. In this situation, the ice equilibrium line is sensitive to temperature possibly more than to precipitation variations. As glaciers reach down to relatively low elevations and vegetation limits are high, there are as a rule good prospects for C-14 dating of moraines. In the more arid outer tropics the climatic environment at the equilibrium line is drastically different. In the hyper-arid Southern tropical Andes in particular, there are mountains of more than 6000 m without perennial snow. The ice equilibrium region is cold, ablation is exclusively through the energetically expensive sublimation, and the ice extent is sensitive to variations in precipitation but not in temperature. The vertical separation between glaciers and vegetation is large, so that C-14 dating of moraines holds little promise.

In addition to the latitudinal control illustrated in Fig. 11.1:2, there are large-scale zonal variations of the ice equilibrium line. Thus in the equatorial zone of the South American Andes, where the lower- to mid-tropospheric easterlies provide the major moisture supply, the equilibrium line is distinctly lower in the Eastern than in the Western Cordillera. Proceeding towards the mid-latitude westerly wind regions of either hemisphere, the westward rise of the equilibrium line gives way to the inverse slope. In the latitude of Northern Chile and Argentina, where the modern ice equilibrium line rises westward, geomorphic evidence indicates an eastward rise of the equilibrium line for some time during the Pleistocene (Hastenrath, 1984, p. 296), thus suggesting a shift in zonal circulation belts.

Apart from the these large-scale meridional and zonal patterns there are azimuth asymmetries of ice distribution on individual mountains. Most readily understood are the North–South differences. Planetary radiation geometry results in larger insolation on the equatorward facing slopes, thus favoring stronger glaciation on the poleward faces. This effect increases away from the Equator. Such meridional asymmetries are, for example, found on Kilimanjaro in East Africa, on the Mexican volcanoes, and in the Ecuadorian Andes (Hastenrath, 1981, pp. 19–24; 1984, pp. 293–296). More remarkable are zonal asymmetries. In the more humid regions there is a tendency for lower-reaching glaciation in the azimuths of major moisture supply, so that zonal asymmetries on individual mountains broadly agree with the large-scale pattern described above. In the Ecuadorian Andes, for example (Hastenrath, 1981, pp. 52–60), lower glacier limits rise from the Eastern to the Western Cordillera, and in both mountain chains the eastward slopes of individual volcanoes are more heavily glaciated than those facing West. In the drier mountain regions, the direction of atmospheric moisture transports gives way to local circulations (Section 2.3) as major controlling factor of zonal asymmetries in ice extent. As discussed in Sections 2.3 and 2.4, the mountain wind system leads to enhanced cloudiness in the afternoon, which reduces insolation on the westward facing slopes, and thus favors glaciation there. The observed tendency for lower-reaching glaciation at westerly azimuths in the more arid mountain regions appears plausible from this mechanism. On the perennially clouded mountains of the humid tropics, such processes would be less effective in controlling the azimuth distribution of ice extent. Lower glacier limits to the West than to the East are, for example, found for the modern and Pleistocene glaciation of Kilimanjaro, for the Pleistocene glaciation of Ethiopia, for the Mexican volcanoes, and the Southern tropical Andes (Hastenrath, 1984, pp. 50–142, 293–296). Apart from certain topographic factors, the spatial distribution of ice thus strongly reflects the predominant climatic patterns.

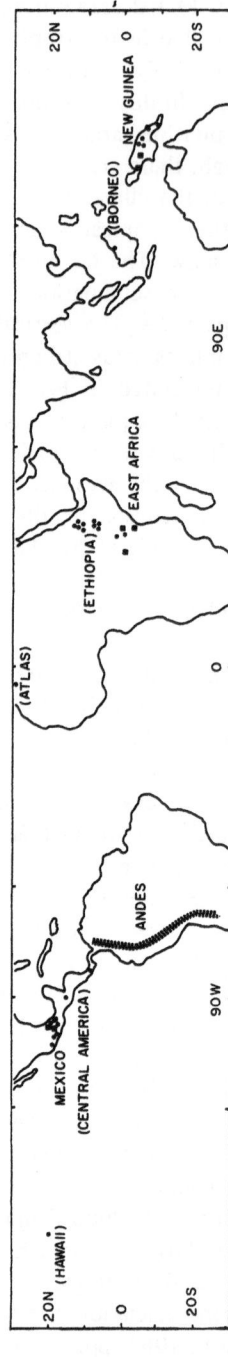

Fig. 11.1:1. Orientation map of tropical glacier regions. Names in parentheses and crosses refer to formerly glaciated sites, all other entries to the modern glaciation. From Hastenrath (1985).

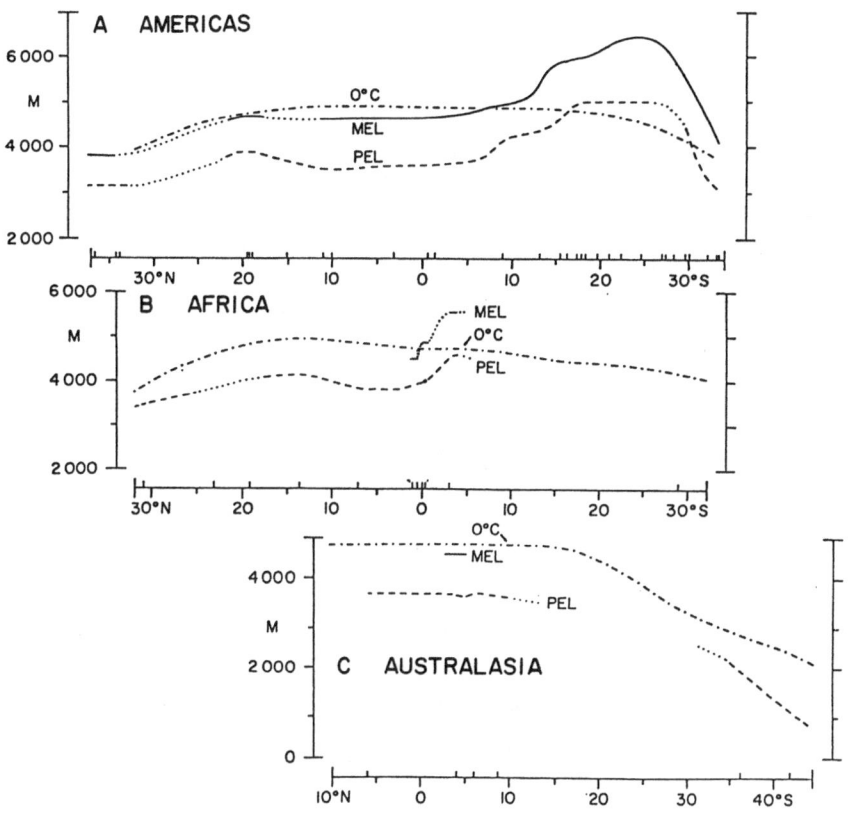

Fig. 11.1:2. Schematic meridional profiles of mean annual isothermal surface, °C, dash-dotted; modern ice equilibrium level, MEL, solid; Pleistocene ice equilibrium level, PEL, broken line. (a) Western Cordillera of the Americas along 75°W; (b) Africa along 30°E; (c) Australasian sector along 145–120°E. From Hastenrath (1985).

11.2. History of Glacier Variations

This section reviews the evidence of tropical glacier variations from half a million years ago to the present, based primarily on three recent publications (Hastenrath, 1981, 1984, 1985). The information is also pertinent to the discussion of past climates of the tropics (Section 12.4). Almost all of the field evidence summarized in Table 11.2:1 has been published since the 1970's, and the present preliminary synopsis of tropical glacier variations is just only now possible. Reference is also made to Clapperton's (1983) recent review for the Andes. The material is here discussed successively for three large time intervals, namely earlier than 100 000 years B.P., 100 000 to 10 000 years B.P., and 10 000 years B.P. to present.

For the era prior to 100 000 years B.P., evidence is summarized in Fig. 11.2:1. Glaciations are indicated around 500 000 years B.P. on Kilimanjaro in East Africa, and around 380 000 years B.P. in New Guinea. Thereafter there is some coincidence in the timing of glaciations between continents. Thus, a glaciation is reported around 300 000 years B.P. for Kilimanjaro, around 290 000 years B.P. for New Guinea, and around 250 000 years B.P. for Hawaii. Furthermore a glaciation at >100 000 years B.P. is evidenced for both Kilimanjaro and the Colombian Andes, broadly coincident with a glaciation around 135 000 years B.P. in Hawaii. This field evidence is in Fig. 11.2:1 compared with an index of oxygen isotope variations of seawater in the low-latitude ocean, which is thought to reflect changes in the global volume of glacial ice (Imbrie *et al.*, 1984). The glaciations >100 000 years B.P. in Colombia and on Kilimanjaro and at ~135 000 years B.P. in Hawaii broadly agree with an extreme positive value of the oxygen isotope index at >100 000 years B.P. Glaciations at 250 000 years B.P. in Hawaii, at 290 000 years B.P. in New Guinea, and at 300 000 years B.P. on Kilimanjaro may correspond to another positive extreme of the oxygen isotope index around that time. These correlations appear reasonable within the tolerance of K–A

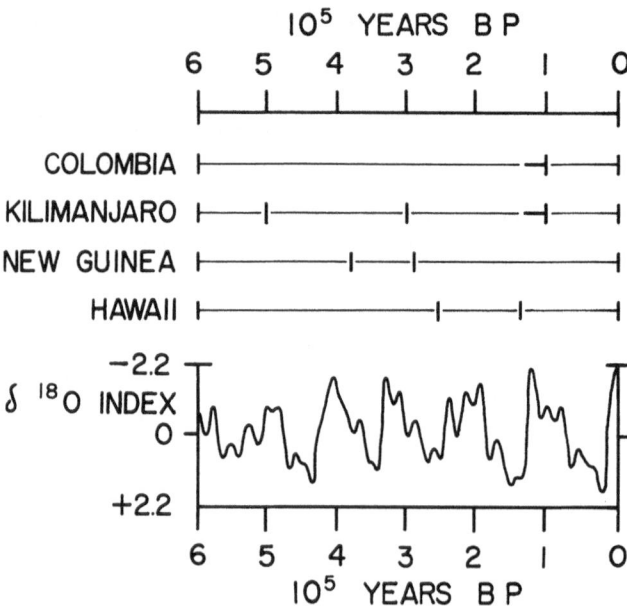

Fig. 11.2:1. Comparison of older glaciations in Colombia, on Kilimanjaro, in New Guinea, and in Hawaii (top; source: Table 11.2:1.) with index of $\delta^{18}O$ of seawater in the low-latitude ocean (bottom; deep-sea cores, variations normalized to zero mean and unit standard deviation; source: Imbrie *et al.*, 1984). From Hastenrath, (1985).

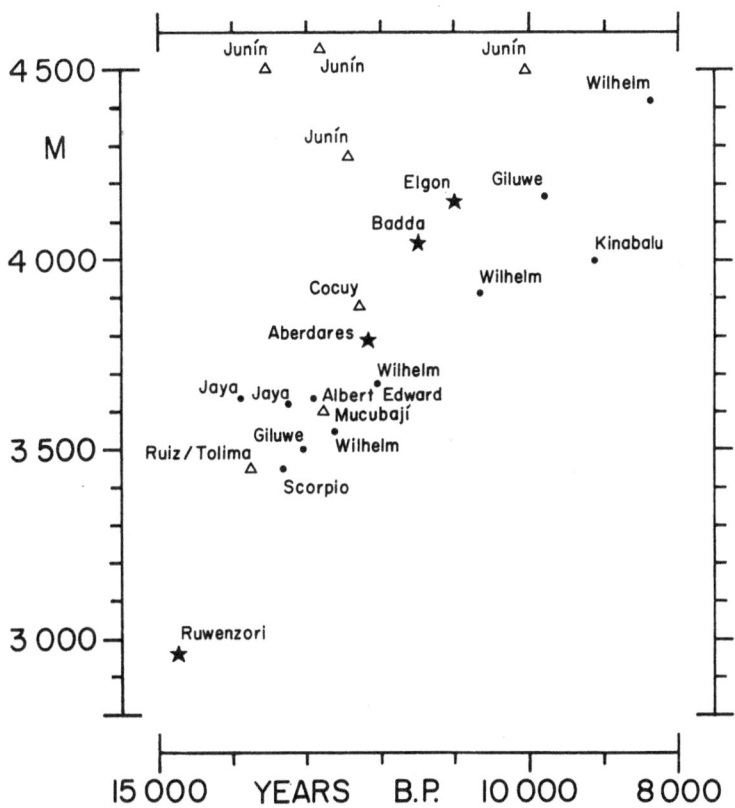

Fig. 11.2:2. Variation of minimum deglaciation age with elevation. (a) Africa, stars; (b) Americas, triangles; (c) Australasia, dots. Data for Kinabalu, Borneo are from Flenley and Morley (1978), for Junín, Peru, from Wright (1983). For other sources refer to Hastenrath (1984). From Hastenrath (1985).

age determinations. The matching of events appears more ambiguous for the glaciation at 380 000 years B.P. in New Guinea and especially for the glaciation at 500 000 years B.P. on Kilimanjaro.

Turning to the younger Pleistocene (Table 11.2:1), there are indications of a glaciation 28 000 – 14 000 years B.P. and minor advances at >12 240 and <10 200 years B.P. in the East Peruvian Andes, and events at >10 000 B.P. in Venezuela, and at 34 000–32 000, respectively at >27 000 and >25 000 years B.P., and at 12 100, 10 000, and 9000 years B.P. in Mexico. Glaciations around 55 000 and 20 000 years B.P. are reported for Hawaii. Particularly noteworthy is the lack of any glaciation dates for the Southern tropical Andes, where C-14 dating is intrinsically difficult, and the apparent absence of major glaciation during 25 000–15 000 years B.P. in Mexico. As discussed in Section 11.1, it is suggested that in this semi-arid environment the ice equilibrium line would be more sensitive to precipitation than temperature changes. Beyond this limited documentation for onset and duration of glaciations in the tropics, considerable evidence has accumulated for the timing of deglaciation in the three glacier regions of the tropics. Dates range between 15 000 and 8000 yeas B.P. This wide time span and the occurrence of various glacier advances during it, at first sight make for a confused picture of deglaciation chronology. However, a plot of elevation versus time (Fig. 11.2:2) reveals a remarkably systematic behavior, in good consistency between the three regions. Deglaciation progressed from around 3000 m after about 15 000 yars B.P. towards the 4500 m level at 8000 years B.P., thus showing a lag of about a millenium per 200 m elevation.

TABLE 11.2:1

Synopsis of evidence of Pleistocene to Holocene glacier events in the tropics. From Hastenrath (1985).

	Americas	Africa	Australia	Hawaii
20th century moraines	Ecuador Peru (Hastenrath, 1981, 1984a, b)	Mt. Kenya (Hastenrath, 1984)		
moraines III ~1400–1800 B.P.	Mexico (Heine, 1975) Ecuador (Hastenrath, 1981) Peru (Mercer et al., 1975; Mercer/Palacios, 1977)	Kilimanjaro, Ruwenzori Mt. Kenya (Hastenrath, 1984)		
moraines	Mexico 2000 B.P. (Heine, 1975) Colombia 2300 B.P. (Gonzalez et al., 1965) Peru >2830 B.P. (Mercer et al., 1975; Mercer/Palacios, 1977)		Irian Jaya 3500 B.P. >1600 B.P. (Peterson/Hope, 1972; Galloway et al., 1973; Hope et al., 1976)	

glaciation deglaciation	Mexico, glac. 34 000–32 000 B.P. >27 000, >25 000 B.P. 12 100, 10 000, 9000 B.P. (Heine, 1975; White/Valastro, 1984) Venezuela, glac., deglac. >13 600 B.P. >10 000 B.P. (Schubert, 1972a, b, 74; Salgado-Labouriau et al., 1977) Colombia, glac., deglac. <13 760 B.P. (Herd/Naeser, 1974; Gonzales et al., 1965) Peru, glaciation 28 000–14 000 B.P. advances >12 240, <10 900 B.P. deglaciation 13 540, 12 800, 11 945, 10 050 B.P. (Mercer et al., 1975; Mercer/Palacios, 1977; Wright, 1983)	deglaciation Ruwenzori >14 700 B.P. (Livingstone, 1962) Aberdares >12 200 B.P. (Perrott, 1982) Badda >11 000 B.P. (Gasse/Descourtieux, 1979) Elgon >11 500 B.P. (Hamilton/Perrott, 1978)	deglaciation New Guinea, Borneo >14 000 B.P., etc. (Peterson/Hope, 1972; Hope et al., 1976) deglaciation Borneo 9200 B.P. (Flenley/Morley, 1978)	glaciation ~20 000, 550 000 B.P. (Porter et al., 1977)
glaciation	Colombia >100 000 B.P. (Herd/Naeser, 1974)	Kilimanjaro >100 000 B.P., 300 000 B.P., 500 000 B.P. (Downie/Wilkinson, 1972)	New Guinea 290 000 B.P. 380 000 B.P. (Löffler, 1972, 76, 82)	~135 000, 250 000 B.P. (Porter et al., 1977)

After this major deglaciation episode, glacial events show less spatial consistency. There are dated moraines of several millenia ago in both the Americas and New Guinea. From elevation and spatial arrangement, certain moraines of the mountains of Eastern Africa may fall broadly within this era.

Progressing into historical times, certain moraine complexes in the Americas can be ascribed to the time span since the Spanish colonization ($\sim$1500–1800 years A.D.). Some moraines on the mountains of Eastern Africa may represent corrollaries. Finally, moraine formations of the 20th century can be identified both in the South American Andes and in East Africa.

Observations of ice extent span more than four centuries in certain regions of the Americas, but date back to only the latter part of the 19th century in East Africa, and to the 1930's in New Guinea. For the Ecuadorian Andes, historical records and field observations pinpoint the onset of ice recession to the middle of the 19th century, at the latest. Inference from numerical modelling (Section 11.3) yields 1880 as onset date of the climatic forcing which led to the retreat of Lewis Glacier on Mount Kenya, and a date around the middle of the 19th century for the Carstensz Glacier on Mount Jaya in New Guinea. In all three regions the glacier recession is continuing to the present.

11.3. Climatic Forcing and Terminus Response

The causality chain from climatic forcing to glacier response is schematically illustrated in Fig. 11.3:1. The climate governs the heat and ice mass budgets of the glacier, which are reflected in the vertical net balance profile. In turn, the distribution of net balance with elevation along with the bedrock configuration control the dynamics of ice flow, which lead to variations in ice extent. It is through this chain of causality that climate variations are ultimately reflected in glacier advance or retreat. Understanding of this causality chain is needed for quantitative translation of the well documented and conspicuous glacier response into the unknown and small climatic forcing.

A method being developed to this end involves the numerical modelling of the ice dynamics (Allison and Kruss, 1977; Kruss, 1981, 1983, 1984). The intent is to reconstruct the climatic forcing from the glacier response. The input to the computer simulation includes (a) the record of terminus positions; (b) bedrock configuration; (c) modern mass budget conditions. An equilibrium glacier of maximum extent, as indicated by moraines, is generated and the corresponding net balance profile is ascertained. Then a retreat is simulated in such a way that the various observed terminus positions are reproduced. This then allows quantitative inference on the history of the climatic forcing.

The modelling of Carstensz Glacier in New Guinea by Allison and Kruss (1977) produced inference on the timing of climatic change and onset of glacier retreat as well as the time sequence

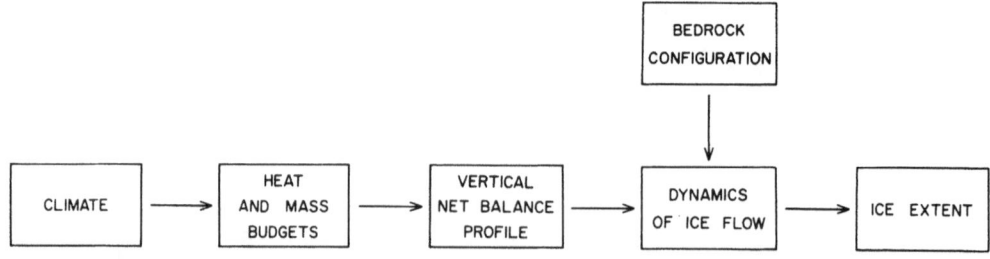

Fig. 11.3:1. Causality chain of climatic forcing and glacier response.

of changes in the vertical net balance profile. Based on this earlier work, Kruss (1981, 1983, 1984) pursued a novel approach for Lewis Glacier on Mount Kenya, by modelling in unison the entire causality chain sketched in Fig. 11.3:1 from climatic forcing to terminus response. Basic to this effort is a multi-annual field investigation of glacier morphology, kinematics, and heat and mass budgets, as well as a reconstruction of Lewis Glacier variations since the later part of the 19th century from a variety of historical sources (Hastenrath, 1984).

Kruss' (1981, 1984) analysis of frequency response yields for net balance variations on the scale of decades to centuries a lag in terminus response of the order of 10–20 years; the amplitude response of the terminus increases rapidly with period on the scale of decades, further amplitude increase with period being small on the scale of centuries. Climatic forcings are considered in terms of precipitation, albedo, and cloudiness decrease, and temperature increase (Kruss, 1981, 1983). The onset of ice retreat from the innermost large moraine of Lewis Glacier is reconstructed at 1890, in response to climatic change beginning around 1883 and continuing to the turn of the century. This is followed by two decades with little variation in climatic conditions, pronounced change from the 1920's to the early 1930's, a reversal in the 1940's, and a recommencement of the earlier climatic tendency in the 1950's. The observed ice retreat could have been produced by any of the aforementioned climatic forcings, above or in various combinations. However, useful constraints are available from independent observations of lake level and temperature variations.

On this basis, the following major forcings emerge as causes of the continued glacier recession since the 19th century: (i) a precipitation decrease of the order of 150 mm a^{-1}, concomitant with a small decrease of cloudiness and surface albedo during the last two decades of the 19th century; and (ii) a temperature increase of a few tenths of °C during the 20th century, concentrated in the 1920's. The drastic precipitation decrease in the latter part of the last century is also reflected in a drastic drop of the water level of East African lakes. The climatic forcings responsible for the conspicuous and well documented glacier response are as a rule too small to be ascertained by conventional sensing techniques. The Lewis Glacier study thus points the way for the quantitative assessment of climate variations from glacier terminus observations.

11.4. Ice Cores

Extraordinary climatic records have been reconstructed from ice cores in the polar regions of both hemispheres (Dansgaard, 1964; Dansgaard et al., 1969; Sharp and Gow, 1970; Johnsen et al., 1972). Indicators evaluated include microparticle concentration, total beta radioactivity, and oxygen isotope ratios. Application of climatic ice core techniques to the tropics is novel. In the tropics, particular difficulties must be considered that are not encountered in the well-established polar glaciology. Aside from logistic considerations, the following criteria guide the choice of an ice body for coring and paleoclimatic studies in the tropics: (i) the ice cap should be at very high elevations and thus low temperature to preclude significant melting and percolation; (ii) for an extended ice plateau with gentle topography, the effects of flow dynamics on stratigraphy are minimized; (iii) location in the outer tropics allows some seasonality in the stratigraphy; (iv) the thickness and net balance are limiting factors for the length of the climatic record to be expected.

These conditions are ideally met for the Quelccaya Ice Cap (Thompson et al., 1979, 1984). This vast ice plateau in the Eastern Andes of Peru has been the object of a decade-long (1974–83) field program, with the central objective of retrieving ice cores for analysis of microparticle and isotope analysis and climate reconstruction. The study of the present climate, heat and mass budget, is an important component of the project, as this is to put paleoclimatic interpretations on a firm footing.

The ice stratigraphy at Quelccaya is characterized by a conspicuous annual layering, the June/July dry season being marked by an ice horizon and concentration of microparticles and total beta radioactivity. The annual net balance is obtained from the mass contained between annual horizons. On the Quelccaya summit plateau net balance is essentially equal to precipitation, because virtually no energy is available for ablation. This is a direct consequence of the large surface albedo (Hastenrath, 1978). The net balance series thus obtained for the Quelccaya summit compares well with precipitation at stations outside the ice cap. Accordingly, the microparticle and beta profiles offer a good prospect for establishing a net balance or precipitation chronology for Quelccaya. Inasmuch as precipitation departures in the interior tend to run inverse to the Ecuador – Peru coast (Hastenrath, 1976), an inverse relation can be expected between the net balance on the Quelccaya Ice Cap and El Niño (Sections 8.2 and 8.4).

The oxygen isotope ratios in the Quelccaya profiles offer a surprise when compared to the classical records from Greenland and Antarctica. In the polar regions (Johnsen et al., 1972), the largest negative values (order $-40^{\circ}/_{\circ\circ}$) are found during the ice age, some 50 000 to 15 000 years B.P., with a change to less negative values (-35 to $-30^{\circ}/_{\circ\circ}$) at present. By contrast at Quelccaya the $\delta^{18}O$ values are on the average less negative (around $-20^{\circ}/_{\circ\circ}$); the annual range (order $-10^{\circ}/_{\circ\circ}$) is about as large as the change from ice age to modern conditions in the polar regions; and the largest negative values occur in the warmer time of the year, or the Southern summer, whereas in the polar regions values are largest negative in the colder ice age. While the ice cores from Greenland and Antarctica have been discussed primarily in terms of a temperature history, such an interpretation is evidently not acceptable in the tropics. It appears that the history of atmospheric water vapor along its transport trajectory will be essential in the climatic interpretation of $\delta^{18}O$ profiles.

After a decade-long effort and the design of a multi-component, portable, solar-powered drill, two 160 m cores to bedrock were retrieved on the Quelccaya summit (5,650 m) in 1983, representing a climate history of one and a half millenia. Shorter cores have also been obtained from the North Peruvian and Ecuadorian Andes. Elsewhere in the tropics, the prospects for climatic ice core studies appear more limited, as melting and percolation at lower elevations contaminate the ice profile. These complications were also recognized at Lewis Glacier on Mount Kenya (Hastenrath, 1984). The summit ice fields of Kilimanjaro appear most nearly suitable for climatic ice core studies in East Africa.

11.5. Synthesis

Tropical glaciers have never amounted to more than a negligible fraction of the total land surface. The ice equilibrium line stands relatively low in the humid inner tropics, where it responds to temperature variations, and highest in the hyper-arid regions of the Southern tropical Andes, where it is most sensitive to precipitation changes. Ice extent tends to be largest on the poleward and least on the equatorward facing slopes, as a result of planetary radiation geometry. Zonal asymmetries of glaciations are not only related to the direction of the atmospheric water vapor transport, but also to diurnal circulation systems which, through the predominant afternoon cloud cover, favor glaciation in the westerly quadrants. The latter effect is most pronounced in the arid high mountain regions.

Field evidence published since the 1970's permits a preliminary synopsis of glacier variations in the tropics. Concerning the era prior to 100 000 years B.P., glaciations have been reported for the epochs 500 000, 380 000, around 250 000–300 000, and somewhat before 100 000 years B.P. Except for the earlier event, these tropical glaciations broadly coincide with maxima of global ice volume as inferred from deep-sea cores. Within the interval from 100 000 to 10 000 years B.P., the time span 25 000–15 000 years B.P. deserves particular attention because this is recognized as an epoch of major glaciation in various parts of the globe. While there are indications of enhanced ice extent during this time in some tropical regions, on the Mexican volcanoes no glaciations are documented. For the hyper-arid Southern tropical Andes, absolute dates are lacking altogether. Deglaciation progressed from around 3000 m after about 15 000 years B.P. towards 4500 m at 8000 years B.P. However, glacier advances also occurred within the time span 14 000–10 000 years B.P. in some regions. About 500 years ago the ice cover was more abundant than now in various tropical mountain areas. Ice retreat began in the 19th century in all three glacier regions, but with remarkable differences in timing. This recession continues.

Concerning the recent glaciation, two methods are being developed to infer climate variations from glacier observations. Modelling of the complete causality chain from climate to net balance, ice dynamics, and terminus variations, permits a quantitative reconstruction of the unknown climatic forcing from the observed glacier response. This method has been applied on the scale of decades to a century. A separate line of quantitative reconstruction of climatic history is based on the analysis of ice cores. The largest record that can be obtained from this method in the tropics spans one and a half millenia.

References

Allison, I., Kruss, P. D., 1977: 'Estimation of recent climate change in Irian Jaya by numerical modeling of its tropical glaciers'. *Arctic and Alpine Research*, 9, 49–60.

Clapperton, C. M., 1983: 'The glaciation of the Andes'. *Quaternary Science Reviews*, 2, 83–155.

Crowley, T. J., 1983: 'The geologic record of climatic change'. *Rev. Geophys. Space Phys.*, 21, 828–877.

Dansgaard, W., 1964: 'Stable isotopes in precipitation'. *Tellus*, 16, 436–468.

Dansgaard, W., Johnsen, S. J., Moller, J., Langway, C. C., Jr., 1969: 'One thousand centuries of climatic record from Camp Century on the Greenland ice sheet'. *Science*, 166, 377–381.

Downie, C., Wilkinson, P., 1972: *The geology of Kilimanjaro*. Geology Department, University of Sheffield, 253 pp.

Flenley, J. R., 1979: 'The late Quaternary vegetational history of the equatorial mountains'. *Progress in Physical Geography*, 4, 488–509.

Flenley, J. R., Morley, R. J., 1978: 'A minimum age for the deglaciation of Mt. Kinabalu, East Malaysia'. *Modern Quaternary Research in Southeast Asia*, 4, 52–61.

Galloway, R. W., Hope, G. S., Löffler, E., Peterson, J. A., 1973: 'Late Quaternary glaciation and periglacial phenomena in Australia and New Guinea'. *Palaeoecology of Africa*, 8, 125–138.

Gasse, F., Descourtieux, C., 1979: 'Diatomées et évolution de trois milieux Éthiopiens d'altitude differente, au cours du quaternaire supérieur'. *Palaeoecology of Africa*, 11, 117–134.

Gonzalez, E., Van der Hammen, Th., Flint, R. F., 1965: 'Late quaternary glacial and vegetational sequence in Valle de Lagunillas, Sierra Nevada del Cocuy, Colombia'. *Leidse Geologische Mededelingen*, 32, 157–182.

Hamilton, A., Perrott, A., 1978: 'Date of deglacierisation of Mount Elgon'. *Nature*, 273, 49.

Hastenrath, S., 1976: 'Variations in low-latitude circulation and extreme climatic events in the tropical Americas'. *J. Atmos. Sci.*, 33, 202–215.

Hastenrath, S., 1978: 'Heat budget measurements on the Quelccaya Ice Cap, Peruvian Andes'. *J. Glaciol.*, 30, 85–97.

Hastenrath, S., 1981: *The glaciation of the Ecuadorian Andes*. Balkema, Rotterdam, 160 pp.

Hastenrath, S., 1984: *The glaciers of equatorial East Africa*. Reidel, Dordrecht, Boston, Lancaster, 353 pp.

Hastenrath, S., 1985: 'A review of Pleistocene to Holocene glacier variations in the tropics'. *Zeitschrift für Gletscherkunde und Glazialgeologie*, 21, 183–194.

Heine, K., 1975: 'Studien zur jungquartären Glazialmorphologie Mexikanischer Vulkane'. Mexiko Projekt DFG, vol. 7, Steiner Verlag, Wiesbaden, 178 pp.

Herd, D. G., Naeser, C. W., 1974: 'Radiometric evidence for Pre-Wisconsin glaciation in the Northern Andes'. *Geology*, 2, 603–604.

Hope, G. S., Peterson, J. A., Allison, I., Radok, U., eds., 1976: *The equatorial glaciers of New Guinea*. Balkema, Rotterdam, 244 pp.

388   CHAPTER 11

Imbrie, J., Hays, J. D., Martinson, D. G., McIntyre, A., Mix, A. C., Morley, J. J., Pisias, N. G., Prell, W. L., Shackleton, N. J., 1984: 'The orbital theory of Pleistocene climate: support from a revised chronology of the marine δ ^{18}O record'. pp. 269–305, part 1, in: Berger, A., Imbrie, J., Hays, J. D., Kukia, G., Saltzman, B., eds.: *Milankovitch and climate; understanding the response to orbital forcing*. Reidel, Dordrecht, Boston, Lancaster, 2 parts, 895 pp.

International Association of Hydrological Sciences – UNESCO, 1977: *Fluctuations of glaciers, 1970–75*. Paris, 269 pp. plus maps.

Johnsen, S. J., Dansgaard, W., Clausen, H. B., Langway, C. C. Jr., 1972: 'Oxygen isotope profiles through the Antarctic and Greenland ice sheets'. *Nature*, 235, 429–434.

Kruss, P. D., 1981: 'Numerical modelling of climate change from the terminus record of Lewis Glacier, Mount Kenya'. Ph.D. Dissertation, Department of Meteorology, University of Wisconsin, Madison, 128 pp.

Kruss, P. D., 1983: 'Climate change in East Africa: numerical modelling from the 100 years of terminus record of Lewis Glacier, Mount Kenya'. *Zeitschrift für Gletscherkunde und Glazialgeologie*, 19, 43–60.

Kruss, P. D., 1984: 'Terminus response of Lewis Glacier, Mount Kenya to sinusoidal net balance forcing'. *J. Glaciol.*, 30, 212–217.

Livingstone, D. A., 1962: 'Age of deglaciation in the Ruwenzori range, Uganda'. *Nature*, 194, 859–860.

Löffler, E., 1972: 'Pleistocene glaciation in Papua and New Guinea'. *Zeitschrift für Geomorphologie*, suppl. vol. 13, 32–58.

Löffler, E., 1976: 'Potassium-Argon dates and pre-Würm glaciations of Mount Giluwe volcano, Papua New Guinea'. *Zeitschrift für Gletscherkunde und Glazialgeologie*, 12, 55–62.

Löffler, E., 1982: 'Pleistocene and present-day glaciations', pp. 39–55, in: Gressit, J. L., ed., *Biogeography and ecology of New Guinea*, Monographiae Biologicae, vol. 42, Junk, The Hague, 983 pp.

Mercer, J. H., Palacios, O. M., 1977: 'Radiocarbon dating of the last glaciation in Peru'. *Geology*, 5, 600–604.

Mercer, J., Thompson, L. G., Marangunič, C., Ricker, J., 1975: 'Peru's Quelccaya Ice Cap, 1975: glaciological and glacial geological studies'. *Antarctic J. of the U.S.*, 10, 19–24.

Perrott, A. R., 1982: 'A postglacial pollen record from Mt. Satima, Aberdare Range, Kenya', p. 153 in Amer. Quat. Assoc., seventh Biennial Conference, Seattle, June 1982, Program and Abstracts, 188 pp.

Peterson, J. A., Hope, G., 1972: 'Lower limit and maximum age for the last major advance of the Carstensz Glacier, West Irian'. *Nature*, 240, 36–37.

Porter, S. C., Stuiver, M., Yang, I. C., 1977: 'Chronology of Hawaiian glaciations'. *Science*, 195, 61–63.

Salgado-Labouriau, M. L., Schubert, C., Valastro, S., 1977: 'Paleoecologic analysis of a Late-Quaternary terrace from Mucubají, Venezuelan Andes'. *J. Biogeogr.*, 4, 131–325.

Schubert, C., 1972a: 'Geomorphology and glacier retreat in the Pico Bolivar area, Sierra Nevada de Mérida, Venezuela'. *Zeitschrift für Gletscherkunde und Glazialgeologie*, 8, 189–202.

Schubert, C., 1972b: 'Late glacial chronology in the Northeastern Venezuelan Andes'. *24th International Geol. congr. Montreal*, Sect. 12, pp. 103–109, Montreal.

Schubert, C., 1974: 'Late pleistocene Mérida glaciation, Venezuelan Andes'. *Boreas*, 3, 147–152.

Servant, M., Fontes, J. C., 1978: 'Les lacs quaternaires des hauts plateaux des Andes Boliviennes: premières interpretations paléoclimatiques'. *Cahiers de l'ORSTOM, Ser. Geol.*, 10, 9–23.

Sharp, S. E., Gow, A. J., 1970: 'Antarctic ice sheet: stable isotope analyses of Byrd Station cases and inter-hemispheric climatic implications'. *Science*, 168, 1570–1572.

Street, F. A., and Grove, A. T., 1979: 'Global maps of lake-level fluctuations since 30 000 B.P.'. *Quat. Res.*, 12, 83–118.

Temporary Technical Secretariat for World Glacier Inventory, UNESCO–UNEP–IASH, ICSI, 1977: 'Instructions for compilation and assemblage of data for a World Glacier Inventory'. ETH Zürich, 29 pp.

Thompson, L. G., Hastenrath, S., Morales Arnao, B., 1979: 'Climatic ice core records from the tropical Quelccaya Ice Cap'. *Science*, 203, 1240–1243.

Thompson, L. G., Thompson, E. M., Grootes, P. M., Pourchet, M., Hastenrath, S., 1984: 'Tropical glaciers: potential for ice core paleoclimatic reconstructions'. *J. Geophys. Res.*, 89, No. D3, 4638–4646.

UNESCO, 1970: 'Perennial ice and snow masses. A guide for compilation of data for a World Inventory'. Technical Papers in Hydrology, No. 1.

United Nations Environment Programme, 1979: 'The Environment Programme, report of the Executive Director'. UNEP Na 79–0075, Nairobi, 147 pp.

White, S., Valastro, S., 1984: 'Pleistocene glaciation of volcano Ajusco, Central Mexico, and comparison with the standard Mexican glacial sequence'. *Quaternary Research*, 21, 21–35.

World Meteorological Organization-ICSU, 1975: *The physical basis of climate and climate modelling*. GARP Publication Series, No. 16, 265 pp.

Wright, H. E., Jr., 1983: 'Late-Pleistocene glaciation and climate around the Junín Plain, Central Peruvian Andes'. *Geografiska Annaler*, 65A, 35–43.

CHAPTER 12

PAST CLIMATES OF THE TROPICS

The functioning of the tropical climate system was throughout this book considered from various perspectives, but always with emphasis on the modern time scale. This final chapter is a preliminary synopsis of climate and circulation conditions in the tropics in the geological past. The available field evidence limits this attempt mainly to the last 30 000 years. For the more distant past refer to the comprehensive review of Crowley (1983).

For some time now there has been a broad consensus (review in Kutzbach, 1976) on the possible causes of climatic change (Section 12.1). However, it is primarily through field research in the course of the last 15 years that a coherent picture is emerging of the climate variations in low latitudes from the late Pleistocene into the Holocene. The pertinent field methods lie outside the traditional expertise of meteorology. Important contributions to the reconstruction of a climatic history have been made through investigations of past vegetation, lakes, and glaciers, and more recently the study of deep-sea cores. Refer to Peterson *et al.* (1979) for an extensive bibliography on the continental record. Establishment of an absolute time scale is crucial in all of the aforementioned approaches. An overview over this varied evidence is presented in Sections 12.2 to 12.4, 11.2, and 12.5. These field results are in Sections 12.6 to 12.9 considered in causal perspective, for various large domains of the tropics, Section 12.10 reviews numerical simulation experiments, and Section 12.11 is a final synthesis.

12.1. Causes of Climate Variations

It is now widely recognized that climatic variations can result from both extra-terrestrial and terrestrial causes (Kutzbach, 1976; Robock, 1978). Fig. 12.1:1. illustrates the characteristic time scales of extra-terrestrial and terrestrial forcings and of the responses of the climate system. Of the extra-terrestrial forcings, solar variability spans all time scales. This affects the global heat budget and hence atmospheric motions and other processes. Orbital variations, with their consequences for insolation, are concentrated on the scale of 10^4 to 10^5 years. Of the terrestrial causes of climatic variations, continental drift is shown active at scales beyond 10^6 years, mountain building and sea-level changes in the range of 10^5 to 10^8 years, and volcanism at nearly all time scales. The latter would affect radiation. All of these are lithospheric processes. As Kutzbach (1976) points out, nonlinear feedback processes could produce oscillation, or autovariation of the climate system. Fig. 12.1:1 shows examples of increasing complexity: autovariations of atmosphere, of atmosphere-hydrosphere, of atmosphere-hydrospere-cryosphere, and of atmosphere-hydrosphere-cryosphere-lithosphere-biosphere, ranging from less than 10^2 to 10^9 years. The effect of human activity extends beyond 10^2 years but possibly much further into the past (Section 10.1). Climatic variations, which are to be considered as the results of the aforementioned processes, are shown ranging from interannual variability on the scale of less than a decade to more than 10^8 years.

The scheme Fig. 12.1:1 is greatly simplified and intended as a general orientation only. The problem of climate variations is complicated not only by the multitude of possible forcings not

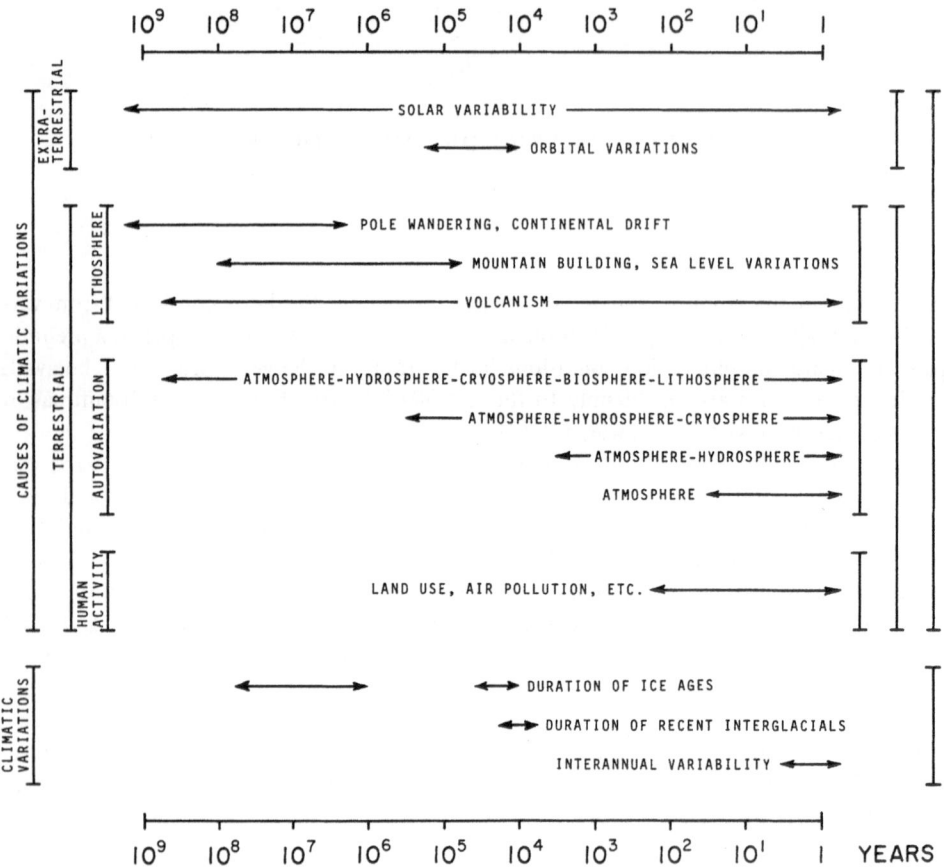

Fig. 12.1:1. Characteristic time scales of climatic Variations and their causes. Sources: Kutzbach (1976), Mitchell (1968, 1976).

detailed in Fig. 12.1:1, but also because diverse processes may act simultaneously, and climatic responses may depend on initial states. As explained by Kutzbach (1976) these in turn depend on previous climatic states because of the inertia of hydrosphere, cryosphere, lithosphere, and biosphere. Moreover, Lorenz (1970) introduced the concept of intransitivity, according to which alternative climatic states may be compatible with a given set of boundary conditions. It is then difficult to unambiguously relate cause and effect in the climatic record.

Of the various forcings indicated in Fig. 12.1:1 the Earth's orbital variations are known with great precision. A response to these periodic forcings is strongly indicated in the record of deep-sea cores (Section 12.5; Hays *et al.*, 1976; Imbrie and Imbrie, 1980; Crowley, 1983). Accordingly, the hypothesis of orbital perturbation influence on climate advanced by Milankovitch (references in Crowley, 1983) received new attention in recent years (Berger, 1978; Schneider and Thompson, 1979; Kutzbach, 1981, 1983; Kutzbach and Otto-Bliesner, 1982; Crowley, 1983; Berger *et al.*, 1984; Imbrie *et al.*, 1984; Start and Prell, 1984; Pisias and Shackleton, 1984).

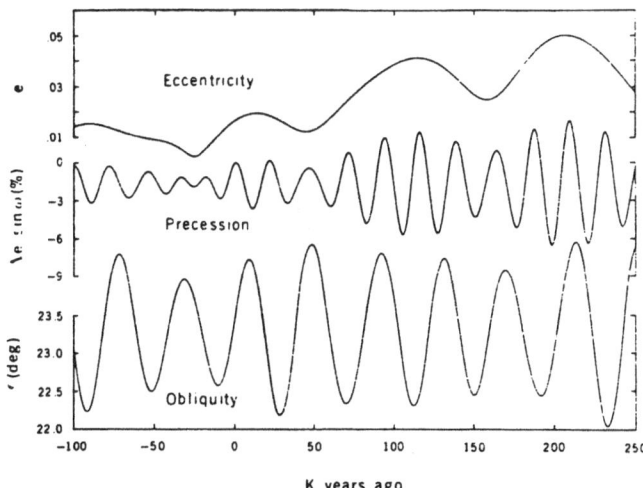

Fig. 12.1:2. Variations in orbital geometry: eccentricity; precession (recorded as index, Δe sin ω, approximately equal to the deviation from its 1950 value of the June earth–sun distance, expressed as fraction of the semi-major axis of the earth's orbit); obliquity. From Imbrie and Imbrie, 1980 (*Science,* **207**, 943–953; copyright 1980 by American Association for the Advancement of Science).

Fig. 12.1:2 illustrates the variations of orbital geometry as a function of time. Of interest are effects on three features of the earth's orbit (Imbrie and Imbrie, 1980; Crowley, 1983): (i) the tilt of the axis of obliquity, period of 41 000 years; (ii) the longitude of perihelion (precession effect, periods of 19 000 and 23 000 years); and (iii) the eccentricity of the orbit, periods of 100 000 and 400 000 years. The first two effects result in varying spatial and seasonal distribution of solar radiation at the top of the atmosphere. Variations of eccentricity produce changes of only 2% in total radiation received, but can strongly modulate the amplitude of the precession effect, with insolation anomalies during large eccentricity being 3–4 times larger than during small eccentricity. As a result of these orbital variations, 9000 years ago perihelion was on 30 July rather than 3 January, and global average insolation in July was 7% higher than now (Kutzbach, 1981). Combined with the concentration of land masses in the Northern hemisphere, especially in the Africa – Indian Ocean sector, this is conducive to enhanced interhemispheric contrasts in surface heating, and stronger monsoon circulations.

12.2. Vegetation

A wealth of evidence on the vegetation history of the tropics has been gathered through the analysis of pollen profiles. The varying composition of pollen assemblages is taken as indication of changes in the environment and indirectly of climate. Three particularly long records, exemplifying Australasia, Africa, and South America, are shown in Figs. 12.2:1 to 12.2:3.

The core from Lynch's Crater at 760 m in tropical Northeastern Australia (Fig. 12.2:1) covers more than 120 000 years. Four distinct periods are recognized (Kershaw, 1974, 1978). Zones E and A are dominated by pollen of rainforest angiosperms. Zone D shows strong contributions by the rainforest gymnosperms Araucaria and Podocarpus and a significant sclerophyll component. In Zone B the sclerophyll taxa Casuarina and Eucalyptus prevail. Zone C is transitional between rainforest and sclerophyll assemblages. The total number of dry land taxa, as a measure of species diversity, is largest in the zones dominated by rainforest angiosperms and lowest in Zone B where sclerophyll taxa prevail.

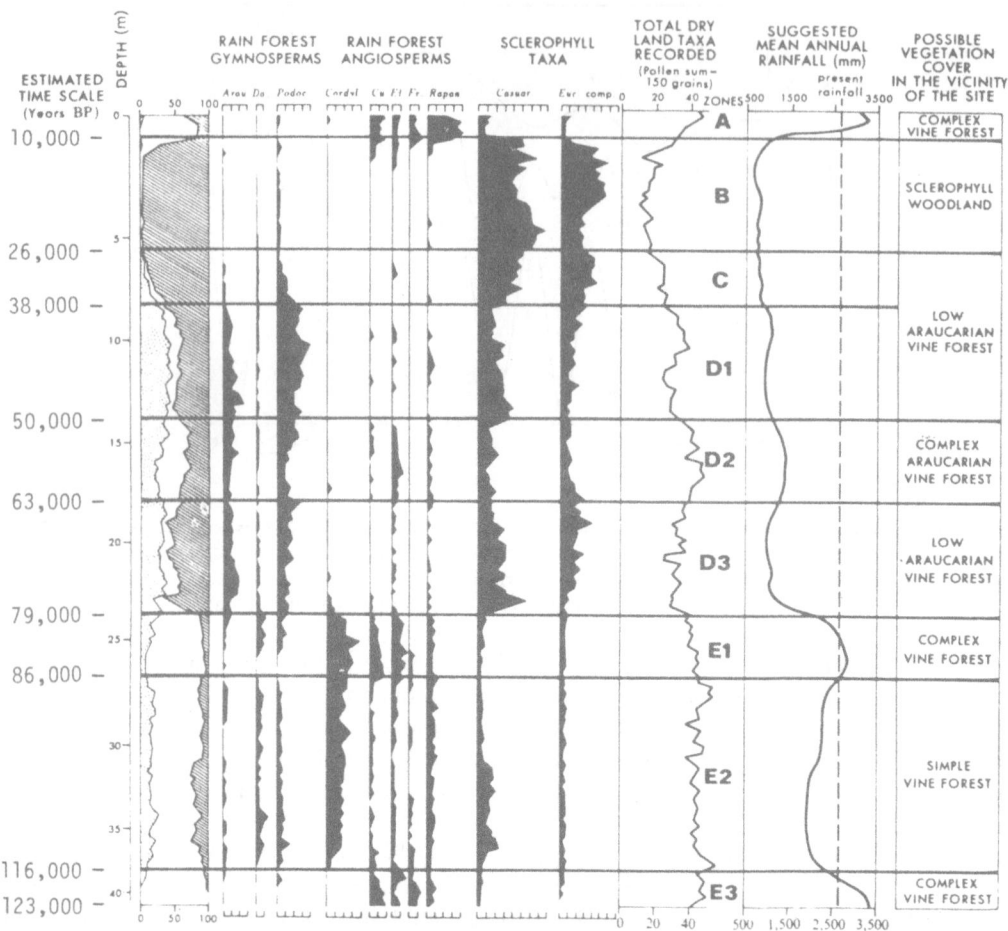

Fig. 12.2:1. Pollen diagram from Lynch's Crater, Queensland, Northeastern Australia, at 760 m. The column on the left shows the percentage of rainforest gymnosperms (stippled), rainforest angiosperms (white), sclerophyll taxa (hatched). The frequencies of pollen of all taxa are shown as percentages of the dry land plant pollen total, each division representing 10 percent of the pollen sum. Abbreviations for taxa: Araucaria Arau.; Dacrydium Da.; Podocarpus Podoc.; Cordyline comp. Cordyl.; Cunoniaceae Cu.; Elaeocarpus El.; Freycinettia Fr.; Rapanea Rapan; Casuarina Casuar; Eucalyptus comp Euc. comp. From Kershaw, 1978 (reprinted by permission from *Nature*, **272** (9 March 1978), 159–161, copyright 1978, Macmillan Journals Limited).

Interpretation of the pollen profile relies on comparisons with the modern vegetation in Northeastern Australia (Kershaw, 1978). The prevalence of rainforest in zones E and A indicates abundant rainfall. A sharp reduction of rainfall is suggested for the base of zone D. The vegetation changes within zone C are believed to reflect further precipitation decrease and burning by aboriginal man. It is suggested that in zone A at the beginning of the Holocene, a pronounced increase in rainfall allowed rainforest to expand from isolated retreats maintained through period B. The pollen profile is thus interpreted primarily in terms of rainfall variations. Comparatively dry conditions are inferred for the period from around 75 000 to 10 000 years B.P., with an extremum towards the end of this interval, followed by a marked increase of rainfall into to Holocene.

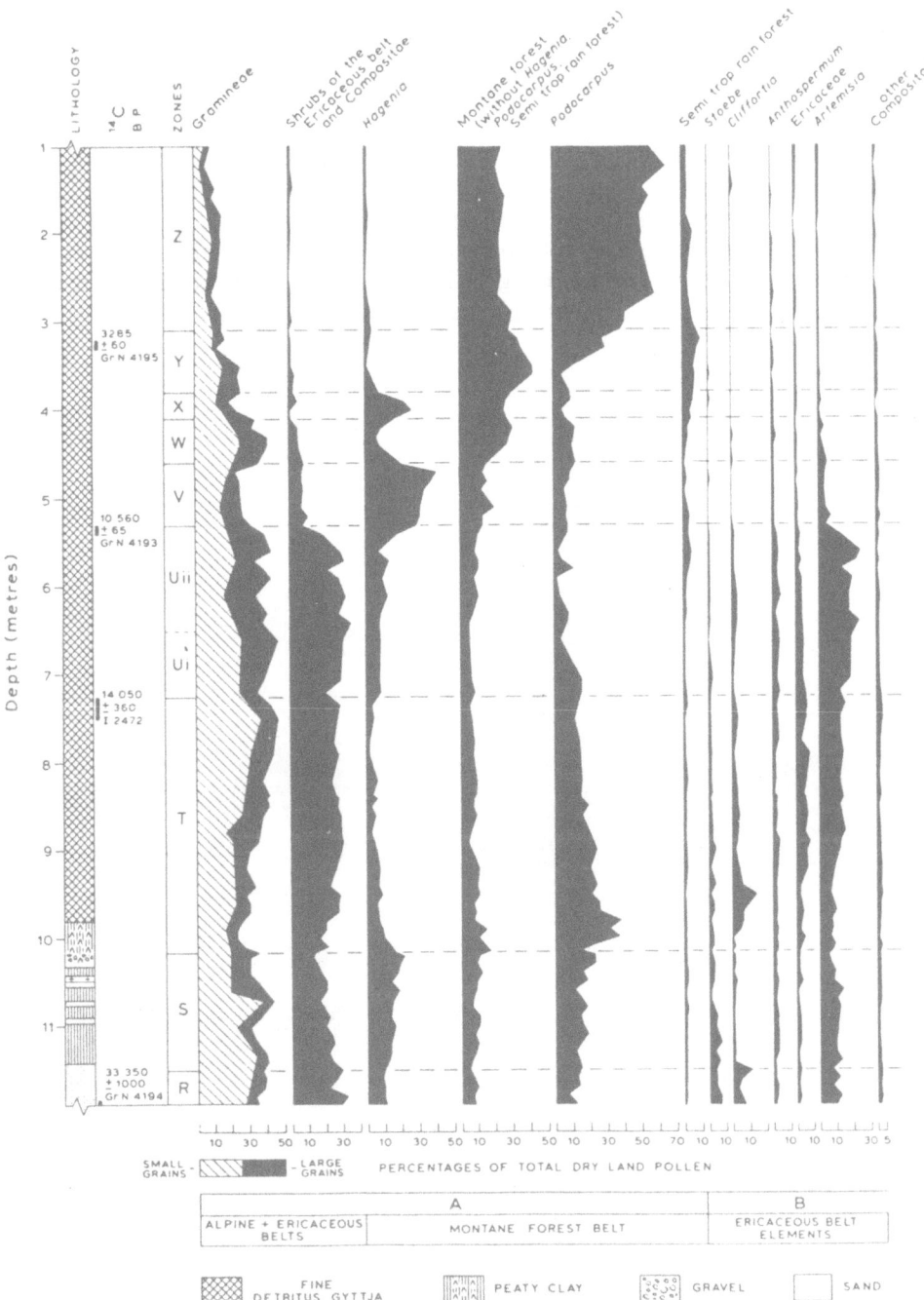

Fig. 12.2:2. Summary pollen diagram from Sacred Lake on Mount Kenya, East Africa, at 2400 m. Values are percentages of total dry land pollen. From Coetzee (1967).

The pollen profile from Sacred Lake on Mount Kenya at 2400 m (Fig. 12.2:2) encompassing more than 30 000 years (Coetzee, 1967), still appears to be the longest core so far obtained in tropical Africa. Although Livingstone (1975) and Flenley (1979a, b) pointed out shortcomings in the absolute chronology and a stratigraphic change at the base, this core provides an internally consistent vegetation sequence. Around 33 000 years B.P. shrub vegetation appears to have prevailed around the lake, similar to the present Ericaceous Belt just above the forest limit of Mount Kenya. Then the forest limit seems to have risen, reaching the elevation domain of the lake around 26 000 years B.P. A drastic change took place thereafter. The forest limit dropped and shrubby vegetation mixed with grasses dominated near the lake until 14 000 years B. P. Arid conditions are indicated until around 10 500 years B.P. Then the forest limit moved drastically upward, and it appears that the climate became both warmer and moister. Cores taken at lower elevations indicate the effect of human activities during the Holocene (Flenley, 1979b). For a comprehensive account of the vegetation history of East Africa refer to Hamilton (1982).

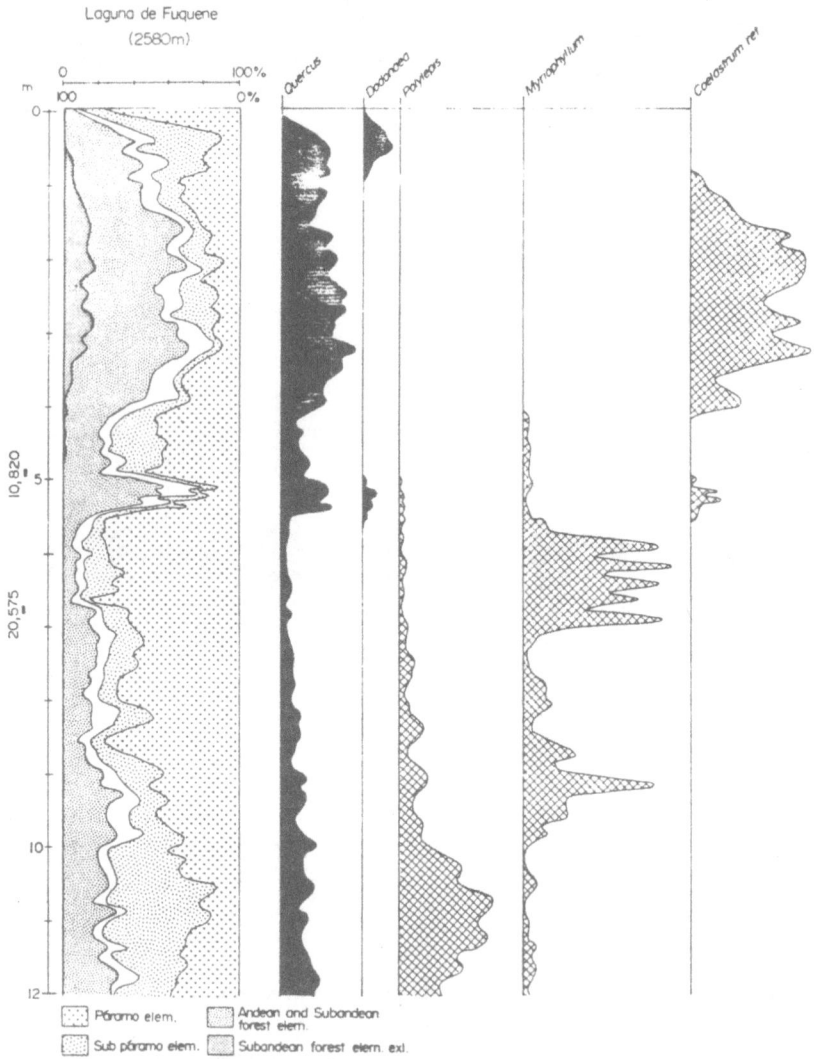

Fig. 12.2:3. Pollen diagram from Laguna de Fuquene, at 2580 m in Eastern Andes of Colombia. The white area represents the percentage of Alnus. From Van der Hammen (1974).

The pollen diagram from the Laguna de Fuquene at 2580 m in the Eastern Colombian Andes (Fig. 12.2:3) provides a vegetation history of the last 32 000 years (Van der Hammen, 1974). Fig. 12.2:4 is a summary interpretation of the core evidence. About 30 000 years ago, the area was dominated by scrub vegetation, similar to the modern subparamo above the forest limit. This is interpreted as reflecting colder and drier conditions, culminating during 20 000–14 000 years B.P. This interval is also characterized by low lake stand. 13 000 years B.P. marks the invasion of Quercus forest into the Fuquene plateau and the surrounding mountains. The forest limit rose until around 9500 years B.P. After 3000 years B.P. forest elements decline and grass pollen increase, presumably reflecting the destruction of forest by man (Flenley, 1979a, b).

The high mountain environment is particularly sensitive to climatic variations in various respects. Of immediate interest here is the vertical displacement of vegetation belts as manifestation of changes in the climatic environment. Figs. 12.2:5 to 12.2:7 illustrate the variation in the altitudinal zonation of vegetation during the last 30 000 years in Australasia, Africa, and the Americas. These

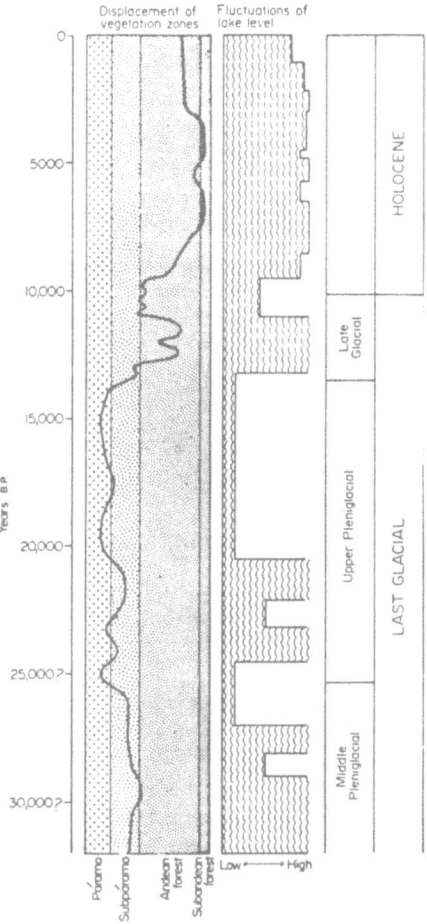

Fig. 12.2:4. Displacement of vegetation zones and lake level fluctuations. Laguna de Fuquene, at 2580 m in the Colombian Andes. From Van der Hammen, 1974 (*J. Biogeography*, Blackwell Scientific Publications).

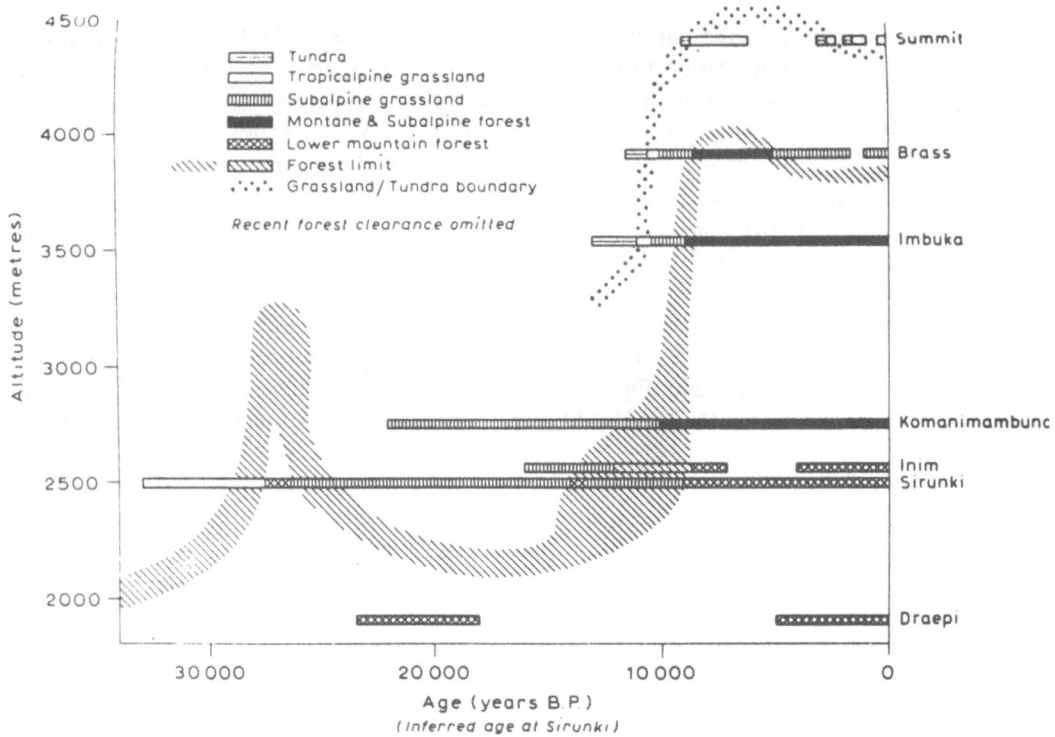

Fig. 12.2:5 Summary diagram of Late Quaternary vegetation changes in the New Guinea highlands. From Flenley, 1979b (*Progress in Physical Geography*, Edward Arnold Publishers).

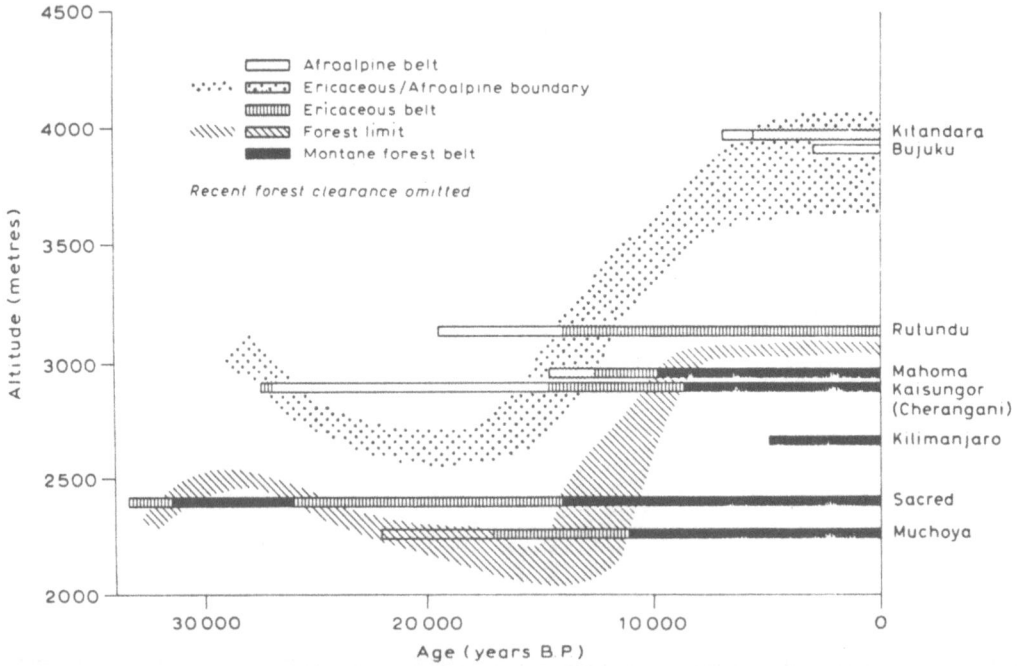

Fig. 12.2:6 Summary diagram of Late Quaternary vegetation changes in the East African mountains. From Flenley, 1979b (*Progress in Physical Geography*, Edward Arnold Publishers).

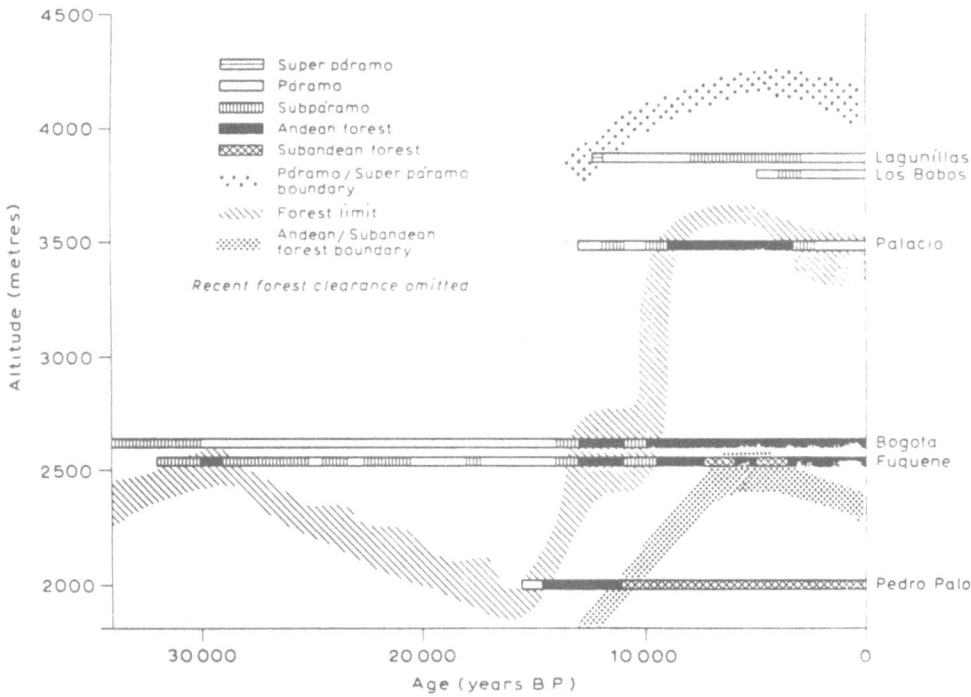

Fig. 12.2:7. Summary diagram of Late Quaternary vegetation changes in the Colombian Andes. From Flenley, 1979b (*Progress in Physical Geography*, Edward Arnold Publishers).

summary diagrams are based on numerous pollen cores such as exemplified by Figs. 12.2:1 to 12.2:3 in accordance with the elevation of sampling sites (Flenley, 1979a, b). While various vegetation belts are described in these diagrams, attention is here focused on the upper forest limit as a particularly conspicuous and indicative feature.

In the New Guinea highlands (Fig. 12.2:5), the forest limit stood low prior to 30 000 years B.P., rose to a maximum elevation around 27 000 years B.P., and then stayed low from around 25 000 to 15 000 years B.P. A rapid rise is indicated thereafter to an absolutely highest elevation around 9500 years B.P., and then a dropoff to the present. Within the overall upward shift of vegetation limits from around 15 000 to 9500 years B.P. the interval 14 000—10 000 years B.P. stands out as an episode with indistinct patterns of vegetation changes. These '4000 years of hesitation' are also characteristic of field evidence from other tropical regions to be reviewed below. Reference to this phenomenon is also found in Roberts *et al.* (1981).

In East Africa (Fig. 12.2:6), the forest limit stood at a maximum elevation around 29 000 years ago, and then dropped to a minimum near 15 000 years B.P. This was followed by a rapid rise until around 9000 years B.P. and comparatively little variation from then to the present. Particularly noteworthy are again the '4000 years of hesitation' in vegetation change, from around 14 000 to 10 000 years B.P.

In the Colombian Andes (Fig. 12.2:7), the forest limit reached a maximum elevation around 29 000 years B.P. and then dropped to a minimum somewhat before 15 000 years B.P. This was followed by a rapid rise to a maximum elevation later than 8000 years B.P. and a small drop to the present. Within the overall rise from around 15 000 to 8000 years B.P., the '4000 years of hesitation' from 14 000 to 10 000 years B.P. are again most conspicuous.

12.3. Lakes

Geomorphic evidence of former water levels in closed lake basins can be informative of past hydrometeorological conditions. Basins with interior drainage are characteristic of the less humid regions of the World, and are found in a zone extending from North Africa across the Middle East into Central Asia, in the semi-arid regions of Southern Africa and Australia, as well as in the American Southwest and in the outer tropics and subtropics of South America.

The relation of lake level variations to climate has been the subject of numerous studies (Street and Grove, 1976; Churchill *et al.*, 1978; Street, 1979, 1980, 1981; Kutzbach 1980; Hastenrath and Kutzbach, 1983, 1985; Swain *et al.*, 1983; Street-Perrott and Harrison, 1984, 1985). Butzer *et al.* (1972) found for the East African lakes and Lake Chad marked high stands during 10 000– 8000 years B.P. (Fig. 12.3:1). In an attempt at a global synopsis, Street and Grove (1979) compiled extensive field evidence of lake level stands in the last 30 000 years in various parts of the World. Fig. 12.3:2 summarizes their results for the American Southwest, tropical Africa, and Australia. The histograms show for both tropical Africa and Australia a tendency for high lake stands prior to about 25 000 years B.P., and a change to prevailingly low stands until around 13 000 years B.P. Thereafter, high lake stands become prevalent again, with most extreme conditions being reached between 10 000 and 7000 years B.P. Proceeding further into the Holocene, this is followed by a change to lower lake stands. The American Southwest shows a pattern almost inverse to the two aforementioned regions: high lake stands are most prevalent during 24 000–21 000 years B.P., but continue until 10 000 years B.P. Thereafter an abrupt change takes place towards predominantly low lake stands.

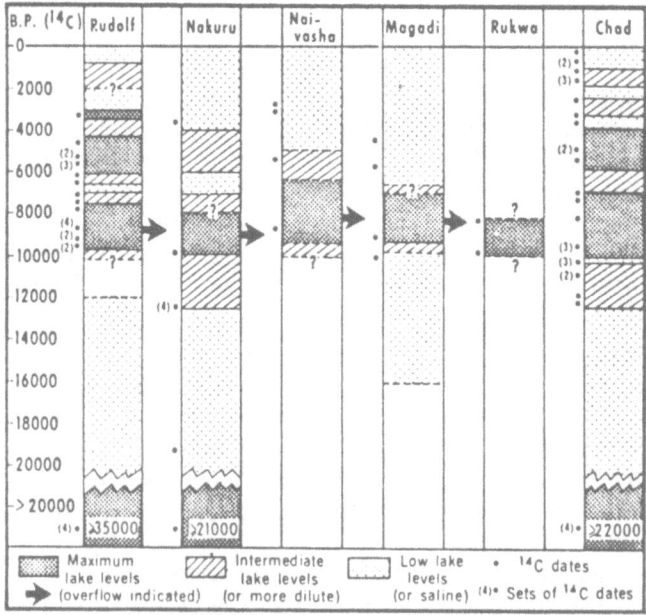

Fig. 12.3:1. 'Fluctuations of African lakes that presently lack outlets. From Butzer *et al.*, 1972 (*Science*, **175**, 1069–1076; copyright 1972 by American Association for the Advancement of Science).

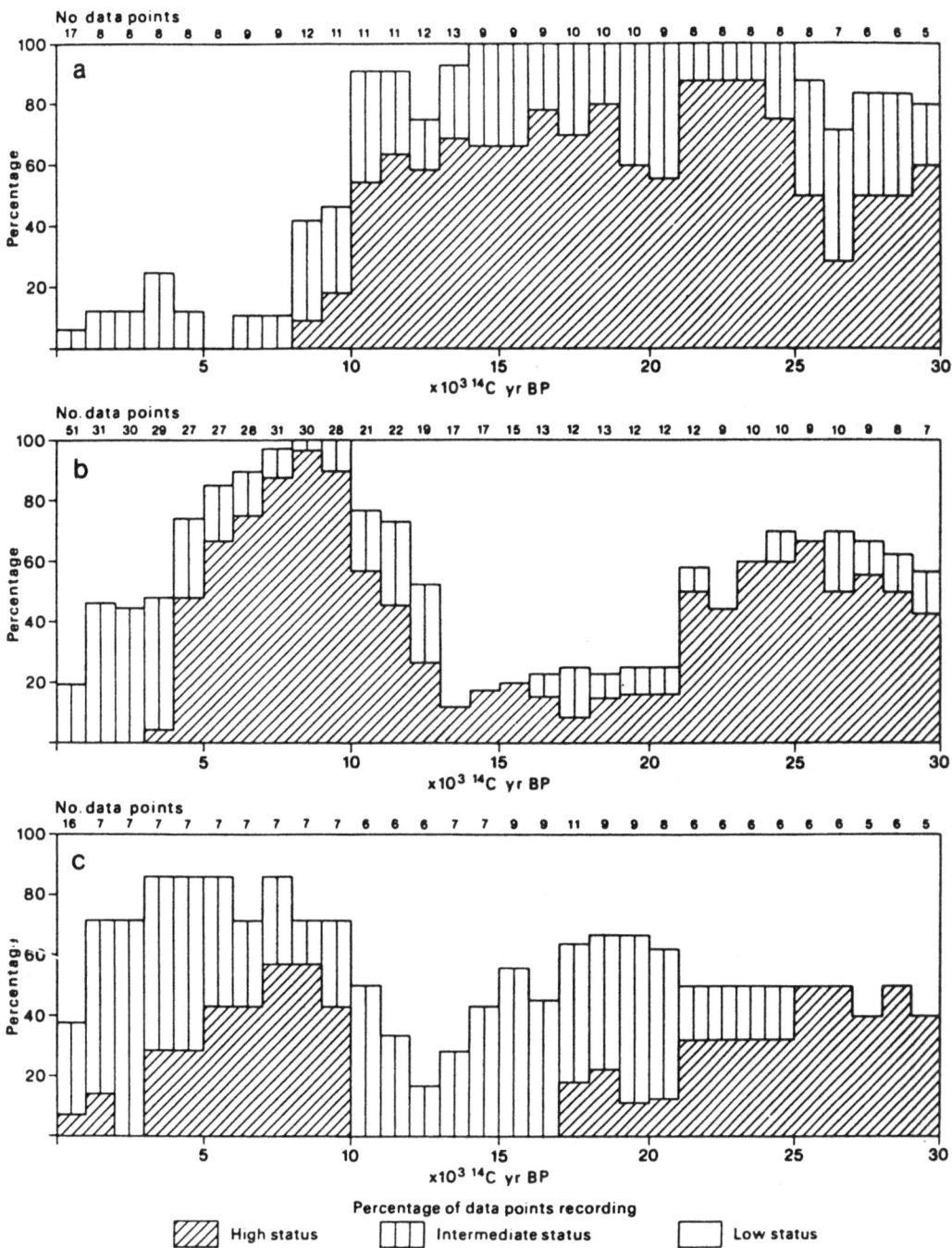

Fig. 12.3:2. Histograms of lake level status for 1000 year time periods from 30 000 years B.P. to present for (a) Southwestern United States, (b) tropical Africa, and (c) Australia. From Street and Grove (1979).

Complementing Street and Grove's (1979) compilation in Fig. 12.3:2, the observations from the Colombian Andes displayed in Fig. 12.2:4 are of interest. In remarkable consistency with tropical Africa and Australia (Fig. 12.3:1), the Laguna de Fuquene had a high water level until sometime prior to 20 000 years B.P., followed by a low stand until around 13 000 years B.P., and higher water levels thereafter.

For the South American Altiplano there is evidence for greatly enlarged lakes during two episodes of the past (Servant and Fontes, 1978; Hastenrath and Kutzbach, 1985). The paleo-lake Minchin is dated at prior to 28 000, and the paleo-lake Tauca at 12 500—11 000 years B.P. Lake Minchin apparently preceded to coincided with a major glaciation in the Eastern Peruvian Andes while the Tauca episode is enclosed by two glacier advances (Table 11.2.1, Fig. 12.4:1 and falls within the '4000 years of hesitation' (Section 12.2).

Field observations of former lake stands provide in the first place a qualitative picture of past hydrometeorological conditions. However, reconstruction of the water and heat budgets (Kutzbach, 1980; Hastenrath and Kutzbach, 1983, 1985; Swain et al., 1983), also permits quantitative rainfall estimates of past episodes.

12.4. Glaciers

The history of glacier variations in the tropics is summarized in Section 11.2. In particular, refer to Figs. 11.2:1 and 11.2:2, and Table 11.2:1. In the present section the glacial evidence is placed in perspective with the vegetation and lake variations in the last 30 000 years (Sections 12.2 and 12.3). Fig. 12.4:1 compares the evidence for three major domains of the tropics. Deglaciation age and elevation is entered from Fig. 11.2:2, the upper forest limit from Figs. 12.2:5 to 12.2:7, and the high water level status of lakes from Fig. 12.3:2.

Part C of Fig. 12.4:1 referring to Australasia shows for New Guinea that the interval between about 15 000 and 8000 years B.P. is characterized by a broadly parallel upward displacement of lower ice limit and upper forest boundary. The variations in the New Guinea forest limits furthermore resemble the long-term changes of lake stands in Australia, with both featuring maxima prior to 25 000 and after 10 000 years B.P., and a broad minimum in the intervening time span.

[on p. 401]

Fig. 12.4:1. Synopsis of glacier, vegetation and lake level changes in (A) Americas, (B) Africa, (C) Australasian sector. Heavy broken line denotes deglaciation age and elevation (from Fig. 11.2:2), solid curve upper forest limit (from Figs. 12.2:5 to 12.2:7), and block-histogram percentage of lakes with high water level status by millenium intervals (from Fig. 12.3:2), with percentage scale given to the right. In part (A) deglaciation curve is for the Northern Andes, forest limit for Colombia. Major glaciation and minor glacier advances in the Quelccaya-Vilcanota region of Peru are entered from Table 11.2:1, datings of the paleo-lakes Minchin and Tauca on the Peruvian-Bolivian Altiplano are from Servant and Fontes (1978). In part (B) deglaciation curve includes one Ethiopian site in addition to East Africa, the forest limit is of East Africa, and the lake histogram refers to tropical Africa at large. In part (C) deglaciation curve includes one Borneo site in addition to New Guinea, the forest limit is of New Guinea, and the lake histograms refer to Australia. Part (D) shows the long-term variation of atmospheric CO_2 levels (in ppm) reconstruction from polar ice cores (Neftel et al., 1982; Crowley, 1983), solid line; and an index of δ ^{18}O of seawater in the low-latitude ocean (ref. caption to Fig. 11.2:1; source: Imbrie et al., 1984), broken line. From Hastenrath (1985).

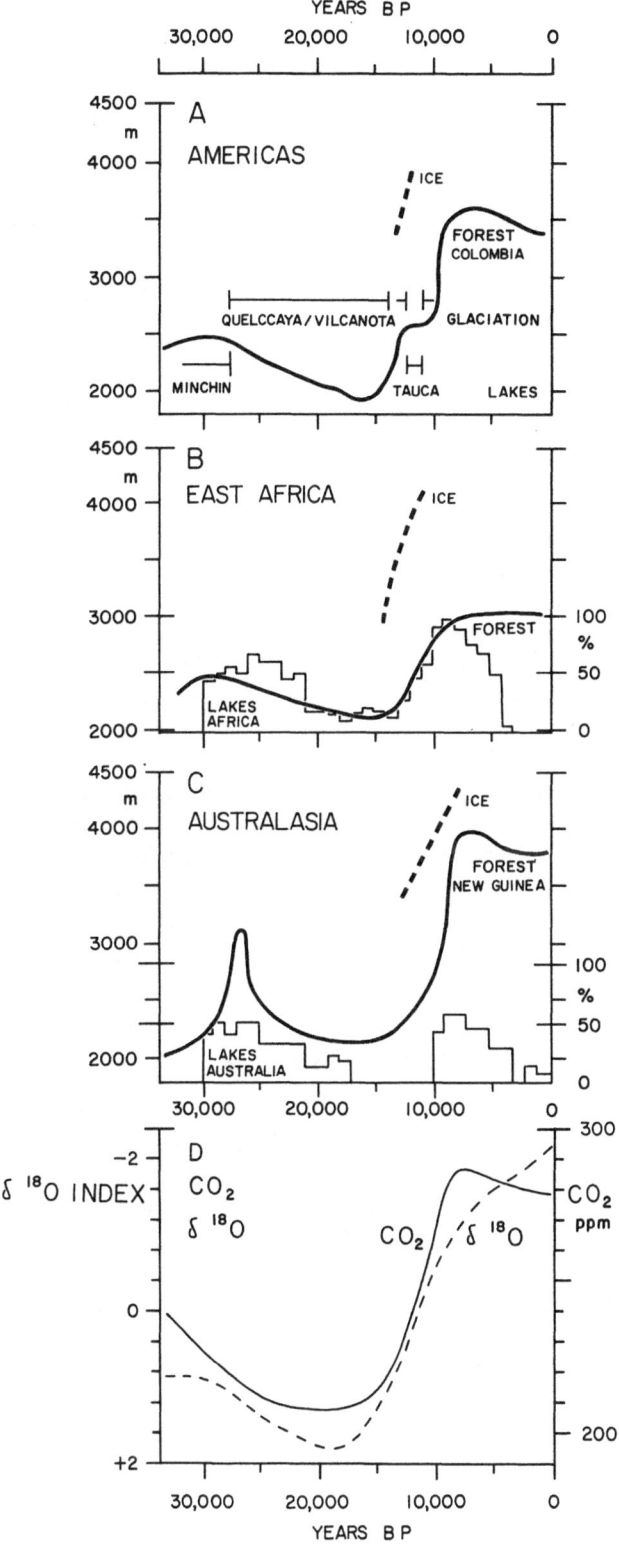

Part B of Fig. 12.4:1. describing tropical Africa indicates after 15 000 years B.P. a drastic rise of the lower ice limit broadly paralleling the upward shift of the forest boundary. Particularly striking is the concordance in the long-term variations of forest limit and lake stands. Both drop from around 25 000 to a broad minimum at 20 000–14 000 years B.P. For the subsequent several millenia lake stands increase and upper forest limit and lower ice limit rise, all broadly in unison.

Part A of Fig. 12.4:1 pertains to the tropical Americas. After 15 000 years B.P. both the lower limit of ice and the upper limit of forest moved up in a broadly parallel fashion. The '4000 years of hesitation' mentioned in Section 12.2 (Fig. 12.2:7) encompass also minor glacier advances in the East Peruvian Andes (Table 11.2:1) and the relatively short-lived paleo-lake Tauca episode on the Peruvian-Bolivian Altiplano. Details of moraine morphology in the mountains of Costa Rica and Guatemala (Hastenrath, 1985) are further suggestive of 'climatic unrest' in this epoch. While the tropical Americas are not included in Street and Grove's (1979) compilation, pertinent lake evidence is contained in Fig. 12.2:4 for the Colombian Andes. Broadly paralleling the varia-tions in forest limit (Fig. 12.2:7, and part A of Fig. 12.4:1), this indicates a high water level prior to 20 000, a low stand from then to about 13 000 years B.P., and higher water levels thereafter. In synthesis a remarkable agreement is emerging between long-term glacier, vegetation and lake variations, not only within but also between the various great domains of the tropics.

Part D of Fig. 12.4:1 illustrates the long-term variation of atmospheric CO_2 levels as deduced from ice cores in Greenland and Antarctica (Neftel *et al.*, 1982; Crowley, 1983). A decrease is apparent to a broad minimum between about 20 000 and 15 000 years B.P., followed by a rapid rise to a maximum around 9000 years B.P. The CO_2 curve thus parallels remarkably closely the long-term variations of glaciation, vegetation, and lakes, as displayed in parts A to C of Fig. 12.4:1. The index of oxygen isotope variations of seawater in the low-latitude ocean (Fig. 12.2:1) also plotted in part D of Fig. 12.4:1 shows extreme positive values between about 25 000 and 15 000 years B.P., followed by a rapid change to negative values, a trend continuing monotonically to the present. Finally, a comparison of the three World regions presented in Fig. 12.4:1 reveals major inter-continental similarities of long-term environmental change. The most conspicuous parallels are the rise of the lower ice limit and upper forest boundary after about 15 000 years B.P., and the maxima of the elevation of the forest limit before 25 000 and after 10 000 years B.P., enclosing a broad minimum around 20 000 to 15 000 years B.P. in all three regions. Moreover, the long-term variations of lake stands are similar and concordant with the vegetation changes.

12.5. Deep-Sea Cores

Cores retrieved from the ocean floor have in the past two decades opened an exciting perspective for the reconstruction of global climatic history. The kinds of evidence and potential of climatic interpretation are manifold. Large rivers cast deltas on the coastal shelf, and the varying nature of these deposits may be indicative of the continental hydrology (Pastouret *et al.*, 1978; Kolla *et al.*, 1979, 1981). Dust from the continental deserts (Parkin and Shackleton, 1973; Sarnthein *et al.*, 1981, 1982a, b; Nicholson, 1982; Tetzlaff *et al.*, 1982), as well as pollen from land areas (Rossignol-Strick and Duzer, 1979; Van Campo *et al.*, 1982), are preserved in the ocean floor sediments and may contain clues on atmospheric transport patterns.

Perhaps even more intriguing is the study of marine microfossils (Bé and Hutson, 1977; Duplessy *et al.*, 1981a, b; Moore *et al.*, 1981). Comparison of fossil assemblages with the behavior patterns of the modern fauna leads to inference of past temperature and salinity conditions (MuIntrye *et al.*, 1972, 1976; Imbrie *et al.*, 1973; CLIMAP project numbers 1976, 1981; Ruddiman and McIntyre, 1979, 1981a, b; Prell and Hutson, 1979; Prell *et al.*, 1980; Ruddiman *et al.*, 1980a, b; Moore *et al.*,

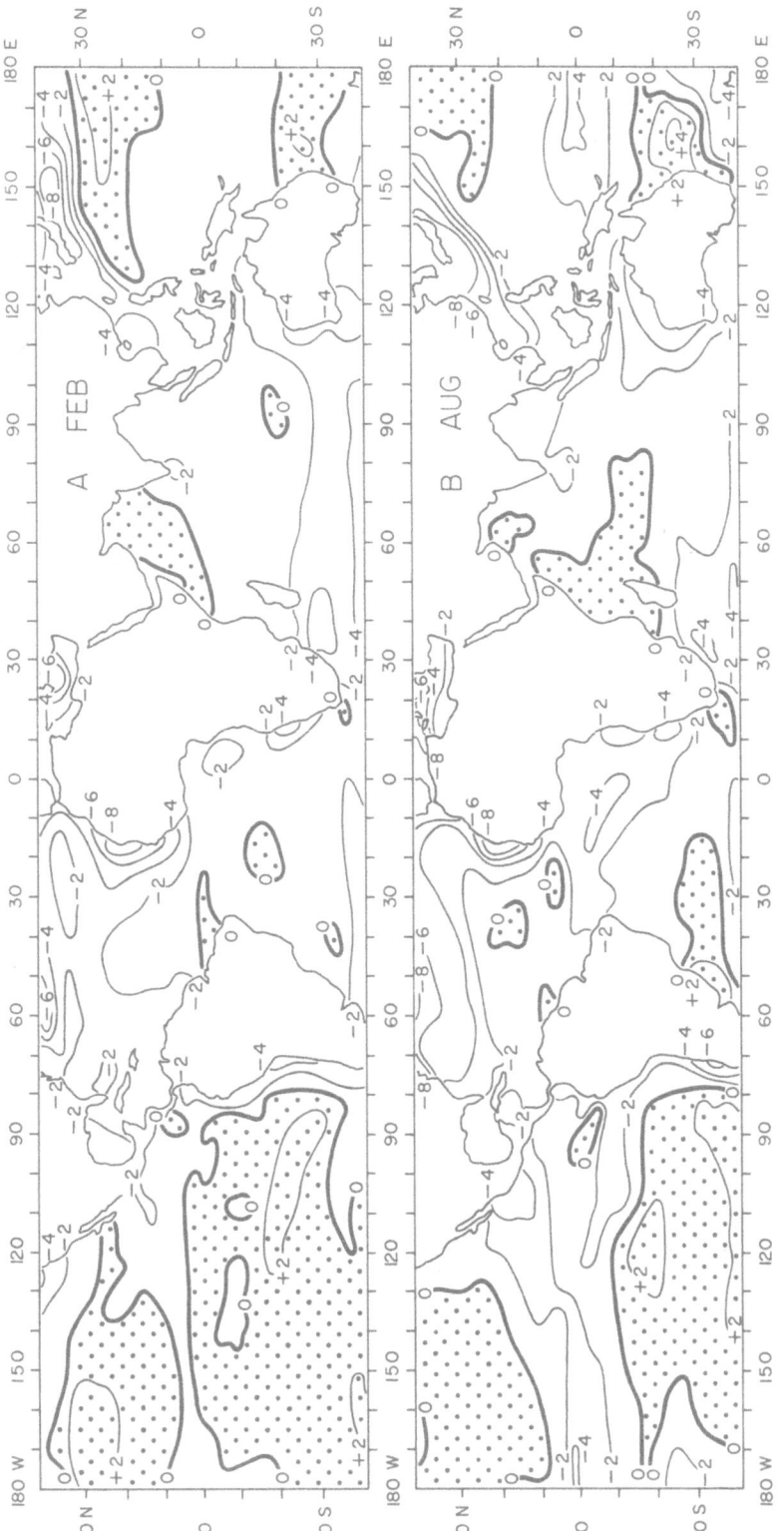

Fig. 12.5:1. Maps of sea surface temperature patterns at 18 000 years B.P., minus modern conditions, in °C. Dot raster denotes positive areas. (a) February, (b) August. Source: CLIMAP project members, 1981 (Geological Society of America Map and Chart Series MC-36).

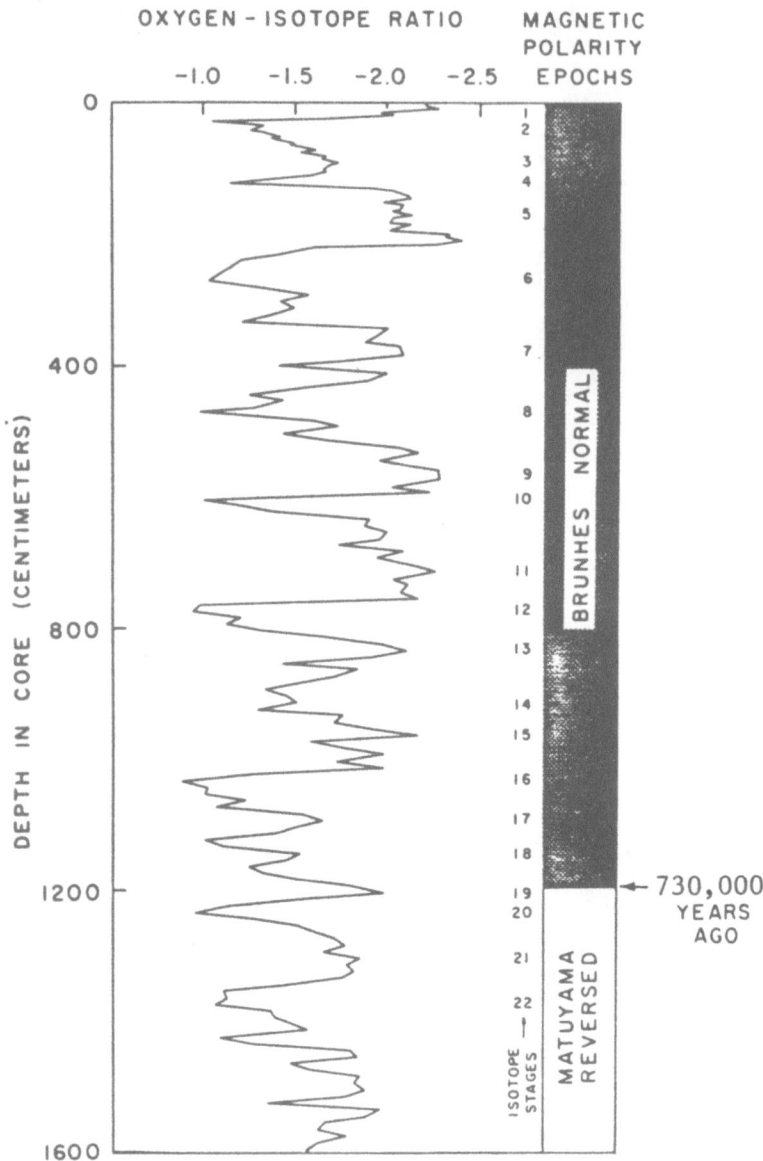

Fig. 12.5:2. Graph of the isotopic and magnetic measurements made on a Pacific deep-sea core (V28–238). (These observations, which established that isotope stage 19 occurs at the boundary between the Brunhes and Matuyama epochs, provided the first accurate chronology of late Pleistocene climate; data from Shackleton and Opdyke, 1973). From Imbrie and Imbrie, 1979 (*Ice ages: solving the mystery*, by John Imbrie and K. P. Imbrie, Enslow Publishers, Box 777, Hillside, NJ, 07205, USA).

1980; Crowley, 1981; Romine and Moore, 1981; Prell and Curry, 1981; Cullen, 1981, Duplessy *et al.*, 1981a; Duplessy, 1982; Molfino *et al.*, 1982; Brunner, 1982). DeVries and Pearcy (1982) estimated past ocean temperature conditions from fish debris in sediments of the upwelling zone off Peru.

A triumph of paleo-oceanography was the reconstruction of sea surface temperature conditions in the global ocean at the height of the last glacial maximum 18 000 years B.P. (CLIMAP project members, 1981). Maps of the temperature difference from modern conditions during the extreme seasons, February and August, are reproduced in Fig. 12.5:1a and b.

$\delta^{18}O$ values of seawater derived from the analysis of calcareous shells (Fig. 11.2:1) are believed to reflect primarily variations in global ice volume (Shackleton and Opdyke, 1973; Emiliani and Shackleton, 1974; CLIMAP project members, 1976; Morley and Hays, 1981; Jones and Ruddiman, 1982; Mix and Ruddiman, 1984; Imbrie *et al.*, 1984).

As with the evidence reviewed in Sections 12.2 to 12.4, the establishment of an absolute chronology is also crucial in the evaluation of deep-sea cores. Fig. 12.5:2 illustrates the oxygen isotopic and magnetic measurements made on a core from the Pacific (V28 − 283). The down core variations of the oxygen isotope ratio should be compared with Fig. 11.2:1. The right-hand portion illustrates the magnetic polarity epochs. The reversal of the Earth's magnetic field at 730 000 years B.P. as ascertained from continental evidence serves as a cardinal time mark for the dating of deep-sea cores (Shackleton and Opdyke, 1973; Hays *et al.*, 1976; Morley and Hays, 1981, Crowley, 1983; Imbrie *et al.*, 1984). The construction of a depth-age scale above this mark involves assumptions on sedimentation rate, correlation of extrema in the $\delta^{18}O$ profiles between various cores (Morley and Hays, 1981, Imbrie *et al.*, 1984), and deliberate mating of the core records with the known orbital variations, as mentioned in Section 12.1 (Imbrie *et al.*, 1984). This method provides a tentative time scale for comparison with other evidence (Fig. 12.4:1). The records from deep-sea cores are in fact instrumental in reconstructing the past climate and circulation patterns in the various large domains of the global tropics (Sections 12.6 to 12.9).

12.6. Pacific and Australasia

The greater Pacific basin occupies a key role in the study of tropical circulations because it appears to contain the major flywheel of the Southern Oscillation (Sections 8.1, 8.2, 9.3), entailing modulations of the atmospheric Walker cell (Section 6.9), variations in the upper hydrosphere of the equatorial Pacific (Section 4.3), and inverse rainfall anomalies at the Eastern Pacific and Australasian dipoles of the Southern Oscillation (Sections 8.1 to 8.5, 9.2). Accordingly, paleoclimatic evidence is of interest with this perspective.

At the height of the last glacial maximum, 18 000 years B.P., much of the presently shallow shelf areas in the Indonesia − North Australia region became land, as a result of the lowered sea level (Quinn, 1971; Bowler *et al.*, 1976; Webster and Streten, 1978). This would prevent cooler waters associated with stronger Southeast trades and equatorial upwelling, characteristic of the positive phase of the Southern Oscillation, from entering the Malay − Indonesia area. Quinn (1971) suggests this as a cause for the prevalence of conditions characterized by a strongly positive mode of the Southern Oscillation (Sections 8.1 and 8.2), a vigorous Walker Circulation (Section 6.9), and pronounced upwelling in the Eastern Equatorial Pacific (Sections 4.3, 8.1, 8.2), and a westward extension of the Equatorial Pacific Dry Zone (Section 6.7.4). He infers these features from deep-sea core evidence of enhanced marine biological activity and steepend sea surface temperature gradients in the Eastern to Central Equatorial Pacific, as well as from phosphate deposits indicative of large guano bird populations.

Following up on Quinn's (1971) ideas, Colinvaux (1972) proposes a climatic interpretation of a core taken from the crater of El Junco at 650 m on the Galápagos Islands, in the eastern portion of the Equatorial Pacific Dry Zone. At more than 34 000 years B.P. El Junco held a small lake. For the next 24 000 years the crater remained dry, and a new lake formed around 10 170 years B.P. and persisted to the present. While conditions on El Junco may be complicated by volcanic activity, it is interesting to compare this evidence with Van der Hammen's (1974) finding for Laguna de Fuquene in the Colombian Andes (Section 12.2, Fig. 12.2:4), where water levels were low from before 20 000 to around 13 000 years B.P. Colinvaux (1972) believes that the drier conditions at El Junco between about 34 000 and 10 000 years B.P. are due to a more northerly position of the Intertropical Convergence Zone (Section 6.7.2) during this era of major glaciation.

Conversely, Newell (1973) argues for a more southerly position of the Intertropical Convergence Zone (Section 6.7.2), on the grounds that for 20 000 years B.P. polar ice core data (Johnsen *et al.*, 1972) indicate a greater steepening of meridional temperature gradients for the Northern as compared to the Southern hemisphere. Newell (1973) concedes, however, that changes in the latitude position of the Intertropical Convergence Zone may differ substantially with longitude. At any rate, his suggestion of a greater steepening of meridional temperature gradients in the Northern as opposed to the Southern hemisphere is supported by the most recent reconstructions of global sea surface temperature patterns at 18 000 years B.P. (CLIMAP Project members, 1981; see also Fig. 12.5:1).

Continuing the controversy, Houvenaghel (1974) suggests that the dry climate at El Junco during 34 000–10 000 years B.P. could be explained from a cooling of the sea surface, without latitude shifts of the Intertropical Convergence Zone. He visualizes a cooling of surface waters in the Eastern Equatorial Pacific as resulting from enhanced Southeast trade winds. Colinvaux (1972), Newell (1973), and Houvenaghel (1974) do not address the issue of an altered intensity of the Walker cell during the last glacial maximum.

Salinger (1980a, b, 1981, 1984) studied the distribution of modern precipitation and temperature anomalies on New Zealand in relation to the broadscale circulation, and evaluated the spatial patterns of fossil pollen records from New Zealand and the Southwest Pacific. On this basis he proposes a weaker Walker cell for 18 000 years and a greater intensity for 9000 years B.P. (Salinger, 1984). Thus Salinger's (1984) conclusions for the last glacial maximum contradict those of Quinn (1971) arrived at from independent evidence and different reasoning; Quinn's (1971) earlier paper is not mentioned.

The controversy over the preferred mode of the Southern Oscillation and the intensity of the Walker Circulation at 18 000 years B.P. is addressed further in Section 12.10, in the context of sea surface temperature patterns reconstructed from deep-sea cores (Fig. 12.5:1) and numerical simulations of the surface wind field (Fig. 12.10:1).

Turning to the western extremity of the Pacific Walker cell (Section 6.9), the paleoclimatic evidence for the greater Australasian region has been discussed by Bowler *et al.* (1976), Rognon and Williams (1977), Webster and Streten (1978, 1981), Prell *et al.* (1980), and Prell (1982). Bowler *et al.* (1976) report conditions drier than at present between 60 000 and 40 000 years B.P. in northeastern Queensland, southern New South Wales, and southeastern South Australia. During 40 000–30 000 years B.P. dryness became even more accentuated in northeastern Queensland, but not in South Australia. There is some indication of colder conditions in New Guinea. Rognon and Williams (1977) report for Southern Australia abundant rainfall and high lake stands during 40 000–20 000 and 11 000–5000 years B.P., and drier conditions during 17 000–12 000 years B.P. As already discussed in Sections 12.2 and 12.3 and illustrated in Figs. 12.2:5, 12.3:2 and 12.4:1, the upper forest limit in New Guinea dropped after 27 000 years B.P., reaching a minimum around 17 000 years B.P., and concurrently there was a tendency for lower lake stands in Australia. This was also an epoch of extensive glaciation in Tasmania and of small glaciers in the Snowy Mountains of Southeastern Australia. As shown in Figs. 12.2:5, 12.3:7 and 12.4:1, about 15 000 years B.P. marks the beginning of the ice retreat and the upward shift of the forest limit in New Guinea, as well as of the rise of water levels of Australian lakes.

Drawing on Bowler et al.'s (1976) review, Webster and Streten (1978) attempted a general circulation interpretation for the Australasian sector during the last glacial maximum. Noting the evidence of preferred equatorward extension of sea ice (CLIMAP project members, 1976) they postulate a pattern characterized by only small annual change, an enhanced Indian Ocean trough, marked ridging over eastern Australia, and another trough further East. They consider such a basic flow pattern consistent with reduced rainfall in southeastern Australia, frequent cold outbreaks favoring glaciation in New Guinea, and enhanced precipitation to nourish the ice caps of Tasmania, and the New Zealand Alps. Webster and Streten (1978, 1981), Prell et al. (1980), and Prell (1982) discuss the apparent discrepancy of temperature estimates based on marine microfossil assemblages (CLIMAP project members, 1976) and various evidence from the New Guinea highlands (Bowler et al., 1976); the former method yielding the smaller values for the temperature lowering against modern conditions. Webster and Streten (1978, 1981) consider frequent and short-lived cold air invasions into the New Guinea highlands as a partial explanation for the apparent discrepancy. However, in the light of plausibly different lapse rates in dry as compared to moist climatic regions, it appears difficult to meaningfully compare temperature around the 5000 m level with sea surface conditions. Moreover, in the light of the widely spaced coring stations and other factors, the sea surface temperature reconstructions from deep-sea cores may be open to revision. The issue of varying intensity of the Walker cell is not addressed in the aforementioned papers.

The various papers summarized in this section in context illustrate the excessive degrees of freedom in paleoclimatic interpretation, where field evidence is limited.

In attempts at inferring circulation characteristics in the Pacific domain during the last glacial maximum, the CLIMAP reconstructions of sea surface temperature patterns (CLIMAP project members, 1981) will be a major source. Refer to Figs. 12.5:1 and 6.1:1. The modern sea surface temperature pattern (Fig. 6.1:1) especially in Northern summer is characterized by a tongue of cold water immediately to the South of the Equator and extending from the South American West coast far into the open Pacific. As discussed in Section 6.7.2.4, this is the result of wind-stress forcing by the cross-equatorial surface wind from the Southern hemisphere. It is noteworthy that in this region the maps of temperature difference 18 000 years B.P. minus modern (Fig. 12.5:1) show a pattern very similar to the modern sea surface temperature distribution (Fig. 6.1:1). This suggests that the wind-stress pattern responsible for the modern 'cold tongue' was active and enhanced during the last glacial maximum. A similar state of affairs will be discussed in Section 12.7 for the tropical Atlantic.

12.7. The Indian Ocean and Surrounding Continents

The paleoclimate of the monsoon ocean has been greatly elucidated through the extensive study of deep-sea cores (Bé and Hutson, 1977; Kolla and Biscaye, 1977; Webster and Streten, 1978, 1982; Prell and Hutson, 1979; Prell et al., 1980; Prell and Curry, 1981; Cullen, 1981; Duplessy et al., 1981a; CLIMAP project members, 1981; Kolla et al., 1981; Duplessy, 1982; Van Campo et al., 1982; Sarnthein et al., 1982a; Prell, 1982; 1984a, b). These pioneering investigations in the past ten years concentrate on the last glacial maximum.

The modern sea surface temperature patterns (Fig. 6.1:1) exhibit some straightforeward relations to the ocean circulation. During Northern winter (Fig. 6.1:1), relatively cold waters are found adjacent to the Asian continent with isotherms extending far southward along the African coast, thus reflecting the wintertime southward directed Somali Current (Fig. 4.4:4) and the effect of the Northeast monsoon winds. The comparatively warm waters in the Mozambique Channel and along the Indian Ocean coast of South Africa are related to the southward flowing Agulhas Current. During the Northern summer (Fig. 6.1:1), cold waters prevail along the coasts of Somalia and Arabia in the realm of the northward flowing Somali Current (Fig. 4.4:4) as a result of pronounced upwelling processes. Isotherms also reflect the gyre-type circulation in the equatorial Indian Ocean (Fig. 4.4:4). The West Australian Current (Fig. 4.4:4) finds a signature in the equatorward bulging of isotherms in the Eastern South Indian Ocean.

Sea surface temperature patterns at the time of the last glacial maximum, 18 000 years B.P., (Fig. 12.5:1) are to be interpreted with reference to the modern temperature distributions, ocean circulation, and wind patterns. Fig. 12.5:1a. shows for February at the last glacial maximum somewhat colder waters in the Western Arabian Sea than presently. This can be interpreted as result of an intensified wintertime southward directed Somali Current and stronger Northeast monsoon winds. Waters off the South African coast appear colder than presently, perhaps indicating a weaker Agulhas Current. Negative departures from modern sea surface temperature conditions extend from the Southern tropical Indian Ocean along the coast of Australia far equatorward. This isopleth arrangement bears some resemblance to the modern sea surface temperature pattern during the Southern winter (Fig. 6.1:1), and suggests that during the last glacial maximum the West Australian Current may have existed also during the Southern summer. Temperature differences against the modern conditions are small in the central equatorial Indian Ocean (Fig. 12.5:1a). For August during the last glacial maximum Fig. 12.5:1b shows various pronounced differences from the modern conditions. Along the coasts of Somalia and Arabia, waters were warmer then, suggesting reduced upwelling and weaker Southwest monsoon winds. Large negative differences in sea surface temperature are apparent along the Australian coast, indicating an enhanced West Australian Current. Near the center of the large oceanic gyre in the central equatorial Indian Ocean, differences from the modern sea surface temperature conditions are small.

Inasmuch as the large-scale sea surface temperature pattern contains a systematic variation with latitude, other features appear enhanced when values are expressed as departures from zonal averages. This analysis mode was chosen by Prell and Hutson (1979) and Prell et al. (1980). Prell and Hutson's (1979) mappings are reproduced in Fig. 12.7:1. The temperature signatures of the Somali, Agulhas and West Australian Currents, as well as the weak gradients near the center of the gyre in the central equatorial Indian Ocean, as discussed above, are apparent on these maps. Arrows are also entered to emphasize the relation of sea surface temperature patterns to ocean circulation. At 18 000 years B.P. in February the temperature signature appears enhanced for the southward flowing Somali Current, but reduced for the warm Agulhas Current; in August the cold waters in the domain of the northward flowing Somali Current are less conspicuous; and in both extreme seasons there is a strong signature of the West Australian Current, while a pronounced pattern is lacking near the center of the gyre in the central equatorial Indian Ocean. Prell et al.'s (1979) proposed synthesis of ocean circulation conditions at 18 000 years B.P. is presented in Fig. 12.7:2. They suggest the following differences from the modern conditions: an equatorward shift of the polar front by 5–10, and of the subtropical convergence by 2–5 degress latitude; an enhanced West Australian Current due to greater extent of sea ice in the Southern Ocean causing a more equatorward location of the band of westerlies, and the deflection of the West wind drift northward along the coast of Australia, in agreement with the propositions of Webster and Streten

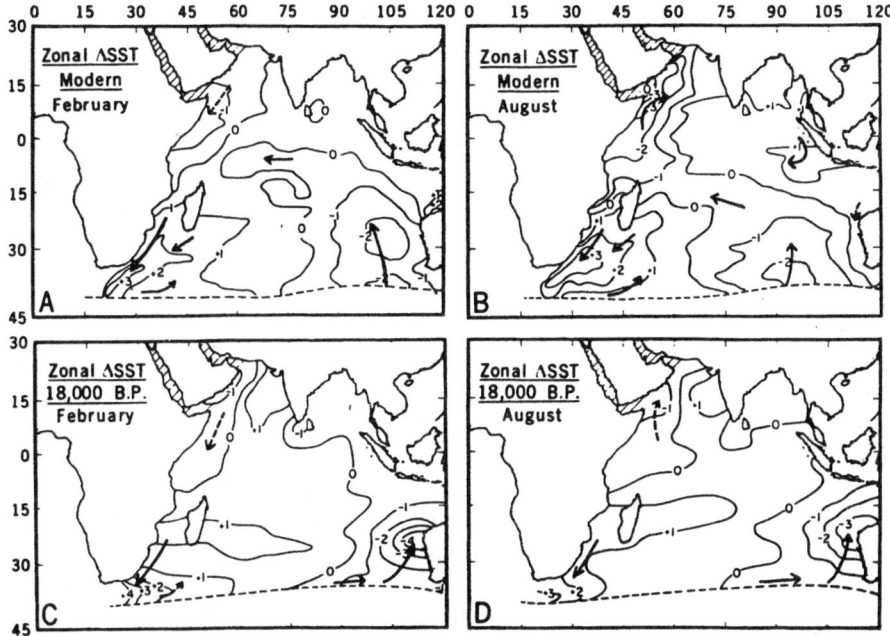

Fig. 12.7:1. Maps of zonal anomalies of sea surface temperature in the Indian Ocean in °C for (a) February and (b) August of modern times, and of estimates of sea surface temperature for (c) February and (d) August at 18 000 years B.P. The broken line gives the location of the subtropical convergence. Solid arrows show major currents and dotted arrows weak or poorly defined currents. From Prell and Hutson, 1979 (*Science*, **206**, 454–456; copyright 1979 by American Association for the Advancement of Science).

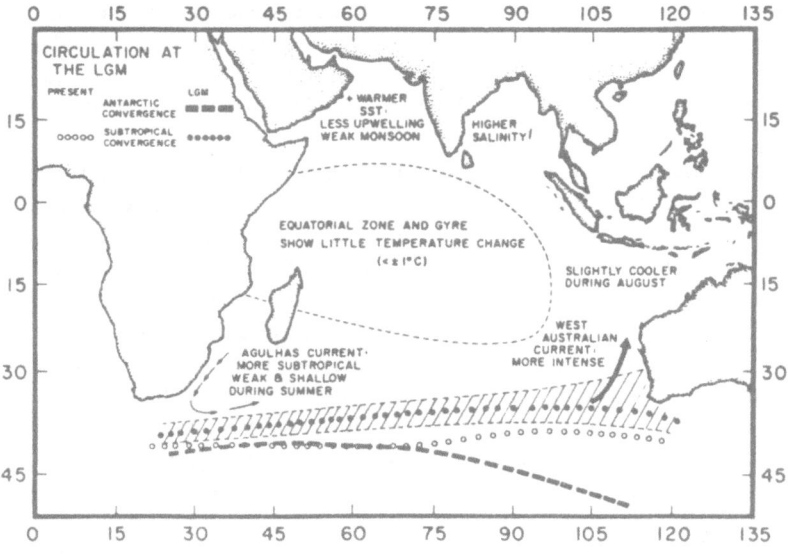

Fig. 12.7.2. Schematic summary of the circulation patterns, temperature patterns, and location of oceanic fronts of the Indian Ocean during the last glacial maximum. From Prell *et al.*, 1980.

(1978); a weaker and cooler Agulhas Current: decreased upwelling in the Arabian Sea and higher salinity in the Bay of Bengal, thought to be related to a weaker Southwest monsoon; a sea surface temperature decrease of the Indian Ocean of the order of $2°C$.

Continuing the reconstruction for the last glacial maximum, Prell and Curry (1981) see in the δ ^{18}O composition of surface-dwelling fauna a further indication of decreased upwelling in the Western Arabian sea during the Southwest monsoon. The picture of weaker Southwest monsoon winds suggested from the analyses of Prell and Hutson (1979), Prell et al. (1980), and Prell and Curry (1981), finds further support in the work of Kolla and Biscayne (1977) who interpret the greater abundance of quartz in Arabian Sea sediments as reflecting a weaker Southwest monsoon and more intense northwesterly winds from the Arabian peninsula.

In extension of his earlier work, Prell (1984a, b) infers from records of faunal variations in the Arabian Sea not only a weaker summer monsoon around 18 000 years B.P., but also an enhanced summer monsoon around 9000 years B.P. Moreover, Prell (1984a) claims that the aforementioned faunal records are highly coherent with solar radiation at the frequencies of orbital precession around 22 000 years (Section 12.1), and refers to supporting numerical simulations of paleo-circulations (Kutzbach, 1981; Kutzbach and Otto-Bliesner, 1982) to be discussed in Section 12.10.

The salinity patterns of the Northern Indian Ocean during the last glacial maximum and the transition to the Holocene have been reconstructed by Cullen (1981) and Duplessy (1982). Cullen (1981) analyzes the distribution patterns of salinity-sensitive fauna, while Duplessy's (1982) method is based on the oxygen isotopic composition (δ ^{18}O) of surface-dwelling planktonic foraminifera. Results of these studies are mutually supportive although Cullen's (1981) work is limited to the Bay of Bengal. For the last glacial maximum, salinity appears increased in the northern portions of the Arabian Sea and Bay of Bengal, but reduced during the transition to the Holocene, 12 000–10 500 years B.P. in the Bay of Bengal. These variations are interpreted as reflecting the freshwater input by the large South Asian rivers and hence continental rainfall.

Van Campo et al. (1982) evaluated the pollen profile in a 150 000 year core off the South coast of Arabia with reference to the oxygen isotope record. They conclude that during low sea-level glacials pollen assemblies indicate saline littoral and arid and steppe inland conditions, while high sea-level interglacials are characterized by savanna-type vegetation. Certain pollen are indicative of distant source regions and suggest intensification of lower-level opposite air flows: the summer Southwest monsoon winds during the last interglacial and the winter Northeast monsoon airstream during the last glacial maximum.

This extensive paleo-oceanographic reconstruction for the tropical Indian Ocean is interesting in relation to evidence from the surrounding land areas. For tropical Australia arid conditions are reported for the last glacial maximum (Rognon and Williams, 1977). Prell et al. (1980) suggest that the cold surface waters of the adjacent Indian Ocean associated with the intensified West Australian Current may through reduced evaporation have contributed to the aridity of tropical Australia; increased continentality caused by lowered sea level is considered as a further factor. As discussed in Section 12.3 (Figs. 12.3:1, 12.3:2, 12.4:1), in tropical Africa, lakes stood low during the last glacial maximum and experienced high stands around 10 000–8000 years B.P.

For the Rajasthan area of Northwest India (Singh et al., 1974; Allchin et al., 1978, pp. 305–330; Bryson and Swain, 1981; Swain et al., 1983), pollen, lake geomorphic, and archaeological evidence indicates arid conditions prior to about 10 000 years B.P., maximum precipitation in the millenia thereafter, and a decrease of rainfall to the present.

In their reconstruction for the Great Indian Desert, Allchin et al. (1978, pp. 305–330) contend that the Thar is an ancient phenomenon, rather than resulting from a combination of man-induced desertification and climatic desiccation in the Holocene. This is at variance with the suggestion of

Bryson and Baerreis (1967) referred to in Section 10.3. Allchin *et al.* (1978, pp. 305–330) propose four major climatic phases in the Late Quaternary of Gujarat and Rajasthan. A dry phase prior to 40 000 years B.P. left major sheets of aeolian sand with some slope-wash material containing calcrete and certain artefacts. A wet phase after 40 000 years B.P. is indicated by weathering and decalcification of dunes, coarse river debris, and substantial human occupations including in currently hyper-arid areas. During a subsequent dry phase lasting to around 10 000 years B.P. dunes extended over lake basins and earlier soils, rivers aggraded, and evidence of human activity is limited. A moist phase around 10 000–9500 years B.P. is evident by pollen-rich lake sediments without aeolion component overlying dunes, and extensive human settlement.

This climatic sequence for Northwest India supports, in particular, the inference from the past salinity patterns of the northern Arabian Sea and Bay of Bengal reviewed above. Moreover, a broad agreement is indicated between other land evidence of past climates and paleo-oceanographic reconstructions in the Indian Ocean sector. However, it is remarkable that the paleoclimatological literature reviewed in this section offers no confirmation by modern analog of the proposed causalities between wind, sea temperature, and rainfall departures. This shortcoming appears symptomatic of current paleoclimatic inferences in general. In order to partly redress this deficiency for the Indian Ocean sector, refer to Section 8.5. The interrelations between surface wind intensity and sea surface temperature in the Arabian Sea, and continental rainfall, proposed in this section from paleoclimatic reconstructions for 18 000 years B.P. appear indeed broadly consistent with the climate anomaly mechanisms deduced in Section 8.5 from observations of the 20th century.

12.8. Africa and the Adjacent Tropical Atlantic

On the continent of Africa, paleoclimatic field research has concentrated on the Sahara desert (Geyh and Jäkel, 1974; Wendorf *et al.*, 1976; Rognon, 1976; Maley, 1977; Jäkel, 1979; Petit-Maire, 1979, 1980; Pachur and Braun, 1980; Fryberger, 1980; Messerli *et al.*, 1980; Petit-Maire and Riser, 1981; Hillaire-Marcel *et al.*, 1983; Smith, 1984) and neighboring land areas to the East (Butzer and Hansen, 1968; Butzer, 1971a, b; 1975; Williams and Adamson, 1974; Adamson *et al.*, 1980; Gasse, 1980; Rognon, 1980; Reiss *et al.*, 1980; Isaar and Bruins, 1983; Gillespie *et al.*, 1983) and to the South (Jeje, 1980; Talbot, 1981, 1984; Maley, 1982; Maley and Livingstone, 1983; Talbot *et al.*, 1984). Lake Chad in particular has been the object of numerous studies (Servant and Servant, 1970; Servant and Servant-Vildary, 1972; Maley, 1977, 1981; Servant-Vildary, 1979; Kutzbach, 1980; Tetzlaff and Adams, 1983; Durand *et al.*, 1984). Important contributions to the continental climate record have also been obtained from deep-sea cores off the Atlantic coasts of Northern hemispheric Africa (Parkin and Shackleton, 1973; Pastouret *et al.*, 1978; Rossignol-Strick and Duzer, 1979, 1980; Kolla *et al.*, 1979; Diester-Haass, 1980; Sarnthein and Koopmann, 1980; Sarnthein *et al.*, 1981, 1982a, b; Tetzlaff *et al.*, 1982; Nicholson, 1982; Agwu and Beug, 1984). As already referred to in Section 12.5, deep-sea cores also formed the basis for the reconstruction (CLIMAP project members, 1976, 1981; McIntyre *et al.*, 1976) of the Atlantic ocean climate at 18 000 years B.P. (Fig. 12.5:1), and other events in the North Atlantic (McIntyre *et al.*, 1972; Duplessy *et al.*, 1981b; Ruddiman and McIntyre, 1981b; Crowley, 1981). Exploration of Southern hemispheric Africa is more limited (Van Zinderen Bakker, 1976, 1980, 1982, 1984a, b; Giresse, 1978; Heine, 1978, 1982; Lancaster, 1981; Giresse and Lanfranchi, 1984; Rust *et al.*, 1984). Butzer (1976), Nicholson and Flohn (1980), Sarnthein (1978), and Sarnthein *et al.* (1982a) contain extensive references for the Pleistocene to early Holocene. For environmental changes during the past two millenia refer to Butzer (1971a, 1981), Nicholson (1978), Riehl and Meitin (1979), and Hassan (1981).

The Tibesti mountains of the Central Sahara are at the margins of influences of both midlatitude and tropical circulation systems. Based on the geomorphological study of a river system and C-14 datings, Jäkel (1979) constructed a climatic history from 16 000 years B.P. onward. A humid period marked by the formation of lakes extended from 16 000 to 7300 years B.P., with a maximum around 10 000–8000 years B.P. Conditions were cool until around 9000 years B.P., with a temperature rise thereafter. Comparison with areas on the northern and southern edges of the Sahara indicates that the Tibesti occupies an intermediate position. It is suggested that from about 16 000 to 8000 years B.P. Tibesti was subject to Mediterranean humid conditions, while tropical humid airmasses occurred until 5000 years B.P., with the two effects overlapping during the most humid interval 10 000–8000 years B.P. Maley (1977) infers for the Central Sahara various tropical influences up to 6500 and 4400 years B.P. Based on geomorphological studies, Messerli et al. (1980) report humid Mediterranean influences in the central Saharan mountains during 15 000–8000 years B.P., and a tendency to drier conditions at 5000–4000 years B.P. Geyh and Jäkel's (1974) results complement this sequence. They note a relatively arid phase 11 700–10 500 years B.P., and also arid time spans 7000–6000 and 4700–3700 years B.P.

For the Atlantic coast region of the Sahara Petit-Maire (1979, 1980) studied the history of human occupation sites and fauna over the last 10 000 years. After about 6000 years B.P. a warming is inferred, indicative of diminished coastal upwelling. Human and animal life along the coast developed significantly after about 4000 years B.P. and aridity is believed to have decreased during 4000–3000 years B.P. Fryberger (1980) finds that various periods of dune formation in Mauritania were not associated with any drastic changes of the surface wind regime, and calls attention to potential upsetting effects of modern human activities on the balance of the arid land system.

For Northeastern Mali Petit-Maire and Riser (1981) infer from faunal evidence relatively humid conditions during 7520–6130 years B.P. Likewise based on investigations of the fossil fauna of Northwestern Mali, Hillaire-Marcel et al. (1983) note a lacustrine period 9500–6500 years B.P., and a drying up of lakes 6500–5500 years B.P. A second lacustrine period is dated at 5300–4500 years B.P. and would correspond to numerous Neolithic settlements in the area. Arid conditions set in again around 4500 years B.P. to continue to the present.

The human prehistory of the Egyptian and Lybian deserts has been studied by Wendorf et al. (1976) and Pachur and Braun (1980). Wendorf et al. (1976) note a drying up of ponds more than 44 000 years ago. For the subsequent period of over 30 000 years they find no traces of any occupation, spring or lacustrine sediments, and suggest that during this long time span, the Western Desert of Egypt was devoid of surface water and any sign of life. A precipitation increase is indicated for 9500–7000 years B.P. Widespread human occupation occurred around 4500 years B.P. Pachur and Braun (1980) conclude that between 10 000 and at least 7100 years B.P. the western part of the Lybian Desert had an environment characterized by widespread, shallow, swamp-like water places. For around 8000 years B.P. they propose a fluvial link between the Tibesti and the Mediterranean.

Butzer and Hansen (1968) studied the environmental changes of Nubia and the Saharan Nile. Of particular interest here are indications of a flood regime similar to the present and with little *wadi* activity 24 000–18 000 years B.P., of enhanced Nile floods prior to 12 000 and during 11 000–10 000 years B.P., of activity in the Red Sea Hills 11 500–8500 years B.P., and of more vigorous summer floods prior to about 5000 years B.P. (Butzer and Hansen, 1968, p. 149, 328, 389). Butzer (1971b, pp. 312–351) also presents a review of late Pleistocene environmental changes in tropical Africa. He concludes (Butzer, 1971b, pp. 325–327; 1975, p. 403) that in Egypt and Nubia *wadi* discharge was more significant between 17 000 and 5000 years B.P., except for three major dry interludes centered around 11 500, 7500, and 6500 years B.P., and that a long arid phase with accelerated dune activity and comparatively low Nile floods began prior to about 4500 years B.P.

Climatic variations are reflected in the river morphology of the Nile. (Butzer, 1971b, pp. 325–327; 1975, p. 405; Butzer and Hansen, 1968, p. 149, 328, 389; Williams and Adamson, 1974; Williams, 1975; Adamson *et al.*, 1980; Williams and Faure, 1980). Adamson *et al.* (1980) find that during the cold phase 20 000–12 500 years B.P. the aggrading Nile was a braided, highly seasonal river. With a change in the headwaters to warmer, wetter conditions, it became an incised, sinuous, suspended load river. The change in the Nile regime began with overflow from Lake Victoria and severe floods in Egypt. Livingstone (1980) details the environmental changes in the Nile headwaters. He reports that just prior to 12 500 years B.P. Lake Victoria and Lake Albert (Mobutu) were not yet connecting to the Nile. Lake Albert possesses the longest known record. It drained to the Nile during 28 000–25 000 years B.P., 18 000–14 000 years B.P., and then since 12 500 years B.P. Lake Victoria became a closed lake again briefly around 10 000 years B.P.

Complementing various findings mentioned above, and broadly consistent with Butzer (1971b, pp. 325–327; 1975, p. 403), Gillespie *et al.* (1983) point out that the widespread early postglacial wet phase in tropical Africa between 12 500 and 5000 years B.P. was penetrated by severe dry episodes during 11 000–10 000, 8500–6500, and 6200–5800 years B.P. Consistent with this overall picture, they present further supporting stratigraphic and dating evidence from the Ziway-Shala lake basin in Ethiopia.

For the Chad basin, Servant and Servant (1970) note from geomorphological and stratigraphic investigations wet phases at 41 000–22 000, > 12 000, and 10 000 years B.P., and dry phases around 7500 and 4000–3500 years B.P. Their work is complemented by Servant-Vildary's (1979) study of the diatoms and paleolimnology, Maley's (1981) palynological investigations, as well as by Durand *et al.*'s (1984) geomorphological and stratigraphic studies of the Chad basin.

Talbot (1981) infers from the study of a fixed aeolian dune ridge at the coast of Ghana that conditions were more arid and southwesterly winds stronger than at present during 4500–3800 years B.P. From a combination of stratigraphic, geomorphic, and palaeontological investigations in Subsaharan Africa Maley (1982) infers a major environmental change around 7000 years B.P., with rivers depositing mainly clay prior to this epoch, and sand thereafter. He believes that a major hydrological change is related to a change in the size of raindrops.

Deep-sea cores off the Atlantic coasts of Northern hemispheric Africa have shed much light on the climatic history of the continent. Pastouret *et al.* (1978) evaluated a core retrieved off the delta of the Niger river. A chronology is constructed from radiocarbon ages of organic matter and the vertical profile of oxygen isotope ratios of foraminifera. The distribution of the pelagic fossil concentrations by the terrigenous material is taken as a measure of freshwater discharge from the Niger. Pastouret *et al.* (1978) infer abundant rainfall in the river drainage area during 13 000–11 800, and 11 500–4500 years B.P., while the intervening intervals are regarded as having been dry as today.

Cores taken along the Senegal coast including off the mouth of the Senegal river have been analyzed for their pollen content by Rossignol-Strick and Duzer (1979, 1980). Chronology is based on the temperature history derived from foraminifera in one of the cores by Pflaumann (1975) and on tentative correlations with the Chad area by Servant and Servant-Vildary (1972). Four major zones are distinguished in the core profiles. A semi-arid interval estimated at 24 000–20 000 years B.P. contains pollen from both North and South of the Sahara. A very arid period believed to be 20 000–12 000 years B.P. is devoid of pollen indicative of the humid regions in the Senegal catchment, has only reduced contributions from the steppic Sahelian element, but an abundance of pollen from North of the Sahara. This is taken as an indication of enhanced trades. The disappearance or desiccation of the Senegal river is mentioned as a further symptom of extreme aridity. Significantly more humid conditions are reported for 13 000–5000 years B.P. The Senegal river again reaches the sea as indicated by pollen from portions of the catchment further inland. The disappearance of pollen from North of the Sahara is interpreted as reflecting a weakening of the trade winds.

Kolla *et al.* (1979) report that a band of maximum concentration of quartz in the surface sediments of the Atlantic at the latitude of the Sahara and Sahel was displaced southward at 18 000 years B.P. They suggest that this was caused not only by the southward expansion of aridity but also by the more southward extent of the Northeast trades.

Sarnthein *et al.* (1981) evaluated the deposition of Saharan dust in the sediments of the Atlantic off Northwest Africa. The chronology is based on oxygen isotope ratios and foraminiferal eco-stratigraphic curves calibrated by C-14 dating. Sarnthein and Koopmann (1980) and Sarnthein *et al.*, (1981) conclude from the inference of aeolic transport patterns that the latitude position of semi-permanent circulation features remained essentially constant throughout glacial and interglacial times. They find this consistent with the inferred constancy in the position of boundaries between major water masses of the Northeastern Atlantic. They further suggest increased meridional wind component at 18 000 years B.P. Complementing this paper, Sarnthein *et al.* (1982a) further state than the grain sizes of near-shore aeolian dust deposits and the correlated meridional wind intensity began to decrease around 17 000–16 000 years B.P. in both hemispheres, reaching a minimum at 13 500 years B.P. McIntyre *et al.* (1976) and Sarthein *et al.* (1982b) report, based on estimates from planktonic foraminiferal counts, that at 18 000 years B.P. the North Atlantic polar front was shifted to 42–45°N and bordered almost on the subtropical gyre. The intensified and colder Canary Current extended further southward, and the productivity of coastal upwelling waters was high. For 6000 years B.P. planktonic foraminiferal counts indicate a warmer Canary Current, and reduced annual temperature range. These features are taken as indications of diminished upwelling during summer, and ascribed to reduced Saharan dust supply and weaker meridional trade wind component.

Grove (1984) suggests that glacial aridity may have been more marked in Africa than in the other tropical continents because of the chilling of the Atlantic related to the extension of the ice sheets in the North Atlantic domain.

A preliminary climatic history of Southern Africa is emerging from the works of Van Zinderen Bakker (1976, 1980, 1982, 1984a, b), Heine (1978, 1982), Giresse (1978), Lancaster (1981), Giresse and Lanfranchi (1984), and Rust *et al.* (1984). Fig. 12.8:1 shows Van Zinderen Bakker's (1976) map of the modern vegetation distribution in Southern Africa (Fig. 12.8:1a) and his reconstructions of glacial (Fig. 12.8:1b) and interglacial (Fig. 12.8:1c) conditions. The modern pattern (Fig. 12.8:1a) shows various types of woodlands receiving summer rainfall to the North of a line extending from 15°S at the Atlantic to 21°S at the Indian Ocean coasts. To the South of this line various drier types of plant cover are found, while the abundant vegetation relying on winter rainfall is confined to the Southern extremity of the continent. During a glacial maximum (Fig. 12.8:1b), the woodlands must have lost much of their area, the Kalahari and Namib deserts shifted northward, an 'arid corridor' extended across the continent in a Southwest-Northeast orientation, and the vegetation types relying on winter rainfall, and now limited to the extreme South, were more extensive. During an interglacial (Fig. 12.8:1c), pattern changes are in part opposite to those described for the glacial maximum. Broadly consistent with Van Zinderen

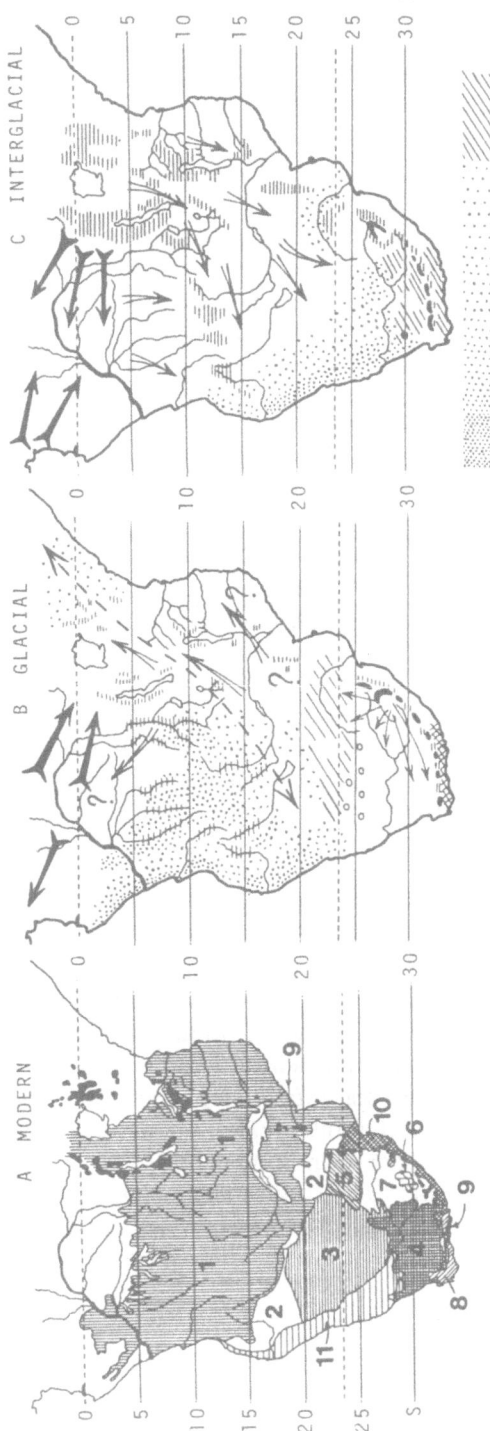

Fig. 12.8:1. Vegetation patterns of Southern Africa in (a) modern (b) glacial, and (c) interglacial times. (a) modern vegetation: 1. different types of woodland; 2. mopane woodland; 3. Acacia – Commiphora woodland; 4. Karoo; 5. Transvaal woodland; 6. Austro afroalpine vegetation; 7. grasslands of the Orange Free State and Natal; 8. Cape macchia; 9. evergreen coastal and montane forest; 10. semi-arid and sub-humid forests and bushveld; 11. Namib desert;

(b) trends of vegetation patterns during a glacial maximum: double shafted arrows = migration of woodland; single shafted arrows = possible routes of alpine grassland; broken line = position of 'arid corridor'; thick arrows in Congo basin = possible migration of humid tropical forest; cross hatched = coastal grassland;

(c) trends of vegetation shifts during an inter-glacial: 1. desert; 2. dry grassland; 3. savanna; 4. Karroo; 5. macchia; 6. coastal and montane forest; 7. temperate and subtropical grassland;

Source: Van Zinderen Bakker (1976).

Bakker's (1976) scheme of climate evolution, Heine (1978) finds for the Southern Kalahari humid conditions 18 000–11 000 years B.P. which he relates to enhanced winter rainfall in the realm of the West wind drift. In the Eastern Central Kalahari pluvial phases are indicated at 30 000–25 000 and around 12 000 years B.P. From the analysis of dune systems, Lancaster (1981) infers arid periods prior to 30 000 and after 20 000–15 000 years B.P. From a combination of continental and marine evidence in the Gabon – Congo sector Giresse (1978) suggests for around 20 000–18 000 years B.P. stronger upwelling and a more equatorward extent of the Benguela Current (Section 4.2) and pronounced aridity. Rust et al. (1984) and Van Zinderen Bakker (1984a) call attention to the difficulties in reconstructing the climatic history of Southwest Africa.

In closing a fundamental discrepancy must be noted concerning the wind patterns inferred from dust transports. Thus, based on the evaluation of deep-sea cores, Kolla et al. (1979) believe that in the West African – Atlantic sector at 18 000 years B.P. the Northeast trades extended further southward, whereas Sarnthein and Koopmann (1980) and Sarnthein et al. (1981) contend that the latitude position of semi-permanent circulation features remained essentially constant throughout glacial and interglacial times. Sarnthein et al. (1981) find their conclusion consistent with Sarnthein's (1978) global study of dune fields during the last glacial maximum. Further pertinent is Fryberger's (1980) suggestion that periods of dune formation in Mauritania were not associated with any drastic changes of the surface wind regime.

The question arises whether a large southward displacement of the surface wind confluence over the Atlantic (Section 6.7.2.1) at 18 000 years B.P. appears plausible. Kolla et al.'s (1979) evaluation in fact suggests a southward shift as large as 8 degrees of latitude. It must first be recalled that even during the most extreme hydrometeorological anomalies of the 20th century in Northeast Brazil and Subsaharan Africa (Sections 8.6 and 8.8), the latitude position of the confluence line between the Northeast trades and the cross-equatorial flow from the Southern hemisphere (Section 6.7.3) departed only by about 1–3 degrees latitude from the long-term average – in fact it can only be ascertained by a fine resolution analysis. It does then not seem obvious that the coarse spatial resolution of deep-sea core networks would warrant discrimination of the latitude position of the wind confluence.

Apart from this reservation, it is interesting to consider the issue of latitude shifts of the surface wind field in relation to the CLIMAP evidence for 18 000 years B.P. (CLIMAP project members, 1981), in part displayed in Fig. 12.5:1. The CLIMAP reconstruction of sea ice cover indicates a substantially greater equatorward extent in the South as opposed to the North Atlantic, this being particularly the case for August. Moreover, the August map (Fig. 12.5:1b) shows a much stronger cooling of surface waters to the South as compared to North of the Equator. Thus both the sea ice distribution and the sea surface temperature departure pattern point, if anything, to a more northerly position of the zone of highest sea surface temperature, which would in turn be of consequence for the latitude position of the Intertropical Convergence Zone and wind confluence (Section 6.7.2). In addition, as mentioned in Section 12.6, Fig. 12.5:1b shows for August at 18 000 years B.P. in both the Pacific and Atlantic a more pronounced 'cold tongue' to the South of the Equator. This suggests a more vigorous cross-equatorial surface airstream from the Southern hemisphere, which in turn would be consistent with a more northerly position of the wind confluence.

These considerations regarding the latitude position of quasi-permanent circulation features in the tropical Atlantic are offered here not as a solution of the problem, but to underline the unforgiving complexity of paleoclimatic interpretations.

12.9. Americas

The exploration of climatic history in the tropical Americas is more limited than for the other large regions of the tropics. Van der Hammen and his collaborators (references in Van der Hammen, 1974, and Van Geel and Van der Hammen, 1973) pioneered in the study of past vegetation in Northern South America. Damuth and Fairbridge (1970) used to advantage the deep-sea core approach to infer variations in the Atlantic domain of the South American continent. Inferences by Tricart (1974) are based on geomorphology. Haffer (1969), Vuilleumier (1971), Simpson (1975), and Simpson and Haffer (1978) studied faunal and floral refuge and speciation patterns. A continent-wide reconstruction of biogeographic distributions is due to Ab'Saber (1977) and Brown and Ab'Saber (1979). Later contributions on the vegetation history include the works of Colinvaux (1972, 1979), Servant and Villaroel (1979), Graf (1981), Markgraf and Bradbury (1982), and Watts and Bradbury (1982). As reviewed in Section 12.4, further important insight into environmental change in the tropical Americas has been obtained from the study of glacier variations.

Van der Hammen's work is epitomized in Figs. 12.2:3, 12.2:4, 12.2:7, and 12.4:1a. Low vegetation limits and relatively cool and dry conditions are indicated for the interval 20 000–14 000 years B.P. The subsequent overall rise of vegetation limits appears halted during 14 000 10 000 years B.P., the '4000 years of hesitation'. The highest forest limit is reached after 8000 years B.P. with some drop towards modern times. For the lowlands of Northern South America, Van der Hammen (1974) also reports long-term vegetation changes. For the Guyanas there are indications for a considerable extension of *savannas* during glacial periods with low sea levels. For the inland *savannas* of Colombia and the Guyanas Van der Hammen (1974) finds alternations between *grass-savanna* and *savanna-woodland* during the Late Pleistocene and Holocene. One of the driest periods in the Guyanas seems to correspond to the time immediately before 13 000 years B.P. Van der Hammen (1972, 1974) concludes that a much drier climate prevailed in large part of South America during the last glacial. This contention is supported by the work of Damuth and Fairbridge (1970) who suggest increased aridity in the lower Amazon basin during the last glacial from the prevalence of arkosic sands in deep-sea cores off the coast of the Guyanas. Moreover, on the basis of satellite and radar imagery over Northern South America Tricart (1974) recognized fossil dune formations, and a dissected land surface near the Amazon, which he believes cannot develop under the dense modern forest cover; the latter formation being coeval with low sea level. This geomorphic evidence is considered indicative of arid conditions in the Late Pleistocene.

For the Mexican plateau, Watts and Bradbury (1982) suggest, based on pollen evidence, relatively wet and dry conditions prior to 10 000 years B.P.; the appearance of maize pollen after 3500 years B.P. is taken as an indication of increased human activity.

For the Venezuelan Andes Salgado-Labouriau *et al.* (1977) deduce, likewise from pollen analysis, a low tree line and cold conditions prior to 12 650 and after 11 960 years B.P., and higher vegetation limits and temperature in the intervening interval. For pollen and other evidence of climatic variations in the Bolivian Andes, reference is made to Servant and Villaroel (1979) and Graf (1981). In particular, Graf (1981) reports relatively cold and dry conditions until after 10 000 years B.P., a moderately humid and warm period 7500–5500 years B.P., and a subsequent increase of humidity until about 3500 years B.P.

Of particular relevance to the climatic history of the tropical Andes and Amazonia are the studies of faunal and floral refuge and speciation patterns by Haffer (1969), Vuilleumier (1971), and Simpson and Haffer (1978). They suggest that the Amazon rainforest was fragmented during periods of relative aridity, which led to the differentiation of floral and faunal elements restricted

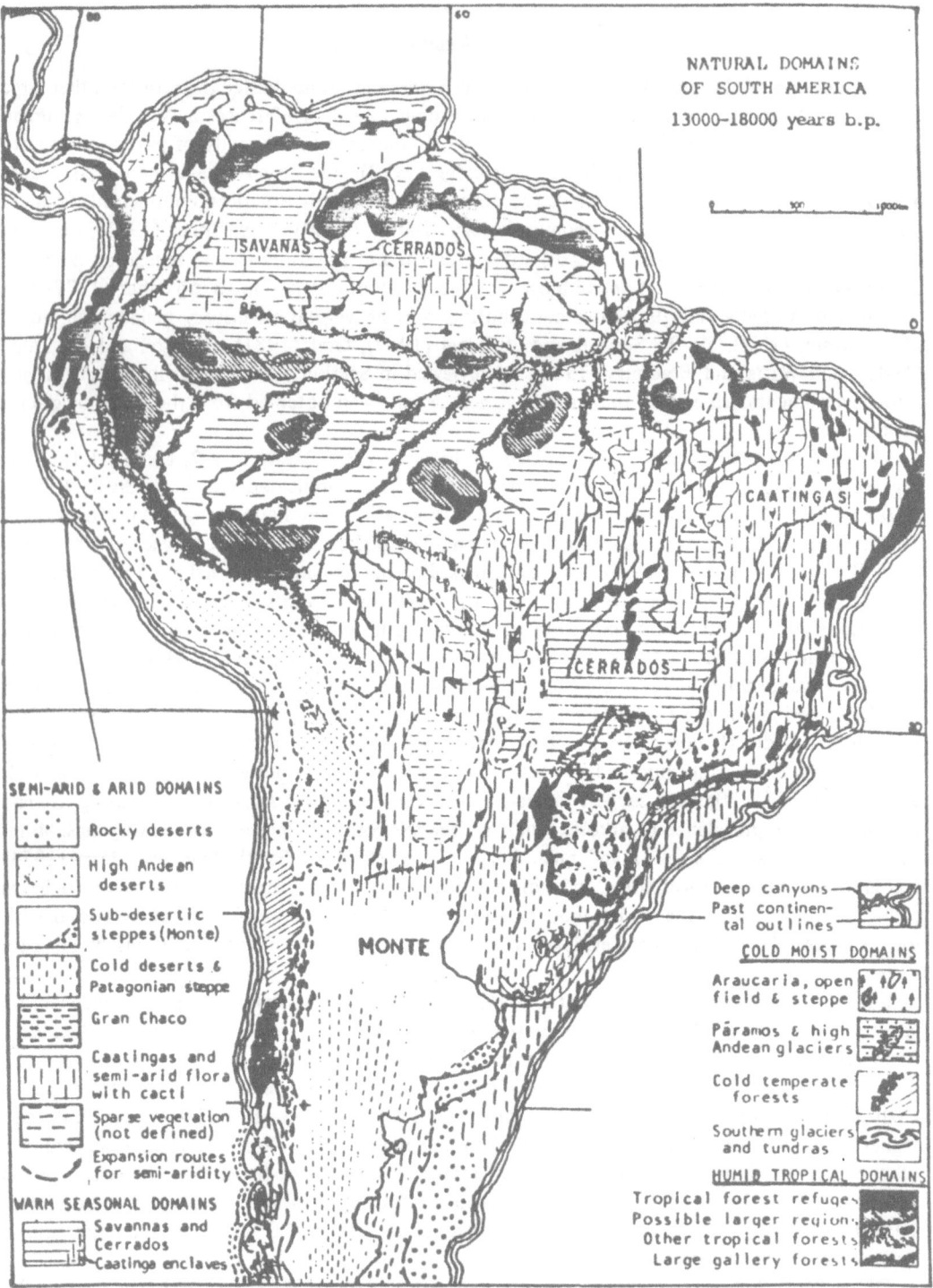

Fig. 12.9:1. Map of South America showing natural domains in which climate and morphological factors favored predominance of characteristic vegetation types, 18 000–13 000 years B.P. From Brown and Ab'Saber (1979), and Ab'Saber (1977).

to humid forest conditions. Refuge core areas deduced from biotic distribution patterns are found to coincide in large part with the regions expected to require sufficient precipitation to maintain rainforest even during arid epochs. Climatic changes and the formation of forest enclaves during the Pleistocene are therefore considered as cause of speciation patterns.

A comprehensive synopsis of vegetation changes in South America has been presented by Ab'Saber (1977) and Brown and Ab'Saber (1979). Their map of the vegetation distribution on the South American continent during 18 000–13 000 years B.P. is reproduced as Fig. 12.9:1. The reconstruction is based on floral, faunal, as well as geomorphic, soil, and other evidence. Ab'saber (1977) calls attention to a macro-enclave of *cerrados* in Goias and Mato Grosso, surrounded by *caatingas* to the West, North, and East, and by *steppes* and open field to the South, forest refuges being rare. A second large core of *cerrados* is noted in the plateaus and hills of the Amazon basin, coexisting with gallery forests and enclaves of *caatingas* and tropical forest refuges. The coastal mountain ranges and plateaus in the northeastern portion of the continent also maintained extensive tropical forest refuges. During the dry glacial episodes, when the drier and open vegetation types reached their greatest extension, the forest core areas are believed to have yielded in part to heterogeneous zones of contact and transition. It appears that extensive and complex transition regions prevailed over the forest core areas. Simpson and Haffer (1978) consider the modern faunal and floral distribution centers as consistent with the forest refuges shown in Ab'Saber's (1977) map for 18 000–13 000 years B.P. (Fig. 12.9:1).

Fig. 12.9:1 merits attention also in the light of the progressive destruction of the tropical rainforest in the Amazon basin (Section 10.1). If the deforestation as such is considered inevitable, it may be important to protect at least certain biologically strategic areas. Brown and Ab'Saber (1979) propose the selective protection of 'genetic banks' to ensure the survival of the greater part of the biological diversity which evolved through the past, for study, breeding, and other uses in the future. In addition to core forest refuge areas, such as shown on the map Fig. 12.9:1, they suggest the need to preserve also nonforest systems, as well as the complex and species-rich transition zones, where biota meet and mix. Brown and Ab'Saber (1979) mention multi-national discussions on conservation planning in the Amazon. Lewin (1984) describes plans for a 30-year experiment in the Amazon basin aimed at ascertaining the minimum critical size of reserves needed to preserve ecosystems within the protected areas. It is interesting to compare Fig. 12.9:1 with the most recent official map of the Brazilian part of the Amazon basin (Ministerio das Minas e Energía, 1983), showing districts under nature conservation and Indian reservations. These categories together make up only a few percent of the total area of the Amazon basin. Moreover, only a small portion of these enclaves coincides with the domains identified by Brown and Ab'Saber (1979) as strategically important for the purpose of 'genetic banks'. In addition to the general concern about the large-scale destruction of the tropical rainforest, the lack of conservation planning is then consequential also in a specific biological sense.

12.10. Numerical Modelling

Experiments have been conducted with various general circulation models (Alyea, 1972; Saltzman and Vernekar, 1973; Williams *et al.*, 1974; Gates, 1976; Manabe and Hahn, 1977; Kutzbach, 1981, 1983; Kutzbach and Otto-Bliesner, 1982; Kutzbach and Guetter, 1984a, b) with the intention of simulating the atmospheric circulation of past epochs, the last glacial maximum being of particular interest. The first two aforementioned experiments were limited to the Northern hemisphere and are not further considered here. As a rule, the input consists of specification of the lower boundary conditions, such as the extent of polar sea ice and continental ice sheets, coast lines, snow cover, and sea surface temperature, based on reconstructions from field evidence (CLIMAP project members, 1976, 1981).

Williams *et al.* (1974) simulated ice age conditions for January and July, and obtained as most significant features a southward displacement of Northern hemispheric storm tracks in January and an appreciable reduction of precipitation over the Northern hemisphere in July.

Gates' (1976) simulation of July at 18 000 years B.P. yielded cooler and drier conditions over the unglaciated continental areas, a southward displacement and strengthening of the westerlies in the vicinity of the Northern hemispheric ice sheets, increased tropical easterlies, and a weakening but no latitude shift of the Hadley cells. The published results allow no estimate of meridional pressure gradients and surface winds over the equatorial Atlantic and Pacific. The continental aridity indicated in both Gates (1976) and Manabe and Hahn's (1977) experiment appears consistent with heat and water budget calculations (Kraus, 1973; Newell *et al.*, 1975), as well as with the field evidence reviewed in Sections 12.2 to 12.9.

Manabe and Hahn (1977) modelled the July—August ice age global circulation, taking into account not only the surface boundary conditions but also the orbital parameters for 18 000 years B.P. They find dry conditions over the continental portions of the tropics, as a result of stronger surface outflow from (or inflow into) the continents, which in turn is related to the larger reduction of atmospheric temperature over land than over sea. The Hadley cells of both hemispheres are intensified, but only due to the contribution of the oceanic tropics, and their latitude positions appear essentially unaltered. Manabe and Hahn (1977) note that the picture of intensified ice age Hadley circulations differs from the model results of Gates (1976) and that causes for this discrepancy have not been identified.

The spatial patterns of circulation differences between ice age and present obtained by Manabe and Hahn (1977) are shown in Figs. 12.10:1a and b. As noted by Manabe and Hahn (1977), the modelled pressure difference pattern (Fig. 12.10:1a) tends to be broadly inverse to that of the prescribed sea surface temperature difference (as approximation see Fig. 12.5.1). Large negative pressure differences appear, in particular, over the equatorial North Pacific and Atlantic. Directly related to the pressure difference map (Fig. 12.10:1a) are the patterns of surface wind vector difference (Fig. 12.10:1b). As Manabe and Hahn (1977) explain, the difference wind vectors point outward from the northern portion of South America and from the West coast of tropical Africa into the relatively warm ocean area around 20°N in the North Atlantic, and similarly outflow is indicated from the East coast of tropical Africa and the Indian subcontinent into the relatively warm areas of the Indian Ocean. Of particular interest in relation to the intensified cold water tongues immediately to the South of the Eastern Pacific and Atlantic Equator (Fig. 12.5:1; discussions in Sections 12.6 and 12.8) is the possibility of enhanced cross-equatorial surface airstreams over the low-latitude Pacific and Atlantic during Northern summer. Fig. 12.10:1b indeed indicates for the ice age a greatly enhanced cross-equatorial flow from the Southern hemisphere over the Eastern Pacific and at least the western half of the Atlantic, whereas the flow conditions mapped over the Gulf of Guinea are more complicated.

For the Eastern tropical Pacific it is interesting to consider the implications of the July—August wind pattern in Fig. 12.10:1b in relation to the February sea surface temperature pattern in Fig. 12.5:1. Over the Eastern Pacific the (modern) confluence line between the Northeast trades and the cross-equatorial flow from the Southern hemisphere varies comparatively little in the course of the year, in contrast to conditions over the Eastern Atlantic. This is substantiated in Hastenrath and Lamb (1977, charts 14—25) and Section 6.7.2:1 (compare parts a and c of Fig. 6.7.2.1:2). Accordingly, the July—August map Fig. 12.10:1b suggests a relatively far poleward position of the wind confluence also for February at 18 000 years B.P. This in turn would entail

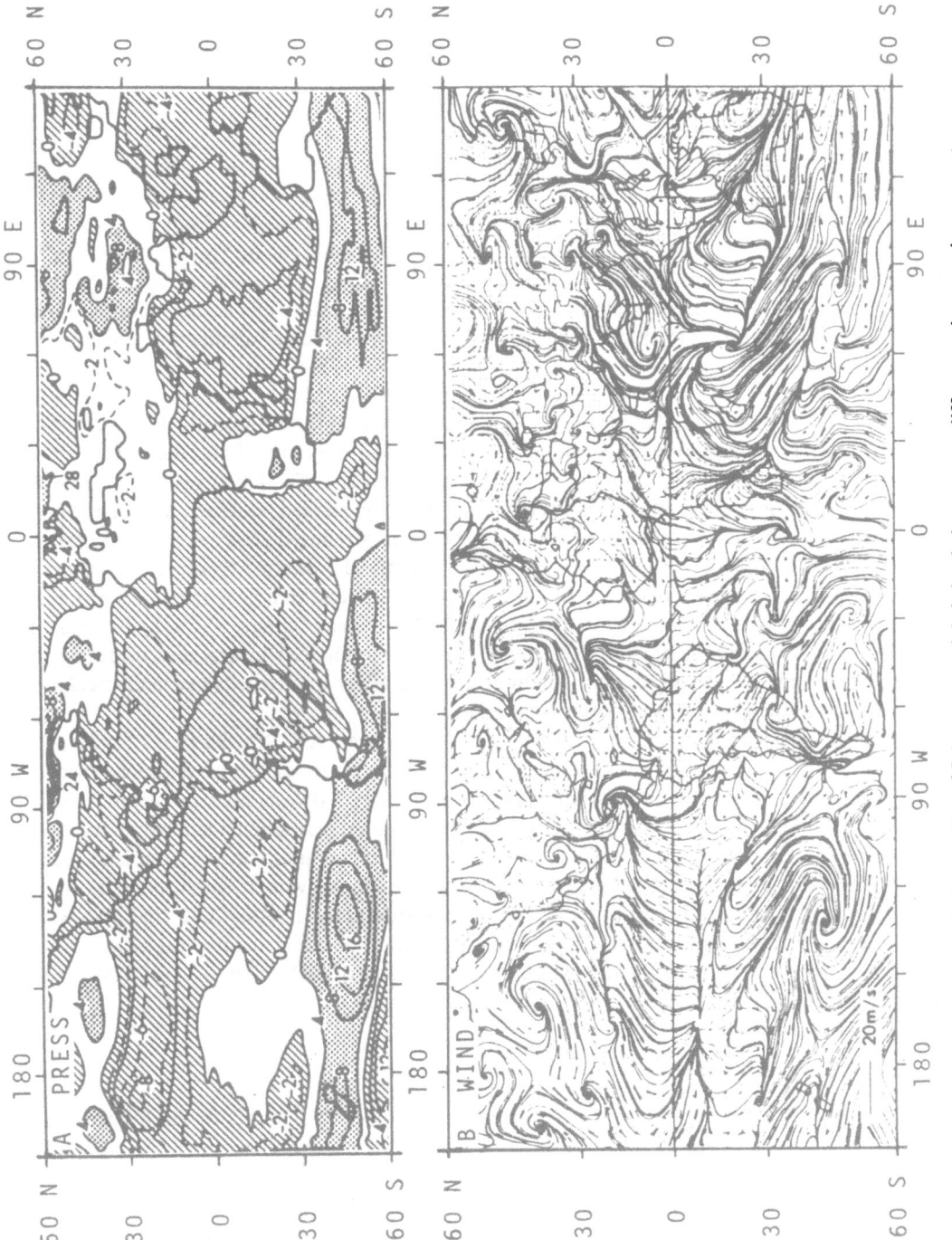

Fig. 12.10:1. Numerical modelling of July–August global circulation; pattern differences, ice age minus present: (a) sea level pressure (in mb), (b) surface wind vector. From Manabe and Hahn (1977).

weaker Northeast trades across the Central American land bridge during Northern winter.

Independent from Fig. 12.10:1b, the jet-like funneling of wintertime winds through the major topographic channels across the Isthmus (ref. also Section 7.9), in particular the Nicaragua Depression, is reflected in the (modern) sea surface temperature pattern of the Pacific waters off Central America, in such a way that a cold water signature downstream from the aforementioned topographic funnels becomes prominent in winter (Hastenrath and Lamb, 1977 charts 50–61; see also the present Fig. 6.1:1a). The positive sea surface temperature difference, 18 000 years B.P. minus present, shown in Fig. 12.5:1a in the Pacific off Nicaragua indicates a weaker wintertime cold water signature and hence a weaker funnel effect and weaker Northeast trades for 18 000 years B.P. Thus the sea surface temperature map Fig. 12.5:1a compiled from deep-sea cores supports the numerical simulation of the surface wind field depicted in Fig. 12.10:1b. The inference of a more northerly position of the wind confluence and of enhanced cross-equatorial flow at 18 000 years B.P. lends support to Quinn's (1971) suggestion of a strongly positive mode of the Southern Oscillation (Sections 8.1, 8.2), and of a vigorous Walker Circulation (Section 6.9), as discussed in Section 12.6.

It is further interesting to consider the implications of Fig. 12.10:1b in the light of the wind stress relaxation mechanisms discussed in Section 8.9 and in relation to Giresse's (1978) suggestion (Section 12.8) of stronger upwelling and a more equatorward extent of the Benguela Current (Section 4.2.) along the coasts of Southwest Africa for around 20 000–18 000 years B.P. The wind pattern displayed in Fig. 12.10:1b, characterized as it is by marked southerly wind departures over much of the equatorial Atlantic, would contribute to a reduced seasonal relaxation (about October to March) of westward wind stress over the equatorial Western Atlantic. As explained in Section 8.9, this would be conducive to cold surface waters and deficient rainfall at the coast of Southwest Africa during the short March–April rainy season. Giresse's (1978) findings are indicative of a cold ocean off Southwestern Africa and pronounced littoral aridity during the last glacial maximum. The above considerations are offered as an alternative general circulation interpretation of the paleo-oceanographic field evidence.

The major objective of Kutzbach's (1981, 1983), Kutzbach and Otto-Bliesner's (1982), and Kutzbach and Guetter's (1984a) experiments is to test the effect of orbital variations (Section 12.1), in particular the different seasons of the perihelion at 9000 years B.P. and present. Among the major results of the 9000 year B.P. simulation are more intense summer and winter monsoon circulations resulting from the larger insolation on the Eurasian and North African land masses in Northern summer and the more pronounced surface cooling in winter. Kutzbach and Guetter (1984b) expanded their earlier work by 'snapshot' simulations for 18 000, 9000, and 6000 years B.P. These numerical modelling results go a long way towards understanding the causes of low stands of tropical lakes during 25 000–15 000 years B.P. and of high stands during 10 000–8000 years B.P. (Figs. 12.3:2 and 12.4:1).

12.11. Synthesis

Research since the late 1960's has shed considerable light on the climates of the tropics during the past 30 000 years. Both internal and external causes of climatic change are recognized, orbital variations being of foremost interest on the scale of 10^5 years and longer. In particular, the change of the perihelion from Northern winter at 18 000 years B.P. to summer at around 9000 years B.P. is considered a factor for enhanced monsoon circulations in the latter epoch. Important field evidence on changes in the climatic environment has been gathered from the study of past vegetation, lakes, glaciers, and deep-sea cores.

Pollen profiles extending over 30 000 years and longer have been retrieved in Australasia, Africa, and the Americas. Abstracting from less important regional differences, drier and cooler environments are suggested up to about 10 000 years B.P. and a change to moister and warmer conditions thereafter. The upper forest limit tended to be low during 25 000–15 000 years B.P., rising to a maximum around 9000 years B.P. A period of indistinct vegetation change around 14 000–10 000 years B.P., the '4000 years of hesitation', is indicated in all three of the aforementioned large domains of the tropics.

Water level variations in closed lake basins over the past 30 000 years have been compiled for tropical Africa and Australia, while field evidence for the Americas is more limited. In both tropical Africa and Australia lakes stood low during 25 000–13 000 years B.P. Thereafter high stands became prevalent, with most extreme conditions being reached between 10 000 and 7000 years B.P. The more limited field evidence for Northern South America and the Caribbean suggests high water levels until prior to 25 000 years B.P., followed by low stand until around 13 000 years B.P., and higher water levels thereafter. For the South American Altiplano greatly enlarged lakes are dated at prior to 28 000 and 12 500–11 000 years B.P.

Long–term glacier variations in Australasia, Africa, and the Americas parallel the vegetation and lake level changes in the past 30 000 years, of particular interest being the upward shift of the lower ice limit from around 15 000 to 8000 years B.P., documented for all three domains.

Deep-sea cores have yielded information not only on past ocean climate but also on the changing climatic environment of the continents. Past sea surface temperature and salinity patterns have been inferred from assemblages of marine microfossils, of particular interest being the reconstruction of sea surface temperature conditions in the global ocean at 18 000 years B.P. The stratigraphy of the deltas built up on the coastal shelf by large rivers may be indicative of the continental hydrology. Dust and pollen originating from land areas and preserved in the ocean floor sediments have been interpreted in terms of atmospheric transport patterns. As with pollen, lake, and glacial field evidence, the establishment of an absolute chronology is also crucial for the evaluation of deep-sea cores. The reversal of the Earth's magnetic field at 730 000 years B.P. serves as a cardinal time mark in the dating of deep-sea cores. This is complemented by assumptions on sedimentation rate, correlation of extrema of the oxygen isotope ratios between various cores, and matching of core records to the known orbital variations.

This diversity of evidence derived from the study of vegetation, lakes, glaciers, and deep-sea cores forms the basis for the reconstruction of past climate and circulation in the various large domains of the tropics. The greater Pacific basin is of particular interest because it appears to contain the most active components of the Southern Oscillation. At 18 000 years B.P. the sea level was lower and much of the presently shallow shelf areas in the Indonesia – North Australia region became land, thus preventing cooler waters of the Equatorial Pacific upwelling regions from entering the Malay – Indonesia area. This has been suggested as a factor for an intensified Walker circulation which would entail enhanced upwelling in the Eastern Equatorial Pacific and a westward extension of the Equatorial Pacific Dry Zone, as inferred from deep-sea core evidence of more vigorous marine biological activity and steepened sea surface temperature gradients, as well as from phosphate deposits indicative of large guano bird populations. However, conflicting conclusions on the intensity of the Walker circulation at 18 000 years B.P. have been drawn from the interpretation of spatial patterns of paleo-vegetation in New Zealand and the Southwest Pacific. Similarly, three mutually incompatible conclusions on the latitude position of the Intertropical Convergence Zone at 18 000 years B.P. have been proposed from the same paleolimnological evidence on the Galápagos Islands. These examples illustrate the importance of field evidence in providing discriminating constraints for paleoclimatic interpretations. The CLIMAP reconstruction

of sea surface temperature patterns at 18 000 years B.P. is an essential source for inferring circulation characteristics at the last glacial maximum. Both in the Eastern Pacific and the Eastern
Atlantic, the cold water tongue immediately to the South of the Equator, so prominent in the
modern sea surface temperature pattern during Northern summer, appears enhanced at 18 000
years B.P. Inasmuch as this seasonal sea surface temperature feature of the present is understood
as the result of wind-stress forcing by the cross-equatorial surface wind from the Southern hemisphere, it is suggested that the causative surface wind pattern was active and enhanced at 18 000
years B.P.

The paleoclimate of the Indian Ocean and surrounding continents has been greatly elucidated
through the study of deep-sea cores. The sea surface temperature pattern at 18 000 years B.P.
has been interpreted with reference to the modern temperature distribution, ocean circulation,
and wind patterns. Thus, colder waters in the Western Arabian Sea, may reflect an intensified
wintertime Somali Current and stronger Northeast monsoon winds. The analysis of pollen in a
deep-sea core off Arabia also indicates intensification of the summer Southwest monsoon winds
during the last interglacial, and a stronger winter Northeast monsoon air stream during the last
glacial maximum. Salinity in the northern portions of the Arabian Sea and Bay of Bengal appears
increased during the last glacial maximum, but reduced in the transition to the Holocene in the
Bay of Bengal. These salinity changes are indicative of varying fresh water input by the South
Asian rivers and thus rainfall on the continent. For 18 000 years B.P., a weaker Agulhas Current
is inferred from lower sea surface temperature off the South African coast. Negative departures
from modern sea surface temperature conditions extending far equatorward along the West coast
of Australia suggest the existence of a vigorous West Australian Current throughout the year.
Warmer waters along the coasts of Somalia and Arabia during Northern summer suggest reduced
upwelling and weaker Southwest monsoon winds. Further inferences from the deep-sea cores
include a greater extent of sea ice in the Southern Ocean, causing a more equatorward location of
the West wind belt, deflection of the West wind drift northward along the West coast of Australia,
the latter feature being the immediate cause of the enhanced West Australian current during
18 000 years B.P. The extensive paleo-oceanographic reconstruction for the Indian Ocean is
complemented by evidence from the surrounding land areas. For the Rajasthan area of Northwest
India, pollen, lake, and other geomorphic evidence indicates arid conditions prior to about 10 000
years B.P., maximum rainfall in the millenia thereafter, and a decrease of precipitation to the
present. For tropical Africa, high lake stands prevailed at 10 000–8000 years B.P., and aridity is
reported for tropical Australia during the last glacial maximum. Thus paleo-oceanographic and
land evidence of past climates in the Indian Ocean sector is mutually supportive.

On the African continent, considerable field research has been carried out in the Sahara desert.
Geomorphological investigations in the Tibesti mountains indicate a humid period from 16 000 to
7300 years B.P., with a maximum around 10 000–8000 years B.P., cool conditions until around
9000 years B.P., and temperature rise thereafter. It is suggested that from 16 000 to 8000 years
B.P. the Central Sahara was subject to Mediterranean humid conditions, and that tropical humid
airmasses prevailed during 8000–5000 years B.P. For the Chad basin, wet phases are indicated at
41 000–22 000, >12 000 and 10 000 years B.P., and dry phases around 7500 and 4000–3500
years B.P. The Nile was braided and highly seasonal during the cold phase 20 000–12 500 years
B.P., but with the change in the headwaters to warmer, wetter conditions it became an incised,
sinuous, suspended load river. The past climates of the continent are further substantiated by
deep-sea cores off the Atlantic coast. From the stratigraphy of a core retrieved in the delta of the
Niger river abundant precipitation is inferred during 13 000–11 800 and 11 500–4500 years B.P.
The pollen profiles in cores taken off the Senegal coast indicate a semi-arid interval 24 000–20 000

and very arid conditions and enhanced trades 20 000–12 000 years B.P. Much more humid conditions are reported for 13 000–5000 years B.P. From the deposition of Saharan dust it has been suggested that the latitude position of quasi-permanent circulation features remained essentially constant throughout glacial and interglacial times, but that the meridional wind component was stronger at 18 000 years B.P. than at present. Moreover there are indications of a more equatorward position of the North Atlantic polar front and an intensified Canary Current during the last glacial maximum. For Southern Africa an equatorward displacement of vegetation zones is suggested for the last glacial maximum, and there are indications of a more equatorward extent of the Benguela Current.

In the tropical Americas, low vegetation limits and relatively cool and dry conditions are indicated for 20 000–14 000 years B.P., and a rise of the upper forest limits thereafter. For the lowlands of South America there is evidence of a considerable extension of savannas during glacial periods with low sea level. Drier conditions during the last glacial maximum are also suggested from the sediment analysis of deep-sea cores. The study of paleo-vegetation in the Amazon basin indicates biological refuge enclaves during the last glacial. These merit attention in the conservation planning of 'genetic banks' aimed at mitigating the adverse biological consequences of large-scale deforestation in the Amazon basin.

Numerical simulations of the atmospheric circulation during the last glacial maximum use as input the specification of lower boundary conditions (extent of polar sea ice and continental ice sheets, coast lines, snow cover, and sea surface temperature) as suggested from field evidence, and altered orbital conditions. Various simulations agree on enhanced ice-age aridity over the tropical continents and little latitudinal variation in the position of the Hadley cells, while there is disagreement concerning their intensity. There is an indication of more vigorous cross-equatorial airstreams from the Southern hemisphere over the Eastern Pacific and part of the Atlantic during Northern summer at 18 000 years B.P., this being pertinent to the origin of enhanced cold water tongues immediately to the South of the Pacific and Atlantic Equator. Of particular interest are experiments with perihelion in Northern winter (9000 years B.P.) and summer (present). These yield more intense monsoon circulations for the former epoch, and appear thus broadly consistent with field evidence from past vegetation, lakes, and deep-sea cores.

The findings on past climatic environments discussed in this final chapter are largely due to research during the past one to two decades. Considerable progress in this area can be expected from the combination of the field disciplines in the earth and ocean sciences with meteorological expertise. The reconstruction of past environments contributes to understanding the functioning of the tropical climate system. The study of climate is most important where it may serve the needs of mankind. In particular, the tropics are above all the lands and seas of the Third World, whose manifold social, economic, and political problems are intricately intertwined with climate.

References

Ab'Saber, A. N., 1977; 'Espaços ocupados pela expansão dos climas secos na America do Sul, por ocasião dos períodos glaciais quaternarios'. Instituto de Geografía, Universidade de São Paulo, *Paleoclimas*, 3, 1–19.

Adamson, D. A., Gasse, F., Street, F. A., Williams, M. A. J., 1980: 'Late Quaternary history of the Nile'. *Nature*, 288, 50–55.

Agwu, C. O. C., Beug, H. J., 1984: 'Palynologische Untersuchungen an marinen Sedimenten vor der West-afrikanischen Küste'. *Palaeoecology of Africa*, 16, 37–52.

Allchin, B., Goudie, A., Hegde, K., 1978: '*The prehistory and paleogeography of the Great Indian Desert*'. Academic Press, London, New York, San Francisco, 370 pp.

Alyea, F. N., 1972: 'Numerical simulation of an ice age paleoclimate'. Department of Atmospheric Science, Colorado State University, Atmospheric Science Paper No. 193, 120 pp.

Bé, A. W. H., Hutson, W. H., 1977: 'Ecology of planktonic foraminifera and biogeographic patterns of life and fossil assemblages in the Indian Ocean'. *Micropal.*, **23**, 369–414.

Berger, A. L., 1978: 'Long-term variations of caloric solar radiation resulting from the earth's orbital elements'. *Quat. Res.*, **9**, 139–167.

Berger, A., Imbrie, J., Hays, J. D., Kukla, G., Saltzman, B., 1984: *Milankovitch and climate; understanding the response to orbital forcing*. Reidel, Dordrecht, Boston, Lancaster, 2 parts, 895 pp.

Bowler, J. M., Hope, G. S., Jennings, J. N., Singh, G., Walker, D., 1976: 'Late Quaternary climates of Australia and New Guinea'. *Quat. Res.*, **6**, 359–399.

Brown, K. S. Jr., Ab'Saber, A. N., 1979: 'Ice-age forest refuges and evolution in the neotropics: correlation of paleoclimatological, geomorphological and pedological data with modern biological endemism'. Instituto de Geografía, Universidade de Saõ Paulo, *Paleoclimas*, **5**, 1–30.

Brunner, C. A., 1982: 'Paleoceanography of surface waters in the Gulf of Mexico during the late Quaternary'. *Quat. Res.*, **17**, 105–119.

Bryson, R. A., Baerreis, D. A., 1967: 'Possibilities of major climatic modification and their implications: Northwest India, a case for study'. *Bull. Amer. Meteor. Soc.*, **48**, 136–142.

Bryson, R. A., Swain, A. M., 1981: 'Holocene variations of monsoon rainfall in Rajasthan'. *Quat. Res.*, **16**, 135–145.

Butzer, K. W., 1971a: 'Recent history of an Ethiopian delta'. University of Chicago, Department of Geography, Research Paper No. 136, 184 pp.

Butzer, K. W., 1971b: *Environment and archaeology; an ecological approach to prehistory*. Aldine-Atherton, Chicago, New York, second edition, 701 pp.

Butzer, K. W., 1975: 'Patterns of environmental change in the Near East during Late Pleistocene and Early Holocene times'. pp. 389–410 in Wendorf, F., Marks, A. E., eds.: *Problems in prehistory: North Africa and the Levant*. Southern Methodist University Press, Dallas, 462 pp.

Butzer, K. W., 1976: 'Pleistocene climates'. *Geoscience and Man*, **13**, 27–43.

Butzer, K. W., 1981: 'Rise and fall of Axum, Ethiopia: a geo-archaeological interpretation'. *American Antiquity*, **46**, 471–495.

Butzer, K. W., Hansen, C. L., 1968: *Desert and river in Nubia*. University of Wisconsin Press, Madison, Milwaukee, London, 562 pp.

Butzer, K. W., Isaac, G. C., Richardson, J. L., Washbourn-Kamau, C., 1972: 'Radiocarbon dating of East African lake levels'. *Science*, **175**, 1069–1076.

Churchill, D. M., Galloway, R. W., Singh, G., 1978: 'Closed lakes and the paleoclimatic record'. pp. 97–108 in: Pittock, A. B., Frakes, L. A., Jenssen, D., Peterson, J. A., Zillman, J. W., eds.: *Climatic change and variability*. Cambridge University Press, Cambridge, London, New York, Melbourne, 455 pp.

CLIMAP project members, 1976: 'The surface of the ice age earth'. *Science*, **191**, 1131–1137.

CLIMAP project members, 1981: 'Seasonal reconstructions of the Earth's surface at the last glacial maximum'. Geol. Soc. Amer. Map and Chart Ser. MC-36, 18 pp., 9 maps, 1 microfiche (also referenced under A. McIntyre as 'coordinator and compilator').

Coetzee, J. A., 1967: 'Pollen analytical studies in East and Southern Africa'. *Paleoecology of Africa*, **3**, 1–146.

Colinvaux, P. A., 1972: 'Climate and the Galapagos Islands'. *Nature*, **240**, 17–20.

Colinvaux, P., 1979: 'The ice-age Amazon'. *Nature*, **278**, 399–400.

Crowley, T. J., 1981: 'Temperature and circulation changes in the eastern North Atlantic during the last 150 000 years: evidence from the planktonic foraminiferal record'. *Mar. Micropal.*, **6**, 97–129.

Crowley, T. J., 1983: 'The geologic record of climatic change'. *Rev. Geophys. Space Phys.*, **21**, 827–877.

Cullen, J. L., 1981: 'Microfossil evidence for changing salinity patterns in the Bay of Bengal over the last 20 000 years'. *Palaeogeogr., Palaeoclimatol., Palaeoecol.*, **35**, 315–356.

Damuth, J. E., Fairbridge, R. W., 1970: 'Equatorial deep-sea arkosic sands and ice-age aridity in tropical South America'. *Bull. Geol. Soc. Amer.*, **81**, 189–206.

DeVries, T. J., Pearcy, W. G., 1982: 'Fish debris in sediments of the upwelling zone off central Peru: a late Quaternary record'. *Deep-Sea Res.*, **A29**, 87–109.

Diester-Haass, L., 1980: 'Upwelling and climate off Northwest Africa during the late Quaternary'. *Palaeoecology of Africa*, **12**, 229–238.

Duplessy, J. C., 1982: 'Glacial to interglacial contrasts in the northern Indian Ocean'. *Nature*, 295, 494–498.

Duplessy, J. C., Bé, A. W. H., Blanc, P. L., 1981a: 'Oxygen and carbon isotopic composition and biogeographic distribution of planktonic foraminifera in the Indian Ocean'. *Paleogeogr., Paleoclimatol., Paleoecol.*, 33, 9–46.

Duplessy, J. C., Delibrias, G., Turon, J. L., Pujol, C., Duprat, J., 1981b: 'Deglacial warming of the northeastern Atlantic Ocean: correlation with the paleoclimatic evolution of the European continent'. *Paleogeogr., Paleoclimatol., Paleoecol.*, 35, 121–144.

Durand, A., Fontes, J.-Ch., Gasse, F., Icole, M., Lang, J., 1984: 'Le Nord-Ouest de Lac Tchad au Quaternaire: étude de paléoenvironnements alluviaux, éoliens, palustres et lacustres'. *Palaeoecology of Africa*, 16, 215–243.

Emiliani, L., Shackleton, N. J., 1974: 'Brunhes epoch: isotopic temperatures'. *Science*, 183, 511–514.

Flenley, J. R., 1979a: *The equatorial rain forest: a geological history*. Butterworths, London, 162 pp.

Flenley, J. R., 1979b: 'The late Quaternary vegetational history of the equatorial mountains'. *Progress Phys. Geogr.*, 3, 488–509.

Fryberger, S. G., 1980: 'Dune forms and wind regime, Mauritania, West Africa: implications for past climate'. *Paleoecology of Africa*, 12, 79–96.

Gasse, F., 1980: 'Late Quaternary changes in lake-levels and diatom assemblages on the south-eastern margin of the Sahara'. *Palaeoecology of Africa*, 12, 333–350.

Gates, W. L., 1976: 'Modeling the ice-age climate'. *Science*, 191, 1138–1144.

Geyh, M. A., Jäkel, D., 1974: 'Spätpleistozäne und holozäne Klimageschichte der Sahara aufgrund zugänglicher ^{14}C-Daten'. *Zeitschrift für Geomorphologie*, N. F., 18, 82–98.

Gillespie, R., Street-Perrott, F. A., Switsur, R., 1983: 'Post-glacial arid episodes in Ethiopia have implications for climate prediction'. *Nature*, 306, 680–683.

Giresse, P., 1978: 'Le contrôle climatique de la sedimentation marine et continentale en Afrique Centrale Atlantique à la fin du Quaternaire – problèmes de correlation'. *Paleogeogr., Paleoclimatol., Paleoecol.*, 23, 57–77.

Giresse, P., Lanfranchi, R., 1984: 'Les climats et les océans de la région Congolaise pendant l'Holocene – bilans selon les échelles et les méthodes de l'observation'. *Palaeoecology of Africa*, 16, 77–88.

Graf, K., 1981: 'Palynological investigations of two post-glacial peat bogs near the boundary of Bolivia and Peru'. *J. Biogeography*, 8, 353–368.

Grove, A. T., 1984: 'Changing climate, changing biomass and changing atmospheric CO_2'. *Progress in Biometeorology*, 3, 5–10. (Section 2.1.)

Haffer, J., 1969: 'Speciation in Amazonian forest birds'. *Science*, 165, 131–137.

Hamilton, A. C., 1982: *Environmental history of East Africa*. Academic Press, London – New York, 328 pp.

Hassan, F. A., 1981: 'Historical Nile floods and their implications for climatic change'. *Science*, 212, 1142–1144.

Hastenrath, S., 1985: 'A review of Pleistocene to Holocene glacier variations in the tropics'. *Zeitschrift für Gletscherkunde und Glazialgeologie*, 21, 183–194.

Hastenrath, S., Kutzbach, J. E., 1983: 'Paleoclimatic estimates from water and energy budgets of East African lakes'. *Quat. Res.*, 19, 141–153.

Hastenrath, S., Kutzbach, J. E., 1985: 'On the Late Pleistocene climate and water budget of the South American Altiplano'. *Quat. Res.*, 24, in press.

Hastenrath, S., Lamb, P. J., 1977: *Climatic atlas of the tropical Atlantic and Eastern Pacific Oceans*. University of Wisconsin Press, 112 pp.

Hays, J. D., Imbrie, J., Shackleton, N. J., 1976: 'Variations in the Earth's orbit: pacemaker of the ice ages'. *Science*, 194, 1121–1132.

Heine, K., 1978: 'Jungquartäre Pluviale und Interpluviale in der Kalahari (Südliches Afrika)'. *Palaeoecology of Africa*, 10, 31–40.

Heine, K., 1982: 'The main stages of the Late Quaternary evolution of the Kalahari region, Southern Africa'. *Palaeoecology of Africa*, 15, 53–76.

Hillaire-Marcel, C., Riser, J., Rognon, P., Petit-Maire, N., Rosso, J. C., Soulie-Marche, I., 1983: 'Radiocarbon chronology of Holocene hydrologic changes in Northeastern Mali'. *Quat. Res.*, 20, 145–164.

Houvenaghel, G. T., 1974: 'Equatorial Undercurrent and the climate in the Galápagos Islands'. *Nature*, 250, 565–566.

Imbrie, J., Hays, J. D., Martinson, D. G., McIntyre, A., Mix, A. C., Morley, J. J., Pisias, N. G., Prell, W. L., Shackleton, N. J., 1984: 'The orbital theory of Pleistocene climate: support from a revised chronology of the marine δ ^{18}O record'. pp. 269–305, part 1, in: Berger, A., Imbrie, J., Hays, J. D., Kukla, G., Saltzman, B., eds.: *Milankovitch and climate; understanding the response to orbital forcing*. Reidel, Dordrecht, Boston, Lancaster, 2 parts, 895 pp.

Imbrie, J., Imbrie, J. Z., 1980: 'Modeling the climatic response to orbital variations'. *Science, 207*, 943–953.

Imbrie, J., Imbrie, K. P., 1979: *Ice ages: solving the mystery*. Enslow, Hillside, N. J., 224 pp.

Imbrie, J., van Donk, J., Kipp, N. G., 1973: 'Paleoclimatic investigation of a late Pleistocene Caribbean deep-sea core: comparison of isotopic and faunal methods'. *Quat. Res., 3*, 10–38.

Isaar, A. S., Bruins, H. J., 1983: 'Special climatological conditions in the deserts of Sinai and Negev during the latest Pleistocene'. *Palaeogeogr., Paleoclimatol., Paleoecol., 43*, 63–72.

Jäkel, D., 1979: 'Run-off and fluvial formation processes in the Tibesti mountains as indications of climatic history in the Central Sahara during the late Pleistocene and Holocene'. *Palaeoecology of Africa, 11*, 13–44.

Jeje, L. K., 1980: 'A review of geomorphic evidence for climatic change since the late Pleistocene in the rainforest areas of Southern Nigeria'. *Palaeogeogr., Paleoclimatol., Paleoecol., 31*, 63–86.

Johnsen, S. J., Dansgaard, W., Clausen, H. B., Langway, C. C., 1972: 'Oxygen isotope profiles through the Antarctic and Greenland ice sheets'. *Nature, 235*, 429–434.

Jones, G. A., Ruddiman, W. F., 1982: 'Assessing the global meltwater spike'. *Quat. Res., 17*, 148–172.

Kershaw, A. P., 1974: 'A long continuous pollen sequence from Northeastern Australia'. *Nature, 251*, 222–223.

Kershaw, A. P., 1978: 'Record of last interglacial-glacial cycle from north-eastern Queensland'. *Nature, 272*, 159–161.

Kolla, V., Biscayne, P. E., 1977: 'Distribution and origin of quartz in the sediments of the Indian Ocean'. *J. Sedimentary Petrology, 47*, 642–649.

Kolla, V., Biscaye, P. E., 1977: 'Distribution and origin of quartz in the sediments of the Indian Ocean'. *J. relation to climate'. *Quat. Res., 11*, 261–277.

Kolla, V., Ray, P. K., Kostecki, J. A., 1981: 'Surficial sediments of the Arabian Sea'. *Marine Geol., 41*, 183–204.

Kraus, E. B., 1973: 'Comparison between ice age and present general circulations'. *Nature, 245*, 129–133.

Kutzbach, J. E., 1976: 'The nature of climate and climatic variations'. *Quat. Res., 6*, 471–480.

Kutzbach, J. E., 1980: 'Estimates of past climate at Paleolake Chad, North Africa, based on a hydrological and energy-balance model'. *Quat. Res., 14*, 210–223.

Kutzbach, J. E., 1981: 'Monsoon climate of the early Holocene: climate experiment with the Earth's orbital parameters for 9000 years ago'. *Science, 214*, 59–61.

Kutzbach, J. E., 1983: 'Modeling of Holocene climates'. pp. 271–277 in Wright, H. E. Jr., ed.: *Late Quaternary environments of the United States*, vol. 2, *The Holocene*. University of Minnesota Press, Minneapolis, 277 pp.

Kutzbach, J. E., Guetter, P. J., 1984a: 'The sensitivity of monsoon climates to orbital parameter changes for 9000 years B.P.: experiments with the NCAR general circulation model'. pp. 801–820 of part 2 in Berger, A., Imbrie, J., Hays, J., Kukla, G., Saltzman, B., eds.: *Milankovitch and climate*. Reidel, Dordrecht, Boston, Lancaster, 2 parts, 895 pp.

Kutzbach, J. E., Guetter, P. J., 1984b: 'Sensitivity of Late-Glacial and Holocene climates to the combined effects of orbital parameter changes and lower boundary condition changes: 'snapshot' simulations with a general circulation model for 18, 9 and 6 Ka BP'. *Annals of Glaciology, 5*, 85–87.

Kutzbach, J. E., Otto-Bliesner, B. L., 1982: 'The sensitivity of the African-Asian monsoonal climate to orbital parameter changes for 9000 years B.P. in a low-resolution general circulation model'. *J. Atmos. Sci., 39*, 1177–1188.

Lancaster, N., 1981: 'Paleoenvironmental implications of fixed dune systems in Southern Africa'. *Palaeogeogr., Paleoclimatol., Paleoecol., 33*, 327–346.

Lewin, R., 1984: 'Parks: how big is big enough?' *Science, 225*, 611–612.

Livingstone, D. A., 1975: 'Late quaternary climatic change in Africa'. *Annual Review of Ecology and Systematics, 6*, 249–280.

Livingstone, D. A., 1980: 'Environmental changes in the Nile headwaters'. pp. 339–359 in: Williams, M. A. J., Faure, H., eds.: *The Sahara and the Nile*, Balkema, Rotterdam, 607 pp.

Lorenz, E. N., 1970: 'Climatic change as mathematical problem'. *J. Appl. Meteor., 9*, 235–239.

Maley, J., 1977: 'Palaeoclimates of central Sahara during the early Holocene'. *Nature, 269*, 573–577.

Maley, J., 1981: 'Études palynologiques dans le bassin du Tchad et paléoclimatologie de l'Afrique Nord-tropicale de 30 000 ans à l'époque actuelle'. *Palaeoecology of Africa, 13*, 45–52.

Maley, J., 1982: 'Dust, clouds, rain types, and climatic variations in tropical North Africa'. *Quat. Res., 18*, 1–16.

Maley, J., Livingstone, D. A., 1983: 'Extension d'un élément montagnard dans le Sud du Ghana (Afrique de l'Ouest) au Pleistocène supérieur et à l'Holocène inférieu: premières résultats poliniques'. *C. R. Acad. Sci. Paris 1*, **296**, 2 Mai 1983, Série II, 1282–1292.

Manabe, S., Hahn, D. B., 1977: 'Simulation of the tropical climate of an ice age'. *J. Geophys. Res.*, **82**, 3889–3911.

Markgraf, V., Bradbury, J. P., 1982: 'Holocene climatic history of South America'. *Striae*, **16**, 40–45.

McIntyre, A., Kipp, N. G., with Bé, A. W. H., Crowley, T., Kellogg, T., Gardner, J. V., Prell, W., Ruddiman, W. F., 1976: 'The glacial North Atlantic 18 000 years ago: A CLIMAP reconstruction'. pp. 43–76 in: Cline, R. M., Hays, J. D., eds.: *Investigation of Late Quaternary Paleoceanography and Paleoclimatology.*, Geol. Soc. Amer, Mem., 145, pp. 43–76.

McIntyre, A., Ruddiman, W. F., Jantzen, R., 1972: 'Southward penetrations of the North Atlantic polar front: faunal and floral evidence of large scale surface water mass movements over the past 225 000 years'. *Deep-Sea Res.*, **19**, 61–77.

Messerli, B., Winiger, M., Rognon, P., 1980: 'The Saharan and East African uplands during the Quaternary'. pp. 82–132 in: Williams, M. A. J., Faure, H., eds.: *The Sahara and the Nile*, Balkema, Rotterdam, 607 pp.

Ministerio das Minas e Energía, Brazil, 1983,: 'Amazonia Legal, Projeto RADAMBRASIL, map scale 1:2 500 000, two sheets, Rio de Janeiro.

Mitchell, J. M. Jr., 1968: 'Concluding remarks'. *Meteorological Monographs*, **8**, 155–159.

Mitchell, J. M. Jr., 1976: 'An overview of climatic variability and its causal mechanisms'. *Quat. Res.*, **6**, 481–493.

Mix, A. C., Ruddiman, W. F., 1984: 'Oxygen-isotope analyses and Pleistocene ice volumes'. *Quat. Res.*, **21**, 1–20.

Molfino, B., Kipp, N. G., Morley, J. J., 1982: 'Comparison of formainiferal, coccolithophorid, and radiolarian paleotemperature equations: assemblage coherency and estimate concordancy'. *Quat. Res.*, **17**, 279–313.

Moore, T. C. Jr., Burckle, L. H., Geitzenauer, K., Luz, B., Molina-Cruz, A., Robinson, J. H., Sachs, H. M., Sancetta, C. A., Thiede, J., Thompson, P., Wenkam, C., 1980: 'The reconstruction of sea surface temperatures in the Pacific Ocean of 18 000 B.P.'. *Marine Micropal.*, **5**, 215–247.

Moore, T. C. Jr. Hutson, W. H., Kipp, N. G., Hays, J. D., Prell, W., Thompson, P., Bock, G., 1981: 'The biological record of the ice-age ocean'. *Palaeogeogr., Paleoclimatol., Paleoecol.*, **35**, 357–370.

Morley, J. J., Hays, J. D., 1981: 'Towards a high-resolution, global, deep-sea chronology for the last 750 000 years'. *Earth Planet. Sci. Letters*, **53**, 279–295.

Neftel, A., Oeschger, H., Schwander, J., Stauffer, B., Zumbrunn, R., 1982: 'Ice core sample measurements give atmospheric CO_2 content during the past 40 000 yr'. *Nature*, **295**, 220–223.

Newell, R. E., 1973: 'Climate and the Galápagos Islands'. *Nature*, **245**, 91–92.

Newell, R. E., Herman, G. F., Gould-Stewart, S., Tanaka, M., 1975: 'Decreased global rainfall during the past ice age'. *Nature*, **253**, 33–34.

Nicholson, S. E., 1978: 'Climatic variations in the Sahel and other African regions during the past five centuries'. *J. Arid. Environment*, **1**, 3–24.

Nicholson, S. E., 1982: 'Pleistocene and Holocene climates in Africa'. *Nature*, **296**, 779.

Nicholson, S. E., Flohn, H., 1980: 'African environmental and climatic changes and the general atmospheric circulation in last Pleistocene and Holocene'. *Climatic Change*, **2**, 313–348.

Pachur, H. J., Braun, G., 1980: 'The Paleoclimate of the Central Sahara, Libya and the Libyan Desert'. *Palaeoecology of Africa*, **12**, 351–363.

Parkin, D. W., Shackleton, N. J., 1973: 'Trade wind and temperature correlations down a deep-sea core off the Saharan coast'. *Nature*, **245**, 455–457.

Pastouret, L., Chamley, H., Delibrias, G., Duplessy, J. C., Thiede, J., 1978: 'Late Quaternary climatic changes in Western tropical Africa deduced from deep-sea sedimentation off the Niger delta'. *Ocean. Acta*, **1**, 217–232.

Peterson, G. M., Webb, T. III, Kutzbach, J. E., van der Hammen, T., Wjimstra, T. A., Street, F. A., 1979: 'The continental record of environmental conditions at 18 000 B.P.: an initial evaluation'. *Quat. Res.*, **12**, 47–82.

Petit-Maire, N., 1979: 'Prehistoric paleoecology of the Sahara Atlantic coast in the last 10 000 years: a synthesis'. *J. Arid Environm.*, **2**, 85–88.

Petit-Maire, N., 1980: 'Holocene biogeographical variations along the Northwestern African coast (19–28° N); palaeoclimatic implications'. *Palaeoecology of Africa*, **12**, 365–377.

Petit-Maire, N., Riser, J., 1981: 'Holocene lake deposits and paleoenvironments in Central Sahara, Northwestern Mali'. *Palaeogeogr., Paleoclimatol., Paleoecol.*, **35**, 45–61.

Pflaumann, U., 1975: 'Late Quaternary stratigraphy based on planktonic foraminifera off Senegal'. *Meteor Forschungsergebnisse, Berlin-Stuttgart*, C23, 1–46.

Pisias, N. G., Shackleton, N. J., 1984: 'Modelling the global climate response to orbital forcing and atmospheric carbon dioxide changes'. *Nature*, 310, 757–759.

Prell, W. L., 1982: 'Reply to comments by P. J. Webster and N. A. Streten regarding "Surface circulation of the Indian Ocean during the last glacial maximum, approximately 18 000 YBP"'. *Quat. Res.*, 17, 128–131.

Prell, W. L., 1984a: 'Monsoonal climate of the Arabian Sea during the late Quaternary: a response to changing solar radiation'. pp. 349–366 of part 1 in Berger, A., Imbrie, J., Hays, J., Kukla, G., Saltzman, B., eds.: *Milankovitch and climate*. Reidel, Dordrecht, Boston, Lancaster, 2 parts, 895 pp.

Prell, W. L., 1984b: 'Variation of monsoonal upwelling: a response to changing solar radiation'. pp. 48–57 in: Hansen, J. E., Takahashi, T., eds.: *Climate processes and climate sensitivity*, Geophysical Monograph 29, Maurice Ewing Volume 5, Amer. Geophys. Union, 368 pp.

Prell, W. L., Curry, W. B., 1981: 'Faunal and isotopic indices of monsoonal upwelling: western Arabian Sea'. *Ocean. Acta*, 4, 91–98.

Prell, W. L., Hutson, W. H., 1979: 'Zonal temperature-anomaly maps of Indian Ocean surface water: modern and ice-age patterns'. *Science*, 206, 454–456.

Prell, W. L., Hutson, W. H., Williams, O. F., Bé, A. W. H., Geitzenauer, K., Molfino, B., 1980: 'Surface circulation of the Indian Ocean during the last glacial maximum, approximately 18 000 yr. B. P.'. *Quat. Res.*, 14, 309–336.

Quinn, W. H., 1971: 'Late Quaternary meteorological and oceanographic developments in the equatorial Pacific'. *Nature*, 229, 330–331.

Reiss, Z., Luz, B., Almogi-Labin, A., Halicz, E., Winter, A., Wolf, M., 1980: 'Late Quaternary paleoceanography of the Gulf of Aquaba (Eilat), Red Sea'. *Quat. Res.*, 14, 294–308.

Riehl, H., Meitín, J., 1979: 'Discharge of the Nile river: a barometer of short-period climate variation'. *Science*, 206, 1178–1179.

Roberts, N., Street-Perrott, F. A., Perrott, A., 1981: 'The "Late-Glacial in the tropics"'. *Quaternary Newsletter*, 35, 1–5.

Robock, A., 1978: 'Internally and externally caused climate change'. *J. Atmos. Sci.*, 35, 1111–1122.

Rognon, P., 1976: 'Essai d'interpretation des variations climatiques au Sahara depuis 40 000 ans'. *Revue de Géographie Physique et Géologie Dynamique*, 18, 251–282.

Rognon, P., 1980: 'Une extension des déserts (Sahara et Moyen-Orient) au cours du Tardiglaciaire (18 000–10 000 years B.P.)'. *Révue de Géologie Dynamique et de Géographie Physique*, 22, 313–328.

Rognon, P., Williams, M. A. J., 1977: 'Late Quaternary climatic changes in Australia and North Africa: a preliminary interpretation'. *Paleogeogr., Paleoclimatol., Paleoecol.*, 21, 285–327.

Romine, K., Moore, T. C. Jr., 1981: 'Radiolarian assemblage distributions and paleoceanography of the eastern equatorial Pacific Ocean during the last 127 000 years'. *Palaeogeogr., Palaeoclimatol., Palaeoecol.*, 35, 281–314.

Rossignol-Strick, M., Duzer, D., 1979: 'A late Quaternary continuous climatic record from palynology of three marine cores off Senegal'. *Palaeoecology of Africa*, 11, 185–188.

Rossignal-Strick, M., Duzer, D., 1980: 'Late Quaternary West African climate inferred from palynology of Atlantic deep-sea cores'. *Palaeoecology of Africa*, 12, 227–228.

Ruddiman, W. F., and CLIMAP project members, 1979: 'The last interglacial ocean'. *Geol. Soc. Am., Ann. Mtg. Abs.*, pp. 507–508.

Ruddiman, W. F., McIntyre, A., 1979: 'Warmth of the subpolar North Atlantic Ocean during Northern hemisphere ice-sheet growth'. *Science*, 204, 173–175.

Ruddiman, W. F., McIntyre, A., 1981a: 'The mode and mechanism of the last deglaciation: oceanic evidence'. *Quat. Res.*, 16, 125–134.

Ruddiman, W. F., McIntyre, A., 1981b: 'The North Atlantic Ocean during the last deglaciation'. *Palaeogeog., Palaeoclim., Palaeoecol.*, 35, 145–214.

Ruddiman, W. F., McIntyre, A., Niebler-Hunt, V., Durazzi, J. T., 1980a: 'Oceanic evidence for the mechanism of rapid northern hemisphere glaciation'. *Quat. Res.*, 13, 33–64.

Ruddiman, W. F., Molfino, B., Esmay, A., Pokras, E., 1980b: 'Evidence bearing on the mechanism of rapid deglaciation'. *Climatic Change*, 3, 65–87.

Rust, U., Schmidt, H. H., Dietz, K. R., 1984: 'Palaeoenvironments of the present day arid Southwestern Africa 30 000–5000 B.P.: results and problems'. *Palaeoecology of Africa*, 16, 109–148.

Salgado-Labouriau, M. L., Schubert, C., Valastro, S., 1977: 'Paleoecologic analysis of a Late-Quaternary terrace from Mucubají, Venezuelan Andes'. *J. Biogeogr.*, 4, 313–325.

Salinger, M. J., 1980a: 'New Zealand climate: 1. precipitation patterns'. *Mon. Wea. Rev.*, 108, 1892–1904.
Salinger, M. J., 1980b: 'New Zealand climate: 2. temperature patterns'. *Mon. Wea. Rev.*, 108, 1905–1912.
Salinger, M. J., 1981: 'Paleoclimates North and South'. *Nature*, 291, 106–107.
Salinger, M. J., 1984: 'New Zealand climate: the last 5 million years'. pp. 131–150 in Vogel, J. C., ed.: *Late Cainozoic palaeoclimates of the Southern hemisphere*; Proceedings of symposium by South African Society for Quaternary Research, Swaziland Aug–Sept 1983. Balkema, Rotterdam, 520 pp.
Saltzman, B., Vernekar, A. D., 1973: 'A solution for the Northern hemisphere climatic zonation during a glacial maximum'. *Quat. Res.*, 5, 307–320.
Sarnthein, M., 1978: 'Sand deserts during glacial maximum and climate optimum'. *Nature*, 272, 43–45.
Sarnthein, M., Erlenkeuser, H., Zahn, R., 1982a: 'Termination I: the response of continental climate in the subtropics as recorded in deep-sea sediments. Actes Colloque International CNRS, Bordeaux, Sept. 81: *Bulletin Inst. Géol. Bassin Aquitaine*, 31, 393–407.
Sarnthein, M., Koopmann, B., 1980: 'Late Quaternary deep-sea record on Northwest African dust supply and wind circulation'. *Palaeoecology of Africa*, 12, 239–253.
Sarnthein, M., Tetzlaff, G., Koopmann, B., Wolter, K., Pflaumann, U., 1981: 'Glacial and interglacial wind regimes over the eastern subtropical Atlantic and North-West Africa'. *Nature*, 293, 193–196.
Sarnthein, M., Thiede, J., Pflaumann, U., Erlenkeuser, H., Fütterer, D., Koopmann, B., Lange, H., Seibold, E., 1982b: 'Atmospheric and oceanic circulation patterns off Northwest Africa during the past 25 million years. pp. 545–604, chapter 24, in: von Rad, U., Hinz, K., Sarnthein, M., Seibold, E., eds.: *Geology of the Northwest African continental margin*, Springer Verlag Berlin, Heidelberg, New York, 709 pp.
Schneider, S. H., Thompson, S. L., 1979: 'Ice ages and orbital variations: some simple theory and modeling'. *Quat. Res.*, 12, 188–203.
Servant, M., Fontes, J. C., 1978: 'Les lacs quaternaires des hauts plateaux des Andes Boliviennes: premières interpretations paléoclimatiques'. *Cahiers de l'ORSTOM, Ser. Geol.*, 10, 9–23.
Servant, M., Servant, S., 1970: 'Les formations lacustres et les diatomées de Quaternaire recent du fond de la cuvette Tchadienne'. *Revue de Géographie Physique et de Géologie Dynamique*, 12, 63–76.
Servant, M., Servant-Vildary, S., 1972: 'Nouvelles données par une interprétation paléoclimatique de séries continentales du bassin Tchadien (Pleistocène recent, Holocène)'. *Palaeoecology of Africa*, 6, 87–92.
Servant, M., Villaroel, R., 1979: 'Le problème paléoclimatique des Andes boliviennes et leurs piedmonts amazoniens au Quaternaire'. *C.R. Acad. Sc. Paris, 288, 19 Février 1979, Serie D*, 665–668.
Servant-Vildary, S., 1979: 'Paléolimnologie des lacs du bassin Tchadienne au Quaternaire recent'. *Palaeoecology of Africa*, 11, 65–78.
Shackleton, N. J., Opdyke, N. D., 1973: 'Oxygen isotope and paleomagnetic stratigraphy of equatorial Pacific core V28–239: oxygen isotope temperatures and ice volume on a 100 000 and 1 000 000 year scale'. *Quat. Res.*, 3, 39–55.
Simpson, B. B., 1975: 'Pleistocene changes in the flora of the high tropical Andes'. *Paleobiol.*, 1, 273–294.
Simpson, B. B., Haffer, J., 1978: 'Speciation patterns in the Amazonian forest biota'. *Ann. Rev. Ecol. Syst.*, 9, 497–518.
Singh, G., Joshi, R. D., Chopra, S. K., Singh, A. B., 1974: 'Late Quaternary history of vegetation and climate of the Rajasthan Desert, India'. *Phil. Trans. Roy. Soc. London, Ser.*, B267, 467–501.
Smith, A. B., 1984: 'The origins of food production in Northeast Africa'. *Palaeoecology of Africa*, 16, 317–324.
Start, G. G., Prell, W. L., 1984: 'Evidence for two Pleistocene climatic modes: data from DSDP site 502'. pp. 3–22 in: Berger, A. L., Nicolis, C., eds.: *New perspectives in climatic modelling*. Developments in Atmospheric Science No. 16, Elsevier, Amsterdam, Oxford, New York, Tokyo, 403 pp.
Street, F. A., 1979: 'Late Quaternary precipitation estimates for the Ziway-Shala Basin, Southern Ethiopia'. *Palaeoecology of Africa*, 11, 135–143.
Street, F. A., 1980: 'The relative importance of climate and local hydrogeological factors in influencing lake-level fluctuations'. *Palaeoecology of Africa*, 12, 137–158.
Street, F. A., 1981: 'Tropical palaeoenvironments'. *Progress Phys. Geogr.*, 5, 157–185.
Street, F. A., Grove, A. T., 1976: 'Environmental and climatic implications of late Quaternary lake-level fluctuations in Africa'. *Science*, 261, 385–390.
Street, F. A., and A. T. Grove, 1979: 'Global maps of lake-level fluctuations since 30 000 BP'. *Quat. Res.*, 12, 83–118.
Street-Perrott, F. A., Harrison, S. P., 1984: 'Temporal variations in lake levels since 30 000 yr B.P. – an index of the global hydrological cycle'. pp. 118–129 in: *Climate processes and climate sensitivity*, Geophysical Monograph No. 29 (Maurice Ewing, vol. 5), 368 pp.

Street-Perrott F. A., Harrison, S. P., 1985: 'Lake-level fluctuations'. pp. 1–80, chapter 7, in: Hecht, A. D., ed.: *Paleoclimate analysis and modeling.* Wiley, New York, 445 pp.

Swain, A. M., Kutzbach, J. E., Hastenrath, S., 1983: 'Estimates of Holocene precipitation for Rajasthan, India, based on pollen and lake-level data'. *Quat. Res., 19,* 1–17.

Talbot, M. R., 1981: 'Holocene changes in tropical wind intensity and rainfall: evidence from Southeast Ghana'. *Quat. Res., 16,* 201–220.

Talbot, M. R., 1984: 'Late Pleistocene rainfall and dune building in the Sahel'. *Palaeoecology of Africa, 16,* 203–214.

Talbot, M. R., Livingstone, D. A., Palmer, P. G., Maley, J., Melack, J. M., Delibrias, G., Gulliksen, S., 1984: 'Preliminary results from sediment cores from Lake Bosumtwi, Ghana'. *Palaeoecology of Africa, 16,* 173–192.

Tetzlaff, G., Adams, L. J., 1983: 'Present-day and early-holocene evaporation of Lake Chad'. pp. 347–360 in: A. Street-Perrott *et al.,* eds.: *Variations in the global water budget,* Reidel, Dordrecht, Boston, Lancaster, 518 pp.

Tetzlaff, G., Sarnthein, M., Wolter, K., 1982: 'Pleistocene and Holocene climates in Africa, reply'. *Nature, 296,* 779–780.

Tricart, J., 1974: 'Existence de périodes sèches au quaternaire en Amazonie et dans les régions voisines'. *Revue de Géomorphologie Dynamique, 23,* 145–158.

Van Campo, E., Duplessy, J. C., Rossignol-Strick, M., 1982: 'Climatic conditions deduced from a 150-kyr oxygen isotope-pollen record from the Arabian Sea'. *Nature, 296,* 56–59.

Van der Hammen, T., 1972: 'Changes in vegetation and climate in the Amazon Basin and surrounding areas during the Pleistocene'. *Geologie en Mijnbouw, 51,* 641–643.

Van der Hammen, T., 1974: 'The Pleistocene changes of vegetation and climate in tropical South America'. *J. Biogeography, 1,* 3–26.

Van Geel, B., van der Hammen, T., 1973: 'Upper Quaternary vegetational and climatic sequence of the Fuquene area (Eastern Cordillera, Colombia)'. *Palaeogeogr., Palaeoclimatol., Palaeoecol., 14,* 9–92.

Van Zinderen Bakker, E. M., Sr., 1976: 'The evolution of late Quaternary palaeoclimates of Southern Africa'. *Palaeoecology of Africa, 9,* 160–202.

Van Zinderen Bakker, E. M., Sr., 1980: 'Comparison of Late Quaternary climatic evolutions in the Sahara and in the Namib-Kalahari region'. *Palaeoecology of Africa, 12,* 381–394.

Van Zinderen Bakker, E. M., Sr., 1982: 'African palaeoclimates 18 000 yrs B.P.'. *Palaeoecology of Africa, 15,* 77–100.

Van Zinderen Bakker, E. M., Sr., 1984a: 'Aridity along the Namib coast'. *Palaeoecology of Africa, 16,* 149–162.

Van Zinderen Bakker, E. M., Sr., 1984b: 'A Late and Post-glacial pollen record from the Namib Desert'. *Palaeoecology of Africa, 16,* 421–428.

Vuilleumier, B. S., 1971: 'Pleistocene changes in the fauna and flora of South America'. *Science, 173,* 771–780.

Watts, W. A., Bradbury, J. P., 1982: 'Paleoecological studies at Lake Patzcuaro on the West-Central Mexican Plateau and at Chalco in the basin of Mexico'. *Quat. Res., 17,* 56–70.

Webster, P. J., Streten, N. A., 1978: 'Late Quaternary ice age climates of tropical Australasia: interpretations and reconstructions'. *Quat. Res., 10,* 279–309.

Webster, P. J., Streten, N. A., 1981: 'Comments on: "surface circulation of the Indian Ocean during the last glacial maximum, approximately 18 000 yr. B.P." '. *Quat. Res., 16,* 421–423.

Wendorf, F., Schild, R., Said, R., Haynes, C. V., Gautier, A., Kobusiewicz, M., 1976: 'The prehistory of the Egyptian Sahara'. *Science, 193,* 103–114.

Williams, J., Barry, R. G., Washington, W. M., 1974: 'Simulation of the atmospheric circulation using the NCAR global circulation model with ice age boundary conditions'. *J. Appl. Meteor., 13,* 305–317.

Williams, M. A. J., 1975: 'Late Pleistocene tropical aridity synchronous in both hemispheres'. *Nature, 253,* 617–618.

Williams, M. A. J., Adamson, D. A., 1974: 'Late Pleistocene desiccation along the White Nile'. *Nature, 248,* 584–586.

Williams, M. A. J., Faure, M., eds., 1980: *The Sahara and the Nile: Quaternary environments and prehistoric occupation in Northern Africa.* Balkema, Rotterdam, 607 pp.

AUTHOR INDEX

433

SUBJECT INDEX